FOREWORD

The sixth edition of the **Agricultural Research Index** was published in 1978. The present edition of this established worldwide reference source continues to list relevant organizations, their staff and work but in considerably greater detail. The publisher and the editor agreed that a more accurate description of the scope and contents of the book is given by the title **Agricultural Research Centres: a world directory of organizations and programmes.**

This directory seeks to provide a comprehensive guide to the organizations throughout the world which conduct or finance research in any aspect of agriculture, veterinary medicine, horticulture, fisheries and aquaculture, food science, forestry, zoology, botany, and relevant ancillary disciplines such as plant production, animal production, soil science, drainage and irrigation, and land use.

Research is defined in broad terms and includes basic and applied research, investigative and development work. Organizations listed include official research centres, educational organizations with research capabilities and activities, industrial firms and their research and development centres, and independent research centres. Purely professional, advisory or educational organizations are not generally included, with the exception of certain major bodies which influence the funding of research.

Agricultural Research Centres gives for each organization listed, where available, the full current address, the type of organization and its affiliation, the names of its Director or Head and senior staff, the general scope of its activities with its major current projects and their leaders, the total number of graduate or equivalent research staff, and information on the titles and frequency of any annual or progress report it publishes.

The present edition includes details on over 11000 laboratories – an increase of 15 per cent on the previous edition. It also contains substantially more information on the current work of organizations listed and on the annual or progress reports they issue. In addition, it contains, for the first time, a subject index as well as an index of establishments. The subject index allows readers to identify organizations in different countries conducting research on particular problems under such headings as fish pathology, pesticides and soil erosion.

It is hoped that this expanded and improved edition will serve as a useful source of worldwide reference for research workers active in the subject range of this directory. It should also serve those who finance, plan and administer such work, members of ancillary industries and professions, advisers, educationalists, librarians and information officers. In addition, it provides information not readily available elsewhere for those in other fields, such as economists and journalists.

No book of this size and complexity could be published without the cooperation of a multitude of international and national organizations throughout the world. Our thanks are due to the numerous bodies which have completed our questionnaire forms and provided the information included in this guide. I particularly wish to thank the following: the Librarian and staff of the Library of the Ministry of Agriculture, Fisheries and Food, London; Dr D. W. Immelman, Director General, Department of Agriculture and Fisheries, South Africa; Anders Fredholm, Head of Information, Sveriges Lantbruksuniversitet; Vehbi Kesici, Director General, Turkish Ministry of Agriculture; Herrn Steinigeweg, Tierärztliche Hochschule Hannover; and Dr Wolfram, Faculty of Agriculture, Rheinische Friedrich-Wilhelms Universität Bonn.

Every effort has been made to make this directory as comprehensive as possible and to present the information in a clear and concise form. It is regretted that in some cases we have been unable to obtain adequate information but we hope that in future editions our efforts will meet with more success. In particular vigorous enquiries on and in the USSR and the People's Republic of China have failed to obtain substantial information. For both countries this directory lists such information as has been received, and for USSR it includes relevant known all-union research centres.

We appreciate that no reference work of this type can hope to be definitive and would be glad to hear from users who can provide information which should be considered for inclusion in future editions, point out any errors of omission or commission, or make any other suggestions for improvement.

London
December 1982

Nigel Harvey, MA, ARICS
Consultant Editor

A NOTE ON THE EDITOR

Nigel Harvey, MA, ARICS spent thirty-two years on the staff first of the UK Ministry of Agriculture, Fisheries and Food, then on the Agricultural Research Council, including one year's secondment as a Kellogg Fellow at Purdue University, Indiana, and three years' secondment to the UK Department of the Environment. For most of this time he was concerned with the collection and collation of scientific and technical information.

In his official capacity, he has been responsible for a number of technical publications and bibliographies of research literature, among them the first edition of the Ministry of Agriculture, Fisheries and Food bulletin on *The housing of pigs,* the Agricultural Research Council's *Bibliography of farm buildings research,* which he edited for twelve years, and the Department of the Environment's *Index of current Government and Government-supported research in environmental pollution in Great Britain,* of which he was co-author, as well as selective bibliographies on *Farm livestock wastes and silage effluent* and *Pesticides* in the Department's Library series. In his private capacity he has written various books on agriculture and agricultural history, including *Farm work study: a history of farm buildings in England and Wales,* and *The industrial archaeology of farming in England and Wales.*

He retired as a Principal Scientific Officer in 1976 and is now Honorary Librarian of the Royal Agricultural Society of England.

AGRICULTURAL RESEARCH CENTRES

REFERENCE BOOKS AVAILABLE FROM LONGMAN GROUP LIMITED

Longman Reference on Research Series

Aerospace Research Index
Agricultural Research Centres
Directory of Scientific Directories
European Research Centres
European Sources of Scientific and Technical Information
Industrial Research in the United Kingdom
International Medical Who's Who
International Who's Who in Energy and Nuclear Sciences
Materials Research Centres
Medical Research Centres
   (supersedes Medical Research Index)
Pollution Research Index
Who's Who in Science in Europe
Who's Who in World Agriculture
World Energy Directory
World Nuclear Directory

Longman Guide to World Science and Technology

  Series editor: Ann Pernet

Science and Technology in the Middle East
  by Ziauddin Sardar
Science and Technology in Latin America
  by Latin American Newsletters Limited
In preparation
Science and Technology in the USSR
  by Vera Rich
Science and Technology in Eastern Europe
  by Vera Rich
Science and Technology in the UK
  by Anthony P. Harvey and Ann Pernet
Science and Technology in China
  by Tong B. Tang
Science and Technology in Japan
  by Alun M. Anderson
Science and Technology in South-East Asia
  by Ziauddin Sardar
Science and Technology in Australasia, Antarctica and the Pacific
  by Jarlath Ronayne

AGRICULTURAL RESEARCH CENTRES

a world directory of organizations and programmes

Consultant editor: Nigel Harvey

Volume 2
Madagascar to Zimbabwe, Indexes

Reference on Research

Longman

AGRICULTURAL RESEARCH CENTRES: a world
directory of organizations and programmes

Fifth edition 1970
Sixth edition 1978
Seventh edition 1983

Published by Longman Group Limited, Professional and Information
Publishing Division, Sixth floor, Westgate House, The High, Harlow,
Essex CM20 1NE, UK
Telephone: Harlow (0279) 442601

Distributed exclusively in the USA and Canada by Gale Research
Company, Book Tower, Detroit, MI 48226, USA

British Library Cataloguing in Publication Data

Agricultural research centres. – (Longman reference on research series)
   1. Agricultural research – Directories
   I. Harvey, Nigel
   630'.72   SZ77

set
ISBN 0 582 90014 X

Printed in Great Britain by
Butler and Tanner Ltd, Frome, Somerset

PUBLISHER'S INTRODUCTION

Agricultural Research Centres: a world directory of organizations and programmes, seventh edition, is the successor to **Agricultural Research Index,** 6th edition published in 1978. This directory is typeset and includes for the first time a subject index. **Agricultural Research Centres** provides details of relevant laboratories and organizations which are carrying out or funding research and development. In it are listed laboratory and research station titles, addresses, names of senior staff, activities, programmes and major current projects, and publications of industrial organizations, research associations, government institutions, and academic departments in over 125 countries in the world.

The directory is intended to be a reference for practising scientists and agriculturalists, information scientists, research administrators, industrial market researchers, job seekers – indeed for all those who want to know about national and international contact organizations in agriculture and related subjects, the names of senior personnel, and research programmes.

The subject coverage is agriculture, veterinary medicine, horticulture, fisheries and aquaculture, food science, forestry, zoology, botany, and relevant ancillary disciplines such as plant production, animal production, soil science, drainage and irrigation, and land use. Pure agricultural engineering is omitted from this volume.

Agricultural Research Centres is arranged by countries in alphabetical order. Organizations spanning more than one country are listed in the opening chapter entitled 'international establishments'. This chapter as well as providing details of research centres carrying out r&d projects substantially funded by more than one country, includes some international bodies influential in recommending research expenditure because of their close links with international projects. National chapters incorporating entries on each research centre or research-funding organization follow.

To locate a particular establishment or laboratory, the reader can either refer directly to the country chapter, or use the Titles of establishments index. The Subject index, which completes the book allows the reader to identify where specific research activities are being conducted. Thus the reader can identify research and organizational details of establishment by country, by title, and by subject matter.

Each country chapter comprises research establishments and research funding organizations arranged in alphabetical order in the language of that country. For countries which do not use the roman alphabet, the title is given in English translation. Universities are entered under their proper title.

Each centre of research has been given a separate entry. Some centres of research have subsidiary bodies administered by them, and these form part of their listing. Examples are universities with several relevant departments or laboratories, and official research centres with several field stations. An example of the weights of the headings used is shown below.

Centre de Recherche d'Antibes
AGRONOMY AND PLANT PHYSIOLOGY STATION
BOTANY AND PLANT PATHOLOGY STATION
FLOWERING PLANT IMPROVEMENT STATION
Seed and Variety Control Study Group

In the above the stations are controlled by the research centre, Centre de Recherche d'Antibes, and the Seed and Variety Control Study Group is administered by the Flowering Plant Improvement Station. A fourth level heading may occasionally be used and appears as in the following example of a university carrying out research:

University of Newcastle Upon Tyne
FACULTY OF AGRICULTURE
Department of Agricultural Economics
AGRICULTURAL ADJUSTMENT UNIT

Overall the chains are made as simple as possible. If the reader cannot immediately locate a particular organization or laboratory in the chapter entry he can use the Titles of establishments index.

Each entry is introduced with the full title of the organization accompanied by its acronym if used. If the title is in a language other than English, a translation into English is supplied. It should be noted that the translated title is not a transliteration of the original language but is an English version intended to give an indication of the work of the particular body. Introduced by the word or phrase typeset in bold below, the following details, where available, are given in an entry: title; acronym if regularly used; English translation of title; full postal **address**; **telephone** number; **telex** address; **status** indicates the type of body; **affiliation** or parent body indicates its administrative links or parentage; the name of the research director, **director** or head is given preceded by his official job title, which is usually given in its English version; the **sections,** departments or divisions provide the scope of the section followed by the name of the person leading that section or that line of research; number of **graduate** (or equivalent) **research staff**; **annual expenditure** may be given in terms of a cash sum or within a particular cash range; the general scope of **activities** of the research centre or unit, indicating the areas of current and proposed research; **publications** gives the titles and frequency of progress reports, and publications issued on a regular basis such as annual reports, year books and research reports; the major current **projects** usually followed by the name of the project leader printed in parentheses.

Overall the entry information is intended to indicate the size of the organization, its major personnel, an overview of its research activities and programmes, and its direct links with other organizations.

An asterisk (*) appearing after the title indicates that a reply was not received in time for inclusion in the book. However the consultant editor believes these organizations are conducting activities which fall within the scope of this directory, so they have been included. In these cases the entry contains data available from **European Research Centres: a directory of organizations in science, technology, agriculture, and medicine** and other public sources.

The Titles of establishments index includes titles of all establishments listed in this reference book, and directs the user by country and entry number to the full information on that establishment. An establishment with a title in a language other than English is entered both under its original language title and its English translation, as given in the entry. Translated titles are printed in italic type. The acronyms of major establishments listed are included in this index.

The Subject index is compiled primarily from the 'activities' and 'projects' sections of each entry and allows the reader to identify in which locations specific research activities are being carried out. The subject terms are based on the controlled vocabulary of the British Standards Institute's published ROOT (1981) thesaurus.

Much of the information in this directory has been elicited by questionnaire and we offer our sincerest thanks to all who made this project possible by completing the form sent to them. The success of this directory has depended on organizations responding to our request for current details.

This publication is the third Longman Reference on Research directory to be typeset. The first was **European Research Centres: a directory of organizations in science, technology, agriculture, and medicine.** The second was **Materials Research Centres: a world directory of organizations and programmes in materials science.** All three publications have been typeset directly from our data base, which incorporates the most recent information obtained by questionnaire on r&d organizations and programmes. This data base allows us to typeset all entry details quickly so that changes discovered as little as three months prior to publication can be readily incorporated.

The continuing development of research programmes and investigative work means that any reference guide of this nature needs regular updating. Consequently it is hoped to produce updated versions of this directory at regular intervals in the future. The publishers would appreciate hearing from users who may have suggestions to make for the improvement of future editions, or who are able to point out any errors of omission or commission.

Colin P. Taylor
Publisher

December 1982

CONTENTS AND INDEX ABBREVIATIONS

MADAGASCAR

Departement de Recherches Agronomiques de la République Malgache* 1

[Department of Agronomic Research of the Malagasy Republic]
Address: BP 1444, Antanarivo
Status: Official research organization
Director: J. Velly

Institut de Recherches Agronomiques Tropicales et des Cultures Vivrières* 2

– IRAT
[Tropical Agriculture and Food Crops Research Institute]
Address: BP 438, Bamako
Status: Official research centre
Director: M. Arraudeau

Universite de Madagascar* 3

[University of Madagascar]
Address: BP 566, Antananarivo
Status: Educational establishment with r&d capability
Administrative director: Guy Rakotovao

DEPARTMENT OF AGRICULTURE* 4

Head: G. Ravelojaona

MALAWI

Baka Agricultural Research Station   1

Address: PO Box 43, Karonga
Parent body: Ministry of Agriculture, Agricultural Research Department
Status: Official research centre
Head: E.M. Nyirenda

Bvumbwe Agricultural Research Station   2

Address: PO Box 5748, Limbe
Parent body: Ministry of Agriculture, Agricultural Research Department
Status: Official research centre
Head: Dr J.H.A. Maida
Activities: Primarily a horticultural research station where investigations are carried out on various fruit, tree nut and vegetables crops including deciduous fruits, tropical fruits, soft fruits, tung, macademia nuts, coffee, essential oils, spices and European potato; major crop protection unit and plant quarantine facilities; research on soils.
Projects: Grain storage.

Chitala Agricultural Research Station   3

Address: Private Bag 13, Salima
Parent body: Ministry of Agriculture, Agricultural Research Department
Status: Official research centre
Head: W.C.C. Mughogho
Activities: Chitala is a sub-station for Chitedze looking at the requirements of the central region lakeshore areas. Variety evaluation, agronomy and plant protection of the following crops are undertaken: maize, sorghum, legumes, cotton, tree crops (production of budded citrus stocks), oil seed crops.

Chitedze Agricultural Research Station   4

Address: PO Box 158, Lilongwe
Parent body: Ministry of Agriculture, Agricultural Research Department
Status: Official research centre
Head: Dr H.K. Mwandemere
Activities: Main research station in the country: its activities include research into variety evaluation, agronomy and plant protection in maize, groundnuts, and other grain legumes such as cowpea, ground beans, and pigeon peas. Research is also carried out on pastures, soil fertility, smallholder appropriate technology, and livestock breeding and management. Studies have been made on rhizobium requirements for nitrogen fixation in legumes with specific rhizobium strains being produced commercially. There is a crop storage training and research centre. Studies on farming systems have recently been initiated.

Fisheries Research Station 5

Address: Box 27, Monkey Bay, Mangochi
Status: Official research centre
Director: B.J. Mukoko
Graduate research staff: 5
Activities: Fresh-water fisheries: fish production.
Publications: Annual report.
Projects: Taxonomy of cichlid fishes of Lake Malawi (Dr D.S.C. Lewis); assessment of stocks of traditional fisheries (S. Alimoso); biology of cichlid fishes (D. Tweddle); plankton of Lake Malawi (S. Mapila).

Forestry Research Institute of Malawi 6

– FRIM
Address: PO Box 270, Zomba
Telephone: (50) 2951-2
Affiliation: Forestry Department, Ministry of Forestry and Natural Resources
Status: Official research centre
Acting senior forestry research officer: C.L. Ingram
Sections: Silviculture, C.L. Ingram; tree breeding, vacant; pathology, N.W.S. Chipompha; entomology, vacant; mensuration, J. Clark; soils, L.A. Sitaubi; ecology, T.M. Darwin; wood products, vacant
Graduate research staff: 9
Activities: Forestry and forest products research aimed at the improvement of the yield and quality of timber and fuelwood plantations, the maintenance and improvement of indigenous protected forests, the monitoring of forest insects and diseases, protection measures, the evaluation of management practices and new equipment and techniques, the provision of a forest seed service, classification of forest soils, and the promotion of the efficient utilization of timber.
Publications: Annual report.
Projects: Plantation silviculture, (C.L. Ingram); indigenous silviculture, (Dr I.D. Edwards); dry zone fuelwood species selection; mycorrhizal fungi, (N.W.S. Chipompha); termite control; primate damage in plantations, (T.M. Darwin); tree selection and seed orchards; species-soil relationships, (L.A. Sitaubi).

Kasinthula Agricultural Research Station 7

Address: PO Box 28, Chikwawa
Parent body: Ministry of Agriculture, Agricultural Research Department
Status: Official research centre
Head: E.S. Mwafulirwa
Activities: This is mainly an irrigatedcrops station although some studies are undertaken on rainfed crops. The main crop research is on rice, maize, horticultural crops and cotton. Basic rice seed is produced under a smallholder scheme.

Lifuwu Agricultural Research Station 8

Address: PO Box 102, Salima
Parent body: Ministry of Agriculture, Agricultural Research Department
Status: Official research centre
Head: A.S. Kumwenda
Activities: Rice research centre covering breeding and agronomy of both irrigated and rainfed rice; agronomic trials on mango, cashew and sugar.

Lunyangwa Agricultural Research Station 9

Address: PO Box 59, Mzuzu
Parent body: Ministry of Agriculture, Agricultural Research Department
Status: Official research centre
Head: R.N.F. Sauti

Makhanga Agricultural Research Station 10

Address: PO Box 20, Chiromo
Parent body: Ministry of Agriculture, Agricultural Research Department
Status: Official research centre
Head: A.E.D. Muwowo
Activities: Dry land and irrigated crops of rice, maize, groundnuts , cotton, cassava, mango and citrus.

Makoka Agricultural Research Station 11

Address: Private Bag 3, Thondwe, Zomba
Telephone: (533) 251
Parent body: Ministry of Agriculture, Agricultural Research Department
Status: Official research centre
Assistant chief agricultural research officer: G.K.C. Nyirenda
Department: Cotton productivity research, G.K.C. Nyirenda; biometrics, A.B. Chirembo
Graduate research staff: 10
Activities: The centre is the headquarters of cotton productivity research in Malawi. Plant breeding: main aims are to develop varieties for smallholder farmers. The varieties should be pest and disease resistant. Yield seed cotton, giving percentage, compact habit, boll size and fibre quality are some of the factors being looked into. Breeding for cold tolerance and earliness is also a major aim. Cotton agronomy: to investigate the use of fertilizers including trace elements such as boron; use of herbicides and growth regulators, and picking techniques by smallholder farmers. Entomology: assessment of insecticides against cotton insect pests both in the laboratory and field; assessment of application equipment and development of application techniques for smallholder farmers; investigation of use of sex pheromones for detection of pest infestations and use of the same for armyworm (Spodoptera exempta) for forecasting; study of the biology of various cotton pests.
Publications: Annual report.
Projects: Biometrics (A.B. Chirembo); cotton breeding (C.F.B. Chigwe); cotton agronomy (M.B.F. Moyo); cotton entomology; armyworm forecasting (G.K.C. Nyirenda).

Mbawa Agricultural Research Station 12

Address: PO Box 8, Embangweni
Parent body: Ministry of Agriculture, Agricultural Research Department
Status: Official research centre
Head: P.M. Malia

Ministry of Agriculture 13

AGRICULTURAL RESEARCH DEPARTMENT 14

Address: PO Box 30134, Lilongwe, 3, Central Region
Telephone: 731 300
Telex: 4113 External M1
Status: Official research centre
Chief agricultural research officer: Dr J.T. Legg
Graduate research staff: 71
Activities: The policy of the department is to conduct applied research. In keeping with the national policy of fast agricultural development, agricultural research in Malawi is carefully orientated towards immediate application. There is no pure or academic research. Where basic research is required to provide a basis for the applied research this is normally reduced to a minimum by studying and applying data from basic investigations which have already been done elsewhere in the world. Administratively, the activities are organized under research stations which in turn have sub-stations and district posts or sites. The main research stations are Chitedze and Lifuwu in the Central Region; Bvumbwe, Makoka, Makhanga and Kasinthula and Ngabu in the Southern Region; and Mbawa, Lunyangwa and Baka in the Northern Region. Chitala is a big sub-station in the Central Region, while Bembeke, Tsangano and Lisasadzi are minor sub-stations. Limphasa, Bolero, Meru, Hara, Wovwe, and Kafikisila are small sub-stations or experimental sites in the Northern Region; and Thuchila in the Southern Region. There are also many shifting experimental sites in the districts where trials are conducted on farmers' gardens.
Publications: Annual report
Projects: Rotations and soil fertility (Dr H.K. Mwandemere, Dr J.H.A. Maida); agricultural chemistry (Dr H.K. Mwandemere, Dr J.H.A. Maida); soil acidity and liming (Dr J.H.A. Maida, Dr H.K. Mwandemere); maize agronomy (L.D.M. Ngwira); maize breeding (B.T. Zambezi); cotton entomology (G.K.C. Nyirenda); cotton breeding (C.F.B. Chigwe, H. Soko); cotton agronomy (M.B.F. Moyo, L.L. Sauti); groundnuts breeding (A.J. Chiyembekeza); groundnuts agronomy (C.E. Mailro); groundnuts pathology (C.T. Kisyombe); soyabean (D.J. Khonje); herbicides and weed control (P.M. Mnyenyembe); tung (W.T. Gondwe); coffee (M.N. Msanjama); cocoa (A.E.D. Muwowo); sugarcane (A.S. Kumwenda); essential oils; vegetables (B.A.M. Malunga); fruits and nuts (E.H.C. Chilembwe); burley tobacco, dark fired tobacco, oriental tobacco (Tobacco Research Authority); sorghum, bulrush, pearl millet (E.M. Chinthu, J.K.T. Kumwenda); finger millet (D.R.B. Manda); fodder crops, pasture (H.D.S. Msiska); phaseolus beans (Dr O.T. Edge); wheat, barley (P.H. Mnyenyembe); rice (A.S. Kumwenda); minor oil crops (P.H. Mnyenyembe); mono grain legumes (C.S.M. Chanika); European potatoes (W.T. Gondwe); other root crops (R.N. Sauti); irrigation (E.S. Mwafulirwa); farm machinery and tools (P.M. Matthews); animal

husbandry: small stock (A.P. Mtukuso); cattle husbandry (A.P. Mtukuso); cattle breeding (M.B.B. Kasowanjete); pigs (Dr P.E. Makhambera); poultry (Bunda College); flue cured tobacco (Tobacco Research Authority); spice crops (B.A.M. Malunga); pharmaceutical crops (Dr J.D. Msonthi); pasture legume, microbiology (D.J. Khonje); unit farms (R. Chikwana); Makande soil research (P. Panje); economics (R. Chikwana); phytosanitary (G.M. Chapola); entomology (A.D. Gadabu); nematology (Dr V.M. Saka).
See separate entries for: Baka Agricultural Research Station
Bvumbwe Agricultural Research Station
Chitala Agricultural Research Station
Chitedze Agricultural Research Station
Kasinthula Agricultural Research Station
Lifuwu Agricultural Research Station
Lunyangwa Agricultural Research Station
Makhanga Agricultural Research Station
Makoka Agricultural Research Station
Mbawa Agricultural Research Station
Ngabu Agricultural Research Station

Ngabu Agricultural Research Station   15

Address: Private Bag, Ngabu
Parent body: Ministry of Agriculture, Agricultural Research Department
Status: Official research centre
Head: J.D.T. Kumwenda
Activities: Agronomy and variety evaluation of sorghum, millet, guar beans, maize, groundnuts, cotton and cowpea.

Tea Research Foundation of Central Africa   16

Address: PO Box 51, Mulanje
Telephone: (486) 218; 224; 255
Status: Independent research centre
Director: Dr R.T. Ellis
Sections: Agronomy, W.J. Grice; horticulture, I.P. Searborough; biochemistry and technology, Dr J.B. Cloughley; plant improvement, V.E.T. Ridpath; plant pathology, Dr P.S. Rattan; entomology, B. Mkwaila
Graduate research staff: 14
Activities: Promotion of research into all matters concerned with the production of tea in the region of central Africa and other areas with similar climatic conditions: fertilizers, pruning, shade, irrigation, planting and establishment, plucking, soil and water conservation,

plant breeding, horticultural work, chemistry, biochemistry, small scale factory, physiology, zinc, economics, plant pathology and entomology.
Publications: Annual report; quarterly newsletter.
Projects: Selection of clones and their methods of production (V.E.T. Ridpath); investigating the potential of tea seed oil; investigating the biochemical factors affecting the quality of tea in the manufacturing process (Dr J.B. Cloughley); investigating the factors affecting the more effective use of fertilizers by growers with below-average yield (W.J. Grice); investigating pest control with special reference to isolated small tea gardens (Dr P.S. Rattan). fertilizers; irrigation works; soil conservation; water resources; plant genetics; horticulture; chemistry; biochemistry; plant anatomy; agricultural economics; zinc; insects; tea; tea pests; tea diseases; tea seed oil; cloning; chemistry agricultural; biochemistry agricultural.

University of Malawi   17

Status: Educational establishment with r&d capability

BUNDA COLLEGE OF AGRICULTURE   18

Address: PO Box 219, Lilongwe
Telephone: 721455
Principal: Dr N.F. Lungu
Departments: Agricultural engineering, Professor Boshoff; crop production, Professor O.T. Edje; livestock production, Professor M. Butterworth; rural development, Professor R. Billingsley
Activities: Plant production: beans, tobacco, maize, groundnuts, oil seeds, plant breeding, crop husbandry, plant protection, development of orchards and pastures; animal production: cattle, goats, pigs, rabbit, sheep, chickens, Muscovy ducks, livestock husbandry, nutrition, veterinary medicine; agricultural engineering: building food grain storage structures, minimum tillage and appropriate technology; agricultural economics: economics of cotton production and marketing in Malawi; economics analysis of factors affecting labour supply in Malawi smallholder agriculture.
Publications: Research bulletin (annual).
Projects: Bean research (Professor O.T. Edje); pig research (Dr P.E. Makhambera); rabbit research (Dr J. McNitt, Dr P.E. Makhambera); nutritive value and protein quality (Dr T. Ngwira, Professor M.H. Butterworth); comparison of extension services in two agricultural settings (Dr I.U. Chaudhry, R.M. Mkandawire); evaluation of the groom equipment and Eicher tractor for Malawi conditions (S.J. Temple, M.L. Mwinjilo).

MALAYSIA

Forest Research Centre 1

Address: Kuching, Sarawak
Telephone: (082) 22161
Affiliation: Department of Forestry
Status: Official research centre
Assistant director (research): H.S. Lee
Research units: Botany, P.P.K. Chai; entomology, A.A. Hamid; pathology, F.H. Chin; soils, C. Phang; silviculture, K.K. Lai
Graduate research staff: 15
Activities: Increased productivity of logged areas in the hill, peat-swamp and mangrove forests; restoration of productive and protective functions of areas deafforested by shifting cultivation. Ecology and silviculture are the main areas of research and these are aimed at an understanding of the ecological effects of human disturbances (logging and deafforestation) on forest ecosystems and therefore the identification of appropriate forest management and regeneration techniques.
Publications: Annual report.
Projects: Silviculture of hill forests; agro-forestry (H.S. Lee); silviculture of swamp (peat and mangrove) forests (K.K. Lai); forest botany and ecology (P.P.K. Chai); plantation silviculture (A. Naruddin); soils survey and plant nutrition (C. Phang); entomology and pest management (A.A. Hamid); diseases in forests and nurseries (F.H. Chin).

Forest Research Institute 2

Address: Kepong, Selangor
Telephone: (03) 662633
Affiliation: Forestry Department, Peninsular Malaysia
Status: Official research centre
Director: Dr Salleh Bin Mohd Nor
Deputy Director Forest Research: Dr Francis Ng Say Pink
Deputy Director Forest Products: Peh Teik Bin
Activities: The Forestry Research Division of the institute carries out research to develop better ways of growing trees and managing forests. The core activity is silviculture supported by a number of disciplines: botany, plant physiology, ecology, entomology, plant pathology, tree improvement, soil science, biometrics and hydrology.
The Forest Products Division's aim is more efficient and diversified utilization of the forest resources. To achieve optimal utilization of timber it is supported by disciplines such as engineering, physics, chemistry, anatomy, entomology and mycology which look into the mechanical, physical and chemical utilization of wood.
Projects: Forestry economics (Abd. Rauf Salim); forestry product economics (Abd. Rauf Salim).

ADMINISTRATION AND RESEARCH 3 SERVICING

Assistant director: Henry van S. Foenander

FOREST BIOLOGY AND PROTECTION 4 BRANCH

Assistant director: Dr Francis Ng Say Pink
Projects: Forest biology, conservation and protection.

FOREST PLANTATION BRANCH 5

Head (acting): Dr Salleh Mohamed
Projects: Plantation forest production.

NATURAL FOREST BRANCH 6

Assistant director: Johari Baharuddin
Projects: Natural forest production; community forest production.

WOOD CHEMISTRY BRANCH 7

Assistant director: Wong Wing Chong

WOOD PROPERTIES AND MACHINING BRANCH 8

Assistant director: Wong Tuck Meng
Projects: Wood waste utilization (Wong Tuck Meng); timber technology development (Wong Tuck Meng); utilization of rubberwood, coconut and oil palm stems as a supplement to natural timber supply (Wong Tuck Meng).

WOOD PROTECTION BRANCH 9

Assistant director: K. Daljeet Singh
Projects: Minor forest products (K. Daljeet Singh).

Institute Penyelidikan dan Kemajuan Pertanian Malaysia 10

– MARDI
[Malaysian Agricultural Research and Development Institute]
Address: Bag Berkunci 202, Unipertama, Serdang, Selangor
Telephone: (03) 356601-12
Status: Official research centre
Director-General: Dato Haji Mohd Tamin bin Yeop
Deputy director-general (research): Dr Mohd Yusof b. Hashim
Deputy director-general (administration): Kamaruzzaman b. Abd. Halim
Graduate research staff: 412
Activities: The overall objective of MARDI is to intensify its research activities according to the research priorities identified. Specifically the objectives of the institute are: increased food productivity and crop diversification; new technologies in farming which can be easily adopted at the local farm level without causing undue financial or other constraints on the farmers so that they can gain maximum return from their inputs; intensified research in food processing and preservation to maximise utilization of agricultural products and reduce wastage in food and non-food production arising from handling, storage and transportation; publication and distribution of research information to extension agencies, farmers' associations, relevant agencies and farmers as a whole in an effort to disseminate new technologies and modernize Malaysian agriculture; establishment of machinery at state and federal level to bring about effective coordination in research, extension and development.
Publications: Annual report; *MARDI Research Bulletin* (biannual).

AGRICULTURAL PRODUCT UTILIZATION DEPARTMENT 11

Director: Hashim b. Hassan
Activities: Plant products, animal products, applied engineering, quality control, advisory services.

ANIMAL PRODUCTION DEPARTMENT 12

Director: Tuan Syed Ali b. Abu Bakar

ANNUAL CROP PRODUCTION DEPARTMENT 13

Director: Hashim b. Abdul Wahab
Activities: Paddy, field crops, tobacco, vegetables.

CENTRAL SERVICES DEPARTMENT 14

Director: Omar b. Abd. Razak
Activities: Information, library, publications, central workshop, training.

DEVELOPMENT RESEARCH DEPARTMENT 15

Activities: Northern, central, southern and eastern region.

FUNDAMENTAL RESEARCH DEPARTMENT 16

Director: N.T. Arasu

PERENNIAL CROP PRODUCTION DEPARTMENT 17

Director: Dr Shariff Ahmad
Activities: Cocoa and coconuts, fruits, tea, pepper.

RESEARCH SERVICES DEPARTMENT   18

Director: Dr Mohd Hashim b. Mohd Noor
Activities: Production economics, sociology and agribusiness, statistics, analytical services, data processing services.

Institut Penyelidikan Getah Malaysia   19

– IPGM
[Rubber Research Institute of Malaysia]
Address: PO Box 150, Kuala Lumpur, 01-02, Selangor
Telephone: (03) 467033
Telex: RRIM MA30369
Status: Independent research centre
Director: Dr Ani bin Arope
Deputy director (research): Dr E.K. Ng
Deputy director (administration): Abdul Wahab bin Abdullah
Graduate research staff: 200
Activities: Research into all aspects of natural rubber cultivation and latex production, the development of new forms of rubber and rubber consumption, as well as technological and end use research in the processing and manufacture of natural rubber products. Advisory and information services extend the benefits of this research and development to all the industry.
Publications: Annual report; *Journal Sains* (annual); journal (3 per year); quarterly journals, including *Planters' Bulletin*; technology bulletin (biannual); monthly statistical bulletin.

DEPARTMENT OF BIOLOGY   20

Assistant director: Dr Abdul Aziz bin Sheikh Abdul Kadir

Crop Protection and Microbiology Division   21

Head: Dr C.K. John
Sections: Pollution and microbiology, Dr Ahmad bin Ibrahim; Crop protection, Dr Lim Tow Ming
Activities: Diseases and pests that attack the rubber tree; microbiological studies of latex, dry rubber, factory effluents and soil.

Plant Science Division   22

Head: Dr Yoon Pooi Kong
Sections : Fundamental physiology, Dr J.B. Gomez; applied physiology, Dr Leong Sook Kwai; breeding and selection, Dr Tan Hong
Activities: Breeding and planting methods; tapping systems; fundamental physiological problems of latex production; botanical aspects of ground cover and other plants associated with Hevea cultivation.

Soils and Crop Management Division   23

Head: Dr E. Pushparajah
Sections: Crop management, Dr K. Sivanadyan; soil survey and lland use, Chan Heun Yin; soil chemistry, Dr Soong Ngin Kwi; international SALB unit (Brazil), Dr Chee Kheng Hoy
Activities: Surveying, mapping classification and evaluation of capability, fertility and physical characteristics of soils under rubber; crop management, including investigations on nutrition of Hevea in relation to clone growth, production and soils and improvement in agro-management practices to improve crop productivity.

Tapping and Exploitation Physiology Division   24

Head: Dr P.D. Abraham
Sections: Tapping, Ismail bin Haji Hasim; exploitation physiology, Dr S.W. Pakianathan

DEPARTMENT OF CHEMISTRY AND TECHNOLOGY   25

Assistant director: Dr Sekaran Nair

Engineering and Testing Services Division   26

Head: Dr Mohinder Singh
Sections: Analytical, Chin Hong Cheaw; physical, Dr Yeoh Oon Hock; instrumentation and engineering services, Dahlan bin Hassan Basri

Polymer Research and Process Division   27

Head: Dr A. Subramaniam
Sections: Non-rubbers biochemical reactions, S.J. Tata; physics, Dr Abdul Kadir bin Mohamad; special rubbers, Dr Sidek bin Dulngali; modified rubbers/chemical reactions, Dr Loo Cheng Teik; characterization, Dr Esah Yip; factory/process development, Ong Chong Oon; latex, Wong Niap Poh

Product Development Division 28

Head: Sin Siew Weng
Sections: Tyres and retreading, Loh Pang Chai; non-tyres, Lim Hun Soo; industrialization TAS, Dr Wan Idris bin Wan Yaacob

Specifications and Quality Control Division 29

Head: Amlir bin Aziz
Sections: Standards and testing, Dr Loke Kum Mun; advisory inspectorate, Ong Chin Teck
Activities: Maintenance of the quality of Standard Malaysian Rubber; provision of testing standards for Standard Malaysian Rubber; provision of technical aids to Standard Malaysian Rubber laboratories; coordination of testing for technically-specified natural rubber between international laboratories.

DEPARTMENT OF RESEARCH AND 30 DEVELOPMENT SUPPORT SERVICES

Assistant Director: Dr Ariffin bin Mohammed Nor
Visiting manager: Chen Kim Koy

Applied Economics and Statistics 31 Division

Sections : Economics, Mohd. Ghazali bin Mohd. Nor; statistics, G.C. Iyer
Activities: Resource allocation: projects undertaken include factors affecting differences in smallholder productivities, and relative merits of replanting with different planting materials and techniques; processing and marketing; rural development; service to the industry: costing and management studies, commercial registration of clones and planting materials; applied statistics and methodology; computer application: survey and experimental results are analysed.

Experiment Station 32

Address: Sungei Buloh, Selangor
Acting manager: Hamidy bin Mohd.
Activities: Experiments on breeding and selection, small-scale trials of hand pollinated seedling families and large-scale trials of selected clones.

RRIM Estate 33

Address: Bukit Ibam
Acting manager: Wong Tee Kiang

RRIM Station 34

Address: Kota Tinggi, Johore
Acting manager: Abdullah bin Hassan
Activities: Experiments were first set up in early 1972. They are mainly on selection of parent mother trees from the large population of clonal seedling trees available. Since then, more trials have been established, including experiments on tapping and yield stimulations, immaturity reduction, planting methods, root disease investigations, weed control and manuring.

DEPARTMENT OF SMALLHOLDERS 35 EXTENSION AND DEVELOPMENT

Assistant director: Dr Samsudin bin Tugiman

Project Development and 36 Implementation Division

Head: Mohd. Shariff bin Kudin
Sections: Replanting and newplanting technology, Tan Peng Hua; exploitation, Dr Najib Lofty Arshad; project implementation, Mohd. Yusof Azaldin bin Taat; processing and marketing, P.S. Rama Rao; training, Cho Shue Nam; project/development evaluation, Lim Bock Tong; extension publication/information, vacant

Pusat Penyelidikan 37 Kehutanan

– PPK
[Forest Research Centre]
Address: 1407 Sepilok, Sandakan, Sabah
Telephone: (089) 44179
Telex: MA 82016
Status: Official research centre
Senior research officer: A. John Hepburn
Sections: Ecology, C.G Phillipps; plantation, Rita Tang; utilization, Chan Hing Hon; wood technology, Dr Ong Seng Heng; chemistry, Dr T.B. Ponnudurai; entomology, Professor R Yoshii; silviculture; tree improvement; seeds; botany; soil
Graduate research staff: 11
Activities: Botany, ecology and silviculture of natural forest; development of plantations both exotic and indigenous including soil aspects; tree improvement and seed source establishment of both exotic and indigenous species; insect pests of trees in particular in plantations; utilization research with emphasis on the lesser known species.
Publications: Lapuran Penyelidik Hutan, biannual.
Projects: Seed source establishment and tree improvement project (N. Jones); sawmilling, kiln drying and wood preservation research (D.K. Gough).

Universiti Pertanian Malaysia 38

– UPM
[University of Agriculture, Malaysia]
Address: Serdang, Selangor
Telephone: (03) 356101/ 10
Telex: MA 37454 UNIPER MALAYSIA
Status: Educational establishment with r&d capability
Vice-Chancellor: Dr Rashdan bin Haji Baba
Graduate research staff: 729
Activities: Multidisciplinary collaborative and problem-oriented research, both fundamental and applied, to optimize the use of resources and to stimulate interchange of ideas: agricultural engineering, agriculture, fisheries and marine science, forestry, resource economics and agribusiness, science and environmental studies, veterinary medicine and animal science, education.
Publications: Annual report.

FACULTY OF AGRICULTURAL ENGINEERING 39

Dean: Choa Swee Lin
Projects: Low-cost materials of construction (Abang Abdullah, Syed Mansor); harvesting and collection of oil palm (Wan Ishak, Dr Zohadie); power from low-speed wind turbines (Fuad Abas); investigations into trickle irrigation in Malaysia (Amin Mohd Soom, Rashidi); drying constant of rough rice as affected by air temperature, velocity and relative humidity (Johari Endan).

Department of Engineering Science, Processing and Environmental Engineering 40

Head: Hussain bin Mohd. Salleh

Department of Field Engineering, Farm Power and Machinery 41

Head: Dr Mohd. Zohadie bin Bardaie

FACULTY OF AGRICULTURE 42

Dean: Dr Khalid bin Nor
Projects: Storage and deterioration of cocoa seeds (Hor Yue Luan); effects of environment and applied growth substances on the growth and development of winged bean (Wong Kai Choo); evaluation of nutrient status of Malaysian soils (Sharifuddin Hj. Abd. Hamid); biochemistry of post-harvest cocoa (Munusamy Padavatan); integrated studies on groundnut production (Dr E.S. Lim, A. Rajan, Mr Surjit); UPM integrated soil research programme (Anuar Rahim, Dr Wan Wulaiman Wan Harun, Sharifuddin Abdul Hamid, Lim Kim Huan, G. Maesschalck); moisture and thermal regime of Malaysian soils (G.G. Maesschalck, Lim Kim Huan); diseases of mango, durian and cashew (Dr Lim T.K.); seed-borne pathogen of selected grain legumes in Malaysia (Wan Zainun W. Nik); ecology and control of bacteria wilt (Hiryati Abdullah); production of industrial alcohol (Dr M. Salleh Ismail); improved quality of locally-produced salted dried fish and sambal daging (an intermediate moisture food) (Dr Yu Swee Yan); solar drying of food commodities (Dr M. Salleh Ismail); water, fertility and soil management of annual crops in rotation (Dr Mok Chak Kim); nutrient status of some selected rice, soil under different conditions (Alias Husin); role of extracellular enzymes in pathogenesis (Dr Sariah Meon); production of vinegar from waste of fruits and vegetables (M. Ismail Karim); nutritional, varietal, cultural and post-harvest studies with local and newly-introduced vegetable crops (Dr M. Md. Ali).

Department of Agronomy and Horticulture 43

Head: Dr Lim Eng Siong

Department of Food Science and Technology 44

Head: Asiah bte Mohd. Zain

Department of Plant Protection 45

Head: Dr Abdul Rahman bin Razak

Department of Soil Science 46

Head: Dr Wan Suleiman bin Wan Harun

FACULTY OF FISHERIES AND MARINE SCIENCE 47

Dean: Dr Ang Kok Jee
Projects: Larval rearing of Microbrachium resenbergi (Dr Ang Kok Jee); heavy metal distribution and their effect on aquatic biota in Kelang Estuary (Dr Law Ah Theem, Amargit Singh); survey of the coral reefs of Pulau Kapas, Trengganu (Dr M.W.R.N. De Silva, Ahmad Zohri, Mohd. Zaki, Lokman bin Shamsudin, Tuan Hj. Umar); the photosynthetic quotient, carbon to nitrogen ration, and the utilization of different sources of nitrogen to a given algal community (Lokman bin Shamsudin); cage culture of fresh-water fishes (Dr Ang Kok Jee); estimation of population size, growth, mortality and distribution of food fishes in a Malaysian lake/river (Mohd. Azmi Ambak).

Marine Science Centre   48

Address: Kuala Trengganu
Head: Mohd Zaki bin Mohd. Said

FACULTY OF FORESTRY   49

Dean: Professor Dr Abdul Manap bin Ahmad
Projects: Ecology of hill forest timber species with particular reference to Seraya (Shorea curtisii) (Dr Kamis Awang, Mohd. Basri Hamzah); tropical mycorrhiza (Professor Abdul Manap Ahmad, M. Noor Shamsuddin); biomass distribution, productivity and nutrient cycling in three or four major forest types (Lim Meng Tsai, Dr Kamis Awang); growth pattern of Pinus caribaea var. hondurensis in Malaysia (P.B.L. Srivastava, Sheikh Ali Abod); insitu timber consumption studies in rural housing (Razali Abd. Kader, Mohd. Zain Jusoh); natural regeneration of mangrove species (Ashari Muktar, P.B.L. Srivastava); root growth studies of Pinus caribaea seedlings (Sheikh Ali Abod).

Department of Forest Management   50

Head: Ashari bin Muktar

Department of Forest Production   51

Head: Mohd. Zin bin Jusoh

FACULTY OF RESOURCE   52
ECONOMICS AND AGRIBUSINESS

Dean: Dr Mohd. Ariff bin Hussein
Projects: Economics of fertilizer production, distribution and use in a longer term perspective (phase II) (Dr Mohd. Ismail Ahmad); role of fragmentation in determining the rate of utilization and intensity of farming in the agricultural sector of Malaysia (Dr Syed Hamid Aljunid); impact of Padi price subsidy on the marketing and production behaviour of Padi in the Muda Area (Dr Ahmad Mahdzan Ayob, Dr Fatimah Mohd Arshad); agriculture, agribusiness and economic development: a case study of Malaysia (Professor R.T. Shand).

Department of Economics   53

Head: Dr Ahmad Mahdzan bin Ayob

Department of Management Studies   54

Head: Zainal Abidin bin Kidam

Department of Natural Resource   55
Economics

Head: Dr Syed Hamid bin Syed Aljunid

FACULTY OF SCIENCE AND   56
ENVIRONMENTAL STUDIES

Dean: Professor Dr Ariffin bin Suhaimi
Projects: Improvement of protein content of grains in Mungbean (Dr Quah Soon Cheang, Dr Ithnin Bujang); biochemical genetic markers in Malaysian and Malaysia's animal and plant species (Dr Gan Yik Yuen, Dr Tan Soon Guan, Faridah Abdullah); investigation of dry weight measurement of rubber latex using microwave technique (Kaida Khalid, Jamil Suradi, Samsudin Mahmood); electrochemical reduction of organic compound (Yap Kon Sang, Dr Anuar Kasim); effects of environmental stress on gonadotrophins (GnRH) sex steroids and Adrenocorticotrophin (Zolkepli Othman).

Department of Biochemistry and   57
Microbiology

Head: Dr Abdullah bin Sipat

Department of Biology   58

Head: Dr Mohamad bin Awang

Department of Chemistry   59

Head: Dr Badri bin Muhammad

Department of Environmental Studies   60

Head: Dr Mohd. Ismail Yaziz

Department of Mathematics   61

Head: Dr Nawi bin Abd. Rahman

Department of Physics   62

Head: Dr Mohd Yusof bin Sulaiman

FACULTY OF VETERINARY MEDICINE   63
AND ANIMAL SCIENCE

Dean: Professor Syed Jalaludin bin Syed Salim
Projects: Mineral status of livestock in Malaysia (Hew Peng Yew); comparative pharmocokinetics of some antimicrobial agents in normal and febrile animals (Abd. Salam Abdullah); breeding for the development of local strain/breed of cattle for improvement of beef and milk production (Dr Baharin Kassim, Dr Mahyuddin, Mak Tian Kwan); improving the reproductive efficiency of buffaloes (Dr M.R. Jainudeen, Tan Hock Seng, Arief Bongso); infertility studies in local and imported cattle (Tan Hock Seng); utilization of untreated and treated carbohydrates and fibrous residues as animal feeds (A. Razak, Dr R.I. Hutagalung, Zainal Aznam, Dr. M.I. Djafar).

Department of Animal Sciences   **64**

Head: Dr Mohd Mahyuddin bin Dahan

Department of Veterinary Clinical   **65**
Studies

Head: Professor M.R. Jainudeen

Department of Veterinary Pathology   **66**
and Microbiology

Head: Dr Sheikh Omar bin Abdul Rahman

MALI

Centre de Recherches Zootechniques de Sotuba   1

– CNRZ
[Sotuba Animal Husbandry Research Centre]
Address: BP 262, Bamako
Telephone: 22 24 49
Affiliation: Institut National de la Recherche Zootechnique, Forestière et Hydrobiologique
Status: Official research centre
Director: Dr Almoustapha Coulibaly
Departments: Genetic breeding, Dr Adama Traore; food and biochemistry, Dr Daouda De Allo; agrostology, Kalifa Kone; animal health, Dr Amadou B. Cisse
Graduate research staff: 20
Activities: Livestock breeding; livestock husbandry and nutrition; veterinary medicine; food science.
Publications: Annual report.

Centre National de Recherches Fruitières*   2

[National Fruit Research Centre]
Address: BP 30, Bamako
Status: Official research centre

Institut de Recherches du Coton et des Textiles*   3

[Cotton and Textile Research Institute]
Address: Bamako
Status: Official research centre

Institut des Recherches Agronomiques Tropicales et des Cultures Vivrières*   4

[Tropical Agriculture and Food Crops Research Institute]
Address: BP 438, Bamako
Status: Official research centre

MARTINIQUE

Institut de Recherches Agronomiques Tropicales 1

– IRAT
[Tropical Agriculture Research Institute]
Address: BP 427, Fort de France
Telephone: (596) 79 1705
Status: Official research centre
Head: M Daly
Departments: Horticulture, M Daly
Graduate research staff: 1
Activities: Plant production; vegetables; plant breeding; plant protection.
Publications: Annual report.
Projects: Varietal vegetable; weed control, vegetable.

Institut de Recherches sur les Fruits et Agromes 2

– IRFA
[Tropical Fruits Research Institute]
Address: BP 153, Fort de France, 97202
Telephone: (596) 71 9201
Affiliation: Ministry of Research
Status: Official research centre
Director: A. Guyot
Sections: Bananas, Ph Melin, J.L. Lachenaud; pineapples, J.J. Lacoeuiche; tropical trees and citrus, Y. Bertin; economics, Y. Dionis du Sejour; soils laboratory, M. Dormoy
Graduate research staff: 6
Activities: Natural resources and plant production: bananas, pineapples, citrus and other tropical fruits.
Publications: Annual report.

MAURITANIA

Institut de Recherches sur les Fruits et Agrumes   1

– IRFA
[Fruit and Citrus Research Institute]
Address: BP 87, Nouakchott
Status: Official research centre
Director: M. Reypercent
Activities: There are stations at Kaédi (pedology), and at Kankossa (dates)

MAURITIUS

Ministry of Agriculture, Natural Resources and the Environment 1

Address: Agricultural Services, Réduit
Telephone: (54) 1091
Status: Government department
Chief agricultural officer: B.D.N. Roy
Departments: Agronomy, A.L. Owadally; horticulture, I. Rajkomar; plant pathology, G.M. Lallmahomed; entomology, J. Monty; animal production, B. Hulman; dairy chemistry, P. Mardamootoo; land use and environment, M. Ramtohul
Graduate research staff: 48
Activities: Natural resources: soil science; plant production: tobacco, rice, tea, ginger, onion, garlic, chillies, bean, crucifers, sweet potato, sesame, citrus, strawberry, mango, letchi, papaya, anthurium; plant breeding: clonal selection and propagation in tea, selection of vegetable crops for local agro-climatic conditions; crop husbandry: weed control on garlic and onion, use of desuckering agents on tobacco, use of chemical maturing agents on tobacco, cultural practices of tomato, okra, cruciferous and leguminous crops and pepper, trial growth media on anthurium; plant protection: screening of all imported plant material and quarantine for seed, exploitation of disease resistance in tobacco, crucifers, tomato and eggplants, chemical control of diseases affecting tomato, tobacco, ginger, garlic and onion, identity, epidemiology and control of bacterial and virus disease of chilli, tobacco, citrus, leguminous and cruciferous crops, host-parasite relationship of root-knot nematodes affecting tomato, ginger and anthurium; animal production: cattle, pigs, goats and sheep; livestock breeding: cross breeding programmes in cattle suited for local conditions; livestock husbandry and nutrition: applied nutritional research in the field of ruminant and non-ruminant nutrition, optimization of utilization of local products and by-products in animal feeding with particular emphasis on the feeding of sugarcane by-products.

Publications: Annual report; technical bulletin (biannual).
Projects: Agronomy research on economic crops (A.L. Owadally); horticultural research on vegetables, fruits and ornamentals (I. Rajkomar); identity, prevention and phytosanitary control of plant diseases (G.M. Lallmahomed); fruit fly eradication project by the sterile male technique, biological control of stable flies, development of sericulture (J. Monty); milk production and processing MAR/75/004 in collaboration with FAO/UNDP (B. Hulman); study of the incidence of mastitis in Mauritius (P. Mardamootoo).

Ministry of Fisheries, Cooperatives and Cooperative Development 2

Address: Registrar-General's Building, 3rd Level, Port Louis
Status: Official research centre
Permanent secretary: P. Padayachy
Divisions: Research, M. Munbodh; protection, Y. Manuel
Graduate research staff: 6
Activities: To carry out research for the management and improvement of fisheries and aquaculture in Mauritius.
Projects: Fresh-water aquaculture: carps and prawns (D. Goorah); hatchery rearing techniques of prawns (V. Chineah).

Sugar Industry Research Institute 3

– SIRI
Address: Réduit
Telephone: (54) 1061-63
Status: Official research centre
Director: J.D. de R. de Saint Antoine
Principal scientific officer: Dr H.R. Julien
Assistant Directors: Dr C. Ricaud, Dr J.R. Williams
Sections: Plant breeding and biometry, Dr H.R. Julien; plant pathology, Dr J.C. Autrey; entomology, H. Dove; botany, Dr G.C. Soopramanien; cultural operations and weed agronomy, G. McIntyre; agricultural chemistry, J. Deville; food crop agronomy, Dr A.R. Pillay; sugar technology, S. Marie-Jeanne; field experimentation, R. Ng Ying Sheung; Mauritius herbarium, Dr H.R. Julien
Graduate research staff: 27
Activities: Cane production and sugar manufacture; production of foodcrops, particularly potatoes, maize, groundnuts and tomatoes by intercropping with cane.
Publications: Annual report.
Projects: Breeding and selection of cane varieties for general as well as for specific adaptation to environment, time of harvest and resistance to major pests and diseases (R. Julien); connected research projects: improving efficiency of utilization of fertilizers in sugarcane; cataloguing of cane fields for their rational utilization and for maximizing total agricultural production in Mauritius (J. Deville); artificial ripening of sugarcane throughout harvest; studies on the water requirements of sugarcane at different stages of growth for increasing the efficiency of irrigation - with particular emphasis on the development of drip irrigation. (G.C. Soopramanien); spacing in cane plantations for maximizing total agricultural production (G. McIntyre, A.R. Pillay); impact of new harvesting and loading operations on cane growth and yield; economic importance and more efficient control of weeds in sugarcane, with particular emphasis on problem weeds (G. McIntyre); control of gumming, yellow spot and smut diseases of sugarcane; improving local seed potato production; improving methods for testing sugarcane for resistance to major diseases (Dr J.C . Autrey); control of the sugarcane borer by the use of sex pheromones (H. Dove); determination of energy balance in sugar factories with a view to increasing electricity production; removal of suspended solids from mixed juice prior to weighing in order to obtain a more precise assessment of sucrose content; treatment of vinasse to reduce pollution from increased ethanol production (S. Marie-Jeanne); selection of potato varieties with resistance to major diseases and suitable for sale as well as for seed production under local constraints; breeding and selection of maize hybrids and varieties resistant to major pests and diseases for growing in different environments and at different times (Dr A.R. Pillay).

MEXICO

Dirección General de Economía Agrícola   1

[General Office of Agriculture Economics]
Address: Carolina 132, México, Distrito Federal
Telephone: (05) 98-54-90; 4-98-55-08
Telex: 5635463
Affiliation: Secretaría de Agricultura y Recursos Hidráulicos
Status: Official research centre
Director, agricultural economics: Efraín Niembro Carsi
Sub-director, statistics and economic studies: José Luis de la Loma y de Oteyza
Sub-director, commerce and international affairs: Francisco Ramos Cantoral
Graduate research staff: 200
Activities: Soil utilization; plant production (cotton, rice, barley, maize, soyabean, wheat, sesame, saffron, sorghum); breeding and production of cattle, sheep, goat, pig, horse, fowl, bees. The following programmes concern the above-mentioned species: agricultural, cattle farming and forestry statistics and product consumption; supply and demand; guaranteed prices; production costs; seed export; quality control; commercialization.
Projects: Vegetable and fruit exports and commercialization (D.G. Fuentes); study of rubber production and the world markets (P.M. Nava); study of production costs of the principal crops in irrigated and temporal zones (G.C. Gutiérrez).

DEPARTMENT OF CATTLE FARMING PRODUCTION   2

Head: Gilberto Castellanos Gutiérrez

DEPARTMENT OF ECONOMIC INTEGRATION AND BORDER ZONES   3

Head: Miguel Angel Medina Torres
Projects: Consumption of agricultural, cattle and forest products in border zones.

DEPARTMENT OF IMPORTS AND EXPORTS   4

Head: Hector Villalobos Cuellar

DEPARTMENT OF INTERNATIONAL RELATIONS IN CATTLE FARMING   5

Head: David Gómez Fuentes

DEPARTMENT OF INVESTIGATION AND PROGRAMMES   6

Head: Jorge Aragón Silva
Projects: Cultivar yield prediction by means of surveys based on probability studies (L.V. Morales).

DEPARTMENT OF NATIONAL CATTLE FARMING STATISTICS   7

Head: Isaac Gómez León

DEPARTMENT OF NATIONAL CATTLE FARMING STUDIES   8

Head: Alfonso Correa Coss
Projects: Study of supply and demand of the principal cattle products.

DEPARTMENT OF SURVEYS 9

Head: Luis Vazguez Morales
Projects: Impact of governmental stimulus given to agriculture in the temperate zones.

DEPARTMENT OF TECHNICAL INTERNATIONAL COOPERATION 10

Head: Pablo Cuellar Morales

Instituto Nacional de Investigaciones sobre Recursos Bióticos 11

– INIREB
[National Institute of Research on Biological Resources]
Address: Apartado Postal 63, Xalapa, Veracruz
Telephone: (0281) 7-52-94
Telex: 15542 INRBME
Status: Official research centre
General director: Dr Arturo Gómez-Pompa
Departments: Ecological planning and soil use, Silvio Olivieri Barra; technological development and biotic resources, Enrique Pardo Tejeda; scientific investigation services, Dr Lorrain Giddings
Graduate research staff: 39
Activities: Research in the following areas: natural resources, including soil science, and land use; plant production, including plant breeding, plant protection, crop husbandry; freshwater fisheries; forestry and forest products; agricultural, fisheries, forestry, and food economics.
Publications: Biotica (four numbers a year); *Flora de Veracruz*; *Madera y su Uso en la Contrucción.*
Projects: Mexican flora (Dr Arturo Gómez-Pompa); basic ecological studies (Dr Silvia del Amo); use of resources; aquatic biology (Dr Luis Morales Zavala); wood science and technology (Dr Ramón Echenique-Monrique); flora of Veracruz (Victoria Sosa); ornithology (Dr Mario Ramos); herbarium (Sergio Avendaño); fresh-water systems (Pedro Noriega); wood characterization (Raymundo Dávalos, José Erdoiza); regional studies; tropical biological resources; ecological reserves (Dr Artunro Gómez-Pompa).

Instituto Tecnológico y de Estudios Superiores de Monterrey* 12

[Technological and Higher Studies Institute, Monterrey]
Address: Sucursal de Correos J, Monterrey, Nuevo León
Telephone: (083) 582000
Status: Educational establishment with r&d capability
Divisions: Agricultural and marine sciences, L. Robles

Universidad Autónoma Chapingo 13

– UACH
[Chapingo Autonomous University]
Address: Km 38.5 Carrettera México-Texcoco, Texcoco
Telephone: (05) 585-02-40
Telex: UAC - 771230-998
Status: Educational establishment with r&d capability
General director: Rogelio Posadas del Río
Academic Directorship: Alfonso Ríos Angeles
Administrative Directorship: César A. Soto Martínez
Cultural Diffusion Directorship: Professor Juan Pablo de Pina García
Graduate research staff: 600
Publications: Annual report.
Projects: Research on the most frequently used agricultural parasiticides in Mexico to improve their use and efficiency (Antonio Segura); analysis of the structure, dynamic function and management of agricultural ecosystems of associated crops (Dr L. Krishnamurthy); production and certification of virus-free potato seed (Daniel Querol); stochastic models for flood analysis (Dr Enrique Cervantes); in vitro propagation of several species (Dr Remigio Madrigal); standardization of the production process of mushroom (Agaricus bisforus) (Salvador Martinez Romero); industrialization of Zea maize (Dr Pedro Valle Vega); survey of cattle production in six counties in the Veracruz forest region (Carlos de Luna Mata); ten year development programme.

AGRICULTURAL INDUSTRIES DEPARTMENT 14

Head: J. Rafael Campos Arredondo
Activities: The preparation of professionals competent to promote, plan and direct agricultural production of raw materials for industrialization.

AGRICULTURAL PARASITOLOGY DEPARTMENT 15

Head: Guillermo F. Lopez Aceves

AGRICULTURAL PREPARATORY DEPARTMENT 16

Head: Julio Goicoechea Moreno

ARID ZONES DEPARTMENT 17

Head: Maximino Sepulveda Bojorquez
Activities: The preparation of professionals in diverse disciplines related to adequate use of natural resources in the arid zones of Mexico.

ECONOMICS DEPARTMENT 18

Head: Ernesto Escalante Cames
Activities: The preparation of professionals competent in economic problems affecting agricultural growth and planning, taking into account rural and sociological phenomena.

FORESTRY DEPARTMENT 19

Head: Hugo Ramírez Maldonado
Activities: Acquisition of the technical knowledge necessary to utilize, preserve and promote all forestry resources, including wood products, wild fauna, and forage.

IRRIGATION DEPARTMENT 20

Head: Hermes Noyola Isgleas
Activities: Studies of the use of water as a fundamental factor in agricultural production, from four principal aspects: hydrological, hydraulic, structural and biological.

PLANT SCIENCE DEPARTMENT 21

Head: Agustín López Herrera
Activities: To increase agricultural production through efficient use of beneficial plants.

RURAL SOCIOLOGY DEPARTMENT 22

Head: Gerardo Gómez González
Activities: The orientation of teaching, learning and investigation towards knowledge of the laws that regulate the growth of nature and society.

SOIL SCIENCE DEPARTMENT 23

Head: Jorge L. Torar Salinas
Activities: The preparation of agronomists on a technical and social level, in order to achieve a larger and more efficient use of soil resources as important factors in agricultural productivity.

ZOOTECHNICS DEPARTMENT 24

Head: Ranulfo Castro Flores
Activities: The main purpose is to orient students towards attaining useful foods from animal sources.

Universidad Autónoma de San Luis Potosi* 25

[Autonomous University of San Luis Potosi]
Address: Alvaro Obregon 64, San Luis Potosi, SLP
Status: Educational establishment with r&d capability

DESERT LANDS RESEARCH INSTITUTE 26

Telephone: (481) 2-66-04
Director: Fernando Medelline-Leal
Departments: Botany, Sonia Salas De Leon; herbarium, Fernando Gomez-Lorence; zoology, Nicolas Vazquez-Rosillo; soils, Raul Grande-Lopez; hydrology, Crescencio Villalobos; phytochemistry, Aldo Torre; desertification, Refugio Ballin
Graduate research staff: 20
Activities: Research into the following areas: natural resources; soil science; hydrology; land use; plant production; plant protection; forestry and forest products.
Publications: Acta Científica Potosina (four issues a year).
Projects: Agricultural soils of arid lands of San Luis Potosi (Urbana); Ramirez-Ochoa genesis of soils of San Luis Potosi (Raul Grande-Lopez); entomological plagues in arid lands of San Luis Potosi (Nicolas Vazquez-Rosillo); hydrogeochemistry of arid lands in San Luis Potosi (Enrique Diaz De Leon); bacteriology of water (Yolanda Gallegos); erosion as an indicator of desertification (J. Carmen Rodríguez); plants as indicators of desertification (J. Refugio Ballin); phytochemistry of toxic to cattle plants (Aldo Torre); weeds of arid lands crops (Sonia Salas De Leon).

Universidad de Guadalajara* 27

[Guadalajara University]
Address: Avenida Juárez 975, Guadalajara, Jal
Status: Educational establishment with r&d capability

FACULTY OF VETERINARY MEDICINE 28 AND ANIMAL HUSBANDRY AT CIUDAD GRUZMÁN*

Dean: Professor A. Ramírez Alvarez

FACULTY OF VETERINARY MEDICINE 29 AND ANIMAL HUSBANDRY AT GUADALAJARA*

Dean: Professor R. Barba López

Universidad Nacional 30 Autónoma de México*

– UNAM
[National University of Mexico]
Address: Ciudad Universitaria, México, 20, DF
Telephone: (05) 50-5215
Status: Educational establishment with r&d capability

FACULTY OF VETERINARY MEDICINE 31 AND ZOOTECHNICS

Director: Juan Garza Ramos
Graduate research staff: 217
Activities: Research in the following areas: veterinary medicine; livestock breeding; livestock husbandry and nutrition; veterinary surgery; equine medicine and surgery; physiology and pharmacology; genetics; poultry production; porcine production; milk production; beef production; ovine production; caprine production; histology; preventive medicine; parasitology; pathology; animal reproduction; virology; immunology.
Publications: Revista Veterinaria-México (quarterly); Ciencia Veterinaria (annual).

Anatomy 32

Head: Santiago Aja

Animal Nutrition and Biochemistry 33

Head: Humberto Troncoso

Bacteriology and Mycology 34

Head: José López Alvarez
Projects: Resistance plasmids against E. coli.

Centre for Research, Teaching and 35 Extension of Production in the Tropics

Head: Javier Escobar
Projects: Milk and beef production in the tropics (J. Escobar); production systems for the goat (E. Suberbie).

Coordination of Scientific Research 36

Head: Alberto Saltiel Cohen

Economy and Administration 37

Head: Alfredo Aguilar

Equine Clinic 38

Head: Raúl Armendáriz

Genetics and Statistics 39

Head: Hilda Castro
Projects: Genetic polymorphism in cattle (Francisco Ayala).

Histology 40

Head: Nuria de Buen Lladó

National Centre for Teaching, 41 Research and Extension of Zootechnics

Head: Carlos Malagón
Projects: Use of industrial and agricultural by-products in bovine feeding.

Parasitology 42

Head: Ramón Meza Beltrán

Physiology and Pharmacology 43

Head: Javier García de la Peña

Pig Production 44

Head: Gilberto Lobo
Projects: Pig genetics (F. Quintana); production systems for pigs (José Doporto).

Pig Research Station 45

Head: Jorge López

Poultry Production 46

Head: Benjamin Lucio

Poultry Research Station 47

Head: Juan J. Romano

Preventive Medicine and Public Health 48

Head: Jorge Cárdenas

Reproduction 49

Head: Carlos Galina
Projects: Sexual behaviour of cattle in the tropics (C. Galina); reproductive activity of the mare (Alberto Saltiel).

Ruminants Production 50

Head: Enrique Sánchez Cruz

Sheep Station 51

Head: Carlos Barrón

Small Animals Clinic 52

Head: Jorge Padilla

Surgery 53

Head: Ciriaco Tista

Virology and Immunology 54

Head: Aurora Valázquez

Universidad Veracruzana* 55

[Veracruz University]
Address: Lomas del Estadio, Xalapa, Ver
Status: Educational establishment with r&d capability

**FACULTY OF VETERINARY MEDICINE 56
AND ANIMAL HUSBANDRY AT
VERACRUZ***

Director: Dr A. Mansicidor Ahuja

MOROCCO

Institut Agronomique et Vétérinaire Hassan II   1

[Hassan II Institute of Agronomy and Veterinary Medicine]
Address: BP 704, Rabat-Agdal
Telephone: (0717) 58159
Telex: AGKOVET 318737
Affiliation: Ministry of Agriculture
Status: Educational establishment with r&d capability
Director: Bekkali Abdallah
Graduate research staff: 200
Activities: Research in the following areas: soil science; drainage and irrigation; plant production (all crops); plant breeding; crop husbandry; plant protection; animal production (all types); livestock breeding; livestock husbandry and nutrition; veterinary medicine; agricultural engineering and building; fresh-water fisheries and food science; forestry and forest products.

ANIMAL PRODUCTION GROUP   2

Research secretary: Mr Guessons

ENVIRONMENTAL SCIENCES GROUP   3

Research secretary: Mr Yacoubi
Departments: Ecology and plant cultivation; earth sciences; water and forestry; zoology.

FOOD TECHNOLOGY GROUP   4

Research secretary: Mr Senhaji

FUNDAMENTAL SCIENCES GROUP   5

Research secretary: Mr Ettalibi

HYDRAULICS AND ENGINEERING GROUP   6

Research secretary: Mr El Abdellaoui
Departments: Mechanization.

HYGIENE AND CONTROL OF ANIMAL FOODSTUFFS GROUP   7

Research secretary: A. Belemlih
Departments: Pathological anatomy; Hidaoa

HUMAN NUTRITION AND FOOD ECONOMICS GROUP   8

Research secretary: Mr Essatara

TOPOGRAPHICAL SCIENCES GROUP   9

Research secretary: Mr El Ghaziani

TRAINING GROUP   10

Research secretary: Mr Pascon

VEGETABLE PRODUCTION GROUP   11

Research secretary: Mr Benchekroun
Departments : Agronomy; horticulture; plant health

VETERINARY MEDICINE GROUP   12

Research secretary: Mr Lahloukassi
Departments: Parasitology; avian pathology; medical and surgical pathology; reproduction

VETERINARY SCIENCES GROUP　13

Research secretary:　Mr Fassi-Fihri
Departments:　Anatomy; microbiology and contagious diseases; pharmacy and toxicology; physiology

Station de Recherches Forestières　14

[Forestry Research Station]
Address:　BP 763, Rabat-Agdal
Telephone:　(07) 727 59
Affiliation:　Direction des Eaux et Forêts et de la Conservation des Sols
Status:　Official research centre
Chief engineer for rivers and forests:　M. Tamri
Departments:　Entomology, M. El Yousfi; genetics, M. Raggabi; meteorology, Mr Laabdi; silviculture (exotics), C. Knockaert; silviculture (indigenous), A. Zaidi; soil sciences, T. Mandouri; statistics and biometrics, H. El Mazzoudi; wood technology, H. Zernij
Graduate research staff:　15
Activities:　Study of forest ecology; study of forest yield; the increase of forest yield, including genetics, fertilization, silvicultural treatments (thinning, pruning, spacings, length of rotation); wood utilization.
Publications: Annales de la Recherche Forestière au Maroc (annual report).

NEPAL

Forest Survey and Research Office   1

Address: Babar Mahal, Katmandu
Telephone: 14943
Status: Official research centre
Survey chief: E.R. Sharma
Graduate research staff: 34

RESEARCH DEPARTMENT   2

Head: B.P. Lamichhaney
Projects: Silvicultural research; management research.

SOIL AND LAND USE   3

Head: K.P. Prajapati
Projects: Soil and land use survey.

Trivahan University *   4

Address: Tripureswar, Katmandu
Telephone: 15313
Status: Educational establishment with r&d capability

INSTITUTE OF AGRICULTURE AND ANIMAL SCIENCE *   5

Director: Netra Bahadur Basnet

INSTITUTE OF FORESTRY *   6

Director: Mazrul Haq

NETHERLANDS

Akzo Naamloze Vennootschap*   1

Address: Velperweg 76, Postbus 60, 6800 AB Arnhem
Telephone: (085) 664422
Telex: 45204
Status: Industrial company

AKZO RESEARCH AND DEVELOPMENT*   2

Address: Velperweg 76, Postbus 60, 6800 AB Arnhem
Research director: Ir P.J.S.Th. Stehouwer
Activities: Coordination of research and development; responsibility for corporate research.

Bedrijfslaboratorium voor de Landbouw in Noord-Nederland*   3

[Laboratory for Agriculture in North Netherlands]
Address: Snekertrekweg 81, 8912 AB Leeuwarden
Telephone: (05100) 27641
Status: Official research centre
Research director: A. Ensing

Centraal Bureau voor Schimmelcultures   4

[Central Bureau of Mould Cultures]
Address: Oosterstraat 1, Postbus 273, 3740 AG Baarn
Telephone: (02154) 11841
Affiliation: Koninklijke Nederlandse Akademie van Wetenschappen
Status: Official research centre
Director: Dr J.A. von Arx
Graduate research staff: 16
Activities: Taxonomy of fungi; medical mycology/veterinary mycology; chemotaxonomy of fungi and yeasts; soil science (soil fungi); phytopathology; biodeterioration; entomogenous fungi; food science.
Publications: Annual reports, *CBS Studies in Mycology* (series of scientific publications); *List of Cultures* , 29 edition, 1978.
Projects: Collections of living fungus cultures; several projects in the field of taxonomy of fungi and applied mycology.

Centraal Diergeneeskundig Instituut   5

– CDI
[Central Veterinary Institute]
Address: Postbus 65, Lelystad
Telephone: (03200) 46664
Telex: 40227
Affiliation: Ministry of Agriculture and Fisheries
Status: Official research centre
General Director: Drs P.H. Bool
Graduate research staff: 55
Activities: Veterinary research and medicine; microbiological, biochemical and toxicological research (cows, sheep, swine, chickens, horses and mink).

BIOLOGICAL DEPARTMENT* 6

Address: Postbus 6007, 3002 AA Rotterdam
or: 3940 AA Doorn, Oude Rijksstraatweg 43; 8219 PH
Lelystad, Edelhertweg 13
Telephone: (010) 153911; (03430) 3641; (03200) 21914
Telex: 26583
Deputy Director: Dr J.M. van Leeuwen
Activities: Diseases and toxicological problems in animals.

CLINICAL DEPARTMENT* 7

Address: Edelhertweg 17, 8219 PH Lelystad
Telephone: (03200) 21525
Head: Drs F.Ph. Talmon

SPF/MDF BREEDING FARM* 8

Address: De Holle Bilt 12, 3732 HM De Bilt
or: Edelhertweg 14, 8219 PH Lelystad
Telephone: (030) 7606772; (03200) 22854
Heads: Dr J.B. van Dijk, Drs P. de Vrcy
Managers: L.S. Bakker, R. Venema
Activities: To breed SPF poultry and livestock for use
by the research departments.

VIROLOGY DEPARTMENT* 9

Address: Houtribweg 39, 8221 RA Lelystad
Telephone: (03200) 26814
Telex: 47907
Director: Professor Dr J.G. van Bekkum
Activities: Virus infections of livestock, especially those
diseases whose handling involves special precautions.

Centraal Proefdierenbedrijf 10
TNO*

– CPB-TNO
[Central Institute for the Breeding of Laboratory
Animals TNO]
Address: Postbus 167, 3700 AD Zeist
Telephone: (03439) 646
Affiliation: Netherlands Organization for Applied
Scientific Research (TNO)
Status: Official research centre
Research director: Dr H. Zwenk
Sections: Animal husbandry; genetics; microbiology;
pathology
Graduate research staff: 3
Activities: Genetics; husbandry; microbiology and
pathology of laboratory animals.

Centrum voor 11
Agrobiologisch Onderzoek

– CABO
[Centre for Agrobiological Research]
Address: Postbus 14, 6700 AA Wageningen
Telephone: (08370) 19012
Affiliation: Ministry of Agriculture and Fisheries
Director: Dr Ir P. Gaastra
Departments: Plant physiology, Dr H.M. Dekhuijzen;
chemistry, Drs N. Vertregt; Crop science, Dr Ir J.H.J.
Spiertz; weed science, Drs J.C.J. van Zon; vegetation
science, Ir Th. A. de Boer
Graduate research staff: 45
Activities: Basic processes of plants, crops and
ecosystems related to problems of crop husbandry, plant
breeding, plant protection, quality of roughage and
concentrates; quality of environment, and regional
management of land use.

CLO-Instituut voor de 12
Veevoeding 'De
Schothorst'*

[CLO-Institute for Animal Nutrition 'De
Schothorst']
Address: Meerkoetenweg 26, 8218 NA Lelystad
Telephone: (03200) 42194
Affiliation: Cooperative Concentrates Manufacturers'
and Farmers' Union
Status: Research association
General director: Ir J.P. Cornelissen
Director: Ir Tj. Bakker
Graduate research staff: 9
Activities: Nutrition of cattle, pigs and poultry particularly with respect to the composition and application
of concentrates.

Delta Instituut voor 13
Hydrobiologisch
Onderzoek*

[Delta Institute for Hydrobiological Research]
Address: Vierstraat 28, 4401 EA Yerseke
Telephone: (01131) 1920
Affiliation: Koninklijke Nederlandse Akademie van
Wetenschappen
Status: Official research centre
Director: Dr E.K. Duursma
Graduate research staff: 11

Activities: Effects of closing river-mouths and sea-arms; detailed surveys of flora and fauna; functioning of estuarine and lagoon systems in general.
Publications: Annual report.

DMV De Melkindustrie Veghel   14

Address: PO Box 13, 5460 BA Veghel
Telephone: (04130) 84411
Telex: dmv nl 50092
Status: Research centre within an industrial company
Dr F. Brouwer
Departments: Product development, Dr F. Visser, A.H.M. van Gennip; process technology, Drs J. v. Leverink; analytical and quality control department, Ir P. de Hoog
Graduate research staff: 20
Activities: Food science and research.

Duphar B.V. *   15

Address: Postbus 2, Weesp, 1380 AA
Telephone: (02940) 71110
Telex: 14359 duph nl
Parent body: Solvay & Cie, Belgium
Status: Industrial company
Technical manager: Dr J.L.M.A. Schlatmann
Divisions: Human health, E. Loman; animal health, B. Tolud; crop protection, Dr H. Geuens; vitamins and chemicals, W.Th. Van Aalzum.
Graduate research staff: 119
Activities: New human and veterinary health pharmaceuticals and vaccines; crop protection agents; vitamins and industrial chemicals; prefilled and automatic syringes.

Gelders-Overijselse Zuivelbond   16

– GOZ
[Dairy Federation (Dutch provinces)]
Address: 162/Nieuwstad 69, 7200 AD Zutphen, Gelderland
Telephone: (05750) 10727
Status: Independent research centre
Director/Secretary: Ing A.G. Faber
Departments: Chartered accountancy, Drs J. Gorter; taxes and statistics, Dr J van Hierden; organization and efficiency, Dr K. Haven; juridical and social affairs, H. E. Gewin; environment affairs: bacteriology/chemistry laboratories, G.J.M. Wolbers, J. Baltjes
Graduate research staff: 7
Activities: GOZ is a dairy federation in the provinces of Gelderland and Overijsel and all activities are advisory, to dairy factories.
Publications: Annual report.

TECHNOLOGICAL CENTRE   17

Address: Burgemeester Smeetsweg 1, Zoeterwoude
Research director: Dr P. van Eerde
Activities: Brewing technology.

Heidemaatschappij Beheer   18

Address: PO Box 33, 6800 LE Arnhem
Telephone: (085) 778436
Telex: 45623
Status: Independent consultancy
President: H.J. Hellema
Graduate research staff: 30
Activities: Improving enviromental conditions, particularly in the developing countries, by consultancy, design, site supervision, construction, contracting, real estate development, management, administration and trade services. The main fields of activity are soil technology and land development, agriculture, forestry and landscaping, hydraulic engineering, environmental technology, cartography, sports and recreation fields, physical and road planning, building design and construction.
Publications: Annual report, activity report.

H.J. Heinz BV* 19

Address: Stationsstraat 50, 6662 BC Elst, (Gelderland)
Telephone: (08819) 1857
Telex: 48043 hein nl
Affiliation: H.J. Heinz Company, Pittsburgh, USA
Status: Industrial company
General Manager, Research and Development and Quality Control: A. Nip
Sections: Equipment; ingredients; food products and processes
Graduate research staff: 5

Houtinstituut TNO* 20

– HI-TNO
[Forest Products Research Institute TNO]
Address: Schoemakerstraat 97, Postbus 151, 2600 AD Delft
Telephone: (015) 569330
Telex: 31660
Affiliation: Netherlands Organization for Applied Scientific Research (TNO)
Status: Official research centre
Head: Ing A. van der Velden
Departments: Materials, Ir J.F. Rijsdijk; technology and products, Ir B. van Heijningen
Graduate research staff: 14
Activities: Wood and wood-base materials, including analysis of structures and identification; drying properties; durability; wood chemistry; technological research.
Publications: Houtbulletin (two a year), *HouTNOtitie* (eight a year).
Projects: Anatomy and microtechnology, (Ing P.B. Laming); deterioration and conservation, (Ing J.W.P.T.v.d. Drift); wood preservation and chemistry, (J.v.d. Elburg); mechanical properties, (Ir H. Buiten; physical properties, (Ir J.F. Rijsdijk); machine wood-working, (Ing J. Kramer); tool maintenance, (A.K.A.v. Gentevoort); drying and moisture-measurement, (Ing H. Dijkstra); carpentry, (Ing G.N. Ruysch); furniture, (Ing H.W.A.M. Frenken), technical research, (H.O. Hoffmann).

Hubrecht Laboratorium, Internationaal Embryologisch Instituut* 21

[Hubrecht Laboratory, International Embryological Institute]
Address: Uppsalaan 8, 3584 CT Utrecht
Telephone: (030) 510211
Affiliation: Koninklijke Nederlandse Akademie van Wetenschappen
Status: Official research centre
Deputy director: Dr J. Faber
Sections: Embryogeny in amphibians, J.G. Bluemink; organogeny in mammals, K.A. Lawson; morphogenesis in cellular slime moulds, A.J. Durston; regulation of the cell cycle, S.W. de Last
Activities: Developmental biology in animals including cellular slime moulds. Methodologies: experimental morphology, ultrastructure, histo- and cytochemistry, organ culture, developmental physiology, biochemistry, biophysics. Systems: cleavage stages and early embryos of amphibians, foetal organs of mammals, later morphogenetic stages of Dictyostelium, murine neuroblastoma cells in vitro.

Instituut CIVO - Analyse TNO 22

[CIVO -Analysis Institute TNO]
Address: Postbus 360, Zeist
Telephone: (03404) 52244
Telex: 40022 civo nl
Affiliation: Netherlands Organization for Applied Scientific Research (TNO)
Status: Official research centre
Director: Drs W.J. Klopper
Activities: Chemical analyses of foodstuffs, amino acid analysis; aroma investigation.
Projects: Hygiene of food and fodder; environmental hygiene in agriculture and food industries; general analytical quality research of food stuffs; biotechnology.

NATIONAAL INSTITUUT VOOR BROUWGERST, MOUT EN BIER - TNO* 23

– NIBEM-TNO
[National Institute for Malting Barley, Malt and Beer - TNO]
Address: Postbus 109, 3700 AC Zeist
Telephone: (03404) 24688
Telex: 40022 civo nl
Status: Official research centre
Head: Ir J. Hiddema

Activities: Testing of malting barleys for agricultural and technological qualities; malts, hops, beers and yeasts; new production methods; soft drinks; wines; spirits.

Instituut CIVO - Technologie TNO   24

[CIVO -Technology Institute TNO]
Address: Postbus 360, 3700 AJ Zeist
Telephone: (03404) 52244
Telex: 40022 civo nl
Affiliation: Netherlands Organization for Applied Scientific Research (TNO)
Status: Official research centre
Director: Ir J.J. Doesburg
Activities: Composite foods; snacks, ready-to-eat meals, sauces, soups, toppings, sugar products; use of stabilizers, thickeners, proteins, flavourings; pollution; energy consumption.
Projects: Product oriented quality research or food stuffs, and food technology.

INSTITUUT VOOR VISSERIJPRODUKTEN TNO   25

– IVP-TNO
[Institute for Fishery Products TNO]
Address: Postbus 183, 1970 AD IJmuiden
Telephone: (02550) 19022
Telex: 40022 civo nl
Head: Ing H. Houwing
Sections: Chemistry, Dr J.B. Luten; technology, Ir B. Meyboom; processing, Ir J. Bon; microbiology, Ir K.J.A. van Spreekens; engineering, L. van Pel
Graduate research staff: 9
Activities: Improving the quality of fish and fishery products by means of organoleptical, chemical and microbiological evaluation; new products and techniques concerning fishery technology (including shellfish and crustaceae), both aboard and ashore; testing of materials (ropes, steel cables, nets, containers, packing materials) for the fishery industry; chemical and microbiological analyses; the former includes trace elements and biogenic amines. Technical assistance concerning product handling and processing equipment, machinery and design of fish handling and processing plants for the industry in the Netherlands and abroad.
Publications: Annual report (in Dutch).
Projects: Technology of blue whiting and molluscs, (Ir J. Bon); the use of w3 fatty acids for human health, analyses of trace biogenic amines in fish elements in fish, (Dr J.B. Luten); the occurrence of spoilage bacteria in fish (Ir K.J.A. v. Spreekens); cooling fish without ice, (L. van Pel); aid for developing countries (Ir B. Meyboom); repeated loading of ropes etc, (Ing M. Kotte).

NEDERLANDS CENTRUM VOOR VLEESTECHNOLOGIE*   26

[Netherlands Centre for Meat Technology]
Address: Zeist
Head: Ir E.J.C. Paardekooper
Activities: Cooling, freezing, thawing and cold transport of meat; determination of non-meat proteins in meat products; analysis of additives; water management and energy conservation in abattoirs and meat-product establishments.

Instituut CIVO - Toxicologie en Voeding TNO   27

[CIVO - Toxicology Institute TNO]
Address: Postbus 360, 3700 AJ Zeist
Telephone: (03404) 52244
Telex: 40022 civo nl
Affiliation: Netherlands Organization for Applied Scientific Research (TNO)
Status: Official research centre
Director: Dr R. Kroes
Activities: Analysis of pesticide residues in foodstuffs; analysis of substances which may be transferred to foods from packaging materials; instrumental analysis; possible toxic effects of substances inhaled, ingested through the mouth, or that come into contact with the skin; birth defects.
Projects: Medical aspects of nutrition.

Instituut TNO voor Verpakking*   28

– IvV-TNO
[Institute TNO for Packaging Research]
Address: Postbus 169, 2600 AD Delft
Telephone: (015) 569330
Telex: 38071 zptno nl
Affiliation: Netherlands Organization for Applied Scientific Research (TNO)
Status: Official research centre
Director: Ir C.J. Overgaauw
Activities: Industrial, retail, agricultural and horticultural packaging; packaging for dangerous goods; environmental studies; testing equipment.

Instituut voor Bewaring en Verwerking van Landbouwproducten 29

– IBVL
[Storage and Processing of Agricultural Produce Institute]
Address: Bornsesteeg 59, Postbus 18, 6700 AA Wageningen
Telephone: (08370) 19043, (08370) 12854 (organoleptic laboratory)
Telex: 45371 ibvl nl
Affiliation: Ministry of Agriculture and Fisheries
Status: Official research centre
Director: Ir J.C.F. Rynja
Sections: Potato storage and handling, Ir A. Rastovski, Ing P.H. de Haan
Biochemical research on potatoes, Ing A. van Es, Drs K.J. Hartmans
Quality aspects of potatoes for the processing industry, Ir J.C. Hesen, Ing C.P. Meijers, Ing P.S. Hak
Hygienic and quality aspects of processing potato products, Ing J.W. Ludwig, Dr Ir M.J.H. Keijbets
Potato processing technology, Ir P.H. Sijbring, Ing H.H.J. van Remmen, Dr Ir M.J.H. Keijbets
Drying and storing of grains, seeds and pulses, Ing P.H. de Haan, Ir H. Sparenberg, J. Jansen
Processing of straw, Ir G. Hofenk, Drs B.A. Rijkens, Professor Dr Ir W.F. du Bois
Production and processing of vegetable protein, Ir P.H. Sijbring, Ing A. van Es, Ir G.J.M. van Laarhoven
Grass and forage crops, Ir H.J. Leutscher, Ing J.J.M. Ogink, H. Timmers
Fibre crops, Ir H. Sparenberg, Ir H.J. Leutscher Professor Dr Ir W.F. du Bois
Graduate research staff: 31
Activities: Improvement of existing techniques and development of new techniques for storing, handling and processing of agricultural crops: potatoes, grains-seeds-pulses; green fodder crops; fibre crops; protein production from pulses; processing of solid (agricultural) wastes.

Instituut voor Bodemvruchtbaarheid 30

– IB
[Soil Fertility Institute]
Address: Postbus 30003, 9750 RA Haren Groningen
Telephone: (050) 346541
Affiliation: Ministry of Agriculture and Fisheries
Status: Official research centre
Director: Ir C.M.J. Sluijsmans
Graduate research staff: 35

Activities: Research on soils and fertilization for the benefit of crop production, grassland farming, and horticulture, to raise production and improve quality of agricultural and horticultural products.
Publications: Annual reports.

BOTANY SECTION 31

Head: Dr L.K. Wiersum
Activities: This section supports the other departments of the institute by making physiological studies of a more general or fundamental nature, such as the effect of aeration upon root growth, the morphology and activity of root systems, and the mechanisms of supply of nutrients to the root.

EXPERIMENTAL FARM IN THE NORTHEAST POLDER 32

Address: Dr H.J. Lovinkhoeve 12, Vollenhoverweg, Marknesse
Manager: J. Koning
Activities: Experiments and research on soil, plant and fertilizer samples.

FERTILIZATION IN HORTICULTURE DEPARTMENT 33

Head: Dr Ir H. Niers
Activities: Cultivation of vegetables under glass; new types of flowers; the use of artificial substrates; cultivation of nursery plants in containers, and methods of water supply; foliar application of fertilizers and related quality problems; seed bed preparation; fertilizer requirement of vegetables grown on regular agricultural soils.
Fruit: effect of nitrogen on yield and quality, effects of grass strips in or between rows of trees in orchards.

FERTILIZATION OF FIELD CROPS AND GRASSLAND DEPARTMENT 34

Head: Dr Ir K.W. Smilde
Activities: Use of inorganic (nitrogen, phosphorous, potassium, magnesium, minor elements and lime) fertilizers; fertilizer efficiency and effects on crop quality; accumulation and decomposition of organic matter; use of manure surpluses and of waste materials (town refuse, sewage sludge, fly ash); pollution of surface and ground water; heavy-metal problems in plant growth and chemical composition; fertilization of turf grass and of (ornamental) crops grown on peat substrates.

MATHEMATICAL STATISTICS SECTION   35

Head: J.T.N. Venekamp
Activities: Development and application of mathematical and statistical procedures to soil fertility problems; experimental designs.

SOIL BIOLOGY DEPARTMENT   36

Head: Dr H. van Dijk
Activities: Biodegradation of pesticides in the soil; chemistry of organic matter and pesticides; turnover dynamics of organic carbon, nitrogen and phosphorous compounds in the soil; pesticides and other organic micropollutants in the soil environment; biogeochemical aspects of organic heavy-metal complexes; antagonism of the soil microflora with respect to soil pathogens.

SOIL CHEMISTRY DEPARTMENT   37

Head: Dr A.J. de Groot
Activities: Pollution of heavy metals in rivers, lakes, estuaries, flood plains and dredge spoils; uptake of heavy metals by plants, and the possible consequences of this for humans and animals.

SOIL PHYSICS AND TILLAGE DEPARTMENT   38

Head: Dr Ir Th.J. Ferrari
Activities: Soil structure; soil tillage; wind erosion; development of mathematical simulation models for solving physical and chemical soil problems; distribution and function of minerals in the plant; root growth and development; nutrient uptake and availability.

Instituut voor   39
Cultuurtechniek en
Waterhuishouding

– ICW
[Institute for Land and Water Management Research]
Address: Postbus 35, 6700 AA Wageningen
Telephone: (08370) 19100
Telex: 75230 ICW
Affiliation: Ministry of Agriculture and Fisheries
Status: Official research centre
Director: Ir G.A. Oosterbaan
Sections: Land use planning, Ir C. Bijkerk; water quality, Dr P.E. Rijtema; hydrology, Dr J. Wesseling; soil technology, Dr G.P. Wind

Graduate research staff: 35
Activities: Hydrology; hydrogeology; drainage; salinity problems; aeration; land treatment of industrial wastewater; regional water management schemes; ground-and surface-water land uses: reallotment of land by computer; modelling of traffic on rural road network; outdoor recreation; multiple land use; short - and long-term effects of amelioration measures on the economy of farm and region; project evaluation; agricultural evaluation of land layout.
Publications: Research digest.

Instituut voor de   40
Veredeling van
Tuinbouwgewassen

– IVT
[Horticultural Plant Breeding Institute]
Address: Postbus 16, 6700 AA Wageningen
Telephone: (08370) 19123
Affiliation: Ministry of Agriculture and Fisheries
Status: Official research centre
Director: Ir C. Dorsman
Divisions: Vegetables, Dr N.G. Hogenboom; Fruit crops and roses, Dr T. Visser; Ornamentals, Dr L.D. Sparnaaij; Special research, Ir M. Nieuwhof
Graduate research staff: 45
Activities: Research into and development of efficient breeding methods creation and distribution of populations from which better varieties can be selected: vegetables: tomatoes, lettuces, cucumbers, sweet peppers, onions, cole crops, beans, carrots; fruits: apples, strawberries; flowers: roses, chrysanthemums, carnations, freesias, gerbera, gloriosa; bulbs: tulips, lilies, hyacinths, nerines; woody ornamentals: rhododendron, clematis.
Publications: Annual report.

Instituut voor Graan, Meel   41
en Brood TNO*

– IGMB-TNO
[Institute for Cereals, Flour and Bread TNO]
Address: Postbus 15, 6700 AA Wageningen
Telephone: (08370) 19051
Affiliation: Netherlands Organization for Applied Scientific Research (TNO)
Status: Official research centre
Director: Ir D. de Ruiter
Sections: Chemistry and cereals, Dr B. Belderok; technology, Dr A.H. Bloksma
Graduate research staff: 10

Activities: Testing of the quality of cereals and cereal foods; wheat improvement; the chemistry of food preparation from cereals; development and improvement of methods for the preparation of food and feed, particularly from cereals; research into the physical properties of raw materials and products; technical advice to the related industries.

Instituut voor Landbouwkundig Onderzoek van Industriële biologische, biochemische en chemische producten* 42

– ILOB
[Institute for Agricultural Research of Industrial Biological, Biochemical and Chemical Products]
Address: Postbus 9, 6700 AA Wageningen
Telephone: (08370) 19134
Telex: 45938 ilob nl
Affiliation: Netherlands Organization for Applied Scientific Research (TNO)
Status: Official research centre
Director: Dr Ir P. van der Wal
Activities: The Institute is a foundation on the board of which the Agricultural University, industry and TNO are represented. Research is in the field of animal nutrition.

Instituut voor Mechanisatie, Arbeid en Gebouwen 43

– IMAG
[Institute of Agricultural Engineering, Labour and Buildings]
Address: Postbus 43, 6700 AA Wageningen
Telephone: (08370) 19119
Telex: 45330 CTWAG
Affiliation: Ministry of Agriculture and Fisheries
Status: Official research centre
Director: Ir F. Coolman
Senior deputy director: Ir A. Duinker
Sections: Engineering, Ir J.J. Laurs; machinery, Ir G.J. Poesse; labour and work management, Drs K.E. Krolis; buildings, Ir W.H. de Brabander; climate control, Ir W.R. Mulder; crops liaison, Ir M.M. de Lint; horticulture and recreation liaison, Dr Ir G.H. Germing; livestock and environment liaison, Ing G. Postna; experimental farms, Ing C.A. den Otter; experimental gardens, G.F. van't Sant, B. Witteneen

Activities: Work organization, mechanization and automization in agriculture and horticulture; agricultural buildings and greenhouses; pollution and the environment. There are experimental farms at Slootdorp and Duiven; experimental gardens at Mansholten and Grebbedijk.

BUILDINGS DIVISION 44

Head: Ir W.H. de Brabander
Deputy head: Ir A.A. Jongebreur
Departments: Agricultural, Ir A.A. Jongebreur; construction, H. Zilverberg; structures, Ir W.H. de Brabander

CLIMATE CONTROL DIVISION 45

Head: Ir W.P. Mulder
Deputy head: Ir B.J. Heijna
Departments: Climate, Ir B.J. Heijna; control, Ir W.P. Mulder

CROPS LIAISON DEPARTMENT 46

Head: Ir M.M. de Lint

ENGINEERING DIVISION 47

Head: Ir J.J. Laurs
Deputy head: Ing J. Maring
Departments: Design, Ir J.J. Laurs; engineering principles, J. Maring.

HORTICULTURE AND RECREATION LIAISON DEPARTMENT 48

Head: Dr Ir G.H. Germing
Deputy head: Ir C.J. van der Post

LABOUR AND WORK MANAGEMENT DIVISION 49

Head: Drs K.E. Krolis
Deputy head: Drs H.K. Krijgsman
Departments: Management, Dr H.K. Krijgsman; operations research, Dr E. van Elderen; systems, Dr K.E. Krolis

LIVESTOCK AND ENVIRONMENT LIAISON DEPARTMENT* 50

Head: Ing G. Postma
Deputy head: Ir J.A.M. Voermans

MACHINERY DIVISION 51

Head: Ir G.J. Poesse
Deputy head: Ir A.R. Kraai
Departments: Field machinery and materials handling,
A. Bouman; livestock machinery, G.A. Benders; soil
tillage and traction, Ir G.J. Poesse; testing, Ir A.R.
Kraai

Instituut voor Oecologisch 52 Onderzoek*

[Institute for Ecological Research]
Address: Kemperbergerweg 67, 6816 RM Arnhem
Telephone: (085) 432841
Affiliation: Koninklijke Nederlandse Akademie van
Wetenschappen
Status: Official research centre
Director: Dr Ir J.W. Woldendorp
Graduate research staff: 17

Instituut voor 53 Plantenziektenkundig Onderzoek

– IPO
[Plant Protection Research Institute]
Address: Postbus 42, 6700 AA Wageningen
Telephone: (08370) 19151
Affiliation: Ministry of Agriculture and Fisheries
Status: Official research centre
Director: G.S. Roosje
Departments: Entomology, Dr Ir A.K. Minks; Mycol-
ogy and bacteriology, Dr Ir A. Tempel; Nematology, Dr
Ir J.W. Seinhorst; Virology, Drs F. Quak; Air pollution,
Dr A.C. Posthumus
Graduate research staff: 40
Activities: Research into control methods for plant
diseases and pests; crop protection; environmental
hygiene, especially the influence of air pollution on
plants; advancement of agriculture in developing coun-
tries, diagnosis methods in fungal, bacterial and virus
diseases, methods of freeing infected plants from viruses.
Publications: Annual report.

Instituut voor Pluimveeonderzoek 'Het Spelderholt' 54

[Spelderholt Institute for Poultry Research]
Address: Spelderholt 9, Postbus 15, 7361 DA Beek-
bergen
Telephone: (05766) 1808
Affiliation: Ministry of Agriculture and Fisheries
Status: Official research centre
Research director: Ir L.H. Huisman
Departments: Production research, Ir A.R. Kuit;
nutrition research, Ir W.M.M.A. Janssen; products
research, Dr Ir B. Erdtsieck
Graduate research staff: 25
Activities: All aspects of poultry research: incubation
and embryology; husbandry; feeds and feeding; poultry
meat; poultry meat products; eggs and egg products.
Publications: Annual report.
Projects: Microbiology, (Ing N.M. Bolder, Ir R.A.W.
Mulder); poultry feeding, (Ir J. Helder); product
research, (Drs A.R. Gerrits); sensoric research, (Drs
A.W. de Vries); broiler feeding, (Ing J.P. Holsheimer);
statistics, (J.W. van Schagen, Ir P.F.G. Vereijken);
technology (Ir C.H. Veerkamp); zootechnics, (Ir D.A.
Ehlhardt, Ing A. Zegwaard); biochemistry, detrimental
compounds, (Drs C.A. Kan); biochemistry, stress
research, (Dr G. Beuving); chemical product research,
(Drs A.C. Germs); consumption eggs, (Ing A. Ooster-
woud); veterinary research, (Drs U. Haye); egg products, (Ir Th. G. Uijttenboogaart); energy metabolism,
(Ing C.W. Scheele); ethology, (Ir H.J. Blokhuis);
physiology, ca-metabolism, (Dr Ir P.C.M. Simons);
genetics, egg production animals, (Ir W.F. van Tijen);
genetics, broilers, (Ir F. R. Leenstra); analytical chemistry, (Ir A.T.G. Steverink).

Institut voor Rationele 55 Suikerproduktie

[Institute for Rationalized Sugar Production]
Address: van Konijnenburgweg 24, PO Box 32, 3600
AA Bergen OP Zoom, Noord Brabant
Telephone: (01640) 34970
Telex: 78273
Status: Official research centre
Director: B.C. Bos
Departments: Agricultural, Ir J. Jorritsma; biological,
Drs W. Heijbroek; chemistry, Ir A. Huijbregts; statis-
tics, L. Withagen; byproducts, Ing J. Haaksma; docu-
mentation, F. van Gils; administration, A. van
Waadenoijen
Graduate research staff: 7

Activities: Research into sugarbeet (beetgrowing techniques, including mechanization; beet quality; weed control; disease protection; value of by-products).
Projects: Research on cultural value of sugar beet varieties; testing of· suitability for sowing of sugar beet seed; research on properties of sugar beet seed other than suitability for sowing; analysis of pelleted beet seed treated with chemical compounds; establishment of an advisory scheme for nitrogen fertilization of arable land on the basis of soil testing; evaluation of the optimum nitrogen nutrition of sugar beet by means of an estimation of the mineral nitrogen content of the soil; urea as a nitrogen fertilizer for sugar beet; application of lime/waste lime in sugar beet growing; weed control in sugar beet; development and improvement of methods for the yield forecast of sugar beet; application of growth regulating substances in sugar beet growing; dry matter production and distribution in sugar beet plants at different growing practices; the practical application of plant protection products and their effects on sugar beet growing; soil fumigation; mechanization in sugar beet growing; frost protection and storage methods for sugar beet (J. Jorristma); chemical control of beet cystnematodes in sugar beet; breeding research for resistance to beet cyst nematode in sugar beet; research into breeding for resistance to the beet eelworm disease (Heterodera schachtii) in the family Cruciferae; population dynamics of nematodes to determine the maximum frequency of sugar beet in different areas; the protection of sugar beet seed and seedlings against soil insects ie integrated control of Onychiurus armatus and Blaniules guttulatus; observations on behalf of the virus yellows warning service; control of virus yellows in sugar beet by chemical control of aphids; control of virus yellows in sugar beet by cultural measures; research concerning acute occurring pests and diseases in sugar beet (W. Heijbroek).

Instituut voor Toepassing van Atoomenergie in de Landbouw*   56

– ITAL
[Institute for Atomic Sciences in Agriculture]
Address: Postbus 48, 6700 AA Wageningen
Telephone: (08370) 19120
Affiliation: Ministry of Agriculture and Fisheries
Status: Official research centre
Directors: Dr A. Ringoet, Drs W.F. Oosterheert
Activities: Radiation induced changes in plants; food preservation and insect control by means of ionizing radiations; behaviour of specific radionuclides in soil; plants and animals forming the 'food chain'; use of radioactive substances in biological research.

Instituut voor Veeteeltkundig Onderzoek 'Schoonoord'   57

[Animal Husbandry Research Institute 'Schoonoord']
Address: Postbus 501, 3700 AM Zeist
Telephone: (03404) 17111
Affiliation: Ministry of Agriculture and Fisheries; National Council for Agricultural Research (TNO)
Status: Official research centre
Director: Dr W. Sybesma
Teams: Livestock management and care, Drs H.K. Wierenga; Slaughter quality, Dr Ir P. Walstra; Reproduction, Ir A. Meijering; Breeding, Dr Ir E.W. Brascamp
Graduate research staff: 24
Activities: All farm animals except poultry: housing; management; reproduction, breeding and selection. There is close cooperation with other institutes including The Institute of Agricultural Engineering, The Netherlands Centre for Meat Technology (TNO) and the Agricultural University at Wageningen. There are experimental farms at Zeist and near Lelystad and a pig station at Maarterisdijk.
Publications: Annual report.

Instituut voor Veevoedingsonderzoek 'Hoorn'   58

– IVVO
[Institute for Livestock Feeding and Nutrition Research Hoorn]
Address: Postbus 160, Runderweg 2, 8200 AD Lelystad
Telephone: (03200) 22514
Affiliation: Ministry of Agriculture and Fisheries
Status: Official research centre
Director: Ir F. de Boer
Research director: Dr Ir A.J.H. van Es
Departments: Ruminants, Dr Ir Y.S. Rijpkema; pigs, Ir A.W. Jongbloed; physiology and biochemistry, Dr Ir Y. van der Honing
Graduate research staff: 25
Activities: Development and testing of chemical methods, laboratory techniques and technical equipment for animal research; surgical techniques for research on livestock feeding and nutrition; feed intake of ruminants; transfer of heavy metals from feed into milk and meat; feed requirements for pigs; feeding-value assessment based on the chemical composition, in particular of cell walls; protein metabolism in the digestive tracts of ruminants; growth of carcass and composition.

Stimulation of food production in developing countries by feed evaluation.

Publications: Annual report, research reports.

Projects: Developing, testing of chemical methods and laboratory techniques, (J.M. van der Meer); development of technical equipment for animal research, (A. van Beers); surgical techniques for research on livestock feeding and nutrition, (H. Everts, H. de Visser, A.M. van Vuuren); digestibility of roughages and concentrates in ruminants, (A. Steg, B. van Donselaar, Y.S. Rijpkema); energy and requirements of dairy cattle by means of feeding trials, (Y.S. Rijpkena, L. van Reeuwijk); feeding standards for ewes, (H. Everts, H. Kuiper); NPN in feeding fattening young cattle, (F. de Boer, H.J. Wentink); feed intake in grazing cattle in loose housing, (J.A.C. Meijs, L. van Reuwijk, H.J. Wentink); factors determining feed-intake of cattle, (H. de Visser, A.M. van Vuuren, A. de Jong); roughage/concentrate ratios in dairy cattle feeding, (A.M. van Vuuren, L. van Reeuwijk, H. de Visser); development of butyric acid bacteria, (S.F. Spoelstra, K. Vreman, G.J. Voerman); transfer of heavy metals from feed into milk and meat, (K. Vreman, G.J. Voerman); digestibility trials on feeds for pigs or calves, (B. Smith); net energy of feeds in by-products in pigfeeding (B. Smits, A.W. Jongbloed, Y. van der Honing); protein requirements in different breeds of pigs (N.P. Lenis, B. Smits, S.H.M. Metz); manipulating the fatty acid pattern in bodyfat (meat) and milk, (S.H.M. Metz, A.W. Jongbloed, S. Tamminga); optimal P-requirement of pigs in relation to environmental aspects, (A.W. Jongbloed); energy balance trials on feedstuffs, (Y. van der Honing, A.J.H. van Ej); feeding value assessment based on technical composition, in particular of cell walls, (J.M. van der Meer); protein metabolism in digestive track of ruminants, (S. Tamminga, C.J. van der Koelen, A.M. van Vuuren); biochemical/physiological basis and parameters of growth and carcass composition (S.H.M. Metz, R.A. Dekker); transfer of environmental critical substances from the feed through the animal to the saleable animal product, (K. Vreman, G.J. Voeman); stimulation of food-production in developing countries by feed value-evaluation (J.M. van der Meer, A. Steg).

Katholieke Universiteit Nijmegen*   59

[Nijmegen Catholic University]
Address: Comeniuslaan 4, Nijmegen
Status: Educational establishment with r&d capability

FACULTY OF MATHEMATICS AND   60
NATURAL SCIENCES*

Address: Toernooiveld, 6525 ED Nijmegen
Dean: Professor A.G.M. Janner

Botany Laboratory*   61

Chemical Cytology Laboratory*   62

Exobiology Laboratory*   63

Genetics Laboratory*   64

Zoology Laboratory*   65

Koninklijk Instituut voor de Tropen   66

– KIT
[Royal Tropical Institute]
Address: Mauritskade 63, 1092 AD Amsterdam
Telephone: (020) 924949
Telex: 15080 kit nl
Status: Official research centre
President: Ir F. Deeleman

DEPARTMENT OF AGRICULTURAL   67
RESEARCH

Director: Ir H.Ph. Huffnagel
Sections: Research, G.J. Koopman; research coordination, P.J. van Rijn
Graduate research staff: 29
Activities: Rural planning, agricultural development planning, soil science, soil survey, soil and water analysis, land evaluation; production of tropical crops (rice, sugar cane, vegetable crops, coffee, rubber, oil palm, etc.); protection of stored crops, control of insects and weeds, small scale processing of agricultural products; food technology; tropical wood technology and marketing.
Publications: Annual report, abstracts on tropical agriculture (monthly), bulletins, *Landendocumentatie* (geographical studies, in Dutch).
Projects: Crop production and income improvement, (de Vries); monitoring of nutrient cycling in soils under shifting cultivation (Andriesse); sulphur status evaluation in tropical soils, (Blees, Limburg, Schelhaas); the effect of soil fertility on the nutritional value of the tropical leaf vegetables Amaranthus spp and Ipomoea reptans, (Grubben, Pol, Rosmalen, Scheer, Wessel); yield increase on calcareous soils in El Minia, Central Egypt, (El Baradi, Pol, Schelhaas); soil-crop interactions in semi-arid zones in West Africa, (Stoop); the effect of ley-farming with legumes on soil fertility improvement of upland soils in Northern Thailand, (Pol, Stoop);

vegetative multiplication of coffee (Coffea spp.), (Pieterse, Roorda, Wormer); physiology and control of aquatic weeds, (Andriessen, Horneman, Hieterse); the relation between soil fertility and Orobanche development, (Pieterse, Roorda, Worna); insect infestation in stored sorghum and millet at different temperatures and relative humidities (Andriessen, Roorda, Schulten); study on the prevention of delayed incompatibility in grafting of black pepper (Piper nigrum L.) on Phytophthora resistant root stocks, (de Waard, Elk, Magendans); small-scale technology for drying of products from tropical agriculture and horticulture, small-scale technology for the extraction of oil from oilseeds, (Heubers, Korthals, Althes); small-scale extraction of oil from fresh coconuts and use of by-products, (Thio); economic methane gas production from palm oil sludge, (Blaak); research and advice on production of weaning food in Benin, (Korthals, Altes, Heubers, Merx); cooperative study on the possibility of standardization and improvement of soil testing methods for soils of the humid tropics with emphasis on soil phosphorus, (Brook Pol, Schelhaas, Blees); improvement of analytical methods for the benefit of the service laboratory, (Pol, Limburg, Blees, Berg); soil-test data interpretation guide with special reference to KIT (RTI)-methods, (Brook, Pol); development of methods for the analysis of organic material to be used as soil amendments, (Schelhaas, Pol, Limburg, Bakels); international crosschecking of soil analytical methods (for classification), (van Reeuwijk, Pol).

Laboratorium voor Bloembollenonderzoek 68

[Bulb Research Centre]
Address: PO Box 85, Vennestraat 22, 2160 AB Lisse
Telephone: (02521) 19104
Affiliation: Ministry of Agriculture; Bulb Growers Association
Status: Official research centre
Director: Dr E.B. van Julsingha
Graduate research staff: 17
Activities: Wide range of research subjects include phytopathology, bulb propagation, soil water management, choice of varieties for specific purposes, temperature treatment of bulbs for flowering in winter and/or other seasons, efficient use of labour and machines, and the economic evaluation of production systems. The centre maintains close cooperation with several institutes, experimental stations, and university laboratories engaged in agricultural, horticultural, or biological research; (resulting in joint projects and exchange of technical facilities) and with the Bulb Inspection Service and the Plant Protection Service.

Publications: Annual reports, *Rapporten* (series, in Dutch), *Praktijkmededelingen* (series, in Dutch), manual or bulb diseases, handbooks on various subjects.

DEPARTMENT OF HORTICULTURE 69 AND ECONOMICS

Head: vacant
Activities: Production ecology, variety trials, economic evaluation of cultural methods.
Projects: Theoretical production ecology, (Ir M. Benschop); variety trials, (Ir H.P. Pasterkamp); lilies, (J. Boontjes); gladioli, (N.P.A. Groen); tulips, (J. Koster); irises, (J.A. Schipper); hyacinths and narcissi, (Ing P.J. M. Vreeburg); miscellaneous bulbs (J.A.Th. de Winter); influence of soil water management on growth and production (Ir G.G.M. van der Valk); economic evaluation of cultural methods (C.O.N. de Vroomen).

DEPARTMENT OF MECHANIZATION 70 AND RATIONALIZATION

Head: Drs A.F.G. Slootweg
Activities: Labour methods and organization, machine testing.
Projects: Improvement of labour methods and organization, (Ing A.J. Bulsink); testing of machines used in bulb industry, (Ing R.S. Bijl); technical information for customers abroad, (J.C.M. Buschan).

DEPARTMENT OF PHYSIOLOGY 71

Head: Dr. G.A. Kamerbeek
Activities: Biochemistry, physiology, flowering (influences), non-parasitic diseases.
Projects: Biochemistry (internal quality of bulbs; resistance of pathogenic fungi against fungicides), (Dr J.C.M. Beijersbergen); physiology of flowering and periodicity (respiration, flower formation and flower development), (Dr G.A. Kamerbeek); flowering (influence of light and temperature), (Dr Ir U. van Meeteren); non-parasitic diseases, (influence of growth substances; flowerbud blasting), (Dr W.J. de Munk); keeping quality of out flowers, (A. Swart).

DEPARTMENT OF PHYTOPATHOLOGY 72

Head: Dr. Ir G. Weststeijn
Projects: Biology and control of Fusarium oxysporum, (Dr B.H.H. Bergman); biology and control of Rhizoctonia spp, (A.W. Doornik); diagnosis and control of diseases caused by bacteria (ie Xanthomonas hyacinthi), (W. Kamerman); biology and control of root rot pathogens (mainly Pythium) (Dr Ir G. Weststeijn); population dynamics and control of eelworms, (Ir W.A. Windrich); diagnosis of parasitic and nonparasitic diseases, (P.J. Muller); chemical and physical control of diseases and pests; chemical weed control (M. de Rooij, A. Koster).

Laboratorium voor Insekticidenonderzoek* 73

– LIO
[Laboratory for Research on Insecticides]
Address: Marijkeweg 22, 6709 PG Wageningen
Telephone: (08370) 11821
Affiliation: Ministry of Agriculture and Fisheries
Status: Official research centre
Director: Dr Ir A.M. van Doorn
Deputy director: Professor Dr F.J. Oppenoorth
Activities: To increase the efficient and selective application of chemicals used in agricultural practice. Interaction between insects and insecticides; between pesticides and the environment, and the use of sex-attractants.

Landbouw-Economisch Instituut 74

– LEI
[Agricultural Economics Research Institute]
Address: Postbus 29703, Conradkade 175, 2502 LS 's-Gravenhage
Telephone: (070) 614161
Affiliation: Ministry of Agriculture and Fisheries
Status: Official research centre
Director: Professor Dr J. de Veer
Departments Structural research, A.L.G.M. Bauwens; agriculture, Dr L.C. Zachariasee; horticulture, D. Meyaard; fisheries and forestry, Drs R. Rijneweld
Graduate research staff: 150
Activities: Management and general economics of Dutch agriculture and fisheries.

AGRARIAN STRUCTURAL RESEARCH DEPARTMENT* 75

Head: Ir A.L.G.M. Bauwens
Deputy head: Drs L. Douw

AGRICULTURAL DEPARTMENT* 76

Head: Professor Drs J. de Veer
Deputy head: Ir C.J. Cleveringa

FISHERIES AND FORESTRY DEPARTMENT* 77

Head: Drs R. Rijneveld

HORTICULTURAL DEPARTMENT* 78

Head: Ir D. Meijaard
Deputy head: Ir W.G. de Haan

STATISTICS DEPARTMENT* 79

Head: Dr Ir G. Hamming
Deputy head: Dr J.A. Kuperus

Landbouwhogeschool Wageningen 80

[Wageningen Agricultural University]
Address: Salverdaplein 11, 6701 DB Wageningen
Telephone: (08370) 84472
Telex: blhwg nl 45015
Status: Educational establishment with r&d capability
Rector Magnificus: Professor Dr C.C. Ooesterlee
Graduate research staff: 600
Activities: Research focuses on all aspects of agriculture; environmental sciences, basic biological research and biotechnology, food science, economics, usage and management of natural resources. There is a botanical gardens; computing centre; experimental farm; water sampling station.
Publications: Mededelingen Landbouwhogeschool (15 per year), miscellaneous papers.

FACULTY OF AGRICULTURAL SCIENCES 81

Dean: Professor Dr Ir J. Schenk

Department of Agricultural Economics   82

Address: Leeuwenborch, Hollandseweg 1, Wageningen
Telephone: (08370) 82949
Senior staff: Professor Dr Ir P.C. van den Noort, Professor Dr Ir J. de Hoogh
Activities: Dairy industry; cost-benefit analysis of agricultural policies; income distribution of Dutch farmers; economic aspects of forestry; world food situation.

Department of Agricultural   83
Engineering

Address: Mansholtlaan 12, 6708 PA Wageningen
Telephone: (08370) 19119/82980
Sections: Agricultural machinery, including tropical mechanization, Professor Ir A. Moens; mechanical engineering, Professor Ir G.J. Quast; construction of agricultural and horticultural buildings, Ir G. Pothoven
Activities: Agricultural machinery; appropriate technology for developing countries; ergonomics.
Projects: Development of agricultural technical systems; farm management; ergonomics, work physiology; appropriate technology; agricultural buildings.

Department of Air Pollution   84

Address: Binnenhaven 12, 6709 PD Wageningen
Telephone: (08370) 82683/82684
Head: Professor Dr E.H. Adema
Activities: Dispersion of air pollutants (model and tracer studies); chemical composition and size distribution of urban aerosols; atmospheric chemistry; effects of air pollution on human health and on vegetation.

Department of Animal Nutrition   85

Address: Haagsteeg 4, Wageningen
Telephone: (08370) 89111
Head: Professor Ir S. Boer Iwema
Activities: Feeding of ruminants, horses and pigs; technology of feedstuffs.
Projects: Feeding and nutrition of ruminants; landscape survey with an emphasis on vegetation; terrestial ecosystems; syntaxonomy and synecology; investigation of microcommunities of bryophytes, vascular plants and fungi in relation to the micropattern of soil and microclimate.

Department of Animal Physiology   86

Address: Haarweg 10, 6709 PJ Wageningen
Telephone: (08370) 84136
Head: Professor Dr P.W.M. van Adrichem
Laboratories: General animal physiology and physiology of domestic animals; General animal physiology and comparative animal physiology
Activities: Energy metabolism of animals; physiology of trace elements; metabolism of proteins; endocrinology; physiology of digestion and absorption; application of isotope techniques in physiology; sensory physiology and neurophysiology; development of analytical methods for use in physiological research.
Projects: Function of the digestive track concerning digestion and resorption; protein and energy metabolism; endocrine regulation of maintenance and production; physiological basis of food-plant relations and food intake regulation in invertebrates.

Department of Animal Production   87

Address: 'Zodiac', Marijkeweg 40, 6709 PG Wageningen
Telephone: (08370) 82335/83120/83959/83801
Sections: Animal breeding, Professor Dr Ir R.D. Politiek; animal husbandry, Professor Dr C.C. Oosterlee; fish culture and inland fisheries, Professor Dr E.A. Huisman; tropical animal production; poultry husbandry, Dr Ir E.H. Ketelaars
Activities: Cattle; sheep; pigs; poultry; dwarf goats; African catfish; carp, evaluation of (re)stocking in fisheries management.

Department of Biochemistry   88

Address: De Dreijen 11, 6703 BC Wageningen
Telephone: (08370) 82868
Head: Professor Dr C. Veeger
Activities: Structure and function of redox enzymes and enzyme systems.
Projects: Structure and function of flavoproteins; mechanism of nitrogen fixation in micro-organisms; model studies of the caralytic centre of flavoproteins; electrotransfer in anaerobic bacteria, structure and function of a-ketodehydrogenase complexes.

Department of Botany   89

Address: Arboretumlaan 4, 6703 BD Wageningen
Telephone: (08370) 82155
Head: Professor Dr M.T.M. Willemse
Activities: Wood anatomy in relation to plant ecology and taxonomy; secondary phloem of Gymnosperms and Angiosperms; fruit development in Solanaceae; development of veins and veinlets; ultrastructure of fertilization and seed development in Angiosperms, Gymnosperms and Ferns.

Department of Civil Engineering and   90
Irrigation

Address: Nieuwe Kanaal 11, 'De Nieuwlanden', 6709 PA Wageningen
Telephone: (08370) 89111
Head: Professor Ir L. Horst
Activities: Irrigation methods and equipment.
Projects: Water balance; countryside roads; structures in irrigation and drainage; on farm irrigation; sediment transport.

Department of Development   91
Economics

Address: Postbus 8130, 6700 EW Wageningen
Telephone: (08370) 89111
Heads: Professor Ir A. Franke, Professor Dr F.P. Jansen
Activities: Agricultural planning in developing countries: project evaluation; problems of farm size in the process of economic transformation.
Projects: Agricultural planning and development problems in developing countries.

Department of Economics   92

Address: Postbus 8130, 6700 EW Wageningen
Telephone: (08370) 84255
Head: Professor Dr Th.L.M. Thurlings
Activities: Common Market economics.
Projects: Methodology of economics; changes in economic order.

Department of Entomology   93

Address: Postbus 8031, 6700 EH Wageningen
Telephone: (08370) 82989
Head: Professor Dr J. de Wilde
Activities: Insect-plant relations; insect biochemistry; bee research; insect histology.

Department of Experimental Animal   94
Morphology and Cell Biology

Address: 'Zodiac', Marijkeweg 40, 6709 PG Wageningen
Telephone: (08370) 83509/83967
Head: Professor Dr J.W.M. Osse
Sections: Zoology, Professor Dr J.W.M. Osse; cell biology, Dr J.F. Jongkind
Activities: Acquisition and digestion of food in fishes; differentiation of cells in vivo and in vitro.

Department of Farm Management   95

Address: Postbus 8130, 6700 EW Wageningen
Telephone: (08370) 84065
Head: Professor Dr Ir J.F. van Riemsdijk
Sections: Horticultural holdings, Drs S. Kostelijk; dairy and mixed farms, Ir G.W.J. Giesen; arable farms, Ir J.H. van Niejenhuis
Activities: Management in relation to agricultural and horticultural undertakings.

Department of Field Crops, Grassland   96
Husbandry, and Weed Science

Address: Haarweg 33, 6709 PH Wageningen
Telephone: (08370) 83040
Head: Professor Ir M.L. 't Hart
Senior staff: Professor Dr P. Zonderwijk
Projects: Production systems and crop ecology; productivity and quality of field crops; production and nutritive value of forages; establishment, utilization and ecology of grass vegetations.

Department of Food Science   97

Address: Postbus 8129, 6700 EV Wageningen
Telephone: (08370) 89111
Laboratories: Process engineering, Professor Dr Ir S. Bruin; Food chemistry, Professor Dr W. Pilnik; Food microbiology and hygiene, Professor Dr E.H. Kampelmacher; Dairy technology and food physics, Professor Ir A. Prins
Activities: Engineering aspects of food production; chemical and biochemical changes in all kinds of food during processing and storage; fermentation; food spoilage, processing of milk products.
Projects: Physical separations; water relations of food; powder technology; biotechnology; hygiene and prevention of spoilage; polysaccharide constituents of plants and their chemical and enzymatic changes; influence on processing and quality of plant food stuffs; proteins; analysis of the daily food; fruit juice technology; ripening of protein-rich milk products; hygienic aspects of milk and milk products; concentration and drying of milk products; structure and stability of disperse foods; transport phenomena in disperse systems, in particular foods.

Department of Forest Management   98

Address: Postbus 342, 6700 AH Wageningen
Telephone: (08370) 82915
Head: Professor Ir A. van Maaren
Activities: Yield and thinning studies; forest mensuration.

Department of Forest Technique 99

Address: Postbus 342, 6700 AH Wageningen
Telephone: (08370) 82120
Head: Professor Ir M.M.G.R. Bol
Sections: Work and efficiency studies in forest operations, Professor Ir M.M.G.R. Bol; operations research in forestry, Ir W. Heij; wood science, Ir N.A. den Hartog J.M. Fundter; ergonomics in forestry, Ir F.J. Staudt
Activities: Mechanical and physical properties of timbers; identification and anatomical description of timber specimens; minor forest products.
Projects: Forest techniques (ergonomics, safety, machines); knowledge of wood products; systems engineering in forest operations.

Department of Genetics 100

Address: Generaal Foulkesweg 53, 6703 BM Wageningen
Telephone: (08770) 83140
Head: Professor Dr Ir J.H. van der Veen
Senior staff: Dr Ir J. Sybenga
Activities: Population genetics; cytogenetics; biochemical genetics; genetics of microorganisms; botanical genetics.
Projects: Cytogenetics; population genetics; microbial genetics; biochemical genetics (transcription); physiological genetics in higher plants; biochemical genetics (metabolism).

Department of Horticulture 101

Address: Postbus 30, 6700 AA Wageningen
Telephone: (08370) 84096
Heads: Professor Dr Ir J.F. Bierhuizen, Professor Dr Ir J. Doorenbos
Activities: Effect of environmental conditions and growth regulators on plant growth, development and regeneration.
Projects: Propagation of horticultural crops; crop production; quality of crops.

Department of Hydraulics and 102
Catchment Hydrology

Address: Nieuwe Kanaal 11, De Nieuwelanden, 6709 PA Wageningen
Telephone: (08370) 82778
Head: Professor Ir D.A. Kraijenhoff van de Leur
Activities: Hydraulics of small structures and vegetated channels; stream flow measurement; long waves; components of the runoff process in catchment areas; deterministic and stochastic modelling in catchment hydrology and in the management of water resources; electronic analogues.

Projects: Coherence between rainfall, evaporation and discharge in catchment areas; flaw concepts in small hydraulic structures in open and closed circuits; surface water management.

Department of Land and Water Use 103

Address: Nieuwe Kanaal 11, De Nieuwelanden, 6709 PA Wageningen
Telephone: (08370) 89111
Senior staff: Professor Dr Ir R.H.A. van Duin, Professor Dr Ir F. Hellinga, Dr Ir H.N. van Lier, Professor Dr Ir W.H. van der Molen
Activities: Water control; drainage and salinity; design of rural road systems; numerical methods in groundwater flow; traffic in rural areas; electric analogue models of groundwater flow; soil management and improvement; soil erosion control; water oriented outdoor recreation; technical criteria for the development of nature reserves.

Department of Landscape Architecture 104

Address: Wilhelminaweg 1, 6703 CC Wageningen
Telephone: (08370) 82050
Head: Professor Ir M.J. Vroom
Activities: Landscape planning and urban design.
Projects: Agrohydrology; land use planning; soil management; recreation.

Department of Mathematics 105

Address: Postbus 8003, 6700 EB Wageningen
Telephone: (08370) 84385
Sections: Pure and applied mathematics, Professor Dr B. van Rootselaar; mathematical and applied statistics, Professor Dr Ir L.C.A. Corsten; operations research, Dr P. van Beek; computer science, Ir M.S. Elzas
Activities: Logic and foundation of mathematics, intuitionistic mathematics; mathematical statistics and its application in research; design and analysis of experiments, mathematical programming, simulation languages and methods, computer-assisted architectural design, parallel processing systems.

Department of Microbiology 106

Address: Postbus 8033, 6700 EJ Wageningen
Telephone: (08370) 89111
Head: Professor Dr Ir E.G. Mulder
Senior staff: Dr Ir C.J.E.A. Bulder
Activities: Soil microbiology; water and wastewater microbiology; DNA composition of bacteria.
Projects: Microbial transformations in soil and/or water; nitrogen fixation; general microbiology; technical microbiology.

Department of Molecular Biology 107

Address: Postbus 8128, 6700 ET Wageningen
Telephone: (08370) 82036
Head: Professor Dr A. van Kammen
Activities: Replication of plant viruses; nitrogen fixation in rhizobium bacteroids in pea root nodules.
Projects: Replication mechanism of cowpea mosaic vitrus; symbiotic nitrogen fixation; structure and regulation of the expression of plant genes.

Department of Molecular Physics 108

Address: Postbus 8091, 6700 EP Wageningen
Telephone: (08370) 82634
Head: Professor Dr T.J. Schaafsma
Activities: Molecular spectroscopy of biologically important compounds and macromolecules; molecular biophysics of plant-water relationships.
Projects: Plant viruses; photosynthesis; water transport and balance in plants; applied molecular phgysics.

Department of Nature Conservation 109

Address: Postbus 8080, 6700 DD Wageningen
Telephone: (08370) 83174
Head: Professor Dr C.W. Stortenbeker
Activities: Scientific and touristic value of nature areas and their function in the landscape; water pollution and water management; wildlife, trees and shrubs in tropical areas; monitoring programme at the Yankari Game Reserve, Nigeria; distribution, ecology and behaviour of the orangutan, the Sumatran rhinoceros and elephant in the Gunung Leuser Reserve, Sumatra, Indonesia.
Projects: Physical planning; land-use planning; management of nature territories and threatened species; nature conservation in the tropics.

Department of Nematology 110

Address: Postbus 8123, 6700 ES Wageningen
Telephone: (08370) 82197
Acting head: Dr J.C. Zadoks
Activities: Population studies; host-parasite relationships; taxonomy and morphology; nematicides; grassland and root knot nematodes.
Projects: Faunistic and taxonomic research on nematodes; biology and ecology of nematodes; physiology of, and control of nematodes.

Department of Organic Chemistry 111

Address: Postbus 8026, 6700 EG Wageningen
Telephone: (08370) 89111
Head: Professor Dr H.C. van der Plas
Senior staff: Dr Ae. de Groot
Activities: Synthesis of natural products; reactivity of carbohydrates and heterocyclic compounds; photochemistry of heteroaromatic compounds; carbon compounds.
Projects: Chemistry of azaheterocyclic compounds; photochemistry of heterocyclic compounds; chemistry of natural products; immobilized enxymes.

Department of Physical and Colloid 112 Chemistry

Address: Postbus 8038, 6700 EK Wageningen
Telephone: (08370) 82279
Head: Professor Dr J. Lyklema
Senior staff: Dr B.H. Bijsterbosch
Activities: Electrochemistry and the adsorption properties of model suspensions; adsorption of polymers and its effect on colloidal stability and emulsification.
Projects: Biopolymers at interfaces; interfacial electrochemistry; colloid stability; thin liquid films; electro-analysis.

Department of Physics and 113 Meteorology

Address: Duivendaal 1 & 2, 6701 AP Wageningen
Telephone: (08370) 82937
Chairman: Dr P.J. Bruijn
Activities: Agricultural and environmental physics, especially in relation to micro-meteorology.
Projects: Glasshouse climates and control of glasshouse climates; turbulent transport; energy balance and transport phenomena; operational weather service; automation and data-processing.

Department of Phytopathology 114

Address: Postbus 8025, 6700 EE Wageningen
Telephone: (08370) 83410
Head: Professor Dr Ir J. Dekker
Sections: Fungicides, Professor Dr Ir J. Dekker; epidemiology and disease resistance, Dr J.C. Zadoks; soil-borne pathogens, Dr Ir T. Limonard; physiology and biochemistry of host-parasite relations, Dr Ir L.C. Davidse
Activities: Physiology and biochemistry of the relation between host plant and parasite; resistance of various crops to pathogenic organisms; systemic fungicides; phytopharmacy; ecology of soil-borne pathogens; epidemiology.
Projects: Systemic fungicides; physiology of parasitism; epidemiology of and resistance to plant diseases; control of pathogenic soil fungi.

Department of Plant Ecology 115

Address: Postbus 8128, 6700 ET Wageningen
Telephone: (08370) 82078
Activities: Description and mapping; structure and functioning of ecosystems; wild annual species; ecology, distribution and taxonomy of cryptogams; micro-habitats and micro-communities.
Projects: Landscape survey with an emphasis on vegetation; terrestrial ecosystems; syntaxonomy and synecology; investigations of microcommunities of bryophytes, vascular plants and fungi in relation to the micropattern of soil and microclimate.

Department of Plant Physiological 116
Research

Address: Generaal Foulkesweg 72, 6703 BW Wageningen
Telephone: (08370) 82800
Head: Dr W.J. Vredenberg
Activities: Bioenergetics; morphogenesis.
Projects: Growth and development; photosynthesis; photomorphogenesis; transport of water and ions.

Department of Plant Physiology 117

Address: Botanical Institute, Arboretumlaan 4, 6703 BD Wageningen
Telephone: (08370) 82146
Head: Professor Dr J. Bruinsma
Activities: Flower physiology; fruit physiology; physiological ecology of seeds; phytogerontology; phototropism; development of hormone determinations.
Projects: Flower physiology; fruit physiology; seed physiology; vegetative development; genetic and hormonal regulation of senescence and pathogenesis phenomena; phybohormone determination methods.

Department of Plant Taxonomy and 118
Plant Geography

Address: Postbus 8010, 6700 ED Wageningen
Telephone: (08370) 83160/82170
Head: Professor Dr H.C.D. de Wit
Sections: Plant taxonomy and plant geography, Professor Dr H.C.D. de Wit; experimental plant taxonomy, Ir J.C. Arends
Activities: Taxonomic botany (Africa, temperate zones); biosystematics; experimental plant taxonomy.
Projects: Revision of African plant clusters; biosystematics.

Department of Public Health 119

Address: Postbus 8128, 6700 ET Wageningen
Telephone: (08370) 82080
Head: Professor Dr K. Biersteker
Activities: Field studies on social and medical aspects of pollution by specific industries in rural areas.

Department of Silviculture 120

Address: Postbus 342, 6700 AH Wageningen
Telephone: (08370) 89111/84426
Head: Professor Dr Ir R.A.A. Oldeman
Activities: Silviculture, architecture and growth dynamics of various tree species; vegetative propagation of forest trees; morphology, identification and cultivation of poplars and willows; provenances and climatological strains of exotic forest trees; ecophysiology of forest trees; natural regeneration of tropical rain forest; forest succession in the tropics; forest architecture and growth dynamics; design of silvicultural systems.

Department of Soil Science and 121
Geology

Address: Postbus 37, 6700 AA Wageningen
Telephone: (08370) 84410
Head: Dr L. van der Plas
Sections: Regional soil science, Professor Dr Ir L.J. Pons; tropical soil science, Professor Dr Ir J. Bennema; geology and mineralogy, Professor Dr J.D. de Jong
Activities: Geomorphological, sedimentological and hydrogeological investigations mainly on Quaternary and Tertiary formations; genesis, chemical composition and agricultural use of soils on sediments; soil formation and classification of Red and Yellow tropical soils; thermodynamic properties of soil minerals and of processes involved in soil genesis; application of aerial photo interpretation and remote sensing techniques for soil science; evaluation for land use planning.
Projects: Red and yellow tropical soils; soils on unconsolidated sediments; ecological land classification; geomorphological-sedimentological research; hydrogeological investigations; model studies and development of methodologies in soil science and mineralogy; ecopedology; soil as an industrial commedity.

Department of Soils and Fertilizers 122

Address: Postbus 8005, 6700 EC Wageningen
Telephone: (08370) 89111
Sections: Soil physics and soil chemistry, Professor Dr Ir G.H. Bolt; soil fertility and fertilizers, Professor Dr Ir A. van Diest; soil fertility and plant nutrition, Dr G.R. Findenegg; soil pollution, Dr Ir F.A.M. de Haan
Activities: Soil fertility; soil chemistry; soil physics; soil pollution.

Department of Surveying 123

Address: Postbus 339, 6700 AH Wageningen
Telephone: (08370) 89111/82130
Head: Ir G.A. van Wely
Activities: Optical distance measurement; application of lasers; electronic survey methods and computations; applications of photomaps; Doppler Satellite Navigation Systems.
Projects: Geodetical observation techniques; photogrammetry processing of survey data and cartographic aspects.

Department of Taxonomy of Cultivated Plants and Weeds 124

Address: Haagsteeg 3, 6708 PM Wageningen
Telephone: (08370) 84390/83377
Head: Dr Ir R.A.H. Legro
Activities: Taxonomy of ornamental plants and their origin; experimental taxonomy (including cytotaxonomy) of ornamental plants and related wild species.
Projects: Taxonomical and experimental-taxonomical research of cultivated plants; fundamental problems concerning the classification and nomenclature of cultivated plants.

Department of Theoretical Production Ecology 125

Address: Postbus 8071, 6700 EH Wageningen
Telephone: (08370) 82141
Head: Professor Dr Ir C.T. de Wit
Activities: Crop production and crop development; population dynamics and integrated pest and disease control.
Projects: Abiotic aspects of plant production; primary productivity; ecology of plant populations; intended and harmful secundary production; agricultural production systems.

Department of Toxicology 126

Address: Biotechnion, De Dreijen 12, 6703 BC Wageningen
Telephone: (08370) 82137
Head: Professor Dr J.H. Koeman
Activities: Cell and genetic toxicology; toxicity studies with trout; effects of toxic compounds on liver morphology and biochemistry; environmental effects of pesticides.
Projects: Fish toxicology; pretoxic effects; cell toxicology; ecotoxicology; mutagenicity and carcinogenicity.

Department of Tropical Crop Science 127

Address: Postbus 341, 6700 AH Wageningen
Telephone: (08370) 83072
Chairman: Dr Ir M. Flach
Senior staff: Professor Dr Ir J.D. Ferwerda
Activities: Consequences of stagnation in early growth of annual crops; ecology of tropical crops.
Projects: Special crop science; general crop science; cropecophysiology.

Department of Virology 128

Address: Postbus 8045, 6700 EM Wageningen
Telephone: (08370) 83090/89111
Chairman: Professor Dr Ir J.P.H. van der Want
Activities: Plant virus infection and multiplication.
Projects: Identification, classification and structure analysis of plant viruses and study of the infection process; the characterization of insect viruses to study their behaviour in the biological control.

Department of Water Purification 129

Address: Postbus 8129, 6700 AB Wageningen
Telephone: (08370) 83339
Head: Professor Dr P.G. Fohr
Sections: Water purification, Professor Dr P.G. Fohr; hydrobiology, Professor Dr C. den Hartog
Activities: Wastewater treatment; structure and function of aquatic ecosystems.
Projects: Biological purification of wastewater; physical-chemical treatment methods of wastewater and drinking water; investigations of surface water.

Department of Zoology 130

Address: De 'Dreyenborch', Ritzema Bosweg 32A, Wageningen
Telephone: (08370) 82973
Acting head: Drs J.H. Kuchlein
Activities: Behaviour of parasites; relationship between prey and its predators; ecology of larval stages of parasites; differentiations among genera and species by electrophoresis.

General and Regional Agriculture 131

Address: Postbus 8130, 6700 EW Wageningen
Telephone: (08370) 84406
Senior staff: Dr Ir J.G. Veldink, Ir G. Blok
Activities: Economic geographical survey of agriculture in Western Europe and other parts of the world, with special reference to the position and role of Dutch agriculture.

Plant Breeding Institute 132

Address: Lawickse Allee 166, 6709 DB Wageningen
Telephone: (08370) 82836
Head: Professor Dr Ir J. Sneep
Activities: Selection methods and natural selection; variation in ploidy: wild species and primitive forms; resistance and breeding for resistance; heterosis and hybrid varieties.

Soil Tillage Laboratory 133

Address: Diedenweg 20, 6703 GW Wageningen
Telephone: (08370) 82066
Head: Professor Ir H. Kuipers
Sections: Plant reaction, Ir F.R. Boone; dry regions, Ir W.B. Hoogmoed; mechanical properties of soil, Dr Ir A.J. Koolen; effects of implements, Ir J.K. Kouwenhoven

Stichting Bloedgroepenonderzoek 134

– SBO
[Foundation for Blood Group Studies in Animals]
Address: 'Zodiac', Marijkeweg 40, Wageningen
Telephone: (08370) 19065
Director: Dr H. van Haeringen
Activities: Correlations between blood groups and other characteristics in animals; genetic composition of populations of domestic animals of the Netherlands; parentage control of animals.

Limnologisch Instituut* 135

[Limnological Institute]
Address: Rijksstraatweg 6, 3631 AC Nieuwersluis
or: De Akkers, 8536 VD Oosterzee
Telephone: (02943) 3251
Affiliation: Koninklijke Nederlandse Akademie van Wetenschappen
Status: Official research centre
Director: Dr S. Parma
Sections: Algology, J. Moed; food chain and production studies, J. Vijverberg; primary and secondary production, R.D. Gulati; mineralization of organic matter, Th.E. Cappenberg
Graduate research staff: 15
Activities: The relation between the chemistry of the aquatic environment and the biochemistry and physiology of the organisms; aquatic ecology and ecological morphogenesis of plants; microbiology, and freshwater biology in general.

Ministerie van Landbouw en Visserij 136

[Ministry of Agriculture and Fisheries]
Address: Postbus 20401, 2500 EK 's-Gravenhage
Telephone: (070) 792110
Telex: 32040; 75044 CTWAG
Status: Government department

DIRECTIE LANDBOUWKUNDIG ONDERZOEK 137

[Division for Agricultural Research]
Address: Postbus 59, 6700 AB Wageningen
Telephone: (08370) 19145
Telex: 75044
General director: Dr Ir D. de Zeeuw
Departments: Natural resources, Dr Ir H.N. Hasselo, Ir J.C.A.M. Bervaes, Drs P. Slot; plant production, plant breeding and plant protection, IR R. Mulder, Ir P.J. Stadhouders, Dr P.A.Th.J. Werry, Ir M. Miedema, Dr Ir P. van Halteren; potato research, Dr Ir D.E. van der Zaag, animal husbandry, Drs A.A.J. van der Leun, Dr P. Leeflang; fisheries, Ir H.G.J. Oudelaar; agricultural engineering and technology, E. Strooker; quality, H.J. Mol; international cooperation, Ir H.H. van der Borg; cooperation with developing countries, Dr Ir J.J. Hardon; Ir W. van Vuure, programming, evaluation and administration of research projects, Ir J. Leeuwangh, Ir B.E.M. Stol; experimental farms, Ing A. Zijlstra
See separate entries for:
Centraal Diergeneeskundig Instituut
Centrum voor Agrobiologisch Onderzoek
Instituut voor Bewaring en Verwerking van Landbouwproducten
Instituut voor Bodemvruchtbaarheid
Instituut voor Cultuurtechniek en Waterhuishouding
Instituut voor de Veredeling van Tuinbouwgewassen
Instituut voor Mechanisatie, Arbeid en Gebouwen
Instituut voor Plantenziektenkundig Onderzoek
Instituut voor Pluimveeonderzoek 'Het Spelderholt'
Instituut voor Toepassing van Atoomenergie in de Landbouw
Instituut voor Veeteeltkundig Onderzoek 'Schoonoord'
Instituut voor Veevoedingsonderzoek 'Hoorn'
Internationaal Agrarische Centrum
International Institute for Land Reclamation and Improvement
Laboratorium voor Insekticidenonderzoek
Landbouw-Economisch Instituut
Proefstation voor de Akkerbouw en de Groenteteelt in de Vollegrond
Proefstation voor de Bloemisterij
Proefstation voor de Boomkwekerij
Proefstation voor de Champignoncultuur
Proefstation voorde Fruitteelt

Proefstation voor de Groenten-en Fruitteelt Onder Glas
Proefstation voor de Rundveehouderij
Rijks Agrarische Afvalwater Dienst
Rijksinstituut voor Het Rassenonderzoek van Cultuurgewassen
Rijksinstituut voor Natuurbeheer
Rijksinstituut voor Onderzoek in de Bos-en Landschapsbouw 'De Dorschkamp'
Rijksinstituut voor Visserijonderzoek
Rijks-Kwaliteitsinstituut voor Land-en Tuinbouwprodukten
Rijksproefstation voor Zaadonderzoek
Sprenger Instituut
Stichting Bureau voor Gemeenschappelijke Diensten
Stichting Laboratorium voor Bloembollenonderzoek
Stichting Technische en Fysische Dienst voor de Landbouw
Stichting voor Bodemkartering
Stichting voor Plantenveredeling

Naamloze Vennootschap DSM* 138

Address: Postbus 65, 6400 AB Heerlen
Telephone: (045) 788111
Telex: 56018
Status: Industrial company
Divisions: Fertilizers; plastics; industrial chemicals; chemical products; energy; plastics processing; building.
Activities: DSM is a group of industrial companies based largely in Western Europe, with the main office at Heerlen. It is owned by the Dutch government.

Naarden International NV* 139

Address: Postbus 2, Naarden-Bussum, 1400 CA
Telephone: (02159) 99111
Telex: 43050 NARD NL
Status: Research association
President: J.P. Guepin
Research director: Dr F. Rykens
Research coordinator: J.G. Spoor
Sections: Flavours and fragrances; enzymes; antioxidants; food preservatives and other additives
Activities: International group of flavour and fragrance producing companies with worldwide organization. Apart from flavours and fragrances, the research programme includes enzymes, antioxidants, food preservatives, and other additives.

Nationale Raad voor Landbouwkundig Onderzoek TNO* 140

– NRLO-TNO
[National Council for Agricultural Research TNO]
Address: Postbus 297, 2501 BD 's-Gravenhage
Telephone: (070) 471021
Telex: 31660 tnogv nl
Affiliation: Netherlands Organization for Applied Scientific Research (TNO)
Status: Official research centre
Chairman: Ir A. de Zeeuw
Sections: Land use planning and management of nature and environment, Ir J. Verkoren; plant production, Professor Dr Ir G.J. Vervelde; animal production, Dr M.J. Dobbelaar; processing and marketing, Professor Dr Ir H.A. Leniger
Activities: The Council is responsible for the planning and coordination of all Dutch agricultural research, and the provision of command facilities which are of importance to such research. Organizations represented on the Council include the Ministry of Agriculture and its research institutes, the Agricultural University Wageningen, the farming industry and the Netherlands Organization for Applied Scientific Research TNO.

Natuurwetenschappelijke Studiekring voor Suriname en de Nederlandse Antillen 141

[Foundation for Scientific Research in Surinam and the Netherlands Antilles]

Address: c/o Zoological Laboratory, Plompetorengracht 9, 3512 CA Utrecht
Telephone: (030) 317241
Affiliation: Zoologisch Laboratorium van de Rijksuniversiteit, Utrecht
Status: Foundation funding research
Secretary: Drs L.J. Weatermann-van der Steen
Activities: Furthering the knowledge and study of the natural sciences concerning the Caribbean region, including the northern part of South America; collection and publication of scientific material, mainly dealing with botany and invertebrates.

Nederlands Instituut voor Zuivelonderzoek* 142

[Netherlands Institute for Dairy Research]
Address: Postbus 20, 6710 BA Ede
Telephone: (08380) 19013
Telex: 37205
Status: Independent research centre
General manager: Dr W.I.J. Aalbersberg
Sections: Analytical chemistry; physical chemistry; food technology; engineering; microbiology; human nutrition physiology
Graduate research staff: 45
Activities: Dairy science and technology; (human) nutrition physiology.
Publications: Annual report.

Nederlandse Organisatie voor Toegepast-Natuurwetenschappelijk Onderzoek* 143

– TNO
[Netherlands Organization for Applied Scientific Research]
Address: Postbus 297, 2501 BD 's-Gravenhage
Telephone: (070) 814481
Telex: 31660 tnogn nl
Status: Official research organization
President of Board of Management: Professor Ir W.A. de Jong
Graduate research staff: 1100
Activities: Industrial development; food and nutrition; defence; health.
Publications: Annual report; newsletter.

HOOFDGROEP VOEDING EN VOEDINGSMIDDELEN TNO 144

– HVV-TNO
[Division for Food and Nutrition Research TNO]
Address: Postbus 360, 3700 AJ Zeist
Telephone: (03404) 52244
Telex: 40022 civo nl
Director: Professor Ir B. Krol
Graduate research staff: 90
Activities: Research, trouble shooting, testing and information with regard to food analysis, food technology, and human nutrition, (quality of foods, food technology, nutrition, metabolic research, toxicology); development and improvement of research procedures, including automation and data processing research directed to support developing contries, dissemination of knowledge.

Publications: Annual report (in Dutch), *Volatile Compounds in Food, Compilation of Man Spectra of Volatile compounds in Food* (Two volumes per year).

Dwarsverband Toxicologie TNO* 145

– DTOX-TNO
[Toxicology TNO]
Address: Postbus 360, 3700 AJ Zeist
Telephone: (03404) 52244
Telex: 40022 civo nl
Chairman: Dr R. Kroes

Researchgroep voor Vlees en Vleeswaren TNO 146

– RVV-TNO
[Research Group for Meat and Meat Products TNO]
Address: Postbus 360, 3700 AJ Zeist
Telephone: (03404) 52244
Telex: 40022 civo nl
Chairman: Dr W. Sybesma
Activities: The group consists of members from the following institutes
Institute CIVO Technology TNO (Ir E.J.C. Paardekooper)
Institute for Animal Husbandry (Dr W. Sybesma)
Institute for Livestock Feeding and Nutrition Research (Ir F. de Boer)
Institute for Food Products of Animal Origin (Professor J.G. van Logtestijn)
Institute for Animal Production (Professor G.J.W.v.d. Mey)
Institute for Animal Breeding. (Professor R.D. Politiek). The group is responsible for the coordination of research in animal production (beef, pork, lamb); livestock breeding; husbandry and nutrition; food science (meat and meat products, and hygiene).

NEDERLANDSE CENTRALE ORGANISATIE TNO 147

[Netherlands' Central Organization for Applied Scientific Research]
Address: Postbus 297, 2501 BD 's-Gravenhage
Telephone: (070) 814481
Telex: 31660
Sections: Industrial research, M.J. Spanraft; food and nutrition research, G. Klein; defence research, H.J. Dirksen; health research, A. Rörsch

Organisch Chemisch Instituut TNO* 148

– OCI-TNO
[Institute for Organic Chemistry TNO]
Address: Postbus 5009, 3502 JA Utrecht
Telephone: (030) 882721
Affiliation: Netherlands Organization for Applied Scientific Research (TNO)
Status: Official research centre
Site manager: Dr A.P.M. van der Veek
Sections: Biochemistry and microbiology, Dr A. Kaars-Sijpesteijn; organometallic and coordination chemistry, Dr J.G. Noltes; physical organic chemistry and chemical analysis, Dr A.P.M. van der Veek
Graduate research staff: 20
Activities: Explorative and applicational research in the field of organic, organometallic and coordination chemistry, biochemistry and microbiology; manufacturing of research chemicals, microanalyses of elements; instrumental analyses; biotechnology; solar energy.

Phytopathologisch Laboratorium 'Willie Commelin Scholten' 149

['Willie Commelin Scholten' Laboratory of Plant Pathology]
Address: Javalaan 20, 3742 CP Baarn
Telephone: (02154) 15654/15655
Affiliation: Rijksuniversiteit van Utrecht; Universiteit van Amsterdam; Vrije Universiteit, Amsterdam
Status: Official research centre
Head: Professor K. Verhoeff
Projects and research leaders: Ecology of pathogenic and saprophytic microorganisms, Dr B. Schippers; mechanisms concerning the susceptibility and resistance to plant diseases, Professor Dr K. Verhoeff; resistance mechanisms of plants to virus infections, Dr D.H. Wieringa Brants
Graduate research staff: 10
Activities: Plant pathology (basic aspects of biological control; mechanisms of disease resistance).
Publications: Annual report (in Dutch).

Primatencentrum TNO* 150

– PC-TNO
[Primate Centre TNO]
Address: Postbus 5815, 2280 HV Rijswijk
Telephone: (015) 140930
Telex: 38191 repgo nl
Affiliation: Netherlands Organization for Applied Scientific Research (TNO)
Status: Official research centre
Director: Professor H. Balner
Sections: Immunology, Professor H. Balner; ethology, Dr C. Goosen; virology, Dr H. Schellekens
Graduate research staff: 6
Activities: Immunogenetics of rhesus monkeys and chimpanzees (histocompatibility, cellular reactivity, immune responses); behavioural studies in macaques and chimpanzees; virology, with emphasis on interferon.

Prins Maurits Laboratorium TNO - Instituut voor Chemische en Technologische Research* 151

– PML-TNO
[Prins Maurits Laboratory TNO - Institute for Chemical and Technological Research]
Address: Postbus 45, 2280 AA Rijswijk
Telephone: (015) 138777
Telex: 38034 pmtno
Affiliation: Netherlands Organization for Applied Scientific Research (TNO)
Status: Official research centre
Director: Dr A.J.J. Ooms
Sections: Technological research, Dr H.J. Pasman; chemical research, Ir M. van Zelm
Graduate research staff: 70
Activities: Determination and evaluation of characteristic properties of noxious substances, in particular chemical warfare agents and pesticides; study of the mechanism of action of noxious substances, in particular chemical warfare agents and pesticides in the human and animal organism; investigation of proper functioning of ammunition and weapons; inherent explosion hazards of industrial products.

Proefstation voor Aardappelverwerking* 152

[Experimental Station for the Utilization of Potatoes]
Address: Rouaanstraat 27, 9723 CC Groningen
Telephone: (050) 130341
Affiliation: Netherlands Organization for Applied Scientific Research (TNO)
Status: Official research centre
Director: Dr T.J. Buma
Activities: Processability of potatoes for starch factories; waste-water of potato-starch factories; fundamental research on starch and its derivatives; enzymatic modifications of starch.

Proefstation voor de Akkerbouw en de Groenteteelt in de Vollegrond 153

– PAGV
[Arable Farming and Field Production of Vegetables Research Station]
Address: Postbus 430, 8200 AK Lelystad
Telephone: (03200) 22714
Affiliation: Ministry of Agriculture and Fisheries
Status: Official research centre
General Director: Ir M. Heuver
Technical Director: Dr Ir J. van Kampen
Research sections: Crop production, Ir G. Liefstingh; farming systems, Ir A. van den Schaaf; farm management, Drs J. Kamminga; regional research, Ir M. van den Beek
Graduate research staff: 50
Activities: Practical research on arable crops and outdoor vegetables: potatoes; sugar beet; cereals and maize; grass seed; oil and fibre crops; pulses and vegetables; crop rotation; soil fertility mechanization; plant protection; farm planning; labour use. There are two experimental farms, at Lelystad and Alkmaar, and the station quides the research of 15 other experimental farms.
Publications: Annual report.

Proefstation voor de Bloemisterij 154

[Floriculture Research Station]
Address: Linnaeurlaan 2a, 1431 JV Aalsmeer
Telephone: (02977) 26151
Affiliation: Ministry of Agriculture and Fisheries
Status: Official research centre
Director: Ir J. van Doesburg
Sections: Plant nutrition and soil science, Dr Ir R. Arnold Bik; economics, efficiency and mechanization, Ir T.H. Edens; plant physiology and selection, Ir J. van Doesburg; cultivation and glasshouse climate, Ir C. Vonk Noordegraaf; pests and diseases, Dr G. Scholten; nursery, Ing J. Bonnyai
Graduate research staff: 13
Activities: Production; climate; plant nutrition; soil science; mycology; entomology; virology; chemical control of pests and diseases; selection; tissue culture; post harvest treatment; efficiency and mechanization.
Publications: Annual. *Bloemisterij Onderzoek in Nederland.*
Projects: Research and introduction of new and less well known cut flowers and potplants (Dr Ir C. Vonk Noordegraaf); cultivar trials for roses, carnation, freesia, (Ing W. van Marsbergen, Ing W. Belgraver; Ing C.V. Nes); soil disinfection trials, (Ir H. Rattink); tissue culture nephrolepis, cordyline, (Dr Ir L. Leffring); growth regulation and keepability, (Dr Ir N. Sytsema); energy saving and glasshouse climate, (Ir G. van de Berg); entomology/acarology, (M. van de Vrie); research on potplants on heated benches, (Ing P. van Weel); virus diseases, (Ir F. Hakkaart).

Proefstation voor de Boomkwekerij 155

– PBB
[Arboriculture Research Station]
Address: Valkenburgerlaan 3, 2770 AC Boskoop
Telephone: (01727) 3220
Affiliation: Ministry of Agriculture and Fisheries
Status: Official research centre
Research director: Ir W.I. Bosch
Departments: Propagation and cultivation, Dr Ir K.M. Joustra; in vitro propagation, Ir F.W. Perquin; breeding, Ir A.S. Bouma; assortment, H.J. van de Laar; crop protection, Ir N.G.M. Dolmans; J.E.A. Caron; work study, W. Schutte; mechanization, Ing P.H.M. Greymans; economics, Ir A.G. van der Zwaan; soil structure and fertility, Ing Th.G.L. Aendekerk
Graduate research staff: 7

Activities: Applied research for the improvement of production methods and quality of nursery stock; applied research to gather information about the behaviour and properties of nursery stock in the urban environment for better planning and management of urban plantings.
Publications: Year book, annual report.

Proefstation voor de Champignoncultuur   156

[Mushroom Experimental Station]
Address: Postbus 6042, 5960 AA Horst
Telephone: (04764) 944
Affiliation: Department of Agriculture and Fisheries
Status: Official research centre
Director: Dr L.J.L.D. van Griensven
Sections: Diseases and pests, Dr Ir A. van Zaayen; physiology of fructification and casing soil, Drs H.R. Visscher; genectis and breeding of new strains, Dr G. Fritsche; mushroom nutrition and composting, Drs J.P.G. Gerrits; mycorrhiza research, Ir G. Straatsma
Graduate research staff: 5
Activities: Plant production, cultivation of mushrooms; diseases and pests.
Publications: De Champignoncultuur, ten times a year.
Projects: Study of the composting process and the nutritional needs of the cultivated mushrooms to achieve the optimum substrate, based on horse manure and other natural manures, (Dr J.P.G. Gerrits); research on behaviour and control of pathogenic and competitive fungi in mushroom culture; research on prevention and control of diseases and pests of the cultivated mushroom; research on bacterial diseases of mushrooms, (Dr A. Van Zaayen); development of the cultivation of compost- and wood-inhabiting mushrooms not belonging to the genus Agaricus; pasteurization and incubation of compost for mushroom growing in bulk, (Dr J.P.G. Gerrits); the function of the casing layer in the culture of edible mushrooms, mainly of the genus Agaricus; studying influences on fructication and development of edible mushrooms, mainly of the genus Agaricus, (Dr H.R. Visscher); the influence of growing and harvesting conditions on quality and shrinkage of preserved mushrooms, (T.G.M. Pampen); breeding with different species of Agaricus; comparison of commercial strains of the cultivated mushroom and research regarding maintenance of strains and aspects of spawn, (Dr G. Fritsche).

Proefstation voor de Fruitteelt   157

– PFW
[Fruit Growing Research Station]
Address: Brugstraat 51, 4475 AN Wilhelminadorp
Telephone: (01100) 16390
Affiliation: Ministry of Agriculture and Fisheries
Status: Official research centre
Director: Ir R.K. Elema
Sections: Pomology, Dr Ir S.J. Wertheim; small fruits, Ir J. Dijkstra; physiology, Dr J. Tromp; soil science, Dr Ir P. Delver; entomology, Drs D.J. de Jong; phytopathology, Drs H.A.T. van der Scheer; economics, J. Goedegebure; experimental orchard Wilhelminadorp, J.J. Lemmens
Graduate research staff: 10
Activities: Fruit growing aspects of: pomology (top fruits and soft fruits), physiology, entomology, mycology, virology, soil science, farm management; improvement of profit ability (research into growing techniques, varieties, pest and disease control, quality, water management).
Publications: Annual report.
Projects: Variety trials for top fruit, (Ing P.D. Goddrie); virus and clonal research top fruit, rootstocks and interstocks, fruit tree nursery research, (Dr Ir H.J. v. Oosten); planting systems and pruning trials, application of growth regulators, testing unusual fruit-species, (Dr Ir S.J. Wertheim); growing techniques and variety trials strawberry and cane fruits, (Ir J. Dijkstra); soil management experiments, significance of N-supply and K-supply for top fruit, bitter pithin apples, trickle irrigation on top fruit, (Dr Ir P. Dlever); relation between shoot growth and fruit production, the uptake of Ca by apple trees, plant growth hormones in apple trees, (Dr J. Tromp); apple cancer, testing fungicides on fruit crops, strawberry root system diseases, (Drs H.A.T. v.d. Scheer); supervised control of fruit pests, protection against minor fruit pests, (Drs D.J. de Jong); economics in fruit growing, (J. Goedegebure).

Proefstation voor de Groenten- en Fruitteelt Onder Glas*   158

[Glasshouse Crops Research and Experiment Station]
Address: Postbus 8, 2670 AA Naaldwijk
Telephone: (01740) 26541
Affiliation: Ministry of Agriculture and Fisheries
Status: Official research centre
Director: Ir E. Kooistra
Sections: Horticulture and glasshouse climate, Ir C.M.M. van Winden; pests and diseases, Dr Ir L. Bravenboer; physiology, Dr Ir P.J.A.L. de Lint; soil science, Ir J. van den Ende; economy, mechanization, labour, Ir J.C.J. Ammerlaan
Activities: Vegetable crops; pests and diseases; growth; soil analysis; horticultural economics.

Proefstation voor de Rundveehouderij   159

– PR
[Cattle Husbandry Research and Advisory Station]
Address: Runderweg 6, 8219 PK Lelystad
Telephone: (03200) 22514
Affiliation: Ministry of Agriculture and Fisheries
Status: Official research centre
Director: Ir L.H. Huisman
Graduate research staff: 55
Activities: Applications of new techniques and systems within complete farms with consideration given to economic and social backgrounds: farm buildings; plant and animal production to contribute towards development of stock farming. Main research is carried out, in 6 experimental farms into grassland production, dairy cattle, beef cattle, and sheep production, agricultural engineering and building.
Publications: Annual report, research reports and publications.

DAIRY CATTLE AND BEEF CATTLE DEPARTMENT   160

Head: M.P. de Jong
Sections: Dairy cattle, A.B. Meijer; beef cattle, D. Oostendorp; veterinary aspects, R. Kommerij
Projects: Research into residuals of medicines and pesticides in milk and meat; improving the health status of cattle by hygienic and preventive measures; influence of housing on health status of dairy cattle; control of ectoparasites in cattle and sheep; management in relation to conception of dairy cattle, (Drs R. Kommerij); influence of housing stock for breeding on health and development; feeding systems for young stock in winter; influence of energy level on growth rate and development of young stock, (Tj. Boxem); application of by-products of the industry in rations for dairy cows; effect of feeding frequency on feed intake, health and performance of dairy cows; efficient methods of feeding concentrates and roughage to dairy cows; intake of different feedstuffs by dairy cattle; interaction of feedstuffs on total feed intake by dairy cows, (A.B. Meijer); indoor finishing of lambs with concentrates, (T. Ruiter); controlling parasites in calves and sheep; housing and well-being of beef bulls; increasing the number of lambs per ewe by heat induction or crossbreeding, (Ir D. Oostendorp); development of breeding schemes for cattle husbandry, (M.C. Verboon); influence of nature, composition and quantity of feeds on growth of bulls for beef, (H.E. Harmsen).

DETACHMENT OF THE AGRICULTURE ECONOMIC INSTITUTE   161

Head: G.J. Wisselink
Sections: Dairy husbandry, P.B. de Boer; beef husbandry, J. Doeksen
Projects: Normative farm models for cattle husbandry, (P.B. de Boer).

DETACHMENT OF THE INSTITUTE FOR MECHANIZATION, LABOUR AND FARM BUILDINGS   162

Head: J.A. Gels

DUTCH FERTILIZER INDUSTRY AGRICULTURAL BUREAU   163

Head: D.J. den Boer
Sections: Advisory farm development, H.M.P. van de Brandt; research farm development, P.B.R. Thiemann; farm equipment, P.J.M. Snijders; farm management systems, Dr Ir A. Kuipers
Projects: Research into the value of straw decomposed with ammonia and application possibilities in livestock feeding.

FARM SYNTHESIS DEPARTMENT   164

Head: Dr. D.C.M. Boonnan
Projects: Simple systems for housing and management of sheep and lambs, (T. Ruiter); model research into the effect of different breeding aims on the profitability of the dairy farm; influence of sprinkling of grassland as a farm system on the utilization possibilities of river basin clay; use of labour with different working methods on cattle farms; milking methods and equipment for milking, (J. van Geneijgen); management on a dairy farm with a limited use of fossil energy; developing and testing of a modern tying stall; orienting research into energy saving methods and systems for cattle husbandry, (P.J.M. Snijders); economic consequences of different grassland management in farming systems, (Drs P.B.R. Thiemann); culling guide for dairy herds, (Dr A. Kuipers); developing grazing systems for young-stock in summer, (Tj. Boxem); technical and economic possibilities of using Piemontese crossbreds for beef production, (D. Oostendorp).

GRASSLAND MANAGEMENT,   165
FODDER HARVESTING AND
CONSERVATION DEPARTMENT

Head: H. Thomas
Sections: Grassland production, H. Wieling; technical problems of cultivation, W. Luten; fodder harvesting and conservation, S. Schukking; restrictions of grassland use, H. Korevaar
Projects: Application of by-products of the industry in rations for beef bulls, (H.E. Harmsen); frequency and causes of urine scorch patches, (J.A. Keuning); protective sheetings on silos without sand cover, (H. Everts); storage and preservation of roughage, (S. Schukking); influence of restrictions of grassland use on gross grassland production, (H. Korevaar); possibilities and difficulties of intensive set stocking systems; control of undesired grasses in grassland, injecting slurry on grassland to prevent stench nuisance; the effects of large amounts of slurry on the growth, yield and composition of maize for silage and on soil and water pollution; the technique of reseeding of grassland: improvement of grassland soils; several aspects of pasture topping; effect of sprinkling grassland on grass growth and utilization of fertilizers, (W. Luten); storage in synthetic silos and separating cattle slurry; influence of sprinkling irrigation of slurry on grassland, on N-utilization and on the environment, (J. van Geneijgen); calculations of the influence of restrictions on land use and technical equipment on grassland management and forage production; research into optimal farming plans for beef and mutton production; standards for forage production; developing and testing labour and grass saving grazing systems for dairy cows, (H. Wieling).

REGIONAL RESEARCH DEPARTMENT   166

Head: M.C. Verboon
Projects: Effects of rotational crossbreeding in cattle, aiming at a combination of milk and beef production.

Rijks Agrarische Afvalwater   167
Dienst *

[Government Agricultural Wastewater Institute]
Address: Kemperbergerweg 67, 6816 RM Arnhem
Telephone: (085) 431245
Affiliation: Ministry of Agriculture and Fisheries
Status: Official research centre
Director: Ir J.H. Voorburg
Sections: Wastes and wastewater from agricultural industries; manure and odours from livestock
Graduate research staff: 4
Activities: Advising dairies, slaughterhouses and other agricultural industries, including farms, on purification of wastewaters, and treatment of manure.

Rijks-Kwaliteitsinstituut   168
voor Land-en
Tuinbouwprodukten

[State Institute for Quality Control of Agricultural Products]
Address: Bornsesteeg 45, 6708 PD Wageningen
Telephone: (08370) 19110
Telex: 75180 RIKIL
Affiliation: Ministry of Agriculture and Fisheries
Status: Official research centre
Director: Dr J.Th. Semeijns de Vries van Doesburgh
Laboratories: Microscopy, Drs W.J.H.J. de Jong; protein chemistry, Drs H.L. Elenbaas; carbohydrate chemistry, Drs B.G. Muuse
Graduate research staff: 15
Activities: Development of quality control criteria for agricultural food products; development of analytical procedures; normalization of analytical procedures; supervision of quality control.

Rijksdienst voor de IJsselmeerpolders 169

[IJsselmeerpolders Development Authority]
Address: Zuiderwagenplein 2, Postbus 600, 8200 AP Lelystad
Telephone: (03200) 99111
Telex: 40115
Affiliation: Ministry of Transport and Public Works
Status: Official research centre
Director: Professor Dr Ir R.H.A. van Duin
Sections: Scientific division, Dr J. de Jong; socio-economic research division, Professor Dr A.K. Constandse; soil mechanics, construction, traffic, Ir P.J.R. Heesterman; operational research division, Ir A. Hagting; research division for public works, Dr J. Nicolai
Graduate research staff: 47
Activities: Preparing new land for agricultural, urban and recreational purposes; settlement and employment; management of government property, natural resources, soil sciences, drainage and land use; plant production (cereals, rope seed and fruit); crop husbandry, plant protection; agricultural engineering, economics and sociology; operational research; management science, forestry.
Publications: Annual report.
Projects: Soil improvement, soil subsidence and ripening of soils, (physical and chemical) (Dr J.J. Ente); water management, drainage techniques and discharge problems, (E. Schultz); farm allocation and allotment, soil classification (J. Bakker); plant production (cereals, colza, apples), (G.J. de Jong); crop protection, (J. Duym); farm management, linear programming, labour and machinery (techniques and processes) for large scale farms, (A. Hagting); forestry, including ecological aspects, (C. Berger).

Rijksinstituut voor het Rassenonderzoek van Cultuurgewassen 170

– RIVRO
[Government Institute for Research on Varieties of Cultivated Plants]
Address: Postbus 32, 6700 AA Wageningen
Telephone: (08370) 19056
Affiliation: Ministry of Agriculture and Fisheries
Status: Official research centre
Director: Ir M.I. Hijink
Deputy director: Ir W. Scheijgrond
Sections: Botanical research on agricultural crops, Ir R. Duyvendak; research on the cultural value of agricultural crops, Ir H. Vos; botanical research on horticultural crops, Ir F. Schneider; performance trials on horticultural crops, Ir J.J. Bakker

Activities: Agricultural, horticultural and forest crops; close cooperation with technical work on testing of potato varieties carried out by the Netherlands Potato Consultative Institute.

BOTANICAL RESEARCH ON AGRICULTURAL CROPS* 171
Head: Ir R. Duyvendak
Departments: Cereals, Ing A.W. den Hartog; Grasses, Ir R. Duyvendak; Potatoes, sugar beet, commercial crops, forage crops, Ir H. Koster; Special research, Ir A. Houwing

BOTANICAL RESEARCH ON HORTICULTURAL CROPS* 172
Address: Postbus 16, 6700 AA Wageningen
Telephone: (08370) 19123
Head: Ir F. Schneider
Departments: Vegetable leaf crops, onion, mushroom and nursery crops, Ir F. Schneider; Cabbage crops, Ir N.P.A. van Marrewijk; Fruit, pulses, fruit vegetables and other vegetable crops, Ir B. Kiès; Bulbs and floricultural crops, Ir C.J. Barendrecht

NETHERLANDS POTATO CONSULTATIVE INSTITUTE TECHNICAL DEPARTMENT* 173
– NIVAA Technical Department
Address: c/o RIVRO, 6700 AA Wageningen

PERFORMANCE TRIALS ON HORTICULTURAL CROPS* 174
Head: Ir J.J. Bakker
Departments: Outdoor vegetables, Ir P. Riepma; Fruit crops, Ing P.D. Goddrie; Glasshouse vegetables, Ir J.H. Stolk

RESEARCH ON THE CULTURAL VALUE OF AGRICULTURAL CROPS* 175
Head: Ir H. Vos
Departments: Cereals, pulses, commercial crops, potatoes, Ir R. Wassenaar; Grasses, forage crops, green manure crops, sugar beets, Ir H.A. ten Velde

Rijksinstituut voor Natuurbeheer 176

– RIN
[Research Institute for Nature Management]
Address: Kemperbergerweg 67, 6816 RM Arnhem
or: Postbus 46, 3956 ZR, Leersum
Telephone: Arnhem (085) 452991; Leersum (03434) 2941
Affiliation: Ministry of Agriculture and Fisheries; Ministry of Cultural Affairs, Recreation and Social Welfare
Status: Official research centre
Director General: Professor Dr A.J. Wiggers
Director, Arnhem Branch: Dr P. Gruys
Director, Leersum Branch: Dr G.J. Saaltink
Sections: Advisory services and general research, D. Kruizinga; botany, Dr C.G. van Leeusun; estuarine ecology, W.J. Wolff; hydrobiology, P. Leentvaar; ornithology, Drs J. Rooth; zoology, Dr A. van Wijngaarden, Dr J.L. van Haaften; soil ecology, Dr D. Barel; physical planning ecology and geography
Graduate research staff: 55
Activities: Applied ecological research on environmental management, in particular on problems of conservation of natural plant and animal communities and of other natural resources, including research on biology and population ecology of insects, birds and game.
Publications: Annual report.

Rijksinstituut voor Onderzoek in de Bos- en Landschapsbouw 'De Dorschkamp' 177

[Dorschkamp Research Institute for Forestry and Landscape Planning]
Address: Postbus 23, 6700 AA Wageningen
Telephone: (08370) 19050
Affiliation: Ministry of Agriculture and Fisheries
Status: Official research centre
Director: Ir A.J. van der Poel
Sections: Silviculture, Ir C.P. van Goor; breeding, Ir R. Koster; forest protection, Dr J. Luitjes; economics, Ir A.P.W. de Wit; landscape planning, Ir P. Tideman
Activities: Silviculture; breeding; forest protection; economics; landscape planning.

Rijksinstituut voor Visserijonderzoek 178

[Netherlands Institute for Fishery Investigations]
Address: Postbus 68, 1970 AB IJmuiden
Telephone: (02550) 19131
Telex: 71044 RIVO
Affiliation: Ministry of Agriculture and Fisheries
Status: Official research centre
Director: K.H. Postuma
Graduate research staff: 23
Activities: Biological, technical, chemical, microbiological, and hydrographic research on behalf of the fishing industry. Interrelations of factors, including behaviour and fluctuations of fish stocks in the sea (especially oysters and mussels), especially the North Sea, and in fresh water. Investigations include control of the marketable product to guarantee wholesomeness.

BIOLOGICAL RESEACH INLAND FISHERY (CYPRINIDS) DEPARTMENT 179

Head: Drs J. Willemsen
Projects: Optimization of the fish stock in Lake Ijssel and related waters, (Drs J. Willemsen); composition of the fish fauna in upper and lower regions of the main Dutch rivers, and the interactions between these populations, relations between environment and characteristics of fish populations and the effects of management, especially for cyprinids, the possibilities for diminishing the deleterious effects of constructions like dams, weirs, sluices on fish populations and migrations, the effects of activities like sand and gravel digging, dumping of mud or similar material, (Drs W.G. Cazemier); optimization of the pike stock, influence of grasscarp on endemic fish species, influence of cooling water discharge upon the fish population, (Drs J. Willemsen).

BIOLOGICAL RESEARCH MARINE FISHERY DEPARTMENT 180

Head: Dr R. Boddeke
Projects: Sole, (Drs F.A. van Beek); plaice, turbot and brill, (Drs A.D. Rijnsdorp); herring, (Drs A.A.H.M. Corten); mackerel, (Drs A.T.G.W. Eltink); crustacean, (Dr R. Boddeke); gadoid fish, (Dr H.J.L. Heessen); interspecific relations between marine fishes, (Dr N. Daan); discard-research, and demersal young fish survey, (Drs F.A. van Beek); young fish surveys, (Drs A.A.H.M. Corten, Dr N. Daan, Dr H.J.L. Heessen).

BIOLOGICAL RESEARCH INLAND FISHERY (EEL) DEPARTMENT* 181

Head: Dr C.L. Deelder
Projects: Investigations on eel larvae, elver immigration metamorphosis, eel-stocks, eel growth, food etc. on behalf of the Dutch eel fisheries, investigations on the bottom fauna, in relation to measures on eel-stocks.

CHEMICAL, RESEARCH DEPARTMENT 182

Head: Drs M.A.T. Kerkhoff
Projects: Investigation into the occurrence and the significance of parasites in fish and shellfish, investigation into the occurrence and significance of diseases in fish and shellfish, (Drs P. van Banning); investigation into the biodegradability of pollutants in the aquatic environment, (R. van den Berg); influence on the environment of the offshore gas and oil industry as well of the sand and gravel industry, harbour dredgings, offshore structures, (Dr S.J. de Groot); investigation into the composition, the distribution, the sequence and the amount of phytoplankton in the Dutch coastal waters in relation with the biological water quality assessment, investigation into the influence of water pollution on the sanitary quality of living shellfish, investigation into the presence of toxic phytoplankton species in shellfish in relation to the occurrence of phytoplankton toxins, (M. Kat); investigation into the influence of aquatic pollution on the contents of PCB and pesticide in fish and shellfish, investigation into the influence of aquatic pollution on the contents of polychlorobenzenes and related compounds in fish and shellfish, investigation into the influence of water pollution on the contents of haloforms in fish and shellfish, (Drs M.A.T. Kerkhoff); hydrography of the Dutch coastal waters, investigation into the influence of water pollution on the contents of halogenated phenols in fish and shellfish, investigation into the influence of water pollution on the content of mercury in fish and shellfish, investigation on the occurrence of natural brominated compounds in marine organisms with natural high bromine contents (Drs H. Pieters).

MOLLUSCAN SHELLFISH RESEARCH DEPARTMENT 183

Address: Postbus 77, 4400 AB Yerseke
Telephone: (01131) 2781
Head: Drs A.C. Drinkwaard
Projects: Biological research concerning oysters and culture of oysters in the Dutch coastal waters and the Grevelingen, biological research concerning cockles and the culture of cockles in the Dutch coastal waters, hydrographic research into the physical and chemicalk composition of the Dutch coastal waters and the Grevelingen as far as relevant to the growth and the storage of shellfish, research into the functioning of the rewatering plots on the Yerseke Bank under the influence of changes as a result of the construction of the stormsurge barrier in the Eastern Scheldt, research into the improvement of the storage for oysters in oyster basins, (Drs R. Dijkema).

Technical Research Department 184

Head: E. J. de Boer
Projects: Application of electrical stimulation of sea and freshwater fish, research and development of a multi-purpose trawler for the year-round operation of low-energy and selective fishing methods, research to improve the operational methods and working conditions on board fishing vessels, research to decrease the energy costs of fishing vessels, research and development of a multi-purpose near-water trawler, (vacant); research to improve fishing with towed gears, (B. van Marlen); research and development of low energy and high selectivity fishing methods, (E.J. de Boer).

Rijksmuseum van Natuurlijke Historie* 185

[National Museum of Natural History]
Address: Raamsteeg 2, Postbus 9517, 2300 RA Leiden
Telephone: (071) 143844
Status: Official research centre
Director: Professor Dr W. Vervoort
Deputy director: Dr P.J. van Helsdingen
Graduate research staff: 16
Activities: Taxonomic research of animal kingdom based on world wide collections; biological oceanography; ecology; earth sciences.

Rijksproefstation voor Zaadcontrole 186

– RPvZ
[Government Seed Testing Station]
Address: Postbus 9104, 6700 HE Wageningen
Telephone: (08370) 19122
Affiliation: Ministry of Agriculture and Fisheries
Status: Official research centre
Research director: Dr Ir G.P. Termohlen
Sections: Purity, identification, cytology and seed cleaning, Ir W.J. van der Burg; germination, moisture and storage, Dr Ir J. Bekendam, Ir J. Vos mycology, Ir C.J. Langerak; bacteriology/virology, Dr J.W.L. van Vuurde; certification; Ing A. Stuurman; cytology, Z. van Dreven
Graduate research staff: 6

Activities: Quality analysis of seed samples; research on seed problems: germination, vigour, moisture content, storage, purity, taxonomy, cytology, pathology, advice on seedhandling, including drying cleaning techniques, disinfestation, storage and package of seeds.
Publications: Annual report.

Rijksuniversiteit Groningen* 187

Address: Postbus 72, 9700 AB Groningen
Telephone: (050) 119111
Telex: 53410 rugro
Status: Educational establishment with r&d capability

FACULTY OF SCIENCE AND TECHNOLOGY* 188

Dean: Dr F. van der Woude
Faculty board: Chairman, Dr F. van der Woude; Deputy Chairman, Professor Dr P.J.C. Kniper; Secretary, Dr A.W.L. Veln
Graduate research staff: 180

Subfaculty of Biology* 189

Activities: Zoology; plant systems; development biology of plants; plant physiology; plant ecology; genetics; microbiology; estuarine biology.

Rijksuniversiteit te Leiden* 190

[Leiden State University]
Address: Stationsweg 46, 2312 AV Leiden
Status: Educational establishment with r&d capability

RIJKSHERBARIUM* 191

[State Herbarium]
Address: Schelpenkade 6, Leiden
Telephone: (071) 13054
Research director: Professor Dr C. Kalkman
Graduate research staff: 25
Activities: Taxonomy of Netherlands and Indo-Malesian flora, including phanerogams, ferns and mosses; algology; mycology; Pacific plant geography; morphology, anatomy and palynology.
Publications: Flora Malesiana Bulletin.

Rijksuniversiteit te Utrecht* 192

[Utrecht University]
Address: Kromme Nieuwe Gracht 29, Utrecht
Status: Educational establishment with r&d capability

FACULTY OF VETERINARY MEDICINE 193

Address: Yatelaar 17, Postbus 80 163, 3508 TD Utrecht
Telephone: (030) 534851
Dean: Professor S.G. van den Bergh

Department of Animal Husbandry 194

Telephone: (030) 53214
Staff: Professor J. Bouw; Professor A. Hoogerbrugge; Professor A.Th. van Klooster; Professor G.J.W. van der Mey; Professor J.A. Renkema; Professor R.A. Prins
Activities: Genetic backgrounds/functional aspects of biochemical diversity in pet and experimental animals (biochemical polymorphism). Hereditary defects in domestic animals (astresia ani, pigs, podotrochleitis, horse, hipdysplasia, dogs, defects of legs and mouth, saddle horses, deafness, white dogs). Reproduction and perinatal mortality (cattle, sheep, pig) oestrus, environmental and hereditary influences). Development and adaptation af behaviour (genetics and environment, circadian rythm, mice, kennel syndrome, dogs). Animal wellbeing (farrowing, rearing and fattening of pigs, social environment and reproduction, pig). Housing and infectious diseases (mastitis, cattle, respiratory diseases, calves, lymphoid leucosis and FAPP housing, poultry). Microbes and digestion (rumen fermentation, domesticated and wild herbivores). Inorganic constituents in animal feed (nitrate poisoning, copper toxicosis). Economics of health programmes (economic significance of diseases).

Department of Bacteriology 195

Address: Biltstraat 172, 3572 BP Utrecht
Telephone: (030) 715544
Head: Professor J.F. Frik
Activities: E. coli enterotoxicosis (K88 adhesion factors, brushborders in pigs); epidemiology of salmonellosis (veal calves); leptospirosis (serodiagnosis); mycobacteria (mycobacterium johnei); dysentery (dysentery doyle, pig). Development of enterococcal resistance (antibiotics, calves); mycology (infectious dermatitis).

Department of Biochemistry   196

Address: Biltstraat 172, 3572 BP Utrecht
Staff: Professor S.G. van den Bergh, Professor L.M.G. van Golde
Activities: Hornonal regulation of hepatic lipogenesis (glycerolipids, ca-ions, hepatic mitochondria during ketosis, acetonaemia); lipid metabolism and surfactant synthesis in the lung (pulmonar type II epithelial cells); lipid metabolism in the brain (lipogenesis, ketone body metabolism); energy metabolism in parasites (liver fluke during maturation).

Department of Food Science and   197
Technology

Address: Biltstraat 172, 3572 BP Utrecht
Telephone: (030) 715544
Staff: Professor B. Krol; Professor J.G. van Logtestijn; Professor D.A.A. Mossel; Professor A. Ruitner; Profesoor D.J. Vervoorn; Professor G.M. Vogely
Activities: Microbiology and microbial hygiene (microbial association, food-borne disease, enteropathogenic bacteria, yeast and moulds, meat and cheese); hygiene of slaughter processes (animal treatment, heat resistance of nonsporeforming bacteria, storage and processing, inspection and quality control).

Department of Functional Morphology 198

Telephone: (030) 534336
Staff: Professor D.M. Badoux; Professor W. Harman; Professor W.A. de Voogd van der Straaten; Professor C.J.G. Wensing
Activities: Morphogenesis of the male and female reproductive system (gubernaculum outgrowth, testicular descent, ovarian development, pigs and dogs); functional morphology of the locomotion apparatus (gastrocnemius and tibialis cranialis muscles, crural mechanism, horse); cell differentiation in vitro (tracheal epithelium, chondrogenic expression, granulosa cells); cytogenetics (chromosomal aberrations in domestic animals, cytotaxonomy, Suidae); cell biology sperm cell (capacitation) .

Department of General Surgery and   199
Surgery of Large Animals

Telephone: (030) 531350
Head: A.W. Kersjes
Activities: General surgery (surgical therapy eyecarcinoma, cattle, cryosurgery sarcoids, horse, BCG (immuno) therapy); disease locomotion apparatus (spavin, horse, hereditary stifle joint, Shetland ponies, anatomy of transphyseal bloodvessels); abdominal surgery (abomasum, pulorotomy, omentopexy, cattle); laryngeal surgery (hemiplegia laryngis, horse).

Subdepartment of Veterinary Anaesthesiology   200

Telephone: (030) 531337
Head: Professor E. Lagerweij
Activities: Clinical anaesthesiology.

Department of Herd Health   201
Programmes and Ambulatory Clinic

Telephone: (030) 531096
Activities: Coliform mastitits in cattle (pathogenesis and treatment); respiratory disturbances in cattle (respiratory virus infections); virus infections in poultry (adenovirus, coronavirus, respiratory diseases).

Department of Immunology   202

Telephone: (030) 532471
Head: vacant
Activities: Clinical immunology (circulation immune complexes, rheumatoid factors, dog, sera, immunoglobulin (sub) classes, horse); regulation immune response (histocomptability systems in ruminants, macrophages and immune response, rat).

Department of Internal Medicine of   203
Large Animals

Telephone: (030) 531234
Staff: Professor H.J. Breukink; Professor J. Kroneman; Professor A.J.H. Schotman
Activities: Respiration and circulation disturbances in horses (allergens, bronchospasmolytics, diagnosis); diarrhoea in horses (pathophysiology); pathophysiology of the digestion (disturbances in digestion, cattle, cattle); diarrhoea in calves (cachexy syndrome in calves, differential diagnosis in neonatal diarrhoea in calves); locomotion disturbances (pig) (osteochondreosis); clinical biochemistry (haemotology, toxicology).

Department of Pathology   204

Telephone: (030) 534303/534365
Staff: Professor: E. Gruys; Professor J.M.V.M. Mouwen; Professor P. Wensvoort; Professor P. Zwart
Activities: Metabolic pathology (secondary amyloidosis congenital goitre, uterine carcinoma, hypertrophic osteodystrophy, hamster/goat/rabbit/dog; gastro-intestinal and hepatic disease (dogs, cats); disease of the locomotion apparatus (osteochondrosis, pig, glycogenosis type II, dog); neuropathology (teratology of central nervous system, eye pathology, domestic animals); renal pathology (proteinuria, glomerulonephritis); pathology of reptiles (skin, kidney, pancreas); varroatose in bees (Varroa jacobsonii); pharmacokinetics of antobiotics in pet birds (canary, pigeon); diseases of fishes (immunization, surface laver of CE-bacterium, Aeromonoas salmonicida complex); toxicity of TCDD (eggs, fry).

Department of Physiology   205

Address: Alexander Numankade 93, 3572 KN Utrecht
Telephone: (030) 715544
Head: Professor G.H. Huisman
Activities: Physiology of steady state exercise and exhaustion (bloodflow distribution in splanchnic area, proteinuria, dog); physiology of brood (O_2-tension, gaseous exchange, embryonic development); physiology of acclimatization (varying ambient temperature and lighting, poultry).

Department of Radiology   206

Telephone: (030) 531264
Head: vacant
Activities: X-ray diagnosis in horses (polyarthritis, podotrochleitis, spavin, myelography); arthography (dogs and horses); enteroclysis (small intestine, dogs).

Department of Small Animal Medicine   207
and Surgery

Telephone: (030) 534021
Staff: Professor A. Rijnberk; Professor H.W. de Vries
Activities: Metabolism and endocrinology (Diabetes Mellitus, Cushing's syndrome, dogs, hypertrophic osteodystrophy); clinical immunology (proteinuria, dog); cardiopulmonology (pumping action of heart,lung perfusion scintigraphy).

Department of Tropical Veterinary   208
Medicine and Protozoan Diseases

Address: Biltstraat 172, 3572 BP Utrecht
Staff: Professor G. Uilenberg; Professor D. Zwart
Activities: Toxoplasmosis (haplo-, diplo- and polyploid of lifecycle stages, Isospora (Toxoplasma) gondii, chemotherapy); ticks and tick-borne disease (Theileriosis, heartwater, Babesiosis); sporozoa infections (Babesia, mice, Eimeria, chicken); trypanosomal infections (immunological reactivity, antypyretics, seroepidemological monitoring of cattle near a wildpark in Mali).

Department of Veterinary   209
Helminthology and Entomology

Telephone: (030) 531117
Head: Professor D. Swierstra
Activities: Strongylata (epidemiology, pathophysiology).

Department of Veterinary   210
Pharmocology, Pharmacy and
Toxicology

Address: Biltstraat 172, 3572 BP Utrecht
Telephone: (030) 715544/531597
Staff: Professor Dr J.M.M. van den Bercken; Professor Dr A.S.J.P.A.M. van Miert
Activities: Experimental fever (inflammatory mediators, rumen motility, body temperature, antipyretics); thermoregulation and fever (central neurotransmission, thermoregulation during fever); pharmacokinetics in healthy and diseased domestic animals (phenyl butazone in dogs, sulfadimethoxine in gestating goats, sulfadimethoxine in goats during endotoxin induced fever); biochemistry and toxicology of foreign compounds (hepatic biotransformation of aromatic nitrocompounds, action mechanisms of phototoxic compounds (porphyrins, phylloerythrin)); biophysics of pyrethroids and DDT-analogues (sensory nervous system, nerve membrane, Na-channels); toxicology of chemical pollutants in aquatic environment (structure-activity relations, accumulation and elimination, toxicity of mixtures, fish); immunotoxicology (tri- and tera-alkyltin compounds, organotin compounds and antitumor activity).

Department of Virology   211

Telephone: (030) 532486
Staff: Professor Dr J.G. van Bekkum; Professor Dr M.C. Horzinek
Activities: Molecular virology (replication coronaviruses, lymphocytic choriomeningitis virus); clinical virology (feline infections peritonitis, coronavirus antigen, parvovirus/dog).

Sprenger Instituut   212

[Institute for Research on Storage and Processing of Horticultural Produce]
Address: Postbus 17, 6700 AA Wageningen
Telephone: (08370) 19013
Affiliation: Ministry of Agriculture and Fisheries
Status: Official research centre
Director: Drs G.J.H. Rijkenbarg
Deputy director: Dipl Ing H.F.Th. Meffert
Sections: Physical engineering, Ir G. van Beek; biology, Chr.E.M. Berkholst; quality research, Ir L. Gersons; chemistry, Dr N. Gorin; economics and statistics, Drs P. Greidanus; biochemistry, Ir W. Klop; microbiology, Ir P.C. Koek; food chemistry, Drs M.A. van der Meer; physics, Ir J.W. Rudolphij; physiology, Drs O.L. Staden; food technology, Ir E.P.H.M. Schijvens; process technology, Drs L.M.M. Tijskens

Activities: The institute is incorporated in the foundation for Agricultural Research in which growers, merchants, industrialists and consumers are represented. Preservation value of new fruit and vegetable types for different processing purposes is researched, for grading, packing, transporting and storage; technological research and quality research on processed products; farm economic and techno-physical research.

Stichting Bureau voor Gemeenschappelijke Diensten   213

– BGD
[Office of Joint Services]
Address: Postbus 33, 6700 AA Wageningen
Telephone: (08370) 19020
Affiliation: Ministry of Agriculture and Fisheries
Status: Official research centre
Director: Ing A. Troost
Sections: Experimental field service, Ing H.E. van Caem; workshop, J.H. van de Weerd, Th. J. Jansen, J. van Beek, R. Jansen; experimental farms, Ing A. van Santen
Activities: Agricultural, technical and administrative services to the Centre for Agrobiological Research (CABO) and the Institute for Research on Storage and Processing of Agricultural Produce (IBVL), including excursions to three experimental farms (at Randwijk, Wageningen, and Nagele).

'DE BOUWING' EXPERIMENTAL FARM*   214

Address: Randwijk
Head: W. de Jager

'DE EEST' EXPERIMENTAL FARM*   215

Head: B. van der Griendt

'DROEVENDAAL' EXPERIMENTAL FARM*   216

Address: Wageningen
Head: Ing P.J. Jochems

Stichting Laboratorium voor Bloembollenonderzoek   217

[Bulb Research Centre]
Address: Postbus 85, 2160 AB Lisse
Telephone: (02521) 19104
Affiliation: Ministry of Agriculture and Fisheries; Bulb Inspection Centre; Plant Protection Service
Status: Official research centre
Director: Dr E.B. van Julsingha
Deputy Director: Dr G.A. Kamerbeek
Sections: Plant physiology and biochemistry, Dr G.A. Kamerbeek; plant pathology, Dr Ir G. Weststeijn; crop science and economics, Ir M.J.G. Timmer; mechanization and rationalization, Irs A.F.G. Slootweg
Graduate research staff: 17
Activities: Research on the problems concerning cultivation, disease control, treatment and handling of bulbs: phytopathology; bulb propagation; soil water management; temperature treatment of bulbs for flowering in winter and/or other seasons; choice of varieties for specific purposes; efficient use of labour and machines; economic evaluation of production systems.
Publications: Annual report.

HORTICULTURE DEPARTMENT*   218

Head: Ir M.J.G. Timmer
Activities: Production and economics in the widest sense: temperatures; growth patterns; soil water management, and the adaptation of results for commercial use.

PHYSIOLOGY AND BIOCHEMISTRY DEPARTMENT*   219

Head: Dr G.A. Kamerbeek
Activities: Biochemistry; physiology of flowers and periodicity; flowering, non-parasitic diseases; keeping quality of cut flowers.

PHYTOPATHOLOGY DEPARTMENT*   220

Head: Ir G. Weststeijn
Activities: Biology; diagnosis and control of disease by bacteria; control of root rot pathogens; population control of eelworms; parasitic and non-parasitic diseases; chemical and physical control of diseases, pests and weeds.

Stitching Landbouwkundig 221 Bureau van de Nederlandse Meststoffenindustrie

– LBNM
[Foundation Agricultural Bureau of the Netherlands Fertilizer Industry]
Address: Thorbeckelaan 360, 15D etage, 2564 BZ 's-Gravenhage
Telephone: (070) 254000
Affiliation: Netherlands fertilizer industry: UKF and NSM
Status: Research association
Director: Dr P.F.J. van Burg
Sections: Arable crop husbandry, Dr K. Dilz; animal husbandry, D.J. den Boer; grassland husbandry, W.H. Prins; agricultural economics, W. Sluiman; agricultural engineering, G.D. van Brakel.
Graduate research staff: 11
Activities: Research directed towards a correct and balanced use of fertilizers in agricultural production systems. Advising industry on future development. Environmental aspects, soil science, crop husbandry, livestock husbandry and nutrition, agricultural economics.
Projects: Nitrogen fertilization of wheat and potatoes, agricultural value of liquid fertilizers, agricultural value of organic manures, (K. Dilz, W.H. Prins); maximizing grass production, intensification and animal health, (J.A. Keuning); intensification and milk quality,' planning and budgeting for optimum grass utilization, (D.J. den Boer); fertilizer handling and application systems, (G.D. van Brakel).

Stichting Nederlands 222 Graan-Centrum*

[Netherlands Grain Centre]
Address: Hamelakkerlaan 40, Postbus 47, 6700 AA Wageningen
Telephone: (08370) 13600
Status: Independent research centre
Secretary: Ir G.E.L. Borm
Graduate research staff: 1
Activities: Short-term research on problems affecting farmers, breeders and millers.

Stichting Technische en 223 Fysische Dienst voor de Landbouw

– TFDL
[Technical and Physical Engineering Research Service]
Address: Postbus 356, 6700 AJ Wageningen
Telephone: (08370) 19143
Telex: 45330 CT Wag
Affiliation: Ministry of Agriculture and Fisheries
Status: Official research centre
Director: Drs A.M.K. van Beek
Deputy director: Ir G. Borel
Sections: Physics, Dr K. Schurer; mechanical engineering, Ing H. Bosch; electronics, Ir G. Borel; calibration, G.J.W. Visscher; buildings, Ing F.J. Levsink; electron microscopy, Ing S. Henstra
Activities: Physical, mechanical and electronic measuring and controlling instruments, machines, tools and technical apparatus; calibration of thermometers, hygrometers etc; design of buildings for agricultural research.

Stichting voor 224 Bodemkartering

– STIBOKA
[Soil Survey Institute]
Address: Postbus 98, 6700 AB Wageningen
Telephone: (08370) 19100
Telex: NL 75230 Stiboka
Affiliation: Ministry of Agriculture and Fisheries
Status: Official research centre
Director: Ir R.P.H.P. van der Schans
Sections: Soil mapping, Ir G.J.W. Westerveld; soil and landscape research, Dr Ir J. Schelling; soil survey applications, Dr Ir J.C.F.M. Haans
Graduate research staff: 32
Activities: Systematic survey of the soils of the Netherlands; interpretation of soil survey data for various types of land use including crop production and forestry; systematic geomorphological survey of the Netherlands; ecological and physiognomic landscape survey of the Netherlands.
Publications: Soil survey papers, *De bodenkartering von Nederland, De boden von Nederland, Bodenkundige Studies, Boor and Spade, Geomortologische Kaart van Nederland* (all with summaries).

DEPARTMENT OF SOIL AND LANDSCAPE RESEARCH* 225

Head: Dr Ir J. Schelling
Sections: Soil chemistry; clay mineralogy; geology; geomorphology; historical geography; soil classification; micromorphology; statistics; information system; physiognomic landscape research

DEPARTMENT OF SOIL SURVEY APPLICATIONS* 226

Head: Dr Ir J.C.F.M. Haans
Sections: Hydrology; soil physical parameters; applied soil physics; arable land and grassland farming; horticulture; forestry; physical planning

DEPARTMENT OF SOIL SURVEYS* 227

Head: Ir G.J.W. Westerveld
Sections: Systematic soil mapping of the Netherlands, scale 1: 50 000; soil surveys and interpretations on contract, scale 1: 5 000 to 1: 100 000; soil cartography

Stichting voor Plantenveredeling 228

– SVP
[Foundation for Agricultural Plant Breeding]
Address: Postbus 117, 6700 AC Wageningen
Telephone: (08370) 19112
Affiliation: Ministry of Agriculture and Fisheries
Status: Official research centre
Director: Dr Ir H. Lamberts
Sections: Potatoes, Ir H.T. Wiersema; cereals, Ir J. Mesdag; flax, Ing M.A. Jongmans; grasses and other fodder crops, Dr G.E. van Dijk; cytogenetics, Dr Ir W. Lange; physiology and chemistry, Dr P. Miedema; biometry, Ir M. Mesken; farm and field trial management, Ing M. van der Weg; Lelystad farm, Ing J. Glerum
Activities: Agricultural plant breeding; the Dutch Breeder's Association, the Industrial Board for Agriculture, the Agricultural University and the Ministry of Agriculture and Fisheries are represented on the foundations board.

Studie- en Informatiecentrum TNO voor Milieu-onderzoek* 229

– SCMO-TNO
[Study and Information Centre on Environmental Research TNO]
Address: Postbus 186, 2600 AD Delft
Telephone: (015) 569330
Telex: 38071 zptno nl
Affiliation: Netherlands Organization for Applied Scientific Research (TNO)
Status: Official research centre
Director: Drs P. Windel

Technisch Adviesbureau van de Unie van Waterschappen BV* 230

– TAUW
[Engineering Consultants to the Union of Drainage and River Boards]
Address: Postbus 479, 7400 AL Deventer
Telephone: (05700) 20711
Telex: 49545 INCON NL
Status: Independent research centre
Director: J.Pieters
Graduate research staff: 120
Activities: Consultancy services on land reclamation, sea defences, drainage and irrigation, roads and bridges, environmental engineering, sewage water treatment plants, sewerage.

Unilever NV* 231

Address: Burgemeester's Jacobplein 1, Rotterdam
Status: Industrial company
Head of Unilever Research: Sir Geoffrey Allen
Activities: See entry for Unilever PLC in the United Kingdom.

UNILEVER RESEARCH VLAARDINGEN LABORATORY* 232

Address: Postbus 114, Vlaardingen
Head: Professor W.J. Beek
Graduate research staff: 300
Activities: A wide range of research in the chemical, phsyical, engineering and biological sciences devoted to the development of foods (edible fats, dairy products, frozen foods, etc), detergents and chemicals.

Vereniging Vleutens Proeftuin   233

[Vleutens Research Garden Association]
Address: Alendorperweg 47, 3451 GL Vleuten
Telephone: (03406) 1326
Director: A.J. Vijverberg
Departments: Garden, A.M. Gresnigt
Graduate research staff: 1
Activities: Plant production and breeding; studies in energy saving production.
Projects: Energy-saving production, production of products in different substratums (eg Rockwool), (A.M. Gresnigt).

Waterloopkundig Laboratorium*   234

[Delft Hydraulics Laboratory]
Address: Rotterdamseweg 185, Postbus 177, 2600 MH Delft
Telephone: (015) 569353
Telex: 38176 hydel-nl
Parent body: Stichting Waterbouwkundig Laboratorium
Affiliation: Netherlands Organization for Applied Scientific Research (TNO)
Status: Official research centre
Director: Ir J.E. Prins
Coordinator of basic research: Ir H.N.C. Breusers
Sections: Delft laboratory, Ir A. Paape;,De Voorst laboratory, Ir J.J. Vinjé; site investigations service, Ir J.E. Mebius
Graduate research staff: 150
Activities: Fundamental and applied scientific research in the fields of hydraulics, hydrodynamics, hydrology, morphology and environment through theoretical studies, mathematical model investigations, hydraulic model investigations and site investigations.
Publications: Quarterly and monthly periodicals.
De Voorst laboratory: Postbus 152, 8300 AD, Emmelord.

NEW CALEDONIA

Office de la Recherche Scientifique et Technique Outre-Mer 1

[Overseas Scientific and Technical Research Office]
Address: Promenade de l'Anse Vata, BP A - 5, Nouméa
Telephone: 26 10 00
Telex: COTRANS 120 NM ATTN ORSTOM
Status: Official research centre
Director: Paul de Boissezon
Graduate research staff: 64
Activities: Terrestrial studies - globe structure, composition and evolution of the global crust, mineral soil and water resources; life sciences natural ecosystems and their alteration, agricultural systems, fungal diseases, arthropod zoology; oceanographic and lacustrine studies - structure and mechanism of the marine environment, optimization and exploitation of resources; continental and brackish waters; human sciences - economic and social development, anthropology, prehistory; hygiene - biological substances.
Publications: Annual report.
Projects: Fertility and evolution of soils under cultivation (B. Bozon, M. Latham); characterization of terrestrial ecosystems; phytogeographic surveys (Ph. Morat); reproduction, germination, growth, photosynthesis (Y. Bailly); soil survey and cartography for cultivation and forestry (M. Latham, A. Beaudou); biological pest control (J. Gutierrez, J. Chazeau, L.O. Brun); mildew in Arabica coffee (B. Boccas, B. Seivert); zones suitable for tunny-fish (J. Marcille, Y. Dandonneau, M. Petit, W. Bour); coastal and lacustrine fisheries (P. Fourmanoir, Ch. Conand, W. Bour); live bait (T. Boely, F. Conand); plant pharmacology (M. Debray, D. Bourret); pharmacological potential of marine substances (C. Levi, A. Clastres, P. Potier, J. Pusset).

AGRONOMY DEPARTMENT 2
Head: B. Bonson

ARCHAEOLOGY DEPARTMENT 3
Head: D. Frimigacci

BOTANY DEPARTMENT 4
Head: P. Morat

ENTOMOLOGY DEPARTMENT 5
Head: J. Guttierez

GEOGRAPHY DEPARTMENT 6
Head: J.C. Roux

GEOLOGY AND GEOPHYSICS DEPARTMENT 7
Head: J. Recy

HYDROLOGY DEPARTMENT 8
Head: J. Danloux

OCEANOGRAPHY DEPARTMENT 9
Head: Y. Dandonneau

PEDOLOGY DEPARTMENT 10
Head: M. Latham

PHARMACOLOGY DEPARTMENT   **11**   **PHYTOPATHOLOGY DEPARTMENT**   **12**

Head: M. Debray   *Head:* B. Boccas

NEW ZEALAND

Agricultural Economics Research Unit 1

Address: Lincoln College, Canterbury
Telephone: (03) 252 811
Affiliation: Lincoln College
Status: Educational establishment with r&d capability
Director: Peter D. Chudleigh
Graduate research staff: 15
Activities: Research, data collection through survey methods, and the sponsoring of seminars and discussions between industry sectors and government on agricultural economics and associated areas.
Publications: Biannual review, research reports and discussion papers.
Projects: Biomass production economics (C. McLeod); economic surveys of farmers (M. Rich); land use and agricultural policy (K. Leathers); agricultural research - priority formation and adoption (A. Beck); rural roading economics (P. Chudleigh); horticultural marketing (R.L. Sheppard); model of world sheepmeat supply and demand (R.L. Sheppard, A.C. Zwart); Japanese agricultural policies (A.C. Zwart).

Forest Research Institute 2

Address: Private Bag, Rotorua
Telephone: 073 82 179
Telex: NZ 21080
Affiliation: New Zealand Forest Service
Status: Official research centre
Director of Research: Dr C. Bassett
Activities: The Forest Research Institute is a multidisciplinary organization where most aspects of forestry and forest products research are undertaken. Aspects of soil science, drainage and irrigation, land use, plant breeding and protection and livestock breeding, husbandry, nutrition and control are also considered when applicable. Many of these aspects are particularly relevant to the Protection Forestry Research Division.
Publications: Annual report.

FOREST PRODUCTS DIVISION 3

Director: A.J. McQuire
Activities: Research into wood quality, drying, preservation, engineering, pulp and paper, adhesives and composite products.
Projects: Wood quality (Dr J.M. Harris); timber drying (Dr J.A. Kininmonth); wood preservation (Dr J.A. Butcher); timber engineering (Dr G.B. Walford); adhesives and composite products (C.I. Hutchinson); pulp and paper (Dr J.M. Uprichard).

PRODUCTION FORESTRY DIVISION 4

Director: E.H. Bunn
Branch heads: Dr G.B. Sweet, J.R. Trustin
Activities: Research into forest establishment, mensuration, pathology, entomology, pine silviculture; genetics and tree improvement, tree physiology, soil and site productivity, social science and harvesting.
Projects: Forest establishment (C.G.R. Chavasse); radiata pine silviculture (Dr W.R.J. Sutton); indigenous forest management (A.E. Beveridge); economics, social science and harvesting (Dr I.J. Bourke); forest mensuration (Dr A.D. McEwen); soils and site productivity (Dr G.M. Will); forest pathology (Dr P.D. Gadgil); forest entomology (Dr G.P. Hosking); tree physiology (Dr D.A. Rook); genetics and tree improvement (I.J. Thulin).

PROTECTION FORESTRY DIVISION 5

Director: J.U. Morris
Activities: Research into geohydrology, plant ecology, animal, and watershed rehabilitation research.
Projects: Geohydrology (Dr C.L. O'Loughlin); plant ecology (Dr U. Benecke); animal research (Dr C.N. Challies); watershed rehabilitation research (A.H. Nordmeyer).

Heavy Engineering Research Association Incorporated 6

– HERA
Address: PO Box 76-134, Auckland
Telephone: (09) 278-9693
Parent body: Department of Scientific and Industrial Research
Director: Dr R. Shepherd
Activities: Research includes hydroelectric project components such as penstocks and gates, elements of buildings, bridges and towers, storage tanks, pressure vessels, process and metalworking machinery, trucks and wagons, crushing plant and concrete mixers, equipment used in the energy industry, in fertilizer manufacture, in pulp and paper manufacture, and in ships.

Invermay Agricultural Research Centre 7

Address: Private Bag, Mosgiel
Telephone: (02489) 5009
Affiliation: Ministry of Agriculture and Fisheries
Status: Official research centre
Regional Director: Dr A.J. Allison
Departments: Agronomy, R.S. Scott; soil chemistry, A.G. Sinclair; animal fertility and production, R.W. Kelly; animal nutrition, K.R. Drew; insect control, K.M. Stewart; weed science, L.D. Bascand
Activities: A wide range of research for this southern south island region includes soils, plants and animals, rate of pasture growth and establishment/renovation by direct drilling, grazing, forestry, cereal cultivar evaluation, mycorrhiza research, irrigation, ecology and weed control, soil chemistry and fertility, biogas production, and deer farming.
Projects: Agronomy: studies of grasslands 'matu' lotus (Dr R.S. Scott); mycorrhiza research (Dr I.R. Hall); legume inoculation and pelleting (Dr W.L. Lowther). Weed science: control of daisy, hawkweed, and sweet briar; tolerance of lucerne to hexazinone and glyphosate; control of weeds in establishing blackcurrants, wheat,

and lotus seed crops (F.A. Meeklah). Entomology: control of grass grub by rolling; control of porina by mob stocking; (Dr K.M. Stewart); field biology; insect pests of clover (Dr B.I.P. Barratt). Soil chemistry and fertility: soil testing and plant analytical services (Dr R. Dolby); studies of soil acidity (J.L. Grigg); effectiveness of sulphur fertilizers (Dr P.D. McIntosh). Energy and environment: biogas production; fluorine survey (Dr D.J. Stewart). Animal production: merino cross breeding evaluation; high fecundity flocks (Dr R.P. Lewer); ram and hoggett reproductive activity (G.H. Davis); control of internal parasites in lambs (K.H.C. Lewis); GnRH hormones in beef cattle (Dr G.W. Montgomery). Animal nutrition: deer farming and winter nutrition (Dr P.F. Fennessy); trace element requirements of young cattle grazing kale (Dr T.N. Barry). Vertebrate pest control: anticoagulant poison toxicity (Dr M.E.R. Godfrey). District-field science: rate of pasture growth; lucerne cultivars evaluation; cereal cultivars, ryegrass and winter hardy grass evaluation; pasture establishment and renovation by direct drilling; winter grazing and grazing forestry; greenfeed maize, potato cultivars (P.B. Greenwood, G.G. Cossens, D.W. Brash, W.H. Risk); irrigation investigations (P.B. Greenwood).

TARA HILLS HIGH COUNTRY RESEARCH STATION 8

Address: Private Bag, Omarama
Officer in charge: Dr A.R. Bray
Sections: Agronomy, B.E. Allan; animal breeding and management, Dr A.R. Bray
Activities: Problems of high country tussock grassland production; grazing and performance of high fecundity sheep.

WOODLANDS RESEARCH STATION 9

Address: Woodlands, RD 1, Invercargill
Telephone: 393-012
Officer in charge: K.F. Thompson
Activities: Research covers studies of grazing management and control of internal parasites in lambs; genetic studies in sheep.
Projects: Pasture allowance studies; control of gastrointestinal parasites in lambs.

Lincoln College 10

Address: Canterbury
Telephone: (03) 252-811
Status: Educational establishment with r&d capability
Principal: Professor J.D. Stewart
Graduate research staff: 117

ANIMAL SCIENCES GROUP 11

Chairman: Professor A.R. Sykes
Activities: Research activities on nutritional diseases and requirements of young ruminants; grazing preferences and intakes; breed selection for adaptation to entirely pastoral systems; control mechanisms on reproduction and growth.
Projects: Growth in young ruminants - parasitism and nutrition (Professor A.R. Sykes); intake and diet selection in young ruminants (Dr D.P. Poppi); calcium metabolism in ruminants - the young lamb, and antler growth in deer (Professor A.R. Sykes); endocrinology - equine reproduction (Professor C.H.G. Irvine); control of antler growth (Dr G.K. Barrell); sheep breed improvement - genetic aspects of tooth wear (D.G. Elvidge); sheep resistance to parasitism, and selection for growth rate and composition (Professor A.R. Sykes).

DEPARTMENT OF AGRICULTURAL ENGINEERING 12

Head: Professor G.T. Ward
Activities: Agricultural engineering and building, water resource development, drainage and irrigation, renewable energy resources.
Projects: Wind energy resource survey of New Zealand (Dr N.J. Cherry); global wind energy assessment (Dr N.J. Cherry); erosion processes in step-pool streams (Dr T.R. Davies); underflow in gravelly river beds (Dr T.R. Davies); slip-metre development (R.E. Chilcott); performance characteristics of wind turbines for advanced farm, rural and remote use (R.E. Chilcott); utilization of solar energy for agricultural processing and building design (Professor G.T. Ward); alternative covering materials for greenhouses (A.L. McLellan); design of surface irrigation structures (Dr D.G. Huber).

DEPARTMENT OF AGRICULTURAL MICROBIOLOGY 13

Head: Professor A.P. Mulcock
Activities: Plant pathology: diseases of agricultural and horticultural crops caused by viruses, fungi and bacteria; integrated pest and disease control and plant disease forecasting; crop loss assessment. Soil and water microbiology: nitrogen fixation by Rhizobium sp; factors affecting the movement of microorganisms through soil and in water; survival and identification of coliform bacteria in water. Food microbiology: survival of pathogens in foods; factors effecting growth of food spoilage microorganisms.
Projects: Ethanol fuel from crops and fermentation and crop storage (Professor A.P. Mulcock); identification of Rhizobium sp; virus diseases of horticultural and agricultural plants, disease control in cereals, peas and lucerne (Dr R.C. Close); causal disease-yield loss rela-

tionships and crop loss assessment in cereals, epidemiology and control of seed-borne diseases in legumes (Dr R.E. Gaunt); studies on the movement of microorganisms in underground water, the microbiology of materials used for holding potable water (Dr M.J. Noonan).

DEPARTMENT OF ENTOMOLOGY 14

Head: Professor G.R. Williams
Activities: Research in applied entomology - especially involving agricultural and horticultural pests, taxonomy and distribution of native insects, island biogeography of native insects and vertebrates.

DEPARTMENT OF FARM MANAGEMENT 15

Head: Professor J.B. Dent
Activities: Management and valuation of both urban and rural land and property, management control systems for farming, land and water use planning, alternative fuels assessment.
Projects: Computer technology application in farming and agricultural industry; information systems development for farmers (Dr Nuthall); alcohol fuels from farm crops; crop protection management (Professor Dent); irrigation planning at regional level (Dr Frengley); water management planning and control at farm level (Mr McGregor) farm machinery replacement (Mr Woodford); economics of farm syndication (Mr Bartholomaeus); recreational land use and value (Mr Jones).

DEPARTMENT OF HORTICULTURE, LANDSCAPE AND PARKS 16

Head: Professor R.N. Rowe
Activities: Research in landscape, parks and recreation, horticultural management and crop production. In the latter area the main thrust is into studies of root growth, shoot growth, flower initiation, fruit training and pruning systems (with special emphasis on mechanical harvesting potential) together with an evaluation of fruit and vegetable cultivars for the local climates and conditions. Studies on the comparative nutrition of nursery plants are also being made, which includes an evaluation of the influence of media and fertilizers during propagation and growing-on, and biological concepts of agricultural production are being demonstrated.
Projects: Nutrition of container-grown plants (Dr M.B. Thomas); biological husbandry (R.A. Crowder); culture and physiology of the grapevine (Dr D.I. Jackson); importance of roots to plant growth (Professor R.N. Rowe).

DEPARTMENT OF PLANT SCIENCE   17

Head: Dr J.G.H. White
Activities: Research into crop and pasture production and experimental approaches adopted go well beyond conventional agronomy and include very detailed measurements of plant performance in the field. Crop physiology features prominently in the research programme, not only in relation to crop/environment interactions but also to growth and development of cereals, grain legumes and pasture plants.
Projects: Physiology of grain production in wheat (Professor R.H.M. Langer); grain development in cereals (Dr W.R. Scott); physiological limitations to crop production (Dr J.N. Gallagher); physiological effects of growth retardants; biology of selected weeds (Dr R.J. Field); agronomy of field crops (Dr J.G.H. White); nutrient uptake by legumes (Dr P. Jarvis); seasonal productivity of pastures (M.L. Smetham); productivity of grain legumes (G.D. Hill); quantitative genetics of wheat (Dr A.G. Fautrier); pasture management (R.J. Lucas); experimental methods (N.S. Mountier, B.G. Love); ecology of woody weeds (Dr G.T. Daly).

DEPARTMENT OF SOIL SCIENCE   18

Head: Professor R.S. Swift
Activities: The integration of a knowledge of soil formation with soil chemical, physical and biological properties to give an understanding of soil fertility and potential land use.
Projects: General research in soil science particularly in areas of pedology, mineralogy, soil chemistry, soil physics and soil fertility.

DEPARTMENT OF WOOL SCIENCE   19

Head (acting): Dr B.R. Wilkinson
Activities: Animal production, qualitative and quantitative wool production, measurement of economically important wool characters, fleece establishment and nature of follicle and fibre populations.
Projects: Prediction of susceptibility to fleece yellowing and fleece rot; seasonal nature of mohair production; efficiency of wool production; fractionation of raw wool (Dr B.R. Wilkinson).

RURAL DEVELOPMENT AND   20
EXTENSION CENTRE

Director: D.B. McSweeney
Activities: Extension teaching and research in the broad area of information/diffusion including related issues of agricultural training and rural social change; research in attitudes of farmers to the state Farm Advisory Service.

Projects: Continuing series of investigations into aspects of the uptake of new information by farmers (G.F. Tate); the long term career prospects for graduates with agriculturally related qualifications (O.M. Wilson).

Lincoln Research Centre   21

Address: Box 24, Lincoln
Telephone: (03) 228 029
Affiliation: Ministry of Agriculture and Fisheries
Status: Official research centre
Regional Director: A.D.H. Joblin
Sections: Agronomy, Dr C.G. Janson; management, J.M. Hayman; pest control, T.E.T. Trought; soil fertility, D.S. Rickard; biometrics, D.J. Saville
Activities: Research programmes are directed to solving this region's agricultural production problems and the development of new production opportunities including investigation into the feasibility of growing beet crops for ethanol production, trials to measure irrigation responses in areas being developed for major irrigation schemes, and development of sheep management strategies that will enhance lamb survival and productivity.
Projects: Agronomy: time of sowing of sugar beet and fodder beet; willows for cellulose production; effect of irrigation and sowing date on cereals (Dr R.J. Martin; fodder beet weed control; wild oats; paraquat formulations; field beans (J.H. Butler); management: relationships between pasture growth and climate; prairie grass in north Canterbury (Dr J.E. Radcliffe); pest control: sitona weevil; Argentine stem weevil biology; grass grub; application techniques (Dr R.A. French, Dr S.L. Goldson, T.E.T. Trought); soil fertility: fertilizers for beets (R.C. Stephenson, Dr B.F.C. Quin).

ADAIR AGRICULTURAL RESEARCH   22
STATION

Address: c/o Ministry of Agriculture and Fisheries, Box 516, Timaru
Telephone: (056) 85-082
Officer in charge: C.C.S. McLeod
Projects: Barley production in south Canterbury; pre-harvest establishment of brassicas in barley; grass cultivar evaluation under mowing; effect of cultivation techniques on wheat yields; angora goat studies.

TEMPLETON RESEARCH STATION   23

Address: Box 33, Templeton
Telephone: (03) 498-165
Scientist in charge: J.M. Hayman
Projects: Evaluation of new grass cultivars under cutting; nitrogen responses on irrigated pasture; effect of soil type on irrigation requirements and dry matter response of pasture and lucerne (J.M. Hayman); influence of nutrition and body composition on ewe milk production; comparison of breeds and crosses for meat production from aged ewes (K.G. Geenty).

WINCHMORE IRRIGATION   24
RESEARCH STATION

Address: Private Bag, Ashburton
Scientist in charge: D.S. Rickard
Projects: Fodder and sugar beet studies; barley irrigation (E.G. Drewitt); nitrogen responses on irrigated pasture; alternatives to superphosphates; chemical and soil physical analyses (Dr B.F.C. Quin); storage losses in hay bales (Dr G.H. Scales); winter and spring nutrition of beef cattle (Dr G.H. Scales); Tasman data loggers (P.D. Fitzgerald).

Massey University   25

Address: Private Bag, Palmerston North
Telephone: (063) 69-099
Status: Educational establishment with r&d capability

AGRICULTURAL AND   26
HORTICULTURAL SCIENCES
FACULTY

Centre for Agricultural Policy Studies   27

Director: Dr E.M. Ojala
Graduate research staff: 4
Activities: The evolving role of the agricultural sector in the New Zealand economy; the evolution of demand, supply and policy factors influencing the international agricultural markets; factors and policies influencing the comparative advantage of New Zealand farming; promotion of discussion on issues of importance to New Zealand agriculture.
Publications: *Agricultural Policy Report Series*; *Agricultural Policy Proceedings Series*.
Projects: New Zealand agriculture in the future world (Dr E.M. Ojala, Professor A.N. Rae, Dr W.R. Schroder, Dr A.C. Lewis); strategic issues in the rural sector (C.W. Maughan); alternative taxation systems for agriculture (Dr A.D. Meister).

Dairy Husbandry Department   28

Director: Dr A.W.F. Davey
Graduate research staff: 13
Activities: The department is interested in increasing animal production of dairy and pig farms through improved livestock husbandry. Emphasis is being placed in the present research on measuring differences in pasture utilization and nutrient metabolism by dairy cows of differing genetic merit.
Projects: Nutrition of the growing pig with emphasis on dietary protein quality (Dr W.C. Smith); studies on the nutritional and physiological differences between dairy cows of high and low breeding index for milk fat yield (Dr A.W.F. Davey); energy partitioning in dairy cows; milking machines (Dr C.W. Holmes); endocrinology of dairy cattle (Professor D.S. Flux); magnesium metabolism in dairy cows (Dr G.F. Wilson); mechanisms of milk secretion (Dr D.D.S. Mackenzie); nitrogen metabolism in ruminants (Dr I.M. Brookes); physiology of overfat sheep (S.W. Peterson).

Seed Technology Centre   29

Director: Dr M.J. Hill
Activities: The aim of the centre is to train both technical and postgraduate students from developing countries in the South-East Asian and Pacific regions in relevant aspects of seed technology. Principle areas of current research include post-harvest technology particularly in the areas of seed drying, seed storage, and seed quality control. Some research, particularly with vegetable, forage and covercrop legumes is also being undertaken.
Projects: Testing (Mrs D.E.M. Meech); processing (C.R. Johnstone); certification (F.W.B. Wilson); production, drying and storage (Dr M.J. Hill).

Sheep Husbandry Department   30

Director: Professor Robert D. Anderson
Graduate research staff: 9
Activities: Animal production (sheep, beef cattle, red deer and goats), livestock breeding, husbandry and nutrition.
Publications: Faculty research report.
Projects: Sire by stocking-rate interactions in Romney sheep (A.L. Rae); response to selection for fleece weight and open faces in Romney sheep (H.T. Blair); performance of half Romney, quarter Cheviot, quarter Booroola-Merino sheep (G.A. Wickham); estimation of variance components in breeding (R.D. Anderson); breeding from ewe hoggets (M.F. McDonald); lamb mortality studies: factors affecting heat loss from the newborn lamb (S.N. McCutcheon); growth and carcass characteristics of Angus steers (R.A. Barton); role of bulls of dairy origin on an intensive fattening farm (S.T. Morris); ultrasonic measurement of fat depth; factors affecting the carcass and meat quality of fallow deer (R.W. Purchas).

Meat Industry Research Institute of New Zealand Incorporated   31

– MIRINZ
Address: PO Box 617, Hamilton, Waikato
Telephone: (071) 56-159
Telex: NZ 21470
Parent body: Department of Scientific and Industrial Research
Status: Independent research centre
Director: Dr C.L. Davey
Departments: Science, Dr P.M. Nottingham; engineering, L.F. Frazerhurst
Graduate research staff: 35
Activities: Research and development on all matters appertaining to meat, meat by-products and their processing, and to implement the results of this work within the New Zealand meat industry. Principal activities comprise basic and applied investigations directed towards promoting improvements in the quality, utilization, and processing of meat and meat by-products, the development and implementation in works' practice of new products and processes, and the efficiency of works' plant, operations and services. These activities include research in chemistry, biochemistry, microbiology, cell structure, home economics and engineering, as well as the application of scientific and engineering principles to processing, refrigeration, packaging, storage, transport, plant design and pollution control. In addition, the institute carries out a comprehensive programme of education and liaison with industry.
Publications: Annual report.
Projects: Construction, design and performance of refrigerated facilities; energy; physics of chilling, freezing, storage and transport of meat (Dr A.K. Fleming); electrical stimulation; meat quality at retail; pre-slaughter stress and meat quality (Dr B.B. Chrystall); handling and storage in cold stores; mechanization of mutton slaughter and dressing (Dr G.R. Longdill); meat processing and microbial storage; microbial ecology of meat; public health and meat hygiene (Dr C.O. Gill); muscle structure and function; stunning and slaughter (Dr C.E. Devine); rendering (T. Fernando); waste treatment methods and waste water analytical techniques (Dr R.N. Cooper).

Ministry of Agriculture and Fisheries   32

Address: Dominion Farmers' Institute Building, PO Box 2298, Wellington
Telephone: (04) 720-367
Status: Government department
Director General: M.L. Cameron
Divisions: Advisory services, J.M. Hercus; dairy, G. McMillan; animal health, Dr G.H. Adlam; fisheries management, B.T. Cunningham

AGRICULTURAL RESEARCH DIVISION   33

Address: Private Bag, Wellington
Telephone: (04) 720 367
Telex: MAFWN (TELEGRAPHIC) 31532
Director: Dr J.B. Hutton
Assistant director (animal research): Dr D.E. Wright
Assistant director (soil and plant research): Dr R.S. Scott
Assistant director (divisional services): B.R. Keenan
Activities: One of ten divisions of the Ministry of Agriculture and Fisheries, the work is mainly concerned with increasing the volume, efficiency and profitability of agricultural and horticultural production. Research and development is carried out on new products and production systems and on agricultural equipment, including evolving methods of measuring and monitoring levels of product and environmental contamination acceptable both within New Zealand and internationally, and providing essential services to producers such as comprehensive soil and plant analysis and associated fertilizer advisory schemes. The division also develops the computerized management information systems which includes the proposed and current research project directory.
Publications: Annual report.
See separate entries for: Invermay Agricultural Research Centre
Lincoln Research Centre
Palmerston North Agricultural Research Centre
Ruakura Agricultural Research Centre
Wallaceville Animal Research Centre
Projects: Improving lambing performance using high fertility strains and breeds of sheep; improving growth of livestock through non-castration; improving ram and ewe meat quality through accelerated conditioning and ageing; developing and implementing improved fertilizer advisory schemes; deer antler growth characteristics and deer veterinary problems; systems modelling, analysis and computer simulation in agriculture; pasture agronomy, soil fertility and irrigation; the application of effective integrated control procedures for pests of pasture and agricultural crops; leaf protein extraction; opossum farming. Horticultural research projects are on

subtropical fruit and citrus production; on supplying quantities of high quality plant material for new and existing crops; viticultural research in the main grape growing regions (Gisborne, Hawkes Bay and Marlborough).

FISHERIES RESEARCH DIVISION 34

Address: Greta Point Box 297, Wellington
Director: G.D. Waugh

National Water and Soil 35
Conservation Organization

Address: Ministry of Works and Development, PO Box 12041, Wellington North
Telephone: (04) 729-929
Affiliation: Ministry of Works and Development
Status: Government department
Director: A.W. Gibson
Activities: Promotion of coordination of water and soil conservation research.

WATER AND SOIL DIVISION 36
RESEARCH AND SURVEY GROUP

Director of Research: Dr M.E.U. Taylor
Activities: The main aim of the group is to develop a quantitative understanding of the technical processes affecting water and soil conservation as a basis for meeting the resource management needs of the National Water and Soil Conservation Organization. The research and survey activities, under the guidance of a small head office group, is concentrated into three science centres. The staff at each centre are organized into groups, each of which is responsible for a defined sector of work. Responsible for all water and soil research and survey programmes, it services the National Water and Soil Conservation Authority, its councils and its Research and Survey Committee. It identifies research needs, evaluates research and survey proposals and guides and coordinates the division's research and survey programmes; coordinates programmes with catchment authorities, other agencies and universities; negotiates and oversees research contracts.
Publications: Technical reports.

Aokautere Science Centre 37

Address: Palmerston North
Scientist in charge: Dr J.G. Hawley
Activities: National centre for land resource studies, plant materials for erosion control, land stability studies, alpine processes and remote sensing techniques, with outstations at Christchurch, Auckland, Napier and Dunedin.
Projects: Preparation of New Zealand land resource inventory and erosion maps; development of remote sensing techniques for land and water resource studies and catchment condition surveys; theoretical studies of land stability under low load conditions, factors governing the stability of earthflows, hill slope instability, earthflow stabilization techniques, role of soil creep in generating mass movement, relationships between geological mass movement, relationships between geological structure, vegetation, rainfall and erosion (with other groups); plant breeding and selection for soil conservation and river protection purposes; tissue culture propagation; plant diseases; roadside stabilization techniques; assessment of effectiveness of plant materials for soil conservation; the reaction of catchments of the Southern Alps to the forces of nature and to land use, alpine climatology, glaciology, land instability, soils and vegetation, erosion dating; sedimentation processes.

District Hydrology Group 38
Group Leader: J. R. Dymond
Activities: Hydrological data are collected by this group as part of the National Hydrological Network Programme. Local data are processed for storage on computer within the national hydrological bank.
Projects: The Taranaki Ring Plain Study concentrates on summarizing water resources for management and planning purposes; regional basin hydrology, data from the ten regional basins in the Wanganui district are being analysed to test their representativeness and assess data extrapolation techniques for these regions; environmental effects of abstraction on river temperature is being studied in detail and methods are being developed to enable catchment boards to carry out similar investigations; long term discharge records from the Wanganui, Rangitikei and Manawatu Rivers will be looked at to quantify the value of additional records when estimating hydrological parameters of significance for management and design purposes.

Land Resources Survey Group 39
Group Leader: G.O. Eyles
Activities: The group collects all data relevant to land use capability planning on a national scale, and ensures maintenance of consistent standards relating to erosion and land use capability assessment.
Publications: Regional bulletins (to provide a guide to more specific use of the New Zealand Land Resource Inventory Survey by water and soil agencies, planning authorities, consultants, and many other groups in the public and private sectors).

Projects: To enable the National Water and Soil Conservation Organization to carry out its responsibilities for development of catchments and the promotion of wise land use, the group has carried out a New Zealand Land Resource Inventory Survey. The survey method has included: a review and collation of available data on soils, geomorphology, vegetation and climate; photogrammetric interpretation and delineation of map units combined with: detailed, comprehensive field assessment; a system of multiple checking involving staff other than the original surveyor; systematic correlation within and between regions; digitization for input into the Vogel Computer Centre data base; urban land resource studies to allow a consistent approach by catchment authorities and the Ministry of Works and Development to assess the physical suitability of areas for urban development; storm damage assessment to provide rapid assessment of storm damage by a combination of aerial photo interpretation and field work.

Land Stability Group   **40**
Group Leader (acting): P. Luckman
Activities: This group studies mass movement (particularly earthflow and soil slip, to develop an understanding of how soil strength and slope stability varies with time and with soil conservation practice) and gully erosion. The land stability laboratories have been equipped to perform a wide range of soil mechanics and soil physics tests in addition to clay separation for mineralogical analysis. The group has an electronics section which is concerned with development of new laboratory and field instrumentation especially for automated data aquisition.

Plant Materials Group   **41**
Group Leader: C.W.S. Van Kraayenoord
Activities: This group develops and selects plant materials for soil conservation and river control and also carries out research into multiplication and establishment techniques. After extensive field testing, the plants are released to catchment authorities and commercial nurseries.
Publications: Advisory leaflets on the most appropriate use and management of soil conservation plants.
Projects: Poplar breeding and selection of a range of improved poplar clones for soil conservation, protection from wind erosion, and for farm forestry; willow breeding and selection to provide an improved range of fast growing willows for hillside stabilization, river control and shelter. Lower growing non-seeding shrub and osier willows with more flexible branches have been developed for river bank protection to replace tree willows; alternative species of trees and tall shrubs suitable for revegetation of dry, exposed and eroded sites (where poplars and willows are less suitable or even unsuitable), including Eucalyptus, Acacia, Platanus, Alnus, Betula and Ulmus; low shrubs, legumes and herbaceous species evaluated for ground cover, soil improvement and as a nursecrop in erosion control;

tissue culture for the rapid propagation of plantlets; establishment and protection techniques for nursery propagation and field establishment; plant diseases and evaluation of new methods of control; stabilization of roadsides and disturbed industrial sites for reclaiming and revegetating roadsides and sites disturbed by industry; river bank protection trials with willows and alders to evaluate both plant species and establishment techniques; biomass production; farm shelter for stock and crop protection.

Remote Sensing Group   **42**
Group Leader: D. Hicks
Activities: This group evaluates the applications of remote sensing to water and soil conservation, and provides an aerial photographic service to other research and survey groups at the three science centres as well as the twenty catchment authorities.
Projects: Remote sensing applications to soil conservation (an aerial photographic method for assessing the condition of slip-prone hill country); aerial photography for the purposes of soil conservation research, water resources research, and the urgent survey of erosion and floods.

Christchurch Science Centre   **43**

Address: Westminster House, Christchurch
Scientist in charge: Dr S.M. Thompson
Activities: National Science Centre for hydrological and groundwater studies, with outstations in Auckland, Hamilton, Napier, Wanganui, Wellington.

Alpine Processes Group   **44**
Group Leader: Dr M.J. McSaveney
Activities: The group is measuring climatic characteristics of the Southern Alps, and quantifying their effects on hydrology and erosion, to enable proposed or unintended impacts of man to be assessed in the context of natural events.
Projects: Studies of selected small and large drainage basins to provide opportunities for developing and testing new techniques for erosion assessment, and for understanding the functions of various erosion processes in maintaining catchment condition; at Shotover River, the effects of storms on a few major sediment sources, and the hydraulics of sediment transport in the stream channel appear to determine the sediment yield; at Cropp River, in the superhumid Western Southern Alps, estimates of flood flow and sediment yield bring into question the adequacy of bridge design standards for rivers draining that region; measurements of rainfall, runoff, sediment sources and sediment yield along a transect across the Southern Alps in the vicinity of Rakaia River, establish storm-induced flood runoff as a dominant influence, along with geomorphic history, on variation in erosion rates from basin to basin within the Southern Alps; the important role of soil erosion processes in maintaining soil fertility over geological time;

dating rock avalanches, scree surfaces, and other screes of bare greywacke rock and the relative stability of screes for process research and for possible future use in planning catchment rehabilitation work; analysis of historical and modern photo-pairs, and marked vegetation plots; longterm changes in plant species composition and the dominating impact of historic storms on the erosion scene, and the markedly faster tempo of erosion and recovery in the wetter regions of the Southern Alps; an inventory of New Zealand glaciers to correct the extent of permanent snow and ice shown on existing maps, and providing information for use in summer low-flow models of alpine drainage basins; measurements of glacier fluctuations and river flow in the Dry Valleys of Antarctica, through annual surveys of the New Zealand snow line, and through studies of recent and Holocene fluctuations of New Zealand glaciers.

Hamilton Science Centre   45

Address: School of Science, University of Waikato, Hamilton
Scientist in charge: vacant
Activities: National centre for developing a source of information and expertise in management, and coastal and estuarine studies.

Estuarine and Coastal Group   46
Group Leader: B.L. Williams
Activities: The broad objective of this group is to be aware of the management oriented estuarine and coastal problems, and to develop or promote expertise in selected areas. Four areas currently under study are: estuarine hydrodynamics and mixing; ocean outfall waste disposal systems; coastal sediment movement; wave climate.
Publications: (In preparation) bibiliographies of published and unpublished information on the hydrology and sedimentology of the important coastline regions; a handbook on techniques for site investigations and design considerations for ocean outfalls, primarily from a water quality perspective.
Projects: (In conjunction with the Auckland Regional Authority) study of the Upper Waitemata Harbour to assess the likely effects of urban development on sedimentation and water quality, involving studies in estuarine hydrology and sedimentology - with subsequent development of mathematical models for predicting dispersion of pollutants entering estuarine systems; hydrological investigations of some proposed combined cycle power station sites to determine the effect of waste heat disposal into estuaries where water availability and temperature impose limitations on thermal discharges; the dynamics and quality of water in coastal and estuarine systems. The group is also involved in coastal erosion studies, for example at Bream Bay and East Clive.

Hydrology Group   47
Group Leader: A. Dons
Activities: This group maintains a local network of automatic raingauges and water level recorders, the recording sites being operated as part of the National Hydrological Network programme that is coordinated by the National Hydrology Group at the Christchurch Science Centre. The group has two field parties, one based in Hamilton, the other at Rotorua. The data obtained are edited and computer archived by the group on a centralized computer located in Wellington. Half of the data acquisition effort is directed toward station design and operation, while the remainder is primarily collected to investigate the effect of land use on hydrology, and the concept of regionalized hydrologic response.
Projects: The group has been involved in the national pilot trial of a network of crest-stage stations as a low cost means of collecting stream flood peak data. To provide a much larger data base of annual maximum stream discharges to supplement the data available from automatic water-level recorder stations, to be used in a flood frequency analysis designed to improve flood estimation procedures in use in New Zealand; investigation of changes in hydrological response of pumice catchments following the planting of pines into pasture (as part of a national programme on hydrological responses to land use).

Laboratory Services Group   48
Group Leader: Dr J.B. Macaskill
Activities: The group works in the fields of analytical chemistry, biology, microbiology, and instrument design. The work carried out by the analytical laboratory is mainly in support of the Hamilton Science Centre operational groups, but services are also provided for other organizations. The group also provides training programmes in methods of water quality assessment. The chemistry laboratory can analyse many of the components found in natural waters. The major techniques include: atomic absorption and emission spectrophotometry; autoanalysis for mineral nutrients; visible and ultraviolet spectrophotometry; gas liquid chromatography.
Projects: The biology section is studying the effects of nutrient addition on algal metabolism and developing short-term bioassays to assess levels of nutrient deficiency in freshwater algae; the section is also developing an algal culture collection for teaching and research purposes throughout New Zealand to assist in setting conditions on water rights; the identification and enumeration of algae for catchment authorities and for specific projects: microbiological investigations are focused on studying processes influencing nitrogenous oxygen demand and denitrification mechanisms in water receiving effluents.

Lakes and Land Use Group   49
Group Leader: Dr R.A. Hoare
Activities: This group has been studying nutrient inputs and their effects on lakes. Their work initially dealt with assessing nutrient loads on Lakes Taupo and Rotorua, partly to give guidance on matters of immediate management concern, but more importantly to become familiar with the problems involved and to develop suitable sampling, analytical and data management strategies. The work is now of two main types - one dealing with the source of nutrients to streams, and the other dealing with the effects of nutrients on lakes. It is intended that in addition to generating scientific reports, this research will be used to guide the writing of handbooks for water quality managers, in which advice can be given to those faced with lake management problems.
Projects: The sources of nutrients are being studied in two projects - one dealing with runoff from a hill-country pasture catchment, and the other dealing with that from a residential urban catchment; the relationships between nutrient concentrations and algal growth; the relationship between algal growth and the occurence of anoxic bottom waters.

Modelling and Systems Group   50
Group Leader: Dr A.G. Barnett
Activities: The group has experience with a range of modelling techniques from simple storage models through time series models to multidimensional hydrodynamic models and has applied these techniques to the water quality of lakes, the dispersion of pollutants in rivers and the response of estuaries to urban pressures. In addition, a close liaison is maintained with the hydrodynamic modelling research contract staff at the University of Auckland (Department of Theoretical and Applied Mechanics). On the systems side the group has taken responsibility for the design work on AQUAL - the Aquatic Quality Unified Archival Library - to cater for water quality data on a systematic national basis. The centre uses a range of sophisticated instruments for field and laboratory measurement of water quality. The systems part of the group is responsible for the development of integrated data management systems based on this data logging equipment.
Projects: Modelling studies have been carried out or are under way on: dissolved oxygen and biochemical oxygen demand in the Waikato, Waipa, Tarawera, Mataura and Manawatu Rivers; vertical transverse and longitudinal dispersion in the Waikato, Waipa, Manawatu (with Manawatu Catchment Board) and Taranaki Ring Plain Rivers (with Taranaki Catchment Commission); nutrient enrichment in Lake Rotorua; trends in water quality in Lake Rotorua and optimal sampling strategies to monitor future changes; deoxygenation in New Zealand lakes based on published data and experimental results gathered by the centre from two small Waikato lakes; nutrient flushing and enrich-

ment of the New River Estuary, Invercargill and the Upper Waitemata Harbour, Auckland; energy spectra of waves at Hicks Bay.

Rivers Group   51
Group Leader: G.B. McBride
Activities: Study of river water quality in terms of oxygen balance, phytoplankton density and bacterial metabolism. River oxygen models and various biochemical methods have been developed as techniques for assessing water quality. The rivers being studied to develop general techniques for river investigations include the Waikato and Upper Clutha.
Publications: (in preparation) *River Dissolved Oxygen Handbook.*
Projects: Longitudinal river surveys have been undertaken each month over a fourteen-month period by the Otago Catchment Board and local Ministry of Works and Development staff; an investigation of the efficiency of modelling techniques for river dissolved oxygen and its production of simple nomographs for the prediction of river assimilative capacity, these being based on the Streeter-Phelps model; more complex models and modelling techniques, including the techniques for defining model parameters and boundary conditions; an investigation of methods of quantitative assessment of river water appearance (especially colour); a study (in association with the Waikato Valley Authority) to examine the response of algal populations in the Waikato River to the cessation of effluent discharge from the Kinleith Pulp and Paper Mill during the industrial stoppage in early 1980.

New Zealand Agricultural   52
Engineering Institute

– NZAEI
Address: Lincoln College, Canterbury
Telephone: (03) 252-811
Affiliation: Lincoln College
Status: Educational establishment with r&d capability
Director: E. M. Watson
Graduate research staff: 18
Activities: Research into agricultural engineering problems and developing new equipment. Water resources for agriculture, irrigation and drainage, crop mechanization, engineering for livestock production, occupational health and safety on farms, farm buildings and environmental control, on-farm processing and storage, agricultural waste management.
Publications: Annual report.
Projects: Non-traditional sources of water irrigation, and irrigation application methods; irrigation in hill country; development of moving sideroll irrigator (T.D. Heiler); fencing development (G.M. Garden); ethanol production from beet (Dr D.J. Painter); direct drilling

and oversowing, mechanical harvest systems, harvesting of berry and pip fruit (J.S. Dunn); spraying methods and improved techniques for horticulture (J.F. Maber); reduced cultivation techniques (J.S. Dunn); fertilizer spreading and application techniques for solid materials (G.M. Garden).

New Zealand Dairy Board   53

Address: PO Box 417, Wellington
Telephone: (04) 724-399
Telex: DAPMARK NZ3348
Status: Government department

FARM PRODUCTION DIVISION   54

Controller of Herd Improvement: J.W. Stichbury
Sections: Research, P. Shannon
Graduate research staff: 5
Activities: Research into agricultural economics and livestock breeding (dairy cattle).
Publications: Annual report.
Projects: Sire and cow evaluation (Dr B.W. Wickham); semen technology (P. Shannon); farm production economics (R.G. Jackson); statistics; genetics (Dr B.W. Wickham).

New Zealand Dairy   55
Research Institute

– DRI
Address: Private Bag, Palmerston North
Telephone: (063) 74-129
Telex: 3960 NZ Daisearch
Parent body: Department of Scientific and Industrial Research
Director: Dr P.S. Robertson
Assistant Research Directors: Dr R.C. Lawrence, Dr K.R. Marshall, Dr W.B. Sanderson
Activities: With fifteen research sections whose programmes cover applied and more basic matters related to the manufacture and composition of dairy products, and includes a wide range of topics relating to new and traditional dairy products, new and improved methods of manufacture, and new uses for milk and its constituents, the institute also provides a comprehensive testing and advisory service to individual companies and plants on a chargeable basis.
Publications: New Zealand Journal of Dairy Science and Technology.

Projects: Lactic streptococci (which play a vital role in the manufacture of almost all varieties of cheese and into the virus-like bacteriophage which can destroy them); butter and milkfat: market evaluation of a butter which spreads easily straight from the refrigerator; low-cost technique for reducing the colour of milkfat which does not have undesirable side effects (such as the development of oxidative flavours); milk powders: commercial adoption of Institute-developed methods for instantizing of fat-containing powders; optimization of yield and energy use.

New Zealand Department   56
of Scientific and Industrial
Research

– DSIR
Address: Charles Fergusson Building, Bowen Street, Wellington, 1
Telephone: (04) 729 979
Telex: RESEARCH NZ 3276
Status: Official research centre
Director General: Dr D. Kear
Publications: Annual report; *DSIR Directory*; *Report of the Department.*
See also entries for: Heavy Engineering Research Association Incorporated
Meat Industry Research Institute of New Zealand Incorporated
New Zealand Dairy Research Institute
New Zealand Fertilizer Manufacturers' Research Association
New Zealand Leather and Shoe Research Association Incorporated
New Zealand Logging Industry Research Association Incorporated
Wool Research Organization of New Zealand Incorporated

APPLIED BIOCHEMISTRY DIVISION   57

Address: Private Bag, Palmerston North
Telephone: (063) 68-019
Telex: PALMSIR 31285
Head: Dr R.W. Bailey
Departments: Biochemistry and microbiology, Dr R.T.J. Clarke; organic chemistry, Dr G.B. Russell; plant chemistry and physiology, Dr P.F. Reay; animal nutrition and physiology, Dr M.J. Ulyatt; services, K.I. Williamson

Activities: Biochemical and microbiological investigations, relevant to plant production and composition, plants as animal feed and both plants and animals as food for man; the study of plant production in relation to nitrogen fixation and biochemical genetics; the study of plant composition in terms of constituents of importance in animal nutrition and disorders, and plant pest resistance; the study of the processes of digestion and nutrient adsorption in ruminants and monogastric animals; the nutritive value of New Zealand foods; processing of biological materials.

APPLIED MATHEMATICS DIVISION   58

Address: PO Box 1335, Wellington
Telephone: (04) 727-855
Telex: NZ 3276
Director: Dr H.R. Thompson
Sections: Mathematical statistics, Dr R.B. Davies; mathematical physics, Dr A. McNabb; operational research, Dr H. Barr; computing, B.K. Campbell
Activities: Mathematical and operational research service work and projects in agricultural production, energy, manufacturing, natural environment, building and construction, transport, sociological and economic projections for DSIR divisions, research associations, government departments, business, industry, statutory organizations, local bodies, universities, and institutes. Provision of statistical advice and assistance. Publicizing of useful techniques.
Projects: Effect of climate on pasture growth; root and leaf growth models; forecasting of milk production; codlin moth emergence times; design and analysis of soil, fertilizer, crop, and grassland trials; production scheduling in the dairy industry; congestion studies in freezing works; analysis of harbour sewage pollution data; population ecology studies; opossum behaviour and feeding trials; effect of soil factors on forest growth; erosion in dispersive clays; statistical methods for earthquake forecasting; Rotorua geothermal monitoring system; computerization of fossil records; computerized bibliography of fresh-water research.

AUCKLAND INDUSTRIAL   59
DEVELOPMENT DIVISION

Address: PO Box 2225, Auckland
Telephone: (09) 34-116
Director: W.R. Beasley
Deputy director: J.B. Cornwall
Sections: Applied physics, R.W. Foster; electronics, Dr R.S. Clist; engineering design and product development, B.H. Hodder; engineering materials, Dr N.A. Miller; vibration acoustics and stress analysis, Dr J.B. Meikle

Activities: Applied research for the manufacturing and processing industries and providing them with a scientific service in the general disciplines of: electronics; mechanical and manufacturing engineering; metallurgy; physics and applied heat; medical and biological engineering. Each section has a balanced programme of applied research, short-term investigations, and servicing.
Projects: Design of instruments to detect multiple pregnancies in sheep, fat thickness in live sheep, and fat thickness in lamb carcasses. Testing of maize, garlic, and onion drying facilities, and cool stores. Investigation of effects of storage and transport conditions on fruit and vegetables.

BOTANY DIVISION   60

Address: Private Bag, Christchurch
Telephone: (03) 252-511
Telex: NZ 4703
Director: Dr H.E. Connor
Sections: Ecology and conservation, Dr P. Wardle; palynology and anatomy, Dr N.T. Moar; taxonomy and herbarium, Dr Elizabeth Edgar
Graduate research staff: 29
Activities: Botanical investigations, botanical surveys, particularly of National Parks and reserves and of representative examples of the different ecological systems; classification of the native, naturalized and horticultural flora of New Zealand and the Pacific Islands and the plant communities in New Zealand - old or new, past or present. The development of a herbarium as the national reference collection. The effects of proposed industrial developments on the botanical environment, and the conservation of endangered habitats and species.
Publications: Triennial report.
Projects: Flora of New Zealand series (Dr Elizabeth Edgar); reproductive systems in New Zealand plants (Dr E.J. Godley); late quaternary vegetation history (Dr N.T. Moar); anatomy of woody plants (R. Patel); synecological and autecological studies of New Zealand vegetation (Dr P. Wardle); botanical surveys of reserves (G.C. Kelly); germination of seeds of native plants (Mrs M.J.A. Bulfin).

CAWTHRON INSTITUTE   61

Address: PO Box 175, Nelson
Telephone: (054) 82-319
Telex: CTG NZ3429
Director: Dr R.H. Thornton
Sections: Research, R.H. Thornton; Cawthron Technical Group: chemical and biological services, A. Cooke; environmental and feasibility services, Dr J. Bamford

Activities: The institute's research programme is directed toward the microbiological degradation of natural and man-made organic materials. Knowledge and understanding of the ways in which microbes degrade and transform organic materials is applicable to study of water and soil quality, nutrient cycling, waste utilization and treatment, and microbial fermentations. The Cawthron Technical Group provides a confidential fully integrated consulting service, where one of more specialist skills are required, for industry, agriculture, local and central Government, and ad hoc authorities.
Projects: The research section are studying biological, chemical, and physical properties of intertidal sediments of Delaware Inlet with particular emphasis on the rates of activity and interrelationships between the microbial processes involved in the transformation and cycling of nitrogen, carbon, and sulphur; comparison of rates of nitrification in intertidal sediments, nearby pasture soils, and in forest soils; collaborative investigations with other organizations, involving measurements of microbial activity and microbial biomass in Westland beech forest soils and in ocean waters off the Westland coast; incidence and degree of transference of antibiotic resistance in coliform bacteria isolated from piggery effluents at environmental temperatures under normal farming practice, with and without antibiotic treatment; investigations to establish whether or not the microbial enzyme urease is carried on plasmid; the extent to which the phosphate in teichoic acids in bacterial cell walls provides a reserve source of phosphate under conditions of inorganic phosphate starvation; 15R36 investigation of the feasibility of linking alcohol fermentation and anaerobic digestion processes for the continuous production of the two fuels ethanol and methane from biomass; the biochemistry of methane production by bacteria in field and laboratory situations. The Cawthron Technical Group are covering forestry; agriculture and horticulture; fishing; manufacturing; local and ad hoc authorities; environmental and feasibility services.

CHEMISTRY DIVISION   62

Address: Private Bag, Petone
Telephone: 666-919
Director: I.R.C. McDonald
Activities: Chemical research in applied chemistry; food; forensic; geochemistry; geothermal; inorganic materials; natural products; organic chemistry; pharmaceutical; physical chemistry; spectroscopy; toxicology; water which is important to community needs and the development of New Zealand resources. The division has district laboratories in Auckland, Christchurch and Wairakei.
Projects: Suitability and safety of containers for food and drugs; toxic elements and compounds, nutritional composition of foods; composition of New Zealand marine fishes; honey quality; therapeutic residues in

animal products; monitoring pesticide levels in food; determination of pesticide residues in fruit and vegetables; persistence of lindane insecticide residues in pasture soils; mycotoxins and nitrosamine residues in foodstuffs; composition of wines; the structure of lignin and condensed tannins; a study of the colour and astringency of fruits and wines; the yield and composition of New Zealand seed oils; chemistry of mammalian pheromones; the classification of native plants by chemical methods; phytoalexins; synthesis of insect pheromones; photochemistry of heterocycles; general aid to government and industry in organic chemistry (analysis and spectroscopy); investigation of water-supply quality and pollution incidents; analytical methods development and interlaboratory testing (the CHEMAQUA programme); chemistry of the Lower Hutt and North Canterbury Plains aquifers; land disposal of effluents; disinfection of swimming pools; geochemistry of spring waters.

CROP RESEARCH DIVISION   63

Address: Private Bag, Christchurch, Canterbury
Telephone: (03) 252511
Telex: LINSIR 4703
Director: Dr Harvey C. Smith
Groups: Environmental, Dr J.W. Sturrock; field crops, Dr M.W. Dunbier; horticultural, T.P. Palmer; genetics, Dr H.C. Smith; industrial crops, G.M. Wright
Graduate research staff: 33
Activities: Studies of the productivity and diversification of agricultural and horticultural crops. Breeding new and improved varieties; effect of shelter, irrigation, augmenting the breeding programme by studies in genetics, seed production, crop agronomy, and processing characteristics.
Publications: Crop Research Division Report Series; Crop Research News; Field Crop News; Cereal News; Wheat Review.
Projects: Effect of wind and water on crop growth (J.W. Sturrock); lucerne (alfalfa) breeding and agronomy (M.W. Dunbier); asparagus breeding (P.G. Falloon); glasshouse tomato breeding (M.T. Malone); strawberry rubus breeding (I.K. Lewis); potato breeding (R.A. Genet); cereal breeding (G.M. Wright); process crop breeding and agronomy (J. Lammerink).

ECOLOGY DIVISION   64

Address: Private Bag, Lower Hutt
Telephone: (04) 694-859
Telex: Geo Surv NZ 31348
Director: Dr Richard M.F. Sadler
Sections: Fresh-water, Dr E. White; agricultural ecology, Dr M.R. Rudge; biometrics, environment and recording, Dr D.G. Dawson; carnivores, rodents, forest birds and invertebrates, Dr B.M. Fitzgerald; South Island forests and reserve management, R.H. Taylor; possum and forest botany group, Dr R.E. Brockie
Activities: Study of the ecology of the fauna and flora of New Zealand and the functioning of terrestrial and fresh-water ecosystems: evaluation of the factors affecting the distribution and abundance of birds and mammals living in the wild, and the effects of introduced birds and mammals on native fauna and flora, on natural vegetation, and on agricultural production; biological surveys particularly of National Parks and reserves, and of other sites of special scientific interest; the causes and effects of the eutrophication of fresh-waters; use of ecological knowledge in land and water management.

ENTOMOLOGY DIVISION   65

Address: Mount Albert Research Centre, 120 Mount Albert Road, Auckland 153, Auckland
Telephone: (09) 893-660
Activities: Research aimed at the solution of applied entomological problems (particularly those affecting New Zealand agriculture, horticulture and the natural environment) and to classify and characterize and develop collections of New Zealand arthropods, nematodes and insect pathogens.

New Zealand Fertilizer   66
Manufacturers Research
Association

– FMRA
Address: Box 23637, Hunters Corner, Auckland
Telephone: (09) 274-7184
Parent body: Department of Scientific and Industrial Research
Status: Independent research centre
Director: D.J. Higgins
Departments: Agronomy, R.C. Stephen; chemical engineering, J.V. Lawlor; fertilizer chemistry, S.J. Gallaher; physical chemistry, A.C. Braithwaite
Graduate research staff: 14
Activities: The association is principally engaged in the production technology of nitrogen and phosphate fertilizers through raw materials, process chemistry effluent control, energy conservation, new process, pro-

duct changes, quality control, and experimental production for agronomic work (involving field and greenhouse testing of nitrogen and phosphorus fertilizers concentrating on new products and relating to other workers).
Publications: Quarterly research reports; annual report; periodic special reports.
Projects: Comparison of phosphate fertilizers (R.C. Stephen); kinetics of solution of fertilizers (S. McConnell); economics of phosphate fertilizers (T.R. Butler); practical fluoride chemistry (A.C. Charlston); liquid effluent processes (K.R. Laing); milling studies (J.V. Lawlor); nitrogen process studies (A.C. Braithwaite); high-analysis fertilizer production (D.J. Higgins).

New Zealand Leather and   67
Shoe Research Association
Incorporated

– LASRA
Address: Private Bag, Palmerston North
Telephone: (063) 82-108
Parent body: Department of Scientific and Industrial Research
Status: Independent research centre
Director: Dr G.W. Vivian
Deputy Director: A. Passman
Sections: Leather technology, A. Passman; footwear technology, A.J. Harvey; effluent, I.G. Mason
Activities: The association is the central research organization of the fellmongering, hide and skin curing, tanning, and footwear manufacturing industries; it determines chemical and physical properties of pelts, hides, and skins, tannery processing liquors and chemicals, leathers and all types of materials and components used in footwear manufacture, and it examines and reports on consumer footwear complaints.
Projects: An investigation into the microstructure of leathers and skins; pathological examination of preserved hides to find the cause of scuff marks on bovine leather, a study of the factors causing white spots on lamb pelts; an assessment of slipe wool yield and leather quality from skins of exotic sheep breeds, and an assessment of sheep breeds in relation to double face shearling quality; investigation of the effect of liming and of bating on pelts and on leather produced from the pelts; evaluation of thickening agents and auxiliaries for depilatory paints; pancreas glands in bating; hide, skin, and pelt preservation; bactericides and fungicides. an investigation into variations of pallet packing methods and into the causes of shipping company and tanner criticism of pallet packed pelts; an examination of works processing vessels, paddles, drums, and hide processors on pickled pelt properties. Leather: rapid wet-blue processing; the use of various syntans/aluminium systems is

to be examined for potential application in woolskin production; woolskin washability and lustre; effect of oxazolidines and other agents; finish drying rate and leather properties, design of sulphide oxidation plant; an investigation of the relationship between sulphide removal rate in liming liquors and oxygen transfer coefficient to tap water in the same apparatus; hair recovery from unhairing liquors; solid waste utilization and the possibilities for small scale utilization of trimmings, shavings, etc, presently dumped.

New Zealand Logging Industry Research Association Incorporated

68

– LIRA
Address: PO Box 147, Rotorua
Telephone: (073) 87-168
Parent body: Department of Scientific and Industrial Research
Status: Independent research centre
Director: J.J.K. Spiers
Graduate research staff: 6
Activities: Forestry and forest products: research and development in the logging industry and extension in the fields of tree harvesting and log transportation.
Publications: Annual report; *Digest*; *Machinery Evaluation*; newsletter; project reports.
Projects: Steep country logging (V.F. Donovan); smallwood harvesting (G.C. Wells); felling and delimbing technique (J.E. Gaskin); log transportation (G.P. Coates); protective equipment (R.L. Prebble).

Noton (New Zealand) Limited

69

Address: 4-10 Alma Street, Newmarket, Auckland
Telephone: (09) 50-4486
Affiliation: Motor Traders (New Zealand) Limited
Status: Industrial company
Manager: P.J. Yarham
Graduate research staff: 1
Activities: Manufacturers and distributors of agricultural dairy farm accessories.
Projects: Calf feeders; sealed milking units; electric fencing; jet spray shed wash nozzles, industrial hoses and teat wash nozzles.

Palmerston North Agricultural Research Centre

70

Address: Private Bag, Palmerston North
Telephone: (063) 68 079
Affiliation: Ministry of Agriculture and Fisheries
Status: Official research centre
Regional Director: Dr J.N. Parle
Sections: Agronomy, I.M. Ritchie; biometrics, D.F. Wright; entomology, W.M. Kain; soil fertility, G. Smith; weed science, A.I. Popay
Activities: This southern North Island region covers the research activities of the Ministry of Agriculture and Fisheries in pastoral and horticultural research in the lower half of the North Island. Research is concerned with field studies (pasture pests, weed control, pasture and crop agronomy, soil zoology and fertility). Studies are also carried out on growth performance of young stock, and on fencing design.
Projects: Agronomy: effects of defoliation frequency and intensity on the production of dairy pasture; production and management of phalaris/lotus swards; performance of lotus on hill country (I.M. Ritchie). Biometrics: evaluation of phalaris/white clover pasture; farm management in the east coast district; trace elements in calves; electric fencing (D.F. Wright). Entomology: stem weevil resistance in ryegrass; pest monitoring and population dynamics; pest/host plant relationships of grass grub; integration of pest management programmes (Dr W.M. Kain); chemical control of grass grub; grass grub diseases; parasites of lucerne aphids (A. Main); effects of gregarines on moth fecundity; effects of baculoviruses on porina populations; physical ecology of eggs; porino moth activity; behaviour of hatching porina larvae (A. Carpenter). Soil fertility: clover responses to nitrogen; nutrient status of east coast hill country soils; effects of lime on plant availability movement of phosphate fertilizer; potassium requirements of dairy pasture; side-dressed nitrogen on maize (G. Smith, G. Mansell). Weed science: gorse; nodding thistle; chemical control of Californian thistle, perennial nettle, and of chamomile in pasture; effects of scrub-weed herbicides on clover content of pastures (Dr A.I. Popay, Dr M.J. Hartley).

FLOCK HOUSE RESEARCH AREA

71

Address: Private Bag, Bulls
Telephone: 439
Officer in charge: W. Stiefel
Activities: The provision of a focal point for research into a wide range of sand country problems.
Projects: Re-establishment of lucerne; evaluation of lucerne cultivars; pasture rate of growth on putepute black sand; national barley cultivar evaluation; national testing of spring wheat cultivars.

HASTINGS HORTICULTURAL RESEARCH STATION 72

Address: Lawn Road, RD 2, Hastings
Officer in charge: J.L. Burgmans
Pastoral group: G. Crouchley
Activities: The station concentrates on vegetable crops of local importance, particularly those grown for processing. A wide range of trial crops has been grown to select varieties suited to local conditions.
Projects: Asparagus and tomatoes (J. Burgmans); dwarf beans, vining peas, potatoes and other crops (B.T. Rogers).

LEVIN HORTICULTURAL RESEARCH CENTRE 73

Address: Private Bag, Levin
Telephone: (069) 87 059
Director: Dr W.M. Kain
Sections: Berryfruits, L.A. Porter; disease and pest control, K.G. Tate; ornamental and greenhouse crops, R.A.J. White; plant and soil analysis, M. Prasad; vegetables, W.T. Bussel; weed control, T.I. Cox
Activities: Improving horticultural production, with emphasis on berryfruit, vegetable and ornamental crops both in open and in greenhouses. Research on husbandry aspects is supported by work on pests, disease and weed control nutrition, plant and soil analysis and tissue culture for all horticultural crops. Production of vegetable crops for processing, diversification of process crop research at Hastings and reorganization of resources for research on ornamental crops have been brought about by the need for research on crops grown for export.

TAKAPU RESEARCH STATION 74

Address: PO Box 63, Waipukurau
Telephone: 864
Technical Officer in charge: M. Slay
Activities: Grassgrub control and pasture plant selection.

TARANAKI AGRICULTURAL RESEARCH STATION 75

Address: PO Box 8, Normanby
Telephone: 28 011
Officer in charge: N.A. Thomson
District Scientist: G.J. Rys
Activities: Grass grub control on dairy farms (station dairy production, research into fertilizers aimed at monitoring soil phosphate levels, value of winter active lucerne, grasses tolerant to grass grub, uses of nitrogen).

Ruakura Agricultural Research Centre 76

Address: Private Bag, Hamilton
Telephone: (071) 62-839
Affiliation: Ministry of Agriculture and Fisheries
Status: Official research centre
Regional Director: Dr J.P. Joyce
Activities: All aspects of horticulture and agriculture research in the Northern North Island region are carried out by the area research stations listed below.
Projects: Biometrics: analysis of ovulation rate data; culling strategies (Dr G.W. Winn); autoanalyser digitizer (Dr R.B. Jordan); environmental data logger; automation of flame photometer system (P.N. Scharre).

NORTHLAND REGION 77

Address: Private Bag, Whangarei
Telephone: (089) 87-179
Sub-Regional Director: Dr G.C. Everitt
Sections: Agronomy and crop production, G.J. Piggot; animal production, I.P.M. McQueen; horticultural production, R.F. Barber; insect control, R.H. Blank; soil fertility, P.W. Shannon; weed control, E.N. Honore
Graduate research staff: 7
Activities: Research projects are carried out in the broad fields of agronomy and crop production, animal production, horticulture production, insect control, soil fertility and weed control.
Projects: Agronomy and crop production: soya beans, summer brassica forages and energy crops (G.J. Piggot); animal production: copper, cobalt and selenium status of sheep and cattle, cobalt and vitamin supplements, autumn lambing (I.P.M. McQueen); horticultural production (R.F. Barber); insect control: blackbeetle and black field cricket population and damage assessment; control in pasture; controlling soldier fly (Dr R.H. Blank); soil fertility (P.W. Shannon); weed control: blackberry control with herbicides; effects of winter application of phenoxy herbicides on white clover and pasture production (E.N. Honore).

ROTOMAHANA RESEARCH STATION 78

Address: Okara Road, RD 3, Rotorua
Officer in charge: T.G. Harvey
Activities: Established in 1979 as an animal breeding research centre to assess the genetic resources now available within the country, and to develop breeding strategies by which they can be best exploited to improve the national flock.
Projects: Assessing genetic performance potential; measuring the influence of merino genes on performance; development of criteria to select sheep which will produce leaner slaughter progeny (H.H. Meyer).

RUAKURA ANIMAL RESEARCH STATION 79

Address: Private Bag, Hamilton
Telephone: (071) 62 839
Station Director: K.E. Jury
Sections: Animal disease control, A.G. Campbell; animal management, Dr J.F. Smith; biophysics, M.W. Woolford; chemistry, Dr E. Payne; dairy science, Dr A.M. Bryant; genetics, Dr J.N. Clarke; meat, Dr A.H. Kirton; mycotoxic diseases, P.H. Mortimer; physiology, Dr K.L. Macmillan
Graduate research staff: 49
Activities: Research into animal breeding, machinery development, dairy and pig production, animal behaviour and protein extraction; the management and utilization of pasture, pasture by-products and forage crops; control of certain animal diseases, particularly those related to animal nutrition and those occurring in Northern New Zealand.
Projects: Animal disease control: metabolic diseases; hypomagnesaemia (A.G. Campbell). Animal management: milking, nutritional and reproductive management (Dr J.F. Smith); biophysics (M.W. Woolford); chemistry (Dr E. Payne). Dairy science: agronomy, protein extraction and engineering (Dr A.M. Bryant). Genetics: sheep breeding, including artificial breeding, and trends in distribution; cattle breeding (Dr J.N. Clarke). Meat: stunning methods; measuring carcasses (Dr A.H. Kirton); mycotoxic diseases (P.H. Mortimer). Physiology: hormonal influences, fertility, udder development, lactation, animal behaviour (Dr K.L. Macmillan).

Wairakei Research Station 80

Address: RD 1, Taupo
Telephone: (074) 48-069
Technical Officer in charge: L.F.C. Brunswick

RUAKURA SOIL AND PLANT RESEARCH STATION 81

Address: Private Bag, Hamilton
Telephone: (071) 62-839
Station Director: N.A. Cullen
Sections: Plant science, J.A. Douglas; soil fertility, Dr W.M.H. Saunders; soil chemistry and microbiology, Dr J.H. Watkinson; plant and analytical chemistry, Dr J.S. Cornforth; insect control, Dr R.P. Pottinger
Graduate research staff: 54
Activities: Research on laboratory and field problems concerned with soil science, chemistry, microbiology, fertility, plant production, weed and insect control in the northern North Island, a large amount of which is carried out on farmers' properties.

Projects: Improving pastures on pumice soils, sugar beet, sainfoin and lucerne, forest farming, agronomy, crop evaluations; plant science: weed control, herbicide residues in soil, aquatic weeds, grass grub (J.A. Douglas); soil fertility - effectiveness of biosuper granules, calibration of Olsen P, potassium fertilizer requirements, nitrogen and other fertilizer requirements, responses to urine, lime and phosphate, pulverized fuel ash, top dressing with magnesium oxide, surveys of trace elements, environmental physics (Dr W.M.H. Saunders); soil chemistry and microbiology - organic matter cycling, soil enzyme activities, ecology, antibodies to tremorganic penicillia, mycorrhizal research in eroded soils, pasture, cropping and fungal selection trials (Dr J.H. Watkinson); plant and analytical chemistry - soils fertilizer, plant and spectrochemical laboratory tests (Dr I.S. Cornforth); insect control - soldier fly, black beetle, stem weevil, slugs, insecticides, instrumentation an analytical organic chemistry, high performance liquid chromatography, spectrometry (Dr R.P. Pottinger).

Manutuke Research Station 82

Address: Box 18, Manutuke
Technical Officer: B.R. Brown

Pukekohe Horticultural Research Station 83

Address: Cronin Road, RD 1, Pukekohe
Scientist in charge: G.J. Wilson
Activities: Problems of vegetable production in the South Auckland area, covering field trials on herbicides, nutrition, cultivar evaluation and crop culture (including kiwifruits, orchids and greenhouse culture).
Projects: Onion spacing, cultivars and herbicides (G.J. Wilson); greenhouse tomato fruit setting (D.J. Brundell).

Te Kauwhata Viticultural Research Station 84

Address: Box 19, Te Kauwhata
Technical Officer in charge: J.G. Whittles
Activities: Evaluation of cultivars: disease resistance/tolerance, maturation, yield and quality.
Projects: Grape cultivars and clones; rootstocks; grafting; viticulture techniques (J.G. Whittles, F.H. Wood, N.S. Brown); oenology (R.S. Eschenbruch, T. van Dam); disease control (J.G. Whittles, T.E. McCarthy).

WHATAWHATA HILL COUNTY RESEARCH STATION   85

Address: Private Bag, Hamilton
Telephone: (071) 298-789
Station Director: Dr P.V. Rattray
Graduate research staff: 15
Activities: Provision of technical information to improve the productivity and profitability of sheep and cattle production on hill country. The main areas being studied are nutrition and management, agronomy, soil fertility, pasture management, reproductive physiology, wool biology, sheep and cattle breeding. Cooperative trials are also undertaken in forestry farming, carcass composition, diseases, entomology and other fields including forestry.
Projects: Soil fertility, pasture management and forestry farming (Dr A.G. Gillingham); animal nutrition and management (D.C. Smeaton); reproductive physiology (Dr T.W. Knight); wool production biology (Dr M.L. Bigham).

Research Areas   86

Dargaville Field Research Area   87
Address: Box 63, Dargaville
Telephone: 7420

Horo Horo Experimental Area   88
Address: RD 1, Rotorva
Telephone: 76 144
Technician in charge: M.D. Fraser

Te Kuiti Field Research Area   89
Address: RD 1, Te Kuiti
Telephone: 708
Technical Officer in charge: H.R. Foskett

Tikitere Forest Farming Research Area   90
Address: Box 951, Rotorua
Telephone: 79 579
Technical Officer in charge: M.F. Hawke

Tussock Grasslands and Mountain Lands Institute   91

Address: PO Box 56, Lincoln College, Canterbury
Telephone: (03) 252-811
Affiliation: Lincoln College
Status: Educational establishment with r&d capability
Director: (vacant)
Graduate research staff: 7
Activities: The aim is to provide technological information and skills for the conservation and use of natural resources in the tussock grassland and mountain lands of New Zealand. To achieve this the institute attempts to facilitate the coordination of research in the region, to identify problems, indicate research requirements, initiate and participate in new lines of enquiry and disseminate information arising from research. Major subject areas of own research are in use and conservation of natural resources, and in pasture and livestock production systems.
Publications: Annual report.
Projects: Primary production and resource economics, public administration of land, resource planning (I.G.C. Kerr); production-consumption modelling, insect population ecology and dynamics (E.G. White); livestock production, pastoral utilization, livestock management systems (M.A. Abrahamson); grassland agronomy, vegetative improvement, alpine ecology, and revegetation (G.A. Dunbar); computer science, bibliographic and information retrieval systems (K.R. Lefever); publication and extension (B.T. Robertson); nitrogen and mineral cycling, edaphic ecology of grasslands, grazing management, land use (K.F. O'Connor).

University of Auckland*   92

Address: Private Bag, Auckland 1
Telephone: (09) 792-300
Status: Educational establishment with r&d capability
Vice-chancellor: C. Maiden

FACULTY OF ENGINEERING   93

Chemical and Materials Engineering Department   94

Head: B.J. Welch
Graduate research staff: 6
Activities: Food science research: industrial processing of food and feed proteins, particularly physical properties of protein precipitates and enzymic hydrolysis of proteins.
Projects: Enzymic hydrolysis of meat proteins; precipitation and dewatering of dairy proteins; physical properties of leaf protein precipitates (P.A. Munro).

FACULTY OF SCIENCE   95

Department of Zoology 96

Head: Professor E.C. Young
Graduate research staff: 22
Activities: Research into plant protection fresh-water fisheries, forestry and forestry protection, marine aquaculture, especially of mollusca.
Projects: Applied entomology (the pests of horticultural crops and of nuisance insects, flies and mosquitoes); the genetics and evolution of Pacific mosquitoes; teaching and research in plant protection, especially of insectan damage and its assessment, in fresh-water fisheries; some studies on the interrelation of native and introduced species, especially of trout and rudd; forestry population and behavioural studies on the possum; the economics of possum management and of their impact on forests.

University of Waikato 97

Address: Private Bag, Hamilton
Telephone: (071) 62-889
Status: Educational establishment with r&d capability
Vice-Chancellor: Dr D.R. Llewellyn

BIOLOGICAL SCIENCES DEPARTMENT 98

Head: Professor J.G. Pendergrast
Activities: Research into insect behaviour, ecology, limnology, autecology and hydrology and related biological sciences.
Projects: Dentrification by rhizobia (Dr R.M. Daniel, Dr K.W. Steele); use of stable enzymes in food processing (Dr R.M. Daniel, Dr H.W. Morgan); ecology of sod webworms in hill country pastures, ecology and economic importance of lucerne flea in pastures (Professor J.G. Pendergrast, Dr R.P. Pottinger, Ruakura Research Centre, Hamilton); ecology and management of peatlands in New Zealand, management of swamps and peatlands on African continent (K. Thompson); study of insects as food in traditional societies of melanesia, habits of entomophagy amongst Australian aborigines and Papua New Guineans (Dr V.B. Meyer-Rochow).

Waitaki New Zealand Refrigerating Company 99

Address: 58 Kilmore Street, PO Box 1472, Christchurch
Status: Industrial company
Sections: Research and development department
Activities: Commercial company that is agriculturally based.

Wallaceville Animal Research Centre 100

Address: Private Bag, Upper Hutt
Telephone: (04) 86-089
Affiliation: Ministry of Agriculture and Fisheries
Status: Official research centre
Director: Dr W.A. Te Punga
Sections: Apiculture, P.G. Clinch; bacteriology, T.M. Skerman; biometrics, L.M. Morrison; epidemiology, D.W. Kane; experimental biology, E. O'Connell; experimental pathology, D.C. Thurley; experimental serology, D.R. Ris; hydatids, D.D. Heath; immunology, W.E. Jonas; meat monitoring and pesticide residues, S.R.B. Solly; membrane biology, J.C. Turner; metabolic diseases, T.F. Allsop; parasitology, R.V. Brunsdon; photography, A.W. Barkus; trace element biochemistry and radiochemistry, K.R. Millar; virology, D.H. Davies
Activities: Research into diseases of livestock affecting production and markets, including study of infectious diseases of farm animals and an intensified programme relating to disease transmitted by cats and dogs, covering foot rot in sheep, salmonellosis, pleurisy and pneumonia in sheep, sarcocystiasis, toxoplasmosis, hydatid diseases, nematodes, anthropods, and immunology.

HYDATID RESEARCH UNIT 101

Address: PO Box 913, Dunedin
Scientist: M.A. Gemmell

Wool Research Organization of New Zealand Incorporated* 102

– WRONZ
Address: Private Bag, Christchurch
Telephone: (03) 228-009
Parent body: Department of Scientific and Industrial Research
Status: Independent research centre
Director: Dr W.S. Simpson
Divisions: Wool and textile, Dr D.A. Ross; chemistry and engineering, R.G. Stewart
Activities: The organization undertakes research on all matters relating to the utilization of New Zealand wool, including chemical and physical properties, marketing, scouring, processing, and end-use performance.

NIGER

Institut National de Recherches Agronomiques du Niger

1

– INRAN
[Niger National Institute of Agronomic Research]
Address: BP 429, Niamey
Telephone: 72 21 44
Telex: INRANI 5201 NI
Affiliation: Ministry of Rural Economy and Climate
Status: Official research centre
Director-General: Moussa Saley
Graduate research staff: 46
Activities: INRAN's role is to organize research and lend technical and scientific assistance in the areas of rural development: ecology, agriculture, zootechnics, forestry, rural economics and teaching. Research is carried out in stations at Tarna, Kolo, Agadez, Tillabery and Grabougoura. There are laboratories for soil analysis, growth, biochemistry, phytogenetics, plant protection, biological pest control, mycotoxin and oleaginous studies.
Publications: Annual report.
Projects: Soil cartography (Quattara Mamadou); green mantle (Idrissa Daouda); groundnut seed production (Mounteila Amadou); cultivation improvement (Issoka Maga).

DEPARTMENT OF AGRICULTURAL RESEARCH

2

Head: I. Maga

DEPARTMENT OF ECOLOGICAL RESEARCH

3

Head: Quattara Mamadou

DEPARTMENT OF FORESTRY RESEARCH

4

Head: Moussa Hassane

DEPARTMENT OF RESEARCH IN AGRICULTURAL ECONOMICS

5

Head: Ly Samba Abdoulaye

DEPARTMENT OF VETERINARY RESEARCH AND ZOOTECHNICS

6

Head: Tahirou Abdou

NIGERIA

Agricultural Extension and Research Liaison Service   1

Address: PMB 1044, Zaria, Kaduna
Telephone: (069) 32589
Affiliation: Ahmadu Bellow University; Federal Ministry of Science and Technology
Status: Official research centre
Director: Alhaji Imrana Yazidu
Sections: Agronomy, Dr E.A. Salako; livestock, Dr S.O. Ogundipe; irrigation, A. Ramalan; farm management and crop protection, Dr N.B. Mijindadi
Graduate research staff: 36
Activities: Linkage of research institutes and state extension services; agricultural extension and advisory services; interpretation and dissemination of research.
Publications: Annual report, half yearly report.
Projects: Adaptation of research recommendations to village situations (Alhaji Imrana Yazidu); home economics for rural women (D.N. Maigida); production and technology transfer in sorghum millet and wheat (Dr E.A. Salako); field problem documentation and feedback (Dr N.B. Mijindadi); agricultural shows, radio and television films and slide production (J.B. Igunnu).

AERLS-ABU OUTSTATION   2

Address: State Liaison Office, Ilorin

Cocoa Research Institute of Nigeria   3

Address: PMB 5244, Ibadan, Oyo State
Telephone: (022) 412430
Telex: CRIN, IBADAN
Affiliation: Federal Ministry of Science and Technology
Status: Official research centre
Director: Dr S.T. Olatoye
Assistant director: Dr M.O.K. Adegbola
Departments: Research, Dr M.O.K. Adegbola; Production, Dr J.A. Williams; Monitoring and Evaluation, vacant.
Graduate research staff: 36
Activities: Production improvement of cocoa, coffee, kola, cashew and tea; investigation of new uses for the products and by-products crops. Plant breeding, agronomy, soil science, biochemistry, plant pathology, entomology, economics, extension research, pilot production.
Publications: Annual reports.
Projects: Breeding of improved cultivars of cocoa, cashew, coffee, kola and tea (Dr E.B. Esan); growth and yield of the above crops (S.A. Adenikinju); soils and nutritional requirements of the above crops (N.E. Egbe); control of diseases associated with above crops (Dr O.A. Olunloyo); control of pests associated with above crops (Dr A. Ojo); economics of production of the above crops (O. Ajobo); utilization of products and by-products of the above crops (Dr M.E. Ukhun); extension research liaison project (Dr M.M. Omole); pilot production projects (Dr J.A. Williams).

BENDE SUBSTATION   4

Address: Bende, Imo State

IKOM SUBSTATION   5

Address: Ikom, Cross River State

MAMBILLA SUBSTATION   6

Address: Gembu, Gongola State

OCHAJA SUBSTATION   7

Address: Ochaja, Kwara State

OWENA SUBSTATION   8

Address: Owena, Ondo State

UHONMORA SUBSTATION   9

Address: Uhonmora, Bendel State

Federal Department of   10
Agriculture

Address: 17-27 Moloney Street, PMB 12613, Lagos,
Lagos State
Telephone: (01) 632387; 630134
Status: Official research centre
Director: Mr Aeoyemi
Graduate research staff: 150
Activities: Formulation and implementation of govern-
ment policy.
Publications: Annual report, quarterly report.

AGRICULTURAL ENGINEERING   11
PROGRAMMES SECTION

Head: Mr Talabi

AGRICULTURAL SERVICE   12
PROGRAMMES SECTION

Head: Dr Anojulu

GOVERNMENT AGRICULTURAL   13
POLICY PLANNING SECTION

Head: A. Babaloia

NATIONAL SEED SERVICES   14

Head: Dr Joshua

PLANT QUARANTINE SERVICES   15

Head: Dr Aluko

Federal Institute of   16
Industrial Research, Oshodi

– FIIRO
Address: PMB 21023 Ikeja, Lagos, Lagos State
Telephone: (01) 962296
Affiliation: Federal Ministry of Science and Technol-
ogy
Status: Official research centre
Director: Dr O.A. Koleoso
Graduate research staff: 108
Activities: Aiding industrialization of Nigeria through
research into industrial use of local resources, development of indigenous technologies, and through routine
services to industries.
Grains, roots and tubers; fruits and vegetables, flavours
and essences; protein research and nutrition; food waste
utilization; materials development; library, information
and documentation (computerized information and
documentation systems); engineering; technoeconomic
studies.
Publications: Annual reports.

ENGINEERING DEPARTMENT   17

Head: S.C.O. Onyekwelu
Projects: Design and fabrication of fish smoke dryer;
design of 500kg/day gari processing machine (O.
Adeoye).

FOOD WASTE UTILIZATION   18
DEPARTMENT

Head: Dr A.B. Oniwinde
Projects: Production of pectin from citrus waste (M.O.
Oresanya).

FRUITS, VEGETABLES, FLAVOURED   19
ESSENCES DEPARTMENT

Head: Dr F.A.O. Osinowo
Projects: Upgrading the local technology of condiment
preparation, iru-fermented African locust beans (Dr
F.A.O. Osinowo); development of flavour concentrates
from ginger for drinks and other uses (A.B. Meadows).

GRAINS, ROOTS AND TUBERS   20
DEPARTMENT

Head: O.O. Onyekwere
Projects: Dry milling of local cereals (Dr O. Olatunji).

LIBRARY, INFORMATION AND DOCUMENTATION DEPARTMENT 21
Head: R.O. Sodipe

MATERIALS DEVELOPMENT DEPARTMENT 22
Head: Dr I. Aladeselu
Projects: Modification of cassava starch for textile industry (L.L. Akerele).

PRODUCTION OF MALT FROM LOCAL CEREAL GRAINS DEPARTMENT 23
Head: Dr O.A. Olaniyi
Projects: Malting of Nigerian cereal grains.

PROTEIN RESEARCH, NUTRITION AND SENSORY EVALUATION DEPARTMENT 24
Head: C.C. Edwards
Projects: Utilization of shea butter (Dr O.A. Koleoso).

TECHNOECONOMIC STUDIES DEPARTMENT 25
Head: O. Olunloyo
Projects: Upgrading the technology of Pito (A.C. Jibogun).

Federal Ministry of Science and Technology 26

Address: Ibadan Office, Moor Plantation, PMB 5382, Ibadan
Telephone: (022) 412202; 412222; 412457
See separate entries for:
Agricultural Extension and Research Liaison Service
Cocoa Research Institute of Nigeria
Federal Institute of Industrial Research, Oshodi
Forestry Research Institute of Nigeria
Institute for Agricultural Research
Institute of Agricultural Research and Training
Kainji Lake Research Institute
Lake Chad Research Institute
National Animal Production Research Institute
National Cereals Research Institute
National Horticultural Research Institute
National Root Crops Research Institute
National Veterinary Research Institute
Nigerian Institute for Palm Oil Research

Nigerian Institute for Trypanosomiasis Research
Nigerian Stored Products Research Institute
Rubber Research Institute of Nigeria

Forestry Research Institute of Nigeria 27

– FRIN
Address: PMB 5054, Ibadan, Oyo State
Telephone: (022) 414441; 414022; 414073
Telex: 312307
Affiliation: Federal Ministry of Science and Technology
Status: Official research centre
Director: Professor P.R.O. Kio
Graduate research staff: 82
Activities: Forestry, wildlife management and forest products utilization and forestry education.
Publications: Progress report, annual; annual report.

ECONOMICS AND MANAGEMENT DEPARTMENT 28
Head: J.O. Abayomi

EXTENSION AND RESEARCH LIAISON DEPARTMENT 29
Head: G.O.B. Dada

FOREST ECOLOGY DEPARTMENT 30
Head: Z.O. Gbile
Projects: Conservation of natural resources.

FOREST PATHOLOGY AND ENTOMOLOGY DEPARTMENT 31
Head: M.O. Akanbi

FOREST PRODUCTS RESEARCH DEPARTMENT 32
Head: S.A.O. Giwa
Projects: Timber utilization (E.O. Ademiluyi); pulp and paper research (S.A.O. Giwa).

FORESTRY EDUCATION AND TRAINING DEPARTMENT 33
Head: I.I. Ero

MANGROVE RESEARCH DEPARTMENT — 34

Head: E. Oddo

SAVANNA FORESTRY RESEARCH DEPARTMENT — 35

Head: A.B. Momodu

SHELTERBELT RESEARCH DEPARTMENT — 36

Head: M.A. Ogigirigi

SOILS AND TREE NUTRITION DEPARTMENT — 37

Head: O. Kadeba

TREE CROP PRODUCTION AND TREE IMPROVEMENT DEPARTMENT — 38

Head: G.O.A. Ojo
Projects: Establishment and improvement of indigenous tree species (G.O.A. Ojo); establishment and improvement of exotic tree species (P.E. Okorie).

Eastern Research Station — 39

Address: PMB 1011, Umuahia

Moist Forest Research Station — 40

Address: PO Box 476, Sapeba, Sapele

Savanna Forestry Research Station — 41

Address: PMB 1039, Samaru, Zaria

School of Forestry, Ibadan — 42

Address: PMB 5054, Ibadan

School of Forestry, Jos — 43

Address: PMB 2019, Jos

Shelterbelt Research Station — 44

Address: PMB 3239, Kano

Institute for Agricultural Research — 45

Address: PMB 1044, Zaria, Kaduna
Telephone: (069) 2571-4
Telex: AGRISEARCH ZARIA
Affiliation: Ahmadu Bello University; Federal Ministry of Science and Technology
Status: Official research centre
Director: J.H. Davies
Graduate research staff: 200
Activities: Natural resources, soil science, drainage and irrigation, land use; plant production (maize, millet, groundnuts, corn); animal production (cows, goats, sheep, chickens, pigs); agricultural engineering and building (harvesters, spraying machines), food science.
Publications: Annual report; *Samaru Agricultural Journal.*
Projects: Socioeconomic research (Professor Abalu); fibre crops research (Dr Ogunlela); irrigation research (Dr Nwa); cereals research (Dr Egharvba); agricultural mechanization (Professor Kaul); horticultural research (Dr Erinle); cropping systems (Baker); oil seed research (Professor Harkness); soil and crop environment (Dr Nuadi); grain legume research (Professor Emechebe); food science technology (Professor Fewster).

AGRICULTURAL ECONOMICS AND RURAL SOCIOLOGY DEPARTMENT — 46

Head: Professor Abalu

AGRICULTURAL ENGINEERING DEPARTMENT — 47

Head: Professor Kaul

ANIMAL SCIENCE DEPARTMENT — 48

Head: Professor Akinola

CROP PROTECTION DEPARTMENT — 49

Head: Dr Erinle

PLANT SCIENCE DEPARTMENT — 50

Head: Professor Olugbemi

SOIL SCIENCE DEPARTMENT — 51

Head: Dr Ojanuga

STATIONS AFFILIATED TO THE   52
INSTITUTE*

Agricultural Research Station, Kano   53
Address: PO Box 1062, Kano

Agricultural Research Station, Mokwa   54
Address: PMB 101, Mokwa via Jebba

Institute of Agricultural   55
Research and Training

Address: PMB 5029, Ibadan, Oyo State
Telephone: (022) 412861
Affiliation: Federal Ministry of Science and Technology, Official research centre
Director: E.A. Olaloku
Sections: Farming systems, Dr M.O. Omidiji; food legumes, K.A. Ayoade; industrial crops, Dr E.A. Ogunremi; soil/water management, Dr R.A. Sobulo; livestock improvement, Dr A. Akinyemi
Graduate research staff: 40
Activities: Improvement of yield and genetic potential, nutritional and utilization qualities of major food crops of south-western Nigeria; soil science, irrigation, plant breeding, crop husbandry, plant protection including post-harvest problems; livestock breeding and management problems.
Publications: Annual reports.
Projects: Improvement of the genetic potential, yield and nutritional qualities of legumes (K.A. Ayoade); improvement and evaluation of S.W. Nigeria farming systems (Dr M.O. Omidiji); improvement of industrial crops such as cotton, sugarcane (Dr E.A. Ogunremi); evaluation of vegetable based farming systems (Dr B.O. Adelana); effect of confinement on the performance of West African cross breeds (Dr Akinyemi).

AGRICULTURAL RESEARCH   56
STATION, ERUWA
Address: Eruwa, Oyo State

AGRICULTURAL RESEARCH   57
STATION, FASHOLA
Address: Fashola, Oyo State

AGRICULTURAL RESEARCH   58
STATION, IKENNE
Address: Ikenne, Ogun State

AGRICULTURAL RESEARCH   59
STATION, ILORA
Address: Ilora, Oyo State

SCHOOL OF AGRICULTURE, AKURE   60
Address: Akure, Ondo State

SCHOOL OF AGRICULTURE, IBADAN   61
Address: Moor Plantation, Ibadan, Oyo State

SCHOOL OF ANIMAL HEALTH   62
Address: Moor Plantation, Ibadan, Oyo State

Kainji Lake Research   63
Institute

– KLRI
Address: PMB 666, New Bussa, Kwara State
Affiliation: Federal Ministry of Science and Technology, Official research centre
Director: Dr H.M. Yesufu
Graduate research staff: 32
Activities: Research into man-made lakes and major rivers in Nigeria.
Publications: Annual report, technical report, progress report.

AGRICULTURE AND IRRIGATION   64
DEPARTMENT
Head: E.C. Erinne
Activities: Development of draw-down farming, drainage and irrigation, soil and agricultural land use.
Projects: Cereals production and improvement: rice, maize and wheat (R.C. Ogo); legumes production and improvement: cowpea, ground nut and winged beans; vegetable crops production and improvement: tomatoes, okro, and leaf vegetables; root crops production and improvement: yam and cassava (Dr A.B. Chaudhry); irrigation development (E.C. Erinne).

FISHERIES DEPARTMENT 65

Head: E.O. Ita
Activities: Abundance, distribution and other biological characteristics of species of fish and practical methods of their rational exploitation in Kainji Lake and other man-made lakes in Nigeria.
Projects: Biology and population dynamics of the economically important species of Kainji, and Tiga and Shiroro lakes; equipment designs and trials; cage culture trials (E.O. Ita); feed technology and nutrition trials in aquarium cage and ponds; chemical composition and nutritive value of Nigerian inland fishes, microbial investigations of spoilage organisms of fresh fish (A.A. Eyo); design and development of mechanized and non-mechanized crafts for lakes and rivers (S.O. Omorodion).

LIMNOLOGY DEPARTMENT 66

Head: H.A. Adeniji
Activities: Limnology, behaviour and characteristics of the Kainji and other man-made lakes in Nigeria and their effects on the fish and other aquatic life.
Projects: Physical, chemical, biological and geolimnological studies; plankton, bacteria, and benthos organism ecology; primary and secondary production of Kainji Lake (H.A. Adeniji).

PUBLIC HEALTH DEPARTMENT 67

Head: Dr E.O. Adekolu-John
Activities: Public health problems arising from the construction of dams and the resettlement of people.
Projects: Health status of some selected villages with particular reference to the following areas: health care delivery facilities, vector and parasitic diseases studies (Dr E.O. Adekolu-John).

SOCIOECONOMIC DEPARTMENT 68

Head: Dr F.P.A. Oyedipe
Activities: Effects of the construction of the lakes on the socioeconomic conditions of the rural populations around Kainji Dam and Jebba Dam.
Projects: Resettlement and social change at New Bussa (Dr F.P.A. Oyedipe); migrant fishermen of Kainji Lake Basin, variables influencing migration patterns and trends; resettlement of Jebba Dam relocatees, a pre-impoundment socioeconomic study (K.M. Arungbemi); cooperative food crop marketing (J.O. Ayanda).

WILDLIFE AND RANGE MANAGEMENT DEPARTMENT 69

Head: Dr J.S.O. Ayeni
Activities: Behaviour and characteristics of wildlife and their management in Kainji Lake National Park; animal production with specific reference to guineafowl and cattle.
Projects: Biology of wildlife species in the Kainji Lake National Park; domestication of selected wildlife species, particularly guineafowl and kob (Dr J.S.O. Ayeni); pasture improvement around Kainji Lake Basin (E. Obot).

Bin Yauri Field Station 70

Gafara Field Station 71

Address: Gafara, Niger State

Papiri Field Station 72

Address: Papiri, New Bussa

Shagunu Field Station 73

Yashikira Field Station 74

Lake Chad Research Institute 75

Address: PMB 1293, Maiduguri, Borno State
Affiliation: Federal Ministry of Science and Technology
Status: Official research centre
Director: V.O. Sagua
Assistant director: Dr B.B. Wudiri
Sections: Crops research, Dr M.L. Jerath; fisheries, N.I. Azeza; livestock, Dr R.K. Pandey; wildlife and forestry, V.O. Sagua; water resources and public health, V.O. Sagua
Graduate research staff: 30
Activities: Crop improvement under irrigated and rainfed conditions (wheat, rice, maize, cotton, millet, sorghum); fisheries, fish resources management and processing on Lake Chad; public health, waterborne diseases associated with irrigation; livestock, pasture improvement; wildlife afforestation and desert encroachment.

Publications: Annual report.
Projects: Dryland farming improvement (Dr L.I. Okafor); irrigated farming improvement (Dr M.L. Jerath); Lake Chad fisheries resources improvement (N.I. Azeza); fish culture in Lake Chad area (V.O. Sagua); livestock improvement (Dr R.K. Pandey); water resource and public health, wildlife studies and arid zone afforestation (V.O. Sagua).

CROP RESEARCH STATION, GAJIBO   76
Address: Gajibo, Borno State

CROP RESEARCH STATION, NGALA   77
Address: Ngala, Borno State

FISHERIES RESEARCH STATION, BAGA   78
Address: Baga, Borno State

FISHERIES RESEARCH STATION, MALAMFATORI   79
Address: Malamfatori, Borno State
Projects: Cattle production (J.E. Umoh); poultry production (J.M. Olomu); sheep and goats production (V. Buvanendran); swine production (M.B. Olayiwole); pasture production (E.C. Agishi).National Animal Production Research Institute, PMB 1096, Zaria, Kaduna State
Telephone: (069) 32596
Affiliation: Federal Ministry of Science and Technology
Status: Official research centre
Director: Professor Saka Nuru
Deputy director: Dr M.B. Olayiwole
Graduate research staff: 50
Activities: Animal production (cattle, sheep, goats, pigs, poultry) and animal products, pasture production and socioeconomic factors that influence production at rural level.
Publications: Animal Production Research Report, annual; bulletin, quarterly; *Journal of Animal Production Research*, biannual.

ANIMAL HUSBANDRY DEPARTMENT   80
Head: J.M. Olomu
Activities: Cattle management and production, sheep and goat production, poultry production, breeding and genetics.

ANIMAL REPRODUCTION DEPARTMENT   81
Head: A.O. Osinowo
Activities: Physiology, artificial insemination research.

LIVESTOCK ECONOMICS AND RURAL SOCIOLOGY DEPARTMENT   82
Head: P. Okaiyeto

National Cereals Research Institute   83
– NCRI
Address: PMB 5042, Ibadan, Oyo State
Telephone: (022) 400920
Affiliation: Federal Ministry of Science and Technology
Status: Official research centre
Director: C.O. Obasola
Graduate research staff: 135
Activities: Production and products of rice, maize, grain legumes and sugarcane; plant breeding, agronomy, soil science, plant protection, storage and processing, economics, rural sociology and agricultural engineering, farming systems.
Publications: Annual report.
Projects: Rice improvement (S.O. Fagade); maize improvement (Dr S.U. Remison); grain legume improvement (S.O. Dina); sugarcane improvement (F.O. Obakin); farming systems (Dr B.A. Olunuga).

AGRICULTURAL EXTENSION AND RESEARCH LIAISON DEPARTMENT   84
Head: G.E.C. Ohiaeri

DEVELOPMENT AND MANAGEMENT DEPARTMENT   85
Head: S.O. Boboye

PLANNING DEPARTMENT   86
Head: Dr A. Abidogun

RESEARCH AND TRAINING DEPARTMENT   87
Head: Dr O.A. Ojomo

STATIONS AFFILIATED TO THE INSTITUTE* — 88

Nifor Station — 89
Address: PMB 1030, Benin City

Rice Research Station, Badeggi — 90
Address: PMB 8, Biga, Niger State

Rice Research Station, Birnin Kebbi — 91
Address: PMB 22, Birnin Kebbi, Sokoto State

Rice Research Substation, Nigala — 92
Address: PMB 1293, Maiduguri, Borne State

Umudike/Amakama Substation — 93
Address: Umudike, PMB 1026, Umahia

Uyo Research Station — 94
Address: PMB 1032, Uyo, Cross River State

National Horticultural Research Institute — 95

Address: PMB 5432, Ibadan, Oyo
Affiliation: Federal Ministry of Science and Technology
Status: Official research centre
Director: S.A. Adeyemi
Graduate research staff: 87
Activities: Plant production: vegetables, fruits such as guava, mango, pineapples, apples, pawpaw, and citrus; plant breeding, agronomy, plant protection, soil science and food science.
Substations are being developed at Bagauda, Kano State, Kbate and Imo State.
Projects: Vegetable improvement (Dr O.A. Denton); fruit improvement (Dr J.C. Obiefuna); citrus improvement (C.A. Amih); commercial production (I.O.A. Emiola).

RESEARCH DEPARTMENT — 96
Head: Dr W.E. Eguagie

SUPPORT SERVICES DEPARTMENT — 97
Head: S.A. Adeyemi

National Root Crops Research Institute — 98

Address: PMB 1006, Umudike, Umuahia, Imo State
Telephone: 220188
Telex: Agrisearch Umuahia
Affiliation: Federal Ministry of Science and Technology
Status: Official research centre
Director: Dr L.S.O. Ene
Programmes: Cassava, Dr J.E. Okeke; yam, Dr O.O. Okoli; Irish potato, Dr O.P. Ifenkwe; other root crops (cocoyam, sweet potato, etc), Dr O.B. Arene
Graduate research staff: 88
Activities: Production and products of yams, cocoyams, cassava, sweet potatoes, Irish potatoes and other root crops of economic importance.
Publications: Annual reports, technical bulletins.
Projects: Cassava improvement (Dr J.E. Okeke); yam improvement (Dr O.O. Okoli); Irish potato improvement (Dr O.P. Ifenkwe); sweet potato improvement, cocoyam improvement (Dr O.B. Arene); irrigation management and root crops (P.M. Nwokedi); germ plasm collection, maintenance and evaluation (Dr N. Ukpa); transfer of agricultural technology by extension workers, foundation seed production and distribution (H.E. Okereke); development of school of agriculture (M.U. Ibe); pilot farm food production testing (C.C. Chinaka); headquarters development and maintenance, development of field studies (Dr L.S.O. Ene, H.E. Okereke).

AGRICULTURAL EXTENSION RESEARCH LIAISON AND TRAINING — 99
Assistant Director: H.E. Okereke
Sections: National accelerated food production programme, Dr S.O. Odurukwe; Subject matter specialist department, Dr T. Enyinnia; School of Agriculture, M.U. Ibe; Planning, monitoring and evaluation department, A.W. Iloka; Livestock unit, S.C. Ujo; Poultry production unit, G.A. Njoka; Crop production unit, M.C. Onwuazor
Graduate research staff: 35
Activities: Collection and dissemination of agricultural information through the activities of Extension Research Liaison Service; mass involvement of farmers in the production of cassava; activities cover the four eastern states of Nigeria, Anambra, Cross River, Imo and Rivers States.
Publications: Annual report.
Projects: Bench-mark survey, training of extension workers and farmers in herbicide technology of rice and cassava/maize production, rice production survey (R.P.A. Unamma); training of extension workers and farmers, field days and agricultural shows, rice production campaigns (S.N. Anyaegbu); mass communication,

anthracnose control in water yam (D. alata), establishment of state liaison offices (H.E. Okereke); yam extension programme, production testing of cassava/maize intercrop (C.C. Chinaka); crop protection campaigns (Dr T. Enyinnia); promotion of sweet potato production and utilization (A.C. Okonko); pig production survey, pig breeding (S.C. Ujo); poultry industrial survey, hatchery project, pullet production testing, broiler and cockerel production, feed mill (A.M. Agbakoba); production and distribution of vegetable seeds, ornamentals, fruit tree seeds and seedlings (C.R.A. Ogbuehi).

Substations *   100

Kuru Substation   101
Address: National Veterinary Research Institute, Vom, Jos, Plateau State

Otobi Substation   102
Address: Divisional Agricultural Office, Oturkpo, Benue State

AGRONOMY DIVISION   103
Head: Dr M.C. Igbokwe

BIOCHEMISTRY DIVISION   104
Head: Dr S.O. Alozie

ENGINEERING RESEARCH DIVISION   105
Head: P.M. Nwokedi

GERMPLASM COLLECTION AND MAINTENANCE DIVISION   106
Head: Dr N. Ukpa

PHYSIOLOGY DIVISION   107
Head: Dr N.H. Igwilo

PLANT BREEDING DIVISION   108
Head: Dr O.O. Okoli

PLANT PROTECTION DIVISION   109
Head: Dr E.C. Nnodu

SOILS DIVISION   110
Head: B.O. Njoku

STATISTICS RECORDS AND AGROMET DIVISION   111
Head: B.I. Anyaegbunam

National Veterinary Research Institute   112

Address: Vom, Plateau State
Affiliation: Federal Ministry of Science and Technology
Status: Official research centre
Director: Dr A.G. Lamorde
Divisions: Bacteriology, Dr O. Onoviran; virology, Dr E.N. Okeke; livestock investigation, Dr I. Umo; poultry, Dr M.C. Njike; biochemistry, Dr A.O. Ikwuegbu; diagnostics, Dr T.O. Osiyemi; parasitology, Dr J.P. Fabiyi
Animal Health Programme: Dr B.Y. Cwolodun
Graduate research staff: 150
Activities: Research into all aspects of diseases of animals and their control; all aspects of animal nutrition and standardization and quality control of animal feeds; production of vaccines and sera; introduction of exotic breeds of animals to improve meat, milk and egg production; training of livestock assistants and superintendents, and medical laboratory technicians and technologists; diagnostic services in the states.
Publications: Annual report.
Projects: Immunology of contagious bovine pleuropneumonia (Dr O. Onoviran); reproductive performance of Friesian and Fulani cows (Dr I. Umo); response of Brucella antigens in guinea-pigs and cattle (Dr E. Eze); development of cell culture rabies vaccine (A.N.C. Okeke); development of Gomboro vaccine (Dr E.N. Okeke); protein requirements of Nigerian poultry (Dr M.C. Njike); efficacy of anthelminthics for inhibited larvae of nematodes of cattle (Dr J.P. Fabiyi); aetiology of caprine vulvovaginitis (Dr J.C. Chima); replacement value of mineral licks (Dr O.A. Ikwuegbu).

VETERINARY DIAGNOSTIC LABORATORY, ILORIN *   113
Address: Ilorin, Kwara State

VETERINARY DIAGNOSTIC LABORATORY, KADUNA *   114
Address: Kaduna, Kaduna State

VETERINARY DIAGNOSTIC LABORATORY, KANO *   115
Address: Kano, Kano State

VETERINARY DIAGNOSTIC LABORATORY, MAIDUGURI* 116

Address: Maiduguri, Borno State

VETERINARY DIAGNOSTIC LABORATORY, UMUDIKE* 117

Address: Umudike, Imo State

Nigerian Institute for Palm Oil Research* 118

Address: PMB 1030, Benin City, Bendel State
Affiliation: Federal Ministry of Science and Technology
Status: Official research centre

ABAK SUBSTATION* 119

Address: PO Box 31, Abak

AGBARHO OUTSTATION* 120

Address: PO Box 57, Orho-Agbarho, via Warri

BADAGRY SUBSTATION* 121

Address: Ikoga Farm, Badagry, Lagos State

DUTSE SUBSTATION* 122

Address: Kano

OBOTME OIL PALM PLANTATION* 123

Address: PMB 32, Arochukwu, Imo

ONISHERE OUTSTATION* 124

Address: PO Box 188, Ondo State

UMUABI OUTSTATION* 125

Address: Udi, Anambra

Nigerian Institute for Trypanosomiasis Research 126

– NITR
Address: PMB 2077, Kaduna, Maduna State
Telephone: (062) 210292; 210271
Affiliation: Federal Ministry of Science and Technology
Status: Official research centre
Director: Y.A. Magaji
Departments: Entomology, Dr J.A. Onyiah; pathology, Dr H. Edeghere; epidemiology, Dr O.E. Uche; parasitology, Dr R.A. Joshua; biochemistry and chemotherapy, J.K. Emeh; veterinary and livestock studies, Dr W.E. Agu
Graduate research staff: 30
Activities: Trypanosomiasis (sleeping sickness) and onchocerciasis (river blindness) research. Pathology of immunology and treatment methods; vector ecology and life-cycle and disease transmission; clinical, biological and other methods of control; socioeconomic effects of the disease on rural populations.
Publications: Annual report; research bulletin.
Projects: Vector biology and control of tsetse and black flies in Nigeria (Dr W. Gregory); diagnosis, treatment and epidemiology of sleeping sickness and river blindness (Dr H. Edeghere); development and production of vaccines against sleeping sickness and river blindness in man and domestic animals (Dr R.A. Joshua); resistance to insecticide, drug resistance and deposit in animal tissue (Dr W.E. Agu).

VOM SUBSTATION 127

Address: Vom, near Jos, Plateau State

Nigerian Stored Products Research Institute 128

– NSPRI
Address: PMB 12543, Lagos, Lagos State
Telephone: (01) 862653; 862789
Affiliation: Federal Ministry of Science and Technology
Status: Official research centre
Director: S.D. Agboola
Assistant Director: Dr S.A. Adesuyi
Departments: Entomology, F.O. Mejule; microbiology, Dr J.O. Oyeniran; biochemistry, J.S. Opadokun; engineering, A.O. Akinnusi; physiology, Y.A. Oluwole; extension/economics, Dr M.A. Adesida
Graduate research staff: 42

Activities: Bulk storage problems of Nigerian export commodities and local food crops, such as cocoa, groundnuts, palm produce, cereals, grain pulses, tubers, etc.
Publications: Annual report.
Projects: Cereals and cereal products (O. Sowunmi; grain legumes and legume products (J.S. Opadokun); roots, tubers and their products (Dr S.A. Adesuyi); fruits and vegetables (Dr J.O. Oyeniran); oilseeds and cakes, fish and meat products, vegetable oils (W.I. Okoye); storage extension and research liaison services (Dr J.O. Oyeniran); economic analysis of recommended storage practices (Dr M.A. Adesida); beverage crops (F.O. Mejule); miscellaneous storage studies (vacant).

IBADAN OUTSTATION 129
Address: PMB 5044, Ibadan

KANO OUTSTATION 130
Address: PMB 3032, Kano

PORT-HARCOURT OUTSTATION 131
Address: PMB 5063, Port Harcourt

SAPELE OUTSTATION 132
Address: PMB 4065, Sapele

Rubber Research Institute of Nigeria* 133

Address: PMB 1049, Benin City
Affiliation: Federal Ministry of Science and Technology
Status: Official research centre

AKWETE OUTSTATION* 134
Address: Akwete, via Aba, Imo State

University of Ibadan 135

Address: Ibadan, Oyo State
Status: Educational establishment with r&d capability

FACULTY OF AGRICULTURE AND FORESTRY 136
Telephone: 400550
Telex: 400626
Dean: Professor G.M. Babatunde

Agricultural Biology Department 137
Head: N.O. Adedipe
Graduate research staff: 18
Activities: Crop physiology, ecology, entomology, genetics and phytopathology.
Publications: Annual report.
Projects: Role of boron in cowpea performance and the assessment of nutrient uptake and translocation profiles in rice (Professor N.O. Adedipe); effects of various factors of production on the growth, development and yield of three grain legumes - cowpea, soyabean and pigeon pea - in the lowland tropics (Dr T.O. Tayo); ecology and control of vertebrate pests (mainly rodent and avian pests) in peasant and plantation rice crops (Dr O. Funmilayo); ecology, biology and control of weeds of arable crops (Dr Sola Ogunyem); damage by thrips on cowpea performance, bioeconomics of pod-sucking bugs on cowpea plant and control of these insects (Dr E.O. Osisanyo); ecology and control (chemical and biological) of pests of okra and cotton and assessment of crop loss by population dynamics (Dr J.A. Odebiyi); identification and control of the seed-borne fungi of economic crops and assessment of host-pathogen relationships in plant disease development (Professor O.F. Esuruoso); microbiological deterioration of basic Nigerian foodstuffs - maize, rice, gari, yam flour; evaluation of nematode-fungal interactions in disease syndromes (Professor M.O. Adeniji); biological control of pests and microorganisms in agriculture and evaluation of the chemical basis of tuber crop resistance to infection by plant pathogens (Dr T. Ikotun); ecology and significance of plant parasitic nematodes under the traditional mixed-cropping systems and assessment of the efficacy of certain soil amendments in reducing nematode infestation of maize (Dr O.A. Egunjobi); ecology, population dynamics and the economic significance of rootknot nematodes in Nigeria and assessment of the significance of migratory ectoparasitic nematodes in Nigerian agriculture (Dr B. Fawole).

Agricultural Economics Department 138
Head: Professor O. Ogunfowora
Publications: Journal of Rural Economics and Development, biannual; occasional reports.
Projects: Cost ration analysis of pig production and broilers with emphasis on cassava, guinea corn and maize as energy sources; beef cattle production supplementary feed economics; plantation agriculture economics; interregional grain production and consumption; minimum price determination for crops and

livestock products in Nigeria (Professor O. Ogunfowora).

Water resources development, role of rural infrastructural facilities in rural development (Professor R.O. Adegboye).

Agricultural price analysis particularly food and livestock prices; effects of agroclimatic factors on Nigeria's food crop production and yields; optimal replacement period analysis for cocoa trees in Nigeria; growth and fluctuations in Nigeria's food crop output (Professor J.K. Olayemi).

Nigerian food policy, dynamic investment, and Nigerian rural-urban migration; economics of information demand by Nigerian farmers; agricultural input subsidy economics; food production resource productivity; macroeconomic policy; Nigerian rural poverty; human capital investment in rural areas; institutional minimum urban wages impact; rural infrastructures (Dr F.S. Idachaba).

Nigerian women in agriculture; agricultural cooperatives and rural development; food and agriculture technology (Dr T.O. Adekanye); women's activities in rural areas of Aniocha, Bendel State; rural development institutions in Oyo State; government agricultural development institutions (Dr S.G. Nwoko); job monitoring analysis, and agriculture and forestry development; bank lending for agriculture; agricultural policy analysis (Dr A.J. Adegeye).

Economic analysis of agribusiness enterprises and food processing industries; organization and management cooperatives evaluation (Dr J.A. Akinwumi).

Fisheries cooperatives structure and management; fish farming economics; input subsidies; domestic food price stabilization through importation; utility analysis of urban working class housewives in Nigeria; monopoly and market concentration in economic development; industrialization; economic effort and earning analysis in Nigerian artisanal fisheries; Nigerian poultry industry cost structure, interregional analysis (Dr A.F. Mabawonku); food products marketing; minimum wage structure determination; investment levels and rural development; Nigerian rice industry economics; economics of amusement parks in Nigeria (Dr A.E. Ikpi); rural development and farm income in Oyo State; markov chains (J.T. Atobatele).

Agroclimatic factors and Nigerian crop economy; Nigerian sugar industry structure; Oyo State expected rainfall limits; impact of land use decree on agricultural production; agricultural financing; palm wine; economic analysis of gari production, processing and marketing (J.O. Akintola).

Acceptable human diets determination models in Oyo State; rice and maize supply response and marketable surplus; Nigerian livestock economy spatial structure (Dr Chris Onyenweaku); performance and impact evaluation of agricultural development projects in Nigeria; public policy impact on rural development (Aja Okorie).

Agricultural Extension Services Department 139

Head: Professor S.K.T. Williams
Graduate research staff: 8
Activities: Operation of pilot projects on rural development, in-service training for farmers and advisory service to State Governments.
Projects: Difficulties faced by agricultural extension agents in bringing service to farmers, subject matter specialists' roles in agricultural extension, impact of agro-service centres on agricultural productivity in Oyo State (S.K.T. Williams); training needs survey of agricultural managers/supervisors, adoption of maize storage practices by farmers in Oyo division, sheep/goat husbandry practices by farmers in Cross River State (A.U. Patel); extension system evaluation of Funtua/Gusau agricultural development projects (A.U. Patel, Dr Ogunfiditimi); feeding patterns and social life of rural farmers in selected villages around Ibadan, soyabean akara acceptability by rural/urban panelists, profitability of cassava cultivation by selected rural women in making gari in southern states of Nigeria, technological innovation in palm oil processing by rural women (Carol E. Williams); post-harvest preservation of grains (M.F. Ivbijaro); adoptive trials on three new varieties of cowpea; intercropping of maize and cassava, maize and cowpea; growth and development of three cowpea varieties; growth and development of two varieties of okra (Hibiscus esculentrum) (E.O. Lucas, Dr M.O. Fawusi).

Agronomy Department 140

Head: Professor H.R. Chheda
Projects: Soils: microbiology - nitrogen cycle, pesticides in soils, environmental pollution by crude petroleum (Professor Odu, Dr Udo, Dr Wahua); physics - soil-water-plant relationship, tillage and non-tillage practices, soil temperature, physical conditions, and root activities of crops (Dr Babalola, Dr Vine); chemistry and fertility - phosphorus chemistry in tropical soils, micronutrients, potassium availability and fixation, organic matter, soil testing and fertilizer recommendations, thermodynamics of cation exchange, chemistry of hydromorphic soils, soil acidity and liming, herbicide-soil interactions, weed-fertility interactions (Professor Agboola, Dr Udo, Dr Omueti, Dr Obigbesan); mineralogy - sand mineralogy, silt and clays fractions of Nigerian soils, geochemistry of fluorine in tropical soils, opaline microfossils (Dr Omueti, Dr Udo, Professor Agboola); survey and pedology - land capability classification, aerial photo interpretation and remote sensing techniques. (Dr Fagbami, Dr Omueti, Dr Udo).
Crops: production - nutrient requirement of tropical food crops, farming systems, weed control (Professor Dr Fayemi, Professor Agboola, Dr Obigbesan, Dr Wahua); plant breeding breeding of tropical forage crops, tropical

vegetables (Professor Chheda, Dr Aken'Ova, Dr Fatokun); pasture agronomy - forage production in lowland tropics, digestibility of grass, legumes and browse plants (Professor Chheda, Dr Saleem); horticulture - citrus rootstock-scion studies, stress studies on indigenous Nigerian vegetables, plant propagation, herbicide residues and weed control (Dr Fawusi, Dr Oputa); crop nutrition and physiology - physiological responses of crops to salinity and herbicides; sources of sugar substitutes, mineral nutrition, yield and quality of improved and local root and tuber crop species, translocation and partitioning of assimilates in selected plantation crops, cassava edaphology (Professor Fayemi, Dr Obigbesan, Dr Oputa, Dr Vine); agricultural mechanization - rural life conveniencies, development of small farm tools (Mr Fawole); weed science - weed biological control, herbicides in the environment (Professor Fayemi, Professor Agboola, Dr Wahua, Dr Oputa).

Animal Science Department   141

Head: Professor A.U. Mba
Projects: Genetic parameter and potentialities of indigenous species of poultry including game birds; genotype-environment interaction in poultry production in Nigeria; nutritional requirements of poultry under tropical conditions; local product utilization in poultry diets in the tropics.

Litter management and types - comparison; management systems comparison, effects on physiology and performance; management and nutritional effects on physiological development.

Cassava utilization by different classes of livestock; nutrient requirements of domesticated game animals; energy and protein requirements of cattle and sheep of different liveweight and age groups.

Comparative studies on the rate of growth of German X N'Dama steers and White Fulani (Zebu) steers of comparable age groups, grazed on pastures only or pastures supplemented with cotton seed cake or palm kernel meal; recycling of cattle wastes as livestock feeds; carbohydrate complex of palm kernel meal utilization by poultry; carbohydrate fractions characterization of local feeds and lesser known tree crops; energy values of poultry feeds; nutrient requirements of farm animals; utilization of byproducts by farm animals; browse plant utilization by small ruminants.

Forest Resources Management   142
Department

Head: Professor D.U.U. Okali
Graduate research staff: 14
Activities: Field, greenhouse and laboratory studies in silviculture, forest biometrics, wood science, forest economics, management of manmade forests of exotic and indigenous species and the natural tropical rainforest.
Publications: Annual report.

FACULTY OF VETERINARY   143
MEDICINE*

Dean: Professor G.O. Esuruoso

University of Ife   144

Address: Ile-Ife, Oyo State
Status: Educational establishment with r&d capability

FACULTY OF AGRICULTURE   145

Telephone: 2291-2299
Dean: Professor A.O. Adenuga
Graduate research staff: 70
Activities: Soil science and land use, crop production, protection and breeding (tomatoes, cowpeas, yams, maize); livestock breeding (poultry), animal production (sheep and goats), livestock husbandry and nutrition (pigs, turkeys and rabbits). Food economics (cowpeas, fishery), land tenure, agricultural marketing, agricultural cooperatives, farm management.
Publications: Faculty research reports, *Ife Journal of Agriculture.*
Projects: Tomato research (Dr Tunde Fatunla); cowpea research (Dr E.A. Akingbohungbe); maize breeding (Dr A.M.A. Fakorede); soil fertility (Dr J.A. Adepetu); yam production (Professor I.C. Onwueme); biological control of insect pests (Dr B.A. Matanmi); goat husbandry and management (Professor A.A. Ademosun); nuclear techniques in animal production (Professor A.A. Adegbola).

Agricultural Economics Department   146

Head: Professor C.A. Osuntogun

Animal Science Department   147

Head: Professor A.A. Ademosun

Extension Education and Rural Sociology Department   148

Head: Professor J.A. Alao

Plant Science Department   149

Head: Professor I.C. Onwueme

Soil Science Department   150

Head: Professor I.I. Ashaye

University of Nigeria   151

Address: Nsukka
Status: Educational establishment with r&d capability

FACULTY OF AGRICULTURE   152

Telephone: 6251-2
Dean: Professor F.O.C. Ezedinma
Graduate research staff: 70
Activities: Soil science.
Management and production of cereals, legumes, roots and tubers, ornamental vegetables, oil fruits, and plantation crops; plant breeding, plant propagation, crop protection.
Animal breeding, nutrition and management involving beef, cattle, swine, goats, poultry and rabbits.
Food science and technology, home economics, dietetics and nutrition.
Agricultural cooperatives, farm management, agricultural projects; planning and evaluation, agricultural extension, education, rural social organizations.

Agricultural Economics Department   153

Head: Professor M.O. Ijere

Agricultural Extension Department   154

Head: Dr L.O. Obibuaku

Animal Science Department   155

Head: Dr F.C. Obioha

Crop Science Department   156

Head: Dr E.U. Okpala

Food Science and Technology Department   157

Head: Professor P.O. Ngoddy

Home Science and Nutrition Department   158

Head: D.O. Nnanyelugo

Soil Science Department   159

Head: Professor W.O. Enwezor

NORWAY

Christian Michelsens Institutt for Videnskap og Åndsfrihet* 1

– MICHFOND
[Christian Michelsen Institute of Science and Intellectual Freedom]
Address: Fantoftvegen 38, 5000 Fantoft
Telephone: (0475) 284410
Telex: 40 006 CMI N
Status: Independent research centre
Director: Arvid Johnsen
Graduate research staff: 55
Activities: Engaging and creating positions for researchers who have shown outstanding abilities, and the furthering of tolerance and forbearance between classes in society, nations and races.

SCIENCE AND TECHNOLOGY DEPARTMENT* 2

Director: Dr Frode L. Galtung
Activities: Marine instrumentation; instrumentation for precision measurements of gas and liquid flow; dynamics of floating structures; microprocessor systems; automatic quality inspection; payloads for scientific rockets; biomedical engineering; petroleum economy; fisheries economy; newspaper economy; gas explosions; dust explosions; powder technology.

Applied Systems Analysis Section* 3

Head: Leif K. Ervik
Activities: Petroleum economics; energy; regional planning; fisheries economy.

Direktoratet for Vilt og Ferskvannsfisk* 4

[Directorate for Wildlife and Freshwater Fish]
Address: Elgesetergaten 10, N-7000 Trondheim
Telephone: (075) 37020
Affiliation: Ministry of Environment
Status: Official research centre
Director: Helge Vikan
Graduate research staff: 10
Activities: Research is aimed at problems concerning wildlife management; the main study objects being: tetraonids; seabirds; great mammalian predators; cervids.
Publications: *Viltrapport; Meddelelser fra Norsk Viltforskning.*

GAME RESEARCH DIVISION* 5

Research director: Svein Myrbeget

SALMON AND FRESHWATER FISHERIES SECTION 6

Address: Postboks 63, 1432 Ås-NLH
Telephone: (02) 941060
Head: Kjell W. Jensen
Activities: Management of freshwater fisheries, including fisheries for anadromous salmonids.
Publications: Rapport fra Fiskeforskningen ; annual reports.

WILDLIFE SECTION* 7

Head: Svein Myrberget

Felleskjøpet* 8

– FK
[Farmers Cooperative Agricultural, Experimental and Seed Growing Station]
Address: Bjørke forsøks- og stamsaedgard, 2344 Ilseng
Telephone: (065) 95411
Telex: 17156 FO N - OSLO
Status: Official research centre
Research director: Stein Frogner
Activities: Testing, increase, release and distribution of new varieties of field crops and maintenance of already released varieties; plant breeding in small grains (barley, oats, wheat) in which field FK is cooperating with Svaløf AB, Sweden.

Fiskeridirektoratets Vitamininstitutt 9

[Fisheries Directorate Vitamin Institute]
Address: Postboks 4285, N-5013 Bergen, Nygårdstangen
Telephone: (05) 230300
Telex: 42 151
Affiliation: Ministry of Fisheries, Directorate of Fisheries
Status: Official research centre
Research director: Professor Olaf R. Braekkan
Sections: Proteins, L.R. Njaa, J.W. Jebsen; vitamin B research, E. Lied; vitamins, marine oil and fats, G. Lambertsen; minerals and trace elements, salt water pollution, K. Julshamn; fish-breeding, nutrients and fodder, F. Utne
Graduate research staff: 12
Activities: Nutrition research comprising vitamins, marine oil and fats as well as proteins, minerals and trace elements; research related to fish-breeding; participation in developing new technical processes for nutrients and fish-fodder products.

Fiskeriteknologisk Forskningsinstitutt* 10

– FTFI
[Institute of Fishery Technology Research]
Address: Postboks 677, N-9001 Tromsø
Telephone: (083) 86586
Affiliation: Norwegian Fisheries Research Council
Status: Official research centre
Director of Administration: Dr Arne Bredesen
Information Manager: Odd Viktor Karlsen
Sections: Processing, Dr Terje Strøm; economics, Terje Vassdal
Graduate research staff: 37
Activities: The activities of the Institute are concentrated on the following research programmes: studies of fish behaviour; catching technology; instrumentation and data processing; development of new types of fishing vessels; safety and working conditions; energy efficiency; processing of capelin; production of feedstuff from fish waste; improvement of working conditions and production techniques in the processing industry; catch handling; processing of blue whiting; operation analysis of fishing and fish processing.
Publications: Annual report.

FISHING GEAR AND METHODS DIVISION* 11

Address: Postboks 1964, N-5011 Nordnes
Telephone: (05) 213773
Research director: Professor Steinar Olsen

VESSEL AND MARINE ENGINEERING DIVISION* 12

Address: Håkon Håkonsonsgt 34, N-7000 Trondheim
Telephone: (075) 95650
Research director: Professor Anders Endal

Havforskningsinstitutt* 13

[Marine Research Institute]
Address: Postboks 1870, N-5011 Nordnes
Telephone: (05) 217760
Telex: 42297 OCEAN N
Affiliation: Ministry of Fisheries, Directorate of Fisheries
Status: Official research centre
Director: Gunnar Saetersdal
Sections: Physical-chemical oceanography; biological oceanography; demersal fish; pelagical fish - south; pelagical fish - north; aquaculture
Activities: Research programmes cover a wide range of topics in fisheries science with an emphasis on studies of the abundance and distribution of the fish resources which support Norway's commercial fisheries. Among the most important species are cod, capelin, mackerel, haddock, saithe and herring, together with shellfish, seals and whales. For these and other species studies are conducted on life history: spawning, growth, age, migrations and food, as this basic knowledge is essential for further investigations of the fish stocks. Several research programmes concentrate on evaluation of the quantity of various fish species present in an area. This work is largely dependent on acoustic instruments: echo sounders, echo integrators and sonar.
Other studies include ecological aspects with assessments of the effects of temperature, ocean currents, pollution and fishing upon the distribution and number of fish.
More recently research on aquaculture and fish farming in the sea have been taken up. Emphasis is also placed on the development of acoustic techniques and methods for scientific investigations of fish.

Hermetikkindustriens Laboratorium 14

– HL
[Norwegian Canning Industry Research Laboratory]
Address: Niels Juelsgate 50, Postboks 68, Stavanger
Telephone: (04) 529044
Telex: 33180 Hermen
Status: Research centre within an industrial company
Research manager: Svenn Rasch
Departments: Bacteriology, Karl Håkon Skramstad; chemistry, Rolv Ragård; engineering and development, Jørg Hviding; packaging, John M. Lunde
Graduate research staff: 7
Activities: Seawater fisheries, food science, canning techniques, canning plant projection, packaging material research.

Publications: Annual report, *Hermetikkindustriens Laboratorium - info* , three monthly.
Projects: Conservation and sterilization procedures (Karl Håkon Skramstad); fish product development (Rolv Ragård); machinery development (Jørg Hviding); packaging materials, cans and plastics (John M. Lunde).

Kongelige Norske Videnskabers Selskab, Museet* 15

[Royal Norwegian Society of Sciences and Letters, the Museum]
Address: Erling Skakkes gaten 47, N-7000 Trondheim
Telephone: (075) 92200
Status: Official research centre
Director: Professor Dr Gunnar Sundnes
Sections: Archaeological Department, Kalle Sognnes; botanical department, Egil I. Aune; zoological department, Svein Haftorn; biological station, Dr Gunnar Sundnes
Graduate research staff: 25
Activities: Studies in biosystematics, biogeography, biosociology, and biocartography; extensive excavations are part of the research programme, many of which are a result of road constructions, building activities, and cultivation.
Publications: Report series in archaeology, botany, and zoology.

Kongelige Selskap for Norges Vel* 16

[Royal Society of Rural Development in Norway]
Address: Hellerud, Postboks 115, N-2013 Skjetten
Telephone: (02) 740610
Status: Independent research centre
Director: Kristian Kaus
Sections: Seed multiplication, Ragnar Hillestad; Hellerud Research Station, Knut Wølner
Graduate research staff: 3

Landbruksteknisk Institutt   17

[Norwegian Institute of Agricultural Engineering]
Address: Postbox 65, N-1432 Ås-NLH
Telephone: (02) 949320
Affiliation: Ministry of Agriculture
Status: Official research centre
Director Professor Kristian Aas
Departments: Testing, Gunnar Weseth; operational techniques, Lars Sjøflot
Graduate research staff: 20
Activities: Research and development in the following areas: farm mechanization, management techniques, ergonomics, automation, materials handling and processing of agricultural products; horticultural techniques and equipment for dispersal of seed, insecticides, herbicides and fungicides; mechanization of field experiments; mechanization of land reclamation and ditching. The Institute has two district stations, one in Western Norway (Voss), and one in Northern Norway (Sortland). The testing and research activities at the Voss station are directed towards the special problems of mechanization of small hill farms; those at Sortland are concerned with the problems of mechanization of special interest for the agriculture in Northern Norway.
Publications: Annual report, test reports, research reports, bulletins.
Projects: Agricultural machinery testing (G. Weseth); managing techniques and work simplification, human engineering (L. Sjøflot); horticultural techniqes (A. Nordby); field experiment mechanization (E. Øyjord); land reclamation and ditching mechanization (H. Aamodt).

Norges   18
Almenvitenskapelige
Forskningsråd*

[Norwegian Research Council for Science and the Humanities]
Address: Munthesgaten 29, Oslo, 2
Status: Official research organization
Chairman, Executive Board: Fredrik Mellbye
Director: Anders Omholt
Chairman, Council for Research for Societal Planning: Hans Chr. Bugge
Chairman, Council for Research in the Humanities: Dosent Dr John Herstad
Chairman, Council for Medical Research: Professor Dr Jon Bremer
Chairman, Council for Natural Science Research: Dosent Per Maltby
Chairman, Council for Social Science Research: Professor Dr Asbjørn Aase

Activities: Promotion of basic research and scholarship, primarily at the universities and colleges but to some extent also at other scientfic-scholarly institutions. Grants are awarded to scientists or institutions for specific research projects and for the publication of scientific literature.

Norges Landbrukshøgskole   19

– NLH
[Agricultural University of Norway]
Address: Postboks 3, N-1432 Ås-NLH
Telephone: (02) 940060
Telex: AGRIUNIV
Status: Educational establishment with r&d capability
President: Professor Dr Ola M. Heide
Vice-President: Professor Dr Arnold Hofset
Graduate research staff: 230
Publications: Annual report; scientific reports.

AGRICULTURAL ECONOMY   20
DEPARTMENT

Director: P.H. Vale
Graduate research staff: 20
Activities: General economics, agricultural and horticultural economics, rural sociology, administration theory, work methods, agricultural history.
Projects: Structural relations within production and marketing of eggs, economics of greenhouse production, the multiplier effect of agriculture, forestry and related industries in a rural area, govermental innovations in the rural area, importance of the agricultural policy to the local rural economy, consumption and self-supply of agricultural products, development and member influence of agricultural organizations.

AGRICULTURAL HYDROTECHNICS   21
DEPARTMENT

Director: Associate Professor E. Myhr
Graduate research staff: 7
Activities: Investigations in run-off, drainage and irrigation as part of a water balance study in areas with different topography, soils and vegetation cover.

AGRICULTURAL STRUCTURES DEPARTMENT 22

Director: Associate Professor A. Nygaard
Graduate research staff: 19
Activities: Studies of various types of agricultural buildings and installations; materials and structures; work organization; animal environment as climate, floor, space, equipment; crop and manure storage; greenhouses; building costs, cost index; standard drawings.
The department has its own field station with various types of experimental buildings.

ANIMAL GENETICS AND BREEDING DEPARTMENT 23

Director: Professor H. Skjervold
Graduate research staff: 16
Activities: Theoretical and practical breeding research in cattle, pigs, sheep and goats.
The department has a laboratory for model experiments with mice and cooperates closely with animal husbandry and breeding associations to apply new breeding and selection methods.
Projects: Cattle - genetic variation in roughage intake and feed efficiency, protein content of milk, connection between hormonal secretion and production traits; swine - selection for fast growth and thin backfat versus slow growth and thick backfat, development of specialized sire and dam lines, genetic variation in constitution and carcass quality; sheep - crossbreeding and selection aiming at an increased number of lambs; goats - genetic variation in the intensity of flavour of goat's milk and the connection between flavour and chemical content; salmonids - comparison of salmon strains from various rivers, crosses between various species of salmonids, genetic variation in tolerance to acidic water.

ANIMAL NUTRITION DEPARTMENT 24

Director: Professor O. Saue
Graduate research staff: 17
Activities: Research relates to conservation of forage, feeding of ruminants and pigs, feed evaluation studies. The department has a biochemistry laboratory which is concerned mainly with analytical work on biochemical materials, such as feeds, milk, fats, blood and urine.
Projects: Conservation of forages; ruminant feeding; pig feeding; feed evaluation.

BEEKEEPING DEPARTMENT 25

Address: 1370 Asker
Director: Professor E. Villumstad
Graduate research staff: 2
Activities: Research in beekeeping economy and management, bee breeding, honey and bee botany, bee diseases. The Norwegian bee control stations, the control of honey and bee diseases on a laboratory scale range under this department.

BOTANY DEPARTMENT 26

Director: Professor E. Dahl
Graduate research staff: 8
Activities: Research in plant physiology and ecology.
Projects: Mineral nutrition of bog plants; nitrogen metabolism and protein production in cereals and legumes; effect of environmental factors on germination, growth, development and hormone balance, mainly in cultivated plants; chemotaxonomy of Cyperaceae; climate demands of thermophilous species; heat exchange and lethal temperatures in plants; plant sociology and vegetation mapping; relationship of vegetation and soil factors; pollution ecology; East African flora.

CHEMISTRY DEPARTMENT 27

Director: Associate Professor A. Stabursvik
Graduate research staff: 8
Activities: Research relates to natural organic compounds (especially carbohydrates) of agricultural interest; chromatography and spectroscopy of natural products.

DAIRY RESEARCH DEPARTMENT 28

Director: Professor G. Syrrist
Sections: Chemistry, bacteriology and fluid milk products; dairy technology; dairy economy; dairy machinery.
Activities: Chemistry, bacteriology and fluid milk products - biochemical and physical aspects of milk, milk quality; milk on the farm, during storage, transport and manufacturing; ordinary fluid milk products and experiments with new products; milk proteins and enzymes from milk and bacteria.
Dairy Technology - dairy product research, cheese, melting cheese, whey cheese, butter, ice cream, dried milk and the influence of various factors in the production of these products; cheese ripening studies; new product development; research with pure cultures for cheese and butter production.
Dairy Economy - demand for fluid milk and other dairy products, marketing, transport, storage of milk and dairy products; product yield and expense analysis; product steering and simulation; structure analysis and

long range planning; work analysis.
Dairy Machinery - machinery used in dairy technology; heating, cooling and freezing, machinery; need and utilization of energy by different processes used in thee dairy industry; testing of dairy machines and other technical equipment.

DENDROLOGY AND NURSERY MANAGEMENT DEPARTMENT 29

Director: Associate Professor M. Sandved
Graduate research staff: 6
Activities: Studies of trees and shrubs, annual and perennial flowers.
Projects: Investigation of ornamental trees and shrubs, introduction of new or little known plants throughout the country, and the collection and testing of ecotypes of some native species; breeding and variety trials in garden roses and turfs, variety trials in several genera of ornamental trees, shrubs and herbaceous perennial flowers; propagation of trees, shrubs and perennial flowers, container growing, fertilizing and soil treatment in nurseries, and indoor storing and packing of nursery crops; revegetation of industrial wasteland, testing the effect of salt to soil and plants bordering roads.

FARM CROPS DEPARTMENT 30

Director: Professor E. Strand
Graduate research staff: 10
Activities: Research into breeding of the most important farm crops - cereals, potatoes, rootcrops, timothy and other grass species for hay and pastures; techniques and methods of growing, harvesting, etc, variety testing.

FARM POWER AND MACHINERY DEPARTMENT 31

Director: Associate Professor A. Nordby
Graduate research staff: 6
Activities: The Department is linked with the State Institute of Agricultural Engineering, which does research work in the various fields of agricultural engineering.

FLORICULTURE AND GREENHOUSE CROPS DEPARTMENT 32

Director: Associate Professor R. Moe
Graduate research staff: 8
Activities: Effect of enviromental factors, particularly temperature, light and daylength, on growth and development of glasshouse crops with emphasis on ornamental plants of commercial significance; chemical control of plant growth; growing media and growing techiques of glasshouse crops; meristem and tissue culture as a means of obtaining disease-free stocks for vegetatively propagated ornamental plants; automatic control of the glasshouse climate.

FOREST ECONOMY DEPARTMENT 33

Director: Associate Professor R. Saether
Graduate research staff: 7
Activities: Cost analysis; accounting; studies of the impact of various financing methods and taxation systems, and the relative value of different tree dimensions as industrial raw material; forest evaluation; market analyses for forest products; studies of the role of forestry in regional economic planning, forestry and recreation, and the economic background of forest policy.

FOREST ENGINEERING DEPARTMENT 34

Director: Associate Professor R. Skaar
Activities: The Department is an integrated part of the Division of Forest Operations and Techniques at the Norwegian Forest Research Institute. Forest engineering comprises the input of people and machines required to carry out the varied aspects of logging, transportation, stand establishment and the necessary support and service functions within the prescriptions of proper forestry and engineering practice.
The subject of forest engineering consists of research work on ergonomics and work physiology, forest terrain, machines and equipment, forest road planning, work and method studies, transport means, planning and control of forest operations.

FOREST MENSURATION DEPARTMENT 35

Director: Associate Professor A. Fitje
Graduate research staff: 6
Activities: Research on accuracy and time consumption for different forest inventory methods and instruments for forest mensuration; applications of aerial photos in inventory work are analyzed; investigations on forest planning, including methods for estimation of long term prospective yield, existing management plans and designation of plans, use of various tools in planning, eg aerial photos and maps; application of operational analysis methods and the electronic computer in forest planning and mensuration.

GENETICS AND PLANT BREEDING DEPARTMENT 36

Director: Assistant Professor M. Bragdø-Aas
Graduate research staff: 6
Activities: Research on the induction of mutations in plants, recombination studies on artificially induced mutations, population studies in grasses, polyploidy breeding in clover, rye, perennial ryegrass and meadow fescue; resistance of clovers to Sclerotinia trifoliorum and Ditylenchus dipsaci.

GEODESY DEPARTMENT 37

Director: Associate Professor Ø. Andersen
Graduate research staff: 5
Activities: Research into methods of surveying and mapping, problems of the theory of errors; photogrammetry, analytical plotters, camera calibration and remote sensing.

GEOLOGY DEPARTMENT 38

Director: Professor P. Jørgensen
Graduate research staff: 7
Activities: Geological mapping, mainly stratigraphical - tectonic investigations of the Eocambrian area in south-eastern Norway in cooperation with the Geological Survey of Norway; detailed mapping of the Precambrian in the neighbourhood of Ås; Quaternary investigation and mapping in south-eastern Norway with type areas in Østerdalen and the Ås region; hydrogeological research.

LAND REALLOCATION AND CONSOLIDATION DEPARTMENT 39

Director: Professor H. Sevatdal
Graduate research staff: 10
Activities: Evaluation, investment and planning in rural districts in connection with land reallocation, appropriation and regional analysis, legal aspects of land reallocation cases; land use and planning in rural regions.

LANDSCAPE ARCHITECTURE DEPARTMENT 40

Director: M. Eggen
Graduate research staff: 10
Activities: Open space planning in residential areas, zoning, planning and development of outdoor recreation areas and urban open spaces; landscape of roads; landscape reclamation; methodology of landscape planning and landscape management.

MATHEMATICS AND STATISTICS DEPARTMENT 41

Director: Associate Professor Å. Lima
Graduate research staff: 7
Activities: Development of statistical and mathematical methods with particular emphasis on agricultural research work; studies of problems in economics and theoretical physics.

MICROBIOLOGY DEPARTMENT 42

Director: Professor T.A. Pedersen
Activities: Occurrence and biochemical basis of antagonistic relationships between bacteria; mechanism of sporulation of fungi in submerged culture; epiphytic flora on grass and development and establishment of an adapted microflora during ensiling; effect of acid precipitation on soil microbiological processes; microbial purification of effluents from ensilage combined with single cell production; conditions for microbial decomposition of organic materials, particularly by means of composting; performance of a test programme of composting toilets as alternative to WCS; biological clogging processes in sand filter beds.

NATURE RESOURCE CONSERVATION 43
AND MANAGEMENT DEPARTMENT

Director: Associate Professor S. Huse
Graduate research staff: 5
Activities: Research in the field of conservation and management of natural resources and environment - criteria of conservation value, methods of resource inventory and planning, techniques of preservation management; wildlife and inland fish resources - main game species, especially moose and grouse; inland fish resources - biology of Norwegian inland fish species and the optimum management of fish resources of lakes and rivers; habitat management and evaluation.

PEDAGOGICS AND SUPPLEMENTARY 44
STUDIES DEPARTMENT

Address: N-1370 Asker
Director: Associate Professor O. Søbstad
Graduate research staff: 8
Activities: Research in the fields of agricultural vocational education and agricultural extension work.

PHYSICS AND METEOROLOGY 45
DEPARTMENT

Director: Associate Professor Vidar Hansen
Graduate research staff: 12
Activities: Research work on agrometeorological and microclimatical problems, such as mapping of the local climate of Ås, investigations of horizontal and vertical differences of soil and air temperature in crops, evapotranspiration and energy balance problems, registration of different meteorological elements, including sun and global radiation; comparison of different meteorological instruments; observations on airglow, aurora, twilightglow and nightglow and spectroscopic studies of discharges in gases, relevant to upper atmospheric problems, are carried out.

POMOLOGY AND FRUIT INDUSTRY 46
DEPARTMENT

Director: Associate Professor S. Vestrheim
Graduate research staff: 4
Activities: Experiments with hard and soft fruits; fertilizer effects on yield and quality, chemical analysis of leaf and fruit samples, surveys of the nutritional status of commercial orchards; cultivar investigations of common marketable fruit species; effects of storage on postharvest life of fruit; bud dormancy and winter injury to fruit plants; pollen viability, incompatibility between cultivars and fruit set, particularly in raspberries.

POULTRY AND FUR ANIMALS 47
DEPARTMENT

Director: Associate Professor N. Kolstad
Activities: Research into feeding, breeding and environmental factors concerning poultry, mink, foxes, rabbits and fish.
Poultry - feeding values of different quality grains and protein feeds, effects of synthetic amino acids, energy, mineral and vitamin metabolism, housing conditions, breeding and selection methods, artificial insemination, factors influencing egg quality and composition.
Mink and blue foxes - energy requirements and levels of protein, fat and carbohydrates, digestibility and balance trials with individual feedstuffs as well as complete diets, possibilities of using dry diets, factors influencing the anemia-causative effect of fur animal diets, different fur characteristics with special emphasis on the effects of genetics and light control
Rabbits - different levels of feeding and age at slaughtering, feed utilization and meat yield, performance testing of breeds and crosses for meat production.
Fish - the effect of using different contents and qualities of fat protein and carbohydrates in the diets for salmonids, the value of some individual feedstuffs as ingredients in the diet.

SILVICULTURE DEPARTMENT 48

Director: Associate Professor O. Haveraaen
Graduate research staff: 10
Activities: Research concentrates on natural and artificial regeneration, growth and yield, provenance trials and tree breeding with aspen and poplars, trials with foreign tree species, yield studies of conifers and broad leaved trees, plant ecology (vegetation types), and draining and nutritional problems of peat land.

SOIL FERTILIZATION AND 49
MANAGEMENT DEPARTMENT

Director: Professor A. Njøs
Graduate research staff: 11
Activities: Research into plant nutrition, fertilization, liming and soil tillage.
Projects: Macronutrients and micronutrients, leaching of plant nutrients by heavy amounts of manure or fertilizer, utilization of sewage sludge and other organic residues, effect of heavy metals on plants, crop rotation and longterm productivity, effect of binding agents on the stability of road slopes against surface splash and run-off.

SOIL SCIENCE DEPARTMENT   50

Director: Associate Professor A. Øien
Graduate research staff: 7
Activities: Research into soil formation, soil mapping, and productivity of soils; testing of laboratory methods; analysis of soil samples.

VEGETABLE CROPS DEPARTMENT   51

Director: Professor A. Persson
Activities: Vegetable production in the open and in glasshouses; research into varieties and methods best suited for various seasons and parts of the country; breeding technique and practical plant breeding with cabbage as the main crop; club root resistance; quality and quantity improvement of tomato yield; means of climate improvement during propogation and field cultivation; variety selection and cultivation of high quality produce for the processing industry; postharvest research and storage experiments; effects of environmental factors on biological quality of vegetables.

WOOD TECHNOLOGY DEPARTMENT   52

Director: Associate Professor R.O. Ullevaalseter
Graduate research staff: 5
Activities: Research work in the various fields of wood utilization.

ZOOLOGY DEPARTMENT   53

Director: Professor Ragnhild Sundby
Graduate research staff: 5
Activities: Research into population dynamics and biological control of insect pests.
Projects: Parasites and predators on greenhouse aphids; practical use of beneficial insects and their efficiency at different temperatures; competition among parasites; mineral balance and adaptation to the environment in reindeer, mineral pools and mineral balance studies using isotope techniques.

Norges Slakterilaboratorium   54

[Norwegian Meat Research Laboratory]
Address: Lörenveien 37, Postboks 96, Refstad, Oslo, 5
Telephone: (02) 150510
Telex: 71302-meat-N
Affiliation: Norwegian Farmers Meat Marketing Organization
Status: Research centre within an industrial company
Director: Ole Sigmund Braathen
Departments: Microbiology, Knut Grösland; chemistry, Arild Andresen; special analysis, Alf Damm; Technology, Jack Nilsen; sensory, Anne-marie Østtveit
Graduate research staff: 20
Activities: Research is carried out on what is happening to slaughter animals from leaving the farm, through the slaughtering, cutting, packing and sausage-making, to the customer processes through retail outlets.
Publications: Annual reports.
Projects: Electrical stimulation of prerigor boned meat; electrical stimulation of sheep and beef carcasses.

Norges Veterinaerhøgskole   55

[Veterinary College of Norway]
Address: Ullevålsveien 72, Box 8146, Oslo, 1
Telephone: (02) 693690
Status: Educational establishment with r&d capability
Director: Professor Dr Weiert Velle
Director of Administration: Arvid Siljan
Graduate research staff: 83
Publications: Annual report.

DAL RESEARCH STATION FOR   56
FURBEARING ANIMALS

Address: Dal Forsøksgard for Pelsdyr, Rustadveien 131, N-1380 Heggedal
Telephone: (02) 797034
Director: Dr Ordin Moeller
Sections: Reproductive physiology and pathology; genetic studies; infectious diseases
Graduate research staff: 2
Activities: Areas of research include: hormonal variations in fur-bearing animals during oestrus and pregnancy; the relationship between different numbers of chromosomes and fertility in the blue fox; epizootiologic features of plasmacytosis, and the specific role of genetic factors in plasmacytosis in mink.
Projects: Reproductive physiology and pathology in furbearing animals (Dr Ordin Moeller); encephalitozoonosis in foxes (Dr Svein Fr. Mohn).

[Sem Research Farm for Animal Husbandry]
Address: Semsveien 168, 1370 Asker
Telephone: (02) 790665
Head: Dr Sverre Tollersrud
Graduate research staff: 1

Norsk Institutt for   72
Naeingsmiddelforskning

[Norwegian Food Research Institute]
Address: Postboks 50, N-1432 Ås-NLH
Telephone: (02) 940860
Status: Independent research centre
Managing Director: Anton Skulberg
Research Director: Tore Høyem, Kjell I. Hildrum
Sections: Chemistry/biochemistry,   Hellmut
Russwurm; food technology, Dr Kiell Ivar Hildrum;
microbiology, Dr Reidar Skjelvåle; packaging research,
vacant; food science, Tore Høyem
Graduate research staff: 32
Activities: To contribute to the development of the
branches of the food industry which base their activities
on meats, vegetables, fruits, berries and eggs through
research, information service and consultative guidance.

   Publications: Annual reports, *NINF-information*
bi monthly.
Projects: Importance of porcine stress syndrome in the
meat industry (Terje Frøystein); quality criteria in
vegetables (Magni Martens); functional properties of
food ingredients (Ole Harbitz); quality control in the
food industry (Bjørn Eidstuen); effect of heat treatment,
water content and storage on the nutritional value of
foods (Bjødne Eskeland); automation of fermented
sausage production (Tore Høyem), pathogenic organisms in food (Reidar Skjelkvåle); microbial conditions
during storage of fresh vegetables (Per Einar Granum).

Norsk Institutt for Skogforskning   73

– NISK
[Norwegian Forest Research Institute]
Address: Postboks 61, 1432 Ås-NLH
Telephone: (02) 941160
Status: Official research centre
Research director: Toralf Austin
Sections: Forest ecology, Professor Kristian Bjor; forest protection, Professor Alf Bakke; forest pathology, Associate professor Kåre Venn; forest zoology, Professor Alf Bakke; forest regeneration, Professor Peder Braathe; forest management and forest yield studies, Associate Professor Helge Braastad; forest genetics and free breeding, Professor Jon Dietrichson; wood science and wood technology, Associate Professor Torbjørn Okstad; forest operations and techniques, Professor Ivar Samset; national forest survey, Associate Professor Torgeir Løvseth; forest biology, Associate Professor Asbjørn Løken; forest yield studies and forest management, Professor Eivind Bauger
Graduate research staff: 60
Activities: NISK has as its main assignment to investigate and clarify the conditions which are of significance to Norwegian forests, including regeneration, production, operations, economy, and protection of nature and the public utilization of forest areas; forest establishment and management on the West Coast.
Publications: Tidsskrift for Skogbruk, Meddelelser fra Norsk institutt for skogforskning, Rapporter fra NISK

DIVISION OF FOREST ECOLOGY*   74

Activities: Studies of climate, soil science, and afforestation of moorland.

DIVISION OF FOREST GENETICS AND TREE BREEDING*   75

Activities: Research into improvement of plant material, including provenance experiments and progeny tests.

DIVISION OF FOREST MANAGEMENT AND YIELD STUDIES*   76

Activities: Preparation of yield tables for various tree species; studies of forest management problems including investigations into thinning; experiments with fertilizing upland sites.

DIVISION OF FOREST PROTECTION*   77

Botanical Section*   78

Activities: Forest pathology.

Zoological Section*   79

Activities: Forest entomology.

DIVISION OF FOREST REGENERATION*   80

Activities: Research in connection with natural regeneration, experiments with forest culture, and nursery experiments.

DIVISION OF LOGGING ENGINEERING AND FOREST WORK SCIENCE*   81

Activities: The Division deals with the analysis of the secondary production apparatus of forestry. This includes: work studies, including work physiology; implement and machine studies; transport studies.

DIVISION OF WOOD SCIENCE AND WOOD TECHNOLOGY*   82

Activities: Investigations into the properties of wood, its protection and storage damage, and scaling of logs.

NATIONAL FOREST SURVEY*   83

Norsk Institut for Skogforskning-Bergen   84

– NISK-Bergen
[Norwegian Forest Research Institute, Bergen]
Address: N-5047 Stend
Telephone: (05) 276370
Affiliation: Norsk institutt for skogforskning - Ås
Status: Official research centre
Director: Associate professor Asbjørn Løken
Division: Forest biology, Asbjørn Løken; forest yield studies, Eivind Bauger
Activities: Plant production, conifers, broadleaf trees; forestry and forest products.
Publications: Meddelelser fra Norsk institutt for skogforskning
Projects: Seed physiology and regeneration, (A. Løken); provenance (S. Magnesen); forest management (E. Bauger).

Norsk Institutt for Vannforskning* 85

– NIVA
[Norwegian Institute for Water Research]
Address: Postboks 333, Blindern, Oslo, 3
Telephone: (02) 235280
Affiliation: Norges Teknisk-Naturvitenskapelige Forskningsråd
Status: Official research centre
Director: Kjell Baalsrud
Head of Research: J.E. Samdal
Information and Public Relations Officer: Knut Pedersen
Sections: Biological analysis; chemical analysis; electronic data processing; investigations in lakes and rivers; investigations in fjords and coastal waters; hydrophysiology; effluents from mining and industry; technical installations (sanitary engineering); special projects
Graduate research staff: 50
Activities: The Institute's research activity comprises the study of water quality management and water pollution of lakes, river systems, estuaries, fjords and coastal waters, and includes transportation and purification of drinking water and waste water.
Specific topics of research include: investigations and quality evaluations of lake, river and fjord waters; methods for chemical, bacteriological and biological analysis of water; experimental biological research; field investigations and development of sampling and testing equipment; humus problems; corrosion; purification of drinking water, water for industrial operation and effluent water from factories and mines; purification methods for sewage water; technical evaluations regarding drainage plants (pipelines, water and sludge treatment plant); operation of water purification plants.
Publications: Yearbook; newsletters.

Norsk Undervannsinstitutt* 86

[Norwegian Underwater Research Institute]
Address: c/o NTNF, Soynsveien 72, Postboks 70, Tåsen, Oslo, 8
Affiliation: Norges Teknisk-Naturvitenskapelige Forskningsråd

Norske Jord- og Myrselskap* 87

[Norwegian Land and Peat Association]
Address: Hellerud i Skedsmo, Postboks 116, N-2013 Skjetten

Norske Potetindustrier, Hveem Forsøksgard* 88

[Norwegian Potato Industries, Hveem Research Station]
Address: N-2856 Bilitt
Telephone: (061) 63363
Status: Research association
Research director: Anton Letnes
Section: Seed potato growing foundation, Egil Gjestvang
Graduate research staff: 3
Activities: Research work with potatoes - meristem culture; testing for virus, sepidonicum etc; storage research; varieties trials; fertilizer and artificial watering trials.
Publications: Annual report, Norwegian.

Norske Videnskaps-Akademi* 89

[Norwegian Academy of Science and Letters]
Address: Drammeusveien 78, Oslo, 2
Telephone: (02) 444296
Chairman: Professor J. Andenaes
Secretary-General: Professor Arne Semb-Johansson
Activities: The Section of Mathematics and Sciences is divided into twelve groups: Pure and Applied Mathematics; Physics; Geophysics; Chemistry; Astronomy; Geodesy and Geography; Mineralogy and Geology; Botany; Zoology; Anatomy; Anthropology and Physiology; Medical Sciences; Technical and Applied Economic Sciences; Agriculture.

Sildolje- og Sildemelindustriens Forskningsinstitutt* 90

[Norwegian Herring Oil and Meal Industry Research Institute]
Address: Bjørgevein 220, N-5033 Fyllingsdalen
Telephone: (05) 164570
Telex: 40087 forsk n
Status: Independent research centre
Research director:. Gudmund Sand
Head of Research: Nils Urdahl
Sections: Science and technology, Nils Urdahl; nutrition, Johannes Opstvedt
Graduate research staff: 15
Activities: Preservation and handling of fish, processing of fish meal, fish oil and fish protein concentrates. Development of new products based on fish.

Statens Biologiske Stasjon Flødevigen* 91

[State Biological Station Flødevigen]
Address: N-4800 Arendal
Telephone: (041) 20580
Affiliation: Ministry of Fisheries, Directorate of Fisheries
Status: Official research centre
Director: Per T. Hognestad
Graduate research staff: 8

Statens Forskingsstasjoner i Landbruk 92

[Norwegian State Agricultural Research Stations]
Address: Postboks 100, N-1430 Ås
Telephone: (02) 942060
Affiliation: Ministry of Agriculture
Status: Official research organization
Graduate research staff: 54
Activities: Natural resources: soil science; drainage and irrigation. Plant production: plant breeding, crop husbandry, plant protection. Animal production: livestock breeding (sheep); livestock husbandry and nutrition (sheep, cattle).
Publications: Research in Norwegian Agriculture.

APELSVOLL AGRICULTURAL RESEARCH STATION 93

Address: N-2858 Kapp
Telephone: (061) 60055
Head: Magnus Jetne
Graduate research staff: 5
Activities: Soil science; crop husbandry (grass, cereals, potatoes).

FURENESET AGRICULTURAL RESEARCH STATION 94

Address: N-6994 Fure
Telephone: (057) 30100; Fure 6
Head: Kristen Myhr
Graduate research staff: 3
Activities: Soil science; soil drainage; plant breeding, grass; crop husbandry, animal crops.

HOLT AGRICULTURAL RESEARCH STATION 95

Address: Postboks 100, N-9001 Tromsø
Telephone: (083) 84875
Head: Ivar Schjeldrup
Graduate research staff: 8
Activities: Soil science; plant breeding, (grass); crop husbandry, (animal crops, vegetables, potatoes, berries); plant protection.

Flaten Division* 96

Address: N-9500 Alta

KISE AGRICULTURAL RESEARCH STATION 97

Address: N-2350 Nes På Hedmark
Telephone: (065) 52300
Head: Johannes Thorsrud
Graduate research staff: 4
Activities: Soil science; drainage and irrigation; crop husbandry, vegetables, berries, fruit.

KVITHAMAR AGRICULTURAL RESEARCH STATION 98

Address: N-7500 Stjørdal
Telephone: (076) 94088
Head: Jon Olav Furunes
Graduate research staff: 11
Activities: Soil science; plant breeding, (grass, cereals); crop husbandry, (cereals, animal crops, vegetables, potatoes, flowers).

Maere Division* 99

Address: N-7710 Sparbu
Telephone: (077) 43341
Manager: Rolf Celius
Graduate research staff: 1
Activities: Research into agricultural plant production on bogs, all round, but mainly fodder production (leys), experiments in draining of bogs; fertilizing experiments, variety trials, etc.

Moldstad Division* 100

Address: N-6577 Nordsmola

LANDVIK AGRICULTURAL RESEARCH   101
STATION

Address: N-4890 Grimstad
Telephone: (041) 42266
Head: Gunvald Jonassen
Graduate research staff: 4
Activities: Plant breeding, (vegetables); crop husbandry, (grass, vegetables).

LØKEN AGRICULTURAL RESEARCH   102
STATION

Address: N-2942 Volbu
Telephone: (061) 50100; Heggenes 59
Head: Erling Olsen
Graduate research staff: 2
Activities: Plant breeding (grass); Crop husbandry (animal crops); mountain husbandry; fertilizer experiments; variety trials in farm crops.

NJØS AGRICULTURAL RESEARCH   103
STATION

Address: N-5840 Hermansverk
Telephone: (056) 53611
Head: Rolf Nestby
Graduate research staff: 2
Activities: Plant breeding (berries); crop husbandry (fruit, berries, vegetables).

SAERHEIM AGRICULTURAL   104
RESEARCH STATION

Address: N-4062 Klepp St
Telephone: (04) 420333
Head: Markus Pestalozzi
Graduate research staff: 4
Activities: Soil science; crop husbandry (grass, cereals, vegetables); livestock husbandry and nutrition (cattle); plant protection.

SAETER AGRICULTURAL RESEARCH   105
STATION

Address: N-2592 Kvikne
Telephone: (063) 61100; Kvikne 3794
Head: Arne Bekken
Graduate research staff: 1
Activities: Livestock breeding (sheep); livestock husbandry and nutrition (shbeep)

TINGVOLL AGRICULTURAL   106
REASEARCH STATION

Address: N-6630 Tingvoll
Telephone: (073) 31372
Head: Johs. Eri
Graduate research staff: 1
Activities: Livestock husbandry and nutrition (sheep).

TJØTTA AGRICULTURAL REASEARCH   107
STATION

Address: N-8860 Tjøtta
Telephone: (086) 46320
Head: Arne W. Våbenø
Graduate research staff: 3
Activities: Crop husbandry (animal crops, vegetables, potatoes); livestock breeding (sheep); livestock husbandry and nutrition (sheep).

ULLENSVAG EXPERIMENTAL FARM   108

Address: N-5774 Lofthus
Telephone: (054) 61105
Head: Jonas Ystaas
Graduate research staff: 3
Activities: Research is concerned with all aspects of fruit husbandry-apples, pears, plums, sweet cherries and small fruits; soil science. The main emphasis is placed on research of fruit quality, fruit tree nutrition, and orchard pests.
Publications: Annual report.

VAAGØNES AGRICULTURAL   109
RESEARCH STATION

Address: Postboks 3010, N-8001 Bodø
Telephone: (081) 25586
Head: Birger Volden
Graduate research staff: 3
Activities: Plant breeding (grass); crop husbandry (animal crops, vegetables), variety trials; livestock husbandry and nutrition; fertilizers and manure.

VOLL AGRICULTURAL RESEARCH   110
STATION*

Address: Postboks 1918, N-7001 Trondheim
Activities: Agronomy; fertilizer and manure studies; variety trials in farm crops; crop husbandry; plant breeding in small grain.
The activities of this station were transferred to Kvithamar (see above) in 1982.

Statens Frøkontroli*   111

[Norwegian State Seed Testing Station]
Address: Box 68, N-1432 Ås-NLH
Affiliation: Ministry of Agriculture

Statens Institutt for   112
Folkehelse*

– SIFF
[National Institute of Public Health]
Address: Geitmyrsveien 75, Oslo
Status: Official research centre
Director: Dr Chr. Lerche
Deputy director: Arne Løkken

ANIMAL DEPARTMENT*   113

Head: Dr Stian Erichsen
Deputy Head: Annelise Lyngset
Activities: Research in laboratory animal science.

TOXICOLOGY DEPARTMENT*   114

Head: Professor Dr Johannes A.B. Barstad
Deputy Head: Dr Erik Dybing
Activities: Areas of research include: the role of metabolism in relation to the toxic effects of environmental chemicals; toxic effects of paraquat; possible sources of tularaemia contamination; insects and ticks serving as vectors for arbovirus.

Statens Institutt for   115
Forbruksforskning

[State Institute of Consumer Research]
Address: Ringstabekkveien 105, N-1340 Bekkestua
Telephone: (02) 533970
Status: Official research centre
Research director: Helga Kringlebotn Emanuelsen
Departments: Food science, Gudrun Rognerud
Graduate research staff: 5
Activities: Research and investigations concerning properties of food, which are of importance to the consumer; nutritional and sensoric quality of food as presented to the consumer through retail markets and catering establishments; studies on effect of methods of processing and preparation on the quality of food.
Publications: Consumer research reports, annual report.

Projects: Nutritional and sensoric properties of bread and other cereal products (Gudrun Rognerud); quality of food from catering establishments (Kirstin Færden); determination of dietary fibre in Norwegian grown potatoes, vegetables and fruit (Ragnhild Reinstad).

Statens Kornforretning   116

[Norwegian Grain Corporation]
Address: Postboks 1367 Vika, Stortingsgt. 28
Telephone: (02) 414500
Telex: 11 608
Status: Official research centre
Head of research department: Lars Sogn
Activities: Work on matters which are of significance for improving the quality of Norwegian grain, and of significance for the best utilization of the grain, crop husbandry.
Projects: Testing new cultivars of cereals and oil seed both domestic and foreign breeds; trials with varieties of malt barley; qualitative experiments with spring and winter wheat; investigations of enzyme activity in grain - wheat, barley and oats; nitrogen fertilizer experiments with spring wheat and spring barley; irrigation experiments with spring wheat, barley and oats; drying and preservation experiments in cereal grain; combine and threshing experiments.

Statens Plantevern*   117

[Norwegian Plant Protection Institute]
Address: Box 70, N-1432 ÅS-NLH
Affiliation: Ministry of Agriculture

ENTOMOLOGY DIVISION*   118

Activities: Areas of research include: insects, mites, nematodes, etc as pests of crops; agro-ecology and control methods.

HERBOLOGY DIVISION*   119

Activities: Research into weeds in horticulture, agriculture and forestry.

PLANT PATHOLOGY DIVISION*   120

Activities: Research into plant diseases caused by bacteria, fungi and viruses.

Statens Veterinaere Laboratorium for Nord Norge   121

[State Veterinary Laboratory for Northern Norway]
Address: PO Box 652, N-9401 Harstad, troms
Telephone: (082) 63155
Status: Official research centre
Research director: Knut Kummeneje
Activities: Subjects of research include: mastitis and milk quality in cows and goats; corynebacterium pseudotuberculosis infection in goats; eradication programme against echinococcus granulosus in reindeer and dog; sarcosporidia in reindeer; pathology, microbiology and parasitology of animal diseases; preventive veterinary medicine including mastitis control and prevention.
Projects: Cow and goat (O. Østerås); goat caseous lymphadenitis, reindeer echinococcosis (K. Kummeneje).

Ullensvang Forsøksgård   122

[Ullensvang Research Station]
Address: 5774 Lofthus, Hardanger
Telephone: (054) 61105
Affiliation: Department of Agriculture
Status: Official research centre
Director: Dr Jonas Ystaas
Graduate research staff: 5
Activities: Plant production - fruit growing, apples, pears, plums, sweet cherries; small fruits, strawberries, raspberries, black currants.
Publications: Annual report.
Projects: Regulation of fruit set and thinning, postharvest physiology of pears, influences of environmental and cultural pratices on fruit quality (Dr A. Kvåle); mineral nutrition of fruit trees, testing of sweet cherry varieties, intensive planting systems and rootstocks (Dr J. Ystaas); investigation of the criteria of fruit quality (E. Vangdal); investigation of root weevils of strawberries (K. Hesjedal).

Universitetet i Bergen*   123

[Bergen University]
Address: Postboks 25, N-5014 Bergen-Universitetet
Telephone: (05) 210040
Status: Educational establishment with r&d capability

FACULTY OF NATURAL SCIENCES*   124

Department of Botany*   125
Head: Professor Peter Emil Kaland
Graduate research staff: 15

Arboretum*   126
Head: Poul Søndergaard
Graduate research staff: 3

Botanical Garden*   127
Head: Peter Magnus Jørgensen
Graduate research staff: 1

Department of Fisheries Biology*   128
Senior staff: Professor Egil Lund, Professor Olav Drageshund, Professor Steinar J. Olsen, Professor Gunnar Soetersdal, Professor Kristian Fredrik Wiborg
Graduate research staff: 12

Department of Marine Biology*   129
Head: Professor John Matthews
Graduate research staff: 9

Department of Microbiology and Plant Physiology*   130
Senior staff: Professor Gjert Knutsen, Professor Ian Dundas, Professor Per Nissen, Professor Jostein Goksøyr
Graduate research staff: 17

Department of Nutrition Biology*   131
Head: Professor Olaf R. Broekkan
Graduate research staff: 4

Zoological Laboratory*   132
Head: Professor Hans Jørgen Fyhn
Graduate research staff: 11

Zoological Museum*   133
Senior staff: Professor Sven Axel Bengtson, Professor Ole Anton Saether
Graduate research staff: 25

Universitetet i Oslo* 134

[Oslo University]
Address: Postboks 1071, Blindern, Oslo, 3
Telephone: (02) 466800
Status: Educational establishment with r&d capability
Rector: Professor Dr Bjarne A. Waaler

FACULTY OF MATHEMATICS AND NATURAL SCIENCES* 135

Dean: Professor Dr Per S. Enger

Animal Physiology Department* 136

Address: Postboks 1051, Blindern, Oslo, 3
Head: Kjell Fugelli

Botanical Laboratory* 137

Address: Postboks 1045, Blindern, Oslo, 3
Director: Birger Grehager

Drøbak Biological Station* 138

Address: N-1440 Drøbak
Director: Finn Walvig

General Genetics Department* 139

Address: Postboks 1031, Blindern, Oslo, 3
Head: Professor Per Oftedal

Marine Biology and Limnology Department* 140

Address: Blindern, Oslo, 3
Director: Professor Eystein Paasche

Limnology Division* 141
Address: Postboks 1027, Blindern, Oslo, 3
Head: Dag Klaveness

Marine Botany Division* 142
Address: Postboks 1069, Blindern, Oslo, 3
Head: Fredrik Beyer

Marine Zoology and Marine Chemistry Division* 143
Address: Postboks 1064, Blindern, Oslo, 3
Head: Fredrik Beyer

Tøyen Botanical Garden and Museum* 144

Address: Tøyen, Oslo, 5
Director: Elmar Marker

Zoological Department* 145

Address: Postboks 1050, Blindern, Oslo, 3
Head: Per Bergan

Zoological Museum* 146

Address: Sarsgaten 1, Tøyen, Oslo, 5
Telephone: (02) 686960
Director: Professor Rolf Vik
Sections: Parasitology, Professor Rolf Vik; insects, Albert Lillehammer; invertebrates I, Marit Christiansen; invertebrates II, Tor A. Bakke; mammals, Jørgen A. Pedersen; fishes, amphibians and reptiles, Per Pethon; freshwater ecology, Svein J. Saltveit; bird/aeroplane problems, Professor Rolf Vik
Graduate research staff: 11
Activities: Cover systematics, animal geography, parasitology and freshwater ecology.

Universitetet i Tromsø* 147

[Tromsø University]
Address: Postboks 635, N-9001 Tromsø
Status: Educational establishment with r&d capability

BIOLOGY AND GEOLOGY DEPARTMENT* 148

Marine Biological Station* 149

Address: Postboks 2550, Sør Tromsøya, N-9001 Tromsø
Activities: Research and development in the following areas: marine biology with special reference to fishery biology, physical oceanography in northern (including Arctic) waters and fjord systems, total production in selected northern Norwegian fjords, sea birds.

FISHERIES DEPARTMENT* 150

Universitetet i Trondheim* 151

[Trondheim University]
Address: Kåkon Magnussons gt 1B, N-7000 Trondheim
Telephone: (075) 15100
Status: Educational establishment with r&d capability

COLLEGE OF ARTS AND SCIENCES*   152

Address: N-7055 Dragroll

Faculty of Science*   153

Dean: Professor K. Mork

Department of Botany*   154

Department of Zoology*   155

Veterinaerinstituttet   156

[National Veterinary Institute]
Address: Postboks 8156, Dep, Oslo, 1
Telephone: (02) 463900
Status: Official research centre
Director: Professor Dr Olav Sandvik
Assistant Directors: Nils Koppang, Dr Bjørn Naess
Graduate research staff: 35
Activities: Diagnostics, control and eradication of animal diseases.
Projects: Taxonomical studies of bacteria (Dr Sandvik); ceroid lipofuscinosis in dogs, toxic effects of nitrosodimetylamins (Dr Koppang); vaccination of animals against ringworm (Dr Naess, Dr Sandvik); nosematosis in the blue fox (Dr Mohn); copper, zinc and molybdenium in sheep, swine and cattle (Dr Norheim); selen in animals (Dr Moksnes); pathogenic mechanisms in swine dysentery (Dr Teige); sarcocystis-infection in cattle and sheep (Dr Bratberg); environment and udder health in dairy farms (Dr Bakken); nuclease of Staphylococcus epidermidis (Dr Gudding); vaccination of swine against atrophic rhinitis (Dr Norberg); paratuberculosis and related mycobacterial diseases (Dr Fodstad); classification of mycobacteria (Dr Saxegaard); maedi, progressive interstitial pneumonia in sheep (Dr Krogsrud); heavy metals, DDE and PCB in Norwegian game (Dr Holt); vaccination of fish against vibriosis (Dr Håstein).

DEPARTMENT I - BACTERIOLOGY,   157
SERODIAGNOSTICS

Head: Dr Bjørn Naess

DEPARTMENT II - POULTRY   158
PATHOLOGY

Head: Dr Finn Kristiansen

DEPARTMENT III - CHEMISTRY,   159
TOXICOLOGY

Head: Dr Arne Frøslie

DEPARTMENT IV - PATHOLOGY   160

Head: Dr Nils Koppang

DEPARTMENT V - PRODUCTION   161
DISORDERS IN CATTLE, MASTITIS
LABORATORY

Head: Dr Gudbrand Bakken

DEPARTMENT VI - VACCINES, SERA   162

Head: Dr Leif Sørum

DEPARTMENT VII - TUBERCULIN   163
PRODUCTION, TUBERCULOSIS
DIAGNOSIS

Head: Dr Saxegaard

DEPARTMENT VIII - VIROLOGY   164

Head: Dr Johan Krogsrud

FEED HYGIENE LABORATORY   165

Head: Dr Liven

FISH DISEASES LABORATORY   166

Head: Dr Tore Håstein

GAME DISEASES LABORATORY   167

Head: Dr Gunnar Holt

PAKISTAN

Irrigation Research Institute*   1

Address: The Mall, Lahore
Status: Official research centre
Administrator: Saad Haroon

Nuclear Institute for Agriculture and Biology*   2

Address: Jhang Road, Lyallpur
Official research centre

Pakistan Animal Husbandry Research Institute*   3

Address: GPO Peshawar Cantt, Peshawar, NWFP
Status: Official research centre
Head: M. Ansari

Pakistan Forest Institute   4

Address: Peshawar, NWFP
Telephone: (0521) 8320
Telex: PAKFI
Affiliation: Peshawar University
Status: Official research centre
Director General: Dr M.N. Malik
Graduate research staff: 50
Activities: Forestry and forest products.
Projects : Forest economics; watershed management (Abdul Aleem); silviculture (M.I. Sheikh); forest mensuration (Raja Willayat Hussain); range management (Dr Sultan Maqsood Khan); wildlife management (Ashiq Ahmad); forest genetics (Dr Shamsur Rehman); forest products (Dr K.M. Siddiqui); forest botany (Dr A.R. Baig); forest chemistry (Fazle Wahid Khan); medicinal plants (Anwar Ahmad Khan); forest pathology (Ch. Zakaullah); forest entomology (M. Ismail Chaudhry).

BIOLOGICAL SCIENCES RESEARCH DIVISION   5

Director: Dr M.N. Malik

Forest Botany Branch   6

Head: Dr A.R. Beg

Forest Chemistry Branch   7

Head: Fazli Wahid Khan

Forest Entomology Branch   8

Head: M. Ismail Chaudhry

Forest Pathology Branch   9

Head: Mr Zakaullah

Medicinal Plants Branch   10

Head: Amwar Khan

FOREST PRODUCTS RESEARCH   11
DIVISION

Director: Dr K.M. Siddiqui

Forest Genetics Branch   12

Head: Dr Shamsur Rehman

FORESTRY RESEARCH DIVISION   13

Director: M.I. Sheikh

Forest Economics Branch   14

Head: Muhammad Amjad

Forest Mensuration Branch   15

Head: Raja Walayat Hussain

Range Management Branch   16

Head: Dr Sultan Maqseed Khan

Silviculture Branch   17

Head: M.I. Sheikh

Watershed Management Branch   18

Head: Muhammad Haif

Wildlife Management Branch   19

Head: Ashiq Ahmad

Pakistan Institute of Cotton   20
Research and Technology*

Address: New Queens Road, Karachi
Status: Official research centre

Punjab Veterinary Research   21
Institute*

Address: Lahore, 13
Status: Official research centre

Rice Research Institute*   22

Address: Dokri
Status: Official research centre

Sind Agriculture University   23

Address: Tandojam, Sind
Telephone: 26881
Status: Educational establishment with r&d capability
Vice-Chancellor: Dr Abdul Qadir Ansari
Graduate research staff: 155
Activities: Advancement of agricultural sciences and dissemination of relevant knowledge to rural workers: agricultural science, soil science, biochemistry, plant breeding and genetics, plant protection, plant pathology, entomology, horticulture, agricultural engineering and building structures, irrigation and drainage, farm power, machinery, veterinary science, pharmacology, livestock management, animal production, physiology, animal husbandry, forestry, fisheries, agricultural economics, land utilization, rural sociology, food technology, cooperative systems.
Publications: Annual report of the University.
Projects: Economic analysis of small animal farming in Sind Province (Sharif Ahmed Siddiqui, Dilawar Hussain); efficacy of various agricultural machinery (Bherulal Deverajani); estimation of consumptive use of water for major crops; national forage and fodder research programme (Ghulam Nabi Kalwar, Minhoun Khan Abbasi); role of predacious arthropods in mite pest control (Abdul Hayee Soomro); incidence and epizootology of ecto- and endoparasites of poultry (Dr Shah Nawaz Buririo).

FACULTY OF AGRICULTURE   24

Head: Karamullah H. Agha

FACULTY OF AGRICULTURAL   25
ENGINEERING

Head: Imam Bux Kundhar

FACULTY OF ANIMAL HUSBANDRY   26
AND VETERINARY SCIENCES

Head: Dr Shah Nawaz Buririo

FACULTY OF SCIENCES   27

Head: Shamsuddin Soomro

PAPUA NEW GUINEA

Appropriate Technology Development Institute 1

– ATDI
Address: Box 793, Lae, Morobe Province
Telephone: 42 4999
Telex: NF 42428
Affiliation: University of Technology; South Pacific Appropriate Technology Foundation
Status: Official research centre
Director: Paul R. Warpeha
Departments: Agriculture, Nett Orimjou; food technology, Rashimah New; energy and shelter, Lukis Romaso; water supply, Charles Nakau; small industry development, Mursalin New
Graduate research staff: 6
Activities: ATDI is a joint venture of the South Pacific Appropriate Technology Foundation (SPATF), the Papua New Guinea University of Technology (Unitech) and the Lik Lik Buk Information Centre. It is dedicated to assisting the people of Papua New Guinea to develop technologies which utilize local skills, materials, and finance, serve to improve health and well-being, create employment, and maintain or renew a sense of self-sufficiency and cultural identity. With ready access to Unitech technical expertise, SPATF's socioeconomic development programmes, and the information resources of Lik Lik Buk, ATDI is able to conduct projects which not only deal with the research of tools, devices, or techniques, but also examine the social and economic processes which are the most important aspect of technological development.
Publications: Annual report.
Projects: Sustained gardening techniques (Nett Orimjou); food processing (Rashimah New); village sawmills (Mursalin New); charcoal production (Lukis Romaso); firewood conserving stoves (Paul Warpeha).

Beef Cattle Research Centre 2

Address: ERAP PO Box 1434, Lae, Morobe Province
Telephone: 42 2274
Parent body: Development of Industry
Status: Official research centre
Head: J.H. Schottler
Sections: Ruminant nutrition and ruminant breeding, J.H. Schottler; pastures, G. Tupper
Graduate research staff: 7
Activities: The centre's aim is to carry out research work on ruminants (mainly cattle and water buffalo), to increase productivity in the Papua New Guinea environment. To achieve this aim, research work is carried out in the fields of nutrition, breeding, behaviour and management of ruminants. Pasture research covers management, species evaluation, nutrition and weed control.
Projects: Mineral nutrition of ruminants (J.H. Schottler); cattle breeding (U. Rova); by-product utilization (K. Wenge); buffalo management and nutrition (W. Cwaiseuk); pasture management (G. Tupper); pastures species evaluation (M. Raurela); weeds in pastures (F. Kamit).

Central Veterinary Laboratory 3

Address: PO Box 6372, Boroko
Telephone: 25 3588
Telex: NE 22147
Parent body: Department of Industry
Status: Official research centre
Chief Veterinary Research Officer: Dr M.J. Nunn, acting
Sections: Veterinary pathology, Dr M.J. Nunn; veterinary microbiology, Dr U. Wernery; veterinary parasitology, Dr I.L. Owen
Graduate research staff: 16
Activities: Animal production and veterinary medicine: veterinary diagnostic laboratory for Papua New Guinea, research into applied/practical problems of animal production and health, advice centre on animal production and health, and training in these fields; food science: microbiological examination of foods (notably seafoods) prior to export.
Projects: Comparison of sheep and goat parasitological status; buffalo fly eradication using insecticide impregnated ear tags; laevamisole resistance in sheep; survey of Salmonella; survey of Leptospira (Dr U. Wernery).

Department of Industry 4

Status: Government department
Publications: Annual report
See separate entries for: Beef Cattle Research Centre
Central Veterinary Laboratory

Office of Forests 5

Address: PO Box 5055, Boroko
Telephone: 25 4022
Telex: NE 22360 'FORESTS'
Status: Official research centre
Director: A.M.D. Yauieb
Sections: Research and training division, (acting) G.P. Samol; forest management research branch, (acting) J.R. Luton
Graduate research staff: 1
Activities: Coordination of silvicultural research by Forest Management Research Branch and Provincial Forest Offices; supply of seed for internal plantations and research, and for international research.
Projects: Statistical advice and analysis of trials (J.R. Luton); seed supply for overseas research (L. Agi).

University of Papua New Guinea 6

Status: Educational establishment with r&d capability
Publications: Annual research report.

FACULTY OF AGRICULTURE 7

Address: PO Box 4820, Port Moresby
Telephone: 25 3900
Head: Dr V. Kesavan
Graduate research staff: 12
Activities: Soil science; plant breeding; crop husbandry; livestock breeding, husbandry and nutrition; agricultural engineering. The scope of studies is problem oriented, and the aim is to assist in developing suitable farming systems in the humid tropics, with special reference to subsistence farmers.
Projects: Agronomy and breeding of winged bean-legumes (Dr V. Kesavan); soil fertility and multiple cropping (Dr K. Thiagalingam); agronomy - yams (A. Pais); crop physiology (Dr G. Browning); parasites of farm animals (Dr J. Varghese); farm economics (Dr D. Das).

Lae Campus 8

Address: c/o University of Technology, PO Box 793, Lae
Telephone: 42 4999
Telex: NF 42428
Administrative head: Professor A.R. Quartermain
Graduate research staff: 5
Activities: Land use; soil conservation; crop husbandry: aroids, banana, cassava, sweet potato; livestock breeding: sheep, goats; livestock husbandry and nutrition: sheep, goats, pigs, ducks; agricultural economics; rural sociology; mechanization in the smallhold sector.
Projects: Growth, behaviour and management of sheep and goats (P. Kohun); sheep and goat breeding; pigs management systems: use of cassava, sweet potato and green legumes (Professor A. Quartermain); effect of topography and land management on soil and water loss (A.R. Williams); factor productivity and income in agriculture in Morobe Province (Dr D.K. Das); agronomy and physiology of Taro and Tannia; collecting germplasm from Papua New Guinea; cassava, banana (Dr A.M. Gurnah).

PERU

Centro Nacional de Patología Animal*   1

[National Centre of Animal Pathology]
Address: Apartado 1128, Lima
Status: Official research centre

Estación Altoandina de Biología y Reserva Zoo-Botánica de Checayani*   2

[High Andes Biology Station and Zoo-botanical Reserve of Checayani]
Address: Checayani, Azangaro (Punto)
Status: Official research centre

Estación Experimental Agrícola de la Molina*   3

[Molina Agricultural Experimental Station]
Address: Apartado 2791, Lima
Status: Official research centre

Estación Experimental Agrícola del Norte*   4

[Northern Agricultural Experiment Station]
Address: Atahualpa 211, Lambayeque
Status: Official research centre

Estación Experimental Agropecuaria de Tulumayo*   5

[Tulumayo Agricultural Research Station]
Address: Apartado 78, Tingo María, Huánuco

Estación Experimental Cajamarca   6

[Cajamarca Experimental Station]
Address: Jr Lima 560, CP 169, Cajamarca
Telephone: 2350
Status: Official research centre
Director: Luis Narro
Departments: Crops, Santiago Franco; forage and livestock, Miguel Barandiaran
Graduate research staff: 9
Activities: Crop and livestock (mainly dairy cows) productivity improvement in the northern sierra of Peru; maize, wheat, barley, lupins, beans, pasture; livestock husbandry and nutrition.
Publications: Annual report.
Projects: Maize (Luis Narro); wheat and barley (Santiago Franco); potato (Victor Vasquez); bean and lupins (Jesus de la Cruz); pasture (Miguel Barandiaran); livestock (Julio Gamarra).

Estación Experimental de 'Vista Florida'* 7

– EEVFL
[Vista Florida Experimental Station]
Address: Apartado 116, Chiclaye
Status: Official research centre
Activities: Agricultural research in rice, grain, sorghum, soyabean, bean, chickpea, cotton, alfalfa, cattle, hogs, and goats.

Instituto de Biología Andina* 8

[Institute of Andean Biology]
Address: Apartado 5073, Lima
Status: Official research centre
Activities: The institute studies biological problems and behaviour of organisms at high altitudes.

Instituto de Zoonosis e Investigación Pecuaria* 9

– Izip
[Institute of Zoonosis and Research of Animal Diseases]
Address: Camilo Carrillo 402, Apartado 1128, Lima
Status: Official research centre

Instituto del Mar de Peru* 10

– IMARPE
[Peruvian Marine Institute]
Address: Esquina General Valle y Gamarra, Apartado 22, Callao
Status: Official research centre
Activities: Technical and scientific oceanographic and fisheries research. In 1982 the Inter-American Development Bank approved a loan of $14.5m to improve fish output in Peru and to strengthen the institute.

Instituto Nacional de Investigación Agraria* 11

[National Agricultural Research Institute]
Address: Sinchi Roca 2728, Oficina 802, Lima, 14
Status: Official research centre

Universidad Nacional Agraria 12

[National Agriculture University]
Address: Apartado 456, Avenida La Universidad s/n, La Molina, Lima
Telephone: (35) 2035
Status: Educational establishment with r&d capability
Rector: Mario Zapata Tejerina
Graduate research staff: 393
Activities: Natural resources; plant production; animal production; food technology; forestry and forest products. The three principal aims are food production, conservation of natural resources and technology use and creation.
Publications: Forestry journal, quarterly; scientific annals, biannual.
Projects: Enriched foods; aviculture; meats; pigs; cereals; floriculture; fruits and native fruits; tropical cattle; horticulture; milk; maize; animal improvement; cotton; sheep and American Camelidae; potatoes.

AGRICULTURAL MECHANIZATION DEPARTMENT 13
Head: Manuel Lecca Rodríguez

ANIMAL HYGIENE DEPARTMENT 14
Head: Mario Bendezú Albela

ANIMAL PRODUCTION DEPARTMENT 15
Head: Manuel Carpio Pino

BIOLOGY DEPARTMENT 16
Head: Dr Carlos López Ocaña

CHEMISTRY DEPARTMENT 17
Head: Pedro Cueva Martín

ECONOMY AND PLANNING DEPARTMENT 18
Head: Dr César Delgado Barreto

ENTOMOLOGY DEPARTMENT 19
Head: Dr Fausto Cisneros Vera

FISH CULTURE AND   **20**
OCEANOGRAPHY DEPARTMENT
Head: Afranio Livia

FISHING TECHNOLOGY   **21**
DEPARTMENT
Head: Julia Olórtegui Vela

FOREST MANAGEMENT   **22**
DEPARTMENT
Head: Dr Augusto Tovar Serpa

HORTICULTURE DEPARTMENT   **23**
Head: Luis Delgado de la Flor

INDUSTRIAL FORESTRY   **24**
DEPARTMENT
Head: Victor Gonzáles Flores

LAND AND CATTLE ECONOMICS   **25**
DEPARTMENT
Head: Carlos Lescano Anadón

LAND AND WATER RESOURCES   **26**
DEPARTMENT
Head: Lorenzo Chang-Navarro

MATHEMATICS DEPARTMENT   **27**
Head: Oscar Manzur Salomón

METEOROLOGY AND PHYSICS   **28**
DEPARTMENT
Head: María del Carmen Cassano

NUTRITION DEPARTMENT   **29**
Head: Arturo Carrasco

PHYTOPATHOLOGY DEPARTMENT   **30**
Head: Ricardo Mont Koc

PHYTOTECHNICS DEPARTMENT   **31**
Head: Salomón Helfgott

RURAL CONSTRUCTION   **32**
DEPARTMENT
Head: César Bellido Peralta

SOILS AND FERTILIZERS   **33**
DEPARTMENT
Head: Dr Sven Villagarcía Hermoza

STATISTICS DEPARTMENT   **34**
Head: S. Alfredo García G.

PHILIPPINES

Bicol Experiment Station   1

Address: San Agustin, Pili, Camarines Sur
Affiliation: Bureau of Plant Industry, Ministry of Agriculture
Status: Official research centre
Regional plant research coordinator and superintendent: Dr Eugenio S. Sabalvoro
Graduate research staff: 36
Activities: The station undertakes agricultural research involving improvement of crop plants; training of technicians and farmers; and dissemination of agricultural information.
Publications: Annual report; *Bicol Agriculture* (quarterly).

AGRICULTURAL ENGINEERING AND   2 CROP UTILIZATION DIVISION

Head: P.G. Salcedo
Projects: Engineering and small tools development.

AGRICULTURAL PRACTICES   3 DIVISION

Head: E.M. Imperial
Projects: Plant physiology and farming systems (Dr E.S. Sabalvoro); fertilizer and weed control (E.M. Imperial).

CROP IMPROVEMENT DIVISION   4

Head: C.N. Abigay
Projects: Rice varietal improvement (C.N. Abigay); vegetable improvement (F.A. Perello); legume improvement (C.B. Imperial).

CROP PRODUCTION DIVISION   5

Head: W.J. Villezar
Projects: Floriculture research (E.M. Aspe).

CROP PROTECTION DIVISION   6

Head: Dr F.D. Laysa
Projects: Plant pathological research (Dr F.D. Laysa); entomological research (M.J. Dancel).

FIELD TRIAL SERVICES DIVISION   7

Head: J.R. Florin
Projects: On-farm trials project.

SEED QUALITY CONTROL SERVICES   8 DIVISION

Head: R.M. Torres
Projects: Seed technology.

Bicol River Basin   9 Development Programme

Address: San Jose, Pili, 4730, Camarines Sur
Affiliation: National Council for Integrated Area Development (NACIAD)
Status: Government agency sponsoring research
Acting programme director: Camilo A. Balisnomo
Departments: Programme planning, Carmelo R. Villacorta; programme management, Onofre R. Alajor
Graduate research staff: 15

Activities: Research in the following areas: soil science; irrigation, drainage and flood control; land use; agro-afforestation; watershed protection; crop production; medicinal plants study; inland and offshore fisheries; health and nutrition; hydrometeorology.

Publications: BRBDP Annual Report; Basin Reporter (monthly).

Projects: Libmanan-Cabusao integrated area development (Ramon Caceres); Pili-Bula land consolidation (Gregorio Beluang); Bicol secondary and feeder roads (Vicente Lopez); cut-off channel (Alfonsito Padua); Bicol river basin irrigation development (Naga-Calabanga, Rinconada) (Graciano Labayog); integrated health, nutrition and population (Salve Tongco); Rinconada-Buhi Lalo development (Padro Brusas); area development programme (Onofre Alajor); rural waterworks development (Roberto Castañeda); medicinal plants (Alicia Vitor); development communication (Dominador Alarkon); agribusiness development and industrialization (Nicolas Beda Priela); Baliwag San Vicente integrated area development; agro-industrial integrated development (Felix Lositaño); Quinali integrated area development (Felix Lositaño); Lake Bato-Pantao bay diversion channel (Marcelo Samson); Naga-Calabanga integrated fisheries development and feasibility study (Nicolas Beda Priela); Sorsogon integrated area development (Pete Jumamil); Caramoan-Partido integrated area development (Cesar Paita); farming systems development (Perfecto J. Bragais); hydrometeorological programme (Elmo Bombase); Bicol multipurpose survey (Teresa Javier).

Central Luzon State University* 10

Address: Munoz, Nueva Ecija
Status: Educational establishment with r&d capability

COLLEGE OF AGRICULTURE* 11
Dean: Marcelo M. Roguel

Agri-Management Department* 12
Head: Diogenes Antonio

Animal Science Department* 13
Head: Dionisio O. Orden

Crop Protection Department* 14
Head: Carlos Alagad

Crop Science Department* 15
Head: Dr Josue A. Irabagon

Soil Science Department* 16
Head: Dr Juliana B. Dacayo

COLLEGE OF EDUCATION* 17
Dean: Nathaniel Lapitan

Agricultural Education Department* 18
Head: Deogracias Ponce

Agricultural Extension* 19
Head: Pastora Coloma

COLLEGE OF ENGINEERING* 20
Dean: Gaudencio Villaroman

Agricultural Engineering Department* 21
Head: Dr Honorato Angeles

Agricultural Mechanics Department* 22
Head: Isaias Lacson

COLLEGE OF INLAND FISHERIES* 23
Dean: Dr Catalino R. De la Cruz

Aquaculture Department* 24
Head: Dr Renato Recometa

Aquatic Biology Department* 25
Head: Professor Luzriminda Guerrero

Inland Fisheries Management Department* 26
Head: Dr Rudolpho G. Arce

COLLEGE OF VETERINARY SCIENCE AND MEDICINE* 27
Dean: Dr Faustino S. Mensalvas

Basic Veterinary Sciences* 28
Head: Dr Oscar D. Quines

Clinical Sciences and Extension Services*   29

Head: Dr Jesus S. de la Rosa

UNIVERSITY ATTACHED INSTITUTES*   30

Central Luzon Agricultural Research Centre, CLSU*   31

Director: Dr Filomena F. Campos

Freshwater Aquaculture Centre, CLSU*   32

Director: Dr Catalino R. de la Cruz

Central Mindanao University   33

– CMU
Address: Musuan, Bukidnon
Affiliation: Independent but linked with National Research Network (Philippine Council for Agricultural Resources Research, PCARR)
Status: Educational establishment with r&d capability
University president: Isabelo S. Alcordo
Graduate research staff: 30
Activities: Research in order to achieve the following objectives to: increase food production, generate export products; develop cheap sources of energy; improve instruction; improve extension approaches; improve the quality of life; develop gainful industries.
Publications: Annual Report of the President; CMU Journal of Science, Education and Humanities (six monthly); *CMU Journal of Agriculture, Food and Nutrition* (quarterly); *NOCEMCARRP Newsletter* (quarterly).
Projects: Pastures and pasture improvement for livestock (Lorenzo Curayag, Prudencio Magadan); livestock breeding and improvement (Lamberto Boloron); plant breeding, evaluation and cultural management (Jose Escarlos); coffee, corn and sorghum breeding (Herminio Pava); agro-forestry and forest conservation (Jesus Manubag); crop protection (Nonito Franje, Lolito Capili); plantation crops management (Arturo Blancaver, Edilberto Flauta); animal health and veterinary related studies (Wilfredo Tamin); utilization of energy sources for agricultural production operations by rural communities and human settlements; systems studies pertaining to software and organizational phase in agricultural engineering (Rizalino Gregorio); breeder sciences (biology, chemistry, mathematics and physics) (Arturo Blancaver); macroeconomics Constancio Cañete); resources development (Raymundo Fonollera); technollogy transfer (Tomasito Redoble).

AGRICULTURE DEPARTMENT   34

Head: Dr Constancio Cañete
Activities: Research in the following areas: soil science, drainage and irrigation, and land use; plant breeding (corn, sorghum, rice, coffee, vegetables); crop husbandry (fruit and tree plantation crops, corn, sorghum, rice, coffee, vegetables, sugarcane); animal production (cattle, carabao, horses, goats, poultry, swine); livestock breeding, livestock husbandry and nutrition; fresh-water fisheries and aquaculture.

ARTS AND SCIENCES DEPARTMENT   35

Head: Dr Arturo Blancaver
Activities: Research on applied chemistry and nutrition, technology development and transference.

ENGINEERING DEPARTMENT   36

Head: Dr Rizalino Gregorio
Activities: Research on agricultural engineering, including energy sources, and the fabrication of small farm tools and implements.

FORESTRY DEPARTMENT   37

Head: Dr Damaso Figarola
Activities: Research on forestry and forest products, including watershed protection, Chincona culture and production, forest conservation.

HOME ECONOMICS DEPARTMENT   38

Head: Dr Lydia Mercado
Activities: Research on food technology, clothing and textiles, family development and sociology.

VETERINARY MEDICINE DEPARTMENT   39

Head: Wilfredo Damin
Activities: Research on parasite control and livestock diseases.

Central Philippine University*   40

Status: Educational establishment with r&d capability

COLLEGE OF AGRICULTURE   41

Address: Iloilo City, 5901
Telephone: 7-34-73
Telex: CPUCA Iloilo City
Dean: Professor Enrique S. Altis
Farm and nursery research coordinator: Erlinda Famoso
Poultry and livestock research coordinator: Julie Cusa
Student research coordinator: Blanquita Garcia
Graduate research staff: 5
Activities: Research in the following areas: soil science; drainage and irrigation; crop husbandry; plant protection; livestock husbandry and nutrition.
Publications: Ang Tuburan (semi-annual); *South East Asia Journal* (semi-annual).
Projects: Philippine rice seed board varietal trial; international mungbean nursery varietal trial (Erlinda Famoso).

Department of Agriculture   42
and Natural Resources*

Address: Diliman, Quezon City, 3008
Status: Official research organization
Activities: Implementation of policies bearing on food and agriculture through the Bureaux of Lands, Forestry and Plant Industry.

Food and Nutrition   43
Research Institute

– FNRI
Address: PO Box EA-467, Manila
Telephone: 595113
Affiliation: National Science Development Board
Status: Official research centre
Director: Josefina Bulatao-Jayme
Divisions: Food consumption surveys, Gracia M. Villavieja; food research, Estelita M. Payumo; nutrition research, Benigna V. Roxas; food management research, Patrocinio E. de Guzman; medical and applied nutrition Asunción C. Baltazar
Graduate research staff: 181
Activities: The general purpose of the institute is to undertake research and provide technical services that will help improve the nutritional status of the population, and it includes the following specific objectives: to analyze and encourage utilization of wholesome Philippine foods and food combinations which are good sources of nutrients; to establish/update nutritional standards that are needed in measuring/attaining good health among Filipinos; to define/monitor malnutrition problems as to causes, types, magnitude and distribu-

tion; and to evaluate/recommend measures that will prevent/correct identified malnutrition problems.
Publications: Annual report; Report of Activities (annual); *Abstracts of Food and Nutrition Researches* (annual); *Summary Report of National Nutrition Surveys* (special report published one year after data collection).
Projects: Physiological standards and nutrient requirements: radioisotope studies, iron stores, nutrition and productivity of sugarcane and construction workers, metabolic rate of pregnant women, liver stores of vitamin A in, and protein requirements of young children, weight growth relationship (P.E. de Guzman, A.C. Baltazar, M.D. Kuizon). Nutrition surveys and ecology of malnutrition: regional food patterns and food wastage, socioeconomic influences on nutrition, clinical surveys for anaemia, parasitism, vitamin A and goitre (G. M. Villavieja, T.E. Valerio). Nutritional aspects of food: amino acids, electrolytes and trace elements in Philippine food, meat analogues from locally available legumes, piloting of snack or dehydrated foods, effect of processing on protein quality, anti-nutritional factors in beans, nitrogen digestibility of protein-rich foods, establishment of recommended daily allowances, development of a short method of dietary analysis, menu guides, supplementary foods, nutritive value of seaweed (A.V. Lontoc, A.D. Santos, P.E. de Guzman, E.M. Payumo). Malnutrition and related conditions: growth patterns, role of RBC protoporphyrin and serum copper cerruloplasmin in the aetiology of pernicious anaemia, nutritional assessment of tuberculous patients, effect of vitamin A on blood in young children, early detection of xeropthalmia (L.E. Villanueva). Community projects (O.C. Valdecañas).

Forest Products Research   44
and Industries
Development Commission*

– FORPRIDECOM
Address: College, Laguna, 3720
Affiliation: National Science Development Board
Status: Official research centre
Commissioner: Francisco N. Tamolang
Activities: Basic and applied research on wood and forest products to increase utility, value, quantity and serviceability; develop new products and improved techniques, provide technical assistance, manpower training and information services.
Publications: Forpride Digest, quarterly, *FORPRIDECOM Technical Note,* monthly.

Forest Research Institute 45

Address: College, Laguna, 3720
Telephone: 2229; 2269; 3320; 3221; 2509
Affiliation: Ministry of Natural Resources
Status: Official research centre
Director: Dr Filiberto S. Pollisco
Assistant director: Dr Florentino O. Tesoro
Sections: Research specialist group, Dr Mario A. Eusebio, Dr Saturnina Halos, Dr Sebastian Quiniones; Technical consultants, Dr Quitolio Viado (forest entomology), Dr Phan-Quang Vinh (wildlife research management), Martin R. Reyes (forest research management), Dr Hoang Hoanh Nguyen (forest soils); Field operations, Warlito R. Natividad, Vicente Cabrera (special project coordinator)
Total research staff: 353
Activities: The principal objective of the institute is to carry out and maintain a programme of research on forest production, on parks, watershed, range and wildlife management and in socioeconomics, so that the forest resources may be perpetuated, and contribute a maximum share to the improvement of the environment, the economic development of the country, and the general well-being of the people. The institute seeks also to maintain an effective outreach programme, the results of which will satisfy the needs of continuously improving the management, development and utilization of forest resources.
Projects: Establishment, utilization and economics of leucaena and albizia species for small tree-farm holders; factors affecting the quality and yield of almaciga resin (Manila Copal); karyotypic analysis of Philippine Dipterocarpaceae; national research programme for fuelwood species; comparative study on mass production of forest tree seedlings using imported Finnish technology (paperpot method) and local production technology; effects of irradiation on callus tissues of Pterocarpus indicus; ex-situ conservation of forest genetic resources (Dr S. Halos). Diseases affecting Pinus species and their control (R. Zamora); studies of diseases affecting Dipterocarpus species and their control (Dr M.A. Eusebio); isolation and identification of rhizobia in leguminous trees used in reafforestation (Dr S. Quiniones); national research programme for premium hardwoods (Dr A. Cornejo).

ADMINISTATIVE SERVICES DIVISION 46

Chief: Rolando L. Metin

AGROFORESTRY RESEARCH CENTRE 47

Coordinator: Maximino L. Generalao
Projects : National research programme for premium hardwood species; effects of storage on the germination of palasan and limuran seeds; effect of different cutting regimes and slope on survival and growth of residual stands; effects of spacing and interplanting high premium species with fast-growing species; relationships of height growth to soil texture depth index in Pagbilao, Quezon; spacing study on the growth and development of guijo; fertilizer studies on molave, teak, acacia and mahogany; fungi associated with forest tree seed in storage and their effects on germination; spacing of balled and bareroot rattan wildlings under different forest covers; germination, survival, growth and development of palasan and limuran species at Pagbilao, Quezon; effects of fruit maturity, storage and pre-treatment germination of fuelwood species; fertilization and liming of leucaena for adaptation to low pH at different climatic conditions.

CONIFER RESEARCH CENTRE 48

Address: Loakan, Baguio City
Coordinator: Vicente P. Veracion
Projects: Timber resources management of tropical conifers; ecological investigations on the hydrological and nutritive values of Pinus Kesiya forest ecosystem; determination and evaluation of emergency measures for quick rehabilitation of newly burned watershed areas in the pine forest; vegetational changes in natural Benguet pine forest after a seed-tree method of logging; reclamation of mine wastes/tailings and surface mined areas through vegetation establishment; hydrological response of Benguet pine at different stocking levels; growth of Benguet pine in areas overseeded with nitrogen-fixing species; provenance trial of various tree species (pines) in Benguet; surface run-off, infiltration rates and sediment yield of different land-use types in Benguet; comparative cost and survival rates of different seedlings, planting germinants, and conventional seedling planting of denuded Benguet; protection and rehabilitation of conifers watersheds.

DIPTEROCARP FOREST RESEARCH 49 CENTRE

Address: Mangagoy, Bislig, Surigao del Sur
Coordinator: Bonifacio A. Apura
Projects: Fertilization of established plantations of fast-growing hardwoods; assessing the developments and effects of moluccan sau planted under coconut groves in Surigao del Sur.

FOREST REGULATION AND UTILIZATION DIVISION 50

Chief: Felizardo Virtucio
Projects: Establishment, utilization and economics of leucaena and albizia species for small tree-farm holders; yield prediction model for Tectona grandis plantations; determination of the appropriate selective logging prescription for the Philippine dipterocarp forests; effect of seasons and different logging methods on log production efficiency and damage to residual trees and reproduction; determination of taper and bark thickness of commercial tree species by site and stand density classes; growth, structure and composition of logged-over dipterocarp and pine forest using CFI plots; growth, yield and economic rotation of giant ipil-ipil and native ipil-ipil at different spacing and site classes for various end-uses; development of pilot scale plantation of selected bamboo species in Rizal and Quezon province for cottage industries; studies on the production and harvesting of some erect bamboo species; determination of the extent and types of natural defects occuring in commercial types in Gebar in Concession; production and utilization of bamboos at the barangay level; evaluation of forest tree resin-producing species as hydrocarbons sources; survey of some erect bamboos stands in the Philippines.

MANGROVE RESEARCH CENTRE 51

Address: Bo Talipan, Pagbilao, Quezon
Coordinator: Maximino L. Generalao
Projects: Phytosociology and development of Palsabangon mangrove forest in Pagbilao, Quezon; rehabilitation of LAICOR fishpond dikes, riverbanks and shoreline protective vegetation.

MULTIPLE-USE RESEARCH CENTRE 52

Address: Palbi, Calapan, Oriental Mindoro
Coordinator: Gregorio A. Reyes
Projects: Planting of fast-growing reafforestation species by cuttings in clusters; trial planting of some reafforestation species by cuttings on beach areas in Oriental Mindoro; turbidity study of important rivers in the province of Mindoro Oriental.

NON-TIMBER FOREST PRODUCTS RESEARCH CENTRE 53

Address: Lantawan, Pasonanca, Zamboanga City
Coordinator: Bayani de Castro
Projects: Effects of various methods of selection cutting on natural regeneration of mangrove forest.

OUTDOOR RECREATION AND WILDLIFE RESEARCH DIVISION 54

Chief: Dr Manuel Bravo
Projects: Ecology and biology of Mindoro Imperial pigeon; census of game birds in Quezon Province; exploratory studies on the type habitat of the tamaraw in relation to population dynamics; performance and profitability of Rana magna macrocephala pond culture method; flora, fauna and outdoor recreation potential of Palawan.

OUTDOOR RECREATION RESEARCH CENTRE 55

Address: BFD, Legaspi City
Coordinator: Isidro T. Zamuco

PALAWAN WILDLIFE AND FOREST RESEARCH CENTRE 56

Address: Tiniguiban, Puerto Princesa City, Palawan
Coordinator: Levi V. Florido

PLANNING AND MANAGEMENT SERVICES DIVISION 57

Chief: Marcelino V. Dalmacio

SILVICULTURE AND FOREST PROTECTION DIVISION 58

Chief: Jesus Benzon
Projects: National research programme for premium hardwood species; national research programme for fuelwood species; collection, treatment and storage of leucaena seeds; survey and control of insects infesting seeds of ipil-ipil; effects of site preparation, fertilization and weeding on the establishment of plantations; asexual propagation of some long-fibred hardwoods; survey of insect pests of long-fibred hardwood species; enrichment planting in inadequately stocked logged-over Philippine dipterocarp forests at Taggat Industries, Incorporated; floristics of the mossy and pine forests in the proposed Luzon's Highest Mountain National Park; control of Ips Calligraphus: experiment with synthetic pheromones and insecticide spraying of standing trees; mass propagation of fast-growing hardwoods through tissue culture and hybridization by cell fusion technique; effects of site preparation and post-planting treatments on morphological grades of leucaena; dates after logging and frequency of timber stand improvement in a residual dipterocarp forest; characterization of selected logged-over areas as to slope, elevation vegetation and soil type; field survival of morphologically graded narra and

mahogany; effect of electrogenic machine and nitrazyme on the germination and development of molave, lumbang and rattan.

SOCIOECONOMIC RESEARCH DIVISION   59

Chief: Celso P. Diaz
Projects: National research programme for premium hardwood species; psychological aspects of forest fire problems; economic study of enrichment planting in selected inadequately stocked logged-over forests in Mindanao; socioeconomic profile of human settlements in selected frontier areas; ethnographic research in the Mount Pulog region with emphasis on shifting cultivation.

TECHNICAL SERVICES DIVISION   60

Chief: Elizio Baltazar

WATERSHED AND RANGE RESEARCH DIVISION   61

Chief: Jemuel Perino
Projects: Hydrological and physical characteristics of improved and unimproved forest plantation watersheds under different rehabilitation of Quiaoit watersheds; erosion control and water yield improvement for Agusan River basin; hydrology of major forest vegetation types; adaptability of seven forage species at different slopes and climatic types for soil protection; production and moisture storage of litter of fast-growing and premium hardwood species; revegetation of mine tailings - covered areas in the Abra River Watershed; different engineering structures to control gully formation in the pine forest watershed; evaluation of some selected plants (legumes and grasses) for prompt stabilization of roadbanks and fill slopes; establishment of Schofield stylo and centrosema in Angat grassland for forage production evaluation under different nitrogen-phosphorus-potassium levels; establishment of Schofield stylo and centrosema around newly outplanted tree seedlings as a biological fertilizer; hydrological and physical characterization of the Lake Bato Majore tributary for watershed management purposes; physical resources characterization and problem analysis for Nagacalabanga IV-B Watershed Management Development Plan; hydrologic and physical resources characterization of the Muleta-Manupali watershed.

WATERSHED RESEARCH CENTRE   62

Address: 6 St Domingo Street, San Jose City, Nueva Ecija
Coordinator: Remilio C. Atabay
Projects: Establishment and cultural treatment of vegetative cover on critical watershed in the Watershed Experimental Forest; effect of different cover crops on the growth and establishment of Enthocephalus chinensis, Gmelina arborea and Endospernum peltatuem in Caranglan rangelands.

Antigue Forest Research Station   63

Address: Barangay 6, San Jose, Antigue
Supervisor: Ernesto Arevalo

Babatngon Forest Research Centre   64

Address: Block 8, Lot 10, PHHC, Tacloban City
Supervisor: Arthur S. Garcia
Projects: Vegetative propagation of fast-growing species: Eucalyptus deglupta and Gmelina arborea.

Batac Forest Research Station   65

Address: c/o Lord Hamilton Tailoring, Batac, Ilocos Norte
Supervisor: Silverio Tolentino

Baybay Forest Research Station   66

Address: c/o Tacloban Forest Research Station, Block 8, Lot 10, PHHC, Tacloban City
Supervisor: Arturo Sazon

Bicol Forest Research Station   67

Address: Bo Laniton, Basud, Camarines Norte
Supervisor: Pepito R. Garcia

Butwan Forest Research Centre   68

Address: 1116 R. Calo Street, Puyohon, Butwan City
Supervisor: Danilo C. Cacanindin
Projects: Timber stand improvement in logged-over areas; production of site indicator plants and lesser-known tree species for the rehabilitation of cogonal areas; determination of appropriate techniques to rehabilitate kaingin areas; regeneration methods of molucan sau plantation; volume table, growth and yield of Kaatoan bangkal; plantation establishment of rattan species.

Cabagan Forest Research Station 69

Address: ISU, Cabagan, Isabela
Supervisory: Roberto C. Apigo
Projects: Comparative study on survival growth and development of molave, tindalo, kalantas and akle planted in grassland areas.

Carranglan Forest Research Station 70

Address: Maringalo, Carranglan, Nueva Ecijan
Supervisor: Remilio C. Atabay
Projects: Grassland survey in the proposed Luzon's highest mountain national park.

Cebu Forest Research Station 71

Address: 1011 F. Llamas Street, Basak, Cebu City
Supervisor: Felimon V. Nañagas
Projects: Establishment of plantation and growth of Rhizophora apiculata and R. mucronata in eastern Negros Oriental; vegetative establishments on mine tailings areas and waste dumps; regeneration of dipterocarp species in newly burned, abandoned kaingin and logged-over areas in two climatic types of Negros Island; effects of thinning and the diameter and height growth of bakauan species in Negros Oriental Province.

Claveria Forest Research Station 72

Address: Claveria, Cagayan
Supervisor: Angelito R. Valencia
Projects: Phenological studies of fast-growing and high-yielding species; seed storage and viability, seed germination and pregermination treatment and sowing density of fast-growing and high yielding species in Taggat Industries, Incorporated.

Davao Forest Research Centre 73

Address: Tagum, Davao del Norte
Supervisory: Generosa C. Aumentado

Hinobaan Forest Research Station 74

Address: c/o ILCO Phils. Inc, Hinobaan, Negros Occidental
Supervisory: Virgilio de la Cruz
Projects: Study on the causes and circumstances of accidents in various logging phases.

Magat Forest Research Station 75

Address: Magat, Diadi, Nueva Viscaya
Supervisor: Marcelino M. Maun
Projects: Growth and development of Acacia suriculaeformis at various spacing; coppicing of kaatoan bangkal; establishment of seed orchard (teak, yemane, and mahogany); effects of thinning and nitrogen-phosphorus-potassium fertilizers on the growth and seed production of yemane for reafforestation.

Malaybalay Forest Research Station 76

Address: Malaybalay, Bukidnon
Supervisor: Constante B. Serna
Projects: Survival and growth of cinchona species in newly opened and previously stocked areas; effects of spacing and bole development of malapapaya for matchwood; effects of thinning on the development of different age Benguet pine plantations in Bukidnon; effects of spacing and planting stock on survival, growth and yield of muzizi for reafforestation in Malaybalay.

Mambusao Forest Research Station 77

Address: MATEC, Mambusao, Capiz
Supervisor: Elvero Eusebio
Projects: Underplanting narra and dao in giant ipil-ipil stand; provenance trial of Pinus caribaea var hondurensis.

Masinloc Forest Research Station 78

Address: Bo. Sta Rita, Masinloc, Zambales
Supervisor: Alberto de los Santos
Projects: Silvicultural characteristics of Mindoro pine in Norther Zambales; growth and development of Mindoro pine regeneration in logged-over dipterocarp forest.

Mindoro Forest Research Station 79

Address: Murtha, San Jose, Occidental Mindoro
Supervisor: Petronilo S. Munez
Projects : National research programme for fuelwood species.
Weight-volume and solid stacked volume relationship of fuelwood species in two climatic types (P. Munez);
Growth, yield and economic rotation of fuelwood species in relation to volume and calorific values in two climatic types (P. Munez);
Effects of potting media on the growth and development of dipterocarp wildings(P. Munez).

Palawan Forest Research Station 80

Projects: Interplanting and spacing effects of fast-growing species on the growth and development of apitong for poles and piles; effect of thinning intensities on the growth and development of apitong for poles and piles; effect of different kinds of manuring on the growth and root development of Rauwolfia serpentina.

Isabela State University 81

– ISU
Address: Echague, Isabela
Status: Educational establishment with r&d capability
Research director: Dr Toribio B. Adaci
Graduate research staff: 40
Activities: Help define research objectives for individual and group researchers, check designs and methodologies, supervise project implementation, redefine local priorities.
Publications: Farmer's Research Journal.
Projects : Post-harvest technology system in Isabela village (Jose Lorenzana); fertilizer interaction in yield of garlic (Henry Lobo); pasture management studies in Cagayan Valley (Professor Pedro Guzman); tobacco priming (Haime Malvar); savings and investment practices of Cagayan Valley farmers (Dr Toribio Adaci); all-Philippine vegetable trials (Lilia Macutay); field trials on cassava and other root crops (Aurelio Baquiran); quantification of factors causing poor performance of selected cropping patterns in farmer's field in Region 02; response of cigar-filler tobacco on different methods of water application (Raul Palaje); control of cercospora leaf spot on cigar-filler tobacco (Alex de Paz).

AGRICULTURAL ENGINEERING DEPARTMENT 82
Head: Jose Lorenzana

CROPS RESEARCH DEPARTMENT 83
Heads : Professor Antonio B. Rocha, Jr, Professor F.T. Agbisit

LIVESTOCK DEPARTMENT 84
Heads: Rogelio Eustaquio, Professor Pedro Guzman

SOCIOECONOMICS DEPARTMENT 85
Head: Dr Fredelita Malvar

Maligaya Rice Research and Training Centre* 86

Address: Munoz, Nueva Ecija
Affiliation: Bureau of Plant Industry
Status: Official research centre
Activities: Research, training and seed production related to rice.

Mindanao State University* 87

– MSU
Status: Educational establishment with r&d capability

COLLEGE OF AGRICULTURE 88
Address: Marawi City, 9014, Lanao del Sur
Dean: Dr Clenio T. Dumlao
Graduate research staff: 72
Activities: The functions of the college are instruction, research and extension. Areas of research include: crop production, animal/livestock production, and dairy production (current research); agricultural engineering, and soil and water sciences (proposed research).

Agricultural Education, Extension and Agribusiness Department 89
Head: Professor Tindugaranao Dayondong
Projects: Agricultural extension services and community laboratory (Joseph Sanguilla).

Agricultural Engineering Department 90
Head: Professor Elpidio Octura
Projects: MSU-New Zealand dairy development project (E.C. Mituda).

Animal Science Department 91
Head: Professor Hernie Tiamting
Projects: Buffalo, swine, and poultry production projects.

Plant Science Department 92
Head: Professor Ottingue Masnar
Projects: White potato research and production (Professors Jenny Agustin, Abobacor Isra).

Northern Luzon State College of Agriculture* 93

Address: Piat, Cagayan
Status: Educational establishment with r&d capability
Activities: Provides professional, technical and special instruction and promotes research, extension service, and progressive leadership in agricultural education, agricultural engineering, home economics and other fields.

Palawan National Agricultural College* 94

Address: Aborlan, Palawan
Status: Educational establishment with r&d capability
President: Miguel P. Palao
Activities: The role of the PNAC Research Department is to assume leadership in the agricultural and agri-industrial development of the province of Palawan. The fields of interest are: coconut, corn and sorghum, field legumes, plantation crops, rice and other cereals, root crops, beef chevon, carabeef, forage, pasture and grasslands, pork, poultry, applied rural sociology, macroeconomics, farming systems, soil resources, and water resources.

Philippine Sugar Commission 95

– PHILSUCOM
Address: PO Box 70, North Avenue Diliman, Quezon City
Telephone: 97-32-88
Telex: 64058 PHILSUCOM PN
Status: Official research centre

RESEARCH AND DEVELOPMENT 96

Research and development director: Rodolfo E. Medina
Research and development assistant directors: Vicente G. Castro (Luzon), Romeo S. Palmares (Visayas and Mindanao).
Activities: Research is directed towards developing high-yielding varieties of sugarcane, and is pursued on an interdisciplinary basis. Over 100 research projects are currently conducted, grouped under a 10 point programme: for details see below.

Agricultural Engineering 97

Head: Mauricio Marcelo (Luzon), Francisco Mercado (Visayas and Mindanao)
Projects: Energy conservation and development of tools and labour saving devices (F. Mercado); packaging of sugarcane production technologies (I. Jimenez).

Agronomy 98

Heads: Silvino M. Samiano (Luzon), Vicente Dosado (Visayas and Mindanao)
Projects: Improvement of ratoon management practices (V. Dosado); development of feasible practices to minimize low sugar content of early - and late-milled canes (R. Tapay); intercropping/diversification (I. Bombio).

Crop Protection 99

Heads: Juliet Recuenco (Luzon), Bienvenido Estioko (Visayas and Mindanao)
Projects: Development of an integrated pest management in sugarcane (B. Estioko).

Soils and Plant Nutrition 100

Heads: Genaro V. Urgel (Luzon), Eduardo Hombrebueno (Visayas and Mindanao)
Projects: Improvement of techniques/concepts towards increasing sugar productivity and procedures for efficiency in research (N. Divinagracia); post-harvest technology with emphasis on sugar losses (E. Hombrebueno); feasible practices to improve alcohol production in sugarcane (G. Urgel).

Sugarcane Breeding and Genetics 101

Heads: Artemio M. Galvez (Luzon), Ernesto Lapastora (Visayas and Mindanao)
Projects: High-yielding variety development (E. Lapastora).

Philippine Tobacco Administration 102

Address: Elliptical Road, Diliman, Quezon City
Telephone: 97-59-16
Status: Official research centre
Chairman and general manager: Demetrio P. Tabije
Research Director: Dr Ricardo C. Briones
Extension Director: Alfredo B. Sulicipan
Trade and Marketing Director: Arnold M. Lewis
Graduate research staff: 50

Activities: The main area of research is tobacco production, including plant breeding; crop husbandry; plant protection; chemical analysis; waste product utilization. There are five experiment stations at Ilagan, Isabela; Tumauini, Isabela; Gattaran, Cagayan; Pasquin, Ilocos Norte; Tubao, La Union.

Projects: Cultural management (agronomy, water management, soil fertility, soil fertilization) (Cosme Aggabao, Vicente C. Rodriguez); varietal improvement projects (Mercedes V. Lopez): adaptation and perpetuation of some tobacco varieties and F_1 hybrids under conditions at Tumauini, Isabela; evaluation of inherent resistance of some tobacco varieties.

EXPORT MARKET DEVELOPMENT DEPARTMENT 103

Head: Florencio T. Telan

INDUSTRIAL AND CHEMICAL RESEARCH DIVISION 104

Address: Chemical Research Laboratory, Diliman, Quezon City

Projects: A survey of the chemical composition of soils in tobacco growing areas.

Studies on the utilization of tobacco wastes in the cultivation of tropical mushrooms (Volvariella volvacea and Auricularia sp.): comparative study on the production of tropical mushrooms in tobacco midribs and other bedding materials (Piedad R. Tolentino).

Philippine Virginia Tobacco Administration 105

Address: Consolacion Building, Cubao, Quezon City, Metro-Manila

Telephone: 78-69-51 to 53

Telex: PHILVITA

Status: Official research centre

Publications: *Tobacco Research Studies* (annual); *Tobacco Abstract.*

RESEARCH DEPARTMENT 106

Assistant manager: Aristides C. Castro

Graduate research staff: 44

Activities: The aims and general scope of activities of our research studies and projects are to improve, develop and stabilize Flue-cured Virginia Tobacco, Light Air-cured (Burley) Tobacco and Oriental (Turkish) tobacco.

Projects: Crop husbandry (Rafael T. Cabanawan); crop protection (Filomeno Gasmen); agricultural engineering (Cecilio P. Costales); socioeconomics (Jose

V. Sim); extension (Reynaldo B. Rabe); cultural management (Rodrigo T. Pagtulingan); soils and plant nutrition (Danilo M. Sumague); irrigation and drainage (Ruben C. Tabili); farm system (Walderico P. Elveña).

Twin Rivers Research Centre 107

– TRRC

Address: PO Box 305, Davao City, 0-404, Davao Del Norte

Telex: TRRC- DAVAO NORTE

Parent body: Twin Rivers Plantation, Incorporated

Status: Research centre within an industrial company

Research manager: Nerius I. Roperos

Departments: Crop protection, Dr Mario O. San Juan; soils, Bonifacio Azucena; agronomy, Nerius I. Roperos

Graduate research staff: 28

Activities: TRRC undertakes three phases of work: research, laboratory analyses and field services. In research, its primary objectives are to develop appropriate production technology for export banana, cacao, coconut and vegetables, and to provide technical expertise in the fields of crop protection, cultural management and plant nutrition on the aforementioned tropical crops. In addition to its purely research activities, TRRC also extends laboratory analyses for soils, tissues and fertilizers. It also provides for diagnosis for pests and diseases either as part of its technical assistance to client companies, or upon specific requests by interested parties.

Publications: Annual research report; termination report of research studies; research monthly progress reports.

Projects: Coconut breeding (Canesio C. Basio); soil fertility monitoring and irrigation studies (Bonifacio Azucena); survey and control of major diseases of banana, cacao and coconut (Dr Mario O. San Juan); pest management in banana and cacao (Nicolas M. Dawi); seed production in vegetables (Jaime B. Rebigan); yield improvement in banana (Daniel D. Eliarda).

University of Eastern Philippines* 108

Status: Educational establishment with r&d capability

UNIVERSITY RESEARCH CENTRE 109

Address: University Town, Northern Samar
Research coordinator: Nestor L. Rubenecia
Departments: Crop sciences, Norman T. Diaz; animal sciences, Nestor L. Rubenecia; agricultural enconomics, Julio M. Baldo; crop protection, Leticia C. Adorro; agro-forestry, Wilfredo A. Baya; fisheries, Adriano A. Salvador; food sciences, Concepcion C. Balanon; social sciences, Elbie Y. Baldo
Graduate research staff: 21
Activities: The centre conducts research in tandem with regional development and establishes links with other research agencies. Main areas of research include: livestock husbandry and nutrition; socioeconomic studies; plant production (vegetables, legumes, cereals, coconut, root crops, fibre crops); agricultural engineering; plant protection; veterinary medicine; agro-forestry; inland fisheries; marketing; food science.
Publications: The Researcher (annual).
Projects: Evaluation of root crops as feed for swine (Nestor L. Rubenecia); selective thinning, intercropping and processing studies on coconuts; evaluation of selected abaca cultivars under established coconut trees (Norman T. Diaz); marketing practices, facilities, channels and flows of selected agricultural products and farm inputs (Julio M. Baldo); socio-cultural and economic profile of farmers in northern Samar (Elbie Y. Baldo); verification research on vegetables (Efren A. Galo); studies on rice production (Fedencio I. Abuke).

University of Mindanao* 110

Status: Educational establishment with r&d capability

FORESTRY DEPARTMENT 111

Address: University of Mindanao, Colton Street, Davao City
Head: Alfredo Bayudan
Activities: Limited research on reafforestation; survey of forest products in Davao area.

University of Southern Mindanao* 112

SOUTHERN MINDANAO AGRICULTURAL RESEARCH CENTER 113

– SMARC
Address: University of Southern Mindanao, Kabacan, 9311, North Cotabato, Mindanao
Affiliation: Philippine Council for Agriculture and Resources Research (PCARR)
Status: Official research centre
Research director: Dr Jaman S. Imlan
Deputy director: Juan Albert Soria
Research staff: 34 (full-time); 99 (part-time)
Activities: SMARC is actively engaged in agricultural research designed to back up the development of its service areas Mindanao and Sulu. It is a national research centre for rubber, coffee, cacao, fibre crops, fruit crops, and corn and sorghum. It is also a regional cooperating centre for legumes, rice, carabeef, swine, cattle, poultry, vegetables, root crops, sugarcane, tobacco, socioeconomics, water management, soil resources and applied rural sociology. It is also recognized by PCARR as the training centre for research and research management in the region.
Publications: USM Research Journal (semi-annual); *SMARC Monitor* (quarterly).

Animal Research Division 114

Head: Professor Ceferino O. Olivo

Crops Research Division 115

Head: Dr Pablito P. Pamplona
Projects: NPK nutrition studies of coffee grown under rubber and in the open field and as affected by mulching materials; cacao breeding (Ruben P. Cabangbang); post-harvest and storage mycoflora of coffee berries and their control (Edna M. Jover); effects of pruning on the yield and quality of chico and rambutan (Vivian Muñasque); stock-scion relationship studies in rambutan and chico (Nicolas Turnos); on-farm demonstration test of cacao under coconut (Ariel Garcia); biology and control of a cacao pod borer (Acrocercrops cramerella Snellen) (Abraham Castillo); study on the sequential occurrence and development of legumes diseases in Mindanao as affected by the time of planting and weather factors (Juan A. Soria); root crops varietal screening (Eugenio Alcala); upland crops countryside action programme (yield trials for corn and sorghum) (P. Pamplona).

Engineering Research Division 116

Head: Felipe D. Vinluan

Management Information Services Division 117

Head: Dr Angelina G. Bautista.

Socioeconomics Research Division   **118**

Head: Professor Lydia P. Oliva

University of the   119
Philippines at Los Baños

– UPLB
Address: Los Baños, 3720, Laguna
Telephone: (3585) 2567
Telex: 2435 UPLB PV
Status: Educational establishment with r&d capability
Chancellor: Dr Emil Q. Javier
Vice-Chancellor for Administration: Dr Domingo M. Lantican
Vice-Chancellor for Academic Affairs: Dr Higino A. Ables
Vice-Chancellor for Planning and Development: Dr Manuel L. Bonita
Director of Extension: Dr Obdulia F. Sison
Director of Instruction: Dr Cristina D. Padolina
Director of Research: Dr Edilberto D. Reyes
Activities: Research at the university reflects the needs of the country, especially in agriculture, forestry and related sciences: results of research are meant for application in the improvement of the socioeconomic conditions of the rural population.

In order to focus the direction of research work, research thrusts were identified, giving emphasis to the problems confronting small farmers and landless labourers. Approaches to the problems are planned to be interdisciplinary and to involve both applied and fundamental research, as well as reinforce the university's functions of instruction, research and extension.

Development, utilization and conservation of energy: it is planned to develop alternative indigenous energy sources, which include, apart from some fossil fuels, geothermal, solar, wind, ocean wave, and biomass sources, for application in agricultural operations, such as the drying of crops, irrigation, cold and air-conditioned storage of fresh produce, and low power generation. Projects in this area of research would include: solar energy application; harnessing wind power for pumping irrigation water and low-power electricity generation; harnessing other local energy sources such as hot springs and marsh gas; utilization of farm biomass as agricultural wastes for fuel and fertilizers; utilization of dendro-thermal energy; alcohol production from organic sources (sugarcane, cassava and/or forest products).

Energy research is also directed towards the development of crop production systems which would emphasize energy conservation, and utilization of non-conventional sources minimizing dependence on fossil fuels. Projects in this area would include: development of crop cultivars which thrive under minimum energy input;

adoption of reduced tillage techniques; utilization of farm waste and other alternative sources of fertilizer; development of small tools for crop production which can be drawn by men or animals; studies on the biological control of leaf hoppers and plant hoppers injurious to to rice plant; nitrogen fixation by legume-rhizobium symbiosis and its contribution to the nitrogen status of Philippine agricultural soils; utilization of Azolla and other nitrogen fixing algae in rice production; studies on conservation and management of irrigation waters in relation to sunshine-based farming.

Appropriate production technology for marginal lands: marginal lands cover an area of about 17.5m hectares, including ill-drained areas, as well as most hilly regions which were formerly rain forests but which logging and 'kaingin' practices have reduced into grassland. The productivity of these areas is very low compared to their productivity when they were newly opened: an initial analysis of upland condition indicates the presence of several interrelated constraints to crop production such as erosion, low nitrogen, low phosphorus, low pH and high water stress. New production technologies adapted to hilly lands are being developed, which are low cost, require less fossil-based energy and will lead to ecological stability. Major research activities will include the following: screening of crops or plant ecotypes for tolerance to water stress and acid conditions; exploitation of rhizobia-legume relationships with the aim of improving the nitrogen status of the soil; investigation of mycorrhizal phenomena and the availability of soil phosphorus; continuing studies of minimum tillage concept in relation to land preparation; studies of the socioeconomic perception of hilly land farmers; studies of physical and biological processes in hilly lands such as erosion and nutrient cycling.

Agricultural diversification and development of non-traditional export products: Research is planned for the development of minor Philippine crops, such as cashew, jackfruit, ornamental or medicinal plants, spices and the like, which have high export potential; research activities would include studies on the production of non-conventional protein sources (vegetable proteins) or agronomic studies on cashew, jackfruit and arrowroot.

Rural income and productivity: demographic, and socio-psychological studies on the unemployed; migratory patterns of the unemployed from rural to urban and vice-versa; culture of the unemployed, unemployment rates; rural labour displacement; role of rural and/or urban institutions in rural unemployment; socio-psychological causes and effects of unemployment; work careers of occupational dropouts.

Post-harvest handling and processing of food system: Research is planned on the development of techniques for increasing processed food production, as well as development of new food products of acceptable quality, good stability and high nutritional value. Research areas suggested are: proper handling, packing, storage and transport of fresh farm produce; development of new food products; studies on thermal processing of local food; food irradiation for food preservation; identification and development of new feed ingredients from indigenous materials.

Environmental management and pollution control: Research is planned to minimize the environmental stresses of rapid industrialization: suggested research areas include: bio-degradation of fresh waste pollutants; recycling of domestic, agricultural and industrial wastes; small watershed management; soil erosion control; agro-afforestation.

Management of rural institutions: Policy research will be undertaken by UPLB researchers to provide solutions to the credit and financial problems of small farmers, to complement the development programmes set up by the Ministry of Agrarian Reform.

Publications: Annual report.

COLLEGE OF AGRICULTURE 120

Dean: Dr Cledualdo B. Perez
Associate Dean: Dr Romulo S. Davide
Departments: Agricultural education, Dr Jaime B. Valera; agronomy, Dr Elpidio L. Rosario; animal science, Dr Vicente G. Momoñgan; development communication, Dr Nora C. Quebral; entomology, Dr Edwin D. Magallona; food science and technology, Dr Elias E. Escueta; horticulture, Dr Eufemio T. Rasco; plant pathology, Dr Tiburcio T. Reyes; soil science, Dr Santiago N. Tilo; sugar technology, Dr Ramon L. Samaniego

ASEAN - Post-harvest Horticulture 121
Training and Research Centre

Telephone: (2444) 3259
Telex: 2435
Acting director: Dr Ofelia K. Bautista
Activities: Research, development and extension in post-harvest handling of fruit, vegetables and ornamental crops. Major objectives include the development of awareness, in the sectors of the industry concerned, of the importance of appropriate handling; and the reduction of losses after harvest by generating relevant technologies.
Publications: ASEAN-PHTRC Biennial Report.
Projects: Technical and economic assessment of the present practices of handling, transporting and storing of white potato; post-harvest physiology and storage of fruits and vegetables, Phase II semi-commercial test on vegetables; ASEAN packinghouse operations and quality control (Dr Ofelia K. Bautista); ASEAN mango and rambutan project/study (Dr Doroteo B. Mendoza).

Central Experiment Station 122

Manager: Telesforo M. Laude

Institute of Plant Breeding 123
Director: Dr Ricardo M. Lantican

La Granja Research and Training 124
Station
Manager: Justino J. Walawala

National Crop Protection Centre 125
Director: Dr Fernando F. Sanchez

University of the Philippines Rural 126
High School
Principal: Professor Arsenia L. Lagasca

COLLEGE OF ARTS AND SCIENCES 127
Dean: Dr Edalwina C. Legaspi
Departments: Botany, Dr Enrique P. Pacardo; chemistry, Dr Carlito R. Barril; mathematics and physics, Rolando Panopio; social science, Dr Corazon V. Lamug; statistics, Dr Ann Inez Gironella; zoology, Professor Pablo J. Alfonso

COLLEGE OF DEVELOPMENT 128
ECONOMICS AND MANAGEMENT
Dean: Dr Pedro R. Sandoval
Departments: Management, Dr Rogelio V. Cuyno; agricultural economics, Dr Narciso R. Deonampo; economics, Dr Tirso B. Paris; agrarian and cooperative studies, Dr Willie C. Depositario

COLLEGE OF FORESTRY 129
Dean: Dr Celso B. Lantican
Associate dean: Dr Adolfo V. Revilla
Departments: Forest biological sciences, Professor Lucio L. Quimbo; forest resources management, Dr Severo Saplaco; silviculture and forest influences, Professor Domingo V. Jacalne; social forestry, Dr Felix M. Estava; wood science and technology, Dr Virgilio A. Fernandez

Centre for Forestry Education 130
Research and Development for the
Asia/Pacific Region
– CFERDAP
Director: Dr Lucrecio L. Rebugio

Forestry Development Centre 131
Director: Dr Adolfo V. Revilla

Forestry Research and Extension Centre 132
Director: Dr Armando A. Villaflor

Makiling Botanic Garden 133
Head: Dr Reynaldo E. dela Cruz

Makiling Experimental and Demonstration Forest 134
Head: Jose G. Sargento

INSTITUTE OF AGRICULTURAL ENGINEERING AND TECHNOLOGY 135
Dean: Dr Reynaldo M. Lantin
Secretary: Professor Maximino G. Villanueva
Departments: Agricultural machinery engineering and technology, Dr Carlos J. del Rosario; agricultural process engineering and technology, Dr Ernesto P. Lozada; agrometeorology, Dr Virgilio G. Gayanilo; land and water resources engineering and technology, Dr Danielito T. Franco

Agricultural Mechanization Testing and Evaluation Centre 136
Director: Dr Reynaldo M. Lantin

UNIVERSITY CENTRES AND INSTITUTES 137

Agrarian Reform Institute 138
Director: Dr Jesus M. Montemayor

Agricultural Credit and Cooperative Institute 139
Director: Dr Rodolfo M. Matienzo

Centre for Policy and Development Studies 140
– CPDS
Telephone: 2595; 3455
Executive director: Dr Ramon L. Nasol
Graduate research staff: 15
Activities: The centre was envisioned as a research and extension centre to help develop the capabilities to organize and manage data, knowledge and creativity to support national development. It serves to integrate human expertise and the scientific and technological outputs of the university in addressing directly or ind-irectly policy and developmental problems confronting the following: rural and agricultural development; regional development; food production and distribution; resource utilization, distribution and conservation; energy development; employment and income.
Projects : Demand and supply analyses for selected crops and livestock in the Philippines (Agnes E. Recto, Duce D. Elazegui); action research for an integrated credit delivery system (Dr Amando M. Dalisay); data bank project (Priscila A. Alcaide, Elvira E. Dumayas); employment structure and potentials in selected provinces in the Philippines (Dr Amando M. Dalisay).

Dairy Training and Research Institute 141
– DTRI
Telephone: 2201-3; 2460; 2441; 2497
Telex: 2435
Director: Assistant Professor Alberto Y. Robles
Divisions: Dairy and forage production, Dr Le Trong Trung; dairy breeding and physiology, Professor Orlando A. Palad; dairy nutrition, Professor Leticia P. Palo; dairy technology, Dr Clara L. Davide; dairy training and information, Dr Antonio L. Ordoveza; action research, Dr Teofilo A. Dulay
Total research staff: 43
Activities: The principal objective of the institute is the acceleration of dairy development to attain self-sufficiency in milk, with research concentrating on dairy production, technology and socioeconomics. Formal, specialized and in-service training in dairy production and processing is also provided, in addition to dairy science courses and post baccalaureate diploma programmes on dairy production and processing. The institute is active in the following subject areas: breeding, physiology, nutrition, health and management of cattle, goats and buffalo; dairy technology; dairy socioeconomics; forage and crops husbandry.
Projects: Studies on the potential non-conventional supplementary feeds to selected forage and legume species for carabao/cara Murrah cow breed; quality evaluation of Philippine pasture and range grasses and legumcs as feed for livestock (R.R. Lopez); preparation and characterization of a potential calf rennet substitute from adult goat, carabao and cattle and its utilization on cheese manufacture (C.L. Davide); goat management, nutrition and reproduction studies in the Philippines: reproductive and artificial breeding studies for improved goat production (O.A. Palad); housing and management studies for improved goat production (A.Y. Robles); utilization of important non-competitive crops and by-products as feed for dairy animals (L.T.Trung).

National Institutes of Biotechnology 142
and Applied Microbiology

– BIOTECH
Telex: 2435 UPLB PU
Director: Dr Emil Q. Javier
Deputy director: Dr William G. Padolina
Activities: The institute conducts mission-oriented research, technology development, extension and planning required to develop small and large-scale industries based on biotechnology and microbiology; and develops and undertakes training programmes for the biotechnological and microbiological industries. In addition, the institute seeks to link research and industrial operations, and to facilitate commercial application of laboratory-tested biotechnological and microbial processes; to extend scientific advice to government and private entities; and to build a national culture collection of microorganisms for research and teaching.
Projects: Biofuels from agricultural crops and residues (Dr Ernesto J. del Rosario); establishment and improvement of rural-based food fermentation processes (Professor Priscilla C. Sanchez); enhancement of nitrogen fixation and soil nutrient availability to crops and reafforestation species (Dr Ruben B. Aspiras); microbial genetics (including microbial gene bank) (Dr Saturnia C. Halos); plant biomass conversion (Dr William G. Padolina); antibiotics, vaccines and microbial insecticides production (Dr Asuncion K. Raymundo, Dr Helen A. Molina); microbial biomass production (Professor William L. Fernandez); hydrocarbon-like oils from plants (Dr Elvira C. Fernandez).

National Training Centre for Rural 143
Development

Director: Dr Florentino Librero

Regional Training Programme on Food 144
and Nutrition Planning

Director: Dr Narciso R. Deonampo

Victorias Milling Company 145
Incorporated

– VICMICO
Address: 6037, Negros Occidental
Telephone: 1-31
Telex: RCA 72222264
Status: Industrial company
Activities: Sugarcane agriculture breeding and selection; tillage; cropping concepts; pest and disease control; soils and fertilizers; irrigation and drainage; chemical ripeners; extension services; crop estimation; block farming; farm and crop management.

SUGARCANE RESEARCH DIVISION 146

Director: Antonio P. Tianco
Graduate research staff: 15
Activities: The general objective is to increase the present level of farm productivity in the Victorias milling district, through the use of high-yielding varieties and improved cultural practices. Main areas of current and proposed research on sugarcane include: natural resources, soil science, and drainage; cane production, (breeding, husbandry, protection); agricultural engineering.
Publications: VMC Sugarcane Research Division Annual Report; animal research reports.

Soils Department 147

Head: vacant
Sections: Soils and nutrition, Manuel Y. Gonzales; soil testing and plant analysis, Magdalena C. Huervas
Projects: Evaluation of parent varieties and their crosses; ecological testing of VMC hybrids in four mill districts; cultural requirements of some promising VMC hybrids (F.C. Barredo); screening cultivars for resistance to sugarcane smut, downy mildew, and leaf scorch; effect of chemicals and organic soil amendments on nematodes affecting sugarcane (R.J. Serra).

Varietal Improvement Department 148

Head: Federico C. Barredo
Sections: Breeding and selection, Aurora T. Barredo; sugarcane pathology, Romeo J. Serra; pest control, Pedro H. Porquez
Projects: Effect of distillery slops on the growth and yield of cane; effect of some organic fertilizers on cane yield (M.Y. Gonzales); evaluation of several chemical methods of assessing available soil Phosphorus to predict phosphorus requirements (M.C. Huervas).

Xavier University* 149

Status: Educational establishment with r&d capability

COLLEGE OF AGRICULTURE 150

Address: Cagayan De Oro City, 8401
Telephone: 3133
Director: William F. Masterson
Departments: Life sciences, Dr Francisco Aclan; animal science, Thelma A. Zablan; soil science, Renato Siong; agricultural economics, Dr Ismael Getubig
Graduate research staff: 15
Activities: Research is more 'applied' than 'basic' and is directed towards an improved production and a more equitable return for the primary producer. Field associations are very much concerned with the small farmer and the economics of production. Another area of research is concerned with getting new technology accepted by the majority of the people and a twenty man team carries out field work with these aims in mind. There is also a Rural Communications Centre with three people involved in media work.
Projects: Weed science (Dr Francisco Aclan); crop protection (Enrico Imperio, Tita Cayme); soil nutrients (Renato Siong); animal nutrition - pasture improvement, feed substitutes (Thelma A. Zablan); agricultural economics - market studies (Dr Ismael Getubig, Dr Eduardo Canlas, Isidro Lico, Claro Cagulada, Wilfredo Yacapin); plant growth stimulants (Dr Erasmo Sagaral); adult/extension education (Anselmo Mercado); rural communications (Genara Banzon); cooperative formation, production, marketing and credit (Dr Eduardo S. Canlas).

POLAND

Akademia Rolnicza   1

[Agricultural Academy]
Address: Akademicka 13, Lublin
Telephone: 028-332-51
Telex: 064-3176 AR-PL
Affiliation: Ministry of Science and technology
Status: Educational establishment with r&d capability
Rector: Professor Edmund Prost
Graduate research staff: 597

FACULTY OF AGRICULTURAL ENGINEERING   2

Dean: Dr Helena Lis

FACULTY OF AGRICULTURE   3

Dean: Professor Czesław Tarkowski
Publications: Annual report.

FACULTY OF ANIMAL SCIENCE   4

Dean: Dr Józef Zięba

FACULTY OF HORTICULTURE   5

Dean: Dr Jósef Nurzyński

FACULTY OF VETERINARY MEDICINE   6

Dean: Professor Stanisław Wołszyn
Publications: *Excerpta Veterinaria,* annual; annual report.

Akademia Rolnicza im Hugona Kołłątaja w Krakowie*   7

[Hugon Kołłątaj Academy of Agriculture, Cracow]
Address: Al Mickiewicza 21, 31-120 Kraków
Telephone: 094-313 36
Telex: 0322469
Status: Educational establishment with r&d capability
Rector: Professor Dr Tomasz Janowski

EXTERNAL FACULTY OF PRODUCTION ECONOMICS AND AGRICULTURAL TRADE*   8

Address: Rzeszów-Zalesie
Dean: Assistant Professor Dr Eugeniusz Machowski

Agricultural Economics and Rural Commerce Organization Institute*   9

Director: Professor Dr Kazimierz Zabierowski

Principles of Agricultural Production and Technology Institute*   10

Director: Assistant Professor Dr Władysław Dubiel

FACULTY OF AGRICULTURAL TECHNIQUES AND ENERGETICS*   11

Address: Ulica Balicka 104, Kraków
Dean: Professor Dr Rudolf Michałek

Agricultural Mechanization and Energetics Institute*   12

Director: Professor Dr Rudolf Michałek

Agricultural Buildings Department* 37

Head: Professor Dr Edward Komarnicki

Hydraulic and Agricultural Structures Institute* 38

Director: Professor Dr Stanisław Polak

Sub-Faculty of Geodesy and Agricultural Installations* 39

Geodesy Institute* 40
Director: Professor Dr Anna Łoś

Planning and Organization of Farmland Department* 41
Head: Assistant Professor Dr Krzysztof Koreleski

INSTITUTE OF TROPICAL AND SUBTROPICAL AGRICULTURE AND FORESTRY* 42

Address: Kraków-Prusy
Director: Assistant Professor Dr Jerzy Solarz

Akademia Rolnicza w Posnaniu* 43

[Academy of Agriculture in Posnan]
Address: Ulica Wojska Polskiego 28, 60-637 Poznań
Telephone: 06-403 34
Telex: 0413322 pl
Status: Educational establishment with r&d capability
Rector: Professor Dr Tadeusz Czwojdrak

FACULTY OF AGRICULTURAL FOOD TECHNOLOGY* 44

Dean: Dr Mieczysław Jankiewicz

Animal Derived Food Technology Institute* 45

Director: Professor Dr Wincenty Pezacki
Activities: Technology of meat and meat products; organization of meat manufacturing processes; microbiology of meat and meat products; quality defects of meat products.

Human Nutrition Institute* 46

Director: Professor Dr Stanisław Stawicki
Activities: Agricultural technology; potatoes and fermentation industry; storage of agricultural products and silage; biochemistry of food.

Plant Derived Food Technology Institute* 47

Director: Professor Dr Kazimierz Szebiotko
Activities: Technology of auxiliary materials in food processing; chemistry of cereals and pulses; technology and engineering of grain processing; technology of fruits and vegetables.

FACULTY OF AGRICULTURE* 48

Dean: Associate Professor Dr Mieczysław Rutkowski

Agricultural Economics Institute* 49

Director: Associate Professor Dr Roman Skoczytas
Activities: Agrarian policy; organization and economics of agricultural enterprises.

Agricultural Engineering Institute* 50

Director: Professor Dr Witold Wogke
Activities: Farm machinery; mechanical engineering; processing and technology; technical farm equipment.

Biochemistry Department* 51

Director: Professor Dr Jerzy Pawełkiewicz
Activities: Biosynthesis of proteins in an isolated plant system; transfer of ribonucleic acids.

Genetics and Plant Breeding Institute* 52

Director: Professor Dr Julian Jaranowski
Activities: Inheritance and variability of nitrogen compounds in forage plants; mutation in flowering plants; polyploids and radiomutations. Cytogenetics and breeding problems of tomatoes, red pepper, beetroot, blueberry, strawberry and wild strawberry.

Plant Husbandry and Soil Cultivation Institute* 53

Director: Professor Dr Kazimierz Piechowiak
Activities: General soil cultivation and plant growing. Research concerning the chemical composition and technological properties of plant material; growing of oil, fibre, potato, cereal and fodder root crops. Meadow and pasture cultivation; grassland phytosociology.

Soil Science and Agrochemistry Institute* 54

Director: Professor Dr Marcelli Andrzejeuski
Activities: Physical properties of soils; soil fertility; soil absorbing properties and soil colloids; soil cartography and monography. Application of trace elements in agriculture; agricultural chemistry and synopsis of pedology; synopsis of fertilization; fertilization in crop rotation. Biosynthesis of active substances by microorganisms; microorganisms within soil ecosystems.

FACULTY OF ANIMAL HUSBANDRY* 55

Dean: Professor Dr Michal Iwaszkiewicz

Animal Breeding and Production Technology Institute* 56

Director: Professor Dr Antoni Kaczamzrek
Activities: Cattle, pigs, horses, sheep, poultry and fur-bearing-animals husbandry. Growing and rearing of colts, piglets and lambs. Crossbreeding of cattle and pigs for commercial production.

Animal Nutrition and Food Management Institute* 57

Director: Professor Dr Kazimierz Gawecki
Activities: Physiology and biochemistry of animals; animal nutrition and feeding stuffs science; zoohygiene and veterinary medicine.

Applied Zoology Institute* 58

Director: Dr Ryszard Graczyk
Activities: Zoology with parasitology; applied ornithology; animal anatomy; fish biology and physiology; apiculture.

FACULTY OF FORESTRY* 59

Dean: Professor Dr Jan Meixna

Forest Organization and Management Institute* 60

Director: Professor Dr Mieczysław Podgorski
Activities: Macro- and microeconomy of forestry; technical progress in forest husbandry; dendrology; science of tree and forest increments.

Forest Protection Institute* 61

Director: Professor Dr Alfred Szmidt
Activities: Forest protection and entomology; problems of insect ecology; forest pathology: game management.

Forest Utilization and Engineering Institute* 62

Director: Associate Professor Dr Tadeusz Cybulko
Activities: Mechanization of transport in forests; technical properties of wood; physiology of woodcutting work; forest engineering and building; water improvement, top dressing of forestways.

Natural Elements of Forestry Institute* 63

Director: Professor Dr Witold Mucha
Activities: Forest botany, phytosociology, silviculture, forest typology, genetics and selection; forest renewal.

FACULTY OF HORTICULTURE* 64

Dean: Professor Dr Tadeusz Hołubowicz

Horticultural Production Institute* 65

Director: Professor Dr Karol Duczual
Activities: Pomiculture; vegetable growing; dendrology; ornamental plant breeding; seed farming, biology, storage and reproduction.

Natural Elements of Plant Production Institute* 66

Director: Professor Dr Stanisław Krol
Activities: Botany and plant physiology; phytosociology, carposociology, plant anatomy, floristics.

Plant Protection Institute* 67

Director: Wiktor Kadlubouski
Activities: Phytopathology, agricultural and horticultural plant diseases. Entomology and plant protection; breeding of useful insects. Methods of plant protection; chemicals for plant protection.

FACULTY OF WATER RECLAMATION AND IMPROVEMENT* 68

Dean: Professor Dr Henyk Mikolajczak

POLAND

Hydraulic Engineering Institute* 69

Director: Associate Professor Dr Bocumił Lewandowski
Activities: Soil engineering; hydraulic constructions; river engineering; water supply and drainage of villages; geodesy, silvicultural geodesy and photogrammetry.

Land and Forest Melioration Institute* 70

Director: Dr Andrzej Kosturkiewicz
Activities: Land and forest melioration; hydrology; plant ecology; meteorology; agrometeorology; climatology.

Akademia Rolnicza w Szczecinie* 71

[Academy of Agriculture, Szczecin]
Address: Ulica Janosika 8, 71-424 Szczecin
Status: Educational establishment with r&d capability
Rector: Professor Dr Idzi Drzycimski

FACULTY OF AGRICULTURE* 72

Dean: Dr Mieczysław Pawlus

Agricultural Mechanization Institute* 73

Director: Jan Wojdak

Chemistry and Crop Preservation Institute* 74

Director: Professor Dr Jerzy Piasecki

Ecology and Environmental Protection Institute* 75

Director: Professor Dr Saturnin Borowiec

Plant Breeding and Seed Production Institute* 76

Director: Dr Mirosław Lapiński

Plant Cultivation and Soil Husbandry Institute* 77

Director: Dr Mieczysław Pawlus

Soil Science and Water Control and Exploitation Institute* 78

Director: Professor Dr Zygmunt Chudecki

FACULTY OF ANIMAL HUSBANDRY* 79

Address: Ulica Dr Judyma 12, Szczecin
Dean: Dr Stanisław Baranow-Baranowski

Animal Breeding and Technology of Animal Production Institute* 80

Director: Dr Arkadiusz Kawecki

Biological Foundations of Animal Husbandry Institute* 81

Director: Professor Dr Marian Kubasiewicz

FACULTY OF SEA FISHERIES AND SEA FOOD TECHNOLOGY* 82

Address: Ulica K. Krolewicza, Szczecin
Dean: Professor Dr Aleksander Winnicki

Exploitation and Protection of Biological Resources of the Sea Institute* 83

Director: Dr Rajmund Trzebiatowski

Fisheries Oceanography and Protection of the Sea Institute* 84

Director: Professor Dr Idzi Drzycimski

Ichthyology Institute* 85

Director: Professor Dr Aleksander Winnicki

Sea Food Technology Institute* 86

Director: Dr Edward Kołakowski

Akademia Rolniczo-Techniczna* 87

[Agricultural and Technical Academy]
Address: Blok 21, 10-957 Olsztyn-Kortowo
Telephone: 027-23310; 28330
Telex: 0526419
Status: Educational establishment with r&d capability
Rector: Professor Dr Andrzej Hopfer
Research director: Professor Dr Stanisław Wajda
Senior staff: Professor Dr S. Grzesiuk; Professor Dr Z. Tomaszewski, Professor Dr W. Niewiadomski, Professor Dr M. Koter, Professor Dr T. Mazur, Professor Dr H. Panak, Professor Dr C. Lewicki, Professor Dr T. Krzymowski, Professor Dr S. Wajda,

Professor Dr P. Znaniecki, Professor Dr S. Poznanski, Professor Dr J. Kisza, Professor Dr H. Kozłowska, Professor Dr Maria Brylińska, Professor Dr T. Januszkiewicz, Professor Dr T. Wojno, Professor Dr R. Fitko, Professor Dr S. Tarczyńsk, Professor Dr H. Janowski, Professor Dr E. Kossowski, Professor Dr J. Kwiatkowski, Professor Dr Z. Pancewicz, Professor Dr W. Baran, Professor Dr A. Hopfer, Professor Dr Cz. Szafranek

Graduate research staff: 800

Activities: Physiology, breeding and technology of plant cultivation and economics of agriculture; new methods for soil fertilization; technology of feed production and multiherd animal feeding, industrial production farms; physiological and biochemical bases of animal production, mainly reproduction and feeding; methods of selection and technology of rearing and breeding stock in industrial management, estimation of animal products; application of cryogenic liquids in storage and transport of chosen animal materials; methods of gaining vegetable and animal proteins, biosynthesis of enzymes and vitamins, byproducts refinery in agricultural-food industry, feed refining, new technologies in food industry; organizational problems in agricultural-food industry; inland fishery and water pollution; prophylaxis and control of livestock diseases; exploitation of machines and agricultural devices, construction and exploitation of machines and devices in the food industry; agricultural and agricultural-food building engineering; aviation in agriculture and in environmental pollution; satelite geodesy and photogrammetry; agricultural management, country planning problems connected with staff education for the needs of agriculture, social and organizational problems concerning agricultural and production complexes and the role of education in introducing technical progress.

Publications: Zeszyty Naukowe ART Olsztyn.

Akademia Techniczno-Rolnicza im Jana i Jedrzeja Sniadeckich w Bydgoszczy* 88

[Technical and Agricultural College, Bydgoszcz]
Address: Jana Olszewskiego 20, 85-225 Bydgoszcz
Telephone: 052-31450
Telex: 0562292
Status: Educational establishment with r&d capability
Rector: Associate Professor Dr Jerzy Roszak
Activities: Apart from the activities connected with educating students in fields corresponding to the Academy faculty structure, basic research is carried out as well as work related to the solution of regional industrial works problems. The essential research orientations of the Academy include: technology and new

building materials examination; technology of organic semi-products; metal-cutting tools technology; gadget optimization, development of telecommunication networks, and measurement apparatus construction; formation of solid body surface properties; technology of cultivation and its ecological foundations; and technology of production and feedstuff maintenance.

FACULTY OF AGRICULTURE* 89
Dean: Professor Dr Jerzy Sypniewski
Senior staff: Professor Dr A. Błażejewska, Associate Professor Dr W. Cieśla, Associate Professor Dr S. Grabarczyk, Professor Dr J. Rogozińska, Professor Dr S. Sadowski, Professor Dr O. Stefaniak, Professor Dr B. Wawrzyniak

FACULTY OF ZOOTECHNOLOGY* 90
Dean: Associate Professor Dr Witold Podkówka
Senior staff: Associate Professor Dr H. Bieguszewski, Professor Dr F. Błażejewski, Professor Dr H. Chmielnik, Associate Professor Dr J. Kluczek, Professor Dr B. Rak, Professor Dr S. Seniczak, Associate Professor Dr J. Załuska, Associate Professor Dr K. Załuska

Centrum Genetyki* 91
[Genetics Research Centre]
Address: Skierniewice, Osada Pałacowa
Status: Official research centre
Director: Professor E. Malinowski

Instytut Badawczy Leśnictwa 92
[Forest Research Institute]
Address: Ulica Wery Kostrzewy 3, skr poczt 61, 00-973 Warszawa
Telephone: 22 32 01
Telex: 812476 ibl pl
Status: Official research centre
Director: Professor Dr Zygmunt Patalas
Graduate research staff: 180
Activities: Research into forest exploitation and conservation, including improvement of technological processes of soil cultivation, and tending of plantations and stands, the production of planting stock, of nursery managing, and the principles of forest amelioration; rendering productive devastated sites, and post-industrial wastelands; recognition of the intraspecific variability of trees, and of the silvicultural values of

various native and introduced species; increasing the production of genetically-improved seed stock; determination of the role of limiting agents, especially of industrial pollution, and of the possibility of forest production in a polluted environment; working out methods of stand reconstruction, especially of coniferous stands, and selecting tree and shrub species for this reconstruction; examining the role of forests in development planning of regions and macroregions; influence of forests on water circulation; biological methods for forest protection; prevention and control of forest diseases and forest fires; improving methods of game damage prevention; optimizing resin-tapping methods; mechanization of main technological processes in nursery operations, silviculture, forest protection, forest utilization and wood transport.

Publications: Prace Instytutu Badawczego Leśnictwa (Proceedings of the Forest Research Institute); *Biuletyn Instytutu Badawczego Leśnictwa* (Bulletin of the Forest Research Institute); *Przegląd Dokumentacyjny Leśnictwa* (Documentation Review of Forestry Literature); *Nowości Piśmiennictwa Leśnego* (News of Forestry Literature).

ADMINISTRATION AND ECONOMY*   93

Telephone: 23 41 08
Deputy Director: Stanisław Wąsikowski

EXPERIMENTAL FORESTS*   94

Address: Instytut   Badawczy   Leśnictwa, Lasy Doświadczalne, Ulica Bohaterów Porytowego Wzgórza, 23-300 Janów, Lubelski
Telex: 062477 ibl pl
Director: Janusz Basiak
Sections:
Experimentation and production, Jerzy Piotrowski
Technology, Leszek Llimek
Economy, Marian Kuta
Activities: Carrying out the research experiments designed by the scientific sections of the Institute; testing of new technologies and new methods of forest management; verification of experimental results; staff training; forest production.

INTRODUCTION INTO PRACTICE*   95

Telephone: 46 20 19
Deputy Director for Technology: Dr Andrzej Gembarzewski
Sections:
Vocational improvement, Seweryn Marck
Experimental production stations, Bronisław Szukiel
Independent laboratory for research effectiveness, Stanisław Zajcac
Economic activity of experimental forests, vacant

SCIENTIFIC RESEARCH I*   96

Deputy-Director for Science and Research: Dr Wiesław Strzelecki

Ecology and Environmental Protection   97
Section, Warsaw*

Telephone: 022 23 45 65
Head: Dr Janusz Wolak
Activities: Research into ecology of economically-important components of forest associations; structure, development dynamics, self-regulation mechanisms, productivity and spatial variation of forest ecosystems; influence of the forest on the climatic regime and the role of the climate in forest development; ecological consequences of various technical and management measures in forest environments.

Forest Management in Industrial   98
Regions Section, Katowice*

Address: Instytut   Badawczy   Leśnictwa,   Zakład Gospodarki Leśnej Rejonów Przemysłowych, Ulica Huberta 35, 40-542 Katowice
Telephone: 032-51 25 47
Head: Dr Janusz Olszowski
Laboratories:
Forest protection in industrial regions, Józef Chłodny
Polluted soils in industrial regions, Świetłana Widera
Stand transformation in industrial regions, Eryk Latocha
Forest recultivation in industrial regions, Zbigniew Hawryś
Activities: Research into the influence of industrial air pollution on forest ecosystems; structure and functioning of ecosystems formed under conditions of pollution stress; evaluation of environment degradation under the influence of industrial emissions; determination and classification of plant communities formed under the influence of industrial emissions; forest recultivation on various post-industrial waste lands; transformtion of coniferous stands damaged by industrial emissions and selection of tree species for various zones of tree damage; methods for protecting forest stands and tree plantations in industrial regions; the influence of environmental pollution on the population dynamics and structure of various insect groups; increasing the resistance of trees to air pollution by agro-amelioration and fertilization.

Forest Management in Mountain Regions Section, Kraków* 99

Address: Instytut Badawczy Leśnictwa, Zakład Gospodarki Leśnej Regionów Górskich, Ulica Prądnicka 80, 31-202 Kraków
Telephone: 094-398 18
Head: Professor Dr Zenon Capecki
Laboratories:
Silviculture in mountain forests, Eugeniusz Jewuła
Mountain forest protection, Professor Dr Zenon Capecki
Activities: Site science and typology of mountain forests; methods of afforestation of bare areas threatened by water erosion, of waste lands and other mountain areas adverse to afforestation; natural and artificial forest regeneration, forms of felling and duration of regeneration periods suitable for mountain conditions; tending of plantations, thickets and stands; rendering them resistant to damage caused by atmospherical and biotic agents; the regression process of silver fir, its natural and man-induced causes and consequences; silviculture procedure in mountain forests utilized as recreation and tourism resorts.

Forest Protection Section, Warsaw* 100

Telephone: 022-22 49 42
Head: Professor Dr Jerzy Burzyński
Laboratories:
Diagnoses and prognoses, Zbigniew Sierpiński
Pesticides, Jerzy Kamiński
Entomology and biological methods, Andrzej Kolk
Forest pathology, Stefan Łukomski
Activities: Research into population dynamics of forest pests; regionalization of forests from the viewpoint of their sanitary condition; the participation of insects and pathogens in the appearance and spread of diseases of forests; biology, ecology and economic importance of harmful and useful forest insects and fungi; progress of disease, studies across a broad range of the biological pectrum, at the level of organism, population and coenobiosis; elaboration and application of new pesticides.

Game Management Section, Warsaw* 101

Telephone: 022-22 49 30
Head: Professor Dr Ryszard Dzięciołowski
Activities: Methods of game rearing and protection in open hunting grounds and in enclosures; census of game populations and their growth rate in relation to geographical and climatic conditions; ecology of some game animals; study and prevention of forest damage due to game animals.

Independent Laboratory for Electronic Data Processing, Warsaw* 102

Telephone: 022-22 49 41
Head: Krzysztof Przyborski
Activities: Research and design of mathematical and statistical methods and models for determined research work carried out at the sections of the Institute; programming; consultation; laboratory activities and seminars.

Nature Protection Section, Białowieża* 103

Address: Instytut Badawczy Leśnictwa, Zakład Ochrony Przyrody, 17-230 Białowieża, Osada Parkowa
Head: Professor Dr Aleksander Sokołowski
Activities: Research into the ecological foundations of the protection of forest animals; studies on the peculiarity of formation of the species composition, structure, organization and performance of natural forest ecosystems under various physiographical conditions; geobotanical studies in natural forest associations.

Scientific, Technical and Economic Information in Forestry Section, Warsaw* 104

Telephone: 022-23 41 28
Head: Leopold Rossakiewicz
Sections: Library, Ewa Bielecka-Fördös
Documentation and publications, Danuta Śledzińska
Activities: Documentation and publication; cooperation with foreign countries within the international information system AGRO INFORM; research in the field of scientific, technical and economic information.

Silviculture Section, Warsaw* 105

Telephone: 022-22 49 43
Head: Dr Ryszard Sobczak
Laboratories:
Stand tending, Jan Zajqczkowski
Stand establishing, Mieczysław Tuszyński
Forest nurseries, Andrzej Gorzelak
Activities: Research into production of planting stock in forest nurseries; soil cultivation for regeneration and afforestation; establishment of forest plantations; natural regeneration of the forest; choice of most suitable fast-growing forest tree species of foreign origin.

Water Economy Section, Warsaw* 106

Telephone: 022-23 49 38
Head: Dr Feliks Białkiewicz
Laboratories:
Water amelioration in forests, Tadeusz Krajewski
Sewage utilization in forests, Konrad Tomaszewski
Activities: Research into water and nutrient requirements of trees and stands; principles of water amelioration in forests and tree plantations and in waste lands with water excess or deficiency; irrigation and fertilization of nurseries; regulation of water conditions in fish ponds and green crop grounds, situated in forests; possibilities of treatment and utilization of sewage, mainly of organic origin, in the forest.

Forest Hydrology Station, Puczniew* **107**
Address: Instytut Badawczy Leśnictwa, Zakład Gospodarki Wodnej, Stacja Hydrologiczno-Leśna, 95-063 Puczniew
Head: Maria Kieruzal

SCIENTIFIC RESEARCH II* 108

First Deputy Director: Professor Dr Józef Stajniak

Economics and Organization of Forest 109 Enterprises Section, Warsaw*

Telephone: 022-22 49 37
Head: Dr Tadeusz Partyka
Activities: Research into the economic principles of the development of forestry, in particular of industries of primary wood processing; assessment of the value of forests and forest grounds, of forest damage, and damage compensation; organization of forest production; principles of planning in forestry; systems of management of forest economy; application of information systems in forestry; geography and history of forestry.

Forest Fire Control Section, Sękocin* 110

Address: Instytut Badawczy Leśnictwa, Sękocin, 05-550 Raszyn
Head: Dr Tytus Karlikowski
Laboratory:
Fire prevention, Zygmunt Santorski
Activities: Research into the prevention of forest fires; influence of fires on the forest environment; increasing the fire resistance of stands; evaluation of forest fire danger; methods of forest fire detection; pirology of forest fires; application of new extinguishing means; new equipment for forest fire control.

Experimental Station, Krzystkowice* **111**
Address: Instytut Badawczy Leśnictwa, Zakład Ochrony Przeciwpożarowej Lasu, Stacja Doświadczalna w Krzystkowicach, Ulica Dzierżyńskiego 11, 68-310 Krzystkowice

Experimental Station, Niedzwiady* **112**
Address: Instytut Badawczy Leśnictwa, Zakład Ochrony Przeciwpożarowej Lasu, Stacja Doświadczalna w Niedźwiadach, 77-220 Koczała
Telephone: 48

Forestry Management and Prognoses 113 Section, Warsaw*

Head: Professor Dr Tadeusz Trampler
Laboratories:
Stand productivity studies, Tadeusz Pirogowicz
Prognoses, Bogdan Łonkiewicz
Activities: Research into methods of forest management; prognoses of the development of forest resources and their utilization; mechanization and automation of work in forest management; classification of forest sites.

Laboratory of Forest Site Classification* **114**
Address: Instytut Badawczy Leśnictwa, Zakład Urządzania Lasu i Prognoz, Pracownia Klasyfikacji Siedlisk Leśnych, Ulica Podhalańska 22, 80-322 Gdańsk, Oliwa
Telephone: 52-39-51
Head: Kazimierz Mąkosa

Labour Protection Section, Sękocin* 115

Address: Instytut Badawczy Leśnictwa, Sękocin, 05-550 Raszyn
Head: vacant
Laboratories:
Labour psychology and accident studies, Maria Kanecka
Technical labour protection, Piotr Jasnos
Activities: Analysis of technological processes in forestry, with the aim of increasing safety at work and reducing its arduousness; physiology of physical effort; prevention methods of occupational and para-occupational diseases; ergonomic evaluation of machines, appliances and devices; evaluation of personal protection appliances; measurement of physically-destructive factors, especially noise and vibration, and development of methods of their elimination.

Mechanization in Forest Utilization 116 Section, Warsaw and Sękocin*

Telephone: 022-23 40 40
Head: Professor Dr Józef Stajniak
Laboratories:
Logging, Jacek Komorowski
Wood transport and timber yards, Wiesław Dziubak
Machine design, Wiktor Lisowicz

Activities: Research into the technology of logging, skidding and transport of wood, as well as work technology and organization in timber yards; processes and theoretical foundations of logging mechanization and timber yard organization; design and construction of machines and appliances for wood working plants.

Mechanization of Silvicultural Operations Section, Sękocin* 117

Address: Instytut Badawczy Leśnictwa, Sękocin, 05-550 Raszyn
Head: Józef Rybczyński
Activities: Improvement of technology and construction of machines and appliances used for silvicultural operations.

Minor Forest Products Section, Warsaw* 118

Telephone: 022-23 40 68
Head: Dr Ryszard Ostalski
Laboratories:
Resin tapping, vacant
Minor forest products analysis, Mirosława Butenko
Activities: Research into stimulants of resin biosynthesis, and their influence on individual trees and on stands; organization and technology of resin-tapping operations, improvement of tools and appliances for resin tapping; chemical composition and quality of gum resin; utilization of bark and foliage; studies on resources of forest ground vegetation and shrubs; evaluation of yield of edible forest fruits and fungi; methods of enlargement and enrichment of supply bases of minor forest products.

Pedology and Fertilization Section, Sękocin* 119

Address: Instytut Badawczy Leśnictwa, Sękocin, 05-550 Raszyn
Telephone: 50 02 14
Activities: Research into the properties, classification, cartography andgeography of forest soils and inventory methods; soil fertility, microbiological indicators and methods of evaluation of the biological activity of forest soils; application of mineral and organic fertilization as means of increasing soil productivity; influence of mineral fertilization and other agro-ameliorative measures on the microbiological processes in soils; influence of industrialization and urbanization on the forest environment; physical, chemical and biological soil analyses.

Raw Wood Section, Warsaw* 120

Head: vacant
Laboratory
Wood measurements, Paweł Cichowski
Activities: Research into the technical usefulness of wood, and individual bases of wood supply; economic importance of defects in wood; elaboration of assortment systems; transportability of wood assortments.

Seed Science and Selection Section, Sękocin* 121

Address: Instytut Badawczy Leśnictwa, Sękocin, 05-550 Raszyn
Head: Stefan Kocięcki
Laboratories:
Forest tree selection, Włodzimierz Dutkiewicz
Forest seed science, Zdzisław Antosiewicz
Physiology of woody plants, Lucjan Janson
Tree plantations, with experimental stations in Prędocin and Chełm Lubelski, Witold Chmielewski
Laboratory of shelter belts, with the experimental station in Sójki, Kazimierz Zajączkowski
Activities: Research into the breeding value of tree populations for planting forests, plantations and shelter belts; studies on the photosynthesis and nutrient requirements of forest tree species; methods of vegetative reproduction of trees; investigation of role of shelter belts in the landscape and for agroclimatic conditions; selection of seed-stands and high-grade trees.

Instytut Biochemii i Biofizyki PAN* 122

[Biochemistry and Biophysics Institute]
Address: Ulica Rakowiecka 36, 02-532 Warszawa
Telephone: 022-49 04 03
Parent body: Polska Akademia Nauk - PAN
Status: Official research centre
Research director: Professor Dr Kazimierz L. Wierzchowski
Departments:
Biophysics, K.L. Wierzchowski
Molecular biology, D. Shugar
Protein biosynthesis, P. Szafrański
Genetics, W. Gajewski
Comparative biochemistry, J. Szarkowski
Plant biochemistry, K. Kleczkowski
Microbial biochemistry, T. Kłopotowski
Lipid biochemistry, T. Chojnacki
Graduate research staff: 150
Activities: The main lines of research concern the structure, function and metabolism of nucleic acids; the biosynthesis of proteins, regulation of gene function in prokaryotes and eukaryotes (yeast, fungi).

Instytut Botaniki PAN* 123

[Botanical Institute - PAN]
Address: Ulica Lubicz 46, 31-512 Kraków
Telephone: 094 161 44
Parent body: Polska Akademia Nauk - PAN
Status: Official research centre
Director: Professor Adam Jasiewicz

Instytut Budownictwa, 124
Mechanizacji i
Elektryfikacji Rolnictwa

[Institute for Buildings, Mechanization and
Electrification in Agriculture]
Address: Ulica Rakowiecka 32, 02-532 Warszawa
Telephone: 022-49 32 31
Telex: 814886 IMER-PL
Affiliation: Ministry of Agriculture and Food Control
Status: Official research centre
Director: Professor Jerzy Tymiński
Deputy directors: Professor Augustyn Fąfara,
Professor Janusz Krzemiński, Professor Zdzisław
Witebski, Professor Zdzisław Wójcicki
Departments: Mechanization and organization of farm
buildings, Dr T. Antolak; agricultural progress exten-
sion, Professor J. Biłowowicki; management and
economics of agricultural engineering, Dr T. Borek; land
reclamation, water and sewage management, Dr S.
Kanafa; mechanization of root crop cultivation and
harvesting, Professor T. Karwowski; material enegineer-
ing, Professor H. Nieborowski; cybernetics, Professor T.
Kostecki; machine operation, Dr T. Olszewski; mechanization of horticulture, Professor J. Pabis; development
of quality and reliability, Professor A. Pawlik; mechanization of livestock raising systems, Dr W. Romaniuk;
farm building design foundation, Professor S. Tomczyk;
agricultural electrification, Dr A. Wejher
Graduate research staff: 220
Activities: Research into farm buildings, mechanization and electrification in agriculture - technical and
technological developments for buildings and
agricultural mechanization; operation of power machinery for cultivation, transport and buildings; assessment
of technical advances and potential of agricultural
machinery, equipment, materials, building units and
constructions; recording, processing and distribution of
scientific, technical and economic information;
electrification in agriculture.
The institute has eleven machine assessment stations:
Bonin, Grodków, Kąty-Wrocławskie, Kraków, Kry-
nica-Bradowiec, Lublin, Płock, Sanok, Stalówka Sroda
Wielkopolska, Tczew. The institute also has nine
demonstration stations for farm buildings: Białystok,
Bydgoszcz, Kielce, Kraków, Łódź, Olsztyn, Szczecin,
Wrocław.

Publications: Annual report, information bulletin, docu-
ment review.
Projects: Energetics in agriculture (Professor J.
Tyminski); crop production mechanization (Professor
Janusz Krzemiński); animal production mechanization
(Professor Augustyn Fąfara); tractor and farm machine
operation (Professor Zdzisław Wójcicki); farm buildings
(Professor Zdzisław Witebski)..

KŁUDZIENKO BRANCH 125

Manager: Dr A. Tabiszewski
Departments: Mechanization of tillage and fertilizing,
Dr E. Kamiński; measurement techniques, Dr T.
Marszałek; motorization and transport, Dr J. Mazur;
mechanization of grain cultivation and harvesting, green
and technical crops, Dr J. Szyszło

POZNAN-STRZESZYN BRANCH 126

Address: Ulica Biskupińska 67, 60-463 Poznań
Manager: Dr S. Borowski
Vice manager: Assistant Professor S. Przygórzewski
Departments: Farm building mechanization, W. Eicke;
feedlot mechanization for cattle, Z. Kowalewski; farm
devices in agriculture, Z. Pankowski; feedlot mechanization for sheep and fur-bearing animals, Dr Z. Pater;
feedlot mechanization for poultry, Dr T. Waligóra

Instytut Dendrologii PAN* 127

[Dendrology Institute - PAN]
Address: Ulica Parkowa 5, 62-035 Kórnik
Telephone: 164; 166
Telex: 041-2948 txgmpl
Parent body: Polska Akademia Nauk - PAN
Status: Official research centre
Research director: Professor Dr Władysław Bugała
Deputy director: Dr R. Siwecki
Departments:
Genetics, Maciej Giertych
Physiology, Zofia Szczotka
Introduction and acclimatization, Władysław Bugała
Systematics and geography, Kazimierz Browicz
Resistance to biotic and abiotic factors, Ryszard Siwecki
Seed biology, Bolesław Suszka
Graduate research staff: 42
Activities: Basic research in various fields of tree biol-
ogy, in particular genetics of conifers and poplars, in-
cluding racial variation, isozyme studies and breeding;
physiology of mycorrhizal symbiosis, hormonal regula-
tion, flowering senescence; selection, collection and nursery techniques for woody ornamentals including gene
banks; taxonomy and chorology of Polish eastern Mediterranean and south west Asian woody plants; host-

pathogen interactions in tree diseases, resistance to low temperatures, to atmospheric and soil pollution; seed storage and stratification techniques.
Publications: Arboretum Kórnickie (Kórnik Arboretum) annual; *Atlas of Distribution of Trees and Shrubs in Poland*; *Nasze Drzewa Leśne* (Our Forest Trees) monographs by species; *Index seminum*, annual.

Instytut Ekologii PAN* 128

[Ecological Institute - PAN]
Address: Dziekanów Leśny, 05-150 Łomianki
Telephone: 34 96 66
Parent body: Polska Akademia Nauk - PAN
Status: Official research centre
Director: Professor Romuald Klekowski

Instytut Ekonomiki Rolnej 129

[Agricultural Economics Institute]
Address: Koszykowa 6, Warszawa, 00564
Telephone: 022-216 271
Affiliation: Ministry of Agriculture and Food Control
Status: Official research centre
Director: Professor Augustyn Woś
Departments: General agricultural economics, Professor Z. Grochowski; economics and organization of socialist agricultural enterprises, Dr L. Wiśniewski; agricultural cooperation and production services, Professor J. Czyszkowska-Dąbrowska; social and economic structures in agriculture and countryside, Professor A. Szemberg; planning and allocation of agricultural production, Assistant professor P. Dabrowski; international comparisons, Assistant professor Z. Smoleński; agricultural bookkeeping, Assistant professor J. St. Zegar; organization and management in agriculture, Assistant professor J. Gajewski
Graduate research staff: 129
Activities: Development of agricultural economics theory and research methods, scientific foundations for planning social and economic development of agriculture and rural communities, socialist production relations in agriculture, development of agricultural policy, spatial planning and allocation of production; recommendations for agricultural policy concerning organization, specialization, concentration; economic analysis of agricultural production, management of farm enterprises; coordination of work on links between agriculture and the national economy, mathematical methods in system analysis of the food economy.
Publications: Annual reports.

Projects: Social and economic reconstruction of agriculture and countryside in Poland, agricultural and food economic system, (Professor A. Woś); research of the production and economic situation in agriculture (Professor Z. Grochowski); agricultural bookkeeping of private farms (Assistant professor J. St. Zegar); elaboration of a management model and informatics system for agriculture, (Assistant professor J. Gajewski).

Instytut Fizjologii i Żywienia Zwierząt PAN 130

[Animal Physiology and Nutrition Institute - PAN]
Address: Ulica Instytucka 3, 05-110 Jabłonna n Warszawy
Telephone: 74 34 22
Telex: 815696 pl
Parent body: Polska Akademia Nauk - PAN
Status: Official research centre
Director: Professor S. Buraczewski
Departments: Endocrinology and neurophysiology, Dr B. Barcikowski; physiology of nutrition, Professor S. Buraczewski; nutrition and physiology of ruminants, Dr T. Zebrowska; pig nutrition, Dr M. Kotarbińska-Urbaniec; environmental physiology and meat science, Professor P. Poczopko; reproduction physiology, Professor R. Stupnicki; analytical laboratory, Dr A. Rymarz
Activities: Nutrition of growing pigs, cattle, sheep and poultry; efficiency of feed protein utilization at different energy to protein ratios and essential amino acid levels; protein digestion, amino acid absorption and utilization in pigs and ruminants; nonprotein nitrogen utilization in ruminants; isolation and characterization of rumen bacteria; digestion in the rumen; evaluation of foodstuffs quality, biological value of proteins and amino acid availability; influence of cold on energy and protein metabolism in new-born animals; biochemical processes and histology of pigs under stress and with deteriorated meat quality; hormonal and biochemical indices connected with tissue growth; modes of action of hypothalamic neuropeptides; hormonal receptors in the sheep ovary; pituitary and ovarian functions in heifers and post-partum cows; development of hormone radioimmunoassay techniques.
Projects: Digestion, absorption and metabolism of nutrients in the digestive tract of animals (Dr T. Zebrowska); protein and amino acid metabolism, amino acid requirements and nutritive value of feedstuffs (Professor S. Buraczewski); utilization and metabolism of nutrients in monogastric animals (Dr M. Kotarbińska-Urbaniec); utilization and metabolism of energy and nitrogen in ruminants (Dr A. Ziołecka); environmental factors influencing meat quality (Professor P. Poczopko); biosynthesis, release and mode of action of

releasing hormones and other hypophysical neuropeptides (Dr K. Kochman); endocrine control of ovarian
function in cattle (Professor R. Stupnicki).

Instytut Genetyki i   131
Hodowli Zwierząt PAN*

[Genetics and Animal Breeding Institute]
Address: Jastrzebiec, 05-551 Mroków
Telephone: 50 05 11
Telex: 814749
Parent body: Polska Akademia Nauk - PAN
Status: Official research centre
Director: Professor M. Zurkowski

Instytut Hodowli i   132
Aklimatyzacji Roślin

[Plant Breeding and Acclimatization Institute]
Address: Radzików, 05-870 Błonie, Warszawskie
Telephone: 55 26 11; 55 26 14
Telex: 812919 ihar pl
Affiliation: Ministry of Agriculture and Food Control
Status: Official research centre
Research director: Professor Dr Stanisław Starzycki
Departments: Genetics, Associate professor Zygmunt
Staszewski; biophysics and plant physiology, Dr Jan
Ciepły; plant immunology, Associate professor
Stanisław Góral; national resources of genes, Associate
professor Julian Jakubiec; seed control and production
methology, Dr Grzegorz Rytko; biological evaluation of
plant products, Professor Maria Rakowska; radiology,
Professor Stanisław Starzycki; maize, Associate
professor Jan Bojanowski; seed biology and storage,
Associate professor Zbigniew Urbaniak; research orga-
nization and extension, Krystyna Zalewska
Independant laboratories: Biochemistry, Professor
Konstancja Raczyńska-Bojanowska; soybean, Professor
Jerzy Szyrmer; electronic computation technique, Jerzy
Serwiński; plant breeding and seed production
economics, Dr Ryszard Cholewa
*Plant breeding and acclimatization experiment sta-
tions:* Bartążek; Bąków; Borowo; Grodkowice;
Konczewice; Małyszyn; Oźánsk; Olésnica Mała;
Przebedowo; Radzików; Smolice; Strzelce
Graduate research staff: 400
Activities: Theoretical bases of breeding and seed pro-
duction of field crops; broad research in the fields of
plant genetics, physiology, biochemistry, phytoim-
munology and seed testing; breeding of new varieties of
main field crops; food science.

*Publications: Hodowla Roślin Aklimatyzacja i Nasien-
nictwo; Biuletyn IHAR; Index Seminum, Delectus
Seminum.*
Projects: Breeding high-yielding cultivars of cereals
and fodder crops of high useful value and working out
their agrotechnics (Professor Stanisław Starzycki);
breeding cultivars of increased useful value and improvement of production methods of sugar, fodder beets
and other root cops (Professor Zdzisław Szota); maize
breeding and improvement of cultivation and utilization
methods (Associate professor Jan Bojanowski); improvement of methods of evaluation and storage of crop
seeds (Dr Grzegorz Rytko); collection and elaboration
of initial materials for plant breeding (Associate
professor Julian Jakubiec); breeding of oil, sunflower,
poppy and linseed and improvement of their cultivation
(Associate professor Andrzej Horodyski); protein production and consumption, improvement of cultivars of
high-protein crops for feeding and consumption purposes (Associate professor Julian Jakubiec).

BYDGOSZCZ BRANCH DIVISION   133

Director: Professor Zdzisław Szota
Departments: Genetics and root crop breeding,
Professor Alfons Kuździowicz; root crops production
technology, Associate professor Izydor Gutmański; con-
trol of root crops pests and diseases, Professor Edward
Berbeć; botanical garden, Dr Bolesław Osiński

KRAKÓW BRANCH DIVISION   134

Departments: Cereals, Professor Stanisław Węgrzyn;
fodder crops, Associate professor Regina Lutyńska

POZNAN BRANCH DIVISION   135

Director: Professor Jan Krzymański
Departments: Oilcrops, Associate professor Andrzej
Horodyski; grasses, Dr Jadwiga Stuczyńska; ecology,
Professor Henryk Szukalski

Instytut Leków*   136

[Drug Research and Control Institute]
Address: Ulica Chełmska 30, 00-725 Warszawa
Telephone: 022-41 29 40
Affiliation: Ministry of Health and Public Welfare
Status: Official research centre
Director: Professor Dr Bogusław Borkowski
Deputy Director: Professor Dr Tadeusz Lesław
Chrusciel
Research Secretary: Assistant Professor Dr Julia
Tyfczyńska

Activities: Coordination of scientific research on biologically-active compounds at national level; problems connected with the quality of drugs - chemical, biological, pharmacological and microbiological analysis; coordination of clinical evaluation of drugs produced in Poland and those which are imported; methodology of studies in the field of clinical pharmacology, rationalization of pharmacotherapy and investigations on the consumption of drugs; preparations of materials for drug registration. Cooperation with the World Health Organization in the international monitoring system of the side effects of drugs.
Publications: In 25 years the Institute has produced 1600 publications in scientific journals, 24 books and 20 popular publications.

DEPARTMENT OF GALENIC DRUGS* 137

Head: Dr Roman Danielak
Laboratories: Drug forms; pharmacognostics; phytochemistry
Activities: Research into methods of drug testing (with particular reference to drugs of plant origin) and auxiliary materials used in pharmacy; chemical investigation of disinfectants and determination of the level of pesticides in herbal raw materials and drugs, in cooperation with the Committee of Polish Pharmacopoea.

Instytut Melioracji i Użytków Zielonych 138

– IMUZ
[Land Reclamation and Grassland Farming Institute]
Address: Falenty, 05-090 Raszyn
Telephone: 28 37 63
Telex: 815494
Affiliation: Ministry of Agriculture
Status: Official research centre
Research director: Professor Czesław Somorowski
Sections: Natural fundamentals of land reclamation, Professor H. Okruszko; land reclamation and operation of structures, Dr G. Nazaruk; grassland farming, Professor S. Grzvb; hydrotechnical and reclamation engineering, Dr W. Mioduszewski; rural water supply and canalization systems, S, Zyliński; economics, J. Prokopowicz
Graduate research staff: 240
Activities: The principal aim of the Institute is to satisfy the needs of agriculture in scientific research and technical progress in the fields of: natural fundamentals of land reclamation; grassland farming; peat science and water supply, and canalization of rural settlements; economics of land reclamation investments projects.

Publications: Wiadomości IMUZ (News of the Institute for Land Reclamation and Grassland Farming); *Biblioteczka Wiadomości IMUZ* (Library News of the Institute for Land Reclamation and Grassland Farming).
Projects: Theoretical fundamentals of water and soil resources management in chosen valleys and river basins with regard to natural environment advantages (Henryk Okruszko); regulation of water relationships of arable lands by means of drainage (Zdzisław Stąpel); techniques and technology of sprinkler irrigation, organization of plants production under irrigation (Stanisław Drupka); land reclamation techniques, exploitation of installations and land reclamation systems (Grzegorz Nazaruk); agricultural utilization of municipal and rural waste waters (Jan Kutera); new constructions and principles of calculation of hydraulic structures (Waldemar Mioduszewski); extension of a pilot water management system in the Upper Notéc river catchment area (Edward Wichłacz); intensification of green forage production from grasslands (Stanisław Grzyb); technologies of grassland fertilization with regard to agrochemical and ecological aspects (Leon Doboszyński); effect of land reclamation on the evolution of hydrogenic sites (Henryk Okruszko).

Instytut Nawozow Sztucznych 139

[Chemical Fertilizers Institute]
Address: 24-110 Puławy
Telephone: 64 44
Telex: INS 0642609
Status: Official research centre
Director: Bolesław Skowronski
Departments: Gas synthesis T. Wąsała; nitrogen compounds, K. Kozłowski; phosphoric and mixed fertilizers, Dr J. Wojcieszek; catalysts, Dr A. Gołębiowski
Graduate research staff: 30
Activities: Analytical and physico-chemical research on gas synthesis, nitrogen compounds, phosphoric and mixed fertilizers, and catalysts.
Projects: Fertilizer production improvement (Dr J. Sas); fertilizer granulation (F. Czornik); bulk transport of fertilizers (Dr T. Gucki); liquid fertilizers (Dr M. Dankiewicz).

POLAND

Instytut Parazytologii PAN* 140

[Parasitology Institut - PAN]
Address: Pasteura 3, skr poczt 153, 00-973 Warszawa
Telephone: 22 25 62
Parent body: Polska Akademia Nauk - PAN
Status: Official research centre
Director: Professor Dr Marian Swietlikowski
Research director: Alicja Guttowa
Sections: Morphology and systematics of parasites, Dr
K. Niewiadoska; evolution of parasites, Professor Dr J.
Dróżdż; immunology, Professor Dr M. Swietlikowski;
epizootiology and pathology of parasitic infections,
Professor Dr A. Malczewski; environmental parasitol-
ogy, Professor Dr K. Kisielewska
Graduate research staff: 48
Activities: Biological, physiological, immunological
and ecological mechanisms forming host-parasite
systems.
*Publications: Acta Parasitologica Polonica; Polska
Bibliografia Parazytologiczna.*

Instytut Podstaw Inżynierii 141
Srodowiska PAN*

[Environmental Engineering Institute - PAN]
Address: Ulica M. Curie-Skłodowskiej 34, 41-800
Zabrze
Telephone: 032-71 70 40; 71 69 50
Telex: 036401 ipis pl
Parent body: Polska Akademia Nauk - PAN
Status: Official research centre
Deputy director: Associate Professor Dr Stefan
Jarzębski
Sections: Atmospheric pollution protection, Dr B.
Raczynski, Professor Dr S. Jarzębski; water pollution
protection and water-sewage management, Associate
Professor Dr Z. Pociecha; reclamation of soil affected by
industry, Associate Professor Dr Z. Harabin, Associate
Professor Dr Z. Strzyszcz; effect of air pollution on
plants and soil, Associate Professor Dr S. Godzik,
Associate Professor Dr M. Warteresiewicz; prognostica-
tion and direction. environmental protection processes,
Dr A. Wrona
Graduate research staff: 150
Activities: Research on atmospheric pollution protection, including the basis of aerosol filtration, the mechanism of dust pollution origin, optimization of collectors,
methods of denitrification and desulphurization of exhaust gases, and measurement of pollution and models
of its dispersion. Water pollution protection and water-
sewage management, including biological and chemical
clearance of sewage, utilization of slurry, clearance and
utilization of highly polluted and mineralized waters,
and the effect of coal mine tips on water environment.

Reclamation of soil affected by industry, including
reclamation of areas affected by underground and opencast mining, by extraction and production of iron ores,
by extraction and flotation of zinc and copper ores and
by opencast sand-pit mining; studies of soil degradation
caused by industrial emissions, industrial mining activity, and oil pollution; biological reclamation of dumps
left after coal fired power plants. Effect of air pollution
on plants and soil, including the intake, localization and
translocation of SO^2 and heavy metals by plants, structural and ultrastructural changes of foliage caused by air
pollutants, effects of pollutants on growth and yield of
agricultural and horticultural plants, usefulness of bioin-
dicators and test plants in regional mapping in environmental trends, and the role of vegetation in absorbing
pollutants around large air pollution sources.
*Publications: Archives of Environmental Protection;
Prace i Studia.*

Instytut Przemysłu 142
Cukrowniczego*

[Sugar Industry Institute]
Address: Ulica Rakowiecka 36, 02-532 Warszawa
Telephone: 022-49 04 24
Telex: 813845ips pl
Status: Official research centre
Director: Dr Wiktor Fornalek
Research director: Professor Dr Edmund
Walerianczyk
Divisions: Sugar beet, Dr Jan Malec; sugar technology,
Dr Józef Marczyński; mechanics and power, Professor
Dr Henryk Dobrowski; analyses, Dr Stefan Eydel
Graduate research staff: 43
Activities: Research into all problems of interest to the
sugar industry: evaluation of sugar beet (varieties and
industrial preservation); technological problems con-
nected with processing sugar beet into sugar; steam
power, power supply, water and waste-water in sugar
factories; mechanization and automation of operations
and apparatus; analysis and microbiology of sugar;
economics and technical progress in the sugar industry.
*Publications: Prace Instytutów i Laboratoriów
Badawczych Przemysłu Spożywczego; Biuletyn Instytutu
Przemysłu Cukrowniczego in Gazeta Cukrownicza; Infor-
macja Ekspresowa.*

Instytut Przemysłu Fermentacyjnego 143

[Fermentation Industry Institute]
Address: Ulica Rakowiecka 36, Warszawa
Telephone: 022-490171
Telex: 813845 ips pl
Affiliation: Ministry of Agriculture and Food Control
Status: Research centre within an industrial company
Director: Professor Tadeusz Gołębiewski
Departments: Protein products technology, technical microbiology and biochemistry, Olga Ilnicka-Olejniczak; technology of fruit and vegetable products, Barbara Sewer- Lewandowska; beer and malt technology, Tadeusz Mozga; instrumental analysis, Stanisław J. Kubacki; scientific, technical and economic information, Maria Lipiec; research work organization, Irena Tabiszewska
Graduate research staff: 52
Activities: Development of food technology, fruit and vegetable product technology, wine, malt, beer, spirits, yeasts, enzymatic preparations, unconventional proteins and instrumental analysis.
Publications: Prace Instytutów i Laboratoriów Badawczych Przemysłu Spożywczego .
Projects: Development of enzyme preparations and their applications in food processing (Regina Sawicka, Antoni Zakrzewski); storage and improvement of microorganisms for industrial applications (Olga Ilnicka-Olejniczak); analysis and technology of flavours from plant materials (Krystyna Karwowska); studies on analysis of toxic substances in foods (Stanisław J. Kubacki, Bogdan Kędzierski); development of beer production in one device technology (Józef Chrostowski); rational ways of hop extracts production and applications (Łucja Dubiel); development of different juice form technologies (Roman Kwaśniewski); technologies of canned fruit and vegetable products, with particular regards to food products for children (Barbara Sewer-Lewandowska, Danuta Mączyńska).

Instytut Przemysłu Mięsnego i Tłuszczowego* 144

[Meat and Fat Industry Institute]
Address: Rakowiecka 36, Warszawa
Telephone: 49 02 26
Status: Official research centre
Graduate research staff: 110
Publications: Roczniki Instytutu Przemysłu Mięsnego i Tłuszczowego.

Instytut Rozwojuws i Rolnictwa PAN* 145

[Rural Development and Agriculture Institute - PAN]
Address: Ulica Nowy Swiat 72, 00-330 Warszawa
Telephone: 022 26 94 36
Parent body: Polska Akademia Nauk - PAN
Status: Official research centre
Director: Professor Dyzma Gataj

Instytut Rybactwa Sródlądowego* 146

[Inland Fisheries Institute]
Address: Bl 5, Olsztyn, Kortowo
Status: Official research centre
Director: Professor Jan Szczerbowski

HEATED EFFLUENTS LABORATORY SIEKERKI* 147

Address: Augustowka 1, 02-981 Warszawa
Head: Dr Lidia Horoszewicz

RIVER FIELD STATION* 148

Address: Bytowska 5, 80-328 Gdańsk, Oliwa
Head: Dr Ryszard Bartel

ZABIENIEC CENTRE* 149

Address: Instytut Rybactwa Sródlądowego, Zakłady w Żabincu, 05-500 Piaseczno
Departments: Fish culture, Dr A. Krüger
River fisheries, Dr R. Sych
Graduate research staff: 30
Activities: The Department of Fish Culture conducts research on common and Chinese carp, hatchery problems, feeding of fry, cultivation, fertilization and feeding in ponds, monitoring natural food in ponds, applied biology of cultivated fish. The Department of River Fisheries conducts research on sea trout fisheries, trout culture, monitoring effects of heated effluents on fish, problems of ageing fish, population dynamics, migration of fishes in the Baltic (salmonids) and in rivers.
Publications: Roczniki Nauk Rolniczych; Rybactwo (Fisheries).

Instytut Sadownictwa i Kwiaciarstwa
150

[Pomology and Floriculture Research Institute]
Address: Pomologiczna 18, 96-100 Skierniewice
Telephone: 20 21
Telex: 886659 Insad PI
Affiliation: Ministry of Agriculture and Food Control
Status: Official research centre
Research director: Professor S.A. Pieniążek
Scientific secretary: Professor Z. Soczek
Departments: Fruit breeding, variety evaluation and nurseries, Dr Alojzy Czynczyk; soil management and fertilization of fruit plants, Dr Augustyn Mika; pest and disease control, Dr Zbigniew Suski; small fruits, Dr Kazimierz Smolarz; horticultural mechanization, Dr Zdzisław Cianciara; fruit storage, Dr Edward Lange; fruit processing, Dr Wiesława Lenartowicz; physiology of horticultural plants, Dr Ryszard Rudnicki; biochemistry, Dr Urszula Dzieciol; breeding and seed production of ornamentals, Dr Kazimierz Mynett; cultivation of ornamentals, Janina Grzeszkiewicz; bee breeding, Dr Leon Bornus; beekeeping management, Dr Cyprian Zmarlicki; pollination of plants, Dr Bolesław Jabłoński; extension service, Dr Andrzej Holewiński
Graduate research staff: 340
Activities: Improvement of fruit production; breeding and improvement of cut flower and ornamental plant production; improvement of beekeeping; pollination of entomophilous plants; physiology of fruit and ornamental plants. The research is conducted in the Central Laboratory in Skierniewice (pomology and floriculture), in Puławy (beekeeping) and in 13 experimental field stations.
Publications: Fruit Science Reports, quarterly; *Prace Instytutu Sadownictwa i Kwiaciarstwa* (Proceedings of the Institute of Pomology and Floriculture); *Rosliny Ozdobne* (Ornamental Plants); *Pszczelnicze Zeszyty Naukowe* (Scientific Apiculture).

Instytut Śląski*
151

[Silesian Institute]
Status: Science association

AGRICULTURAL ECONOMY DEPARTMENT
152

Address: Luboszycka 3, PL 45-036 Opole
Telephone: 038-3 64 41; 3 64 42
Head: Dr Zenon Baranowski
Graduate research staff: 8
Activities: Land use, agricultural economics, farm management.
Publications: Studia Społeczno-ekonomiczne .
Projects: Influence of the habitat on the organization of agricultural production and farming (Professor Józef Góralczyk, Dr Józef Makowiecki); agricultural settlements in the Opole province during the process of land socialization (Dr Jan Tkocz); social - economic farming effectiveness of agricultural production cooperatives (Dr Zenon Baranowski).

Instytut Uprawy, Nawoźenia i Gleboznawstwa
153

– IUNG
[Soil Science and Cultivation of Plants Institute]
Address: Osada Palacowa, , 24-100 Puławy, Lublin
Telephone: 028-34 21
Telex: 84410
Affiliation: Ministry of Agriculture and Food Control
Status: Official research centre
Director: Professor Stanislaw Nawrocki
Deputy director: Professor Zdzislaw Gonet
Departments: Research methods and computer facilities, Professor Jerzy Krzymuski; soil science and land protection, Professor D. Tadeusz Witek; agricultural meteorology, Assistant professor Tadeusz Górski; fertilization, Professor Mariusz Fotyma; soil cultivation, Professor Jerzy Sienkiewicz; agricultural microbiology, Assistant professor Wanda Maliszewska; weed ecology and control, Assistant professor D. Jósef Rola; extension service, D. Eugeniusz Polak; physiology and biochemistry of plant nutrition, Assistant professor Michał Płoszyński; cereal crop cultivation, Assistant professor Jan Mazurek; fodder crop cultivation, Professor Anna Jelinowska; tobacco breeding and cultivation, Professor Jan Berbeć; hop cultivation and breeding, Professor Zbigniew Wirowski; economics and crop production management, Assistant professor Kazimierz Bis; regional experiments, Leszek Maj,
Independent laboratories: Plant nutrition, Professor Maria Ruszkowska
Graduate research staff: 412

Activities: Cultivation, fertilization and pedology in science and agricultural practice; cognition of natural condition of agricultural productive areas and their formation methods, investigation and mapping of soil, land protection and reclamation in rural areas, land cultivation and crop rotation systems, fertilization, microbiology, agrometeorology; biology, biochemistry, physiology and cultivation procedures for main crop and fodder plants; breeding and cultivation of tobacco and hops.
Publications: Pamiętnik Puławski, annual; *Zalecenia Agrotechniczne* .
Projects: Improved fertilization (Professor Roman Czuba); methods for increasing crop production in different soil and climatic conditions (Professor Marek Ruszkowski); principles for utilization and protection of agricultural areas (Professor D. Tadeusz Witek); biological and agrotechnical methods for soil fertility improvement (Professor Zdzisław Gonet); breeding principles and hop production method improvement (Professor Zbigniew Wirowski); breeding principles and tobacco production method improvement (Professor Jan Berbeć); development and application of new technologies for fodder production (Professor Stanislaw Nawrocki).

Instytut Weterynarii   154

[Veterinary Research Institute]
Address: Partyzantow Street 57, 24-100 Puławy, Lublin
Telephone: 028-30 51
Telex: 0642401 iwet pl
Affiliation: Ministry of Agriculture and Food Control
Status: Official research centre
Director: Professor M. Truszczyński
Departments: Cattle and sheep diseases, S. Cąkała; swine diseases, Professor S. Tereszczuk; poultry diseases, Professor W. Karczewski; fish diseases, Dr J. Antychowicz; microbiology, Professor M. Truszczyński; virology, Professor Z. Baczyński; pathology, Professor J. Zadura; parasitology, Professor W. Chowaniec; biochemistry, Professor J. Juśko-Grundboeck; pharmacology and toxicology, Professor T. Juszkiewicz; hygiene of animal food products, Dr B. Wojtoń; organization and popularization, Dr A. Jarosz
Laboratories: Cell pathology, Professor M. Grundboeck; radiobiology, Professor S. Kossakowski; tuberculosis immunology, Professor C. Zórawski; diseases of young animals, Professor W. Radomiński; water fowl diseases, Professor A. Cąkałowa; immunoprophylaxis, Dr J. Górski
Graduate research staff: 158

Activities: Development of veterinary sciences; elaboration of prophylactic programmes, diagnostic methods and control measures of diseases of animals; principles of veterinary inspection, hygiene of animal products; microbiology, virology, epizootiology, parasitology, physiopathology of reproduction, hygiene of animal products, zoohygiene, toxicology, drug technology, biochemistry and pathology.
Publications: Biannual bulletin.
Projects: Methods for prevention and control of infectious and parasitic diseases of animals (Professor M. Truszczyński); methods for the prevention and control of non-infectious diseases of animals (Professor S. Cąkała); hygiene and toxicology of food of animal origin and veterinary drugs (Professor T. Juszkiewicz); methods for the control of animal infertility (Professor J. Romaniuk).

BYDGOSZCZ BRANCH   155

Departments: Physiopathology of reproduction and insemination, Professor J. Romaniuk, animal hygiene, Professor J. Wiśniowski; deficiency diseases, Dr A. Lachowski
Laboratories: Horse diseases, Dr E. Wiśniewski

GDANSK BRANCH   156

Laboratories: Brucellosis, Dr M. Królak; salmonid fish diseases, Dr F. Flondro

SWARZEDZ BRANCH   157

Departments: Bee diseases, Professor R. Kostecki; prophylaxis of sterility, Professor K. Rosłanowski; animal ecology, Dr W. Więckowski

WARSZAWA BRANCH   158

Departments: Technology and control of veterinary drugs, Professor Z. Synowiedzki
Laboratories: Microbiology and biochemistry of animal products, Dr J. Maleszewski

ZDUNSKA WOLA BRANCH   159

Departments: Foot and mouth diseases, Dr A. Kęsy

Instytut Ziemniaka   160

[Potato Research Institute]
Address: Bonin, 75-016, Koszalin
Telephone: 26 99 7
Status: Official research centre
Director: Dr Edward Kapsa
Departments: Genetics and parental line breeding,
Professor K. M. Swieźyński; experimental breeding, Dr
E. Werner; virus diseases and seed production, Professor
W. Gabriel; diseases and pests, Dr J. Pietkiewicz;
cultivation, fertilization, mechanization, Dr S.
Roztropowicz; storage, Dr K. Kubicki
Graduate research staff: 86
Activities: Polish potato industry research particularly
breeding and improvement of production techniques.
Projects: Development of parental lines with resis-
tances, starch and protein content, table quality
(Professor K. M. Swieźyński); selection of table and
starch varieties (Dr E. Werner); epidemiology of virus
diseases and virus free seed production (Professor W.
Gabriel); testing storability of new varieties (Dr K.
Kubicki); biology of pathogens and development of
control methods (Dr J. Pietkiewicz); potato production
economics (Dr Stachurski); virus antisera production
(Dr Mierzwa); complex mechanization of potato production, agronomical characterization of new varieties
(Dr S. Roztropowicz).

Instytut Zoologii PAN*   161

[Zoological Institute - PAN]
Address: Ulica Wilcza 64, skr poczt 1007, 00-679
Warszawa
Telephone: 022 29 32 21
Parent body: Polska Akademia Nauk - PAN
Status: Official research centre
Director: Professor Henyk Szelzgiewicz

Instytut Zootechniki w   162
Polsce

[Zootechnics Institute of Poland]
Address: 32-083 Balice near Kraków
Affiliation: Ministry of Agriculture and Food Control
Status: Official research centre
Director: Professor Stefan Wawrzyńczak

DEPARTMENT OF ANIMAL   163
PRODUCTION TECHNOLOGY

Telephone: 012 132 11
Telex: 0325 602 PL
Head Dr Adam Pilarczyk
Graduate research staff: 6
Activities: Improvement of the technological and
environmental function of farm buildings for cattle and
pigs, design of new buildings and modernization of
existing buildings; research into the animal – microclimate relation to the energy balance of organisms. ,
Publications: Roczniki Naukowe Zootechniki, biannual;
Biuletyn Informacyjny Instytutu
Zootechniki, 6 times a year.
Projects: Influence of the environment on the productivity and behaviour of animals; influence of different
beddingless floors on the productivity, behaviour and
state of health of animals (Dr A. Pilarczyk); principles of
connection of the external forms of modern farm buildings with traditional and regional forms in complexion
of landscape protection (Dr J. Witek).

DEPARTMENT OF ANIMAL   164
REPRODUCTION AND ARTIFICIAL
INSEMINATION

Telephone: 012-102 94
Telex: 0325602 izet pl
Head: Professor Stefan Wierzbowski
Graduate research staff: 6
Activities: Animal production, livestock breeding,
veterinary medicine, reproduction and artificial insemination.
Projects: Improving methods of freezing farm animal
embryos (Dr Z. Smorąg); invitro culture of bovine
oocytes (Lucyna Kątska); use of embryo freezing and
transfer methods in genetic reserve programme in Polish
red cattle (Dr E. Wierzchoś); use of embryo freezing and
transfer for genetic progress control in olkuska sheep
(Dr W. Kareta); influence of body and gonad weight on
sexual performance in bull (Assistant professor A.
Laszczka); long-term storage of bull semen in Central
Semen Depot (Dr J. Pilch); bacteriological control of
semen stored in CSD (Professor S. Wierzbowski).

DEPARTMENT OF CATTLE BREEDING   165

Telephone: 012-11 32 11
Telex: 0322304 zoot pl
Head: Professor Jósef Romer
Graduate research staff: 15
Activities: Methods of genetic improvement, preserva-
tion of a native cattle breed, organization and control of
reproduction.

Projects: Genetic improvement in milk and meat productivity of dual purpose cattle breeds, modification of national breeding and selection programme for pure breeding, crossbreeding experiments (J. Romer, Maria Stolzman, K. Nahlik, Z. Pasierbski, T. Krempa, K. Zukowski, Hanna Czaja, P. Cieślar); methods of improving beef performance by means of commercial crossing (J. Romer, J. Trela, J. Rey); estimation of breeding value of bulls in Poland (Hanna Czaja, K. Nahlik, Lidia Lewińska); formation of gene reserves of a native Polish red cattle breed (K. Żukowski); organization and control of cattle reproduction in cooperation with INRA, Nouzilly, France (J. Rey); evaluation of factors affecting milk production in large dairy farms (M. Zaczek, Z. Pasierbski).

DEPARTMENT OF HORSE BREEDING   166

Telephone: 012-11 32 11
Telex: 0325602 PL
Head: Dr Stanisław Deskur
Graduate research staff: 3
Activities: Methods in performance improvement of Polish horse breeds, including rearing, selection, performance testing.
Projects: Evaluation of breeding programme in training units for half-bred stallions, modernization of the Polish racing programme for thoroughbreds, the relationship between the season of birth and the breeding value and performance of a foal.

DEPARTMENT OF POULTRY SCIENCE   167

Telephone: 012-130 30
Telex: 0322304 PL
Head: Professor Stanisław Węzyk
Graduate research staff: 16
Activities: Increasing the quantity and quality of poultry meat and eggs though breeding, rearing, nutrition methods and environmental conditions.
Projects: Genetic improvement of poultry population (Professor Stanisław Węzyk); poultry nutrition (Dr Roman Kaniok); biological aspects of poultry production (Dr Barbara Kaminska); poultry reproduction (Dr Alina Łada Gorzowska); quality of poultry products (Dr Barbara Różycka); poultry behaviour (Dr Eugeniusz Herbut).

LABORATORY OF DAIRY RESEARCH   168

Telephone: 012-130 30
Head: Professor Irena Leonhard-Kluz
Sections: Basic milk research, Irena Leonhard-Kluz; mineral and toxic substances in milk, Franciszek Bielak
Graduate research staff: 8
Activities: Improvement of milk quality; economically important properties and components of milk; breeding, feeding and adaptability in the dairy industry.
Projects: Economically important components and properties of milk of domestic animals such as cows, sheep, goats, mares considering their state of health; new parameters in dairy cow milk production evaluation; effect of feeding on milk quality (Irena Leonhard-Kluz); variability and dependence between components and properties of milk; hygienic quality of milk under different conditions of keeping of cows (Andrzej Gwoździewicz); minerals, nitrates and toxic substances in milk, blood and forage; lowering of toxic substance levels in milk (Franciszek Bielak).

Morski Instytut Rybacki*   169

[Sea Fisheries Institute]
Address: Al Zjednoczenia 1, 81-345 Gdynia
Telephone: 098-21 70 21
Telex: 051-543
Affiliation: Central Fisheries Board, Szczecin
Status: Official research centre
Director: Dr Bohdan Draganik
Research director: Professor Dr Andrzej Ropelewski
Sections: Oceanography, Dr S. Grimm;
Ichthyology, Professor J. Popiel;
Fish processing, Dr P. Bykowski;
Fish economics, Dr Z. Polanski
Graduate research staff: 200
Activities: Research is conducted, principally aboard the marine research vessel 'Professor Siedlecki', in the following areas: fishery, oceanography and biology, including determination of levels and conditions of fish populations; biological studies of economically important species; pollution studies; design of trawling and fishing technology; fish processing and food technology; economic and ecological forecasting; marine engineering; hydroacoustics.
Publications: Report of the Sea Fisheries Institute in Gdynia.

Ogród Botaniczny PAN* 170

[Botanic Gardens - PAN]
Address: Ulica Prawdziwka 2, 02-973 Warszawa, 34
Telephone: 42 79 14
Parent body: Polska Akademia Nauk - PAN
Status: Official research centre
Director: Dr Bogusław A. Molski
Departments: Crop plants genetic resources, Dr W. Łuczak
Gene bank, Dr Ł. Pietrzak
Plant ecology, Dr W. Dumchowski
Experimental taxonomy, Dr J. Puchalski
Dendrology, A. Marczewski
Horticulture, T. Szymański
Graduate research staff: 50
Activities: Plant genetic resources, rye, bioindication of environmental pollution.
Publications: Gardens, botanic journal.

Państwowy Zakład Higieny* 171

[National Institute of Hygiene]
Address: Ulica Chocimska 24, 00-791 Warszawa
Telephone: 022-49 40 51
Status: Official research centre
Acting Director: Associate Professor W. Magdzik
Graduate research staff: 175

HYGIENE DIVISION* 172

Vice-Director: Professor Dr M. Nikonorow

Department of Food and Consumer Articles Examination* 173

Head: Professor Dr M. Nikonorow
Activities: Hygienic evaluation of food, food additives, consumer articles, plastics; food microbiology; toxicological investigation, including evaluation of pesticide residues; physico-chemical analyses.

Politechnika Krakowska* 174

[Kraków Technical University]
Address: Ulica Warszawska 24, 31-155 Kraków
Telephone: 094-303 00
Telex: 0322468 pl
Status: Educational establishment with r&d capability
Rector: Professor Dr Bolesław Kordas

FACULTY OF MECHANICAL ENGINEERING* 175

Dean: Professor Dr Stanisław Rudnik

Building, Road and Agricultural Machines Institute* 176

Director: Associate Professor Dr Kazimierz Szewczyk
Sections:
Theory of mechanisms and machines, Professor Jan Korecki

Politechnika Łodzka* 177

[Łódź Technical University]
Address: Ulica Zwirki 36, 90-954 Łódź
Telephone: 04-655
Telex: 88-61-36
Affiliation: Ministry of Science, Technology and Higher Education
Status: Educational establishment with r&d capability
Rector: Professor Dr hab Edward Galas
Graduate research staff: 1 490

FACULTY OF FOOD CHEMISTRY* 178

Dean: Associate Professor Dr Piotr Moszczyński

INSTITUTE OF CHEMICAL TECHNOLOGY OF FOOD* 179

Address: Ulica Gdańska 162/168, Łódź
Director: Professor Dr Zygmunt Niedzielski
Senior staff: Professor Dr Jan Dobrzycki

INSTITUTE OF FUNDAMENTAL FOOD CHEMISTRY* 180

Address: Ulica Gdańska 162/168, Łódź
Director: Associate Professor Dr Józef Góra

Politechnika Wrocławska* 181

[Wrocław Technical University]
Address: Wybrzeże Wyspiańskiego 27, 50-370 Wrocław
Telephone: 07-20 22 17; 22 73 36
Telex: 0712554 PWr PL
Status: Educational establishment with r&d capability
Vice-President: Professor Dr Wacław Kasprzak
Graduate research staff: 2 440

INSTITUTE OF INORGANIC 182
TECHNOLOGY AND CHEMICAL
FERTILIZERS*

Address: Ulica Smoluchowskiego 25, 50-372 Wrocław
Telephone: 22 74 25; 20 24 84
Director: Professor Jerzy Schroeder
Activities: Scientific research on mathematical modelling of technological processes, methodological basis of the technology of mineral fertilizers and intermediate products used for their production, methodology of electrochemical phenomena and corrosion.

INSTITUTE OF ORGANIC AND 183
POLYMER TECHNOLOGY*

Address: Ulica Lukasiewicza, 50-371 Wrocław
Telephone: 07-20 24 26; 22 45 14
Activities: Scientific research on polymer materials science, methodological basis of the technology of surface-active compounds, plant protection agents and plastics.

Polska Akademia Nauk* 184

– PAN
[Polish Academy of Sciences]
Address: Pałac Kultury i Nauki, 00-901 Warszawa
Telephone: 20 44 70
Telex: 022-813929
Status: Official research organization
President: W. Nowacki
See separate entries for:
Section I - Social Sciences
Instytut Rozwojuws i Rolnictwa PAN
Section II - Biological Sciences
Instytut Biochemii i Biofizyki PAN
Instytut Botaniki PAN
Instytut Dendrologii PAN
Instytut Ekologii PAN
Instytut Parazytologii PAN
Instytut Zoologii PAN
Ogród Botaniczny PAN
Zakład Antropologii PAN
Zakład Badania Ssaków PAN
Zakład Biologii Wód PAN
Zakład Ochrony Przyrody PAN
Zakład Paleobiologii PAN
Zakład Zoologii Systematycznej i Doświadczalnej PAN
Section IV - Technical Sciences
Instytut Podsław Inżynierii Srodowiska PAN
See below for Section V

SECTION V - AGRICULTURE AND 185
FORESTRY SCIENCES

Telephone: 022-20 33 71
Telex: 81 72 57 KOSMO PL
Secretary: Professor Janusz Haman ·
Institute for Agrobiology and Forestry: Professor L. Ryszkowski
Graduate research staff: 281
Activities: Animal genetics; animal physiology; plant genetics; plant physiology; physical properties of soils, plants and animal materials; forest ecology; endocrinology; relationship between agriculture and environment.

Przemysłowy Instytut 186
Maszyn Rolniczych

[Industrial Institute of Agricultural Machinery]
Address: Ulica Starołecka 31, 60-963 Poznań
Telephone: 06-746 41
Telex: 04 13761 pimr pl
Status: Official research centre
Research director: Professor Kazimierz Mielec
Departments: Cultivating and harvesting machinery, Z. Balloniak; machinery for fertilization, seeding and pest control, J. Łącki; reapers and driers M. Liska; animal husbandry, Dr H. Wojciechowski; energetics and road transport B. Koczorowski; quality testing and economics, Professor R. Wiza; technological research, Dr Z. Kiełpiński; measuring apparatus, Professor M. Cywinski; foreign relations, B. Jenek
Branch centres: Standardization, W. Bryl; industrial design, ergonomics and safety of machinery service, T. Banaś; scientific, technical and economical information, Professor A. Górski
Graduate research staff: 85
Publications: Agricultural machinery record review, monthly; quarterly reports.

Szkotł Głowna 187
Gospodarstwa Wiejskiego -
Akademia Rolnicza w
Warszawie*

[Agricultural University of Warsaw]
Address: Ulica Rakowiecka 26-30, 02-528 Warszawa
Telephone: 022 44 22 51
Status: Educational establishment with r&d capability
Rector: Professor Dr M.J. Radomska

FACULTY OF AGRICULTURAL ECONOMICS* 188

Address: Ulica Nowoursynowska 166, 02-766 Warszawa
Telephone: 022-43 11 73
Telex: SGGW-AR 81 47 90
Dean: Assistant Professor Dr Mieczysław Adamowicz
Graduate research staff: 80
Activities: Research on factors influencing the economic effectiveness of peasant farms; evaluation of possibilities of agricultural production increase on individual farms of the central and eastern parts of Poland; management systems on state farms; development trends in world agriculture, and their links with food economy in Poland.
Publications: Agricultural Economics and Rural Sociology, Annals of the Agricultural University of Warsaw.

Agricultural Economics Institute* 189

Director: Professor Dr B. Struzek

Farm Organization and Management Institute* 190

Director: Professor Dr T. Rychlik

World Agriculture Institute* 191

Director: Professor Dr J. Górecki

FACULTY OF AGRICULTURE* 192

Dean: Professor Dr Zygmunt Brogowski

FACULTY OF AGRICULTURE AND FORESTRY ENGINEERING* 193

Dean: Dr Stanistaw Pabris

FACULTY OF ANIMAL SCIENCE* 194

Address: Ulica Przejazd 4, 05-840 Brwinów
Telephone: 58 90 16

Department of Animal Breeding and Technology of Animal Production* 195

Senior staff: Associate Professor Dr Jerzy Chachuła, Associate Professor Dr Feliks Mały, Assistant Professor Dr Stanisław Jankowski, Assistant Professor Dr Maria Soroczyńska, Assistant Professor Dr Franciszek Horszczaruk
Graduate research staff: 17

Department of Animal Nutrition* 196

Senior staff: Professor Dr Franciszek Witczak, Associate Professor Dr Jozef Karaś, Jadwiga Chachuła
Graduate research staff: 16

Department of Biological Basis of Animal Production* 197

Senior staff: Professor Dr Jerzy Woyke, Associate Professor Dr Andrzej Łysak, Assistant Professor Dr Tadeusz Sławiński, Dr Witold Skolasiński
Graduate research staff: 26

Department of Cattle Breeding and Milk Production* 198

Senior staff: Professor Dr Henryk Jasiorowski, Associate Professor Dr Tadeusz Czaplak, Professor Dr Ena Lipińska, Assistant Professor Dr Wojciech Empel, Assistant Professor Dr Barbara Reklewska, Assistant Professor Dr Marek Jurczak, Assistant Professor Dr Zenon Tomicki, Dr Jerzy Mostin
Graduate research staff: 36

Department of Poultry Production* 199

Senior staff: Professor Dr Ewa Potemkowska, Assistant Professor Dr Andrzej Frindt, Assistant Professor Dr Maria Szymkiewicz, Assistant Professor Dr Bogumił Strzyżewski
Graduate research staff: 16

Department of Zoohygiene and Prophylaxis in Animal Production* 200

Senior staff: Associate Professor Dr Jerzy Mazurczak, Assistant Professor Dr Eligiusz Rokicki, Assistant Professor Dr Stefan Marcinkiewicz
Graduate research staff: 21

FACULTY OF FOOD TECHNOLOGY* 201

Address: Ulica Rakowiecka 26/30, 02-528 Warszawa
Telephone: 022-49 66 36
Telex: 81 47 90
Dean: Piotr P. Lewicki
Activities: Research covers the characteristics of raw materials, technological processes, biochemical and microbial problems in food technology; engineering, energy conservation and mechanization and automation of food processing.

Dairy Technology Department* 202

Head: Assistant Professor S. Zmarlicki

Food Engineering Department* 203

Head: Assistant Professor P. Lewicki

Fruit and Vegetable Technology Department* 204

Head: Professor A. Horubata

Industrial Microbiology Department* 205

Head: Assistant Professor E. Subczak

Meat and Oil Technology Department* 206

Head: Professor A. Rutkowski

FACULTY OF FORESTRY* 207

Dean: Professor Tadeusz Marsatek

FACULTY OF HORTICULTURE* 208

Dean: Dr A. Sadowski

FACULTY OF HUMAN NUTRITION AND RURAL HOME ECONOMICS* 209

Dean: Professor S. Berger

FACULTY OF LAND RECLAMATION* 210

Dean: Dr Tadeusz Kicinski

FACULTY OF VETERINARY MEDICINE* 211

Dean: Professor M. Szulc

FACULTY OF WOOD TECHNOLOGY* 212

Dean: Professor Bolestaw Gonet

Uniwersytet Jagielloński* 213

[Jagiellonian University]
Address: Gotebia 24, 31-007 Kraków
Status: Educational establishment with r&d capability

FACULTY OF BIOLOGY AND EARTH SCIENCES* 214

Botanical Institute* 215

Address: Ulica Lubicz 46, 31-512 Kraków
Director: Professor Eugenia Pogan

Zoology Institute* 216

Address: Ulica Krupnicza 50, 30-060 Kraków
Director: Professor Czestaw Jura

Uniwersytet Łódzki* 217

[Łódź University]
Address: Ulica Narutowicza 65, 90-131 Łódź
Telephone: 04 858 12
Telex: 886291
Status: Educational establishment with r&d capability
Rector: Professor Dr Romuald Skowroński
Pro-rector for Research: Professor Dr Halina Mortimer-Szymcsak

FACULTY OF BIOLOGY AND EARTH SCIENCES* 218

Dean: Professor Dr Andrzej Romaniuk

Environmental Biology Institute* 219

Address: Ulica Banacha 12-16, Łódź
Director: Professor Dr Romuald Olaczek
Sections:
Algae, Professor Dr Joanna Kadłubowska
Ichthyology, Professor Dr Antoni Kulamowicz
Plant systems, Professor Dr Romuald Olaczek, Professor Dr Ryszard Sowa
Hydrobiology, Professor Dr Krzysztof Jażdżewski, Professor Dr Franciszek Woitas
Zoology and fish ecology, Professor Dr Tadeusz Penczak, Professor Dr Wanda Susłowska
Entomology and ecology, Professor Dr Tadeusz Penczak, Professor Dr Włodzimierz Romaniszvn
Zoology, Professor Dr Cezarv Tomaszewski
Anthropology, Professor Dr Zdzisław Kapica

Physiology and Cytology Institute* 220

Address: Ulica Banacha 12-16, Łódź
Director: Professor Dr Andrzej Romaniuk
Sections:
Plant physiology, Professor Dr Stanisław Knypl, Professor Dr Wacława Maciejewska-Potapczyk
Animal physiology, Professor Dr Maria Lewińska, Professor Dr Andrzej Romaniuk
Plant cytology and cytochemistry, Professor Dr Barbara Gabara, Professor Dr Maria Kwiatkowska, Professor Dr Eugenia Mikulska, Professor Dr Maria Olszewska

Protozoology, Professor Dr Maria Wolska
Zoology and comparative anatomy, Professor Dr Krystyna Urbanowicz

Uniwersytet Marii Curie-Skłodowskiej* 221

[Marie Curie-Skłodowska University]
Address: Plac Marii Curie-Skłodowskiej 5, 20-031 Lublin
Telephone: 028-37 51 00
Telex: 643223
Affiliation: Ministry of Higher Education, Science and Technology
Status: Educational establishment with r&d capability
Rector: Wiesław Skrzydlo
Research director: Eugeniusz Gąsior

FACULTY OF BIOLOGY AND EARTH SCIENCES* 222

Biology Institute* 223

Activities: Problems concerning the protection and economics of the environment; environmental biology for the blind; phenology of plant embryology; taxonomical, phyto- and zoogeographical investigations, as well as the structure and function of plant organs.

Microbiology and Biochemistry Institute* 224

Activities: Research on the biosynthesis of albumins and the mechanisms of these processes; the genetics of papillose bacteria capable of assimilating free nitrogen; antiviral agents, new enzyme production and technology of organic compounds for the food industry.

Botanical Garden* 225

Activities: Collection of varieties and assessment of resources of Polish plant life; research into original vegetation and acclimatization of new plants; nurturing South European, Asiatic, and American species, ornamental, onion and tropical plants.

Uniwersytet Nikołaja Kopernika w Toruniu* 226

[Nicholas Copernicus University, Torun]
Address: Ulica Gagarina 11, 87-100 Toruń
Telephone: 056-22 69 4
Status: Educational establishment with r&d capability
Rector: Professor Dr Ryszard Bohr
Vice-Rectors: Professor Dr Wiesław Domasłowski, Professor Dr Jerzy Tomala, Professor Dr Jan Winiarz

FACULTY OF BIOLOGY AND EARTH SCIENCES* 227

Dean: Professor Dr Juliusz Nabębski

Biology Institute* 228

Address: Ulica Gagarina 3, 87-100 Toruń
Director: Dr Jan Kopcewicz
Sections:
Animal ecology, Professor Dr A. Gromadska
Animal physiology, Professor Dr L. Janiszewski
Botany, Professor Dr K. Kepczyński
Histology and embryology, Professor Dr J. Czopek
Geomorphology, Professor Dr W. Niewiarowski
Biochemistry, Professor Dr P. Masłowski
Microbiology, Professor Dr E. Strzelczyk
Neurophysiology, Professor Dr J. Narebski
Plant physiology, Professor Dr M. Michniewicz
Plant systematics and ecology; Professor Dr R. Bohr
Soil sciences, Professor Dr Z. Prusinkiewicz

Geography Institute* 229

Address: Ulica Fredry 8, 87-100 Toruń
Director: Professor Dr W. Niewiarowski
Section: Hydrobiology, Professor Dr J. Mikulski

Uniwersytet Ślaski* 230

[Silesian University]
Address: Ulica Bankowa 12, 40-007 Katowice
Status: Educational establishment with r&d capability

FACULTY OF BIOLOGY AND NATURAL SCIENCES* 231

Address: Ulica Jagiellónska 28, 40-032 Katowice

Botany Institute* 232

Zoology Institute*   **233**

Uniwersytet Warszawski*   234

[Warszawa University]
Address: Ulica Krakowskie Przedmieście 26/28, Warszawa
Telephone: 022-20 03 81
Telex: 815439 UW PL
Status: Educational establishment with r&d capability
Pro-rector: Professor Kazimierz Dobrowolski

FACULTY OF BIOLOGY*   235

Address: Ulica Oboźna 8, 00-332 Warszawa
Dean: Professor Ewa Pieczyńska
Graduate research staff: 187

Institute of Biochemistry*   236

Address: Ulica Zwirki i Wigury 93, 00-089 Warszawa
Director: Professor Kazimierz Toczko
Divisions:
General biochemistry, Professor Kazimierz Toczko
Plant biochemistry, Professor Zofia Kasprzyk
Natural compounds biochemistry, Associate Professor Zdzisław Wojciechowski
Enzymatics, Professor Zbigniew Kaniuga
Metabolism regulation, Associate Professor Jadwiga Bryła

Institute of Botanics*   237

Address: Ulica Krakowskie Przedmieście 26/28, 00-927 Warszawa
Director: Professor Stanisław Lewak
Divisions:
General botanics, Dr Jadwiga Tarkowska
Phytosociology and plant ecology, Professor Matuszkiewicz
Genetics, Professor Wacław Gajewski
Plant cell physiology and biology, Professor Jerzy Poskuta
Systematics and geography of plants, Professor Alina Skirgiełło
Plant adaptation and physiology, Professor Stanisław Lewak

Institute of Zoology*   238

Address: Ulica Krakowskie Przedmieście 26/28, 00-927 Warszawa
Director: Professor Andrzej Krzysztof Tarkowski
Divisions:
Cytology, Associate Professor Bohdan Matuszewski
Embryology, Professor Andrzej Krzysztof Tarkowski
Physiology of invertebrates, Associate Professor Bronisław Cymborowski
Physiology of vertebrate animals, Professor Janusz Gill
Hydrobiology, Professor Ewa Pieczyńska
Immunology, Associate Professor Mirosław Stankiewicz
Parasitology, Professor Bernard Bezubik
Zoology and ecology, Professor Kazimierz Dobrowolski

Uniwersytet Wrocławski im Bolesława Bieruta*   239

[Bolesław Bierut University, Wrocław]
Address: Pl Uniewersytecki 1, 50-137 Wrocław
Telephone: 07-402 212
Telex: 0712791
Status: Educational establishment with r&d capability
Rector: Professor Dr Kazimierz Urbanik
Vice Rectors: Professor Dr Lucjan Sobczyk, Associate Professor Dr Marek Mazurkiewicz, Associate Professor Dr hab Hanna Wałkówska, Associate Professor Dr hab Karol Fiedor
Graduate research staff: 900

FACULTY OF NATURAL SCIENCES*   240

Dean: Professor Dr Anna Jerzmanska

Botanical Garden*   241

Director: vacant

Institute of Botany*   242

Address: Ulica Kanonia 6/8, 50-328 Wrocław
Director: vacant
Senior staff: Professor Dr Józef Buczek, Professor Dr Stanisław Marek

Institute of Zoology*   243

Address: Ulica Sienkiewicza 21, 50-335 Wrocław
Director: Associate Professor Dr hab Andrzej Warchołowski
Senior staff: Professor Dr Zbigniew Jara, Professor Dr Anna Jerzmańska, Professor Dr Tadeusz Krupiński

Museum of Natural Sciences* 244

Address: Ulica Sienkiewicza 21, 50-335 Wrocław
Director: Professor Dr Andrzej Wiktor

Zakład Agrofizyki PAN* 245

[Agrophysics Institute - PAN]
Address: Ulica Krakowskie Przedmieście 39, 20-076 Lublin
Telephone: 028-298 57
Telex: 0643101 tur pl
Parent body: Polska Akademia Nauk - PAN
Status: Official research centre
Research director: Professor Dr Ignacy Dechnik
Deputy director: Professor Dr Jan Gliński
Laboratories:
Soil physics, Dr R. Walczak
Soil physiocochemistry, Professor J. Gliński
Basic problems of soil improvement, Professor I. Dechnik
Plant physics, B. Szot
Prototype apparatus, M. Grochowicz
Graduate research staff: 40
Activities: Qualitative study and interpretation of physical and physicochemical phenomena occurring in soils, cultivated plants and agricultural products; creation of physical environment models typical of plant growth in order to develop simulating systems of the structure: soil-plant atmosphere.

Zakład Antropologii PAN* 246

[Anthropology Institute - PAN]
Address: Ulica Kuźnicza 35, 50-951 Wrocław
Telephone: 094 386 74
Parent body: Polska Akademia Nauk - PAN
Status: Official research centre
Director: Professor Tadeusz Bielicki

Zakład Badania Ssaków PAN* 247

[Mammals Research Institute - PAN]
Address: Gen Waszkiewicza 1d, 17-230 Białowicza
Telephone: 34
Parent body: Polska Akademia Nauk - PAN
Status: Official research centre
Director: Professor Dr Zdzisław Pucek
Deputy director: Associate Professor Dr Marek Gebczyński

Sections: Ecological physiology; ecology of ungulates; bioclimatology
Graduate research staff: 18
Activities: Variability of mammals (morphological, physiological, phenetical, biochemical); seasonal, developmental, and population studies; mammal fauna of Poland; ecology of mammalian populations and communities (small mammals - rodents and shrews, ungulates); breeding voles as new models for biomedical research; bioclimatology of forest ecosystems.
Publications: Acta Theriologica.

Zakład Biologii Wód PAN* 248

[Water Biology Laboratory - PAN]
Address: Ulica Sławkowska 17, 31-016 Kraków
Telephone: 094 221 15
Telex: 0322414 pan pl
Parent body: Polska Akademia Nauk - PAN
Status: Official research centre
Research director: Professor Dr Kazimierz Matusiak
Sections: Hydrochemistry, Dr M. Bombówna; microbiology, Dr A. Starzecka; hydrobiology, Dr A. Kownacki; ichthyology, Professor Dr J. Włodek
Graduate research staff: 31
Activities: Investigation of the structure and function of different aquatic ecosystems, both naturally occurring and as subject to transformations as a result of miscellaneous human activity. The investigations which are carried out comprise hydrochemistry, microbiology, phyto- and zooplankton, attached algae, micro- and macrofauna, and ichthyofauna; the trophic relation between selected organism groups will also be determined.
Publications: Acta Hydrobiologica.

FISH CULTURE EXPERIMENTAL STATION, GOŁYSZ* 249

Address: Zakład Doświadczalny Gołysz, 43-422 Chybie
Telephone: Rudzica 81
Telex: 038316 pang
Research director: Jan Szumiec
Sections: Intensive fish pond culture; industrial fish culture; hydrobiology and hydrochemistry; genetics
Graduate research staff: 20
Activities: Biological principles of fish pond culture and methods of its intensification; induced fish breeding and industrial methods of fish fry rearing; hydrobiology of pond ecosystems; selection and genetics of carp.

HYDROBIOLOGICAL STATION* 250

Address: Stacja Hydrobiologiczna, Ulica Jeziorna 80, 43-230 Goczatkowice

Zakład Fizjologii Roślin PAN* 251

[Plant Physiology Institute - PAN]
Address: Ulica Sw Jana 22, 50-156 Kraków
Telephone: 094-279 44
Parent body: Polska Akademia Nauk - PAN
Status: Official research centre
Director: Professor Włodzimierz Starzecki

Zakład Genetyki Roślin PAN* 252

[Plant Genetics Institute - PAN]
Address: Ulica Strzeszyńska 30/36, 60-479 Poznań
Telephone: 06-20 37 41
Parent body: Polska Akademia Nauk - PAN
Status: Official research centre
Director: Professor Ignacy Wiatroszak

Zakład Higieny Weterynaryjnej 253

– ZHW
[Veterinary Hygiene Laboratory]
Address: 21 Lechicka str, 02 156, Warszawa
Telephone: 022-46 01 68
Telex: PL-81 55 28
Affiliation: Stołeczny Zakład Weterynarii
Status: Official research centre
Director: Dr Stefan Samól
Departments: Biochemistry; toxicology; virology; bacteriology; serology; mycology; poultry diseases; fish diseases; swine diseases; cattle diseases; parasitology
Graduate research staff: 25
Activities: Veterinary medicine - laboratory diagnosis of animal diseases; meat and fodder investigation and estimation, macrocomponents and microcomponents, vitamins, toxic components and residues; state veterinary consultancy; preparation and introduction of new diagnostic methods in veterinary virology, bacteriology, biochemistry and toxicology.

Zakład Ochrony Przyrody PAN* 254

[Nature Conservation Institute - PAN]
Address: Ulica Lubicz 46, 31-512 Kraków
Telephone: 094 161 44
Parent body: Polska Akademia Nauk - PAN
Status: Official research centre
Director: Professor Roman Ney

Zakład Paleobiologii PAN* 255

[Palaeobiology Institute - PAN]
Address: Al Zwirki i Wigury 93, 02-089 Warszawa
Telephone: 22 1652
Parent body: Polska Akademia Nauk - PAN
Status: Official research centre
Director: Professor Dr Zofia Kielan-Jaworowska
Graduate research staff: 28
Activities: Morphology and anatomy of fossil plants, invertebrates and vertebrates; palaeoecology, including Jurassic oyster banks, and Cretaceous terrestrial ecosystems of Gobi Desert, Mongolia; paleophysiology of archosaurs; biosedimentation, including diagenetic structures in carbonate deposits and skeletons, and biosedimental significance of algae and bacteria.
Publications: Acta Palaeontologica Polonica; Palaeontologia Polonica.

Zakład Zoologii Systematycznej i Doświadczalnej PAN* 256

[Systematic and Experimental Zoology Institute - PAN]
Address: Ulica Stawkowska 17, 31-016 Kraków
Telephone: 094 264 10
Telex: 032414
Parent body: Polska Akademia Nauk - PAN
Status: Official research centre
Director: Professor Kazimierz Kowalski

PORTUGAL

Aquário Vasco da Gama*   1

[Vasco da Gama Aquarium]
Address: Dafundo, 1495 Lisboa
Telephone: (019) 212338
Affiliation: Ministério da Marinha
Status: Official research centre
Director: M.L. de Mendonça
Research director: M.T. Dinis
Sections: Aquaculture, M.T. Dinis; chemistry, M.A. Madeira
Graduate research staff: 4
Activities: The principal research activities are concerned with aquaculture (sole, seabass, eel) and marine biology/ecology.
Publications: Relatorios de Actividades do Aquário Vasco da Gama.

Centro de Botânica da Junta de Investigações Científicas do Ultramar*   2

[Botanical Centre of the Council for Overseas Scientific Research]
Address: Rua da Junqueira 86, 1300 Lisboa
Telephone: (019) 645518
Parent body: Junta de Investigações Científicas do Ultramar
Status: Official research centre
Research director: E.J. Mendes
Sections: Herbarium; Library
Graduate research staff: 5
Activities: Botanical taxonomy and phytogeography of Africa south of the Sahara, especially of Angola, Mozambique and of the 'Flora Zambesiaca' area.
Publications: Conspectus Florae Angolensis; Flora Zambesiaca; Flora de Moçambique; Garcia de Orta, Série de Botânica.

Centro de Estudos de Defesa Fitossanitária dos Produtos Ultramarinos   3

– CEDFPU
[Centre for Pest Infestation Control of Tropical Stored Products]
Address: Travessa do Conde da Ribeira 9, 1300 Lisboa
Telephone: (019) 64 49 46
Affiliation: Laboratório Nacional de Investigação Científica Tropical
Director: Eng Artur Soares de Gouveia
Sections: Entomology (systematics, bio-ecology), Artur Soares de Gouveia; micology (systematics, bio-ecology), Maria Antonieta Mourato; pest control (safe storage, chemical control, nutritional loss assessment), Maria Elmina Lopes
Graduate research staff: 6
Activities: Ecology and pest control of stored products: fungi, insects, nutritional loss assessment; physical and chemical control.
Projects: Ecological studies on stack and bulk products (Maria Elmina Lopes); bio-ecological studies of fungi (Maria Antonieta Mourato); bio-ecological studies of insects (Artur Soares de Gouveia); nutritional loss assessment (Dr Maria Alice Santos); evaluation of physical factors concerning safe storage (António Maia); physical and chemical control (António Barbosa).

Centro de Estudos de 4
Economia Agrária

[Agricultural Economics Research Centre]
Address: Apartado 14, 2781 Oeiras Codex, Lisboa
Telephone: (019) 24 31 436
Telex: 12345 GULB
Affiliation: Fundação Calouste Gulbenkian
Status: Independent research centre
Director: Mário Pereira
Graduate research staff: 19
Activities: Policy analysis and socioeconomic studies: the agricultural sector and the European Economic Community, alternative technologies and Portuguese agriculture; education and training of personnel; information and documentation.
Publications: Annual report.
Projects: Consequences of the entry of Portugal to the European Economic Community (F. Brito Soares); alternative technologies (Agostinho de Carvalho); Cape Vert: study of the social stratification and basic needs of a rural community in a defavourized ecosystem (A. Trigo de Abreu); part-time agriculture: types, scope and effects in the region of Lisbon (J. Silva Lourenço); modelling the Portuguese agricultural sector (F. Brito Soares).

Centro de Estudos de 5
Pedologia Tropical

– CEPT
[Tropical Pedology Research Centre]
Address: Instituto Superior de Agronomia, Tapada da Ajuda, 1399 Lisboa Codex
Telephone: (019) 63 81 61
Affiliation: Junta de Investigações Científicas do Ultramar
Director: Professor R. Pinto Ricardo
Departments: Soil physics, vacant; soil chemistry, Professor R. Pinto Ricardo; soil genesis, classification, and cartography, E.P. Cardoso Franco; soil mineralogy, A.F.A. Sanches Furtado; geomorphology and geology, M. Monteiro Marques
Graduate research staff: 12
Activities: Soil science: classification of soils; soils and plant production; soil genesis; soil conservation; tropical soils; general soil map of Angola.
Projects: General soil map of Angola (E.P. Cardoso Franco, E.M. Silva da Câmara); soil studies in Madeira archipelago (R. Pinto Ricardo, E.P. Cardoso Franco); pedogenesis of tropical soils (R. Pinto Ricardo); mineralogical and chemical studies of day and other soil fractions; alteration of the rocks and soil genesis (A.F.A. Sanchez Furtado); genesis of laterite formations (M. Monteiro Marques, A.F.A. Sanches Furtado); geo-

morphological studies applied to soil conservation (M. Monteiro Marques).

Centro de Investigação das 6
Ferrugens do Cafeeiro *

[Coffee Rusts Research Centre]
Address: Estação Agronómica Nacional, 2780 Oeiras
Telephone: (019) 2430442
Affiliation: Junta de Investigações Científicas do Ultramar; Laboratório Nacional de Investigação Científica Tropical
Status: Official research centre
Research director: Carlos José Rodrigues Junior
Sections: Genetics and breeding, A.J. Bettencourt; Histopathology, Luisete Rijo; Pathology and physiopathology, C.J. Rodrigues Junior
Graduate research staff: 6
Activities: Coffee rusts; physiological specialization and biology of Hemileia vastatrix; screening of coffee for rust resistance; genetics and breeding for rust resistance; histopathology and biochemistry of the coffee-rust interactions; coffee tissue culture and induction of mutations.

Centro de Zoologia * 7

[Zoology Centre]
Address: Rua da Junqueira 14, 1300 Lisboa
Telephone: (019) 637055
Affiliation: Junta de Investigações Científicas do Ultramar
Status: Official research centre
Research director: Professor Dr João Tendeiro
Sections: Vertebrates, Dr João Crawford-Cabral; Hydrobiology, Dr Emerita Marques; Arachnoentomology, Dr M. Luisa Gomes Alves; Parasitology, Professor Dr João Tendeiro; Biological Control, Dr J. Piedade Guerreiro; Apiculture, Eng J.F. Rosário Nunes
Graduate research staff: 16
Activities: Mammalogy (mainly wild animals); ornithology (mainly wild birds); herpetology; copepodology; conchology; general entomology; parasitology (mainly wild mammals and birds); biological control and apiculture.
Publications: Garcia de Orta, Serie de Zoologica; Estudos Ensaios e Documentos; Memórias da Junta de Investigações Científicas do Ultramar.

Divisão de Produção Piscícola do Norte 8

[Northern Portugal Freshwater Fisheries Division]
Address: 4481 Vila do Conde Codex
Telephone: (022) 63 241
Affiliation: Ministério da Agricultura e Pescas
Status: Official research centre
Head: Eduardo Alberto de Castro Léncastre
Graduate research staff: 6
Activities: The division is less concerned with research than with the restocking of rivers and giving assistance to private fisheries, but some observations are made on the chemical quality of water and the biology of rivers especially with regard to benthonic invertebrates, particularly aquatic insects.

Estação Agronómica Nacional 9

[National Agronomic Station]
Address: Quinta do Marquês, 2780 Oeiras
Telephone: (019) 2430442 and 2431505
Affiliation: Instituto Nacional de Investigação Agrária
Status: Official research centre
Director: Dr J.V. de Carvalho Cardoso
Graduate research staff: 242
Activities: The Station is concerned with scientific research in a number of areas: ecological surveys; cartography; cultivation of vineyards and tomato plantations; pomology; rice growing; cytology and genetics of wheat, maize, barley, etc; arabica coffee beans; horticulture; insects, fungi, viruses, bacteria and nematodes with scientific or economic interest; mathematical methods and taxonomy.
The Station has close contacts with other investigation centres on the Continent and in Africa in addition to links with international scientific centres.
Publications: Agronomia Lusitana (quarterly); *De Flora Lusitana Commentarii; Index Seminum.*

DEPARTMENT OF AGRONOMY 10

Head: A. Mendes Gaspar
Activities: Plant production: studies on Rhizobium - isolation and selection of Rhizobium strains, strain identification by means of antisera, phages, and antibiotics, physiological aspects of dinitrogen fixation by Rhizobium.
Projects: Techniques and systems of crop production.

DEPARTMENT OF CHEMISTRY 11

Head: A. Pereira
Activities: Analytical chemistry - quality of agricultural products; determination of amino acids by electrophoresis; analysis of forage and silage; chromatography of amino acids in turfs; increase of proteins in cereals by fertilizing methods; olive oil yields.
Projects: Quality of production and chemical support for plant breeding.

DEPARTMENT OF ENTOMOLOGY 12

Head: G. Magalhães Silva
Activities: New species or species that have become popular recently; bioecology of species with economic importance; insect control by means of integrated biological and microbiological approaches; study of the lesser known acaro- and entomofauna.
Projects: Survey of entomofauna and crop pests.

DEPARTMENT OF EXPERIMENTAL STATISTICS 13

Head: Augusto Oliveira
Activities: Delineation and analysis of seed density and planting seasons; econometric analysis of annual and long-term planting experiments with special reference to the use of fertilizers, water supply and ground fertility; technoeconomic analysis; electronic computations and calculus for biological, agronomic and economic investigations; econometric analysis of experiments on the feeding of cattle and sheep.
Projects: Econometric studies and experimental statistics.

DEPARTMENT OF GENETICS* 14

Head: Miguel Mota
Activities: Ultrastructure of chromosomes and other cellular elements; production of auto-tetraploidal barley and rye; production of triticum; morphology and fertilization of the olive; high quality wheats; genetic study of the resistance of Coffea arabica to Hemileia vastatrix; identification of wheat genes according to production and locality.

DEPARTMENT OF MICROBIOLOGY 15

Head: Dr Augusto S. Cordeiro Zagallo
Activities: Symbiotic fixation of atmospheric nitrogen-reducing bacteria; herbicides; water quality and hygiene by microbiological means.
Projects: Microbiological studies.

DEPARTMENT OF PEDOLOGY 16

Head: Dr A.J. da Silva Teixeira
Activities: Soil science: fertility and soil technology; macronutrients; micronutrients; technology of irrigated soils; cartography of ecological characteristics.
Projects: Basic pedological studies; fertility, drainage, and salinity of agricultural soils.

DEPARTMENT OF PHYTOPATHOLOGY 17

Head: Oscar Sequeira
Activities: Bacteriology; mycology; nematology; virology; pests and diseases of cotton, sugarbeet, tobacco, corn, sunflower, soyabean, orchard fruits, grapevine, and vegetable garden crops; resistance to Puccinia recondita of wheat; clover rusts.
Projects: Studies of bacteriosis, mycosis, nematosis, and virosis of cultivated plants.

DEPARTMENT OF PHYTOSYSTEMATICS AND PHYTOSOCIOLOGY 18

Head: A.R. Pinto da Silva
Activities: Documentation of current flora; archaeological agricultural flora; inventory, analysis, description and documentation of Portuguese plant life; analysis of productivity of agricultural ecosystems; pollen studies of the atmosphere and of the soil.
Projects: Vegetation maps and ecosystems productivity (A. Teles).

DEPARTMENT OF PLANT BREEDING AND GENETICS 19

Head: M. Vianna e Silva
Sections: Improvement of rice, M. Vianna e Silva; Improvement of forage crops, João R. Marques De Almeida; Improvement of horticultural plants, Dr Alberto H. Garde; Improvement of vineyards, H. Padua De Carvalho
Activities: Cytogenetics of rye, barley and wheat; multiple crosses of maize; breeding of rice, forage, grapevine and olive trees; prebasis seed production of rice and forage; crop production: maize, sorghum and other cereals, oil-seeds, sugarbeet, tobacco and cotton; subtropical fruit trees; chemical study of quality of production and crop breeding support in cereals and olive-oil; physiological control of crops; plant-cold relationships: frost control; physiological control of water in irrigated and non-irrigated crops; physiological diagnosis of the mineral nutritional status of crops; postharvest physiology; fruit cold storage; physiological studies with radioisotopes; vegetation maps and ecosystems, productivity; statistical and econometric analysis of agricultural experiments; computer analysis of data.
Projects: Studies in genetics and plant breeding.

DEPARTMENT OF PLANT PHYSIOLOGY 20

Head: Dr J. de Oliveira Contreiras
Activities: Pomology; sugar beet; detection of radioactivity by biological means.
Projects: Physiological and biochemical support to agriculture.

DEPARTMENT OF POMOLOGY* 21

Head: vacant

Estação de Avicultura Nacional* 22

[Poultry Improvement Station]
Address: Rua Elias Garcia, 38 Venda Nova, Amadora
Status: Official research centre

Estação Nacional de Selecção e Reprodução Animal 23

[National Station for Animal Selection and Reproduction]
Address: Rua Elias Garcia 38, Venda Nova, 2700 Amadora
Telephone: (019) 97 45 05
Affiliation: Ministério da Agricultura e Pescas
Status: Official research centre
Director: Manuel Joaquim Freire
Sections: Reproductive clinical pathology; laboratory for the physiopathology of reproduction; national bank of semen and artificial insemination control; progeny testing.
Activities: Research work on animal genetics and reproduction is carried on intermittently at the station.

Estação Zootécnica Nacional 24

– EZN
[National Husbandry Station]
Address: Fonte-Boa, 2000 Vale de Santarem
Telephone: 76202/3/4
Affiliation: Instituto Nacional de Investigação Agrária
Status: Official research centre
Director: Professor Dr Apolinário J.B.C. Vaz Portugal
Departments: Nutrition; Genetics; Reproduction; Cattle; Nongastrics
Graduate research staff: 50
Activities: Animal production - beef and dairy cattle, sheep, goats, swine, and poultry; livestock breeding; livestock husbandry and nutrition; veterinary medicine. Agricultural engineering and building. Food science. There are three farms with 600 acres, 1 000 cattle, 1 200 sheep and goats, 3 250 poultry and 389 pigs.
Publications: Annual report.

AGRICULTURE DEPARTMENT 25

Head: Eduardo O. Sousa
Projects: crop production in the EZN farms; studies on forage conservation techniques (silage production); maintenance of parks and gardens.

ANIMAL PHYSIOLOGY DEPARTMENT 26

Head: Nuno M.V.B. Potes
Projects: Determination of the circulation parameters of the national breeds; hormonal characterization of local breeds.

BEEF AND DAIRY CATTLE DEPARTMENT 27

Head: Luis A. Cortes Martins
Projects: Milk production: maximization of grass consumption on milk production, studies of different diets and levels of feeding for indoor dairy cows; meat production: studies of beef cattle in 'feed-lot', maximization of grass consumption for beef production; maintenance herd: growth and feeding of heifers and cows; artificial rearing of calves: production with different levels of milk intake.

CARCASS EVALUATION DEPARTMENT 28

Head: Jorge A. Simões
Projects: Carcass evaluation as a complement to the different experiments carried out in EZN.

FEED MANUFACTURING PLANT DEPARTMENT 29

Head: Vasco S. Antunes
Projects: Formulation of diets; production of diets; studies on the technology and processing of diets.

GENETICS DEPARTMENT 30

Head: Gabriel S.N. Barata
Projects: Genetic improvement of milking ewes; studies of the Portuguese breeds of goats; studies in collaboration with other departments, and with the Centro de Produção Cavalar.

NON-RUMINANTS DEPARTMENT 31

Head: José Pires da Costa
Projects: Reproduction: utilization of prostaglandins as farrowing inducers in sows; utilization of hormones inducing heat in gilts and sows; sugar utilization in pregnant sows to reduce the death rate on birth. Feeding: study on energy levels in poultry diets; utilization of different grains in pig feeding; utilization of by-products in poultry feeding; study of different levels of cellulose in rabbit diets; utilization of non-traditional elements in poultry diets; restricted feeding of semi-heavy laying hens. Housing: comparative study on different kinds of accommodation for maternity and weaning of pigs; study of different densities of laying hens in batteries. Products: Phosphate measurements in pig carcasses; studies on the improvement of eggshell resistance.

NUTRITION DEPARTMENT   32

Head: João M.C. Ramalho Ribeiro
Projects: Chemical analysis for work in progress; characterization and nutritive value of Portuguese hays; correlation between chemical and nutritive parameters on hays; protein improvement on maize silage; studies of lipid metabolic disorders affecting production of dairy cows; copper sulphate as feed supplement in pig diets; formulation of diets for experiments in progress (João M.C. Ramalho Ribeiro.

REPRODUCTION DEPARTMENT   33

Head: Nuno M.V.B. Potes
Projects: Studies on reproductive disorders; reproductive implement techniques.

SHEEP HUSBANDRY DEPARTMENT   34

Head: Francisco S.C. Calheiros
Projects: Studies on meat, milk, and goat production; improvement and husbandry of the flock.

STATISTICS AND BIOMETRICS DEPARTMENT   35

Head: Joaquim Vacas de Carvalho
Projects: Statistical support for work in progress; econometric studies on different cost prices; characterization of ecological systems of grass production in the humid region of the north of Portugal.

VETERINARY SURGERY DEPARTMENT   36

Head: José J.A.C. Cardoso
Projects: Prophylaxis for and treatment of EZN farm animals.

Instituto do Azeite e Produtos Oleaginosos*   37

– IAPO
[Institute for Oils and Oleaginous Products]
Address: Avantonio, Augusto Aguiar, 23 P 1000 Lisboa
Status: Official research centre

Instituto do Vinho do Porto*   38

[Port Wine Institute]
Address: Rua de Ferreira Borges, 4000 Porto
Telephone: (029) 26522
Telex: 25337
Affiliation: Ministério do Comércio e Turismo
Status: Official research centre
Director: Eng Eduardo Mendia Freire de Serpa Pimentel
Research director: Eng Jose Viana Marques Gomes
Sections: Scientific research, Eng Agostinho Guimarães; oenotechnological research, Eng Moura Bastos
Graduate research staff: 10
Activities: The analytical control of, and the protection of the genuineness and quality of port wine through oenological research (microfermentation); physico-chemical research (chromatography in gas and liquid phases, spectrophotometry of atomic absorption, spectrometry of scintillation); microbiological research (yeasts, bacteria, and biological methods).
Publications: Anais do Instituto do Vinho do Porto; Cadernos Mensais de Estatística e Informação.

Instituto dos Cerealis*   39

[Cereals Institute]
Address: Apartado 2405, P-1000 Lisboa
Status: Official research centre

Instituto Nacional de Veterinária   40

[National Veterinary Institute]
Address: Lisboa
Status: Official research centre

ÉVORA LABORATORY 41

Address: Rua D. Isabel 8-2, 7000 Évora, Alentejo
Telephone: (069) 22825
Status: Official research centre
Service director: Luís Durval Botelho Borges Ferreira
Departments: Bacteriology, F.J. Alface Reis; bromatology, F.J. Alface Reis; parasitology, L.D.B. Borges Ferreira; pathology, M.M.V. Carrilho Ferreira; sorology, A.J.F.V. Gomes Ferreira; virology, A.L. Duarte Serralha
Graduate research staff: 6
Activities: Diagnosis and analysis of microbial and parasitic diseases occurring in the provinces of Alto e Baixo Alentejo.

Instituto Universitário de Trás-os-Montes e Alto Douro* 42

[University Institute of Trás-of-Montes and Alto Douro]
Address: Apartado 202, 5001 Vila Real Codex
Telephone: (099) 23688
Affiliation: Ministry of Education and Science
Status: Educational establishment with r&d capability
Research director and Rector: Professor Dr Fernando Real
Graduate research staff: 40
Activities: Research activities deal mainly with animal production, agricultural production and forestry. As such, there are experimentation fields, research laboratories and community services working on: documentation; biology (including herbarium); forestry; geosciences; animal husbandry; agriculture; food science; plant pathology; animal pathology. Much of the data from applied research is passed on to the Local Rural Extension Services of the Ministry of Agriculture.
Publications: Informação Bibliográfica; Relatório Anual de Actividade.

Jardim e Museu Agrícola do Ultramar 43

[Garden and Tropical Agricultural Museum]
Address: Calçada Galvão, 1400 Lisboa
Telephone: (019) 63 70 23
Affiliation: Laboratório Nacional de Investigação Cientifica Tropical, Junta de Investigações Cientificas
Status: Official research centre
Director: Cláudio Manuel Bugalho Semedo
Sections: General horticulture, Cláudio Manuel Bugalho Semedo; hydroponics, Rosalina Santos Victor; botany, Maria Cândida Liberato; museum, Rogério Dias Pereira; seeds, Henrique Alberto Almeida Lima
Graduate research staff: 5
Activities: Land use; irrigation; plant protection; taxonomy of agricultural and tropical plants; plastics in agriculture.
Publications: Index-Seminum (annual).
Projects: General horticulture; plastics in agriculture (C.M. Bugalho Semedo); hydroponics in horticulture (Rosalina Santos Victor); taxonomy of tropical and agricultural plants (Maria Cândida Liberato).

Laboratório Nacional de Engenharia e Tecnologia Industrial* 44

– LNETI
[National Laboratory for Engineering and Industrial Technology]
Address: Rua São Pedro Alcântara 79, 1200 Lisboa
Telephone: (019) 368856/8, 321221/3
Telex: 12727 NUCLAB P
Affiliation: Ministério da Indústria e Energia
Status: Official research centre
Research director: Professor José Veiga Simão
Graduate research staff: 360
Activities: LNETI has been given the necessary facilities to modernize national industry and to assure its development, through industrial development, up-to-date technology, training programmes for skilled manpower, and technical information systems for industry.

INSTITUTE OF INDUSTRIAL TECHNOLOGY* 45

Research director: Eng Coelho de Carvalho
Activities: Scientific and technical assistance and training for various industrial sectors, with facilities to promote research, experimental development and transfer of technology in the industrial areas.

Technology Department for Food Industries* 46

Activities: Developing new technologies for the conversion and preservation of food, for the milk, meat and vegetable industries.

QUIMIGAL - Quimica de Portugal, EP* 47

Address: Avenida Infante Santo 2, 1399 Lisboa, Codex
or: Aportado 2026, 1101 Lisboa Codex
Telephone: (019) 604040/609511/609561
Telex: 12301 FABRIL P
Status: Industrial company

CHEMICAL PRODUCTS FOR AGRICULTURE DIVISION* 48

Activities: The Division manages QUIMIGAL's activities in the manufacture and commercialization of agricultural chemicals, especially fertilizers and pesticides. The importance of agriculture in the home economy, and the responsibilities of the Corporation towards it, demand the special attention of the Division to technical assistance, tests and research.

Agricultural Development Centre* 49

Address: Quinta dos Almostéis, 2685 Sacavem
Manager: Eng Vieira De Brito
Graduate research staff: 9
Activities: The Agricultural Development Centre delineates, controls and draws conculsions from tests carried out in the field the laboratory or the greenhouse; develops research and trial studies designed to optimize the cost/efficiency ratio of products and to ensure their legal and technical requirements; maintains continuous actualization and interpretation of data enabling assessment of the potential and opportunities for agricultural development in the different areas of the country, advising on courses of action and cooperating in their respective applications.
Publications: Ao Serviço da Lavoura, monthly.

FEEDSTUFF PRODUCTS DIVISION* 50

Address: Avenida Infante Santo 34, 1399 Lisboa, Codex
Activities: The Division manages QUIMIGAL's activities in the manufactureand commercialization of compound feedstuffs for livestock. Its responsibilities include technical support and experimentation, which is systematically executed in its own animal production units.

OILS AND SOAPS DIVISION* 51

Address: Avenida Infante Santo 34, 1399 Lisboa, Codex
Activities: The Division manages QUIMIGAL's activities in the manufacture and commercialization of consumer goods such as margarines, edible oils and soaps. The characteristics of some of the consumer products gave rise to the creation of a specialized company, Sovena (an association of QUIMIGAL and two other companies).

Sociedade Broteriana* 52

[Broteriana Society]
Address: Instituto Botânico da Universidade, Arcos do Jardim, 3049 Coimbra
Telephone: (039) 22897
Status: Independent research society
President: A. Fernandes
Sections: Taxonomic botany, Rosette Fernandes; cytotaxonomy, A. Fernandes; ecology and phytogeography, J. Barros Neves; genetics, J. Montezuma de Carvalho; plant physiology, Gil Silva Cruz
Activities: Taxonomy of vascular plants from Portugal, Azores, Cabo Verde, Angola and Mozambique; cytotaxonomy of vascular plants from Portugal.
Publications: Annual report; bulletin.

Universidade de Aveiro* 53

[Aveiro University]
Address: Rua Calouste Gulbenkian, P-3800 Aveiro
Telephone: (034) 28391/-2
Telex: 25373
Status: Educational establishment with r&d capability
Rector: Professor Dr José Ernesto de Mesquità Rodrigues

BIOLOGY DEPARTMENT* 54

Head: Professor Dr Gustavo Cardoso Nunes Caldeira
Graduate research staff: 5

ENVIRONMENT DEPARTMENT* 55

Head: Dr Fernando Jorge M. Antunes Pereira
Graduate research staff: 2

Universidade de Évora* 56

[Évora University]
Address: Largo dos Colegiais, 2 and 3, Apartado 94, 7000 Évora
Telephone: (069) 25572/3/4
Telex: 18771
Status: Educational establishment with r&d capability
Rector: Ario Lobo Azevedo

DEPARTMENT OF AGRICULTURAL EXTENSION* 57

Head: Professor Augusto da Silva
Senior staff: Eduardo Alvaro do Carmo Figueira

DEPARTMENT OF ANIMAL AND PLANT PATHOLOGY* 58

Head: Professor Victor Manuel Caeiro Pais
Senior staff: Professor Maria Ivone E. da Clara Henriques

DEPARTMENT OF ANIMAL HUSBANDRY* 59

Head: Professor George Braz Pereira
Senior staff: Dr Nuno Maria Vilas Boas Potes

DEPARTMENT OF BIOLOGY* 60

Head: Professor Carlos Alberto Martins Portas
Senior staff: Professor Jorge Quina Ribeiro de Araújo, Professor Fernando Manuel Santos Ferreira Henriques

DEPARTMENT OF CROP SCIENCE* 61

Head: Professor Carlos Alberto Martins Portas
Senior staff: Eng João Antero Araújo

DEPARTMENT OF ECOLOGY* 62

Head: Professor Eduardo Cruz de Carvalho

DEPARTMENT OF FISHERY SCIENCE* 63

Head: Professor George E. Braz Pereira
Senior staff: Dr Vasco de Almeida Valdez

DEPARTMENT OF LANDSCAPE PLANNING* 64

Head: Professor Gonçalo Ribeiro Telles
Senior staff: Alexandre d'Orey Cancela d'Abreu, Nuno José de Noronha Mendoça

Universidade de Lisboa* 65

[Lisbon University]
Address: Cidade Universitária, 1699 Lisboa, Codex
Telephone: (019) 767624
Status: Educational establishment with r&d capability
Rector: Professor Dr Raul Miguel de Oliveira Rosado Fernandes

FACULTY OF PHARMACY* 66

Address: Avenida das Forças Armadas, 1600 Lisboa
Dean: Professor Dr José do Nascimento Júnior

Department of Pharmacognosy* 67

Head: Professor Dr Joao Adriano Borralho da Graca

FACULTY OF SCIENCE* 68

Address: Rua da Escola Politécnica 58, 1200 Lisboa

Botanical Museum, Laboratory and Garden* 69

Head: Professor Dr J.E. dos Santos Pinto Hopes

UNIVERSITY INSTITUTES AND RESEARCH CENTRES* 70

Zoological and Anthropological Museum and Laboratory (Museu Bocage)* 71

Address: Forte de Nosenhora de Goia, Cascais
Commission: Dr António Augusto Soares; Dr Luis Vieira Caldas Saldanha; Dr José Antunes Serra

Universidade do Minho* 72

[Minho University]
Address: Largo do Paço, 4700 Braga
Telephone: (023) 27021
Telex: 32135

UNIT OF EXACT AND NATURAL SCIENCES* 73

Head: Professor J. Simão

Universidade do Porto* 74

[Oporto University]
Address: Apartado 211, 4003 Porto, Codex
Status: Educational establishment with r&d capability

FACULTY OF SCIENCES* 75

Address: 4000 Porto
Telephone: (029) 31 02 90

'Dr Augusto Nobre' Institute of Zoology* 76

Telephone: (029) 310290
Research director: Professor Dr Amilcar de Magalhaes Mateus
Graduate research staff: 6
Publications: Publicacūes do Instituto de Zoologia 'Dr Augusto Nobre'.

Universidade Tecnica de Lisboa* 77

[Lisbon Technical University]
Address: Rua Gonçalves Crespo 20, 1100 Lisboa
Telephone: (019) 530533
Status: Educational establishment with r&d capability

INSTITUTE OF AGRONOMY* 78

Address: Tapada de Ajuda, Lisboa, 3
Director: Professor Dr Pedro Varennes Monteiro de Mendonça

INSTITUTE OF VETERINARY MEDICINE* 79

Address: Rua Gomes Freira, Lisboa, 1
Director: Professor Dr José Luís da Silva Leitão

PUERTO RICO

Institute of Tropical Forestry 1

– ITF
Address: PO Box AQ, Rio Piedras, 00928
Telephone: (809) 763 3939
Affiliation: United States Department of Agriculture, Forest Service
Status: Official research centre

SOUTHERN FORESTRY EXPERIMENTAL STATION 2

Project leader: Ariel E. Lugo
Graduate research staff: 8
Activities: Forestry research.
Publications: Annual report.
Projects: Timber plantation culture; naturally regenerated forest ecosystems; conservation of forest wildlife, especially the endangered Puerto Rican parrot.

Mayaguez Institute of Tropical Agriculture 3

Address: PO Box 70, Mayaguez, 00709
Telephone: (809) 832 2435
Affiliation: United States Department of Agriculture
Status: Official research centre
Location leader: Dr Antonio Sotomayor-Ríos
Graduate research staff: 5
Activities: Aid to continental United States agriculture through winter nurseries that permit year-round outdoor experiment and development; introduction, evaluation, selection, multiplication, preservation, and development of tropical crop germplasm; cooperative studies; tropical research programmes.
Publications: Annual report.

Projects: Introduction, classification, maintenance, evaluation and documentation of plant germplasm; breeding and production - soyabeans, peanuts and other oilseed crops (F. Vázquez); protein improvement in tropical grain legumes; multiple-disease resistant cowpeas; improvement of beans (Phaseolus vulgaris) for the tropics (G.F. Freytag); breeding of corn for the tropics; improving sorghum germplasm by breeding and genetic research; breeding and seed increase of sorghum, corn and pearl millet (A. Sotomayor-Ríos); breeding and production: tomatoes and sweet potatoes (F.W. Martin).

University of Puerto Rico,* 4

Address: Mayaguez, 00708
Status: Educational establishment with r&d capability

FACULTY OF AGRICULTURAL SCIENCES* 5

Dean: Dr Luis A. Mejia-Mattei

Academic Extension Division* 6

Director: Julio César Pérez

Agricultural Experimental Station* 7

Director: Dr Raúl Abrams

Agricultural Extension Service* 8

Director: Dr Roberto Vázquez-Romero

REUNION

Institut de Recherches Agronomignes Tropicales*   1

– IRAT
[Tropical Institute of Agronomic Research]
Address: Saint Denis, 97487
Status: Official research centre

Institut de Recherches sur les Fruits et Agrumes   2

– IRFA
[Tropical Fruit Research Institute]
Address: BP 180, Saint Pierre, 97455
Telephone: 25 01 24
Status: Official research centre
Director: B. Moreau
Departments: Bacteriology and virology, B. Aubert; tropical fruit research, C. Vuillaume; temperate fruit research, P. Fournier; agroeconomics, M Freyssinel; nursery, A. Sizaret
Graduate research staff: 6
Activities: Citrus irrigation; plant breeding, protection and husbandry. Tropical fruits and nursery. Temperate fruits.
Publications: Annual report.
Projects: Bacteriology and virology (B. Aubert); temperate fruits - apple and peach (P. Fournier); tropical fruits - techniques and pest control (C. Villaume); agroeconomy - citrus, banana, pineapple (M. Freysinnel).

ROMANIA

Academia Republicii Socialiste Romania* 1

[Academy of the Socialist Republic of Romania]
Address: Calea Victoriei 125, Bucureşti
Status: Official research organization
Scientific sections: Mathematical sciences; physical sciences; chemical sciences; technical sciences; biological sciences; agricultural and forestry sciences; geological, geophysics and geographical sciences section; medical sciences
Activities: The Academy is the major national scientific central body, and has a president and several vice-presidents. Each section has several full members, and some corresponding members.

Centrul de Cercetări Biologice Cluj-Napoca* 2

[Biological Research Centre, Cluj-Napoca]
Address: Strada Republicii 48, 3400 Cluj-Napoca
Affiliation: Ministry of Education and Training
Status: Official research centre
Director: Professor Dr V. Preda

Centrul de Cercetări Biologice Iaşi* 3

[Biological Research Centre]
Address: August 11, Iaşi
Status: Official research centre
Director: Professor Dr C.C. Zolyneak

STATIUNEA DE CERCETĂRI STEJARUL* 4

['Stejarul' Research Station]
Address: Pîngăraţi, Neamţ
Telephone: (936) 10809
Director: Dr Ion Băra

Genetics of Plants Department* 5

Head: Dr Ion Băra
Activities: Plant productivity.

Hydrobiology Department* 6

Head: Dr Ionel Miron
Activities: Limnology of dam reservoirs; hydrobiology of rivers; underground sources for water supply; water toxicology; hydrology; hydrochemistry; algae; zooplankton; zoobenthos; ichthyology; microbiology.

Institutul Agronomic Dr Petru Groza 7

[Dr Petru Groza Institute of Agronomy]
Address: Strada Manastur 3, 3400 Cluj-Napoca, Cluj
Telephone: (951) 12440; 12441
Status: Educational establishment with r&d capability
Rector: Professor Ioan Puia
Pro-Rector: Professor Zoltan Nagy
Graduate research staff: 230
Activities: Natural resources: soil science and soil improvement, land use and land improvement, fertilization and soil conditioning, drainage and irrigation; plant production: cereals, pulses, forage, root and tuber, textile, oil, medicinal, aromatic, horticultural plants - plant genetics and breeding, crop husbandry, plant protection; animal production: horses, cattle, sheep, goats, pigs, poultry, bees - animal genetics and breeding, livestock husbandry and nutrition, technology, hygiene and in-

spection of animal products, fisheries, veterinary medicine; agricultural and zootechnical engineering and building.
Publications: Buletinul Institutului Agronomic Cluj-Napoca series on agriculture (annually), and series on animal husbandry and veterinary medicine (annually).
Projects: Ecology and noxious factors in ecosystems; land improvement; production of new varieties of forage, technical and horticultural plants; improvement of cultivation technology and of sanitary plant protection methodology; breed improvement; livestock breeding, increased production, improved cattle, sheep, pig and poultry rearing and exploitation technology; improved hygienic quality of food products and inspection methodology; prevention and control of intensively reared domestic animals' diseases.

FACULTY OF AGRICULTURE AND HORTICULTURE*   8

Dean: Professor Alexandru Salontai

Crop Farming, Plant Protection and Farm Mechanization Department*   9

Head: Professor Ioan Bobes

Farm Economy and Management Department*   10

Head: Professor Iosif Timen

Forage Crops and Soil Science Department   11

Head: Professor Victor Miclăuş

Genetics and Horticulture Department*   12

Head: Professor Costică Panfil

FACULTY OF ANIMAL BREEDING AND VETERINARY MEDICINE*   13

Dean: Professor Ioan Boitor

Anatomy Department*   14

Head: Professor Eugen Muresianu

Animal Breeding and Biochemistry Department   15

Head: Professor Octavian Popa

Livestock Husbandry Department   16

Head: Professor Augustin Pop

Veterinary Pathology Department*   17

Head: Professor Eronim Suteu

Institutul Agronomic 'Ion Ionescu de la Brad'*   18

[Ion Ionescu de la Brad Institute of Agronomy]
Address: Aleea M. Sadoveanu 3, Iaşi
Status: Educational establishment with r&d capability
Activities: There are faculties for agriculture, animal husbandry, horticulture and veterinary medicine.

Institutul Agronomic Timişoara   19

[Timişoara Institute of Agronomy]
Address: Calea Aradului 119, 1900 Timişoara
Telephone: (0961) 43016
Status: Educational establishment with r&d capability
Rector: Professor Ilie Duvlea
Departments: Anatomy and physiology, Professor Valeriu Pintea; general technology, Professor Ştefan Romoşan; infectious diseases, Professor Octavian Popa; special technology, Professor Ioan Fazecaş; State farms' organization and management, Professor Constantin Anderca; zootechnics and nutrition, Professor Marin Miloş
Graduate research staff: 158
Activities: Plant breeding (wheat, maize, rice); crop husbandry; drainage and irrigation; livestock breeding; livestock husbandry and nutrition; veterinary medicine.
Publications: Lucrari, Stiinţifice, annual series on agriculture, annual series on zootechnics, and annual series on veterinary medicine.

Institutul Central de Chimie*   20

[Central Institute of Chemistry]
Address: Splaiul Independentei 202, 77208 Bucureşti
Telephone: (90) 49 32 60
Telex: 10944 ichim
Affiliation: Ministerul Industriei Chimice
Status: Official research organization
Director General: Dipl Eng Maria Ionescu

CENTRUL DE CERCETĂRI PENTRU INGRĂSĂMINTE CHIMILE CRAIOVA* 21

[Research Centre for Chemical Fertilizers]
Address: Centrala Industrială de Îngrăsăminte Chimice Comuna Isalnita, 1100 Craiova
Telephone: (941) 19147
Director: Eng Ion Doca
Activities: Research on fertilizers and catalysts.

INSTITUTUL DE CERCETĂRI PESTICIDE* 22

– ICP
[Research Institute for Pesticides]
Address: Splaiul Independentei 202, 77208 Bucureşti
Telephone: (90) 49 29 86
Director: Professor Dr Nicolae Bărbulescu
Activities: Research on pesticides.

Institutul de Cercetăre şi 23 Inginerie technologica pentru Irigatii şi Drenaje

[Research and Engineering Institute for Irrigation and Drainage]
Address: Baneasa - Giurgiu, 8384 Jud Giurgiu
Affiliation: Academia de Ştiinte Agricole şi Silvice
Director: Dr Gheorghe Hâncu
Departments: Land improvement and agricultural constructions, E. Sârbu
Graduate research staff: 80
Activities: Drainage, irrigation and land use.
Publications: Biennial report.
Projects: Irrigation methods (Dr Gh. Hâncu); irrigation equipment (Dr N. Iga); drainage of agricultural lands (Dr I. Stanciu); management of irrigation and drainage systems (Dr N. Grumeza); optimization of irrigation rates application (Eng Avrigeanu).

Institutul de Cercetăre şi 24 Producţie a Cartofului

[Potato Research and Production Institute]
Address: Strada Fundaturii 2, 2200 Braşov, Braşov
Telephone: (0921) 12620
Director: Dr Tanase Gorea
Sections: Potato breeding and seed production, Dr T. Catelly; potato cropping laboratory, Dr H. Bredt
Activities: Potato crops - soil science, irrigation, land use, potato breeding, husbandry, potato protection, potato economics.

Publications: Annual report.
Projects: Potato genetics, initial material for breeding, mutations, heploids (Dr T. Gorea); potato breeding (Dr T. Catelly); potato seed production (Dr S. Man); potato protection: virology, phytopathology, entomology (N. Cojocaru); potato cropping: tillage, fertilizers, herbicides, crop rotation, mechanization (Dr H. Bredt); potato processing and storage (Dr S. Muresan); potato crop economics (I. Mezabrovszky); potato biochemistry and physiology (G. Ólteanu).

Institutul de Cercetăre şi 25 Producţie pentru Creşterea Bovinelor

– ICPCB
[Research and Production Institute for Cattle Breeding]
Address: PO 8113, Baloteşti, Sectorul agricol Ilfov - Bucureşti
Telephone: 17 77 43
Affiliation: Academia de Ştiinţe Agricole şi Silvice
Status: Official research centre
Director: Dr Smarand Duică
Laboratoires: Genetics and breeding, Dr I. Granciu; physiology and feeding, Dr D. Popovici; technology of bovine management, Ing P. Gheorghe
Graduate research staff: 70
Activities: Breeding and management of cattle, including buffaloes: breed improvement - applied quantitative and qualitative genetics, immunobiochemical genetics, cytogenetics, reproduction, physiology, behaviour; feeding; housing; management; bull production and sire progeny testing in field and special units. ICPCB integrates scientific research with production by the introduction of new methods and technologies in cattle husbandry.
Publications: Annual report.
Projects: Bases for the increase of bovine milk and meat production (Dr Ioan Granciu); increase of milk and meat production in the Romanian spotted breed (Dr Emil Silvaş); increase of milk and meat production in the brown breed (Dr Dumitru Şerban); increases of milk and meat production in Friesian cattle (Dr Dumitru Georgescu).

Institutul de Cercetăre şi Producţie pentru Cultura Pajiştilor

26

– ICPCP
[Grassland Research and Production Institue]
Address: Strada Cucului 5, Braşov
Telephone: (0921) 17620
Status: Official research centre
Director: Dr Vasile Cardaşol
Departments: Animal production on herbage, M. Proca; chemical composition, G. Oprea; genetic resources, Dr J.A. Kovács; herbage breeding, Dr M. Krauss; Dr I. Breazu; management of cultivated grasslands, C. Constantinide; management of natural grasslands, Dr V. Cardaşol; seed production, I. Tănase
Graduate research staff: 25
Activities: Creation and improvement of herbage varieties (poaceae and leguminoasae) for Romanian agriculture; elaboration of technologies for improvement of natural and cultivated grasslands; for seed production, fodder conservation, grazing behaviour, animal production; increased production potential of Romanian meadows and pastures.
Publications: Lucrări ştiinţifice (Scientific Works, annual).
Projects: Breeding and seed production of perennial grasses and legumes (Dr M. Krauss); technologies and management of cultivated and natural grasslands (Dr N. Simtea); mechanization in grassland management (Dr P. Mănişor); economic aspects in grassland agriculture (P. Săbădeanu).

Institutul de Cercetăre şi producţie pentru Pomicultură

27

[Fruit Research Institute]
Address: 0300 Piteşti - Mărăcineni, Arges
Telephone: (0976) 34292
Telex: 18294 ipomi r
Status: Official research centre
Director: Dr Pirvan Parnia
Departments: Biochemistry and physiology, A.P. Bădescu; breeding and genetics, Dr V. Vasilescu; fruit technology, Dr Şt. Chiriac; growth regulators, Dr Sabina Stan; herbicides, Dr St. Coman; orchard management and agrotechnics, Dr A. Şuta; orchard mechanization, Gh. Stan; plant protection, Dr Victoria Suta: soil improvement, Dr M. Iancu; tissue culture, Tatiana Coman
Graduate research staff: 212

Activities: The institute has 26 fruit research stations where work is carried out on a variety of topics
Plant production: reduction of energy consumption in orchard management; concomitant execution of increased fertilization, soil treatment and anti-parasitic treatments; unconventional energy sources especially solar energy. Plant breeding: creation and multiplication of new fruit varieties tolerant of and resistant to plant diseases. Crop husbandry. Plant protection: reduction of number of antiparasitic treatments; application of reduced and ultra-reduced volume of treatments on area fruit.
Publications: Lucrari Stiintifice (Scientific Works, annual).
Projects: Apple, pear and quince culture (Dr A. Şuta); plum culture (Dr St. Coman); apricot, peach and almond culture (Dr V. Cociu); cherry culture (Dr T. Gozob); nut species culture (Dr V. Vasilescu); low bush species culture (Dr M. Botez); strawberry culture (Dr A. Teodorescu); tree propagation (Dr P. Parnia); dendrology and landscape design (T. Chiriţă).

Institutul de Cercetări pentru Cereale şi Plante Tehnice

28

– ICCPT
[Research Institute for Cereals and Industrial Crops]
Address: R-8264 Fundulea, Călăraşi
Telephone: (90) 15 08 05; 13 70 62; 18 36 44; 15 40 40
Telex: R-10489
Affiliation: Academia de Ştiinţe Agricole şi Silvice
Status: Official research centre
Director: Nichifor Ceapoiu
Research director: Dr Cristian Hera
Production director: Dr Nicolae Gumaniuc
Departments: Plant genetics, Viorel Vrânceanu; plant breeding I (self pollinated crops), Lazăr Drăghici; plant breeding II (open pollinated crops), Octavian Cosmin; crop husbandry I (non-irrigated crops), Gheorghe Sipoş; crop husbandry II (irrigated crops), Ion Picu; plant protection, Alexandru Bărbulescu; chemistry and physiology, Victoria Alexandrescu; seed multiplication, Ion Păcurar
Graduate research staff: 175
Activities: Fertilization and plant nutrition; land use and irrigation; plant rotation; cultural practices; plant genetics and physiology, genetic engineering; cell and tissue culture; collection, conservation and study of plant genetic resources; plant breeding (winter wheat, winter barley, rice, maize, sorghum, peas, beans, soyabeans, sunflower, oil, flax, cotton, lucerne, Italian ryegrass); plant protection (disease, insect and weed

control); seed multiplication; seed technology; seed testing; chemistry and biochemistry.

Publications: Annual report; *Probleme de genetică teoretică şi aplicată* , bimonthly; *Probleme de agrofitotehnie teoretică şi aplicată* , quarterly; *Probleme de protecţia plantelor*, quarterly

Projects: Cultivation: wheat and rye (Dr Nichifor Ceapoiu); barley and oats (Lazăr Drăghici); rice (Florica Melachrionos, Ion Albescu); maize and sorghum (Dr Cristian Hera, Octavian Cosmin); field peas (Alexandru Covor, Gheorghe Popa); field beans; soyabeans (Alexandru Covor, Ion Picu); sunflower and other oil crops (Dr Nicolae Gumaniuc, Viorel Vrânceanu); flax (Mircea Doucet); hemp (Oprea Segărceanu); cotton (Traian Cărpinişan); forage crops (Iuliu Moga); hop (Alexandru Salontai); production of biomass for chemical processing (Nicolae Hurduc).

Institutul de Cercetări pentru Legumicultură şi Floricultură   29

– VIDRA
[Vegetable and Flower Growing Research Institute]
Address: Vidra, Giurgiu
Telephone: 13 92 92
Affiliation: Ministerul Agriculturii şi Industrie Alimentare
Status: Official research centre
Director: Dr Ghorghe Vîlceanu
Departments: Agrochemistry and biochemistry, Victor Lăcătuş; genetics and breeding, Dr Orest Năstase; mechanization and growing techniques, Mihail Dumitrescu; plant protection, Dr Marcel Costache; protected crops, Laurian Maier; seed production, Nicolae Gherman
Graduate research staff: 106
Activities: Plant breeding and selection; super-elite and elite seed production; plant protection; soil science; organization, mechanization and economic efficiency; growing techniques; protected vegetable and flower crops; unconventional energy sources.
Publications: Annual report with summaries in English and French.
Projects: Field grown tomatoes (Mihail Dumitrescu); field grown pepper and eggplants (Cornel Tănăsescu); field grown onion and garlic (Viorel Rădoi); field grown cucurbits (Victor Rojancovski); field grown cole crops (Maria Sveatchievici); vegetables under high plastic tunnels (Lucia Luncă); flowers in the field and under plastic (Rodica Tepordei); genetics and plant physiology (Dr Orest Năstase); mechanization (Alexandru Conea).

Institutul de Cercatări pentru Mecanizarea Agriculturii*   30

[Agricultural Mechanization Research Institute]
Address: Bulevardul Ion Ionescu de la Brad 6, Bucureşti 1
Status: Official research centre

Institutul de Cercetări pentru Pedologie şi Agrochimie   31

– ICPA
[Soil Science and Agrochemistry Research Institute]
Address: Bulevardul Mărăşti 61, Bucureşti 71331
Telephone: (90) 17 21 80
Telex: 11394 asas r
Affiliation: Academia de Ştiinţe Agricole şi Silvice
Status: Official research centre
Director: Corneliu Răuţă
Departments: Agricultural chemistry; land evaluation; soil analyses; soil genesis, classification and soil mapping; soil physics and technology; soil pollution control
Graduate research staff: 110
Activities: Soil science; fertilizers; land use; soil physics; soil pollution control.
Publications: Anale ICPA , annual.

Institutul de Cercetări pentru Protecţia Plantelor*   32

– ICPP
[Plant Protection Research Institute]
Address: Bulevardul Ion Ionescu de la Brad 8, 71592 Bucureşti 1
Affiliation: Academia de Ştiinţe Agricole şi Silvice
Status: Official research centre
Director: Dr T. Baicu
Activities: Pesticides; entomology; virology, bacteriology and mycology.

Institutul de Cercetări pentru Viticultură şi Vinificaţie   33

– ICVV
[Vinegrowing and Winemaking Research Institute]
Address: cod 2040 Valea Călugărească, Prahova
Telephone: (971) 11200; (972) 35895; 35351
Telex: 19270; 19288
Affiliation: Economic Production Trust for Viticulture and Oenology
Status: Independent research centre
Director: Dr Leonida Mihalache
Departments: Genetics and amelioration, I. Poenaru; oenology, M. Macici; science, M. Macici; vine protection, Elisabeta Stoian; viticultural agrotechnics, L. Mihalache
Graduate research staff: 51
Activities: Vine genetics and improvement; vine propagation; viticultural improvements; nutrition and fertilization; mechanization; viticultural economy; technologies for alcoholic beverages and wine producing; microbiology and chemistry of wines.
Publications: Annual report.
Projects: Vine genetics and improvement (I. Poenaru); vine propagation (V. Grecu); viticultural improvements (L. Mihalache); nutrition and fertilization (Gh. Condei); vine protection (Elisabeta Stoian); oenology (M. Macici).

Institutul de Cercetări şi Proiectări 'Delta Dunării'*   34

– ICPDD
[Danube Delta Research and Design Institute]
Address: Strada Alexandru Sahia 2, 8800 Tulcea
Affiliation: Academia de Ştiinţe Agricole şi Silvice
Status: Official research centre
Director: A. Volcov
Activities: Investigation into the resources of the Danube Delta; agriculture; pisciculture; forestry.

Institutul de Cercetări si Proiectări pentru Valorificarea si industrializarea Legumelor şi Fructelor   35

[Research and Design Institute for Valorization and Processing of Vegetables and Fruits]
Address: Strada Lînăriei 93-95, PO Box 93, 4 Bucureşti 75162
Telephone: (90) 23 00 99
Affiliation: Ministerul Agriculturii şi Industriei Alimentare
Status: Official research centre
Director: Dr Ion Rădulescu
Sections: Processing of horticultural products, Eng Octavian Burtea; laboratory for conditioning and keeping of vegetables and fruits, Dr Andrei Gherghi; laboratory for packing, transportation, presentation and economics of vegetables and fruit valorization, Dr Ion Mircea; mechanization and automation of valorization processes of vegetables and fruits, Dr Liviu Ionescu
Activities: Horticultural products: steady market supply; maintenance of produce freshness; store pest control; promotion of latest processing methods; economical packaging; minimization of loss during transport; marketing methods.
Publications: Scientific Works .

Institutul de Cercetări Veterinare şi Biopreparate Pasteur   36

– ICVB Pasteur
[Pasteur Institute for Veterinary Research and Biological Products]
Address: Soseauna Giuleşti 333, 7000 Bucureşti 77 826
Telephone: (90) 18 04 90
Telex: 10238 icvbp r
Affiliation: Academia de Ştiinţe Agricole şi Silvice
Director: Dr V. Popovici
Sections: Pathology of ruminants; swine pathology; biochemistry; nonspecific pathology; viral diseases; parasitology; bacterial diseases, immunoprophylactic and diagnosis biologicals and allergens
Graduate research staff: 146
Activities: Studies of infectious, parasitic, nonspecific and reproduction diseases in animals, disease agents, diagnosis, prevention and control methods; elaboration of modern methodologies for the preparation of vaccines against infectious diseases in animals, diagnosis reagents and medicated products for the prevention and control of diseases in animals. Particular investigations are

undertaken into the following: viral diseases; swine pathology; pathology of sheep and horses; sheep pathology and parasitic diseases; chemotherapy - toxicology - disinfection; transfer of genetic information with application in animal pathology; veterinary immunology; improvement of biological products and of vaccination and diagnosis methods.
Publications: Archiva Veterinaria (annual); *Lucrările ICVB Pasteur* (Scientific Papers of the Pasteur Institute, annual).

Institutul de Constructii*   37

[Civil Engineering Institute]
Address: Bulevardul Republicii 176, Bucureşti
Telephone: (90) 424200
Affiliation: Ministry of Education and Training
Status: Educational establishment with r&d capability
Activities: There are departments of civil, industrial and agricultural engineering; railways, bridges and geodesy; hydraulics; building construction equipment.

Institutul Politehnic Bucureşti   38

– 1 PB
[Bucharest Polytechnic]
Address: Splaiul Independenţei 313, Bucureşti 77206
Status: Educational establishment with r&d capability

FACULTY OF FARM MECHANICS   39

Telephone: (970) 31 28 91
Dean: Dr Pavel Babiciu
Departments: Descriptive geometry and technical drawing, Eng Ion Enache; farm machinery, Professor Marcel Segărceanu; mechanics, Professor Dumitru Voiculescu
Graduate research staff: 18
Activities: Agricultural engineering, farm machinery and equipment - primary tillage and planting equipment; spraying, dusting, and fertilizing equipment; hay and forage harvesting equipment; grain and corn harvesting combines; special farm and ranch equipment; automation of farm machinery and equipment; farm power machinery.
Projects: Crop planting machines - grain drill and fertilizers (Dr Pavel Babiciu); grain harvesting combines - threshing and separation mechanisms (Professor Marcel Segărceanu).

Institutul Politehnic din Cluj-Napoca*   40

[Cluj-Napoca Polytechnic]
Address: Strada Emil Isac 15, 3400 Cluj-Napoca
Telephone: (951) 34565 25699
Telex: 31-352
Status: Educational establishment with r&d capability
Rector: Dr Eng Attila Pálfalvi
Pro-Rectors: Eng Th. Alb, Dr Eng Al. Cătărig, Dr Eng D. Comşa
Registrar: V. Ilieş

FACULTY OF CONSTRUCTION*   41
Dean: Dr Eng G. Bârsan

Agricultural Constructions; Constructions in Transportation*   42
Head: D. Marusceas

FACULTY OF MECHANICS*   43
Dean: Dr Eng V. Ionuţ

Heat Engines, Testing and Experimenting on Agricultural Machines*   44
Head: Professor N. Băţagă

Thermo-Technics and Thermal Equipment in Farming*   45
Head: Professor A. Chirilă

POLYTECHNIC ATTACHED RESEARCH CENTRES*   46

Research and Design Team for Agricultural Machines*   47

Institutul Politehnic 48
'Gheorghe Gheorghiu-Dej'
Bucureşti*

[Gheorghe Gheorghiu-Dej Polytechnic, Bucharest]
Address: Splaiul Independenţei 313, 77.206 Bucureşti
Telephone: (90) 37 20 55
Telex: 10 490-PIBUH
Affiliation: Ministry of Education and Training
Status: Educational establishment with r&d capability
Rector: Professor Dr Ing Radu Voinea

FACULTY OF AGRICULTURAL 49
ENGINEERING*

Dean: P. Bibiciu

Institutul Politehnic 'Traian 50
Vuia' Timişoara*

['Traian Vuia' Polytechnic of Timişoara]
Address: Bulevardul 30 Decembrie 2, 1900 Timişoara
Telephone: (961) 34713
Telex: 71347
Status: Educational establishment with r&d capability
Rector: I. Anton
Pro-Rectors: I. DeSabata, H. Theil, N. Bogoevici, A. Saimac

FACULTY OF AGRICULTURAL 51
MECHANICS*

Dean: I. Niţă

Agricultural Machines* 52

Head: Professor St. Căproiu

Tractors and Automobiles* 53

Head: Professor N. Tecuşan

Use of Agricultural Machinery* 54

Head: Professor A. Sandru

Institutul Roman de 55
Cercetări Marine*

[Marine Research Institute of Romania]
Address: Bulevardul Lenin 300, Constanta, 8700
Telephone: (016) 43288;47288
Telex: 14286
Status: Official research centre
Director: Contraamiral Ing C. Tomescu
Scientific director: Dr Ing V. Iordanescu
Sections: Fish research; hydrology and hydrochemistry; sea pollution; marine ecology
Graduate research staff: 100
Activities: Studies in the Black Sea and Atlantic Ocean on physical and chemical oceanography, plankton, invertebrates, fish, water pollution and marine technology.

Staţiuhea de Cercetări 56
pentru Plante Medicinale şi
Aromatice

– SCPMA
[Medicinal and Aromatic Plants Research Station]
Address: Cod 8264, Fundulea, Călăraî
Telephone: 16 68 42
Affiliation: Academia de Ştiinţe Agricole şi Silvice
Status: Official research centre
Director: Dr Emil Paun
Activities: Plants - breeding, production, and protection.
Publications: Herba Romanica , annual.

Universitatea Babes-Bolyai* 57

['Babes-Bolyai' University, Cluj-Napoca]
Address: Strada Kogălniceanu 1, 3400 Cluj-Napoca
Telephone: (951) 16100
Status: Educational establishment with r&d capability
Rector: Professor Dr Ion Vlad
Pro-Rector: Professor Dr Ionel Haiduc
Graduate research staff: 900

COLLEGE OF BIOLOGY, 58
GEOGRAPHY AND GEOLOGY*

Address: Strada Clinicilor 5-7, Cluj-Napoca
Telephone: (951) 14296
Dean: Professor Dr Stefan Kiss

ROMANIA

GRĂDINE BOTANICĂ A 59
UNIVERSITĂŢII DIN CLUJ-NAPOCA*

[Botanical Garden of University of Cluj-Napoca]
Address: Strada Republicii 42, 3400 Cluj-Napoca
Telephone: (951) 24060;21604
Director: Professor Dr Onoriu Ratiu
Sections: Taxonomy; phytosociology; physiology of plants; mycology of soil
Graduate research staff: 6
Activities: Flora and vegetation of Romania; physiology of germination; soil mycology; introduction of new species of plants.

Universitatea din Brasov* 60

[Braşov University]
Address: Bulevardul Gheorghe-Gheorghiu-Dej 29, 2200 Braşov
Telephone: (921) 4 15 80
Telex: 61318
Affiliation: Ministry of Education and Training
Status: Educational establishment with r&d capability
Rector: Professor Dr Eng Flores Dudiţa

FACULTY OF FORESTRY* 61

Graduate research staff: 7

Forest Management and Survey 62
Department*

Senior staff: N. Boş, S. Munteanu, A. Rusu, P. Gatej, Tr. Popovici, I. Tudor, O. Marcu

Logging and Transport Department* 63

Senior staff: R. Bereziuc, H. Furnica, V. Chiru, Gh. Ionaşcu, S. Corlaţeanu

Silviculture and Afforestation 64
Department*

Senior staff: I. Ciortuz, M. Marcu, F. Negruţiu, I. Damian, Gh. Mihai, D. Parascan, I. Florescu, A. Negruţiu, V. Stanescu

FACULTY OF MECHANICS* 65

Graduate research staff: 9

Automobiles, Tractors and Agricultural 66
Engineering Department*

Senior staff: D. Abăităncei, T. Nagy, C. Sălăjan, Gh. Bobescu, S. Năstăsoiu, I. Soare, V. Cîmpian, S. Popescu, Fr. Tanase, E. Ionescu, Gh. Radu, M. Untaru

Universitatea din 67
Bucureşti*

[Bucharest University]
Address: Bulevardul Gheorghe Gheorghiu-Dej 64, Bucureşti
Telephone: (90) 160187
Affiliation: Ministry of Education and Training
Status: Educational establishment with r&d capability

FACULTY OF BIOLOGY* 68

Address: Spl Independenţei 91-93, Bucureşti
Telephone: (90) 151902
Dean: Professor Dr Gh. Zarnea

Universitatea din Craiova* 69

[Craiova University]
Address: 13 Strada Al I. Cuza, Craiova
Telephone: (941) 14398
Affiliation: Ministry of Education and Training

FACULTY OF AGRICULTURE* 70

Address: Piţa Libertăţii 15, Craiova
Telephone: (941) 23475
Dean: Professor Dr N. Marin

FACULTY OF HORTICULTURE* 71

Telephone: (941) 12427
Dean: Professor Dr F. Ene

Universitatea din Galati* 72

[Galati University]
Address: Bulevardul Republicii 47, 6200 Galati
Telephone: (934) 93413602
Telex: 51292
Status: Educational establishment with r&d capability
Rector: Professor Dr Ing Ion Crudu

ST HELENA

Agriculture and Forestry Department

1

Telephone: 202
Status: Official research centre
Agricultural and Forestry Officer: G. Thwaites
Graduate research staff: 2
Activities: Soil science; land use; plant production, potatoes and vegetables; plant breeding; crop husbandry; plant protection; animal production, cattle, sheep, goats, pigs; livestock breeding, husbandry and nutrition; veterinary medicine; forestry and forest products.

ST LUCIA

Windward Islands Banana Growers Association*

1

WINBAN RESEARCH AND DEVELOPMENT

2

Address: Box 115, Castries
Telephone: (16) 255
Telex: 323 WINBAN LC
Status: Independent research centre
Director: Dr Joseph Edmunds
Departments: Nematology, Dr Edmunds; chemistry, J.H.L. Messing; fruit quality, Colin Borton; extension agronomy, Ben Laville; plant pathology, Dr Keith Cronshaw; crop protection, Everton Ambrose; library, Gloria E. Kast
Graduate research staff: 9
Activities: Plant production, fruits, bananas; quality improvement of bananas grown in the West Indies.
Publications: Annual report.
Projects: Intercropping of bananas with various food crops.

ST TOME

Ministério da Agricultura   1
Estaçáo Experimental

[Ministry of Agriculture]
Address: CP 47, Sao Tomé, Sao Tomé e Principe
Telephone: Madalena 15
Telex: MINAGRI 230
Status: Official research centre
Director: Maria da Graca Viegas
Sections: Plant health; cacao; food; industry; agrochemicals; biostatistics
Graduate research staff: 4
Activities: Soil science, land use, plant production and protection, crop husbandry.
Projects: Palm and coconut production; plant quarantine; food science.

SENEGAL

Centre National de Recherches Agronomiques   1

– CNRA
[National Agronomic Research Centre]
Address: BP 53, Bambey, Région de Biourbel
Telephone: 58 63 51; 58 63 52; 58 63 54
Affiliation: Institut Sénégalais de Recherches Agricoles
Status: Official research centre
Director: Mahawa Mbodj
Graduate research staff: 37
Publications: Annual report.

CEREAL IMPROVEMENT DEPARTMENT   2

Head: J. Chantereau
Sections: Wheat genetics and improvement, S.C. Gupta, Mr Nidoye, S. Hanne; wheat physiology, F. Digo; mineral nutrients, T. Diouf, S. Faye; genetics and improvement of sorghum, J. Chantereau, M. Dallo, M. Galiba, S. Samb; genetics and improvement of corn, P.A. Camara, A.A. Coly

DRY SOIL FERTILITY DEPARTMENT   3

Head: R. Oliver
Sections: Chemistry of mineral fertilization, S. Diatta, J.P. Ndiaye, L. Cisse; organic nitrogenous matter, F. Ganry, F. Gueye; rhizobiology, J. Wey, J.L. Alard, B. Ndir, M.Ndiaye; mineral nutrition of plants, vacant; analysis laboratory, R. Oliver, A. Ndiaye

INDUSTRIAL CULTIVATION IMPROVEMENT AND DIVERSIFICATION DEPARTMENT   4

Head: J. Gautreau
Sections: Genetics and improvement of the ground-nut, O. De Pins; ground-nut physiology, J. Gautreau, A. Ndiaye; resistance to flavotoxins, vacant; soyabean phytotechnics, J. Larcher; diversification of species, S. Thiaw

PRODUCTION SYSTEMS AND APPLIED RESEARCH DEPARTMENT   5

Heads: G. Pocthier, M. Fall
Sections: Rural economy, M. Fall; applied research, G. Pocthier, M. Fall; coordination and experimentation, L. Diack

PROTECTION DEPARTMENT   6

Head: Mb. Ndoye
Sections: Wheat entomology, Mb. Nidoye, K. Diop; sorghum entomology, R. Gahukar, S. Diop; wheat pathology, D. Louvel, D. Lassey; sorghum pathology, vacant; phytopharmacy, M. Ly, M. Sarr; integrated control, Mb. Ndoye, G. Pierrard

TECHNIQUES AND SYSTEMS DEPARTMENT   7

Head: S. Hernandez
Sections: Mezereon, S. Hernandez, M. Wade; zootechnics, A. Faye, E.H. Diallo; agricultural machinery and rural engineering, M. Havard, H. Mbengue; post-harvest technology, vacant; exploitation of large farms, S. Manga; seed production, R. Guegan

WATER UTILIZATION DEPARTMENT   8

Head: T.M. Duc
Sections: Bioclimatology, C. Dancette, N. Piton;
agricultural hydraulics, T.M. Duc, J. Sene; soil physics,
J.L. Chopard

Institut Sénégalais de   9
Recherches Agricoles*

– ISRA
[Senegal Institute of Agricultural Research]
Address: BP 3120, Dakar
Affiliation: State Secretariat of Scientific and Techni-
cal Research
Status: Official research centre

Institut de Technologie   10
Alimentaire

[Food Technology Institute]
Address: Route des Pères Maristes, BP 2765, Dakar
Telephone: 22 00 70
Status: Official research centre
Director: Dr Ousmane Kane
Departments: Plant production, Dr B. Diallo; Animal
production, M. Diop; Laboratories, Y. Gaye; Research
support, Dr C. Ndiaye
Graduate research staff: 25
Activities: Stabilization of local food production.
Publications: Annual report.
Projects: Millet processing (Dr B. Diallo); functional
properties of south-west African millet and sorghum (Dr
A.A. Thiam); sun drying of fish, preservation of fish in
refrigerated sea water (N. Diouf); clotted milk produc-
tion (M. Diop).

Laboratoire de l'Élevage et   11
de Recherches Vétérinaires

– LNERV
**[National Laboratory of Animal Health and Pro-
duction]**
Address: BP 2057, Dakar
Telephone: (320) 21 12 75; 21 51 46
Affiliation: Institut Sénégalais de Recherches
Agricoles
Status: Official research centre
Director: Dr Abdel Kader Diallo
Sections: Bacteriology, M.P. Doutre; parasitology,
S.M. Toure; virology, F. Sagna; agrostology and forage
crops, J. Valenza, G. Roberge; nutritional physiology, N.
Mbaye; zootechnology, J. P. Denis; documentation, Ph.
Lhoste; vaccine production, A. Niasse
Graduate research staff: 20
Activities: Animal pathology, virology, microbiology
and parasitology; animal husbandry and production,
nutrition and alimentation, pasture and fodder crops,
sheep and milk production; livestock breeding; veterin-
ary medicine; vaccine production.
Publications: Annual reports.

SINGAPORE

Parks and Recreation Department 1

Address: 19th Floor, Ministry of National Development Building, Maxwell Road, 0106
Telephone: 2220044
Status: Official research centre
Commissioner: Wong Yew Kwan
Activities: Tropical horticulture; plant breeding; protection; introduction; propagation and taxonomy; soil science; tissue culture.
Projects: Standard sampling methods for Angsanas, Rain trees and other common shade trees; liquid feed for Saintpaulias, (Foong Thai Wu); propagation of woody species, shrubs and creepers through tissue culture techniques, (Lim-Ho Chee Len, Lee Sing Kong); effect of soil treatments on the growth of Saintpaulias produced by tissue culture, (Foong Thai Wu, Lim-Ho Chee Len); establishment of clonal plantings of flowering trees, production of new varieties of Cassia by hybridization, (Jennifer Ng); introduction of scent into Vandaceous hybrids using Vanda tessellata, (Ng Siew Yin); common fungi in Singapore, (Chang Kiaw Lan); disease survey on ornamentals, phytotoxicity and efficacy testings of fungicides, hormone treatment on woody species, (Fong Yok King); effect of foliar nutrients on the growth rate of Ficus pumila, (Phua Kin Keng).

DEVELOPMENT AND RESEARCH BRANCH 2

Head: C.H. Kee
Sections: Planning and development, K.Y. Ho; training and research, S.Y. Ng; nursery services, A.S. Kau

MAINTENANCE AND ADVISORY BRANCH 3

Head: G.C. Ang

SOLOMON ISLANDS

Dodo Creek Research Station   1

Address: PO Box G11, Honiara
Telex: PRIMUS HQ 66311
Affiliation: Ministry of Agriculture and Lands
Status: Official research centre
Chief Research Officer: vacant
Departments: Entomology, R. Macfarlane; plant pathology, Dr G.V.H. Jackson; soils and plant nutrition, L.D. Chase; pasture agronomy, H. Siota; horticulture, P. Linton
Graduate research staff: 7
Activities: Productivity increase of land and labour to meet food requirements; investigation of suitable new crops and appropriate management practices; investigation of identified agricultural opportunity area potential; technically and economically sound management practice development for cattle and pastures; pest and disease problem identification and control on crops and livestock.
Soil and land use development; plant production (coconut, cocoa, spices, maize, sorghum, peanuts, yam, taro); crop husbandry; animal production, cattle (Brahman Crosses), goats, ducks, geese; livestock breeding, cattle 50 : 50 Bos Indicus/Bos Taurus; veterinary medicine, cattle disease survey and eradication.
Publications: Occasional reports.
Projects: Control of Scapanes Australis by chemical and biological means; control of major diseases of cassava, sweet potato, taro and yam through cultivar selection and breeding for resistance; control of Phytophthora Palmivora on cocoa and Marasmiellus Cocophilus on coconuts; semi-intensive soil and land suitability surveys for major crops such as coconut, cocoa, oil palm, rice and foodcrops; evaluation of suitable grass and legume pastures on land systems, beef cattle potential, fertilizer requirements and management systems; varietal and fertilizer evaluation of a wide variety of tree crops and foodcrops on differing soils and climatic conditions.

Ministry of Natural Resources Forestry Division   2

Address: PO Box G24, Honiara
Telephone: 521
Status: Official research centre
Chief Forest Officer: Samson Gaviro
Departments: Research, G. Chaplin; planning, A.C. Orr; plantation, J. Sandom; management, E. Kwanairara; utilization, T. Nolan; timber control, E. Dolaiano
Graduate research staff: 1
Activities: Line planting techniques, species selection and site matching.
Publications: Annual reports.

SOUTH AFRICA

AECI Limited 1

Address: PO North Rand, 1645, Transvaal
Telephone: (011) 608-1201
Telex: 8 7886
Status: Research centre within an industrial company
Research manager: Dr D.O. Hughes
Formations group: C.M. Fullarton
Graduate research staff: 4
Activities: Crop protection; pesticide formulation.

Animal and Dairy Science Research Institute 2

Address: Private Bag X2, Irene, 1675, Transvaal
Telephone: (012) 65721
Affiliation: Department of Agriculture and Fisheries
Status: Official research centre
Director: Dr Jan H. Hofmeyr
Divisions: National improvement schemes, B.R. Grobler; nutrition and physiology, Dr J.G. Cloete; poultry, W.C.J. Viljoen; dairy science and technology, Dr J.C. Oosthuizen
Graduate research staff: 108
Activities: The institute provides services in: the planning, development and promotion of production in the livestock and dairy industries; policy formulation, legislation and regulations pertaining to livestock production; conduct of pure and applied research on animal and poultry production, meat and dairy science, and its associated technology; auxiliary services on behalf of national farm livestock improvement schemes and programmes to improve the quality of dairy products, and specialist advice to producers, manufacturers and technicians.
Specifications of livestock: dairy and beef cattle, sheep, goats, pigs and poultry.
Publications: Biannual report.

ANIMAL SCIENCE RESEARCH 3

Deputy research director: Dr P.E. Lombard
Projects: Various projects on: meat production and improvement; breeding of cattle, goat, pig, sheep; milk, milk products, and wool production; animal feeds; animal nutrition.

DAIRY INDUSTRIAL DEVELOPMENT 4

Projects: Dairy products improvement (J.J. du Toit); milk quality payment and dairy extension (J.H. Labuschange, A. Dixon, N.H. Robertson).

DAIRY SCIENCE RESEARCH 5

Deputy research director: Dr J.C. Oosthuizen
Projects: Dairy microbiology (J.F. Mostert, H. Luck, T.J. Britz, J.B. Kriel); dairy chemistry (J.D. de Lange, T.E.H. Downes, J.C. Pelskr, J.H. Labuschagne, J.C. Meurs, J.A. Niewoudt, L.E. Smit).

Botanical Research Institute* 6

Address: Private Bag XlOl, Pretoria, 0001
Telephone: (012) 86-1164/5/6
Affiliation: Department of Agriculture and Fisheries
Status: Official research centre
Director: Dr B. de Winter
Deputy director: Dr D.J.B. Killick
Graduate research staff: 46
Activities: The institute is situated in a 77 ha botanical garden containing 5000 indigenous species, representing over a quarter of the flowering plants of South Africa. The building also houses the National Herbarium, which contains 600 000 specimens, including the national reference collection. There are regional herbaria at Stellenbosch, Grahamstown and Durban. The plant

information and identification service deals with some 32 000 specimens annually.

Projects: There are four main divisions of research: taxonomic research on the angiosperm, gymnosperm, pteridophyte, bryophyte and freshwater algal flora of Southern Africa, which includes taxonomic revisions, taxonomic monographs, preparation of the *Flora of Southern Africa* and regional floras, as well as palaeoflora and plant geography studies, herbarium methods and procedures, processing and retrieval of herbarium data.

Anatomical and cytogenetic studies of the Southern African flora, especially of grasses; comparative anatomy from a taxonomic viewpoint.

Ecological research including the ecology and phytosociology of Southern African terrestrial and freshwater macrophyte vegetation in general reconnaissance; physiognomic, floristic and ecological classification of vegetation; vegetation mapping, remote sensing, air photo interpretation and earth satellite evaluation and interpretation; quantitative analytic and synthetic methods of vegetation study; computer processing, ecological data storage and retrieval systems; autecology; primary productivity, especially of trees and shrubs, ecophysiology and environmental relations of aquatic macrophytes, especially Eichhornia crassipes.

Economic aspects of the indigenous flora, including: plant exploration for medicinal drugs, anti-tumour activity, use in survival situations, horticulture; ethnobotanical studies of present day and cave deposit floras; problem plants containing drug sources or properties poisonous to humans and domestic stock; agronomic, range and aquatic weeds; introduction and cultivation of indigenous flora.

Citrus and Subtropical Fruit   7
Research Institute

– CSFRI
Address: Private Bag X11208, Nelspruit, Transvaal
Telephone: (01311) 24241
Affiliation: Department of Agriculture and Fisheries
Status: Official research centre
Director: Dr J.H. Grobler
Deputy director: A. van Oostrum
Departments: Citrus, Dr G.S. Bredell; soil and chemistry, Dr S.F. du Plessis; entomology, Dr M.A. van den Berg; subtropical fruits, A. Joubert; plant pathology, Dr J. Moll
Graduate research staff: 36
Activities: Plant production including citrus, bananas, avocados, lychees, papaya, pineapple, spices and herbs, tea, coffee, bixa, guava, kiwi fruit, nuts; soil science, drainage and irrigation, land use; food science; plant protection and pathology; entomology.

Projects: Citrus research is carried out at Nelspruit, Friedenheim (near Nelspruit, in the Eastern Transvaal), Rustenburg (Western Transvaal), Citrusdal (near Cape Town) and Addo (near Port Elizabeth), and includes: cross-pollination studies; climatological studies; storage treatments; selection for disease resistance; fruit quality and easy-peeling cultivars; replant problems; investigations into soils and fertilization; irrigation, mulching, pruning and windbreaks; biological and integrated control of insect pests; control of nematodes, viruses, mycoplasms, bacterial and fungal diseases as well as quarantine schedules. At Messina, on the banks of the Limpopo River, virus-free citrus is produced.

Research on subtropical fruit is carried out both at Nelspruit and throughout the country on experimental farms and substations situated in areas with different climatic and soil conditions. At Nelspruit research is concentrated on citrus, mangoes, litchis, macadamia and pecan nuts, papayas, granadillas, guavas and the kiwi fruit. At Levubu (Northern Transvaal) research is conducted on bananas, macadamia nuts and other subtropical crops; at Burgershall (Eastern Transvaal) on these crops as well as on avocados, granadillas, coffee, tea, pepper and ginger; at Makatini research is conducted mainly on cashew nuts and coconuts.

In the Natal Region problems with subtropical crops such as bananas, pineapples and coffee are dealt with at Pinetown. The station at East London (in the Eastern Cape Region) is devoted entirely to pineapples.

Dairy Products   8
Manufacturers Association*

– DPMA
Address: PO Box 14624, Verwoerdburg, 0140
Status: Official research centre

Department of Agriculture   9
and Fisheries

Address: Private Bag X250, Pretoria, 0001
Telephone: (021) 284657
Status: Official research centre
Director-general: Dr D.W. Immelman
Graduate research staff: 1 493
Activities: The Department of Agriculture and Fisheries is active throughout South Africa. Farming conditions and farming practices differ greatly in the various agro-ecological regions of the country, and local research extension services under the department are geared to serve the farmers and the agricultural industry in each region: the department is also responsible for the formal training of farmers at five colleges of agriculture.

The department seeks to create an economically sound agricultural industry, and provides the following services: the formulation of agricultural economic policy; orderly marketing and price stabilization of agricultural products via the various control boards and in consultation with the National Marketing Council; registration of and guidance to cooperative societies and companies; the provision of credit for the purchase of land; granting of financial assistance to farmers under emergency conditions; promulgation of grading regulations and provision of inspection services for most agricultural products; research regarding general agricultural economic conditions, farming practices and planning as well as marketing; provision of statistical services for the determination of current conditions and expected tendencies; and the administration of various Acts relating to these matters.

Note: See also separate entries for:
Animal and Dairy Science Research Institute
Botanical Research Institute
Citrus and Subtropical Fruit Research Institute
Directorate of Agricultural Economics and Marketing
Directorate of Agricultural Technical Services including Fisheries
Division of Agricultural Engineering
Division of Agricultural Marketing Research
Division of Agricultural Production Economics
Eastern Cape Region: Agricultural Research Institute
Fruit and Fruit Technology Research Institute
Highveld Region: Agricultural Reasearch Institue
Horticultural Research Institute
Karoo Region: Agricultural Research Centre
Natal Region: Agricultural Research Institute
National Institute for Grain Crops
Oenological and Viticultural Research Institute
Orange Free State: Agricultural Research Institute
Plant Protection Research Institute
Sea Fisheries Institute
Soil and Irrigation Research Institute
Tobacco Research Institute
Transvaal Region: Agricultural Research Institute
Veterinary Research Institute
Winter Rainfall Region: Agricultural Research Centre
Publications: Agroanimalia (animal sciences); *Agrochemophysica* (soil, chemical and physical sciences); *Botanical Survey Memoirs*; *Bothalia*; *Entomology Memoirs*; *Flora of Southern Africa*; *Flowering Plants of Africa*; *Onderstepoort Journal of Veterinary Research*; *Phytophylactica* (plant protection sciences and microbiology); *Agroplantae* (plant sciences) *Crops and Markets*; *Trends in the Agricultural Sector*; *Abstract of Agriculture Statistics*; *Statistics on Fresh Markets*.

Directorate of Agricultural Economics and Marketing   10

Address: Private Bag X250, Pretoria, 0001
Telephone: (021) 28-4657
Telex: 3757
Affiliation: Department of Agriculture and Fisheries
Status: Official research centre
Deputy director-general: W.J. Treurnicht
Chief director: I.S. Geldenhuys
Activities: The Directorate exists to promote the development of an economically sound agricultural industry.

Directorate of Agricultural Technical Services including Fisheries   11

Address: Private Bag X116, Pretoria, 0001
Telephone: (021) 21/8111
Telex: 3756
Affiliation: Department of Agriculture and Fisheries
Status: Official research centre
Deputy director-general: Dr D.J. Agenbach
Chief Directors: Dr J.C. Strydom (chairman); Professor S.A. Hulme; Dr J.E. Erasmus; Dr E. Strydom
Activities: The Directorate exists to promote the productivity and efficiency of the South African agricultural and sea fisheries industries and to protect their natural resources.

Division of Agricultural Engineering   12

Address: Private Bag X515, Silverton, 0127, Transvaal
Telephone: (012) 861116
Telex: 3756 sa
Affiliation: Department of Agriculture and Fisheries
Status: Official research centre
Director: J.J. Bruwer
Subdivisions: Agricultural mechanization, J. Veenstra; soil conservation works, P. Pienaar; agricultural energy, F.J.C. Hugo; irrigation and drainage, J.F. van Staden; farm development, C.J. van Schalkwyk; winter rainfall complex, I. Scheepers
Graduate research staff: 50
Activities: The division conducts research, development, testing and advisory work in the following subject areas: drainage and irrigation; mechanization in plant production; mechanization in animal production, buildings and facilities for livestock husbandry; soil conservation; mechanization in fruit and wine production; agricultural energy; animal waste management.

Publications: Annual report.
Projects: Soil conservation (C.T. Crosby); alternatives to diesel fuels, plant oil (F.J.C. Hugo); official testing of tractors (J.B. du Plessis); official testing of agricultural machinery (Dr B.V.A. Boshoff).

Division of Agricultural Marketing Research* 13

Address: Private Bag X246, Pretoria, 0001
Telephone: (021) 284985
Affiliation: Department of Agriculture and Fisheries
Status: Official research centre
Director: J.K. Siertsema
Deputy director: C.J. Bester
Activities: The division undertakes agricultural economic research and collects, processes and makes available statistics, in order to promote efficiency in the marketing of agricultural produce and to serve as a basis for the formulation and execution of agricultural policy by the State.
Projects: Research into the local and foreign marketing of agricultural produce not controlled in terms of the Marketing Act, 1968, that is, all types of vegetables and fresh fruit, except deciduous and citrus fruit and bananas, for which there are control boards.

Division of Agricultural Production Economics 14

Address: Private Bag X416, Pretoria, 0001
Telephone: (021) 28-4985
Affiliation: Department of Agriculture and Fisheries
Status: Official research centre
Director: H.S. Hattingh
Deputy director: J.S.G. Joubert
Activities: The principal task of the division is to undertake research into the economics of agricultural production, and to ensure that the research results and general economic principles are brought to the notice of producers and other interested parties. The object is to promote efficiency in the agricultural field and thus to place agricultural production on a sound economic basis.
Projects: Investigations into the profitability of farming systems and farming practices in the various agricultural regions, and an analysis of the factors influencing the profitability of farming in any region, as well as the economic planning of farming enterprises; development of research methods for the analysis and evaluation of production economics data; investigations with regard to labour, mechanization and land ownership, soil conservation, land use, capital financing,

managerial and socio-economic problems and the processing of agroeconomic data to standards and forms which can serve as aids in the economic planning of farming enterprises.

Eastern Cape Region: Agricultural Research Institute 15

Address: Private Bag X15, Stutterheim, 4930
Telephone: 710-714
Affiliation: Department of Agriculture and Fisheries
Status: Official research centre
Director: H.S. Niehaus
Deputy director: Dr A.D. Ventner
Graduate research staff: 14
Activities: Agricultural research, with particular reference to crop and animal production under the soil and climatic conditions of the Eastern Cape region.
Projects: Animal science; agronomy; pasture research; soil chemistry; crop production, including chicory and subtropical fruits; management and utilization systems for cattle and Boer goats in the Valley Bushveld.
The Agricultural Research institute of the region is based at Dohne, and has analytical laboratories, a controlled-environment hothouse, a farm unit, computer terminal, and library; research stations are situated at Bathurst and Adelaide.

Fruit and Fruit Technology Research Institute 16

Address: Private Bag X5013, Stellenbosch, 7600
Telephone: (02231) 2001
Affiliation: Department of Agriculture and Fisheries
Status: Official research centre
Director: Dr P.G. Marais
Deputy director: Dr J.H. Terblanche
Departments: Chemistry/physics, Dr W.A.G. Kotze; horticulture, Dr N. Hurter; plant protection, Dr P.L. Swart; plant physiology, Dr H.J. Van Zyl; technology, Dr B. Nortje
Graduate research staff: 15
Activities: Study of soils and soil problems, plant nutrition and irrigation with special reference to water requirements of deciduous fruit trees and table grapes; breeding of new cultivars of fruit and table grapes for disease resistance and cold requirements; horticultural and viticultural practices such as spacing, trellising, rootstocks, nursery practices, pruning; selection of outstanding clones of existing cultivars, evaluation of imported cultivars and hybrids from own breeding programmes; diseases of fruits and grapes, including post-

SOUTH AFRICA

harvest diseases and their control; spray application techniques; pesticide residue determinations; life cycles of insect pests of fruit and table grapes and their control by various methods; agrometeorology, with special reference to the correlation between weather and growth of plants and effect on diseases; enzymes in fruit maturity; optimum conditions of storage for every cultivar of fruit exported, marketed locally or canned; and canning, drying and freezing of fruit and vegetables, fruit juices and other processed fruit products.

The institute has the following experimental farms: Bien Donné (agrometeorology), Bellevue (grape growing), Robertson (apricots and peaches), Hexvalley (table grape trials), Elgin (apple trials), Drostersnes, (adjuct to Elgin), Citrusdal (stone fruits, table grapes, indigenous red tea, Aspalthus linearis), Langkloof (apple trials); Helderfontein.

Projects: Nutrition/irrigation/soil management (Dr W.A.G. Kotze); plant breeding (Dr N. Hurter); planting and pruning system and crop control - pome fruit (O. Bergh); planting and pruning system and crop control - stone fruit (Dr P.J.C. Stassen); disease control (Dr W.F.S. Schwabe); pest control (Dr P.L. Swart); plant physiology (Dr H.J. Van Zyl); post-harvest technology (Dr B.K. Nortje); cold storage (Dr G.J. Eksteen).

Highveld Region: Agricultural Research Institute   17

Address: Private Bag X804, Potchefstroom, 2520, Transvaal
Telephone: (01481) 4221
Affiliation: Department of Agriculture and Fisheries
Status: Official research centre
Director: Dr J.D. Slabber
Departments: Pasture science, R.H. Drewes; soil science, J.E. Volschenk; animal husbandry, J.W. Cilliers; crop production, M.C. Walters; education, I.E. De Waal; extension, Dr J.J. Coetsee; resource utilization, J.J. Scheepers
Graduate research staff: 50
Activities: The main aim of the institute is the development of production techniques and farming systems, adapted for optimal resource utilization within discrete soil-climatic-topographic units, known as relatively homogeneous farming areas. The development of computer models, systems research, and problem investigation in soil science, land use, crop husbandry, livestock husbandry and nutrition and pasture science are the main fields of activity.

The Potchefstroom complex houses the High veld Region Agricultural Research Institute, three outstations mainly for veld management and animal husbandry research, and the Highveld College of Agriculture; it

is also the base of the Research Centre for Oil- and Protein-rich Seeds and the Summer Grain Centre, both divisions of the National Institute for Grain Crops (please see separate entry).

Projects: Beef production from fertilized veld (J.W. Cilliers); beef production systems for extensive semi-intensive and intensive conditions (J.W. Cilliers and R.H. Drewes); utilization of Digitaria species for sheep production; selection of Digitaria eriantha cultivars for use in the Highveld; selection of adapted annual and perennial legume pasture species for use in veld situations (C.S. Dannhauser); liming requirements for crop production on acid soils (P.E. Haumann); kinetics of applied phosphorus in the soil (O.J. Van der Walt); moisture retention in Highveld soils (A. Lambooy); rearing and early weaning of dairy calves (C.W. Cruywagen); simulation model for beef production (Dr G. Coetsee); delineation of relatively homogeneous farming areas (J.J. Scheepers).

HIGHVELD COLLEGE OF AGRICULTURE*   18

Horticultural Research Institute   19

Address: Private Bag X293, Pretoria, 0001, Transvaal
Telephone: (012) 821694
Affiliation: Department of Agriculture and Fisheries
Status: Official research centre
Director: Dr A.J. Heyns
Deputy director: Dr J.T. Meynlardt
Departments: Potato, F.J. Du Plooy; vegetable, O.J. Olivièr; fruit and ornamental, E.P. Evans; plant protection, B.W. Young
Graduate research staff: 47
Activities: The institute undertakes fundamental and applied research throughout the country on vegetable crops and in floriculture and ornamental horticulture; it is also responsible for research on deciduous fruit in the summer rainfall area. It undertakes the development through breeding and selection, of improved cultivars suitable for South Africa, particularly of crops such as vegetables, potatoes, indigenous flower bulb crops, deciduous fruits and table grapes. At Roodeplaat, approximately 200 ha of irrigable land is used for field experiments. There are substations at Elsenburg (Stellenbosch), George, Glen (Bloemfontein), Ficksburg, Cedara (Pietermaritzburg), Nelspruit and Addo (Eastern Cape).
Projects: Flower bulb crops (Lachenalia and Ornithogalum) (J. Lubbinge); potato (F.J. Du Plooy); peach and table grape (E.P. Evans); tomato (E. Aucamp); bean (A.F. Coertze); onion (J.J.B. Van Zijl); apricot (T. Haulik); sweet potato (C.P. Plooy).

Karoo Region: Agricultural Research Centre

20

Address: Private Bag X529, Middelburg, 5900, Cape Province
Telephone: (04832) 21113
Affiliation: Department of Agriculture and Fisheries
Status: Official research centre
Director: Dr P.W. Roux
Sections: Agronomy, G.C. de Kock; animal production, Dr D. Wentzel; extension and soil conservation, Dr A.J. Siepker; pasture science, M. V. Vorster
Graduate research staff: 45
Activities: Natural resources science and technology: delimitation of reasonable homogeneous farming areas; ecological studies on plant communities; quantitative evaluation of Karoo veld; establishing of norms for grazing capacity; study on animal ratios, animal load and physical impact of stock on veld; development of formulae to estimate grazing capacity and potential. Plant production: breeding of improved drought fodder crops, eg Atriplex, Opuntia; breedinq of lucerne (Medicago sativa) for alkaline, acid, insect and disease resistance; utilization and evaluation of cultivated crops for animal production. Animal production: breeding of Merino, Dorper, Afrino sheep and Angora goats for higher production and fertility; research on quality, weathering and influence of management and feeding regimes on wool and mohair; nutritional evaluation of veld, drought and other fodder crops; research on the nutritional requirements of the Angora goat and other types of small stock with emphasis on grazing animals; supplementation of deficiencies and strategic feeding; study on hormone levels, reproduction of ewes and fertility of rams, artificial changes of oestrus cycle; improvement of reproduction rate.
Projects: Comparison of different selection methods in Merino sheep (J.J. Olivier); half-class fleece mass system as selection method in Merino sheep (J.A.A. Baard); selection for fleece properties and fertility in the Angora goat (L. de W. Louw); meat production and body composition and fertility in Dorper sheep (P.G. Marais); weathering and age effects of wool (Dr J.J. Venter); quality differences between under-crimped, Deurdentrue and over-crimped wool (P.D. Grobbelaar); comparitive study on the digestibility of three Atriplex species (G.A. Jacobs); voluntary intake, digestibility and chemical composition of Karoo veld selected by small stock (P.J.L. Zeeman); influence of energy and protein supplementation with and without phosphorus on production of Merino sheep on four types of Karoo veld (F.C. Hayward); sheep production on irrigated lucerne pasture; fodder utilization of Merino sheep (A.G. Bezuidenhout); influence of cold stress on abortion in the Angora goat; sex hormone levels of the Angora ewe (Dr D. Wentzel); hormone levels of the Angora ram (P.G. Loubser); fodder production; improvement of Oldman saltbush (G.C. de Kock); veld grazing experiment (Dr P.W. Roux); influence of climate on the vegetation of six veld types (E. Sykes); norms for carrying capacity of the biomass; selective grazing behaviour of sheep, cattle and goats (Dr P. Botha); photographic survey technique for the evaluation and measuring of changes in Karoo veld (Dr P.W. Roux); chemical control of misquite (Prosopus) in the north western Karoo (M. Vorster).

GROOTFONTEIN COLLEGE OF AGRICULTURE*

21

Head: J.A.A. Baard

Natal Region: Agricultural Research Institute

22

Address: Private Bag X9059, Pietermaritzburg, 3200, Natal
Telephone: (0331) 33371
Affiliation: Department of Agriculture and Fisheries
Status: Official research centre
Director: Dr P. Hildyard
Assistant director (Extension): Dr D.M. Scotney
Assistant director (research): Dr J.W.C. Mostert
Assistant director (training): U.W. Nänni
Departments: Animal husbandry, S.F. Lesch; biochemistry, J.P. Marais; entomology, Dr P. Joubert; land use, P. Fotheringham; pasture service, Dr P.J. Edwards; plant pathology, G. Nevill; soil science, R. Bennet; summer grains, Dr J. Mallett; agrometeorology, B. Clemence; agricultural engineering, B. Whittal; agricultural economics, N. Whitehead; agronomy, F. Hyam; systems section, R. Jones
Graduate research staff: 50
Activities: Animal production promotion (beef, dairy and sheep) - feeding/management, production potential; pasture utilization research - veld management/planted pastures and fodder production; plant production - maize, annual legumes and fodder crops; pasture grass and legume (clover) breeding; land use; soil science.
Publications: Annual report.
Projects: Animal production promotion - beef/dairy cattle and sheep feeding/management projects (S.F. Lesch); pasture utilization - veld management/ecology, grass/legume breeding, planted pastures, fodder crops (P.J. Edwards); agricultural resources utilization, soil surveys, land use (P.J. Fotheringham); soil/plant/water analysis for extension use (R. Bennet); summer grain research - tillage/moisture studies (J. Mallett); summer grain research - fertility studies (M.P. Farina); systems modelling in agriculture - animal/plant production (R. Jones); fodder crops and legumes (F. Hyam).

DUNDEE RESEARCH CENTRE 23

Address: Private Bag 626, Dundee, 300, Natal
Telephone: (0341) 22479
Officer in charge: A.P. van Schalkwyk
Sections: Animal science, A.P. van Schalkwyk; agronomy, F.P.C. Blamey
Activities: Beef production - fertility and management; production of oil seed crops; veld management and pasture/fodder production.
Projects: Production and reproduction of beef cattle (H.J. Meaker); fertilization of sandy scurveld (cold lands) under grazing (P. du Toit); various projects on sunflower, soyabean, groundnuts, and maize (F.P.C. Blaney, J. Chapman).

ESTCOURT RESEARCH COMPLEX 24

Address: Private Bag X752, Estcourt, 3310, Natal
Telephone: (03631) 3070
Officer in charge: B. Mappeldoram
Activities: Animal production - beef; pasture production, screening, production potential and management.
Projects: Planted pastures - introductions, screening production potential and management (B. Mappeldoram); beef production from veld and pastures; livestock husbandry and nutrition (S.D. Lesch).

GEDARA COLLEGE OF 25
AGRICULTURE*

KOKSTAD RESEARCH STATION 26

Address: Private Bag X501, Kokstad, 4700, Natal
Telephone: (0372) 72
Officer in charge: A.D. Lyle
Graduate research staff: 1
Activities: Animal production promotion - beef and woolled sheep; livestock, husbandry and nutrition; veld management, pastures and fodder crops.
Projects: Veld management and production potential; planted pasture and fodder crops; beef and wool production under different feeding/management regimes.

SUMMER GRAIN SUBCENTRE - 27
MAIZE BREEDING

Address: University of Natal, Private Bag 375, Pietermaritzburg, 3200, Natal
Telephone: (0331) 63320
Officer in charge: Dr H.P. Gevers
Graduate research staff: 2
Activities: Maize breeding. See also entry under National Institute for Grain Crops.

National Food Research 28
Institute

– N Food RI
Address: Private Bag 395, Pretoria, 0001, Transvaal
Telephone: (012) 86-9211
Telex: 3630
Affiliation: Council for Scientific and Industrial Research
Status: Official research centre
Director: Dr L. Novellie
Graduate research staff: 70
Activities: The main aim of the N Food RI is to promote the effective utilization of South Africa's food resources. Its activities include both fundamental and applied research into aspects of food consumption, utilization, preservation, packaging and storage, as well as product and process development. Another function of the institute is to provide information, research and specialized development facilities to government departments, to government sponsored institutions and the South African food industry.

BIOLOGICAL EVALUATION 29
DIVISION

Head: Dr J.J. Dreyer
Activities: All aspects of food evaluation, including physiological, biochemical, histological and biometrical analyses.
Projects: The digestibility and physiological assimilability of foods, as affected by food-technological manipulation; supplementary effects of protein foods upon one another; long term effects upon baboons of supplementation of predominantly cereal diets with proteins, vitamins and minerals from specific sources; effects of antinutrients such as polyphenols in grain sorghum, phytic acid in cereals and enzyme inhibitors in legume seeds; the response of primates to intake of foods of high fibre content; methodology of protein evaluation; the indigestible dry matter content of foods.

BREWING TECHNOLOGY DIVISION 30

Head: P. de Schaepdrijver
Projects: Studies of alcoholic and lactic acid fermentation; studies on the use of microbial enzymes and of selected pure brewery yeast cultures; process and plant engineering; development of brewing control methods; technological information and communications; testing of new plant and techniques on the industrial scale; studies on cooking and conversion by continuous processes; sterilization of beer and production of concentrates.

CEREAL BIOCHEMISTRY DIVISION 31

Head: Dr K.H. Daiber
Activities: Research into grain sorghum, including genetic, environmental, agronomic and seasonal effects, in order to develop malt qualities required by the brewing industry.
Projects: The development of enzymes in the germinating cereal grain.

FERMENTATION TECHNOLOGY DIVISION 32

Head: Dr T.G. Watson
Activities: Investigations of the technology of the production of a wide variety of substances through microbial activities: the substances produced range from simple to complex organic molecules and include a range of biopolymers not yet synthesized.

FOOD CHEMISTRY DIVISION 33

Head: A.S. Wehmeyer
Activities: The division is concerned mainly with the determination of the nutrient composition of foods, including water, fat, ash, crude fibre, protein, carbohydrate, minerals, vitamins and the amino acid composition of proteins.

FOOD TECHNOLOGY DIVISION 34

Head: Dr P.J. van Wyk
Activities: Investigation into food processing technology, and the suitability of new cultivars for processing.
Projects: The evaluation of the milling properties of maize cultivars bred by the Department of Agricultural Technical Services; the evaluation of new sweet-corn varieties with respect to their canning properties; technological aspects of biltong manufacture; the processing of subtropical products like mango, guava, avocado, macadamia nuts and ginger; determination of the suitability of new dry-bean cultivars for canning and a study of the various processes which can be used for processing of dried beans; evaluation of new fruit and vegetable cultivars from the Department of Agricultural Technical Services for their processing properties; an investigation of the fatty acid and glyceride composition of various fats and oils and their suitability for usage in food processing; the extraction, separation and analysis of the flavour components of foods; the correlation of organoleptic evaluation of flavours with instrumental methods of analysis; the use of immobilized enzymes in food processing.

MICROBIOLOGY RESEARCH GROUP 35

Head: Dr J.P. van der Walt
Activities: All aspects of microbiology, including morphology, taxonomy, genetics, biochemistry, ecology, and the interaction of microbial and higher forms of life.
Projects: Investigations aimed at advancing the knowledge of industrially important yeasts, primarily on taxonomic aspects, with special reference to ecology, the application of genetic criteria and the elucidation of life-cycles; investigations into microbial transformations with possible industrial application, including the commercial production of citric acid from locally available carbohydrate sources by specially selected mutant yeast strains; microbiological investigations relating to the requirements of the South African food industries, primarily around the bacteriological and mycological aspects of biltong production on industrial scale.

National Institute for Grain Crops 36

Address: Private Bag X804, Potchestroom, 2520, Transvaal
Telephone: (01481) 25504
Affiliation: Department of Agriculture and Fisheries
Status: Official research centre

RESEARCH CENTRE FOR OIL- AND PROTEIN-RICH SEEDS 37

Director: Dr W.P. Grobbelaar
Leader, research-group: Dr J.W. Snyman
Graduate research staff: 15
Activities: The centre is responsible for all aspects of plant production research related to soyabeans, peanuts (Arachis hypogaea), dry-beans (Phaseolus vulgaris and Phaseolus coccineus) and high-oil sunflower.
Projects: Breeding of high-oil sunflower (W.J. Vermeulen); breeding of high quality peanuts (P.J.A. van der Merwe); breeding of dry-beans (Phaseolus vulgaris) (W.J. Vermeulen); sunflower agronomy (H.L. Loubser); peanut agronomy (Dr C.J. Swanevelder); dry-bean agronomy (A.J. Liebenberg); soyabean agronomy (M.A. Smit).

SMALL GRAIN CENTRE 38

Address: Bethlehem Agricultural Research Station
Activities: Research on feeding, cultivation, evaluation, nutrition and physiology of wheat and barley, and to a lesser extent of oats, rye, durum, and triticale.
Projects: Breeding; participation in and evaluation of international nurseries; screening including chemical methods; cultivar evaluation; nutrition; weed control; production potential under dryland and irrigation; quality evaluation; disease and pest control.

SUMMER GRAIN CENTRE 39

Director: Dr W.P. Grobbelaar
Assistant director: J. De Kock
Departments: Breeding, Dr H.O. Gevers; cultivar evaluation, J.P. Pretorius; soil tillage, Dr J.B. Mallett; plant nutrition, Dr M. Farina; entomology, Dr J.B.J. Van Rensburg; plant pathology, K. Chambers; herbicides, M. Le Court de Billot
Graduate research staff: 45
Activities: The main function of the Summer Grain Centre is to conduct and to coordinate research on all aspects of production and quality of summer grains (maize, grain sorghum, buckwheat, millet and rice). In addition it serves as a centre for technical information regarding all aspects of summer grain production and conveys this information to all strategic points for the use of commodity advisers.
Projects: Population development of maize (Dr H.O. Gevers); improvement of morphology and physiology of maize (A.P. Fourie); production studies in summer grains (maize, grain sorghum, millets, buckwheat) (J.P. Pretorius); the influence of physiological factors on growth and development of maize (Dr C.C. Martin); soil tillage studies in maize (Dr J.B. Mallett); herbicide studies in maize (M. Le Court de Billot); maize soil fertility studies (Dr M. Farina); soil-plant-water relationships (R. Mottram); plant diseases studies in maize (K. Chambers); seedling diseases of grain sorghum (N. Mclaren); entomology studies in maize (Dr J.B.J. Van Rensburg); nematology studies in maize (S. Zondagh).

National Timber Research Institute 40

– NTRI
Address: Private Bag 395, Pretoria, 0001
Telephone: (012) 86-9211
Telex: 3-630 SA
Affiliation: Council for Scientific and Industrial Research
Status: Official research centre
Director: Dr D.L. Bosman

Divisions: Timber engineering, P.A. Bryant; timber economics, D.J.T. van Niekerk; wood processing and products, P. Sorfa; pulp and paper, Dr J.S.M. Venter; wood chemistry, Dr A. Pizzi
Graduate research staff: 30
Activities: The main purpose of the institute is to further timber technology through research and development for the benefit of the first products industry, and its particular aims are: to promote the effective utilization of South African resources; to assist in developing satisfactory products; to assist in developing and improving manufacturing processes; to promote the effective use of timber. Subject areas of research include; timber engineering; timber economics; wood processing and wood products; pulp and paper; wood chemistry.
Publications: Annual report.
Projects: Thermo-mechanical pulping (G.W. Vinopal); energy - pulp and paper (A.B.J. du Plooy); product development - pulp and paper (J.J. Hough); pulping and bleaching processes (A.B.J. du Plooy); drying of timber (Dr H.P. Stöhr); grading of timber (G.W. Vinopal); glulam (F. Louw); energy - sawmills (P. Sorfa); production planning and control (D.J.T. van Niekerk); economic studies (W.E. Lubbe); timber structures (F.R.P. Pinenaar); lignin utilization (K. Psotta); product development - composite materials (P. Sorfa); adhesives (Dr A. Pizzi); producer gas (W.T. Gore); properties of wood (F.S. Malan); preservation of wood (W.E. Conradie).

Oenological and Viticultural Research Institute 41

Address: Private Bag X026, Stellenbosch, 7600
Telephone: (02231) 7-0110
Affiliation: Department of Agriculture and Fisheries
Status: Official research centre
Director: Dr J.D. Burger
Deputy director: Dr J. Deist
Graduate research staff: 32
Activities: The institute serves the wine and dried grape industries through research at Nietvoorbij, the main research farm. There are also research stations at Robertson and Vredendal, and viticultural sections on other experimental farms at Oudtshoorn, Upington and Vaalharts.
Projects: Breeding, selection, evaluation and propagation of cultivars and rootstocks; pruning, trellising and spacing of vines; irrigation and cultivation practices; chemical weed control; nutritional and fertilizer studies; disease and pest control; efficient use of yeasts in winemaking; influence of environmental factors on grape and wine quality; influence of spray residues on fermentation; influence of wine processing techniques on quality; bouquet constituents; distillation of brandy.

Orange Free State: 42
Agricultural Research
Institute

Address: Private Bag X01, Glen, 9360, Orange Free State
Telephone: (05214) 811
Telex: B690
I *Affiliation:* Department of Agriculture and Fisheries
Status: Official research centre
Director: T.E. Skinner
Deputy director: G.W.O. Oosthuysen
Departments: Plant production, C. Engelbrecht; animal production, Dr E.A.N. Engels; pasture science, J. van Niekerk; agricultural education, C. F. van der Linde
Activities: Soil science; drainage and irrigation; land use; breeding of beef cattle; Karakul sheep; mutton sheep; nutrition of cattle and sheep; husbandry of cattle and sheep; protection of crop plants.
Attached to the institute are the agricultural research stations at Vaalharts (mainly for irrigation and large stock research); at Upington, one on the banks of the lower Orange River (for viticulture and cotton research) and another nearby for Karakul research; Armoedsvlakte, near Vryburg (for beef production); Koopmansfontein, north of Kimberley (for research on Dorper sheep and for veld management); and the substations at Sand-Vet and Riet River where wheat, oil seeds, summer cereals and crop rotation systems are investigated. Among the facilities are: abattoirs; barns for sheep and wool, animal nutrition, cattle and calves; dairy research building with dairy pilot plant; adiabatic bomb calorimeter, gas chromatographic and other equipment for analysing organic and inorganic components in soil, plant and animal samples; workships; a library.
Projects: Nutritive evaluation of pastures and forages; increased beef production by purposive cross breeding (Dr E.A.N. Engels); ecological studies on natural pastures (J. van Niekerk); breeding of Karakul sheep for high quality pelt production (A.S. Faure); chemical and physical criteria of different soil series for optimal crop production under different cultivation practices (C. Engelbrecht).

COLLEGE OF AGRICULTURE* 43

Plant Protection Research 44
Institute

– PPRI
Address: Private Bag X134, Pretoria, 0001
Telephone: (012) 21-3111
Telex: Entomologist, Pretoria
Affiliation: Department of Agriculture and Fisheries
Status: Official research centre
Director: Dr I.H. Wiese
Departments: Entomology/zoology, Dr D.P. Keetch; phytosanitary/microbiology, Dr B.W. Strydom; weeds, pesticides and herbicides, Dr H.A. van de Venter
Graduate research staff: 71
Activities: National-level directed and fundamental research on taxonomy, biology and ecology of insects, nematodes, phytophagous mites and spiders, the biological, chemical and integrated control of insects, nematodes, mites and rodents of importance in agriculture and forestry; apiculture; taxonomic mycology; mycotoxins; nitrogen fixation; epidemiology of fungus and bacterial diseases of importance in agriculture and forestry; plant virology; ecology and physiology of weeds; biological, chemical and integrated control of weeds with emphasis on invasive species of veld and catchment areas; pesticide, fungicide and herbicide dynamics. Technical advice on the registration of pesticides, fungicides and herbicides. In addition, with the exception of the inspection services, all phytosanitary aspects of the importation of plant material, including disinfection, disinfestation and virus elimination.
Projects: Forest and timber insects (G. Tribe); National Collection of Insects (Dr G.L. Prinsloo); integrated control of stalk-borers (Dr H. van Hamburg); integrated control of aphids (Dr N.J. van Rensburg); integrated control of American bollworm (N.C. Basson); insects of stored products (J.H. Viljoen); apiculture (Dr R.H. Anderson); plant nematology (Dr E. van den Berg); phytophagous mites (Dr M.K.P. Meyer); spiders (Dr A.S. Dippenaar); rodents (S. Stiemie); taxonomic mycology and mycotoxins (Dr G.C.A. van der Westhuizen); nitrogen fixation (Dr B.W. Strydom); plant virology (D.J. Engelbrecht); weed control (Dr H.A. van de Venter); plant quarantine (H. P. van Heerden); technical advice (pesticides, fungicides, herbicides) (Dr J. Bot); pesticide and herbicide dynamics (Dr L.P. van Dyk).

SA Forest Investments 45
Limited

Address: Private Bag 69, Sabie, 1260, Transvaal
Telephone: (0131512) 81
Telex: 8-4443
Status: Research centure within an industrial company

RESEARCH AND DEVELOPMENT DIVISION   46

Group research and development manager: P.H. Müller
Graduate research staff: 2
Activities: Research into most aspects of forest products; pine and eucalyptus sawmilling and kiln drying; benefits of live pruning on sawn yield.
Projects: Most projects and tasks are of a confidential or semiconfidential nature; current emphasis is placed on computer sawing simulation studies using real log data, drying of pine and eucalyptus timber for maximum throughput and minimum degradation and all aspects of production of sawn eucalyptus grandis.

Sea Fisheries Institute   47

Address: Private Bag X2, Rogge Bay, 8012
Telephone: 21-6290 (Heerengracht); 44-9521 (Sea Point); 41-2361 (Green Point); 21-1480 (Foreshore)
Telex: 57-20796 (Heerengracht); 57-26425 (Foreshore); 56-0061 (Walvis Bay)
I *Affiliation:* Department of Agriculture and Fisheries
Status: Official research centre
Director: G.H. Stander
Total research staff: 164
Activities: The main aim of the institute is to conduct research into the living resources of the sea in order to assure a maximum sustainable yield, and to provide the minister of Agriculture and Fisheries with advice on the management of marine resources according to scientific principles to ensure rational exploitation. The institute is responsible for the administration of the Sea Fisheries Act and other legislation which deals with the catching or capturing, marketing, export and protection of marine species. It also controls and manages 11 proclaimed fishing harbours and enforces fish conservation measures: for the latter purpose 63 full-time sea fisheries inspectors make use of an effective land-based inspection system and 5 highspeed patrol vessels. The institute maintains well-equipped laboratories in Sea Point for planktology, on the Cape Town Foreshore for fisheries biology, in Green Point for physical/chemical oceanography, marine environmental pollution and marine mammals, and at the University of Cape Town for seaweed research. In addition, field laboratories are manned when required in Saldanha, Hout Bay, Gans Bay, Mossel Bay, Lüderitz and Walvis Bay. A fleet of eight research vessels, five of which are based at Cape Town, two at Walvis Bay and one at Lüderitz, provides extensive coverage for inshore and off-shore research surveys, which are conducted throughout the year mainly in the west coast area between Cape Town and the Kunene River. Supplementary research services embrace mechanical and electronic research workshops at Sea Point, an electronic maintenance workshop and net repair shop at Paarden Island, and computer facilities at the Foreshore laboratories. Other facilities include automatic chemical analysis, spectrophotometer, gas chromatograph; infrared spectrophotometer; electronic fish-detecting equipment; shipborne echo-integrator; airborne radiation thermometer; mobile laboratory.
Publications: Annual and investigational reports.
Projects: The research functions entail the collection, processing and analysis of comprehensive information on the biology and population dynamics of exploited stocks, as well as on their relation to the biotic and abiotic environment. Research is conducted on a priority basis according to the economic importance of the various exploited species. Current research is therefore focused on South West African pelagic fish, South African pelagic fish, South African and South West African West Coast rock lobster, and South African and South West African trawlfish. Lesser projects include investigations of marine mammals (whales and seals), seaweed, snoek and other line-caught fish. Oceanographic research covers an extensive area off and along the coasts of South Africa and South West Africa. Special attention is given to upwelling, primary productivity, chemical studies, and pollution by pesticide residues, factory effluents, heavy metals, and mineral oil. South Africa is a member of the International Commission for the South East Atlantic Fisheries, the body which controls fishing activities in international waters in the area, and scientists of the institute are actively involved in the work of this commission, especially in respect of hake conservation. Rock lobster studies include the tagging of fish to study growth and mortality in relation to the maximum sustainable yield which may be obtained in the various commercial fishing areas. Scientists of the institute are furthermore active in the International Whaling Commission and the International Commission for the Conservation of Atlantic Tunas.

Soil and Irrigation Research Institute   48

– SIRI
Address: Private Bag X79, Pretoria, 0001, Transvaal
Telephone: (012) 28-4048
Affiliation: Department of Agriculture and Fisheries
Status: Official research centre
Director: Dr M.C.F. du Plessis
Departments: Pedology and irrigation planning, Dr C.N. MacVicar; soil and water research, G.C. Green; agrometeorology, Dr A.L. du Pisani; analytical services, Dr C.E.G. Schutte

Graduate research staff: 54

Activities: The aims of the institute are to promote the optimal utilization of the country's natural resources, soil, water and climate, by means of comprehensive survey and mapping programmes and supporting research in the fields of soil chemistry, mineralogy, physics, fertility and genetics, irrigation and water quality and agrometeorology. Additional functions are to evaluate land for irrigation development by integrating all available physical, agronomic and economic information, and to provide services, soil surveys, and statutory quality control of fertilizers, other agricultural chemicals and stock feeds.

Projects: Survey and mapping of natural resources (C.N. MacVicar); evaluation of land for irrigation development (J.A. Stofberg); agrometeorological station network and data bank (A.L. du Pisani); soil mineralogical studies (R.W. Fitzpatrick); soil fertility investigations (A.J. van der Merwe); soil-plant-atmosphere water relations and irrigation research (A. Streutker); water quality investigations (H.M. du Plessis); chemistry of acid soils (H.C.H. Hahne); hydraulic and structural properties of soils (J.L. Hutson).

South African Forestry Research Institute   49

− SAFRI

Address: Private Bag 727, Pretoria, 0001, Transvaal
Telephone: (012) 287120
Telex: 3717
Affiliation: Department of Water Affairs, Forestry and Environmental Conservation
Status: Official research centre
Director: Dr C.P. Kromhout
Graduate research staff: 45 L

Activities: Research into effective utilization of national resources, including soil science and land use; plant production, including plant breeding, and plant protection; forestry and forest products research.

Projects: Forest botany, trial plantings and arboriculture (Dr R.J. Poynton); tree improvement (Dr G. van Wyk); conservation forestry (F.J. Kruger); forest protection and fire; silviculture (H.A. van der Syde); forest management (N.C. Loveday); minor forest products and waste wood (W.K. Darrow); national forestry economy (A.A. Arnold).

South African Sugar Association Experiment Station   50

− SASA

Address: PO, Mount Edgecombe, 4300, Natal
Telephone: (031) 591805
Telex: 62215
Affiliation: South African Sugar Association
Status: Research centre within an industrial company
Director: Dr Gerald D. Thompson
Divisions: Agricultural Engineering, Dr A.G. de Beer (mechanization), J.R. Pilcher (design and development), J.P. Fourie (land and water management); Agronomy, Chemistry and Soils, P.K. Moberly, Dr R.A. Wood; Extension, vacant; Plant Breeding and Pathology, Dr J.C.S. Allison, R.S. Bond (selection), K. J. Nuss (crossing), R.A. Bailey (pathology and nematology); Services, H. de Vink; biometry, M.G. Murdoch; entomology, Dr A.J.M. Carnegie
Graduate research staff: 48

Activities: The primary objective of the experiment station is to improve the productivity and profitability of the South African canegrower. This is done by conducting applied research into cane production problems; developing improved machines and cultivars; providing educational courses in sugarcane growing; training agricultural labour and semiskilled operatives; providing a specialist advisory service and an extension service to make all these results and facilities readily available to sugarcane growers. The station is under contract to the Swaziland Sugar Association to provide the same technological services to their members as it does to its own members.

Publications: Annual report.

Projects: Develop new varieties of sugarcane both by introduction and breeding and to propagate and distribute selected varieties to meet the diverse needs of the industry (Dr J.C.S. Allison); investigate the agronomic problems of sugarcane production (P.K. Moberly); test and when necessary develop machines and equipment used for production and harvesting (Dr A.G. de Beer); study soils in sugarcane-producing areas and to relate these to crop management (J.H. Meyer); study the nutritional requirements of sugarcane (Dr R.A. Wood); study and monitor the pests and diseases of sugarcane and so develop appropriate control measures (R.A. Bailey); conduct basic research on the germination of setts, sucrose production, translocation and storage, and on the influences of the environment on these processes (Dr P.L. Greenfield).

South African Sugar Technologists' Association　51

Address: c/o SASA Experiment Station, PO, Mount Edgecombe, 4300, Natal
Telephone: (031) 591805
Telex: 6-2215
Affiliation: South African Sugar Association
Status: Industrial association organizing research
President: Dr G.W. Shuker
Vice-president: Dr A.B. Ravnö
Activities: The interchange of scientific knowledge of, and the discussions and investigation of technical problems related to the production of sugar; improvement in the accuracy and rational standardization of methods of factory chemical control; encouraging and assisting in the improvement of the technical knowledge of persons engaged in the South African sugar industry; encouraging and promoting research into all aspects of sugarcane agriculture and sugar milling practice and in ancillary fields. The association has approximately 1 200 members in the sugar industry and related areas.
See also entry under: Sugar Milling Research Institute.
Publications: Annual proceedings.

Sugar Milling Research Institute　52

– SMRI
Address: c/o University of Natal, King George V Avenue, Durban, 4001, Natal
Telephone: (031) 253241
Telex: SA 60139
Affiliation: Council for Scientific and Industrial Research; South African Sugar Association
Status: Independent research centre
Director: Dr A.B. Ravnö
Divisions: Research and development, Dr J. Bruijn; analytical services, P. Mellet; sugar processing, J.P. Lamusse; sugar engineering, G.N. Allan
Graduate research staff: 16
Activities: The objectives of the institute are to promote, control, conduct, manage, organize and assist research and other scientific work on matters affecting the sugar milling and refining industries, and other allied industries.
Publications: Annual report.
Projects: Cane quality (J.P. Lamusse); factory performance; equipment evaluation (G.N. Allan, J.P. Lamusse); cost reduction (G.N. Allan); analytical developments (P. Mellet); by-products development (J. Bruijn).

Tobacco Research Institute　53

Address: Private Bag X82075, Rustenburg, 0300
Telephone: (01421) 93171
Affiliation: Department of Agriculture and Fisheries
Status: Official research centre
Director: Dr D.J. Rossouw
Deputy director: Dr W.J. Pienaar
Departments: Tobacco breeding, Dr M.P. Lamprecht; plant protection, R.J. van Wyk; biochemistry/technology, J.G. Nel; soil chemistry, vacant; cotton breeding, Dr H.G. van Heerden; agronomy, M.C. Dippenaar
Activities: All aspects of basic and field research concerning the production of tobacco and cotton, from seed production to the on-farm processing of the crop in preparation for marketing, including plant breeding, agronomy, soil science, plant protection and biochemistry.The facilities include: tobacco cutting machines; cigarette machines; critical temperature and moisture control units; lysimeters; phytotron equipment; cotton fibre laboratory; research station at Kroondal; experimental farm at Groblersdal and near Brits; facilities for research into Oriental tobacco production and seed bulking at the Stellenbosch-Elsenburg research station.
Projects: Tobacco; breeding and cultural practices, herbicides (M.P. Lamprecht); water requirements (W.J.J. Roux); nematodes, virus diseases (R.J. van Wyk); diseases (G.C. Prinsloo); pests (E. Eulitz); quality (J.G. Nel); nutrition (H.J. Boshoff). Cotton: breeding (H. G. van Heerden); water requirements and nutrition (M.C. Dippenaar).

Transvaal Region: Agricultural Research Institute　54

Address: Private Bag X180, Pretoria, 0001
Telephone: (021) 268111
Affiliation: Department of Agriculture and Fisheries
Status: Official research centre
Director: D.F. de Wet
Deputy director: S.J. Gericke
Graduate research staff: 38
Activities: Agricultural research with particular reference to crop and animal production under the soil and climatic conditions of the Transvaal region.

University of Fort Hare*　55

Status: Educational establishment with r&d capability

AGRICULTURAL AND RURAL DEVELOPMENT RESEARCH INSTITUTE   56

– ARDRI
Address: Private Bag X1314, Alice, 5700, Cape Province
Telephone: (043522) 281
Telex: 746193
Director: Professor P.J. Burger
Graduate research staff: 5
Activities: The institute is engaged on research relating to agricultural and associated development in Southern Africa's CDCS. There is a strong bias towards a human approach. Areas currently being investigated are: land use; crop production; forestry; beef production and marketing; freshwater fisheries; agricultural and freshwater fisheries; economics; drainage and irrigation.
Publications: Annual report.
Projects: Forest farming (P. Adams); Ciskei essential oils (Professor E. Graven); ecotope project (Professor J. Marais); Amatola basin rural development (Professor P.J. Burger); Kwazulu beef industry; Ciskei fisheries industry; Transkei beef marketing (D.R. Tapson); irrigation design parameters for Makatini flats (C.J. van Rooyen); evaluation of Sheila/Mooifontein maize project (Professor T.J. Bembridge).

University of Natal*   57

Status: Educational establishment with r&d capability

FACULTY OF AGRICULTURE   58

Address: Private Bag 375, Pietermaritzburg, 3200, Natal
Telephone: (0331) 63320
Telex: 63719
Dean: Professor W.J. Stielau
Departments: Agricultural economics, Professor H.I. Behrmann; agricultural engineering, Professor P. Meiring; animal and poultry science, Professor W.J. Stielau; crop science, Professor K. Nathanson; dietetics and home economics, E. Nel; genetics, Professor W.H. Weyers; horticultural science, Professor P. Allan; microbiology and plant pathology, Professor M.M. Martin; pasture science, Professor N.M. Tainton; soil science and agrometeorology, Professor J. De Villiers
Graduate research staff: 50
Activities: Soil science; drainage and irrigation; land use; plant breeding; crop husbandry; plant protection; livestock breeding; livestock husbandry and nutrition; agricultural economics; food science.

WATTLE RESEARCH INSTITUTE   59

Address: University of Natal, Private Bag 375, Pietermaritzburg, 3200, Natal
Telephone: (0331) 62314
Telex: 643719
Director: Professor J.A. Stubbings
Departments: Silviculture, Dr A.P.G. Schönau; Tree breeding and genetics K.M. Nixon; Plant physiology and pathology, vacant; Chemistry, D.C.F. Garbutt; Entomology, R.B. Borthwick; Productivity and mechanization, R. M. de Laborde; Field research, liaison and planning, V.R. Davidson
Graduate research staff: 11
Activities: The institute conducts research into forestry and forestry products, participating in government-sponsored national research programmes; it also provides seed services, insect and fungi identification biometrics, training, plantation planning, promotion and analytical services.
Publications: Annual report.
Projects: Suitability of forest-tree species (acacia, eucalyptus) for planting in South Africa; phenology and factors influencing growth; breeding and hybridization; vegetative propagation; design and management of seed orchards (K.M. Nixon, S.F. Hagedorn); silviculture; soils and climate; seed stocks and raising seedlings (acacia and eucalyptus), regeneration, weeding and cultivation (A.P.G. Schönau, K.M. Nixon); forest protection and fire research: control of insect pests (acacia, eucalyptus, pine); weed control, including chemical control (R.B. Borthwick); forest management: sampling and assessment methods; mensuration studies on trees and stands; site and stand quality; yield data acquisition and processing and the development of mathematical yield models and tables; effects of insect pests on yields (acacia, eucalyptus, pine); valuation; management systems of plantations and exotic species (A.P.G. Schönau, D.I. Boden); wood properties: chemical properties of acacia and eucalyptus timbers related to heredity; roundwood moisture loss; stockpile moisture content of pulpwood chips; wattle bark tannin; cucalyptus bark for fodder; charcoal; acid hydrolysis of plantation timber and waste wood (D.C.F. Garbutt, M.A. Millar); forest engineering and ergonomics: evaluation of methods, equipment and tools and the development, planning and implementation of improved versions of these in the fields of: silvicultural operations, timber and bark harvesting and extraction, fire protection, and motivation of plantation labour (R.M. de Labourde, T.G. Goodricke); national forestry economy: surveys, analyses and projections of wattle bark and timber resources; capital, cost, return and price structre in the timber growing industries (under the aegis of the Federation of Timber Growers' Associations) (J.A. Stubbings).

University of the Orange Free State* 60

Status: Educational establishment with r&d capability

FACULTY OF AGRICULTURE 61

Address: Private Bag 339, Bloemfontein, Orange Free State
Telephone: (051) 70711
Dean: H.A. Kotze
Departments: Agricultural economics, Professor C.S. Blignaut; agronomy, Professor J.J. Humau; agricultural meteorology, Professor J.M. de Jager; agricultural engineering, W.R. van Zyl; biometry, Professor A.J. van den Merwe; animal breeding and small stock breeding, Professor J.A. Nel; animal husbandry, Professor A. Smith; dairy technology, Professor J.C. Novella; plant pathology, Professor G.D.C. Pauer; soil science, Professor R. du T. Burgeo; pasture science.
Graduate research staff: 100
Activities: Soil science; land use; plant breeding; crop husbandry; plant protection; animal production - cattle, sheep, and pigs; agricultural engineering; agricultural food economics; food science.
Projects: General agricultural research, organized according to the departments listed above.

University of the Witwatersrand 62

Address: 1 Jan Smuts Avenue, Johannesburg, 2000,
Telephone: (011) 726 1111
Telex: 8-7330 SA
Status: Educational establishment with r&d capability
Affiliation: Council for Scietific and Industrial Research

DEPARTMENT OF BOTANY AND MICROBIOLOGY* 63

Photosynthetic Nitrogen Metabolism Unit 64

Address: 1 Jan Smuts Avenue, Johannesburg, 2000, Transvaal
Telephone: (011) 39 4011
Telex: 4-22460
Affiliation: Council for Scientific and Industrial Research
Director: Professor C.F. Cresswell
Graduate research staff: 15

Activities: Investigations of plant physiological components associated with crop production, with particular emphasis on carbon and nitrogen metabolism.
Projects: Nitrogen uptake and assimilation in maize, wheat and barley; relationship between nitrogen supply and photosynthetic carbon metabolism in maize and wheat; primary productivity of a savannah area in Northern Transvaal.

Veterinary Research Institute 65

Address: PO, Onderstepoort, 0110
Telephone: (012) 554141
Affiliation: Department of Agriculture and Fisheries
Status: Official research centre
Director: Dr R.D. Bigalke
Deputy director: Dr T.W. Naudé, Dr C.J. Howell, Dr A. Pini
Departments: Bacteriology, Dr C.M. Cameron; virology, Dr B.J. Erasmus; reproduction, Dr C.J.V. Trichard; protozoology, Dr A.J. de Vos; helminthology, Dr J.A. van Wyk; entomology, Dr J.D. Bezuidenhout; food hygiene, Dr W.H. Giesecke; poultry diseases, Dr S.B. Buys; molecular biology, Dr D.W. Verwoerd; biochemistry, Dr H. Huismans; toxicology, Dr T.S. Kellerman; chemical pathology, Dr L.A.P. Anderson; pathology, Dr L. Prozesky; foot and mouth disease, Dr A. Pini
Graduate research staff: 83
Activities: Research on diseases of animals, poultry and fish including wild animals but with the emphasis on food producing animals; the provision of a specialized veterinary diagnostic service; production of vaccines for use in animals; technical advice on the registration of veterinary medicines.
Publications: Ondersteport Journal of Veterinary Research (quarterly).
Projects: Research on the wide variety of infectious and other diseases of livestock occurring in subtropical regions of Africa. The taxonomy, ecology, life cycles and methods of control of such pests are investigated. 46 vaccines against bacterial, viral, rickettsial and protozoan diseases are produced. Other investigations include studies on toxic principles of poisonous plants and their pathogenic effects; poisoning associated with industrialization or with the use of dips or veterinary remedies; malnutrition related to the variable rainfall (and thus pasture condition) or mineral deficiencies; immunology; and molecular biology of viruses and cancer.

Viticultural and Oenological Research Institute*

66

Address: Nietvoorbij, Private Bag X5026, Stellenbosch
Status: Official research centre

Winter Rainfall Region: Agricultural Research Centre

67

Address: Private Bag X5023, Stellenbosch, 7600
Telephone: (02231) 2211
Affiliation: Department of Agriculture and Fisheries
Status: Official research centre
Director: Dr J. Serfontein
Deputy director: Dr P.J.S. Pieterse
Graduate research staff: 39
Activities: Agricultural research, with particular reference to crop and animal production under the soil and climatic conditions of the Winter Rainfall region.

Projects: Optimum number of livestock to balance other types of farming; feeding livestock with locally-produced crops; chemical and physical properties of various types of soil; trace elements; effects of water quality and soil amendments on soil properties; multiplication and evaluation of new pasture plants; plant nutritional requirements; pasture utilization; stabilization of windblown sand; ecology and control of insect pests; extensive studies on the adaptability of winter cereals and pasture crops; and soil cultivation studies for crop production.

ELSENBERG COLLEGE OF AGRICULTURE*

68

Address: Stellenbosch, 7600

SPAIN

Asociación de Investigación para la Mejora de la Alfalfa* 1

[Alfalfa Breeding Research Association]
Address: Barrio de Movera, 165 Dupdo Zaragoza
Telephone: 29 20 88
Status: Research association
Research director: Fernando Hidalgo Maynar
Sections: Breeding; cultivation; seed production
Graduate research staff: 3
Activities: Classification of Spanish ecotypes; conservation of Aragon ecotypes; adaptation of varieties; seed production; fertilization; breeding and genetics.
Publications: Annual reports, research reports.

Asociación de Investigación Técnica de la Industria Papelera Española* 2

– IPE
[Spanish Paper Industry Research Association]
Address: Carretera de la Coruña km 7, PO Box 33045, Madrid, 35
Telephone: 207 09 77
Status: Research association
Research director: Dr José Luis Asenjo Martinez
Sections: Kraft sack paper; offset paper; eucalyptus bleached pulp
Graduate research staff: 4
Activities: Research on raw materials for pulp and paper manufacture.

Asociación de Investigación Técnica de las Industrias de la Madera y Corcho* 3

– AITIM
[Wood and Cork Industries' Research Association]
Address: Flora 3, Madrid, 13
Telephone: 242 78 64
Status: Research association
Research director: Dr Ing Luis Mombiedro
Activities: Problems of interest to the wood and cork industries.

Centro de Edafología y Biología Aplicada de Salamanca 4

[Edaphology and Applied Biology Centre]
Address: Cordel de Merinas, 40 Apartado 257, Salamanca
Telephone: 923–21 96 06
Affiliation: Consejo Superior de Investigaciones Científicas
Status: Official research centre
Director: D. José Manuel Gómez Gutiérrez
Departments: Plant nutrition and soil chemistry, L. Sánchez de la Puente; soils, A. García Rodríguez; clay mineralogy and physicochemistry, M. Sánchez Camazano; animal pathology, E. de Santiago Redel; pasture and bioclimatology, B. García Criado; rural socioeconomy, A. Cabo Alonso; nitrogen fixation and soil biochemistry, C. Rodríguez Barrueco; mineralogy, J. Saavedra Alonso
Graduate research staff: 30 (1981)

Activities: Computing for statistics, animal pathology, animal parasitology, plant ecology, plant nutrition, plant physiology, bioclimatology, exploration geochemistry, mineralogy, agricultural meteorology, soil biochemistry, soil cartography, soil chemistry, soil classification, soil microbiology, soil mineralogy, soil morphology and genesis, soil physics, pasture, soil fertility, agriculture economics, rural sociology.
Publications: Annual report.
Projects: Plant nutrition, wheat, corn, strawberry, sugarbeet (L. Sánchez de la Puente); soil micronutrients (L. Prat Pérez); soil survey and cartography (A. García Rodriguez); chemicals and clay interactions (M. Sánchez Camazano); clay mineralogy (A. Vicente Hernández); animal parasitology (F. Simón Vicente); infectious diseases (E. de Santiago Redel); bioclimatology (A. Blanco de Pablos); rural econometry (L. Jiménez Díaz); rural sociology (F. Sánchez López); nitrogen fixation (C. Rodríguez Barrueco); soil organic matter (J. Gallardo Lancho); mineralogy and or deposits (J. Saavedra Alonso).

Centro de Edafología y Biología Aplicada del Cuarto   5

– CEBAC
[Edaphology and Applied Biology Centre, Cuarto]
Address: Cortijo del Cuarto, Apartado 1052, Bellavista, Sevilla
Telephone: 954–69 07 00; 69 07 04
Affiliation: Consejo Superior de Investigaciones Científicas
Director: Professor José Martin Aranda
Departments: Soil chemistry, Professor Pablo Arambarri; soil physics, Dr Felix Moreno; soil genesis, Dr Juan Olmedo; oil cartography, Dr José Luis Mudarra; soil biochemistry, Professor Francisco Martin; soil fertility, Dr Antonio Troncoso; clay mineralogy, Dr José Luis Perez; plant biochemistry, Dr Luis Catalina
Graduate research staff: 30 (1981)
Projects: Spanish sericite research, mineralogical aspects of art painting erosion (Dr J. L. Perez); urban waste used in agriculture (Professor J. Martin-Aranda); use of waste water filter residues in agriculture (Professor P. Arambarri); irrigation and fertility of olive trees (Dr A. Troncoso); vineyard product biochemistry (Dr A. Catalina); water balance in olive trees (Dr F. Moreno); organic composition and use of lignites (Professor F. Martin).

Centro de Edafología Biología Aplicada del Segura*   6

[Edaphology and Applied Biology Centre, Segura]
Address: Apartado 195, Murcia
Affiliation: Consejo Superior de Investgaciones Científicas
Status: Official research centre

Centro de Investigación y Desarrollo, Santander*   7

[Investigation and Development Centre, Santander]
Address: Avenida de los Castros s/n, Santander
Affiliation: Consejo Superior de Investigaciones Científicas
Status: Official research centre

Centro de Investigaciones Agrícolas*   8

[Agricultural Research Centre]
Address: Virgen de la Soledad 5, Badajoz
Affiliation: Consejo Superior de Investigaciones Científicas
Status: Official research centre

Centro de Investigaciones Biológicas*   9

– CIB
[Biological Investigations Centre]
Address: Velazquez 144, Madrid, 6
Affiliation: Consejo Superior de Investigaciones Científicas
Status: Official research centre
President: Dr Antonio Portolés Alonso

JAIME FERRAN MICROBIOLOGY INSTITUTE*   10

Director: Miguel Rubio Huertos

Bacteriology Department*   11

Head: Professor Román Vicente Jordana
Sections: Symbiotic nitrogen fixation
Direct nitrogen fixation
Bacterial physiology and genetics

General Microbiology Department*   12

Head: M. Pilar Aznar Ortíz
Sections: Yeasts
Technical microbiology
Phytopathological microbiology
Microbe chemistry
Laboratory of aetiopathology of bacterial tumours in plants
Laboratory of applied mycology
Laboratory of fungi biochemistry
Laboratory of algae physiology and biochemistry

Virology Department*   13

Head: Miguel Rubio Huertos
Sections: Plant viruses; animal viruses

SANTIAGO RAMON Y CAJAL   14
INSTITUTE*

Director: Professor Alfredo Carrato Ibañez

Comparative Neuroanatomy Section*   15

Head: Dr Facundo Valverde García

Plant Pathology Section*   16

Head: Dr José Rodrigo García

Centro Nacional de   17
Alimentación y Nutrición

[National Centre for Food and Nutrition]
Address: Majadahonda, Madrid
Telephone: 91-638 11 11
Status: Official research centre
Director: Dr Antonio Borregón Martinez
Departments: Contamination of foods, Da Pinilla; instrumental analysis of foods, Dr Sanchez Saez; microbiology of foods, Dra Pascual Anderson; meat and milk, Dr Jimenez; toxicology, Dr Sanz; nutrition, Dra Santo-Domingo; documentation, Dr Barga
Graduate research staff: 250
Activities: Food sciences.

Project: Additives and their degradation products in foods and in the body; chemical contamination in food and food products (Dra Pinilla); toxicity of additives by animals and cell cultures (Dr Sanz); mutagenic effects of food, food products and additives; food microbiology (Dra Pascual Anderson).

Centro Nacional de   18
Investigaciones Pesqueras
Barcelona

[National Centre for Fisheries Research, Barcelona]
Address: Paseo Nacional s/n, Barcelona, 3
Affiliation: Consejo Superior de Investigaciones Científicas
Status: Official research centre

Centro Nacional de   19
Investigaciones Pesqueras
Cadiz*

[National Centre for Fisheries Research, Cádiz]
Address: Puerto Pesquero, Cádiz
Affiliation: Consejo Superior de Investigaciones Científicas
Status: Official research centre

Centro Pirenaico de   20
Biología Experimental*

[Pyrenean Centre for Experimental Biology]
Address: Apartado 64, Jaca (Huesca)
Affiliation: Consejo Superior de Investigaciones Científicas
Status: Official research centre

Consejo Superior de Investigaciones Científicas* 21

– CSIC
[Higher Council for Scientific Research]
Address: Serrano 117, Madrid, 6
Status: Official research organization
See separate entries for:
Centro de Edafología y Biología Aplicada de Salamanca
Centro de Edafología y Biología Aplicada del Cuarto
Centro de Edafología y Biología Aplicada del Segura
Centro de Investigación y Desarollo, Santander
Centro de Investigaciones Agrícolas
Centro Nacional de Investigaciones Pesqueras, Barcelona
Centro Nacional de Investigaciones Pesqueras, Cádiz
Centro Pirenaico de Biología Experimental
Departamento de Economía Agraria
Estación Agrícola Experimental de León
Estación de Investigaciones Pesquerras de Torre de la Sal
Estación Experimental 'Aula Dei'
Estación Experimental del 'Zaidin'
Estación Experimental La Mayora
INIP Laboratoriao de Barcelona
Instituto de Agroquímica y Tecnología de Alimentos
Instituto de Alimentación y Productividad Animal
Instituto de Botánica 'Antonio José de Cavanilles'
Instituto de Economía y Producciones
Instituto de Edafología y Biología Vegetal
Instituto de Fermentaciones Industriales
Instituto de Industrias Cárnicas
Instituto de Investigaciones Agrobiológias de Galicia
Instituto de Investigaciones en Patología de las Colectividades Ganaderas
Instituto de Investigaciones Pesqueras de Cádiz
Instituto de Investigaciones Pesqueras de Vigo
Instituto de Investigaciones Veterinarias
Instituto de la Grasa y sus Derivados
Instituto de Nutrición
Instituto de Productos Lacteos
Instituto de Zoología 'José de Acosta'
Instituto de Zootecnia
Instituto del Frío
Instituto Español de Entomología
Instituto 'López-Neyra' de Parasitología
Instituto Nacional de Edafología 'José María Albareda'
Sección de Edafología de la Palma
Universidad Autónoma de Madrid, Agricultural Chemistry Department
Universidad Complutense de Madrid, Faculty of Pharmacy, Department of Bromatology, Toxicology and Applied Chemical Analysis

PATRONATO FOR BIOLOGICAL AND MEDICAL SCIENCES* 22

Address: Velásquez 144, Madrid, 6

PATRONATO FOR MATHEMATICS, MEDICINE AND NATURAL SCIENCES* 23

Address: Serrano 117, Madrid, 6

PATRONATO FOR NATURAL AND AGRICULTURAL SCIENCES* 24

Address: Serrano 113, Madrid, 6

Departamento de Economía Agraria* 25

[Agrarian Economics Department]
Address: Serrano 113, Madrid
Affiliation: Consejo Superior de Investigaciones Científicas
Status: Official research centre

Empresa Nacional de Fertilizantes SA* 26

– ENFERSA,
[National Fertilizer Company]
Address: Calle Prim 12, Madrid, 4
Status: Industrial company

Escuela Superior de Técnica Empresarial Agrícola 27

– ETEA
[Agricultural Business Administration College]
Address: Escritor Castilla Aguayo 4, Apartado 439, Córdoba
Telephone: 0957-29 61 33
Affiliation: Universidad de Córdoba
Status: Educational establishment with r&d capability
Director: Jaime Loring
Departments: Social sciences, Gaspar Rul-Lan; mathematics, José Manuel Retenaga; law sciences, Vicente Theotonio; management sciences, Francisco Amador; economics, José Juan Romero
Graduate research staff: 10

Activities: ETEA is part of the Federación Libre de Escuelas de Ciencias Empresariales, jointly with Deusto (Bilbao), ESTE (San Sebastián), ESADE (Barcelona), ICADE (Madrid), and ESCE (Alicante), and conducts locally oriented research and offers a consultancy service in the region. ETEA and the university college has received research contracts from Contratado por la Confederación Nacional de Agricultores y Granaderos, Ministerio de Agricultura, Fondo para la Investigación Económica y Social de la Confederación Española de Cajas de Ahorros, Ministerio de Industria y Energía, Instituto de Estudios Fiscales, and the Asociación de Agricultores y Ganaderos (ASAGA) de Sevilla.
Publications: Annual report.
Projects: Public finance (Vicente Theotonio); finance of agricultural activities (Manuel Delgado); economics of the firm (Manuel Cabanes); accountancy and management control (Jaime Loring); industrial psychology (Alfonso Lopez); cooperative credit institutions (Adolfo Rodero); management of the educational system (Francisco Amador).

Escuela Técnica Superior de Ingenieros Agrónomos   28

[Technical College for Agronomy Engineers]
Address: Ciudad Universitaria, Madrid, 3, Madrid
Telephone: 41-244 48 02
Status: Educational establishment with r&d capability
Director: Juan F. Gálvez
Departments: Mathematics, José L. de Miguel, José M. Antón; physics, Dario Maravall; geometry, Francisco Puerta; statistics, Angel Anós; topography, Fernando Lopez de Sagredo; hydraulics, Faustino Garcia; electrotechnics, José Camacho; agricultural buildings, José R. Marcet, Rafael Dal-Re; agricultural industries, José Carballo, José M. Xandri; agricultural engineering, Jaime Ortiz, Jesus Garcia; chemistry, Segundo Jimenez; biochemistry, Francisco Garcia; soil science, Carlos Roquero; plant physiology, Cesar Gomez; botany, Eduardo Prieto; microbiology, Juan Santamaria; plant breeding, Enrique Sanchez Monge; entomology, Manuel Arroyo; plant protection, Eloy Mateo Sagasta; crop husbandry, José M. Mateo; horticulture, Joaquín Miranda; fruticulture, Fernand Gil Albert; animal nutrition, Juan F. Gálvez; animal production, Carlos Buxadé; animal breeding, Marcos Rico; agricultural economics, Arturo Camilleri, Leovigildo Garrido, Enrique Ballestero; sociology, Juan J. Sanz; agricultural project, José I. Trueba
Graduate research staff: 298
Activities: Soil science, drainage and irrigation, land use; plant breeding, crop husbandry, plant protection; livestock breeding, husbandry and nutrition; agricultural engineering and building; food science.
Publications: Annual report.

Estación Agrícola Experimental de León   29

– EAE
[Agricultural Experiment Station of Léon]
Address: Apartado 788, León
Affiliation: Consejo Superior de Investigaciones Científicas
Status: Official research centre
Director: Professor A. Suárez
Departments: Grassland research, Professor A. Suárez; animal production, Professor E. Zorita; animal parasitology, Professor M. Cordero
Graduate research staff: 14 (1981)
Activities: Crop husbandry, herbage; livestock husbandry and nutrition, sheep; epizoology and control of parasitic diseases.
Projects: Grassland improvement in the mountains (A. Suárez); nutritive requirements of the milk sheep in northern Spain (E. Zorita); parasitic diseases of ruminants, mainly sheep; fish diseases (M. Cordero).

Estación Biológica de Doñana   30

[Doñana Biological Station]
Address: Calle Paraguay 1-2, Sevilla, 12
Telephone: 954-61 52 91; 61 13 41; 61 04 57
Affiliation: Consejo Superior de Investigaciones Científicas; Ministerio de Universidades e Investigación
Status: Official research centre
Director: Dr Javier Castroviejo Bolíbar
Sections: Ecology and ethology, Dr Fernando Álvárez González; vertebrate zoogeography and systematics, Dr José A. Valverde
Graduate research staff: 10
Publications: Doñana Acta Vertebrata, biannual.
Projects: Impact of hervibores on the vegetation of Doñana (for 1982) (Dr Carlos Herrera Maliani); Iberian fauna - vertebrates and mammals (Dr Miguel Delibes de Castro); ornithological research and observation; general research work on ecology, ethology, zoogeography, systematics and conservation, mainly on vertebrates; botanical research.

Estación de Investigaciones Pesqueras de Torre de la Sal* 31

[Torre de la Sal Fisheries Research Station]
Address: Castellón
Affiliation: Consejo Superior de Investigaciones Científicas
Status: Official research centre

Estación Experimental 'Aula Dei' 32

[Aula Dei Experimental Station]
Address: Apartado 202, Montañana 177, Zaragoza
Telephone: 976-29 21 45
Affiliation: Consejo Superior de Investigaciones Científicas
Status: Official research centre
Director: Dr Luis Heras Cobo
Departments: Soil fertility and plant nutrition, Dr Luis Montañés García; edaphology, Dr Manuel Catalán Calvo; pomology, Dr Joaquín Herrero Catalina; genetics and plant production; Dr José M. Lasa Dolhagaray, virology, Dr Félix Martínez Cordón
Activities: Soil science: macro-and micronutrient dynamics; breakdown and extraction methods; soil genesis, cartography and contamination; manuring. Plant production: plant nutrition (olive, grape, fruit); nutrition deficiency; organic and mineral metabolism; plant improvement (winter cereals, forage crops, sugarbeet, corn); cytogenetics; fruit selection; graft incompatibility; effect of climate on fruit plants; virus control; vectors and virus-vector interaction; isolation of vine stocks; virus inhibitors. Agricultural engineering: mechanization and instrumentation; methods of sowing, harvest and treatment; precision sowers; automatization.
Projects Soil fertility – grape and olive (Dr L. Hera Cobo); microelements in soil and plant (Dr L. Montañés García); trisomics of sugarbeet (Dr J. M. Lasa Dolhagaray); fruit selection and experimentation (Dr J. Herrero Catalina); cartography of saline soils (Dr F. Alberto Jimenez); new species of alfalfa and date (Dr H. Hycka Maruniak); properties, dissemination and control of sugarbeet virus (Dr F. Martínez Cordón); chemical treatment of agricultural residues with little commercial value to evaluate their value as ruminant feed (Dr M. Catalán Calvo); effect of climate on fruit plants (Dr Tabuenca Abadía); variety-host relationship (Dr R. Cambra y Ruiz de Velasco) plant metabolism (Dr E. Monge Pacheco).

Estación Experimental de Zonas Áridas 33

[Arid Zone Experimental Station]
Address: Calle General Segura 1, Almeria
Telephone: 951-23 65 00; 23 26 22
Affiliation: Consejo Superior de Investigaciones Científicas
Status: Official research centre
Research director: Dr Guillermo Verdejo Vivas
Departments: Agronomy, Dr G. Verdejo Vivas; entomology, Dr A. Cobos Sanchez; zoology, Dr Francisco Amores Carredano
Graduate research staff: 5
Projects: Agronomic research in arid lands (Dr G. Verdejo Vivas); entomology research (Dr A. Cobos Sanchez, Dr F. Suarez Egea); zoology of arid lands (Dr Francisco Amores Carredano).

Estación Experimental del Zaidin 34

[Zaidin Experimental Station]
Address: Professor Albareda 1, Granada, Granada
Telephone: 958-25 88 00
Affiliation: Consejo Superior de Investigaciones Científicas
Status: Official research centre
Director: Dr Manuel Lachica Garrido
Vice-director: Dr Julio López-Gorge
Departments: Clay minerals, Dr Gonzalo Dios; pedology and botany, Dr José Luis Chicano; applied analytical chemistry, Dr Francisco Girela; agricultural chemistry, Dr Eduardo Estéban; plant physiology, Dr Antonio Leal; plant biochemistry; Dr Julio López-Gorge; microbiology, Dr José Olivares; phytopathology, Dr Pedro Ramos; animal physiology, Dr Juristo Fonolla
Graduate research staff: 39 (1981)
Activities: Natural resources, plant nutrition, plant protection, soil science, livestock nutrition.
Projects: Mineralogy and geochemistry of peridotitic soils (Dr José Linares); adsorption of pesticides on laminar clay minerals (Dr Gonzalo Dias); study of soils and vegetation in eastern Andalucía (Dr José Luis Guardiola, Dr Pablo Prieto); prediction of micronutrient deficiencies in soil and plants (Dr Manuel Lachica); normalization of analytical methods for pesticide analysis (Dr Francisco Sanchez Rasero); use of urban wastes as agricultural fertilizers (Dr Francisco Gallardo); physiology and biochemistry of lipid metabolism in olive (Dr Juan Pedro Donaire); light activation mechanism of CO_2 assimilation (Dr Julio López-Gorge); nitrogen fixation in the symbiosis rhizobium-leguminosae (Dr José Olivares); phytopathology of olive (Dr Pedro Ramos); improve-

ment in the quality of agriculture by-products for their use in animal feeding (Dr Juristo Fonolla); improvement in the quality and feeding conditions of the goat (Granada breed) (Dr Julio Boza).

Estación Experimental La Mayora   35

[La Mayora Experimental Station]
Address: Algar-robo-Costa, Málaga
Telephone: 952-51 10 00
Affiliation: Consejo Superior de Investigaciones Científicas
Status: Official research centre
Director: Dr Antonio Aguilar
Sections: Fertility and nutrition, Dr M. Casado; plant protection, J. Ruiz; plant improvement, Dr J. Cuartero
Graduate research staff: 15
Activities: Thirty per cent of activities are concerned with the cultivation and improvement of avocado. Extensive investigation in horticulture, varieties, pests, greenhouses, etc. Research is also carried out on plant nutrition and sub-tropical plant fertility.
Projects: Under-cover cultivation (Dr M. Casado); horticultural genetic improvement (Dr J. Cuartero); sub-tropical fruits improvement (Dr J. M. Farre); pest control R. Moreno).

INIP Laboratorio de Barcelona*   36

[INIP Laboratory of Barcelona]
Address: Paseo Nacional s/n, Barcelona, 3
Affiliation: Consejo Superior de Investigaciones Científicas
Status: Official research centre

Institut Català de la Carn*   37

[Catalan Meat Institute]
Address: Granja Camps i Armet, Monells, Girona
Telephone: 972-20 67 70
Affiliation: Departament d'Agricultura Ramaderia i Pesca de la Generalitat de Catalunya
Status: Official research centre
Technical director: Dr Josep Monfort Bolívar
Research director: Dr Antoni Nadal Amat
Sections: Analytical Department, Gloria Cassademont; microbiology, Margarita Garriga; technology, Esteve Corominas
Graduate research staff: 8

Instituto Botánico de Barcelona*   38

[Barcelona Botanical Institute]
Address: Avenida de Montañans, Parque de Montjuich, Barcelona, 4, Cataluña
Telephone: 34 3 325 80 50
Status: Official research centre
Research director: Dr O. de Bolós
Sections: Phanerogamy; geobotany
Graduate research staff: 3
Activities: Geobotany and botany including floristics of the western Mediterranean.

Instituto de Agroquímica y Tecnología de Alimentos   39

[Food Technology and Agrochemistry Institute]
Address: Jaime Roig 11, Valencia, 10
Telephone: 96-369 08 00
Affiliation: Consejo Superior de Investigaciones Científicas; Ministerio de Universidades e Investigación
Status: Official research centre
Director: Dr E. Tortosa
Sections: Canning, Dr L. Durán; meat products, Dr J. Flores; fruits and fruit juices, Dr B. Lafuente; cereals, Dr S. Barber; agricultural chemistry, Dr P. Cuñat; microbiology, Dr E. Hernández; instrumental analysis, Dr J. Alberola; biology, Dr J. Flores; soils and plant nutrition, Dr J. Rubio; processing, Dr A. Flors; canning fruit and vegetables, L. Durán; economics and statistics, I. Fernández
Graduate research staff: 50
Publications: Annual report.
Projects: Basic and applied studies to improve the breadmaking process and bread quality (S. Barber); quality studies on fruits and vegetables (L. Durán); quality of citrus fruits for the industry and the fresh market (P. Cuñat); utilization of agricultural residues for energy purposes (A. Flors); solar drying of fruits and vegetables (F. Piñaga); viroid diseases of citrus (R. Flores); economics of spanish citrus production (P. Serra).

Instituto de Alimentación y Productividad Animal   40

[Animal Feeding and Productivity Institute]
Address: Ciudad Universitaria, Madrid, 3
Telephone: 91-449 23 00
Affiliation: Consejo Superior de Investigaciones Científicas
Status: Official research centre
Director: Professor Gaspar Gonzalez
Departments: Physiology of nutrition, Dr R. Viñarás; grassland and roughages, Dr V. González; feeding and productivity, Dr F. Tortuero
Graduate research staff: 14
Activities: Animal production, monogastric feeding; nutritive evaluation of feedstuffs; poultry and rabbit feeding; grassland, grass and forage conservation and ruminant feeding; economy of animal feeding and productivity.
Publications: Annual report.
Projects: New sources of protein in animal feeding (Dr R. Viñarás); influence of feeding upon the carcass quality of pigs; utilization of agricultural by products in ruminant feeding (Dr V. Gonzalez); utilization of roughages and concentrates; yield and nutritive value of legumes (grain and roughages); aminoacidic composition of legume seeds; minerals in poultry and rabbit nutrition (Dr F. Tortuero); different sources of energy for poultry and rabbits; nutrition and cardiovascular disease in animals; alternative models for extensive livestock production.

Instituto de Botánica 'Antonio José de Cavanilles'*   41

[Antonio José de Cavanilles Botany Institute]
Address: Plaza de Murillo 2, Madrid, 14
Affiliation: Consejo Superior de Investigaciones Científicas
Status: Official research centre

Instituto de Economía y Producciones*   42

[Economics and Productivity Institute]
Address: Facultad de Veterinaria, Zaragoza
Affiliation: Consejo Superior de Investigaciones Científicas
Status: Official research centre

Instituto de Edafología y Biología Vegetal*   43

[Edaphology and Plant Biology Institute]
Address: Serrano 115 bis, Madrid, 6
Affiliation: Consejo Superior de Investigaciones Científicas
Status: Official research centre

Instituto de Fermentaciones Industriales*   44

[Industrial Fermentation Institute]
Address: Juan de la Cierva 3, Madrid, 6
Affiliation: Consejo Superior de Investigaciones Científicas
Status: Official research centre

Instituto de Industrias Cárnicas*   45

[Meat Industry Institute]
Address: Serrano 17, Madrid, 6
Affiliation: Consejo Superior de Investigaciones Científicas
Status: Official research centre

Instituto de Investigaciones Agrobiológicas*   46

[Agrobiological Investigation Institute, Galicia]
Address: Apartado 122, Santiago de Compostela
Affiliation: Consejo Superior de Investigaciones Científicas
Status: Official research centre

Instituto de Investigaciones en Patología de las Colectividades Ganaderas*   47

[Pathology Research Institute for the Cattle Industry]
Address: Facultad de Veterinaria, Zaragoza
Affiliation: Consejo Superior de Investigaciones Científicas
Status: Official research centre

Instituto de Investigaciones Pesqueras de Cádiz* 48

[Fisheries Research Institute, Cádiz]
Address: Puerto Pesquero, Cádiz
Affiliation: Consejo Superior de Investigaciones Científicas
Status: Official research centre

Instituto de Investigaciones Pesqueras de Vigo* 49

[Fisheries Research Institute, Vigo]
Address: Muelle de Bouzas, Vigo (Pontevedra)
Affiliation: Consejo Superior de Investigaciones Científicas
Status: Official research centre

Instituto de Investigaciones Veterinarias 50

[Veterinary Research Institute]
Address: Facultad de Veterinaria Universidad Computense, Madrid, 3
Telephone: 91-449 16 00
Affiliation: Consejo Superior de Investigaciones,
Status: Official research centre
Director: Professor C. Sánchez Botija
Departments: Biochemistry, Professor M. Ruiz Amil; physiopathology of reproduction, Professor F. Pérez Pérez; genetics, Professor C. L. Cuenca; surgical pathology, Professor C. García Alfonso; microbiology, Professor G. Suarez Fernández; virology, Professor C. Sánchez Botija
Graduate research staff: 60
Activities: Animal production; livestock breeding; livestock husbandry and nutrition; veterinary medicine; food science.
Publications: Biannual report

Instituto de la Grasa y sus Derivados 51

[Fats and Derivatives Institute]
Address: Apartado 1078, Avenida Padre García Tejero 4, Sevilla, 12
Telephone: 954-61 15 50
Affiliation: Consejo Superior de Investigaciones Científicas
Status: Official research centre

Director: Dr José Manuel Martínez Suárez
Departments: Chemistry and biochemistry of fats, oil-seeds and oil-fruits, especially olive oil, Dr Augustín Vioque Pizarro; analysis and quality control, Dr Jaime Gracian Tous; oil-seeds cultivated in Spain and their technology, with especially protein extraction and utilization, Dr Eduardo Vioque Pizarro, Dr Félix Ramos Ayerbe; detergents, surface active agents, soaps and related products, Dr Carlos Gómez Herrera; olive oil technology, residue profits, and waste treatment, Dr José Manuel Martínez Suárez; chemistry, microbiology, and technology of table olives and other vegetable products Dr Matías José Fernández Díaz
Graduate research staff: 46
Activities: Crop husbandry; food science.
Publications: Grasas y Aceites, bimonthly.

Instituto de Nutrición* 52

[Nutrition Institute]
Address: Facultad de Farmacia, Cátedra Fisiología Animal, Ciudad Universitaria, Madrid, 3
Affiliation: Consejo Superior de Investigaciones Científicas
Status: Official research centre

Instituto de Productos Lacteos* 53

[Milk Products Institute]
Address: Apartado 78, Arganda del Rey, Madrid
Affiliation: Consejo Superior de Investigaciones Científicas
Status: Official research centre

Instituto de Zoología 'José de Acosta'* 54

['José de Acosta' Zoology Institute]
Address: PO de la Castellana 84, Madrid
Affiliation: Consejo Superior de Investigaciones Científicas
Status: Official research centre

Instituto de Zootecnia 55

[Animal Production Institute]
Address: Facultad de Veterinaria, Avenida Medina Azahara 9, Córdoba
Telephone: 57-23 75 89
Affiliation: Consejo Superior de Investigaciones Científicas
Director: Professor Diego Jordano Barea
Sections: Applied biology, Professor D. Jordan Barea; blood groups and biochemical polymorphism, Professor A. Rodero Franganillo; animal nutrition, Professor M. Pérez Cuesta; bromatology, Professor R. Pozo Lora; toxicology and environmental contamination, Professor F. Infante Miranda; plant production, Professor M. Medina Blanco; ethnology, vacant; physio-zootechnics, Professor G. Gómez Cárdenas
Laboratories: Parasitology, Professor F. Martínez Gómez; animal breeding Professor J. Aparicio Macarro
Graduate research staff: 92
Activities: Plant production; animal production; food science.
Publications: Archivos de Zootechnia.
Projects: Neurone cultivation (Professor D. Jordano Barea); improvement of autochthonous breeds of ovidae (Professor A. Rodero Franganillo); Lupinus albus research – nutrition (Professor M. Pérez Cuesta); food contamination (Professor R. Pozo Lora); parasitological map of Spain (Professor F. Martínez Gómez); pest contaminanats in the Guadalquivir river basin (Professor F. Infante Miranda); use of Mediterranean shrubs for ruminants (Professor M. Medina Blanco); productivity of the Iberian pig (Professor J. B. Aparicioo Macarro); ovine fertility (Professor F. Aparicio Ruiz); enzymatic study of birds (Professor G. Gomez Cardenas); craneo-encephalic studies of domestic species (Professor E. Aguera Carmona); study of lesions resulting from African swine disease (Professor A. Jover Moyano); temporary sterility in the rabbit (Professor J. Sanz Parejo).

Instituto del Frío 56

[Refrigeration Institute]
Address: Ciudad Universitaria, Madrid, 3
Telephone: 91-449 61 62; 449 61 66
Affiliation: Consejo Superior de Investigaciones Científicas
Status: Official research centre
Director: Manuel Estada Girauta
Sections: Refrigeration engineering; treatment and freezing of vegetable products; treatment and freezing of animal products
Graduate research staff: 40

Activities: Refrigeration engineering - insulation and insulating materials; physical and mathematical modelling of heat transfer in refrigeration plants; thermophysical properties of food products; technological uses of solar energy.
Conservation of food products by freezing - modern technology and comparative studies of fresh and frozen fruits and vegetables; fungi control; improvement to the commercial value of soft foods subjected to refrigeration processes; problems related to the freezing of meat; milk and dairy products, fish; weight loss in frozen foods.

Instituto Español de Entomología* 57

[Spanish Institute of Entomology]
Address: José Gutierrez Abascal 2, Madrid
Telephone: 91-262 01 36
Affiliation: Consejo Superior de Investigaciones Científicas
Status: Official research centre
Director: Professor Dr Salvador V. Peris
Sections: Taxonomy, E. Mingo; ecology and experimental entomology, Professor Dr J. Templado; soil zoology, Professor Dr D. Selga; nematology, Professor Dr H. Arias
Graduate research staff: 14
Activities: The main subject of research is systematic work on arthropods of which the Institute has a large collection. Other areas of research concern: life of insects and the ecology of some species; plant-insect relations; surveys of the pests affecting the main Spanish crops; biological methods of insect breeding techniques and control; terricolous invertebrates, especially Collembola, mites and earthworms (taxonomy and influence on soil structure); parasitic nematodes affecting plants of economic importance (systematics, distribution, economy).
Publications: EOS Revista Española de Entomología; Graellsia, Revista de Entomólogos Ibéricos.

Instituto Español de Oceanografía* 58

[Spanish Institute of Oceanography]
Address: Calle Alcalá 27, Madrid, 14
Telephone: 91-470 17 11/12/13
Affiliation: Ministerio de Agricultura y Pesca
Status: Official research centre
Director: Miguel Oliver Massutí
Sub-director: Joaquin Ros Vicent
Graduate research staff: 120

Activities: The Institute is responsible for advising the Government on matters of oceanography, for coordinating oceanographic research in Spain and for conducting oceanographic research itself. There are eight central units situated in Madrid itself and seven coastal centres, in addition to two service centres.

Research is carried out in the following areas and in conjunction with parallel topics of major importance such as fisheries biology and technology, marine biology, marine culture, physics, chemistry and pollution:

Applied biology - marine production (especially molluscs); algae investigation and exploitation.

Marine biology - oceanic biology; marine production; importance and application of marine biology; applied plankton studies; applied microbiology; environmental impact on marine life; work in the Mediterranean and the Cantabrian area.

Fisheries biology and technology - demersal fishing; biological research into the West African and the Mediterranean area fishing industries; pelagic fisheries; Cantabrian anchovy; sardines; tunny fish; fishing technology; stock evaluation; crustacean and molluscan fishing.

Marine pollution - effects of pollution on coastal ecosystems and marine organisms; study of hydrocarbons and petroleum waste; heavy metals; organochloride hydrocarbons.

Physical oceanography - oceanographic conditions of the Spanish coastline; setting up a system of oceanographic buoys.

Marine geology - marine geology and geophysics; studies of the continental shelf; bilateral studies with the USA; applied geophysics.

Oceanographic chemistry - elements in both maximum and minimum proportion; gases dissolved in the ocean; particulate and dissolved organic matter; sediments; fish, molluscs and crustacea; algae; new analytical methods.

Publications: Bulletin; various reports.

DEPARTMENT OF APPLIED BIOLOGY*   59

DEPARTMENT OF CHEMISTRY*   60

Head: Jesús Arabio
Activities: Chemical oceanography; analysis of marine species.

DEPARTMENT OF FISHERIES BIOLOGY*   61

Head: Miguel Oliver Massutí
Activities: Fisheries production and fishing technology.

DEPARTMENT OF MARINE BIOLOGY*   62

Head: Jerónimo Corral
Activities: Plankton studies, basic productivity, ecology.

DEPARTMENT OF POLLUTION*   63

Head: Joaquín Ros

Marine Observation Network*   64

Activities: The Network's duties are: to carry out regular checks on coastal waters; the systematic study of certain organic communities (ie phytoplankton); to alert the authorities to the presence of toxic species; to define the parameters which determine the quality of the waters; to follow the evolution of ecosystems in different zones.

Balearic Coastal Laboratory*   65

Address: Muelle de Pelaires, Palma de Mallorca
Head: Miguel Duran
Activities: General oceanography in the area with special reference to the Western Mediterranean; fisheries; plankton studies.

Canary Islands Coastal Laboratory*   66

Address: Jose Antonio 3, Santa Cruz de Tenerife
Head: Jerónimo Bravo
Activities: General oceanographic research with emphasis on the problems of the archipelago and the Saharah bank fisheries.

Coruña Coastal Laboratory*   67

Address: Muelle de las Ánimas, La Coruña
Head: Miguel Torre Cervigon
Activities: Marine ecology; marine dynamics; molluscan production; local fishing problems; plankton studies and pollution.

Málaga Coastal Laboratory*   68

Address: Paseo de la Farole 27, Málaga
Head: Natalio Cano
Activities: Regional oceanography; geological surveys; fisheries biology; marine pollution; Mollusca.

Mar Menor Coastal Laboratory* 69

Address: Magallanes, San Pedro del Pinatar, Murcia
Head: Argeo Rodríguez de León
Activities: The Laboratory has a semi-industrial plant for marine culture. The main research topics are marine culture, especially fishing and pollution.

Santander Coastal Laboratory* 70

Address: Promontorio de San Martín, Apartado de Correos 240, Santander
Head: Orestes Cendrero
Activities: Fishing problems affecting the Cantabrian area; algae exploitation; marine culture - ecology, microbiology, marine pollution.

Vigo Coastal Laboratory* 71

Address: Orillamar 47, Vigo
Head: Rafael Robles Pariente
Activities: Fisheries biotechnology; marine pollution, chemistry and ecology.

Instituto 'Jaime Ferrán' de Microbiología* 72

[Jaime Ferrán Microbiology Institute]
Address: Joaquín Costa 32, Madrid, 6
Telephone: 91-261 72 43
Affiliation: Consejo Superior de Investigaciones Científicas
Status: Official research centre
Director: Angel García-Gancedo
Units: Animal virology, Angel García-Gancedo; plant virology, Miguel Rubio-Huertos; microphycology and protozoology, Maximiano Rodríguez-López; industrial microbiology, Carlos Ramírez; physiology and biochemistry of fungi, Rafael Lahoz; phytobacteriology.
Graduate research staff: 29
Publications: Microbiología Española.

Instituto 'López-Neyra' de Parasitología* 73

[López-Neyra Parasitology Institute]
Address: Calle Ventanilla 11, Granada
Telephone: 958-23 15 85
Affiliation: Consejo Superior de Investigaciones Científicas
Status: Official research centre
Director: Miguel Monteoliva-Hernández

Sections: Biochemistry, M. Monteoliva-Hernández; immunology, V. Gómez; nematology, A. Tobar
Graduate research staff: 15
Publications: Revista Ibérica de Parasitología, annually.

Instituto Nacional de Edafología 'José María Albareda'* 74

[José María Albareda National Institute of Edaphology]
Address: Pinar 25, Madrid
Affiliation: Consejo Superior de Investigaciones Científicas
Status: Official research centre

Instituto Nacional de Investigaciones Agrarias 75

– INIA
[Agricultural Research National Institute]
Address: Calle de José Abascal 56, Madrid, 3
Telephone: 91-442 31 99; 442 22 56
Telex: 23425 AGRIM E
Affiliation: Ministerio de Agricultura y Pescas
Status: Official research centre
Research director: Gerardo García Fernández
Scientific, director Pedro Veyrat; technical director Jesús de la Maza
Graduate research staff: 805
Activities: INIA is responsible for a series of national research programmes with the aim of contributing towards agricultural policy as formulated by the Ministry of Agriculture. Research is carried out in 9 regional centres (Centros Regionales de Investigación y Desarollo Agrario – CRIDA). These centres are in Galicia, Ebro, Cataluña, Duero, Tajo, Levante, Extremadura, Andalucía, Canarias. The national programmes include the following: agricultural research on by-products: citrus fruits, subtropical crops, fruit growing, olive, viticulture and oenology, rice, cereals, horticulture, leguminous grains, pastures and feed crops, sugarbeet; forestry research by-products: forestry production, forest products; livestock research by-products: bovine production, sheep production, swine, domestic bird and other; agricultural research: ecology, agriculture, economics and rural sociology, forest breeding and stand implantation, animal nutrition, animal pathology, forest protection, plant protection, animal reproduction, animal breeding, agricultural pollution, irrigation and drainage, soil fertility, plant nutrition, fertilizers; agricultural technological research: agricultural food industries, agroenergy.

Publications: Agrícola; Ganadera; Forestal; Economía y Sociología Agrarias, all series; technical reports.

Instituto Tecnológico del Tabaco   76

[Tobacco Technological Institute]
Address: Barriada Elcano, Apartado 583, Sevilla
Telephone: 954-61 19 50
Status: Official research centre
Director: Manuel C. Llanos
Departments: Biology, Antonio Izquierdo; agronomy, Vicente Cortes; chemistry, Antonio Olleros; pedology, J. A. Galbis Muñoz
Graduate research staff: 7
Activities: Improvement of the economic production of high quality tobacco; information on Spanish tobacco types.
Projects: Tobacco breeding improvement (Manuel Duran); physical properties of tobacco (Vicente Cortes); chemical properties of tobacco (Manuel Ortiz); tobacco smoke (Antonio Olleros); microbiology and phytopathology of tobacco (Antonio Izquierdo); soil and irrigation for tobacco (J. A. Galbis Múnoz).

Laboratorio Municipal de Barcelona*   77

[Barcelona Municipal Laboratory]
Address: Wellington 44, Barcelona, 5
Telephone: 34-3-309 20 50; 34-3-309 41 00
Status: Official research centre
Research director: Dr Francisco Fernández Pérez
Sections: Water quality control and research, Dr Manuel Subirá Rocamora; air pollution control and research, Dr Pere Conillera Vives; food quality control and research, Dr Mercedes Centrich Sureda; applied microbiology, Dr Maria Dolores Ferrer Escobar; rabies control and research, Dr Manuel Gómez Quintana
Graduate research staff: 17
Activities: Research based on the systematic control of sanitary quality of air, food, water, etc; research into control of rabies.
Publications: Annual reports.

Mision Biológica de Galicia   78

[Biological Mission of Galicia]
Address: Apartado 28, Pontevedra
Telephone: 986-85 48 00
Status: Official research centre
Director: Dr Alfonso Solano
Departments: Applied genetics, Dr Amando Ordas; agricultural chemistry, Dr Benito Sanchez
Graduate research staff: 4
Activities: Soil science; plant breeding; livestock breeding.
Projects: Corn breeding and genetics (Dr Amando Ordas); swine inbreeding (Dr Alfonso Soland); soil fertility and soil chemistry, corn nutrition (Dr Benito Sanchez).

Productos Aditivos SA*   79

Address: Paseo de Cracia 89, Sa Planta, Barcelona, 8
Telephone: 216 06 45
Status: Industrial company
Research director: Pedro Valiere Piño
Activities: Artificial and natural sweeteners; pharmaceuticals.

Real Jardín Botánico de Madrid*   80

[Madrid Royal Botanical Garden]
Address: Plaza de Murillo 2, Madrid, 14
Telephone: 91-468 20 25
Affiliation: Consejo Superior de Investigaciones Científicas
Status: Official research centre
Director: Dr F.D. Calonge
Sections: Systematics and ecology of fungi, Dr F.D. Calonge; systematics and ecology of vascular plants, Dr S. Castroviejo
Graduate research staff: 7
Activities: Systematics and ecology of fungi and vascular plants.
Publications: Anales del Jardín Botánico de Madrid.

Servicio Agrícola de la Caja Insular de Ahorros de Gran Canaria* 81

[Agricultural Service for the Canary Islands Savings Bank]
Address: CP 854, Las Palmas, Canary Islands
Telephone: 928-70 00 35
Status: Independent research centre
Director: Dr G. Pérez Melián
Sections: Agricultural chemistry and hydroponics, Dr G. Pérez Melián; phytopathology, R. Rodríguez Rodríguez; fruticulture, N. Quintana Cabrera; floriculture, J.I. Buxens Barandiarán; horticulture, J. Mantique de Lara Gil; energy, A. Guerra Galván
Graduate research staff: 15
Activities: The need for nitrogen, phosphorus and potassium in plant cultivation; content and development of the fatty acids in avocados; pesticide residues in fruit and vegetables; hydroponic acclimatization of meristem plants; etymology of the main plant diseases in the Canary Islands; study of the distribution and control of Meloidogyne spp in the Canary Islands; study of 'Panama' disease in plants with a view to its control.
Publications: Xoba.

Servico Nacional de Cultivo y Fermentación del Tabaco* 82

[National Service for Tobacco Cutivation and Fermentation]
Address: Zurbano 3, Madrid
Status: Official research centre

Universidad Autónoma de Barcelona* 83

[Barcelona Autonomous University]
Address: Bellaterra, Barcelona
Telephone: 692 00 00
Telex: 52 040
Status: Educational establishment with r&d capability
Rector: Professor Antonio Serra Ramoneda

INSTITUTE OF FUNDAMENTAL BIOLOGY* 84

Address: Casa de Convelecencia del Hospital de la Santa Cruz y San Pablo, Avenida San Antonio María Claret 171, Barcelona

Universidad Central de Barcelona* 85

[Barcelona Central University]
Address: Catalanas 585, Barcelona, 7
Telephone: 93-301 42 36
Status: Educational establishment with r&d capability

AGRICULTURAL CHEMISTRY DEPARTMENT* 86

Affiliation: Consejo Superior de Investigaciones Científicas

Universidad Complutense de Madrid* 87

[Complutense University of Madrid]
Address: Ciudad Universitaria, Madrid, 3
Telephone: 91-243 92 00
Status: Educational establishment with r&d capability
President: Professor Dr Angel Vian Ortuño

FACULTY OF BIOLOGICAL SCIENCES* 88

Telephone: 91-243 95 17
Dean: Dr Carlos Vicente Córdoba

Department of Botany and Plant Physiology* 89

Director: Francisco Bellot Rodríguez

Department of Genetics* 90

Director: Juan Ramón Lacadena Calero

Department of Microbiology* 91

Director: Dimas Fernández-Galiano Fernández

Department of Microscopic Morphology* 92

Director: Alfredo Carrato Ibañez

Department of Zoology and Animal Physiology* 93

Director: Rafael Alvarado Ballester

FACULTY OF PHARMACY*   94

Telephone: 91-244 57 73
Director: Dr Antonio Doadrio López

Department of Animal Physiology*   95

Director: Dr Gregorio Varela Mosquera

Department of Parasitology*   96

Director: Dr Antonio R. Martínez Fernández

Department of Pharmaceutical Botany*   97

Director: Dr Salvador Rivas Martínez

Department of Pharmacognosy and Pharmacodynamics*   98

Director: Dr Manuel Gómez-Seranillos Fernández

Department of Plant Physiology*   99

Director: vacant

FACULTY OF VETERINARY SCIENCES*   100

Dean: Dr Rafael Martín Roldán

Department of Agriculture and Agrarian Economy*   101

Director: Dr Gaspar González González

Department of Anatomy and Embryology*   102

Director: Dr Rafael Martín Roldán

Department of Animal Production*   103

Director: Dr Vicente Serrano Tomé

Department of Animal Surgery and Reproduction*   104

Director: Dr Félix Pérez y Pérez

Department of Biochemistry*   105

Director: Dr Manuel Ruiz Amil

Department of Food Hygiene and Microbiology*   106

Director: Dr Bernabé Sanz Pérez

Department of Food Technology and Biochemistry*   107

Director: Dr Pascual López Lorenzo

Department of General and Medical Pathology and Nutrition*   108

Director: Dr Pedro Carda Aparici

Department of Genetics*   109

Director: Dr Carlos Luis de Cuenca

Department of Histology and Pathological Anatomy*   110

Director: Dr Eduardo Gallego García

Department of Infectious and Parasitic Pathology*   111

Director: Dr Carlos Sánchez Botija

Department of Microbiology*   112

Director: Dr Guillermo Suárez Fernández

Department of Nutrition and Food Science*   113

Director: Dr Gonzalo Díaz Rodríguez-Ponga

Department of Pharmacology and Toxicology*   114

Director: Dr Félix Sanz Sánchez

Department of Physiology*   115

Director: Dr Mariano Illera Martín

Universidad de Córdoba*   116

[Córdoba University]
Address: Calle Alfonso XII, Córdoba
Status: Educational establishment with r&d capability

FACULTY OF VETERINARY SCIENCE*   117

Dean: Professor Amador Jovar Moyano

Universidad de Extremadura* 118

[Extremadura University]
Address: Carretera de Portugal, Badajoz
Telephone: 924-23 88 00
Telex: 28-638
Status: Educational establishment with r&d capability
Rector: Dr Andrés Chordi-Corbo
Secretary-General: Dr Cristóbal Valenzuela Calahorro

FACULTY OF SCIENCES* 119

Address: Carretera de Portugal, Badajoz
Telephone: 924-23 88 00
Dean: Dr José Sotelo-Sancho
Deputy Dean: Dr Guillermo Rodríguez-Izquierdo Gavala

Department of Biology* 120

Sections: Botany, Professor Alberto Romero Lamo; microbiology, Dr Andrés Chordi-Corbo; geology, Dr M. Jesús Liso Rubio

UNIVERSITY COLLEGE OF AGRICULTURAL ENGINEERING* 121

Address: Carretera de San Vicente, Badajoz
Telephone: 942-23 62 04
Director: Enrique Trabadela Gómez
Sections: Genetics and phytopathology, Juan Villalobos Borrachero; phytotechnics and extensive cultivation, Enrique Trabadela Gómez; horticulture and intense cultivation, Leonor Fernández Pizarro

Universidad de León* 122

[León University]
Address: Campus Universitaria La Palomera, León
Status: Educational establishment with r&d capability

FACULTY OF VETERINARY MEDICINE* 123

Dean: J. Burgos González

SCHOOL OF AGRICULTURAL ENGINEERING* 124

Dean: R. Cos Jarlhing

Universidad de Murcia* 125

[Murcia University]
Address: Santo Cristo 1, Murcia
Telephone: 968-24 92 00
Telex: 67058
Status: Educational establishment with r&d capability
Secretary-General: Antonio Bódalo Santoyo

Agricultural Chemistry Department* 126

Head: Dr S. Navarro

Biology Department* 127

Head: Dr F. Sabater

Universidad de Oviedo* 128

[Oviedo University]
Address: San Francisco, Oviedo
Telephone: 985-21 98 85
Status: Educational establishment with r&d capability

FACULTY OF BIOLOGY* 129

Dean: Professor Carlos Hardisson Rumen

FACULTY OF VETERINARY MEDICINE* 130

Dean: Professor J.B. Delgado

Universidad de Sevilla* 131

[Seville University]
Address: Calle San Fernando, Sevilla
Telephone: 954-21 86 00
Status: Educational establishment with r&d capability

FACULTY OF BIOLOGY* 132

Dean: Dr Julio Pérez Silva

Universidad de Valencia* 133

[Valencia University]
Address: Nave 2, Valencia, 3
Telephone: 96-321 73 80
Status: Educational establishment with r&d capability

FACULTY OF BIOLOGY* 134

Dean: Professor José L. Mensua Fernández

Universidad de Zaragoza* 135

[Zaragoza University]
Address: Ciudad Universitaria, Zaragoza
Telephone: 976-35 41 00
Telex: 58064
Status: Educational establishment with r&d capability

FACULTY OF VETERINARY 136 MEDICINE*

Dean: Professor José Antonio Bascuas Asta

Universidad Politècnica de 137 Barcelona*

[Barcelona Technical University]
Address: Avenida Dr Gregorio Marañon, Barcelona, 28
Telephone: 93-249 38 04
Telex: 52821 upbe
Status: Educational establishment with r&d capability
Rector: Gabriel Ferraté i Pascual
Vice-Rector, Research: Manuel Marti Recóber
Graduate research staff: 320

HIGHER TECHNICAL COLLEGE OF 138 AGRONOMIC ENGINEERING - LLEIDA*

Director: Juan Antonio Martín Sánchez

Department of Agrarian Economy* 139

Senior staff: José María Torralba
Activities: Technical and socioeconomic study of agrarian cooperatives in Lérida; repercussions of entry into the EEC.

Department of Agricultural 140 Entomology*

Senior staff: Ramon Albajes
Activities: Whitefly in greenhouses; lilies.

Department of Edaphology, Geology 141 and Climatology*

Senior staff: Jaume Porta, Rafael Rodríguez Ochoa, Pere Ferrer
Activities: Study of soils for hazelnut plantations; study of boron and toxicity levels in soils; use of purines in agriculture - effects on the soil; databank on climate, soil and water for Catalonia; soil desalination.

Department of General and 142 Agricultural Electrotechnics*

Senior staff: José A. Giné
Activities: Use of solar energy for drying alfalfa at low temperatures; use of electric analogues in the movement of liquids through porous media.

Department of General and 143 Agricultural Hydraulics*

Senior staff: Javier Barragán Fernández, Aniceto Casanas Cladellas
Activities: Irrigation levels for corn growing; design and construction of a mini-tractor to operate in rugged and sloping areas.

Department of Physiology and 144 Biology*

Senior staff: Inmaculada Recasens
Activities: Biochemical processes in the growing and maturation of Starking apples.

Department of Plant Genetics* 145

Senior staff: Antonio Michelena Barcena
Activities: Study of esparto grass varieties.

Department of Soil Mechanics* 146

Senior staff: Miquel Llorca Marquis
Activities: Alterations to the physical structure of soils as a result of agricultural mechanization.

UNIVERSITY COLLEGE OF 147 AGRICULTURAL ENGINEERING - BARCELONA*

Director: Jaume Bech i Borrás

UNIVERSITY COLLEGE OF 148 AGRONOMIC ENGINEERING - LLEIDA*

Director: Juan Antonio Martín Sánchez

Department of Chemical and Agricultural Analyses* 149

Senior staff: Xavier Ferran Caldero
Activities: Analytic methods which detect boron in soils.

Department of Zootechnics* 150

Senior staff: Mateu Torrent Mollevi

UNIVERSITY POLYTECHNIC OF GIRONA* 151

Director: Josep Arnau Figuerola

Department of Phytotechnics* 152

Senior staff: Laurea Fuster Berenguer
Activities: Humidifying organic matter according to origin and soil type; improvement of soya bean cultivation.

Universidad Politécnica de Valencia* 153

[Valencia Technical University]
Address: Camino de Vera, Valencia, 14
Telephone: 96-369 67 00
Status: Educational establishment with r&d capability

SCHOOL OF AGRICULTURAL ENGINEERING 154

SRI LANKA

Agrarian Research and Training Institute   1

– ARTI
Address: PO Box 1522, Colombo, 7
Telephone: (01) 94873, 96437, 96981
Affiliation: Ministry of Agricultural Development and Research
Status: Official research centre
Director: T.B. Subasinghe
Deputy director: Dr S.B.D. de Silva
Graduate research staff: 35
Activities: Research in the following areas: agricultural economics; rural sociology; agricultural marketing; land economics; irrigation and water management; agricultural extension and communication.
Publications: Annual report; *Sri Lanka General Agrarian Studies* (biannual); research studies.

AGRICULTURAL LABOUR AND   2
EMPLOYMENT INFORMATION
CENTRE

Head: P.J. Gunawardana
Projects: Agricultural statistics (Dr M.D. Sumanasekera).

AGRICULTURAL PLANNING AND   3
IMPLEMENTATION GROUP

Head: A. Wanasinghe
Projects: Kurunegala: rural development (Dr H. D. Sumanasekera); planning in the agricultural sector (M. Samad); monitoring and evaluation of Kirindioya: irrigation and settlement (A. Wanasinghe).

FARM MARKETING AND   4
MANAGEMENT GROUP

Head: S.M.P. Senanayake
Projects: Preparation of an inventory of animal feed marketing resources in Sri Lanka (R.M.B. Fernando).

FOOD POLICY (MRU) GROUP   5

Head: S.B.D. de Silva

IRRIGATION AND WATER   6
MANAGEMENT GROUP

Head: C.M. Wijeratne
Projects: Farm power water use (F. Abeyratne); water management (C.M. Wijeratne); northwest land water resources development (M.V. Isak Lebbe).

LAND AND AGRARIAN RELATIONS   7
GROUP

Head: G. Wickramasinghe
Projects: Smallholder rubber rehabilitation (W.G. Jayasena).

RURAL ORGANIZATIONS AND   8
COMMUNICATIONS GROUP

Head: S.B.R. Nikehetiya
Projects: Management training for leaders of small farmer groups.

Central Agricultural Research Institute 9

Address: Gannoruwa, Peradeniya
Telephone: (08) 8011
Telex: AGRESEARCH GANNORUWA
Status: Official research centre
Deputy director of agricultural research: Dr Nallini Wickremasinghe
Activities: The institute conducts research in the areas of natural resources and plant production in the Mid Country Wet Zone, in respect of cereals, root crops, grain legumes, vegetables and fruits; it also provides training and support services to other research stations of the Department of Agriculture. The following areas are being currently covered: identification of different soil types, water regimes and climates which exist in the Mid Country Wet Zone; breeding and selecting of varieties of cereals, root crops, grain legumes, vegetables and fruits suited to these different agro-ecological regions; evaluation of the fertility status of soils and determination of the fertiliser requirements of crops: studies on efficient use of inorganic and organic fertilizers; analysis of soil, plant, water fertilizer and pesticides; identification and control of weeds which adversely affect crop growth; studies on the various plant diseases and strategies for their control; identification and control of insect pests causing damage to crops; development of techniques to improve the quality and value of agricultural produce: studies on post-harvest processing and utilization.
Projects: Identifying donors of resistance to rice pests, diseases and tolerance to low phosphorus (Y. Elikewela, Dr S.N. de S. Seneviratne, Velmurugu, in collaboration with local breeders and the International Rice Research Institute); utilization of cheaper forms of nitrogen-phosphorus-potassium (Dr S.L. Amarasiri, K. Wickremasinghe, and M.Gunatilake); upgrading of seeds and plant stock (Dr S.N. de S. Seneviratne, Premala Jeyanandarajah, and M.E.R. Pinto).

AGRONOMY DIVISION 10

Head: Dr C.R. de Vaz
Projects: Development of root and tuber crops for food, livestock feed, and industry.

BOTANY DIVISION 11

Head: V. Velmurugu

CEREAL CHEMISTRY DIVISION 12

Head: C. Breckenridge

CHEMISTRY DIVISION 13

Head: Dr S. Nagarajah

ENTOMOLOGY DIVISION 14

Head: Y. Elikewela
Projects: Integrated pest control project for rice (I.D.R. Peries).

HORTICULTURE DIVISION 15

sri
Head: M.E.R. Pinto
Projects: Development of tropical fruit production in Sri Lanka.

PLANT PATHOLOGY DIVISION 16

Head: Dr S.N. de S. Seneviratne
Projects : Bacterial wilt programme (Srimathi Vdugama).

PLANT QUARANTINE DIVISION 17

Head: Dr P. Shivanathan

SOYA FOOD RESEARCH CENTRE 18

Head: T.D.W. Siriwardena
Projects: Processing and utilization of soyabean.

Coconut Research Institute 19

– CRI
Address: Bandirippuwa Estate, Lunuwila
Telephone: 95 Dankotuwa
Parent body: Coconut Development Authority
Status: Official research centre
Acting director: Dr R. Mahindapala
Activities: Conducting and furthering of scientific research in respect of the growth and cultivation of coconut palms and other crops; animal husbandry on coconut plantations; prevention and cure of diseases and pestilence; conducting and furthering of scientific research in connection with the processing and utilization of coconut products.
Publications: Ceylon Coconut Quarterly (quarterly); *Ceylon Coconut Planters' Review.*

AGRONOMY DEPARTMENT   20

Officer-in-charge: L.V.K. Liyanage
Projects: Utilization of interspace in coconut lands
(D.E.F. Ferdinandez).

BIOMETRY AND AGRICULTURAL   21
ECONOMICS DEPARTMENT

Officer-in-charge: D.T. Mathes
Projects: Irrigation, crop-forecasting and biometrical
studies, agroeconomic studies.

BOTANY AND PLANT BREEDING   22
DEPARTMENT

Head: Dr M.R.T. Wickramaratne
Projects: Genetical improvement, hybridization and
propagation.

COCONUT PROCESSING   23
DEPARTMENT

Head: Dr S. Mohanadas
Projects: Processing of products and by-products.

CROP PROTECTION DEPARTMENT   24

Head: Dr R. Mahindapala
Projects: Pest and disease control.

SOILS AND PLANT NUTRITION   25
DEPARTMENT

Head: M. Jeganathan
Projects: Soils, soil surveys, fertilizers and nutritional
studies.

Department of Animal   26
Production and Health*

– DAPH
Address: Getambe, Peradeniya
Parent body: Ministry of Rural Industrial Development
Status: Official research centre
Activities: The Asian Development Bank has approved
a loan for a livestock development project in Sri Lanka
designed to expand the country's food production. The
objective of the project is to increase the quality and
productivity of cattle, buffalo, pigs and poultry in
selected districts of Sri Lanka. This will be done by
introducing better breeds; improving animal nutrition
and husbandry; strengthening livestock marketing

systems; expanding livestock extension, health services
and training; and encouraging farmers to develop their
livestock enterprises. The project is also designed to
increase the participation of small-scale farmers in
livestock production by expanding integrated farming
techniques in order to improve the economic and social
conditions of rural dwellers through higher incomes and
greater on-farm employment opportunities.

Forests Department   27
Research Branch

Address: PO Box 509, Colombo, 2
Telephone: 32251
Status: Official research centre
Forests Conservator: V.R. Nanayakkara
Research divisions: Silvicultural, Dr K. Vivekanandan;
timber utilization, E.W. Seneviratne
Graduate research staff: 4
Activities: The main areas of research are: species
evaluation; genetic improvement of species; yield and
growth studies; disease and pest control; timber technol-
ogy; seasoning, preservation and utilization research.
Publications: Sri Lanka Forester (biannual); *Administra-
tion Report* (annual).
Projects: Silvicultural (Dr K. Vivekanandan); entomological (Dr N.B. Ratnasiri); timber technology
(A.M.T. Soyza).

Land Use Division   28

– LUD
Address: Irrigation Department, PO Box 1138, Colombo
Telephone: 86427
Affiliation: Irrigation Department, Ministry of Lands
and Land Development
Status: Official research centre
Head: Dr K.A. De Alwis
Graduate research staff: 9
Activities: The division is mainly concerned with soil
surveys, land classification, land suitability evaluation
for agricultural purposes, land use mapping, and basic
investigations on physico-chemical and mineralogical
properties of soils, soil salinity, and trace element contents of soils. The division is also engaged in a programme of on-farm water management research, and
carries out studies on the environmental effects of irrigation.

Projects: Soil survey of Sri Lanka (Dr K.A. De Alwis); on-farm water management research project (S. Dimantha); environmental study of immediate catchment of Victoria reservoir (Dr N.S. Jayawardane); environmental monitoring of Kalawewa irrigation project (T.S.B. Weerasekera); soil survey and land classification of system 'A' Mahaweli project (S. Kumarakulasuriyar); soil studies in Red Latosoi region (A. Senarath).

Ministry of Lands and Land Development * 29

Address: (Bullers Road) Bauddhaloka Mawatha, PO Box 1138, Colombo 7
Status: Official research organization

IRRIGATION DEPARTMENT * 30

Activities: The Kirindi Oya irrigation and settlement project, financed by the Asian Development Bank.

Minor Export Crops Research Project 31

Address: Matale
Telephone: (0662) 391
Affiliation: Department of Minor Export Crops, Kandy
Status: Official research centre
Assistant director (research): Dr W.S. Alles
Activities: The Minor Export Crops Research Project is concerned with research on the following crops: cocoa, coffee, pepper, cinnamon, cardamom, clove, nutmeg, aromatic grasses, pawpaw for papain, and a few other perennial crops of lesser importance. The project also maintains an intensive programme of work on intercropping of some of the above crops on coconut, tea and rubber plantations, as well as in mixed cropping systems. Research work also includes plant breeding, crop husbandry, plant protection, post-harvest technology and economics of the crops and cropping systems.

Rubber Research Institute of Sri Lanka 32

Address: Dartonfield, Agalawatte
Telephone: 26
Status: Official research centre
Director: Dr O.S. Peries
Graduate research staff: 21

Activities: Classification of Sri Lanka soil types; Hevea breeding; agronomy; phytopathology; polymer chemistry - natural polymers.
Publications: Annual report; *Quarterly Research Journal.*

GENETICS AND PLANT BREEDING DEPARTMENT 33

Head: D.M. Fernando
Projects: Studies on production of higher-yielding more vigorous trees.

INTERCROPPING DEPARTMENT 34

Head: L.B. Chandrasekara

PLANT PATHOLOGY DEPARTMENT 35

Head: Dr A. de S. Liyanage
Projects: Studies on the phytopathology of rubber and associated crops.

PLANT SCIENCE DEPARTMENT 36

Head: Dr A.C.I. Samaranayake
Projects: Improvement of planting methods and tapping methods.

RUBBER CHEMISTRY DEPARTMENT 37

Head: S.W. Karunaratne
Projects: Studies on the chemistry of rubber and rubber products.

SOILS AND PLANT NUTRITION DEPARTMENT 38

Head: Dr N. Yogaratnem
Projects: Studies on the nutrition of rubber trees.

University of Peradeniya * 39

Status: Educational establishment with r&d capability

FACULTY OF AGRICULTURE   40

Address: Peradeniya
Telephone: (08) 8041
Dean: Professor Y.D.A. Senanayake
Graduate research staff: 70
Activities: The main objectives of the faculty are to undertake semibasic research which would complement the applied research conducted in other research stations of government planned in accordance with national agricultural priorities.
Projects: Studies on winged bean (Professor Y.D.A. Senanayake); cereal/legume cropping system (Professor H.P.M. Gunasena); soyabean agronomy (Dr R. Clements); poultry nutrition (Professor A.S.B. Rajaguru); straw enrichment (Dr M.C.N. Jayasuriya); reproductive physiology of buffalo (Dr R. Rajamahendran); socioeconomics (Dr S. Pinnaduwage); production economics (Dr H.M.G. Herath); development studies (Professor T. Jorgaratnam); crop physiology (Professor H.M.W. Herath); stress physiology (Dr M. Thiagarajah); plant pathology (Dr J.M.R.S. Bandara); mineral nutrition (Dr V. Pavanasasivam); soil science (Professor M.W. Thenabadu); alternate sources of energy (Dr S. Illangantileke).

Agricultural Biology Department   41

Head: Dr J.M.R.S. Bandara

Agricultural Chemistry Department   42

Head: Professor M.W. Thenabadu

Agricultural Economics Department   43

Head: Dr S. Pinnaduwage

Agricultural Engineering Department   44

Head: Dr S. Illangantileke .

Animal Science Department   45

Head: Professor A.S.B. Rajaguru

Crop Science Department   46

Head: Professor H.P.M. Gunasena

FACULTY OF VETERINARY MEDICINE   47
AND ANIMAL HEALTH*

Dean: Professor Dr S.T. Fernando

Veterinary Research Institute   48

– VRI
Address: Peradeniya
Telephone: (08) 8311
Status: Official research centre
Director: Dr J.A. de S. Siriwardene
Graduate research staff: 30
Activities: Research on all aspects of animal health and production.

ANIMAL BREEDING DIVISION   49

Head: E.F.A. Jalatge
Projects: Cattle breeding and evaluation of crossbreeds (Dr N. Thilekeratne); evaluation of breeding programmes with sheep and goats (E.F.A. Jalatge).

ANIMAL NUTRITION DIVISION   50

Head: Dr G.A.P. Ganegoda
Projects: Fish silage production and utilization (G.A.P. Ganegoda); mineral status and deficiencies in livestock (Dr S.S.E. Ranawana).

BACTERIOLOGY DIVISION   51

Head: Dr P. Kulesekeram
Projects: Haemorrhagic septicaemia vaccine improvement (Dr M.C.L. de Alwis); mastitis surveys and efficacy of control methods (Dr D.D. Wanasinghe); improvement of Newcastle disease vaccines (Dr P. Kulesekeram).

PARASITOLOGY DIVISION   52

Head: H.H.M.L. Wanduragala
Projects: Tick-borne diseases (Dr D.J. Weilgama).

PASTURE AND FODDER CROPS   53
DIVISION

Head: A.B.P. Jayawardene
Projects: Research on non-conventional feedstuffs (Dr K.K. Pathirana); utilization of rice straw and crop residues as animal feedstuffs (Dr K.K. Pathirana).

REPRODUCTIVE DISORDERS   54
DIVISION

Head: G.S. Peiris

VIROLOGY DIVISION 55

Head: Dr W.W.H.S. Fernando
Projects: Epidemiology and control of foot and mouth
disease in cattle.

SUDAN

University of Gezira*   1

Status: Educational establishment with r&d capability

FACULTY OF AGRICULTURAL   2
SCIENCES

Address: PO Box 20, Wad Medani, Gezira Province
Dean: Dr Yousif Mohamed El Tayeb
Graduate research staff: 50
Activities: The main activities of the faculty consist of undergraduate training, graduate training and research. The latter two aim to contribute in improving existing agricultural systems and practices, and to increase regional and national agricultural production. Priority areas are crop protection, plant breeding, irrigation and water relations.

Agricultural Mechanization   3
Department

Head: Mohamed A. Ali
Projects: Effects of various tillage systems.

Animal Production Department   4

Head: Dr Abdullahi H. Eljack

Crop Production Department   5

Head: Professor Osman A. Ali
Projects: Vegetable production research (Dr Hassan A. Izzeldin); pomology research (Dr Osman A. Sidahmed); plant breeding (Dr Abdul Hassan S. Ibrahim).

Crop Protection Department   6

Head: Dr Sami O. Freigoun
Projects: Different aspects of entomological research (Dr Y.M. El Tayeb); phytopathological research (Professor Omer H. Giha).

Environmental Science and Natural   7
Resources Department

Head: Dr Hussein S. Adam
Projects: Plant irrigation, soil, and water relations (Professor O.A. Ali).

University of Juba   8

Address: PO Box 321/1, Khartoum
Status: Educational establishment with r&d capability

COLLEGE OF NATURAL RESOURCES   9
AND ENVIRONMENTAL STUDIES

Address: PO Box 82, Juba
Telephone: 2113-4
Dean: Dr Peter Obadayo Tingwa
Graduate research staff: 20
Activities: The college is primarily a teaching institution, but it is also involved in limited research in natural resources and environmental studies.
Projects: Wheat project (Dr M. Ashraf); soil erosion (Dr Anwarul Haq).

Animal Science Department   10

Head: Dr Joseph Awad Morgan

Crop Science Department   11

Head: Dr M. Ashraf

Fisheries Department   12

Head: Dr V.G. Krishnamurthy

Forestry Department 13
Head: Dr T.O. Davies

Wildlife Science Department 14
Head: Dr M. Sommerlatte

University of Khartoum 15

Status: Educational establishment with r&d capability

FACULTY OF AGRICULTURE 16

Address: PO Box 32, Shambat
Dean: Professor A/Mohsin H. El Nadi
Graduate research staff: 184
Activities: Research on all aspects of agriculture.

Agricultural Botany Department 17

Head: Dr Hamid A. Dirar
Projects: Nile Valley project for improvement of broad beans (Vicia faba) (Professor A/Mohsin H. El Nadi).

Agricultural Engineering Department 18

Head: Dr T.F. Demian
Projects: Development of improved grain handling and storage facilities (Dr Sherif M. Khaeiri).

Agronomy Department 19

Head: Dr A.M.O. El Karuri

Animal Production Department 20

Head: Dr Gaafar A. El Hag
Projects: Promotion of bee culture in the Sudan (Dr M.S. El Sarrag).

Biochemistry and Soil Science Department 21

Head: Dr A.H. El Tinay

Crop Protection Department 22

Head: Dr. Ahmed M. Baghdadi
Projects: Rhizobium inoculum production for leguminous crops (Dr Ahmed Ali Mahdi).

Forestry Department 23

Head: Professor M.A. Loeschan

Horticulture Department 24

Head: Dr Gaffar M. El Hassan
Projects: Biological control of water hyacinths (Dr M.O. Beshir).

Rural Economy Department 25

Head: Professor Farah H. Adam
Projects: Sudan Striga project (Professor M.O. El Khidir); economics of mechanization in the Sudan; self-sufficiency and the insufficiency of inputs (Professor F.H. Adam).

FACULTY OF VETERINARY SCIENCE* 26

Dean: Professor Amir Mohammed Salih Mukhtar

SURINAME

Centre for Agricultural Research 1

Address: PO Box 1914, Paramaribo
Telephone: 60244
Status: Independent research centre
Acting director: B.A. Halfhide
Graduate research staff: 12
Activities: Research in the following areas: soil science; crop husbandry; plant protection; agricultural engineering; weed science; forest ecology; silviculture; hydrology.
Publications: Quarterly Reports (in Dutch); annual reports, by project.
Projects: Human interference in the tropical rainforest (Dr P. Schmidt); permanent cultivation of rainfed annual crops on the loamy soils of the Zanderij formation (Dr J.F. Wienk).

Landpouwproefstation* 2

[Agricultural Experiment Station]
Address: POB 160, Paramaribo
Affiliation: Department of Agriculture, Animal Husbandry and Fisheries
Status: Official research centre
Director: I.E. Soe Agnie
Publications: Suriname Agriculture (biannual); annual report.

Naturwetenschappelijke 3 Studiekring vour Suriname en de Nederlandse Antillen*

[Foundation for Scientific Research in Suriname and the Netherlands Antilles]
See entry under Netherlands.

SWAZILAND

Lowveld Experiment Station* 1

Address: PO Box 53, Big Bend
Status: Official research centre
Senior research officer: J.S. Watson
Departments: Cotton breeding, J.S. Watson; cotton entomology, R.D.S. Clarke; dryland crop agronomy, K.D. Shepherd
Activities: Dryland crops research for variety, fertilization, pest control, agronomy and cultivation techniques on cotton, maize, sorghum, caster, soya bean and sunflower on several soil types in semi-arid areas. Some horticultural research, mainly variety testing, and soil physics and chemistry investigations, mainly irrigated soils, in cooperation with staff at the Malkerns Research Station.

Malkerns Research Station 2

Address: PO Box 4, Malkerns
Telephone: (0194) 83017
Status: Official research centre
Chief research officer: F.M. Buckham
Departments: Agronomy, J. Pali-Shikhulu; pathology, I.S. Kunene; horticulture, D. Gama; entomology (cotton), M.P. Mtshali; breeding (cotton), C.T. Nkwanyana; socioeconomics F.G. Simelane; biometry, S. Matsebula; pasture, P.D. Mkhatshwa; dryland agronomy, A.I. Mamba; weed control, M.V. Mkhonta
Graduate research staff: 150
Activities: Plant production, cotton, maize, sorghum, wheat, legumes, vegetables and fruits; cotton breeding; crop husbandry; plant production.
Publications: Annual report; biennial advisory bulletin.
Projects: Cropping systems research (Dr T.B. King).

Mpisi Cattle Breeding Experiment Station* 3

Address: Mpisi
Status: Official research centre
Director: R.A. John
Manager: I.A. Morley Hewitt
Activities: Improvement of indigenous Nguni cattle; providing multiplication studs of Brahman, Simmentaler and Friesland cattle for beef, milk and crossbreeding.

University College of Swaziland* 4

Address: Private Bag, Kwaluseni
Status: Educational establishment with r&d capability

FACULTY OF AGRICULTURE* 5

Dean: Dr G.T. Magagula

SWEDEN

Alfa-Laval AB   1

FARM EQUIPMENT DIVISION   2

Address: PO Box 39, S-147 00 Tumba
Telephone: (0753) 31100
Telex: 13944; 12310
Status: Research centre within an industrial company
Research director: Johannes Schmekel
Graduate research staff: 11
Activities: Livestock husbandry and nutrition (cattle, sheep, goats, buffalo, pigs); veterinary medicine; agricultural engineering; aquaculture.

Arla   3

RESEARCH AND DEVELOPMENT DEPARTMENT   4

Address: 105 46 Stockholm
Telephone: (08) 22 60 60
Telex: 19844 MCMILKS
Status: Research centre within an industrial company
Graduate research staff: 20
Activities: Food science and technology - mainly milk and milk products.

⟩ AB Cardo*   5

Address: Box 17050, S-200 10 Malmö
Telephone: 040-736 40
Telex: 32416 SOCKER S
Status: Industrial company
Chairman: Ernst Herslow
Managing Director: Per Lindblad
Graduate research staff: 70

HILLESHÖG AB   6

Address: Box 302, S-261 23 Landskrona
Telephone: 0418-260 60
Managing Director: Arne Emanuelsson
Research director: Nils Olof Bosemark
Activities: The Company is one of the world's leading plant breeding enterprises for sugar beet seed. Among its major achievements has been the introduction in 1966 of a monogerm beet seed, which, by producing only one plant per seed, reduced the necessity for time-demanding and expensive hand thinning and led towards wholly mechanized sugar beet growing. More recent research has been into the pelleting of seed, not only to aid mechanical sowing but also to utilize the possibility of including plant nutrient and pest control preparations within the pelleting mass. This technique can also be applied to, for instance, rape and vegetable seeds. The company also pursues a policy of research and plant breeding into forestry, specializing in conifers, and has 5 400 hectares of forest land at the Gisslarbo Forest Experimental Institute in Västmanland, central Sweden. Other research interests lie in the development of rape, cereal and vegetable seeds.

SÄBYHOLM JORDBRUKS AB*   7

Managing Director: Nils Ohlsson
Activities: The Company operates ten highly mechanized farms and continuously tests new rational forms of operation. Although the main interest is not in livestock, several well-known bulls of lowland race have been bred for artificial insemination, and on one farm 3 000 piglets are bred annually in a unit controlled by one man.

AB SORIGONA*   8

Address: Box 139, S-245 00 Staffanstorp
Telephone: 046-25 50 40
Managing Director: Harald Skogman
Activities: Sorigona is a development company specializing in biotechnology. Its main interests are as follows: manufacture of raw dextran and tryptophane for the pharmaceutical industry; manufacture of fruit sugar, yeast protein, polysaccharides, amino acids and citric acid for the food industry; production of yeast protein for use in animal fodder from waste water from potato processing; purification of waste water from the sugar industry.

SVENSKA SOCKERFABRIKS AB*   9

Managing Director: Hans-Erik Leufstedt
Sections: Department of Agricultural Engineering, John Erik Nilsson; Research Laboratory, Rolf Bergkvist
Activities: The Company supplies Sweden's growers of sugar beet with seed. It has a special unit, the Department of Agricultural Engineering at Staffanstorp, which works with agricultural and mechanical development, and is responsible for information and advice to farmers in beet-growing matters.
The research laboratory at Arlöv, outside Malmö, is the centre of the Cardo Group's chemical and microbiological research and development. Development and control of methods for sugar production have been the basis for the laboratory's work, and today biotechnical processes are the most important features of research. The laboratory has divisions for biochemistry, microbiology, organic chemistry, analytical chemistry as well as for process and fermentation technology. Extensive research is carried out in, for instance, the production of various substances, such as amino acids, polysaccharides, enzymes, organic acids, ethanol and protein. New projects are the production of biologically active carbohydrates and natural products from plant material. The development of by-products from the sugar production as well as of new feedstuffs is included in the programme. The whole field of sweeteners is studied, especially with regard to the possibilities of manufacturing alternative products out of starch or other raw materials. The programme also contains diet and nutrition questions and the laboratory takes part in various studies connected with these questions.
A number of processes developed at the laboratory have been exploited within and outside the Group, including processes for the production of tryptophane, citric acid, dextran, fructose, lysine and 'Symba' yeast. The laboratory is in close contact with universities and research institutes.

W. WEIBULL AB*   10

Address: Box 520, S-261 24 Landskrona
Telephone: 0418-780 00
Managing Director: Wider Weibull
Research director: Göran Ewertson
Activities: Breeding and marketing of agricultural seeds as well as vegetable and flower seeds.

Fiskeristyrelsen*   11

[National Board of Fisheries]
Address: Box 2565, S-403 17 Gothenburg
Status: Official research centre

FÖRSÖKSGRUPPEN FÖR   12
FISKEVÅRDANDE ÅTGÄRDER I
KRAFTVERKSMAGASIN

– FÅK
[Research Group for Fishery Management in River Reservoirs]
Address: FÅK Fiskeriintendenten, St Torget 3, S-87100 Härnösand
Telephone: 0611-100 50
Research director: Adam P. Gönczi
Senior staff: Dr Jan Henricson
Graduate research staff: 3

HAVSFISKELABORATORIET   13
[Marine Research Institute]
Address: Box 5, S-453 00 Lysekil
Telephone: 0523-104 58
Director: Dr Armin Lindquist
Sections: Herring and coastal research, west coast, Dr Rutger Rosenberg; herring and cod in the Baltic, Gunnar Otterling; stock assessment, Bengt Sjöstrand; shellfish, Dr Bernt Ingemar Dybern
Graduate research staff: 12
Publications: Reports; *Meddelande.*

Göteborgs Universitet*   14

[Gothenburg University]
Address: Vasaparken, S-411 24 Göteborg
Telephone: 031-81 04 00
Status: Educational establishment with r&d capability
Rector: Professor Georg Lundgren

FACULTY OF MATHEMATICS AND NATURAL SCIENCES*   15

Dean: Professor Bo Malmström

Biology-Geographical Section*   16

Dean: Professor Anders Enemar

Biology Laboratory*   17
Address: Carl Skottsbergs Gata 22, S-413 19 Göteborg
Telephone: 031-41 87 00
Head: Dr Tore Mörnsjö

Botany Department*   18
Address: Carl Skottsbergs Gata 22, S-413 19 Göteborg
Telephone: 031-41 87 00
Head: Dr Uno Eliasson
Senior staff: Physiological botany, Professor Hemming Virgin; marine botany, Professor Erik Jaasund; marine microbiology, vacant; systematic botany, Professor Gunnar Harling; phytobotany, Professor Per Wendelbo

Genetics Department*   19
Address: Stigbergsliden 14, S-414 63 Göteborg
Telephone: 031-14 90 64
Head: Professor Göran Levan

Tjärnö Marine Biology Laboratory*   20
Address: PL 2781, S-452 00 Strömstad
Telephone: 0526-251 16
Head: Lars Afzelius

Zoology Department*   21
Address: Medicinaregatan, Box 25059, S-400 31 Göteborg
Telephone: 031-41 08 00
Head: Stefan Nilsson
Senior staff: Structural and ecological zoology, Professor Anders Enemar; zoophysiology, Professor Ragmar Fänge

Chemistry Section*   22

Biochemistry and Biophysics Department*   23

Institutet för Skogsförbättring   24

[Institute for Forest Improvement]
Address: Box 7007, S-750 07 Uppsala
Telephone: (018) 30 07 16
Status: Independent research centre
Director: Karl-Rune Samuelson
Sections: Forest fertilization, Göran Möller; Southern district, Martin Werner; Central district, Gustaf Hadders; Northern district, Ola Rosvall
Graduate research staff: 13
Activities: Forest tree breeding; forest fertilization.
Publications: Yearbook.
Projects: Forest fertilization on mineral soil (Göran Möller); forest fertilization on peat soil (Göran Möller); general forest tree breeding (Gustaf Hadders); seed orchard research (Gustaf Hadders); testing of native and introduced species (Ola Rosvall); seed extraction research (Ola Rosvall); Norway spruce cuttings (Martin Werner).

Jordbrukstekniska Institutet   25

[Swedish Institute of Agricultural Engineering]
Address: PO Box 7033, S-750 07 Uppsala
Telephone: (018) 30 19 30
Affiliation: Ministry of Agriculture - Foundation for Swedish Agricultural Engineering Research
Status: Official research centre
Director: Sven-Uno Skarp
Graduate research staff: 12
Activities: The institute conducts agricultural engineering research in order to develop farm machinery and improve work methods, monitors the development of mechanical and electrical farm equipment, and provides educational and extension services in the rational use of mechanization in Swedish agriculture. These tasks are carried out in close cooperation with other institutes, the farmers' organizations, the farm machinery industry, agricultural advisers, and practical farmers. At present particular attention is being given to studies on the work environment and energy in agriculture.
Publications: Bulletin; annual report.

Köttforskningsinstitutet*   26

[Meat Research Institute]
Address: PO Box 504, S-244 00 Kävlinge
Telephone: 046-73 22 30
Telex: 32206 Meatres S
Parent body: Swedish Meat Marketing Association
Status: Official research centre

Research director: Kurt Östlund
Sections: Microbiology, Göran Molin; food chemistry, Anita Laser-Reutersvärd; meat quality, Eva Tornberg
Graduate research staff: 7
Activities: The study, ·by using modern physical, chemical, biochemical and microbiological methods, of different processes in meat and meat products in connection with slaughtering, cutting, processing etc, with regard to efficiency, improved quality and shelf life. Specifically research is concentrated on the study of microflora on fresh meat and meat products, the classification of microorganisms, cold shortening and electrical stimulation connected with meat quality.

Kristinebergs Marinbiologiska Station* 27

[Kristineberg Marine Biological Station]
Address: Kristineberg 2130, S-450 34 Fiskebäckskil
Telephone: 0523-22007
Affiliation: Royal Swedish Academy of Sciences
Status: Official research centre
Director: Professor Jarl-Ove Strömberg
Sections: Crustacean deep-sea biology, Professor J.-O. Strömberg; algal physiology, Lennart Axelsson; ecotoxicology, Åke Granmo; hardbottom ecology, Tomas Lundälv; softbottom ecology, Alf Josefson; planktology, Lars Hernroth, Odd Lindahl
Graduate research staff: 6

Kungliga Fysiografiska Sällskapet i Lund* 28

[Royal Physiographical Society of Lund]
Address: Oorganisk Kemi 1, Kemicentrum, PO Box 740, S-220 07 Lund
Telephone: 046-10 81 02
Status: Learned society funding research
Secretary: Professor Sten Ahrland
Activities: The Society is not a research organization but awards donations and grants from various foundations in order to support research in the fields of natural history, mathematics and physics, medicine, practical natural sciences, and - to some degree - humanities.
Publications: Årsbok, (Year book).

Kungliga Skogs- och Lantbruksakademien 29

[Royal Swedish Academy of Agriculture and Forestry]
Address: Box 6806, S-11386 Stockholm
Telephone: 08-30 30 07
Status: Learned society funding research
Permanent secretary: Professor Olle Johansson
Activities: The Academy is a learned society whose purpose is to encourage progress in agriculture and forestry by intellectual and financial support. The Forestry and Agricultural Research Council is closely connected with the academy. Subjects covered include soil science, drainage and irrigation, land use, plant breeding, crop husbandry, plant protection, livestock breeding, livestock husbandry and nutrition, veterinary medicine, agricultural engineering and building, freshwater fisheries, food science, forestry and forest products.
Publications: Bimonthly journal; *Acta Agriculturae Scandinavica* (quarterly).

Kungliga Tekniska Högskolan* 30

– KTH
[Royal Institute of Technology]
Address: S-100-44 Stockholm
Telephone: 08-787 70 00
Telex: 10389 Kthlb S
Status: Educational establishment with r&d capability
President: Professor Anders Rasmuson
Faculty Dean: Professor Sune Berndt
Graduate research staff: 1 100

SCHOOL OF SURVEYING* 31
Dean: Professor E. Carlegrim

Department of Geodesy* 32
Head: Professor A. Bjerhammar

Department of Land Improvement and Drainage* 33
Head: Professor G. Knutsson

Department of Photogrammetry* 34
Head: Professor K. Torlegård

Department of Real Estate Economics* 35
Head: Professor E. Carlegrim

Department of Real Estate Planning*   36

Head: Professor G. Larsson

Kungliga Vetenskaps- och   37
Vitterhets-Samhället i
Göteborg*

[Royal Society of Arts and Sciences of Gothenburg]
Address: Kungsportsavenyn 32, S-411 36 Göteborg
Telephone: 031-18 94 60
Status: Learned society
Secretary: Professor Erik J. Holmberg
Activities: The Society as such does not carry out any research. It does, however, award a number of grants to promote scientific research in various fields.

Kungliga   38
Vetenskapsakademien*

[Royal Swedish Academy of Sciences]
Address: PO Box 50005, S-104 05 Stockholm
Telephone: 08-15 04 30
Telex: 17073
Status: Learned society funding and administering research
President: Gunnar Hoppe
Secretary General: Professor Dr Tord Ganelius
Sections: Biology; astrophysics, mathematics; energy
Graduate research staff: 25
Activities: The Academy maintains national committees, advises the government in scientific matters, awards research grants and is responsible for the operation of several research institutions.

Laxforskningsinstitutet*   39

[Salmon Research Institute]
Address: S-810 70 Älvkarleby
Telephone: 026-727 51
Status: Official research centre
Director: Dr Nils Johansson
Sections: Fish pathology, Dr Nils Johansson; salmon ecology, Dr P.-O. Larsson; salmon physiology, Dr Eva Bergström; salmon rearing, Yngve Ottosson
Graduate research staff: 4

Activities: Exploration of methods for producing young salmon in order to maintain the salmon stock in the Baltic. Fishery biology: basic research on the physiology of salmon and ecological research on Swedish salmon stocks. Fish pathology: study of diseases in salmonids; the laboratory is a centre for disease control in all Swedish salmon rearing plants. Salmon rearing: rearing methods are developed and tested on a full industrial scale; rearing experiments are supplemented by detailed research under laboratory conditions and their efficiency is tested by an extensive tagging programme.
Publications: Swedish Salmon Research Institute Report, series.

Lunds Universitet*   40

[Lund University]
Address: Box 1703, S-221 01 Lund
Telephone: 046-10 70 00
Telex: LUNIVER S-33533
Status: Educational establishment with r&d capability

FACULTY OF MATHEMATICS AND   41
NATURAL SCIENCE*

Dean: Professor Lennart Eberson

School of Biological and Earth   42
Sciences*

Dean: Professor Nils Malmer
Graduate research staff: 326

Botanical Garden*   43
Address: Ostra Vallgartan 20, S-223 61 Lund
Director: Lennart Engstrand

Department of Biology*   44
Address: Helgonarägan 5, S-223 62 Lund
Head: Sten Rundgren

Department of Genetics*   45
Address: Sölvegatan 29, S-223 62 Lund
Head: Professor Karl Fredga

> **Tumour Cytogenetics Division***   46
> *Address:*   Wallenberglaboratoriet, Sölvegatan 33, Box 7031, S-220 07 Lund
> *Senior staff:*   Professor Karl Fredga, Professor Åke Gustafsson, Professor Arne Lundqvist, Professor Arne Müntzing

Department of Limnology*   47
Address: Box 3060, S-220 03 Lund
Head: Dr Wilhelm Granéli
Senior staff: Professor Sven Bjork

Naturhistoriska Riksmuseet* 73

[Swedish Museum of Natural History]
Address: Box 500 07, S-104 05 Stockholm
Telephone: 08-15 02 40
Chairman of the Board: Roland Morell
Director of Museum Department: K. Engström

RESEARCH DEPARTMENT* 74

Head: Professor Alf G. Johnels
Graduate research staff: 48
Publications: Fauna och flora.

Naturvetenskapliga forskningsrådet* 75

– NFR
[Swedish Natural Science Research Council]
Address: Box 6711, S-113 85 Stockholm
Telephone: 08-15 15 80
Telex: 13599 Rescoun S
Head: Dr Mats-Ola Ottoson
Secretary: Professor Ingvar Lindvist
Sections: Biology, Professor Jan-Erik Kihlström; physics, Professor Sven Kullander; geosciences, Professor Kurt Boström; chemistry, Professor Lennart Ebberson; energy, Professor Ingvar Lindqvist
Activities: The essential task of the Council is to promote fundamental scientific research, to encourage cooperation between workers in this field, and to be a contact organization for international cooperation. The Council allocates grants for research to individuals or to institutions, for pursuing specific investigations and for publishing the research results. Grants are also allocated for exploration expeditions and for conferences and symposia. Research groups are established in connection with some of the Council's professorships; these cover the following fields of research: immunobiology, biochemical cytogenetics, ethology, glaciology, molecular cell physiology, and thermochemistry.
Publications: Yearbook; annual report; projects catalogue (in Swedish).

Nordreco AB 76

Address: Box 500, S-26700 Bjuv
Telephone: (042) 70800
Telex: 72 200
Affiliation: Nestlé
Status: Research centre
Director: Professor Erik Von Sydow
Departments: Agriculture, Anders Dahlkvist; culinary products, frozen foods,
Graduate research staff: 8
Activities: Plant production, protection, and storage; testing of harvesting machines.
Projects: Plant breeding - vegetables (Rolf Stegmark); cultivar tests - vegetables, berries (Bernt Bengtsson); weed control (Jan Olofsson); plant protection-insects (Bodil Jönsson); crop scheduling (Stig Nyström); storage (Carl Wikberg); testing of harvesting machines (Hilding Christensson).

Osteologiska Forskningslaboratoriet* 77

[Osteological Research Laboratory]
Address: Ulriksdals kungsgård, S-171 71 Solna
Telephone: 08-85 73 11
Affiliation: Stockholm University
Status: Official research centre
Director: Professor Dr Torstein Sjøvold
Activities: Research covers: archaeo-osteology, human and animal; prehistoric environmental destruction of fauna; forensic osteology; biostatistics.

Patscentre Scandinavia AB* 78

Address: Box 30007, S-104 25 Stockholm
Telephone: 08-13 13 00
Telex: 17837
Affiliation: Patscentre International, UK
Status: Independent research centre
Managing Director: Per F. Tengblad
Sections: Materials; optics; optronics; microelectronics/electronics; mechanical engineering; biotechnology; food technology
Graduate research staff: 150

SIK - Svenska Livsmedelsinstitutet
79

[SIK - Swedish Food Institute]
Address: Box 5401, S-420 29 Göteborg
Telephone: (031) 40 01 20
Telex: 21651 SIK S
Affiliation: Swedish Board for Technical Development
Status: Official research centre
Director: Professor Nils Bengtsson
Sections: Chemical analysis, Jonas Andersson, Ulla Stöllman; biochemistry, Svante Svensson; microbiology, Benkt Göran Snygg; sensory evaluation, Birgit Lundgren; technology, Thomas Ohlsson; contract work, Pär Olsson
Graduate research staff: 35
Activities: Food: raw materials, processing, packaging, storage, distribution and consumption.
Publications: Annual report, Swedish and English.
Projects: Utilization of raw materials (Svante Svensson); distintegration of raw materials and recombination of components (Anne-Marie Hermansson); food preparation (Yngve Andersson); measurment of sensory quality (Birgit Lundgren); nutrients and quality affecting substances (Gunnar Hall); industrial heat processing and chilled storage (Thomas Ohlsson).

Skogs- Och Jordbrukets Forskningsråd
80

[Swedish Council for Forestry and Agricultural Research]
Address: Box 6806, S-113 86 Stockholm
Telephone: (08) 30 21 05; (08) 30 11 50
Affiliation: Royal Swedish Academy of Agriculture and Forestry
Status: Official research organization
Governor: Ragnar Edenman
Sections: Forestry, G. Lundeberg; agriculture, S. Fernros
Graduate research staff: 3
Activities: The council was established in 1967 and reorganized in 1981. It promotes and supports research useful to forestry, agriculture, and horticulture. The chairman and the other ten members of the council are elected by the government. The research supported often takes the form of special projects undertaken at the University of Agricultural Sciences or at other universities.
Publications: SJFR informerar: Projektrererat, quarterly.

Standardiseringskommissionen i Sverige*
81

– SIS
[Swedish Standards Institution]
Address: Box 3295, S-103 66 Stockholm
Telephone: 08-23 04 00
Telex: 17453 sis s
Status: Independent research organization
Research director: Folke Hermanson-Snickars
Sections: SIS-STG (writing of standards), A. Vogel; SIS-U (search of foreign standards and official regulations), Folke Hermanson-Snickars
Graduate research staff: 60
Activities: SIS is an independent national standards body, issuing standards in the following areas: measuring and weighing systems; electrical engineering - power technology and telecommunication and electronics; pneumation and refrigeration; pipelines; drawing and tolerances; welding; machinery components and parts; machines and tools; mining; land and road vehicle engineering; agricultural machinery and products; domestic utensils; administration and data processing; printing; packaging and distribution; transport engineering; chemical industry and petroleum; paints and varnishes; metallurgy; textiles; rubber; plastics; building industry; environment; school equipment.
Publications: Swedish standards, annually.

Statens Lantbrukskemiska Laboratorium
82

[National Laboratory for Agricultural Chemistry]
Address: Box 7004, S 750 07 Uppsala
Telephone: (018) 10 20 20
Official research centre
Research director: Professor Bengt Nygård
Departments: Agriculture, Anders Dahlgren
Activities: Soil science and plant nutrition: soil analyses for estimation of soil fertility, analyses of fertilizers and soil amendments, estimation of food quality; control of pesticide levels in food, environment, and people using pesticide sprays; control of purity of pesticide formulations.
Projects: Active aluminium in soils and lime requirement; (Sten Ståhlberg); determination of selenium in agricultural products; multi-element analysis with plasma spectrometry (Per Jennische); enzymatic analysis of carbohydrates; analysis of food additives (Kjell Larsson); analysis of pesticide residues in vegetables and soils; analysis of pesticide formulations; occupational exposure to pesticides (Malin Åkerblom)

Statens Livsmedelsverk* 83

– SLV
[National Food Administration]
Address: PO Box 622, S-751 26 Uppsala
Telephone: 018-15 22 00
Affiliation: Ministry of Agriculture
Status: Official research centre
Director-General: Arne Engström
Graduate research staff: 19
Publications: Vår Föda (Our Food).

FOOD RESEARCH DEPARTMENT* 84

Head: L.E. Tammelin

Biological Laboratory* 85

Head: G. Nillson
Activities: The Laboratory carries out microbiological, biological and sensory analyses of food. It also has the supervision of about 50 laboratories engaged in local food control (Public Health Committees).

Food Laboratory* 86

Head: B. von Hofsten
Activities: The Laboratory carries out chemical analyses for food contaminants, food additives, and naturally-occurring food constituents. It also elaborates and evaluates methods of analysis.

Nutrition Laboratory* 87

Head: L. Reio
Activities: The Laboratory carries out chemical, microbiological, and biological analyses on the nutritive value and vitamin content of food and on the vitamin content of drugs and animal feeds. It also investigates the food-consumption patterns of different groups within the population and draws up tables showing the nutritional value of foodstuffs.

Toxicology Laboratory* 88

Head: L. Albanus
Activities: The Laboratory is responsible for evaluating the toxicity of food contaminants and food additives and carries out research in the field of food toxicology. A computer-based system for registering additives present in food sold in Sweden (SLIT) has been developed with the laboratory.

Statens Naturvårdsverk* 89

– SNV
[National Swedish Environment Protection Board]
Status: Official research centre

RESEARCH COMMITTEE* 90

Research Secretariat* 91

Head: Jan Nilsson
Graduate research staff: 19 (in research administration)
Activities: The Research Secretariat administers the activities of the Research Committee of the Environment Protection Board. This includes planning and financing of applied research performed at different research institutions within the field of environment protection.

RESEARCH DEPARTMENT* 92

Address: PO Box 1302, S-171 25 Solna
Telephone: 08-98 13 20
Telex: 111 31 ENVIRON-S
Head of Section: Anders Akerblom
Research director: Göran Persson
Graduate research staff: 80 (research and research administration)
Activities: The Department is responsible for applied nature conservancy research and investigations into the effects of pollution and other human impacts on the outer environment. Responsibilities also include the compilation of the knowledge of environmental effects as well as monitoring and reporting of the status and changes in the environment. The Department consists of five principal research sections.

Brackish Water Toxicology Laboratory* 93

Address: Studsvik, S-611 82 Nyköping
Head: Olof Svanberg
Graduate research staff: 4

Special Analytical Laboratory * 94

Address: Wallenberg Laboratory, S-106 91 Stockholm
Temporary head: Göran Sundström
Graduate research staff: 5
Activities: The Special Analytical Laboratory is performing research and development in the field of organic analytical chemistry. Projects are mostly elaborated and carried out in cooperation with other laboratories or sections within the Protection Board or at University departments working in the fields of ecology, organic chemistry, genetics, toxicology etc.

Water Pollution Research Laboratory * 95

Research director: Erik Vasseur
Graduate research staff: 20

Water Quality Laboratory, Uppsala * 96

Sections: Inland water, Dr Thorsten Ahl; coastal water, Dr Ulf Grimås
Graduate research staff: 21

Statens Veterinärmedicinska Anstalt 97

– SVA
[National Veterinary Institute]
Address: Norra Ulture 2, S-750 07 Uppsala
Telephone: 018-15 58 20
Status: Official research centre
Director: Professor Hans-Jörgen Hansen
Bacteriology, Professor B. Hurvell; virology, Professor B. Morein; pathology, Professor N-E. Björklund; parasitology, Professor O. Ronéus; chemistry, Professor K. Erne; epizootiology, Professor G. Hugoson; Advisory, Professor N.O. Lindgren; production (vaccines), Professor K.A. Karlsson
Graduate research staff
Activities: The institute is the central reference veterinary laboratory for the country. Its functions cover diagnostic services, field and laboratory investigations, and extension services. Research is undertaken into methods for the diagnosis of animal diseases, the development of vaccines, and the aetiology and control of major animal diseases. Other concerns of the institute are food hygiene, fodder hygiene, environmental hygiene, and laboratory animal
Publications: Collected papers of the institute (biennial).

Stockholms Universitet * 98

[Stockholm University]
Address: Universitetsvägen 10, S-106 91 Stockholm
Telephone: 08-15 01 60
Status: Educational establishment with r&d capability
President: Professor Staffan Helmfrid
University Director: Rune Lindquist

FACULTY OF NATURAL SCIENCES: 99 BIOLOGY-GEOGRAPHY SECTION *

Dean: Professor Torsten Hemberg

Botany Department * 100

Address: Lilla Frescati, S-106 91 Stockholm
Director: Professor Torsten Hemberg
Activities: Areas of research include: physiological botany (plant physiology, growth hormones and inhibitors, ion uptake and photosynthesis, experimental ecology); morphological botany (plant morphology and taxonomy).

Genetics Department * 101

Address: Box 6801, S-113 86 Stockholm
Director: Professor Karl Gustav Lüning
Activities: Subjects of research include: mutations in Drosophila and mice; wildlife genetics.

Microbiology Department * 102

Address: Roslagsvägen 101, S-106 91 Stockholm
Director: Professor Hans G. Bornan
Activities: Studies of the structure, physiology and genetics of microorganisms.

Radiobiology Department * 103

Address: Wallenberg Laboratory, Lilla Frescati, S-104 05 Stockholm
Director: Göran Löfroth
Activities: Subjects of research include: biological and chemical effects of radiations and genotoxic chemicals; basic radiobiology and its applications, eg, in environmental health and agriculture.

Zoology Department * 104

Address: Box 6801, S-113 86 Stockholm
Director: Tommy Radesäter
Activities: Studies of structural and ultrastructural zoology, ecology, ethology, systematics and evolution.

Wenner-Grens Institute (Life Sciences) *   105

Address: Norrtullsgatan 16, S-113 45 Stockholm
Director: vacant
Activities: Experimental biology, particularly on the borderline of chemistry, physics and medicine; cytology.

Styrelsen för teknisk utveckling *   106

– STU
[Swedish Board for Technical Development]
Address: Box 43200, S-100 72 Stockholm
Telephone: 08-744 51 00
Telex: 10840 Swedstu S
Status: Official research organization
Director General: Sigvard Tomner
Deputy Director General: Jan Olof Carlsson
Sections: Planning Department - cooperative industrial research, Björn Englund; knowledge development programmes, Göran Friborg; budget proposals, Christina Wennmark; long-term planning, Bo Stenviken, Lennart Stenberg, Göran Friborg, Lennart Palm; evaluation, Staffan Håkansson
Project Departments - Information and systems technology, Lennart Lindeborg; industrial processes, Paul Forsgren; medical, welfare and life sciences, Hans-Göran Karlsson; energy technology, Staffan Ulvönäs; manufacturing and product development, Peter Jörgensen
Graduate research staff: 3 000
Activities: STU is the central government agency for the support of technical research and development in Sweden. Its directives from the government are: to follow technical development by keeping in touch with scientists, institutions and companies; to organize and support cooperation in technical research and industrial development and also to encourage contacts between authorities, industry, commerce and research institutions; to take the initiative on technical research of importance to industry, commerce and society and also to further such research and its utilization; to plan and allocate governmental support through loans and grants to technical research, industrial development work and inventions; to follow and monitor the activities of industrial research associations and other cooperative research institutions at which the research work is conducted with governmental support; to give advice to inventors and act as an intermediary when it comes to the commercial utilization of research results; to further international technical cooperation with foreign institutions and international organizations.

Svalöf AB   107

Address: S-268 00 Svalöv
Telephone: 0418-625 10
Telex: 72476 Svaloef S
Affiliation: Swedish Farmers' Association
Status: Official research centre
Managing Director: Göran Kuylenstjerna
Departments: Plant breeding, Gösta Olsson; agricultural production, Lars Olsson; agricultural marketing, Paul Vigre; horticultural marketing, Ulf Arvidsson; Ohlsens Enke A/S Copenhagen, John V. Möller; international relations, Anders-Hugo Silfwerberg
Graduate research staff: 60
Activities: In 1980 Sveriges Utsädesförening (Swedish Seed Association) was amalgamated with its commercial counterpart Allmänna Svenska Utsädes AB (General Swedish Seed Company) into a new enterprise, Svalöf AB. This enterprise is jointly owned by the Swedish State and the Swedish Farmers' Association - SLR.
Research concerns plant breeding and includes the following topics: breeding methods, introduction of new genes from wild species, milling and baking quality in wheat, sprouting resistance in rye and wheat, disease resistance in different crops, winter hardiness and drought tolerance, improved protein quality in barley, changed fatty acid composition in oil crops.
Publications: Sveriges Utsädesförenings Tidskrift (Journal of the Swedish Seed Association' 'quarterly).
Projects: Wheat - breeding (Thore Denward); barley and rye - breeding (Göran Persson); oats - breeding (Bengt Mattsson); triticale - breeding (Arnulf Merker); oil crops - breeding (Roland Jönsson); fodder crops - breeding (Jan Sjödin); potatoes - breeding (Lennart Erjefält); peas - breeding (Nils Johansson); horticultural plants - breeding (Gösta Carlsson); crops for northern Sweden - breeding (Arne Wiberg); plant physiology (Volkmar Stoy); plant tissue cultures (Ove Hall); baking quality and nutritional value (Robert Olered).

PLANT BREEDING AND RESEARCH DIVISION *   108

Research director: Dr Gösta Julén
Graduate research staff: 60
Activities: The main purpose is to carry on plant breeding in order to develop improved plant varieties for Swedish agriculture and horticulture. In connection with this, research work is carried out in the fields of crop physiology, cytogenetics, tissue culture techniques and biochemistry.

Biochemistry Department*   109
Head: Dr Robert Olered

Biological Development Department* **110**
Head: Dr Volkmar Stoy
Sections: Plant physiology, Dr Volkmar Stoy;
cytogenetics, Dr Arnulf Merker; plant pathology, Jan
Meyer; tissue culture, Dr Ove Hall

Forage Crops Department* **111**
Head: Jan Sjödin

Oil Crops Department* **112**
Head: Dr Gösta Olsson

Potato Department* **113**
Head: Dr Lennart Erjefält

Rye and Barley Department* **114**
Head: Dr Göran Persson

Wheat and Oats Department* **115**
Head: Dr Thore Denward

Departments outside Svalöv* **116**

Department for Central Sweden, Ultuna* **117**
Address: S-751 05 Uppsala
Head: Dr Nils Johansson

Department for Northern Sweden* **118**
Address: Röbäcksdalen, Box 720, S-901 10 Umeå
Head: Dr Arne Wiberg

Horticulture Department, Hammenhög* **119**
Address: S-270 50 Hammenhög
Head: Dr Gösta Carlsson

Stations* **120**

Östergötland Branch Station* **121**
Address: S-596 00 Skänninge
Head: Dr Åke Borg

Västergötland Branch Station* **122**
Address: S-460 60 Vargön
Head: Göran Engqvist

AB Svensk laboratorietjänst 123

– Svelab
[Swedish Laboratory Services Limited]
Address: 105 33 Stockholm
Telephone: (08) 14 16 00
Status: Official research association
Managing director: Anders S. Nilsson
Chairman: Hans-Jörgen Hansen
Regional laboratories: Luleå, Klas Johnsson; Borlänge,
Per-Arne Persson; Västerås, Nils Nilsson; Linköping,
Lars Starkhammar; Skara, Rune Bucht; Kalmar, Nils-
Georg Carlsson; Jönköping, Ulf Lagerquist; Halmstad,
Lennart Persson; Kristianstad, Nils Persson;

Helsingborg, Krister Lange; Malmö, Sten Vesterlund
Graduate research staff: 30
Activities: Svelab was founded in 1969 and is a partly
state-owned laboratory company. Ownership is shared
between the government, the agricultural industry, consumer cooperatives, and the private foodstuffs industry,
the respective shares being 52 per cent, 24 per cent, 12
per cent, and 12 per cent. It provides laboratory services
for veterinary diagnostic testing, the testing of
foodstuffs, animal fodder, and water supplies. These
services are currently available at 11 regional laboratories.
Projects: Epizootiology and control of Ascaris suum
intexions of growing pigs (Professor Kjell Martinson,
Olle Nilsson).

Svenska Mejeriernas Riksförening Centrallaboratoriet 124

– SMR
[Swedish Dairies' Association Central Laboratory]
Address: Box 205, S-201 22 Malmö
Telephone: (040) 715 50
Telex: 32 333
Status: Research association
Director: Dr Hans Överström
Departments: Chemistry, Dr Hans Jönsson; microbiol-
ogy, Dr Hans-Erik Pettersson
Graduate research staff: 11
Activities: Dairy research and development: chemical
and microbiological analysis and investigations; studies
in starter microbiology.
Publications: Nytt från SMR Malmö, bimonthly.
Projects: Development of starter cultures for dairy
products (Dr Hans-Erik Pettersson); microbial flora of
raw milk (Gunnar Olsson); bacteriophages against starter streptococci (Jean Dufeu); studies on the ripening of
hard cheese (Sven Borgström); research and development of processes for the utilization of whey (Dr Hans
Jönsson); development of analytical procedures for dairy
purposes (Torsten Nilsson).

Sveriges Geologiska Undersökning* 125

– SGU
[Geological Survey of Sweden]
Address: Villavägen 18, Box 670, S-751 28 Uppsala
Telephone: 018-15 52 80
Status: Official research organization
Director: Gunnar Ekevärn

DEPARTMENT OF QUATERNARY 126
GEOLOGY AND HYDROGEOLOGY*

Sections: Geological; service
Activities: Geological mapping; hydrogeological mapping; groundwater documentation; exploration for groundwater sand and gravel; soil science; pollen analysis.

Sveriges 127
Lantbruksuniversitet

– SLU
[Swedish University of Agricultural Sciences]
Address: Ultuna, S-750 07 Uppsala
Telephone: (018) 10 20 00
Telex: 760 62 Ultbibl S
Affiliation: Ministry of Agriculture
Status: Educational establishment with r&d capability
Rector: Professor Lennart Hjelm
Activities: SLU concentrates mainly on research. The topics covered are agriculture, forestry, veterinary medicine, horticulture, and landscape architecture. The university also has a teaching programme on these subjects, and has an information service on the practical results of the research activities. The governing body - the Board - of the university is also the Board of the National Veterinary Institute. (see separate entry). Basic research activities use about 20 per cent of the university's resources and, in addition, are financed by grants from funds and research councils, such as the Swedish Council for Forestry and Agricultural Research. Basic research ranges over several subjects in the basic natural sciences and veterinary medicine, with application in agriculture, forestry, animal health, food product control, horticulture, landscape architecture, and environmental management. Programmed Applied Research takes about 55 per cent of the university's resources. The activities of the university take place at some thirty places within Sweden. The main locations, at Alnarp, Skara, Garpenberg, Skinnskatteberg, Röbäcksdalen, Umeå, Lund, Bispgården, and Värnamo, are detailed below. In addition, the university cooperates closely with the Swedish International Development Authority (SIDA), and a special centre for international rural development, mainly financed by SIDA, is based at Ultuna, where research is undertaken in agriculture, forestry, veterinary medicine, ecology, and rural development in the tropics.
Publications: Swedish Journal of Agricultural Research, quarterly; *Aktuellt från Lantbruksuniversitetet; Studia Forestalia Suecica*; reports.

FACULTY OF AGRICULTURE* 128

Address: Röbäcksdalen, S-901 10 Umeå
Telephone: 090-13 53 10
Dean: Professor Ulf Renborg
Activities: Research towards increased production of plants and animals, in particular: comparison of cultivars, soil tillage, fertilizers, weed control, etc; reduction of the use of pesticides; replacing chemicals with biological and cultural methods; nutrient seepage and water pollution - optimal nitrogenous fertilizer application; genetic research with cattle, pigs, poultry and sheep, and to a lesser extent fish, game, fur animals, bees, and reindeer; improvements in animal feed preparation and dispensing; disease resistance, storage, and glasshouse and field cultivation techniques in horticulture; reduction of glasshouse energy requirements; cultivation of ornamental plants; function, management and construction of parks and recreational areas; planning and construction of farm buildings; improvements in farm management and machinery; planning of machinery utilization; economic analysis of the farm as a business; marketing; structural development of agriculture; agricultural policy.Agricultural research is mainly concentrated at Ultuna, Alnarp, Lund, and Umeå.

Centre for Horticultural Science 129

Address: S-23053 Alnarp
Telephone: (040) 41 50 00
Activities: Horticulture, landscape architecture, plant protection, farm buildings; experimental work in livestock and horticulture. There are 400 employees.

Department of Agricultural 130
Economics*

Head: Professor K.G.L. Hjelm

Department of Agricultural 131
Hydrotechnics*

Head: Professor A. Håkansson

Department of Agricultural 132
Marketing*

Head: Professor L.-G. Folkesson

Department of Animal Breeding* 133

Head: Professor K. Rönningen

Department of Animal Husbandry* 134

Head: Professor C.O. Claesson

Department of Animal Nutrition* 135

Head: Professor S.S. Eriksson

Department of Animal Physiology* 136

Head: Professor P.G.K. Knutsson

Department of Applied Entomology* 137

Head: Professor J.G.P. Pettersson

Department of Environment Research 138
and Ecology*

Head: Professor E.V. Steen

Department of Farm Buildings 139

Address: Box 624, Mellanvångsvägen, S-220 06 Lund
Telephone: (046) 11 75 10
Status: Official research centre
Head: Professor Rolf Henriksson
Senior staff: Division of Farm Building Construction,
Ingvar Jansson; Division of Economics and Environment, Bengt Gustafsson; Division of Horticultural
Engineering, Bengt Landgren; Division of Administration, R. Henriksson
Activities: The main purpose of research is to adapt
agricultural and horticultural buildings to biological,
technological and economical developments as regards
function, labour, and construction technology together
with a consideration of the internal and external environment. In addition, the work concentrates upon
developing a foundation for building technology requirements and construction principles. Specific topics include: system solutions for farm buildings; development
of suitable housing within agriculture; materials and
construction; fire problems in farm buildings; climate in
animal barns; energy saving methods; building production; improving the environment for animals in stables;
improving work methods and work conditions in farm
buildings; planning and utilization of buildings and
building systems; more accurate basic data for economic
calculations; greenhouse construction; climate in greenhouses; climate in storage and packing areas.
Projects: Agricultural buildings (R. Henriksson);
greenhouses and farm buildings (B. Gustafsson); improvement of conditions in farm buildings and animal
houses (I. Jansson); heating and cooling systems, and
energy systems for greenhouses (B. Landgren).

RESEARCH AND TRAINING DIVISION 140
Address: LBT, Box 7032, S-750 07 Uppsala
Telephone: (018) 10 20 00
Director: Lennart Bengtsson

Department of Farm Mechanization* 141

Head: Professor L.Å. Haraldson

Department of Fruit Breeding* 142

Address: Balsgård, Fjälkestadsvägen 123-1, S-291 94
Kristianstad
Telephone: 044-750 41
Head: Professor I.B. Fernqvist

Department of Genetics* 143

Head: Professor S.G. Östergren

Department of Genetics and Plant 144
Breeding*

Senior staff: Professor N.A. Hagberg, Professor J.F.
Mac Key

Department of Horticultural 145
Economics*

Head: Professor C.M. Carlsson

Department of Inorganic Chemistry* 146

Head: Professor F.I. Lindqvist

Department of Landscape 147
Architecture*

Head: Professor P.-A. Friberg

Department of Microbiology* 148

Head: Professor A.B. Norén

Department of Organic Chemistry and 149
Biochemistry*

Head: Professor O. Theander

Department of Ornamental 150
Gardening*

Head: Professor T. Kristoffersen

Department of Overhead Planning* 151

Head: Professor O.R. Srage

Department of Pedology* 152

Head: Professor N.F.S. Odén

Department of Plant and Forest Protection   153

Address: PO Box 7044, S-750 07, Uppsala
Telephone: (018) 10 20 00
Head: Professor Per Oxelfelt
Graduate research staff: 25
Publications: Växtskyddsrapporter; Växtskyddsnotiser.

EXPERIMENTAL DIVISION OF ENTOMOLOGY   154
Head: Dr Hans von Rosen
Activities: Applied research concerning pests in agricultural and horticultural crops.
Projects: Pest control in cereals (Hans Larsson); insect-pests of oil-seed crops (Christer Nilsson); pest control in sugar-beets (Hans Larsson); testing of pesticides (Johan Mörner); biological control of pests in glasshouses (Barbro Nedstam); insect pests in field grown vegetables (Christer Persson).

FOREST ENTOMOLOGY DIVISION   155
Head: Professor Hubertus Eidmann
Projects: Ecology of pine shoot beetles and their impact on forestry (Bo Långström); insects attacking lodge pole pine (Hans Iacobaeus); insects attacking conifer seedlings; forest protection and population dynamics in forest insects (Hubertus H. Eidmann); insect pathology and microbial control (Einar Olofsson); cone and seed damaging insects, mainly in seed orchards (Nicolaas Wiersma); biology and control of spruce bark beetles (Ips typographus) (Hubertus H. Eidmann); population ecology of leaf-eating insects, including plant/insect interactions (Olle Tenow); studies of the soil fauna (especially microarthropods) in forest soils in southern Norrland; connections between collembola and phytopathogenic fungi (Högni Bödvarsson); bark beetles and forest protection (Bo Långström); soil animals as possible index species of different soil types (Högni Bödvarsson); behavioural chemicals and forest insects (Hubertus H. Eidmann); insect-attacks in raw timber; insect fauna in virgin and managed forest stands (Bengt Ehnström); parasitic hymenoptera: taxonomy and biology (Göran Nordlander).

Research and Training Division of Entomology   156

Head: Professor J. Pettersson
Activities: Plant production and plant protection: occurrence and distribution of plant parasitic nematodes in the country, and their significance in the aetiology of plant disease.
Projects: Population dynamics of pest species and their natural enemies (B.S. Ekbom); host plant relations of aphids (J. Pettersson).

Research and Training Division of Mycology and Bacteriology   157

Head: V. Umaerus
Projects: Basic studies of resistance mechanisms in plants; resistance to gangrene and dry rot in breeding material of potatoes (V. Umaerus); studies of Fusarium root rots of forage legumes (S. Rufelt); studies of cereals and grasses on root growth, root morphology and physiology, and resistance to root diseases; biological weed control; side-effects of herbicides on plant diseases; remote sensing and image analysis at macroscopic and microscopic levels in plant pathology (H.E. Nilsson); microbial and microbial-plant interactions in the root zone (B. Gerhardson); rhizobacteria influencing fungal disease development and plant growth (S. Alström).

Research and Training of Nematology   158

Head: Professor K. Bengt Eriksson
Projects: Faunistic/taxonomic studies of plant parasitic nematodes; interactions between nematodes and fungi in plant disease (K.B. Eriksson); nematodes in monoxenic cultures (nematode bank) and associated studies related to breeding for resistance (Sigrid Bingefors); nematodes as parasites on forest trees (Ch. Magnusson); studies on potato cyst nematodes, Globodera spp, part of an inter-Scandinavian project with K.B. Eriksson as the leader (Marja Leena Magnusson).

Research and Training Division of Virology   159

Head Professor Per Oxelfelt
Projects: Development of diagnostic methods for barley yellow dwarf virus in grasses and cereals; survey of potato virus A and related viruses in seed potatoes (P. Oxelfelt); characterization of gemini virus causing wheat dwarf in Sweden; plant hopper borne viruses of Gramineae in Sweden (K. Lindsten).

Department of Plant Husbandry*   160
Head: Professor P.R. Larsson

Department of Plant Pathology*   161
Head: Professor V.R. Umaerus

Department of Plant Physiology*   162
Head: Professor G. Stenlid

Department of Soil Fertility*   163
Head: Professor S.L.H. Jansson

Department of Soil Management* 164

Head: Professor R. Heinonen

Department of Vegetable Growing* 165

Head: Professor K.L. Ottoson

FACULTY OF FORESTRY* 166

Address: S-901 83 Umeå
Telephone: (090) 16 50 00
Dean: Professor Gustaf von Segebaden
Senior staff: Forest vertebrate ecology, Professor T.I. Ahlén; forest yield research, Professor S.-O. Andersson; reforestation, Professor P.O. Bäckström, Professor G. Sirén; forest entomology, Professor H.H.T. Eidmann; forest management, Professor B. Jonsson; forest genetics, Professor D.G. Lindgren; forest economics, Professor G. von Malmborg; forest biometry, Professor B. Matérn; forest survey, Professor G.O.U. von Segebaden; operational efficiency, Professor A. Staaf, Professor U. Sundberg; forest ecology, Professor C.O. Tamm; forest soils, Professor T. Troedsson; forest mycology and microbial ecology, Professor T. K.-G. Unestami
Activities: Research is centred in five main areas: land use - utilization of forest land for timber production, recreational activities, and hunting; inventory of forest resources - development of methods for collection, documentation and analysis of data describing the condition of the forest and the changes taking place therein (to generate a National Forest Survey); timber yield of a site - effect of site conditions on quantity and quality of timber yield, forest regeneration, protection from damage caused by fungi and insects (eg Pine Weevil), production of biomass as an energy source; techniques concerning stand establishment and timber harvesting and utilization; production, preservation, management, and utilization of other resources and environmental assets such as the gathering of berries and edible fungi, and the overall preservation of species, gene resources, and ecosystems. Forestry research is conducted at Umeå, Garpenberg, Ultuna, and several forest research and experimental stations at different places throughout the country.

Forest Research Centre 167

Address: S-77073 Garpenberg
Telephone: (0225) 22100

Northern Forestry Institute* 168

Address: S-840 73 Bispgården
Telephone: 0696-302 11

School for Forest Engineers* 169

Address: Box 43, S-779 00 Skinnskatteberg
Telephone: (0222) 100 84
Activities: Research into forestry is conducted at this site as well as at Garpenberg. There are 175 employees.

Southern Forestry Institute* 170

Address: Box 1000, S-331 01 Värnamo
Telephone: 0370-155 20

FACULTY OF VETERINARY MEDICINE 171

Dean: Professor Ingmar Månsson
Activities: Subjects of research include diseases which often occur among both humans and animals, such as infectious diseases; problems involving circulation, respiration, digestion, nutrition deficiency diseases, and transplantation; mastitis in cows; disease and high mortality rates among piglets; muscular function, coordination and endurance in horses; canine diseases; animal breeding; environmentally caused diseases; food product control. Veterinary research at Uppsala is concentrated at the Clinical Centre, the Animal Sciences Centre, and the Biomedical Centre, and is also carried out at Skara.

Department of Anatomy and Histology* 172

Head: Professor L. Plöen

Department of Animal Breeding and Genetics* 173

Department of Animal Hygiene* 174

Senior staff: Professor S.G. Bengtsson, Professor A.G.I. Ekesbo

Department of Bacteriology and Epizoology* 175

Address: Biomedicum, Artillerigärdet, S-751 23 Uppsala
Telephone: 018-15 20 00
Head: Professor I.M. Månsson

Department of Clinical Chemistry* 176

Head: Professor L.G. Ekman

Department of Clinical Radiology* 177

Department of Food Hygiene*   178

Head: Professor T.E. Nilsson

Department of Medical and   179
Physiological Chemistry*

Address: Biomedicum, Artillerigärdet, S-751 23 Uppsala
Telephone: 018-15 20 00
Head: Professor U.P.F. Lindahl

Department of Medicine I (non-   180
ruminants)*

Head: Professor S.G.B. Persson

Department of Medicine II   181
(ruminants)*

Head: Professor P.H. Holtenius

Department of Obstetrics and   182
Gynaecology*

Head: Professor S.G. Einarsson

Department of Parasitology*   183

Head: Professor O. Ronéus

Department of Pathology*   184

Head: Professor K.G.A. Nilsson

Department of Pharmacology*   185

Address: Biomedicum, Artillerigärdet, S-751 23 Uppsala
Telephone: 018-15 20 00
Head: Professor C. G. Schmiterlöw

Department of Physiology*   186

Department of Surgery*   187

Head: Professor B.D.E. Funkquist

Department of Virology*   188

Address: Biomedicum, Artillerigärdet, S-751 23 Uppsala
Telephone: 018-15 20 00
Head: Professor Z. Dinter

Veterinary Institute   189

Address: Box 234, S-532 00 Skara
Telephone: 0511-162 20
Activities: The institute has an animal hospital and farriers' school, and carries out livestock research. A minor part of the training of veterinary students takes place at the institute, which has 100 employees.

Umeå Universitet*   190

[Umeå University]
Address: S-109 87 Umeå
Telephone: 090-16 50 00
Status: Educational establishment with r&d capability

FACULTY OF MATHEMATICAL AND   191
NATURAL SCIENCES*

Dean: Professor Arne Claesson

Ecological Botany Department*   192

Head: Professor Bengt Pettersson

Ecological Zoology Department*   193

Head: Professor Karl Müller

Forest Products Department*   194

Head: Professor Mats Hagner

Genetics Department*   195

Head: Professor Bertil Rasmuson

Microbiology Department*   196

Head: Professor Glenn Björk

Physiological Botany Department*   197

Head: Professor Lennart Eliasson

Zoophysiology Department*   198

Head: Professor Søren Løvtrup

Uppsala Universitet* 199

[Uppsala University]
Address: Box 256, S-751 05 Uppsala
Telephone: 018-15 54 00
Telex: 76024 univups s
Status: Educational establishment with r&d capability

FACULTY OF MATHEMATICS AND NATURAL SCIENCES* 200

Biology-Earth Science Section* 201

Biology Laboratory* 202
Address: Villavägen 7, S-752 36 Uppsala
Telephone: 018-10 17 15
Head: Dr Åke Franzén

Botanical Garden* 203
Address: Villavägen 8, S-752 36 Uppsala
Telephone: 018-13 08 18
Director: Dr Örjan Nilsson

Ecological Biology Department* 204
Address: Villavägen 14, Box 559, S-751 22 Uppsala
Telephone: 018-13 99 55
Head: Professor Eddy van der Maarel

Genetics Department* 205
Address: Dag Hammarskjölds väg 181, Box 7003, S-750 07 Uppsala
Telephone: 018-10 20 00
Head: Dr Gunnar Almgård

Gustaf Werners Institute* 206

Physical Biology Division* 207
Address: Thunbergsvägen 5 (Kemicum), Box 531, S-751 21 Uppsala
Telephone: 018-13 94 60
Head: Professor Börje Larsson

Limnology Department* 208
Address: Norbyvägen 20, Box 557, S-751 22 Uppsala
Telephone: 018-12 03 60
Head: Professor Curt Forsberg
Senior staff: Professor Birger Pejler

Meteorology Department* 209
Address: Kyrkogårdsgatan 6, Box 516, S-751 20 Uppsala
Telephone: 018-13 67 58
Head: Dr Biger Rindert
Senior staff: Professor Ulf Högström

Microbiology Department* 210
Address: Biomedicinska Centrum, Hursagatan 3, Box 531, S-751 23 Uppsala
Telephone: 018-15 20 00
Head: Professor Lennart Philipson

Molecular Biology Department* 211
Address: Wallenberglaboratoriet, Dag Hammarskjölds väg 21, Box 562, S-751 22 Uppsala
Telephone: 018-15 16 62
Head: Professor Charles Kurland
Senior staff: Professor Bror Strandberg

Natural Geography Department* 212
Address: Norbyvägen 18B, Box 554, S-751 22 Uppsala
Telephone: 018-12 03 60
Head: Professor John O. Norrman
Senior staff: Professor Åke Sundborg

Palaeobiology Department* 213
Address: Döbelnsgatan 20A, Box 564, S-751 22 Uppsala
Telephone: 018-11 51 60
Head: Professor Anders Martinsson

Physiological Botany Department* 214
Address: Botaniska trädgården, Thunbergsvägen 2-8, Box 540, S-751 21 Uppsala
Telephone: 018-13 22 58
Head: Professor Tage Eriksson

Systematical Botany Department* 215
Address: Botaniska trädgården, Thunbergsvägen 2-8, Box 541, S-751 21 Uppsala
Head: Professor Olov Hedberg

Zoology Department* 216
Address: Villavägen 9, Box 561, S-751 22 Uppsala
Telephone: 018-14 25 82
Head and Professor of Morphological Zoology: Professor Carl-Olof Jacobson
Senior staff: Ecological zoology, Professor Staffan Ulfstrand

Entomology Division* 217
Telephone: 018-14 52 24
Head: Professor Christine Dahl

Klubbans Biological Station* 218
Address: Östersidan, Pl.2535, S-450 34 Fiskebäckskil
Telephone: 0523-221 02
Director: Göran Gezelius

SWITZERLAND

Battelle Centres de Recherche de Genève* 1

[Battelle Geneva Research Centres]
Address: 7 Route de Drize, CH-1227 Carouge-Genève
Telephone: (022) 43 98 31
Telex: 23-472 CH
Parent body: Battelle Memorial Institute, USA
Status: Independent research centre
General director: Dr V. Stingelin
Graduate research staff: 160
Activities: Battelle's Geneva operations perform contract research in various technical areas (engineering, metallurgy, chemistry, biology, physics, electronics etc) and in the field of human sciences (applied economics, social sciences, education and training).

INDUSTRIAL TECHNOLOGY CENTRE* 2

Director: Dr F. Trojer
Activities: Applied research; analytical and experimental investigations; design and development; and basic research in the following fields:
Materials technology - development of improved or new products made of metals, alloys, glass, ceramics and composites; conception and development of new processes in the automotive, construction, oil and gas, power generation, chemical, and related industries; quality control techniques and equipment are developed for specific sponsor's needs.
Computer sciences - analysis, design and implementation of user-orientated systems for government agencies and industry; computer centre operations, basic software development, and mathematical, scientific and economic modelling, software development and implementation of mini- and microprocessors in the engineering and process industries.
Energy and environment - development and improvement of equipment and systems for more effective use of conventional and alternative energy resources in industrial and domestic applications, included are technical and economic studies as well as hardware investigation and development; investigations to recover materials and to treat industrial waste are conducted, specialized chemical analyses of air and water pollution are developed.
Mechanical and process engineering - investigations focus on applications in the textile and automotive industries, including industrial automation. Process engineering is directed towards combustion and incineration technology, fuels, thermal engineering, and chemical processes; included are conceptual design of new equipment and prototypes/pilot development.
Chemistry and food technology - tomorrow's needs are prepared for by pursuing research into areas undergoing considerable evolution because of changing technical and economic factors; biomass chemistry presents itself as a medium to long-term alternative to petrochemistry, and laser photochemistry could provide a new tool for organic synthesis; food chemistry is mainly concerned with developing products for the food and feed industry and includes nutrition, food safety and microbiological studies.
Optical and electronic instrumentation - activities focus on prototype and product development for instrumentation and automation in the industrial, public and consumer sectors; biomedical instrumentation combines precision mechanics, optics, electronics and data processing to develop and improve automation of biomedical and analytical laboratory equipment.

Bundesamt für Landwirtschaft   3

[Department of Agriculture]
Address: Mattenhofstrasse 5, 3003 Bern
Telephone: (031) 61 25 07
Parent Body: Federal Ministry of Public Economy
Status: Official research organization
Research Director: J. Cl. Piot
Deputy Director: Professor Dr J. Von Ah
Graduate research staff: 220

Bundesamt für Veterinärwesen*   4

– BVET
[Swiss Federal Veterinary Office]
Official research centre

MEAT CHEMISTRY SECTION   5

Address: Viktoriastrasse 85, CH 3000 Bern
Telephone: (031) 61 28 78; 61 26 82
Director: Dr Eugen Hauser
Graduate research staff: 2
Publications: Annual reports.

F.J. Burrus & Cie SA*   6

Address: CH-2926 Boncourt
Telephone: (066) 75 55 61
Telex: 34543 TABUR CH
Status: Industrial company
Vice-President, Research: Dr Jost Wild
Sections: Tobacco research
Graduate research staff: 4
Activities: Development of new smoking products; research in filters.

Ciba-Geigy AG*   7

Address: Postfach, CH-4002 Basle
Telephone: (061) 3611 11
Telex: 62991
Status: Industrial company
Divisional r&d: Dyes and chemicals, Dr H. Ackermann; pharmaceuticals, Dr K. Heusher; agrochemicals, Dr E. Knüsli; plastics and additives, Professor H. Batzer; photochemicals, Dr D. Wyrsch
Activities: Research laboratories in Basle, Marly, Regensdorf (Switzerland), Horsham, Duxford, Manchester, Stamford Lodge, Brentwood (United Kingdom), Marienberg, Paris, and India for the development of dyes and pigments, textile products, detergent and paper industry products, photochemicals, and animal health products. In 1978 research and development expenditure totalled 8.5 per cent of total sales.

Dow Chemical Europe*   8

Address: Bachtobelstrasse 3, CH-8810 Horgen
Telephone: (01) 728 21 11
Telex: 54313
Status: Industrial company
Director, Research and Development: Denis Wilcock
Sections: Chemicals and hydrocarbons, E. Dyhrenfurth; speciality chemicals films and foams, W.A. Riese; coatings products, R.F. Sansone; plastics (styrenic/olefinic), H. Schumacher; agricultural products, T. Thomson (Kings Lynn, United Kingdom); pharmaceutical products, F. Leavitt, (Milan, Italy)
Graduate research staff: 450
Activities: Product and process and applications research, development and technical service in support of the Dow Chemical Europe range of products. The Research and Development headquarters are in Horgen, Switzerland with laboratories in Italy, Holland, Spain, United Kingdom and Germany.

École Polytechnique Fédérale de Lausanne/ Eidgenössische Technische Hochschule Lausanne/ Politecnico Federale di Losanna* 9

– EPFL
[Swiss Federal Institute of Technology, Lausanne]
Address: 33 Avenue de Cour, CH-1007 Lausanne
Telephone: (021) 47 11 11
Telex: 24478
Status: Educational establishment with r&d capability
President: Professor Bernard Vittoz
Vice-President: Professor Roland Crottaz

AGRICULTURAL ENGINEERING AND SURVEYING DEPARTMENT* 10

Address: 61 Avenue de Cour, CH-1007 Lausanne
Head: Professor Jean-Claude Piguet

Eidgenössische Anstalt für das Forstliche Versuchswesen/Institut Fédéral de Recherches Forestières/Istituto Federale di Ricerche Forestali 11

[Swiss Federal Institute of Forestry Research]
Address: CH-8903 Birmensdorf
Telephone: (01) 737 14 11
Affiliation: Swiss Federal Institute of Technology, Zürich
Status: Official research centre
Director: Dr Walter Bosshard
Sections: Inventory and productivity, Dr P. Schmid; forest technology and planning, Dr F. Pfister; landscape, Dr K. Ewald; forest protection and timber, Dr O. Lenz; permanent station, Dr H. Flüher; woods and environment, Dr H. Turner; pollution protection, Dr Th. Keller; regional forest inventory, E. Wullschleger; conservation and hydrology, J. Zeller
Graduate research staff: 55
Activities: Production planning; mensuration; growth research; remote sensing; data processing; working techniques and management of forest enterprises; methods of planting and tending; structure and organization of forest enterprises; management of the municipal forests of Bremgarten; forest seed and plants;

forest plant propagation; mycorrhiza; forest and area planning; forest protection and wood; wood quality; entomology; phytopathology; site and water; soil science; hydrology; plant ecology; forest and environment; bioclimatology; ecophysiology; air pollution detection; physiology; biochemistry; national forest inventory; mountain torrent control and slope stabilization.
Contract work: Yes
Publications: Berichte; Mitteilungen; Jahresbericht

AIR POLLUTION DETECTION DEPARTMENT 12

Director: Dr Th. Keller
Sections: Bio-indicators and monitors, Dr Th. Keller; physiology, Dr J. Bucher; biochemistry, Dr W. Landolf

FOREST AND ENVIRONMENT DEPARTMENT 13

Director: Dr H. Turner
Sections: Bioclimatology, Dr H. Turner; ecophysiology, Dr R. Hässler

FOREST PROTECTION AND WOOD DEPARTMENT 14

Director: Dr O. Lenz
Sections: Wood quality, Dr O. Lenz; phytopathology, Dr G. Bazzigher; tree rings and site, quaternary woods, Dr F.H. Schweingruber

LANDSCAPE DEPARTMENT 15

Director: Dr K. Ewald
Sections: Landscape history and planning, Dr K. Ewald; fauna, Dr G. Eichenberger; vegetation and sampling desing, Dr O. Wildi

MANAGEMENT RESERCH UNIT 16

Directors: Dr W. Bosshard, E. Wullschleger
Sections: Fallowland, Dr E. Surber; entomology, Dr J.K. Maksymov; afforestation and treatment of young stands, G. Beda; management of municipal forests of Bremgarten, G. Beda; structure and organization, A. Speich; effects of forest policy, H. Kapser

MECHANIZATION AND LAND USE PLANNING DEPARTMENT   17

Director: Dr F. Pfister
Sections: Forest and land use planning, Dr F. Pfister; nursery practices and stand establishment, Dr M. Hocevar; timber harvesting, B. Abegg

NATIONAL FOREST INVENTORY DEPARTMENT   18

Director: E. Wullschleger
Sections: Objectives of national forest inventory, E. Wullschleger; methods and evaluation in national forest inventory, F. Mahrer

PRODUCTION PLANNING DEPARTMENT   19

Director: Dr P. Schmid
Sections: Mensuration, Dr P. Schmid; growth and yield, Dr W. Keller; planning methods, C. Gadola

PROTECTION WORKS AND HYDROLOGY DEPARTMENT   20

Director: J. Zeller
Sections: Mountain torrent control and slope stabilization consultation, J. Zeller; mountain torrent control and slope stabilization research, H. Geiger; hydrology, Dr H. Keller

SITE STUDIES DEPARTMENT   21

Director: Dr H. Flühler
Sections: Biophysics, Dr H. Flühler; vegetation ecology, Dr N. Kuhn; mycorrhiza, S. Egli; soil science, Dr P. Blaser

SWISS GUEST INSTITUTES *   22

Bureau for Advice on Hedges   23

Director: W. Müller

Laboratory of Soil Physics   24

Director: Professor F. Richard

Swiss Interest Group for Industrial Wood   25

Director: M. Leidig

Eidgenössische Forschungsanstalt für Agrikulturchemie und Unwelthygiene/ Station Fédérale de Recherches en Chimie Agricole et sur l'Hygiène de l'Environment/Stazione Federale di Ricerche per la Chimica Agraria e l'Igiene dell'Ambiente   26

[Federal Research Station for Agricultural Chemistry and Hygiene of the Environment]
Address: Schwarzenburgstrasse 155, CH-3097 Liebefeld-Bern
Telephone: (031) 59 81 11
Parent body: Bundesamt für Landwirtschaft
Status: Official research centre
Director: Dr Ernest Bovay
Sections: Soils and fertilizers, Dr O. Furrer; plant physiology and nutrition, Dr H. Schnetzer; air pollution, Dr R. Zuber; analytical chemistry, Dr R. Daniel
Graduate research staff: 16
Publications: Biannual report.

Eidgenössische Forschungsanstalt für Betriebswirtschaft und Landtechnik/Station Fédérale de Recherches d'Économie, d'Enterprise et de Génie Rural/Stazione Federale di Ricerche d'Economia Aziendale e di Genio Rurale   27

[Federal Research Station for Farm Management and Agricultural Engineering]
Address: CH-8355 Tänikon, Thurgau
Telephone: (052) 472025
Parent body: Bundesamt für Landwirtschaft
Status: Official research centre
Director: Walter Meier
Sections: Project coordination, Fritz Bergman; farm management, Emanuel Dettwiler; labour economy, Albert Schönenberger; agricultural machinery, Rudolf Studer; cultivation techniques, Witold Zumbach; farmstead mechanization, Mathäus Rohrer; farm building, Alex Studer
Graduate research staff: 45
Activities: Research on agricultural engineering and building, farm management and labour economics.
Publications: Annual report.

Eidgenössische Forschungsanstalt für Landwirtschaftlichen Pflanzenbau/Station Fédérale de Recherches Agronomiques/Stazione Federale di Ricerche Agrarie   28

[Federal Research Station for Agronomy]
Address: Reckenholzstrasse 191/211, CH-8046 Zurich
Telephone: (01) 57 88 00
Parent body: Bundesamt für Landwirtschaft
Status: Official research centre
Director: Dr Alfred Brönnimann
Divisions: Plant production, Dr H. Guyer; agricultural chemistry, Dr K. Peyer; plant-breeding, Dr F. Weilenmann; plant protection, Dr W. Meier
Graduate research staff: 45

Activities: Soil science; recultivation and improvement; soil mapping; plant breeding - wheat, maize, grasses and clover; crop husbandry - pasture, meadow, potatoes, sugar beet, rape; seed testing; plant protection - pests and diseases in field crops - bacteria, fungi, viruses, nematoda, homologation of pesticides.
Publications: Biannual report; monthly periodical.
Projects: Improvement of alpine pasture; effects of pesticides on soil; rhizobia and nitrogen fixation; quality of organic matter in soil; influence of various tillage methods on soil structure; mole drainage; soil fertility and biological soil activity; heavy metals in soil; straw and green manuring; influence of sewage on meadows; gülle on arable land; new varieties of cereals and N+CCC; mobilization in soil; percolation of nitrogen in soil (lysimeters); influence of herbicides on soil and weed populations; testing application of herbicides in rivers; growth regulators; herbicides to improve natural pastures; quantification of damage of weeds; soil maps on different scales; resistance breeding (Septoria, Yellow rust, mildew).

Eidgenössische Forschungsanstalt für Milchwirtschaft/Station Fédérale de Recherches Laitières/Stazione Federale di Ricerche dell'Industria del Latte *   29

[Federal Dairy Research Institute]
Address: Schwarzenburgstrasse 155, CH-3097 Liebefeld-Bern
Telephone: (031) 59 81 11
Parent body: Bundesamt für Landwirtschaft
Status: Official research centre
Director: Professor Dr Bernard H. Blanc
Sections: Basic research, Professor Dr B.H. Blanc; research in cheesemaking, Dr Chr. Steffen; hygiene, Dr M. Schällibaum; technology, Dr E. Flückiger; nutrition, Professor Dr B.H. Blanc; biology of bees and honey research, Dr H. Wille
Graduate research staff: 35
Activities: Scientific research in dairying, cheesemaking, milk hygiene, technology, nutrition; testing of dairy products; examination of items subject to authorization, ie detergents and sanitizers for dairy equipment, dairy equipment, refrigerators and milking machines as to their technical operation; advisory service and publicizing of specialized knowledge to cheesemakers and to the regional udder health services; biology and diseases of bees and honey research.
Publications: Biennial Activity Report.

Eidgenössische Forschungsanstalt für Obst- , Wein- und Gartenbau/ Station Fédérale de Recherches en Arboriculture, Viticulture et Horticulture/Stazione Federal di Ricerche in Frutticoltura, Viticoltura e Orticoltura* 30

[Federal Research Station for Fruit-Growing, Viticulture and Horticulture]
Address: Schloss, CH-8820 Wädenswil
Telephone: (01) 780 13 33
Parent body: Bundesamt für Landwirtschaft
Status: Official research centre
Director: Professor Dr Robert Fritzsche
Sections: Fruit growing, Dr Robert Schumacher; viticulture and oenology, Dr Koblet Werner; horticulture, Dr Fritz Kobel; plant protection, Dr Lukas Stalder; chemistry, biology, and technology of beverages, Dr Ulrich Schobinger; fruit and vegetables storage and processing, Dr Karl Stoll; biochemistry, Dr Alfred Temperli
Graduate research staff: 35
Activities: Research in the following areas: plant physiology and plant anatomy; genetics, breeding, and testing of varieties; work rationalization, management; cultivation techniques; entomology, nematology, phytopathology, virology, bacteriology, herbology; microbiology and chemistry of beverages; biochemistry; chemistry of pesticides; food technology.
Contract work: Yes
Publications: Tätigkeitsbericht; Informationsschrift; Schweizerische Zeitschrift für Obst- und Weinbau.

Eidgenössische Forschungsanstalt für Viehwirtschaftliche Produktion/Station Fédérale de Recherches sur la Production Animale/ Stazione Federale di Ricerche della Produzione Animale 31

[Federal Research Station for Animal Production]
Address: Grangeneuve, CH-1725 Posieux
Telephone: (037) 82 11 81
Parent body: Bundesamt für Landwirtschaft
Status: Official research centre
Director: Dr H. Schneeberger
Departments: Ruminant nutrition, F. Jans; pig nutrition, M. Jost; physiology of nutrition, Dr R. Daccord; forage conservation, E. Gallasz; feedstuffs and feed additives, Dr J. Morel; analytics, Dr T. Rihs; veterinary service, Dr M. Wanner
Graduate research staff: 20
Activities: Optimum utilization of home grown roughages, indigestions in the high-yielding dairy cow, fattening of cross-bred animals from dairy farms, use of antimicrobial substances in calf fattening.
Publications: Biannual report.
Projects: Influence of botanical compostion of grasslands on feed intake and milk production of dairy cows (F. Jans); fattening of crossbred animals from dairy farms (E. Lehmann); feed evaluation and nutrient requirements of ruminants and swine (Dr. R. Daccord); use of whey in pig fattening, effect of roughages in the daily ration of pregnant and suckling sows (M. Jost); effect of wilting on the nutritive value of green fodder silage (E. Gallasz).

Eidgenössische Technische Hochschule Zürich/École Polytechnique Fédérale Zürich* 32

– ETHZ
[Swiss Federal Institute of Technology, Zürich]
Address: Rämistrasse 101, CH-8092 Zürich
Telephone: (01) 256 22 11
Status: Educational establishment with r&d capability
President: Professor Heinrich Ursprung
Executive Director: Dr Eduard Freitag
Secretary General: Dr Hans Rudolf Denzler
Information Officer: Dr Rold Guggenbühl

SCHOOL OF AGRICULTURE*   33

Dean: Professor E. R. Keller

Animal Production Institute*   34

Address: Universitätstrasse 2, CH-8092 Zürich
Director: Professor Jakob Landis

Animal Breeding Institute*   35
Address: Universitätstrasse 2, CH-8092 Zürich
Directors: Professor Niklaus Künzi, Professor Gerald Stranzinger
Activities: Population genetics and problems of selection for milk and beef in cattle, food conversion and carcass qualities in pigs, fertility and lamb production in sheep, egg weight and egg qualities in poultry, climatic physiology of farm animals.

Animal Nutrition Institute*   36
Address: Universitätstrasse 2, CH-8092 Zürich
Directors: Professor Alfred Schürch, Professor Jakob Landis
Activities: Experiments on energy metabolism of animals. Determination of the chemical composition, biological value and metabolism of proteins and lipids contained in different feedstuffs. Influence of vitamins, hormones, antibiotics on animal performance. Research on fodder preservation.

Biometry and Population Genetics Laboratory*   37
Address: Gloriastrasse 35, CH-8092 Zürich
Head: Professor Henry Louis Le Roy
Activities: Applied statistics; population genetics; simulation studies.

Physiology and Hygiene of Farm Animals Institute* 38
Address: Universitätstrasse 2, CH-8092 Zürich
Director: Professor Hans Heusser
Activities: Various problems of livestock management and sanitation, especially climatology of stables. Studies on iron metabolism, physiology, parathyroidhormone and calcitonine in cattle.

Chair of Agricultural Economics*   39

Address: Sonneggstrasse 33, CH-8092 Zürich
Professor: Professor Dietmar Onigkeit
Activities: National planning of agricultural production and nutrition. Distribution systems of vitally necessary goods. Interregional model of agricultural structure. Long run investigation for Swiss factories. Applications of operations research and econometrics in agricultural economics. Research of supply and demand for milk and meat in Switzerland.

Food and Nutrition Science Institute*   40

Address: Universitätstrasse 25, CH-8092 Zürich
Director: Professor Jürg Solms

Chemistry, Technology and Agriculture Laboratory* 41
Address: Universitätstrasse 2, CH-8092 Zürich
Directors: Professor Johann Neukomm, Professor R. Bach, Professor J. Solms, Professor Hans Sticher
Activities: Chemistry, biochemistry and technology of foods, especially cereals, fruits and vegetables, chemistry of food flavours, carbohydrates; morphological, chemical, and genetical studies of Swiss soils; chemistry of clay minerals.

Dairy Science Laboratory*   42
Address: Eisgasse 8, CH-8004 Zürich
Directors: Professor Zdenko Puhan, Professor M. Bachmann
Activities: Dairy chemistry, bacteriology, technology and economics. Electrophoretic study of milk proteins, estimation of protein break-down products, metabolic products of lactic cultures, heat inactivation of lactic cultures and contaminants during heat treatment of dairy products, cheese technology, problems of developing countries.

Food Microbiology Laboratory*   43
Address: Universitätstrasse 25, CH-8092 Zürich
Director: Professor W. Schmidt-Lorenz

Food Technology Laboratory*   44
Address: Universitätstrasse 25, CH-8092 Zürich
Director: Professor Franz Emch
Activities: Concentration and drying; pasteurization; sterilization; aseptic processing; different technical aspects in food technology.

Plant Production Institute*   45

Address: Universitätstrasse 2, CH-8092 Zürich
Director: Professor Josef Nösberger
Activities: Ecophysiology of crop plants and meadows (analysis of growth, studies on competition). Development of swards of natural meadows in dependence on fertilization and utilization. Growth and breeding of soya bean and horse bean. Studies on potato haploids. Maintenance of yield potential of the soil under modern techniques of crop production.

Rural Economics Institute*   46

Address: Sonneggstrasse 3, CH-8092 Zürich
Director: Professor W. Meier
Activities: Farm management; integration of production processing and marketing; natural resource economics; economic development of rural areas.

SCHOOL OF FORESTRY*   47

Forestry Research Institute*   48

Address: Rämistrasse 101, CH-8092 Zürich
Head: Professor Felix Richard
Activities: Geological aspects of forestry engineering; sylviculture; forest management and planning; forest policy and economy; wood science and technology; dendrology and genetics; soil science and ecology.

SCHOOL OF NATURAL SCIENCES*   49

Entomology Institute*   50

Address: Clausiusstrasse 21, CH-8092 Zürich
Director: Professor Georg Benz
Activities: Population dynamics of the larch bud moth. Ecology of agricultural insect pests. Microbiological control of the codling moth and basic research on insect viruses. Physiology of insect reproduction, especially hormonal control and pheromones. Systematics and faunistics, especially on microlepidoptera.

General Botany Institute*   51

Address: Universitätstrasse 2, CH-8092 Zürich
Director: Professor Philippe Matile
Activities: Quantitative cytochemistry of DNA and proteins at the light microscope level. Highest resolution electron microscopy of biopolymers. Development of the freeze-etching technique. Characterization of the lysosomal compartment of the plant cells.

Geobotany Institute*   52

Address: Zürichbergstrasse 39, CH-8044 Zürich
Director: Professor E. Landolt
Activities: Plant sociology and ecology; cytology and evolution of ecotypes.

Special Botany Institute*   53

Address: Universitätstrasse 2, CH-8092 Zürich
Director: Professor Heinz Kern
Activities: Plant pathology (toxic metabolic products of plant parasites, mechanisms of disease resistance); taxonomy and biology of fungi (ascomycetes and basidiomycetes), physiology of spore formation; experimental taxonomy of vascular plants.

Eidgenössisches Institut für   54 Schnee- und Lawinenforschung/Institut Fédéral pour l'Étude de la Neige et des Avalanches/ Istituto Federale per lo Studio della Neve e delle Valanghe*

[Swiss Federal Institute for Snow and Avalanche Research]
Address: CH-7260 Weissfluhjoch/Davos
Telephone: (083) 5 32 64
Telex: 74 309
Affiliation: Bundesamt für Forstwesen; Department of the Interior
Status: Official research centre
Director: Dr C. Jaccard
Sections: Snow meteorology, Dr P. Föhn; snow mechanics, Dr B. Salm; snow vegetation, H.R. in der Gand; snow physics, Dr W. Good
Graduate research staff: 15

SNOW AND VEGETATION SECTION   55

Address: Fluelastrasse 9, 7260 Davos Dorf, Graubuenden
Telephone: (083) 5 13 47
Director: H. in der Gand
Departments: Forestry, Dr M. de Coulon
Graduate research staff: 3
Activities: Snow cover and avalanches as ecological factors for forest plants, forest plant protection against negative influences of snow cover and avalanches, forestry and forest products.
Publications: Annual reports.
Projects: Interaction between cover, avalanches, antiavalanche structures and forest plants (J. Rychetnik); interaction between snow duration, snow gliding and forest plants (W. Frey); protection against snow gliding and avalanches in forest zone, snow and snow motion in forests with avalanche protection function (H. in der Gand).

Fabriques de Tabac Réunies SA* 56

Address: Postfach 11, CH-2003 Neuchatel
Telephone: (038) 21 11 45
Telex: 35137
Parent body: Philip Morris Incorporated
Status: Industrial company
Research director: Dr H.W. Gaisch
Sections: Analytical chemistry, Dr W. Fink; biotechnology, Dr D. Schulthess; technical services, Dr C. Jeanneret
Graduate research staff: 15
Activities: Science and technology of cigarette manufacture: tobacco chemistry; smoke chemistry; ingredients and materials; process development; process control.

Genossenschaft UFA* 57

Address: Theaterstrasse 3, Postfach 344, CH-8401 Winterthur
Telephone: (052) 22 35 08
Telex: CH 76 866
Parent body: Union of the Federations of Swiss Agricultural Cooperatives
Status: Independent research centre
Research director: Dr H.P. Pfirter
Management director: Dr P. Schmid
Genetic activities director: Dr A. Schneider
Sections: Animal nutrition research; pig breeding
Graduate research staff: 3
Activities: Management and experiments for all domestic animals, and research in the nutrition and breeding sectors.

F. Hoffman-la Roche et Cie AG* 58

[Hoffmann-la Roche and Company Limited]
Address: CH-8157 Dieksdorf
Status: Industrial company

AGROCHEMICAL RESEARCH AND DEVELOPMENT DEPARTMENT 59

Address: CH-4002 Basle
Telephone: (061) 27 27 93; 27 11 22
Coordinator: Hans Thommen

IMS Aktiengesellschaft* 60

Address: Gartenstrasse 2, CH-6300 Zug
Status: Industrial company
Activities: Market research in the pharmaceutical, medical, veterinary, medical promotion and cosmetic fields. Toxicological research.
Publications: Pharmaceutical news bulletin.

Ingenieurschule Wädenswil für Obst- Wein- und Gartenbau 61

[Engineering School of Wädenswil]
Address: PO Box 434, CH-8820 Wädenswil
Telephone: (01) 780 19 75
Status: Educational establishment with r&d capability
Director: Dr Walter Müller
Graduate research staff: 9
Activities: Variety testing; cultivation methods, economical aspects; technology.
Projects: Viticulture and oenology (Dr Walter Eggenberger); arboriculture (Dr Johann Spichiger); vegetable growing (Dr Rolf Grabherr); floriculture (Dr Theodor Zwygart).

Institut für Ernährungsforschung - Stiftung 'Im Grüne'* 62

[Institute for Nutrition Research Foundation 'Green Meadow']
Address: Seestrasse 72, CH-8803 Rüschlikon
Telephone: (01) 724 25 20
Status: Official research centre
Director: Dr A. Blumenthal
Graduate research staff: 4
Activities: Survey of dietary and living habits of different groups of population; studies of the contents of nutrients in foods and their changes during processing and preparation; consulting the authorities and industry on nutritional questions.

Institut Sérothérapique et Vaccinal Suisse* 63

[Swiss Serum and Vaccine Institute]
Address: Postfach 2707, CH-3001 Bern
Status: Official research centre
Activities: Human and veterinary immunology; bacteriology, virology and specific immunoglobulins.

Interfood Limited* 64

Address: Avenue de Cour 107, CH-1001 Lausanne
Telephone: (021) 27 15 61
Status: Industrial organization
Technical and production manager: Norbert Bucher
Activities: The laboratories and experimental section, incorporated into the headquarters at Lausanne, carry out investigations into new products and new manufacturing methods.
Publications: Annual report.

Kommission zur Förderung der Wissenschaftlichen Forschung/Commission pour l'Encouragement de la Recherche Scientifique* 65

[Commission for the Promotion of Scientific Research]
Address: Wildhainweg 21, Postfach 2338, CH-3001 Bern
Telephone: (031) 61 21 43; (031) 61 21 46; (031) 61 21 49
Status: Official research organization
President: Dr W. Jucker
Secretary: Dr P. Kuentz

Lonza AG 66

Address: Postfach, CH-4002 Basel
Telephone: (061) 55 88 50
Telex: 62323 Lonza Basel
Affiliation: Schweizerische Aluminium AG -Alusuisse
Status: Industrial company
Director: F. Friedli
Department: Agrochemicals, O. Häller
Graduate research staff: 10
Activities: Plant protection (molluscicides), plant nutrition (organic and inorganic fertilizers), silage improvement.
Publications: Annual report.
Projects: Speciality fertilizers (A. Löliger); molluscicides (O. Häller); recycling and granulated slow release fertilizers (M. Bernheim).

Sandoz Ltd* 67

Address: Lichtstrasse 35, S-4002 Basle
Telephone: 061-24 11 11
Telex: 63275
Status: Industrial company
Director, r&d: Dr J. Rutschmann
Divisional r&d leaders: Dyestuffs and chemicals, Dr J. Benz, Dr A. Kaufmann; pharmaceuticals, Dr B. Berde, Dr K. Saameli, Dr S. Guttmann; agricultural products, Dr K. Lutz

CROP PROTECTION RESEARCH DEPARTMENT 68

Telephone: (061) 24 31 09
Manager: H.P. Schelling
Sections: Biology, Paul Roth; synthesis, Fred Kuhnen
Graduate research staff: 25
Activities: Plant protection, insect, control, stored products protection, feedstock ectoparasites.
Publications: Annual report.
Projects: Search for new pesticides (H.P. Schelling).

Schweizerische Geflügelzuchtschule   69

[Swiss Poultry Institute]
Address: Burgerweg 22, 3052 Zollikofen
Telephone: (031) 57 02 22
Status: Educational establishment with r&d capability
Director: Dr Werner Thomann
Sections: Experimental, H.P. Guler
Graduate research staff: 3
Activities: Poultry nutrition and husbandry, ethology.
Publications: Annual reports.
Projects: Zeolite in broiler and layer nutritition, testing of various additives in broiler nutrition (H.P. Guler); development and testing of practical alternatives to the conventional laying batteries (H.P. Gueler, Hans Oester).

Schweizerische Meteorologische Anstalt/ Institut Suisse de Météorologie*   70

– SMA
[Swiss Meteorological Institute]
Address: Postfach, CH-8044 Zürich
Telephone: (01) 252 67 20
Telex: 52202 metzu
Parent body: Federal Department of the Interior
Status: Official research centre
Graduate research staff: 11
Activities: Future research and development activities will concentrate on the following: improvement of weather prediction; development of new systems and methods for a combined treatment of new and classical data (eg radar, satellite automatic network and synoptic observations); mesoscale modelling of the Alpine region; determination of the characteristics of the airflow and mass field over and around the Alpine mountain complex, including upstream effects and special weather situations such as Foehn winds and blockings. (In the frame of the international ALPEX Mountain Sub-Programme of the Global Atmospheric Research Programme of the World Meteorological Organization/ International Council of Scientific Unions); satellite meteorology; turbulent diffusion processes within the exchange layer (up to about 1000 m/ground); heat and momentum transfer in the boundary layer; solar radiation research; special questions in the field of agro-and biometeorology; climatology (climatological atlas, climatology of different weather types, studies in the field of technical meteorology); cloud physics: Radar-meteorology and selected precipitation processes.

Publications: Veröffentlichungen der Schweizerische Meteorologische Anstalt; Arbeitsbericht; Klimatologie der Schweiz.

APPLIED METEOROLOGY AND DATA   71 ACQUISITION DIVISION*

Address: c/o Station Aérologique, CH-1530 Payerne
Deputy Director: Dr A. Junod

Agricultural Meteorology*   72

Address: Postfach, CH-8044 Zürich
Head: Dr B. Primault

Schweizerischer Nationalfonds zur Förderung der Wissenschaftlichen Forschung/Fonds National Suisse de la Recherche Scientifique*   73

[Swiss National Science Foundation]
Address: Wildhainweg 20, Postfach 2338, CH-3001 Bern
Telephone: (031) 24 54 24
Telex: 33 413
Status: Official research organization
Secretary-General: Dr Peter E. Fricker
Graduate research staff: 750 (on projects funded by the Foundation)
Activities: The Foundation provides financial support for basic research in all sciences and the humanities.

Internal

AN FORAS TALÚNTAIS

Memo

To:

Ref.

Subject:

Date

Agric. Dept. of the canton of Fribourg.
181 Place Notre Dame
PO 1700 Fribourg.
Swit 3.

Société d'Assistance Technique pour Produits Nestlé SA 74

– NESTEC
Address: PO Box 88, CH-1814 La Tour de Peilz
Telephone: (021) 51 02 11
Telex: 451 333 nta
Status: Research centre within an industrial company
Research director: Professor Dr Jean Mauron
Sections: Fundamental sciences, Jean-Pierre Bouldoires; biological sciences, Dr M. Horisberger; food sciences, Dr V. Wenner; Biological Laboratory, Orbe (Labior), Dr H.P. Wuerzner
Graduate research staff: 100
Activities: Studies on: food composition; raw materials and processing; nutrition mechanism and physiology; malnutrition; metabolic diseases; modern diseases; food product safety.

Station Fédérale de Recherches Agronomiques/ Eidgenössische Landwirtschaftliche Forschungsanstalt Changins/Stazione Federale di Ricerche Agrarie Changins 75

[Swiss Federal Agricultural Research Station]
Address: Route de Duillier, CH-1260 Nyon, Vaud
Telephone: (022) 61 54 51
Telex: 22 785
Parent body: Bundesamt für Landwirtschaft
Status: Official research centre
Director: H. Alexandre Vez
Departments: Plant protection, Dr Bovey; plant breeding, Dr Badoux; crop husbandry, M. Charles; viticulture and technology, M. Simon; arboriculture, horticulture, Dr Perraudin
Graduate research staff: 60
Activities: Soil science, drainage and plant production; plant breeding; crop husbandry; plant protection; food science.
Publications: Biannual report.

Universität Basel* 76

[Basel University]
Address: Petersplatz 1, CH-4051 Basel
Telephone: (061) 25 73 73
Status: Educational establishment with r&d capability
Rector: Dr Frank Vischer

FACULTY OF PHILOSOPHY AND NATURAL SCIENCES* 77

Address: Klingelbergstrasse 70, CH-4056 Basel
Telephone: (061) 25 04 55
Dean: Professor Dr Walter J. Gehring

Biocentre* 78

Address: Klingelbergstrasse 70, CH-4056 Basel
Telephone: (061) 25 38 80
Head: Professor Dr W. Arber

Microbiology Research Department* 79
Research head: Professor Dr E. Kellenberger
Senior staff: Professor Dr W. Arber

Structural Biology Research Department* 80
Research head: Professor Dr R.M. Franklin
Senior staff: Professor Dr J.N. Jansonius

Botany Institute* 81

Address: Botanischen Garten, Schönbeinstrasse 6, CH-4056 Basel
Telephone: (061) 25 69 15
Head: Professor Dr H. Zoller

Zoology Institute* 82

Address: Rheinsprung 9, CH-4051 Basel
Telephone: (061) 25 25 35
Heads: Professor Dr H. Nüesch, Professor Dr C.H.F. Rowell

Universität Bern* 83

[Berne University]
Address: Hochschulstrasse 4, CH-3012 Bern
Telephone: (031) 65 81 11
Status: Educational establishment with r&d capability
Rector: Professor Dr R. Fankhauser

FACULTY OF PURE SCIENCE* 84

Dean: Professor Dr H. Carnal

Botany Department*   85
Head: Professor K. Erismann

**Experimental Morphology   86
Department***
Head: Professor M. Lüscher

General Biology Department*   87
Head: Professor U. Leupold

Microbiology Department*   88
Heads: Professor R. Braun, Professor V. Leupold

**Systematic Botany and Geobotany   89
Department***
Head: Professor G. Lang

Zoology Department*   90
Heads: Professor P. Tschumi, Professor R. Weber, Professor B. Tschanz

Zoophysiology Department*   91
Head: Professor M. Lüscher

**FACULTY OF VETERINARY   92
MEDICINE***
Dean: Professor H. Gerber

Anatomy Department*   93
Head: Professor W. Mosimann

**Animal Breeding and Hygiene   94
Department***
Head: Professor W. Weber

Bacteriology Department*   95
Head: Professor H. Fey

**Domestic Animal Diseases   96
Department***
Head: Professor H. Gerber

Embryology Department*   97
Head: Professor W. Mosimann

Histology Department*   98
Head: Professor W. Mosimann

**Neuropathology of Domestic Animals 99
Department***
Head: Professor R. Fankhauser

Parasitology Department*   100
Head: Professor H. Fey

Pathology Department*   101
Head: Professor U. Freudiger

Pharmacology Department*   102
Head: Professor H.-J. Schatzmann

Serology Department*   103
Head: Professor H. Fey

Veterinary Pathology Department* 104
Head: Professor H. Luginbühl

Universität Zürich*   105

[Zürich University]
Address: Rämistrasse 71, CH-8006 Zürich
Telephone: (01) 257 11 11
Telex: unizh 54 864
Status: Educational establishment with r&d capability
Rector: Professor Dr Gerold Hilby
Secretary: Dr Franz Züsli-Nicosi

**FACULTY OF PHILOSOPHY II   106
(SCIENCES)***
Dean: Professor Dr Kurt Strebel

**Botanical Garden and Systematic   107
Botany Institute***
Address: Zollikerstrasse 107, CH-8008 Zürich
Director: Professor Dr C.D.K. Cook

Molecular Biology Institute I 108
(Chemical Genetics)*

Address: Hönggerberg, H-8093 Zürich
Director: Professor Dr Ch. Weissmann
Activities: Gene-expression of higher organisms; studies on virus QB; gene isolation and modification.

Molecular Biology Institute II 109
(Genetics-Cell Biology)*

Address: Hönggerberg, Dr M. Birnstiel, CH-8093 Zürich
Director: Professor Dr M. Birnstiel
Activities: Molecular analysis of isolated genes of multicell organisms; gene regulation; genetic engineering; surrogate genetics.

Plant Biology Institute* 110

Address: Zollikerstrasse 107, CH-8008 Zürich
Director: Professor Dr H. Wanner
Activities: Cytology; cellular mould and lichen fungi; physiology and biochemistry of green plants; technical photosynthesis - biological aspects of sun energy utilization; production and utilization of biomass.

Hydrobiological-Limnological Station* 111
Address: Seestrasse 187, CH-8802 Kilchberg
Head: Professor Dr E. Thomas

Zoology Institute* 112

Address: Winterthurerstrasse 190, CH-8057 Zürich
Director: Professor Dr R. Nöthiger
Activities: Developmental biology of biochemical direction; developmental biology and regeneration of lower invertebrates; developmental physiological genetics; neurobiology; applied research on wild animals and the deer problem in the national park and its environment; research on the interactions of primates.

Ethology and Wild Animal Research 113
Department*
Address: Birchstrasse 95, CH-8050 Zürich
Head: Professor Dr H. Kummer

Zoological Museum* 114
Director: Professor Dr H. Burla

FACULTY OF VETERINARY MEDICINE 115

Address: Winterthurerstrasse 260, CH-8057 Zürich
Telephone: (013) 65 11 11
Dean: Professor Dr Konrad Zerobin
Graduate research staff: 70
Activities: Research and teaching in the field of veterinary medicine.

Animal Breeding Institute* 116

Director: Professor Dr K. Zerobin
Activities: Reproduction physiology and pathophysiology; sperm preservation and examination; zootechnical manipulation of sexuality (cattle); keeping and diseases of zoo, domestic and laboratory animals.

Parasitology Institute* 117

Director: Professor Dr J. Eckert
Activities: Intramitochondrial energy studies; mechanism of anthelminthica; cellular immunity reactions; chemotherapy and differentiation of echinococcus; ruminant infestations.

Pharmacology and Biochemistry 118
Institute*

Director: Professor Dr E. Jenny
Activities: Pharmacokinetics; pharmacodynamics; muscle biology; immunology; neurobiology; developmental biology.

Veterinary Anatomy Institute* 119

Director: Professor Dr J. Frewein
Activities: Morphological, morphometric and immuno-cytochemical research; cell dynamics and secretion cycle studies; neonatal development; epithelial substance and diabetes research.

Veterinary Hospital* 120

Gynaecology Clinic* 121
Head: Professor Dr M. Berchtold

Medical Clinic* 122
Head: Professor H. Sutter

Surgical Clinic* 123
Head: Professor Dr A. Müller

Veterinary Hygiene Institute* 124

Director: Professor H.U. Bertschinger
Activities: Infectious diseases of domestic cattle; meat hygiene; milk hygiene; environmental hygiene.

Veterinary Pathology Institute* 125

Director: Professor Dr H. Stünzi
Activities: Comparative histopathological studies of lung carcinoma in domestic carnivora and endometrium of the bitch; spontaneous heart diseases in cats; immunity system development (puppies); pathogenesis of neonatal diarrhoea (calf).

Veterinary Physiology Institute*   126

Director: vacant

Virology Institute*   127

Director: Professor Dr R. Wyler

Université de Fribourg/   128
Universität Freiburg*

[Fribourg University]
Address: Miséricorde, CH-1700 Fribourg
Telephone: (037) 21 91 11
Status: Educational establishment with r&d capability
Rector: Professor Dr Bernhard Schnyder
Director: Sir Hans E. Brülhart

FACULTY OF NATURAL SCIENCES*   129

Dean: Professor André Antille

Botany Department*   130

Head: Professor Hans Meier

Zoology Department*   131

Head: Professor Heinz Tobler
Senior staff: Professor Jean Schowing, Professor Gerolf Lampel

Université de Genève*   132

[Geneva University]
Address: 24 Rue Général Dufour, CH-1211 Genève, 4
Telephone: (022) 20 93 33
Status: Educational establishment with r&d capability
Rector: Professor Justin Thorens
Managing director: Claude Boissy
General Secretary: Bernard Ducret

FACULTY OF SCIENCES*   133

Dean: Professor Hubert Greppin

Biology Section*   134

Address: Quai Ernest-Ansermet 20, CH-1211 Genève, 4
President: Professor Marco Crippa

Animal Biology Department*   135

Plant Biology Department*   136

Université de Lausanne*   137

[Lausanne University]
Address: 4 Place de la Cathédrale, CH-1000 Lausanne, 17
Telephone: (021) 22 00 31
Status: Educational establishment with r&d capability
Rector: Professor Claude Bridel

FACULTY OF SCIENCE*   138

Address: Collège Propédeutique, CH-1015 Lausanne Dorigny
Dean: Professor Oscar Burlet

Animal Biology Institute*   139

Address: Palais de Rumine, CH-1005 Lausanne
Director: Professor Walter Wahli

Animal Ecology and Zoology   140
Institute*

Address: 19 Place du Tunnel, CH-1005 Lausanne
Director: Professor Peter Vogel

Plant Biology and Physiology   141
Institute*

Address: Palais de Rumine, CH-1005 Lausanne
Director: Professor Paul-Emile Pilet

Systematic Botany and Geobotany   142
Institute*

Address: 14b Avenue de Cour, CH-1007 Lausanne
Director: Professor Pierre Villaret

Université de Neuchatel*   143

[Neuchatel University]
Address: Avenue du 1er Mars 26, CH-2000 Neuchatel
Status: Educational establishment with r&d capability

FACULTY OF SCIENCE*   144

Dean: Professor Klaus Bernauer

Botany Institute* 145

Address: Chantemerle 22, CH-2000 Neuchatel
Director: Professor Claude Favarger

Plant Physiology Laboratory* 146

Address: Chantemerle 18-20, CH-2000 Neuchatel
Director: Professor Paul-André Siegenthaler

Zoology Institute* 147

Address: Chantermerle 22, CH-2000 Neuchatel
Director: Professor André Aeschlimann

SYRIA

Cotton Bureau 1

Address: Al-Maydan, Aleppo
Telephone: 47600
Status: Official research centre
Director: Dr M. Ali Deiri
Graduate research staff: 15
Activities: Breeding and creation of new varieties suitable for growing in Syria; modern implement use in cultural practice; fertilization, irrigation, planting dates, plant density.
Publications: Annual report (Arabic), fortnightly report (English).
Projects: Cotton breeding (Jaara Abdul Aziz); cotton farming mechanization (Eid Muhsen); cultural practice improvement (Hakim Ala-Eldine).

University of Aleppo* 2

Address: Aleppo
Status: Educational establishment with r&d capability

FACULTY OF AGRICULTURE 3

Telephone: 54199
Dean: Fayez El-Yassin
Graduate research staff: 120
Activities: Natural resources, range science and forestry (synecology, autecology and management); plant breeding, cereals, food legumes, and range plants; plant protection, cereals, legumes and industrial crops; livestock husbandry and nutrition.
Publications: Arid Zone Ecology Report, annual; Pistachio Unit Report, annual.

Agricultural Economics Department 4

Head: Dr Khaled Sabei El-Najar

Agricultural Engineering Department 5

Head: Dr Souheil Barbara

Agronomy and Range Science Department 6

Head: Professor Dr Mohamed Nazir Sankary
Projects: Germplasm resources of arid zones, potential vegetation maps of the Arab countries, revegetation and desertification control of arid zones.

Animal Production Department 7

Head: Dr Ghassan Ghadri
Projects: Animal improvement and feeding (Dr Fayez El-Yassin).

Food Science Department 8

Head: Dr Adel Mehio

Horticulture and Forestry Association 9

Head: Dr Adhian Hadj-Hassan
Projects: Varieties, pollination, rootstocks and growth regulators in Pistachio trees (Dr Adnan Hadj-Hassan); mediterranean climate and forest trees (Professor Dr Ibrahim Nahal); Syrian vegetable crop varieties (Dr Hassan El-Warreh).

Plant Protection Association 10

Head: Dr Bassam Bayaa
Projects: Food legumes entomology, Syrian fauna survey, insects (Professor Dr Ghazi El-Hariri); fungal plant diseases, olive wilt (Dr Bassam Bayaa).

Soil Science Department 11

Head: Dr M. Khaldon Dormoch
Projects: Soil-plant-water relationships.

TAIWAN

Academia Sinica* 1

Address: Nankang, Taipei, 115
Status: Official research centre

INSTITUTE OF BOTANY 2

Telephone: 02 761 6050
Director: Dr Hong-Pang Wu
Graduate research staff: 34
Activities: Plant sciences, applied botanical studies, promoting academic cooperations in botanical sciences both domestically and internationally.
Publications: Annual report; botanical bulletin, semi-annual; monographs.
Projects: Research on rice plants; cell and tissue culture; microorganisms; plant pathology; plant physiology; biochemistry; molecular biology; ecology.

Chia-yi Agricultural 3
Experiment Station

Address: 2 Min-Chung Road, Chia-yi
Telephone: (052) 271341
Status: Official research centre
Director: L. Li
Departments: Agronomy, W.L. Chang; horticulture, C.K. Chu; plant protection, C.H. Cheng
Graduate research staff: 6
Activities: Improvement of sweet potato varieties, cultural practices, quality improvement, product utilization; improvement of Indica rice varieties, cultural practices, quality improvement; collection, conservation and cataloguing of tropical and subtropical fruit germplasm; variety improvement and cultural practices of subtropical fruits, particularly citrus; ecology and control of diseases and insect pests of rice, upland crops, tropical and subtropical fruits.

Publications: Annual report.
Projects: Breeding sweet potatoes for high starch, carotene, and protein contents and resistance to major diseases (L. Li, H. Wang); breeding methods, selection criteria, and field plot techniques for major characteristics of sweet potato varieties (L. Li, C.H. Liao); breeding rices for resistance to major diseases, insect pests, high yielding ability (W.L. Chang, L.C. Chen); breeding rices for tolerance to low temperature and acceptable grain quality (W.L. Chang, S.C. Yang); breeding citrus, pineapple, passion fruit and other important tropical and subtropical fruits for better horticultural characteristics; collection and conservation of germplasm of tropical and subtropical fruits and observation on the adaptabilities of newly-introduced fruits (C.K. Chu, C.T. Lin, Ch.C. Chang, C.C. Chang, A.S. Hwang); resistance of rice varieties to major diseases and insect pests (C.H. Cheng, W.H. Tsai); integrated control of major diseases and insect pests on rice and other important upland and fruit crops (C.H. Cheng, W.H. Tsai, P.J. Ann, M.L. Chung).

Food Industry Research 4
and Development Institute

Address: PO Box 246, Hsinchu, 300
Telephone: (035) 223191/3
Telex: FOODEVELOP
Status: Independent research centre
Director: Dr Paul C. Ma
Graduate research staff: 76
Activities: Carrying out basic and applied research in the field of science and technology to help the local food industries to develop new products; the introduction of new processing equipment and techniques to improve the production efficiency and the quality of existing processed products; organizing demonstrations and training programmes to upgrade the level of processing technology; disseminating information on recent

developments in food science and technology; carrying out feasibility studies to advise on product development and new processing procedures.

Publications: Research report; *Food Industry*, monthly; seminar reports.

Projects: Can corrosion (Dr W.C. Tsai); double seam improvement (Y. T. Lin); improvement and extension of food processing machinery; bamboo shoots peeling system, automatic control of sterilization process (J.S. Huang); simple detection of frying oil deterioration (Dr T.S. Chu); palm oil application (Dr T.Y. Liu); cold solvent extraction of rice bran oil and wax separation by pilot scale (S.C. Chang); quality of edible oils and their products (Y.H. Fu); flavour chemistry of Chinese food, flavour complex from spent chicken (L.B. Wu).

FOOD SCIENCE DEPARTMENT   5

Head: Dr Wei-Chung Tsai

FOOD TECHNOLOGY DEPARTMENT   6

Head: Chung Ping Huang

National Chung Hsing University   7

Address: Taichung, 400
Telephone: 225911
Status: Educational establishment with r&d capability

COLLEGE OF AGRICULTURE   8

Address: 250 Kuokung Road, Taichung, 400
Dean: Dr Y.H. Han
Graduate research staff: 235
Activities: This college incorporates 14 departments and 10 research institutes. For 1980-81 year, there were 187 projects and US $1.7m of funding from governmental and private agencies. The subject matters include natural resources, plant production, animal production, agricultural engineering and building, food science, and forestry and forest products.

Agricultural Economics Research Institute   9

Head: Dr Chaur-Shyan Lee

Agricultural Machinery Department   10

Head: Dr Yang-Ren Hwang

Agricultural Marketing Department   11

Head: Dr Yi-Chung Kuo

Animal Husbandry Research Institute   12

Head: Dr Hsi-Shan Chang

Entomology Research Institute   13

Head: Dr Chaim-Ing T. Shin

Food Crops Research Institute   14

Head: Dr Fu-Sheng Thseng

Food Science Research Institute   15

Head: Dr Tsu-Han Lai

Forestry Research Institute   16

Head: Dr Shaw-Lin Lo

Horticulture Research Institute   17

Head: Dr Kuo-Chuan Lee

Plant Pathology Research Institute   18

Head: Dr Shih-Tien Hsu

Soil Science Research Institute   19

Head: Dr Tzo-Chuan Juang

Veterinary Medicine Department   20

Head: Dr Happy K. Hsien

Water and Soil Conservation Research Institute   21

Head: Dr Cheng-Ping Yen

Tainan District Agricultural Improvement Station   22

Address: 480 Tong-men Road, Tainan
Telephone: (062) 379111
Status: Official research centre
Director: Dr C.C. Tu
Graduate research staff: 36
Activities: Breeding and multiplication of new varieties, experimentation of the crop improvement, varieties and cultural practice, farm machinery improvement and extension, experimentation on farming and farm management improvement for special regions, research into soil and fertilizers and rendering of services required; research in plant protection and pest and disease forecasting, radio and television programmes; local advisor and farmer training; home economics, regional agricultural development advice and assistance.
Publications: Annual report, research bulletins.

AGRICULTURAL EXTENSION DEPARTMENT   23

Head: H.C. Lu
Projects: Rural economics (H.C. Lu); agricultural extension education and home economics (S.S. Chang, Y.P. Lee); farm management (J. Lee).

CHIAYI BRANCH STATION   24

Head: S.L. Chuang
Project: Rice breeding (S.L. Chuang).

CROP ENVIRONMENT DEPARTMENT   25

Head: Yung-Hsiung Chang
Project: Beneficial effects of Endomycorrhizae on soyabean (Y.H. Cheng).

CROP IMPROVEMENT DEPARTMENT   26

Head: Ching-Sheng Hsu
Projects: Peanut breeding for early crops, large-podded and high-yielding variety (Ching-Sheng Hsu); sunflower varietal improvement-F1 hybrid and oil products (S.Y. Hseih); breeding of Muskmelon hybrids (Cucumis melo) (S.L. Hwang); screening of resistant lines and bio-control of asparagus fusarium wilt; effects of rotation systems on the population change of Rhizoctonia solani in Chianan district (Dr C.C. Tu).

HSINHUA BRANCH STATION   27

Head: M.T. Chang
Project: Management of orchards, integrated crop-livestock farming on sloping area (M.T. Chang).

PUTZE BRANCH STATION   28

Head: S.C. Chang
Project: Breeding and cultivation of corn and sorghum (S.C. Chang).

YICHU BRANCH STATION   29

Head: Fu-Yao Kuo
Project: Asparagus breeding and cultivation (Fu-Yao Kuo).

Taiwan Agricultural Research Institute   30

– TARI
Address: 189 Chung Cheng Road, Wufeng, Taichung
Telephone: (043) 302301
Status: Official research centre
Director: Hsiung Wan
Graduate research staff: 191
Activities: Yield increase, quality improvement, reduction of loss from pests and diseases of crops; breeding; crop husbandry; soil science; plant protection on rice, sweet potato, corn, peanut, soyabean, various vegetables, and subtropical and tropical fruits, such as cabbage, cauliflower, beans, radish, cucumber, gourds, eggplant, pepper, etc, and mango, papaya, guava, pineapple, banana, lichi, longan, citrus, grape, grapefruit, pears and peaches.
Publications: Journal of Agricultural Research of China, quarterly.

AGRICULTURAL CHEMISTRY DEPARTMENT   31

Head: C.F. Lin
Projects: Soil survey in Taiwan (C.F. Lin); microflora in the root-sphere of rice for nitrogen fixation (S.C. Lin).

AGRICULTURAL ENGINEERING DEPARTMENT   32

Head: C.C. Yen

AGRONOMY DEPARTMENT   33

Head: S.C. Hsieh

APPLIED ZOOLOGY DEPARTMENT   34

Head: S.C. Chiu
Projects: Insect fauna in Taiwan, biological control of diamondback moth and vectors on vegetables and fruit trees.

CHIA-YI EXPERIMENTAL STATION   35

Head: L. Li

FENGSHAN HORTICULTURAL   36
EXPERIMENTAL STATION

Head: T.F. Shen

HORTICULTURE DEPARTMENT   37

Head: M.J. Yen
Projects: Development of crop varieties, including rice, corn, peanuts, soyabean, sweet potato, fruit trees and vegetables for high-yielding and pest resistance; hybrid rice development, control of citrus pests (C.H. Huang); potato production by true seeds (S.J. Tsao); pollen and tip culture for rice, asparagus, sweet potato and orchid (H.S. Tsuy); edible mushrooms (K.J. Hu).

PLANT PATHOLOGY DEPARTMENT   38

Head: Y.S. Lin

Taiwan Fertilizer Company   39
Limited *

FERTILIZER RESEARCH DEPARTMENT   40

Address: 63 Yung-Chen Road, Yung-Ho, Taipei, 234
Telephone: (02) 9212670; 9245728
Telex: 5142
Status: Research centre within an industrial company
Director: Dr Ching-Tsun Chiang
Sections: Fertilizer technology, Yu-Yen Liao; soil science, Horng-Shyang Chai; plant nutrition, Dao-Nan Jian
Graduate research staff: 20
Activities: Manufacture and study of new kinds or new formulae of fertilizers and other agriculture materials; formulation of proper NPK compound fertilizers for different crops; fertilizer application technique; soil science and plant nutrition.
Publications: Technical bulletin, annual report.
Projects: Availability and utilization of phosphate rock

and low-acidified phosphorous fertilizer (Ching-Tsun Chiang, Horng-Shyang Chai, Kuang-Sheng Yang); effect of supplemental slag, fly ash and bottom ash of industrial waste material on soil conditioner and fertilizer utilization (Ching-Tsun Chiang, Horng-Shyang Chai, Hwa-Wei Sun); N-15 studies on the transformation of various nitrogen fertilizers in relation to rice uptake (Horng-Shyang Chai, Jong Ching).

Taiwan Forestry Research   41
Institute

Address: 53 Nan-Hai Road, Taipei
Telephone: (02) 3817107
Status: Official research centre
Director: Shen-Chen Liu
Graduate research staff: 107
Activities: Forest sciences in Taiwan, and to help improve applied techniques for forest industries and forest management, by covering a wide range of research dealing with forest biology, the selection of trees, silviculture practices, forest breeding, forest economics, watershed management, utilities and properties of wood and bamboo, forest chemistry, wood cellulose and other forest oriented subjects.
Publications: Bulletins, irregular.

FOREST BIOLOGY DIVISION   42

Head: T.Y. Chen
Projects: Powdery mildew fungi on forest plants in Taiwan (H.J. Hsieh); forest tree seed biology (Y.L. Chung).

FOREST CHEMISTRY DIVISION   43

Head: T.P. Ma
Projects: Essential oil of Ylang Ylang in Taiwan (J.C. Shieh); effectiveness of chemicals in preventing attack on bamboo products by sap-stain (S.F. Wang); forest soils (W.E. Cheng).

FOREST ECONOMICS DIVISION   44

Head: S.C. Liu
Projects: Growth and yield estimation for China fir (Cunningham lameolata) in Taiwan (S.C. Liu); impact of environmental pollution on urban forests (K.C. Lin); shelter-belt and windbreak establishment (W.H. Kan).

FOREST EXTENSION DIVISION   45

Head: Z.J. Kang
Projects: Introduction of exotic Acacia species and symbiotic nitrogen fixation (T.Y. Cheng, J.C. Yang); exotic tropical pine introduction (J.C. Yang).

FOREST MANAGEMENT DIVISION   46

Head: L.P. Hung
Projects: Selective cutting for managing natural forest stands of cypress in Taiwan, effects of thinning in the Luanta-fir plantation (L.P. Hung); height growth and site index curves for mahogany (Swietenia mahagoni) in Taiwan (M.H. Lin).

FOREST UTILIZATION DIVISION   47

Head: T.P. Ma
Projects: Effect of moisture content and carbonization of bamboo material on mechanical and processing properties of laminated bamboo (T.P. Ma); properties and utilization of wood from planted, fast-grown Leucaena (J.L. Tang); minor forest products (T.W. Hu); wood consumption prospects in Taiwan (I.A. Jen).

WATERSHED MANAGEMENT DIVISION   48

Head: B.Y. Yang
Projects: Effect of land use on stream water quality (H.B. King); water relations of native tree species (H.H. Chung).

WOOD CELLULOSE DIVISION   49

Head: Y.C. Ku
Projects: Semichemical pulping of Leucaena wood, rice straw pulping.

SILVICULTURE DIVISION   50

Head: T.W. Hu
Projects: Silviculture studies of economically important tree species (T.W. Hu); cultivation of fast-growing tree species (W.C. Shi); genetic improvement of forest trees (C.M. Lu); bamboo cultivation (W.C. Lin).

Taiwan Sugar Research Institute   51

Address: 54 Sheng Chan Road, Tainan
Telephone: (062) 378101
Affiliation: Taiwan Sugar Corporation
Status: Official research centre
Director: Dr S.C. Shih
Graduate research staff: 128
Activities: Sugarcane breeding, agricultural improvement on sugarcane cultivation, agricultural engineering, agricultural economics; plant nutrition, soil science, plant protection; sugar processing and engineering, by-products utilization.
Publications: Annual report, quarterly report.

AGRONOMY DEPARTMENT   52

Head: C.C. Lo
Projects: Ratoon crop improvement (P.C. Yang, C.C. Wei); high-water-table surface drainage system (P.L. Wang, Y.T. Chang); Sha-Lun plantation water resource development and utilization (Y.T. Fang).

BY-PRODUCT UTILIZATION   53

Head: S.L. Sang
Projects: Glucose isomerase immobilization in whole cells for high fructose syrup production (C.L. Lai); microbial gum improvement (W.P. Chen); microbial protein production from bagasse (L.H. Wang).

PLANT NUTRITION DEPARTMENT   54

Head: C.C. Wang
Projects: Utilization of sugar and nitrogen by sugarcane cells in suspension culture (W.H. Chen); amino acid distribution and variation in cane plants (T.T. Yang); residue effect and recovery of long-term phosphorus and potassium fertilizations on sugarcane (C.C. Wang); sugarcane nitrogen efficacy studies, using N^{15} (Y.Y. Chan).

PLANT PROTECTION DEPARTMENT   55

Head: Y.S. Pan
Projects: Herbicidual residue in sugarcane fields (H.J. Yeh); effect of soil inoculant on sugarcane growth (C.H. Chang); methods of screening sugarcane varieties for disease resistance (C.S. Lee); chlorotic streak disease (C.T. Chen); biological insect pest control (W.Y. Cheng); sugarcane resistance to moth borers (Y.S. Pan).

SUGAR TECHNOLOGY DEPARTMENT   56

Head: H.T. Cheng
Projects: Sugarcane mini-harvester development (T.W. Lin, Y.M. Lei); sugar technology improvement and development, ultrafiltration and electrodialysis juice clarification (H.T. Cheng); new type clarifier development (C.H. Chen).

SUGARCANE BREEDING DEPARTMENT   57

Head: I.S. Shen
Projects: Breeding of prominent sugarcane varieties.

TANZANIA

Kituo Cha Utafiti wa Uzalishaji Mifugo   1

[Livestock Production Research Institute]
Address: Private Bag, Mpwapwa, Dodoma Region
Telephone: MPWAPWA 21
Status: Official research centre
Director: Peter M.J. Katyega
Departments: Animal breeding and genetics, D.F. Msuya; animal nutrition, A.V. Goodchild; pasture, R.N. Mero; disease control, S.M. Das; national livestock registry, S.N. Bitende
Graduate research staff: 14
Activities: Livestock production; animal breeding and genetics; animal nutrition; animal management; pastures; coordination of animal production research in other livestock research centres (Malya, Tanga, West Kilimanjaro, Kongwa Pasture Research Station).
Publications: Progressive Stockman Periodical, quarterly; annual report.
Projects: Breeds and breeding systems in commercial beef production, development of dual purpose Zebu type cattle, hybrid crosslines and backcross of Mpwapwa cattle for dairy purposes (D.F. Msuya); Tanzania shorthorn Zebu breeding, effect of an anobolic growth implant Ralgro on steers (A. Masaoa); Serena-Shorghum dual purpose crops, molasses for growing pigs, early wet season liveweight loss (A.V. Goodchild); stall feeding trials, sheep digestibility (J. Ndosa); evaluation of a range monitoring programme for Tanzania (R.N. Mero).

Malya Livestock Research Centre   2

Address: PO Box 3021, Malya
Telephone: Malya 15
Affiliation: Tanzania Livestock Research Organization
Status: Official research centre
Officer-in-charge: S. Busungu
Departments: Cattle recording, P.K. Saku; sheep and goat recording, S. Rushoke; dairying, B. Mashalla; calf rearing, E.L. Minja; animal health, J.S. Kiaratu
Graduate research staff: 1
Activities: Animal production; livestock breeding, husbandry and nutrition.
Publications: Annual report.
Projects: Development and use of Mpwapwa cattle for milk and meat production; breeding, selection and management of Borna beef cattle; breeding of Tanzania shorthorn Zebu cattle; development of meat producing goats; Blackhead Persian sheep, meat and fat production.

Ministry of Agriculture   3

Address: Pamba House, PO Box 9071, Dar es Salaam
Status: Official research centre

RESEARCH DIVISION   4

Head: Dr J.M. Liwenga
Sections: Manpower development, V.F. Malima; bilateral projects, E.N. Ruzika; plant protection, A.H. Shayo; documentation, Dr Julie Woolman

Tanzania Agricultural Research Organization   5

– TARO
Address: PO Box 9761, Dar es Salaam
Affiliation: Ministry of Agriculture
Status: Official research centre
Director: Dr J. Kasembe
Graduate research staff: 140
Activities: Basic and applied interdisciplinary research into all aspects of crops, soil and farming systems in Tanzania.
Publications: Cotton research annual report.
Projects: Cotton (G. Jones); coffee (D.L. Kessy); sisal (L.M.E. Muliahela); sugar (F.E. Mbemba); cashewnuts (Dr G.B. Assi); oilseeds (P. Naylor); tea (J.S.L. Machaga); tobacco (J.D. Madulu); coconuts (Dr D. Speidel); roots and tubers (M.M. Msabaha); wheat (P. Fehr, A.S. Mosha; maize (Dr J. Deutsch); rice (H. Ching'ang'a); sorghum and millet (C. Mushi); grain legumes (Dr M. Price); phaseolus beans (E.M. Koinange); horticulture (R.D. Kyamba); soil and fertilizers (J.K. Samki).

IFAKARA AGRICULTURAL RESEARCH INSTITUTE   6

Head: M.J. Kapalasula

ILONGA AGRICULTURAL RESEARCH INSTITUTE*   7

Address: PO Post Bag Ilonga, Kilosa, Morogoro
Telephone: (056) 49
Director: P.J. Makundi
Departments: Maize, Dr J. Deustch; grain legumes, Dr Mazo Price; sorghums/millets, C.S. Mushi; agricultural economics and farming systems research, P.J. Makundi; agricultural entomology, Dr D. Lakhani
Graduate research staff: 28
Activities: Plant breeding, disease and pest resistance; maize, grain legumes including cowpease, soyabeans, rice, sorghums and millets; crop husbandry; agricultural economics and farming systems research, production economics; dissemination of technologies to the farming community.
Publications: Annual report.
Projects: Land development and irrigation (E.E. Mlay); maize agronomy (D.M. Mwanjali); maize breeding (Z. Mnduruma); grain legume breeding (Dr B.B. Singh); grain legume agronomy (F.Z. Machange); cotton breeding (Dr D. Lakhani); entomology (Dr D. Hackett, J.C.B. Kabissa), armyworm programme (Dr P.J. Marrett).

LYAMUNGU AGRICULTURAL RESEARCH INSTITUTE   8

Address: PO Box 3004, Moshi
Head: J. Kondera

MARIKITANDA AGRICULTURAL RESEARCH INSTITUTE   9

Head: J.S.E. Machaga

MARUKU AGRICULTURAL RESEARCH INSTITUTE   10

Address: 127, Bukoba, Kagera Region
Telephone: 533
Director: Dunstan Ndamugoba
Departments: Banana, C.N.C. Cumming; coffee, M. Hamisi; tea, D. Ndamugoba; legumes, I.K.K. Ndamucoba; cereals, K.T. Kalemela; soil chemistry, J. Nyonyi
Graduate research staff: 4
Activities: Plant protection, sustained high yields; fertilizers, varieties, management; yields, clones, fertilizers, quality; plant protection, selection, management and spacing; soil and leaf sample testing, mineral contents of soil and plant tissues.
Publications: Annual report, quarterly report.
Projects: Banana crop protection (C.N.C. Cumming); coffee fertilizer trials (D.L. Kessy); tea tipping, fertilizers and environment (D. Ndamugoba); legume variety trials (I.K.K. Ndamucoba); cereals management and variety trials (K.T. Kalemela); soil chemistry (J. Nyonyi).

MLINGANO AGRICULTURAL RESEARCH INSTITUTE   11

Address: Private Bag, Ngomeni, Tanga
Head: J.K. Samki

NALIENDELE AGRICULTURAL RESEARCH INSTITUTE   12

Address: PO Box 509, Mtwara
Head: S.H. Shomari

SELIAN AGRICULTURAL RESEARCH INSTITUTE   13

Head: A.S. Mosha

TUMBI AGRICULTURAL RESEARCH INSTITUTE 14

Address: PO Box 306, Tabora
Head: J.D. Madulu

UKIRIGURU AGRICULTURAL RESEARCH INSTITUTE 15

Address: 1433, Mwanza
Telephone: (068) 40596
Director: A.H. Kishimba
Departments: Cotton breeding, George Jones, M.P.K. Kapinge; cotton pathology, F.S. Ngulu; cotton entomology, A. Treen, B. Nyambo; soil science, I. Mkamba; maize and legumes, H.B. Akonaay; root crops, A.K. Nyango, J. Ndikumana; fibre laboratory, L. Zikaukuba
Graduate research staff: 12
Activities: Plant protection, crop husbandry, cotton, sorghum and millets, maize and legumes, root crops. Development of high yield pest and disease resistance and high-quality yarn cotton varieties; soil science, soil fertility, moisture conservation.
Publications: Annual report.

Tanzania Livestock Research Organization 16

– TALIRO
Address: Box 73, Tabora
Telephone: (062) 2293
Affiliation: Central Veterinary Laboratory
Status: Official research centre
Director: Dr R.K. Bhalikulije
Departments: Histopathology, parasitology, Dr R.K. Bhalikulije; microbiology, pathology, Dr G.M. Nsengwa
Graduate research staff: 2
Activities: Veterinary investigations; animal disease survey; livestock disease intensity; animal disease control; veterinary public health, water supply, animal products, animal feeds; veterinary medicine drug use.
Projects: Fascioliasis epidemiology in western zone, tick control (Dr R.K. Bhalikulije); salmonellosis survey, brucellosis survey (Dr G.M. Nsengwa).

Tanzania Pesticides Research Institute 17

– TPRI
Address: Box 3024, Arusha
Telephone: (057) 3557
Affiliation: Ministry of Agriculture
Status: Official research centre
Director General: Dr B. Simon
Divisions: Agricultural research; medical and veterinary research; chemical and physical research; National Herbarium; plant quarantine section.
Graduate research staff: 17
Activities: Research on the fundamental aspects of pesticide application and behaviour in relation to the control of tropical pests by both ground and aerial spraying techniques.
Publications: Annual research reports.
Projects: Agricultural entomology (J. Saidi); plant pathology (J. Bujulu); bird pests (D. Manyanza); rodent pests (T. Mbise); tsetse entomology (C. Muangirwa); mosquito entomology (H. Qorro); malacology (Dr J. Nguma); ticks (acaricides) (Dr S. Mchinja); pesticide toxicology (A. Ngowi); chemistry (soils) (A. Moshi); physics (Mr Ketau); botany (E. Ritoine); environmental pollution (J. Akhabuhaya).

Tanzania Silvicultural Research Institute 18

Address: PO Box 95, Lushoto, Tanga
Telephone: (053) 32
Affiliation: Ministry of Natural Resources and Tourism
Status: Official research centre
Silviculturist: L. Nshubemuki
Departments: Soils and ecology, S.M. Maliondo; seeds and nurseries, I.M. Shchaghilo; tree breeding, J.A. Mushi; plantation management, S.T. Mwihomeke; botany and herbarium, C.K. Ruffo; indigenous forests, A.G. Mugasha; dry area afforestation, L. Nshubemuki; forest protection, E.N. Mshiu
Graduate research staff: 20
Activities: Forestry and forest product studies.
Publications: Annual report; technical and research notes.
Projects: Interpretation of soil and climatic data in relation to afforestation (S.M. Maliondo); provenance and progeny trials including the establishment of seed orchards (J.M. Wanyangha); tending of plantations, spacing, thinning, yield table preparation (A.G. Kalaghe); edible plants from Tanzania indigenous forests (C.K. Ruffo); silviculture of important Tanzanian indigenous species (A.G. Mugasha); afforestation techniques in dry areas (L. Nshubemuki); biological control of forest pests (E.N. Mshiu).

Tsetse Research Institute 19

Address: PO Box 1026, Tanga
Telephone: 053 2577
Affiliation: Tanzania Livestock Research Organization
Status: Official research centre
Director: C.S. Tarimo
Graduate research staff: 5
Activities: Tsetse control by the sterile insect release method, African trypanosomiasis elimination; livestock production improvement; ecology, population dynamics, distribution, control techniques of different Glossina species.
Publications: Annual reports, quarterly reports, monthly reports.
Projects: Sterile insect release method against the Tsetse fly Glossina morsitans morsitans in an integrated tsetse control programme.

Uyole Agricultural Centre 20

– SKU
Address: PO Box 400, Mbeya
Telephone: (065) 2116; 2117
Affiliation: Ministry of Agriculture
Status: Official research centre
Director: vacant

RESEARCH INSTITUTE 21

Chief research officer: Dr G.H. Semuguruka
Graduate research staff: 32
Activities: Development of high-yielding varieties with disease and insect pest resistance, management practices, adaptive research; dairy cattle improvement in high altitudes of southern Tanzania, breeding genetics, husbandry, nutrition, pasture and forage improvement; agricultural engineering, farm structures, implements, machinery, village technology, draught animals; agricultural economics and rural sociology.
Publications: Annual report.
Projects: Maize improvement for high altitudes (A.E.M. Temu); rice (vacant); sorghum and millet (R.O.F. Mwambene); wheat and triticale (E. Morris); barley breeding (G.A. Mrimi); legumes (C. Madata); oilseeds (C.M. Mayona); roots and tubers, potato improvement (C. Macha); horticulture, vegetable improvement (E.E. Meela); horticulture, temperate and tropical fruit (A. Nyomora); coffee (vacant); pyrethrum (E.Z. Manang); plant protection, pathology (L.H.T. Nsemwa); plant protection, entomology (G.A. Mallya); plant protection, nematology (I. Swai); soil survey (S.P. Msaki); soil fertility (J.A. Kamasho); soil physics (G. Ley); soil science (H.A. Chande).

Agroeconomics Department 22

Head: T.N. Kirway

Agroengineering Department 23

Head: E. Kwiligwa

Agroextension Department 24

Head: C.K.J. Ponjee

Crops Department 25

Head: R.O.F. Mwambene

Food Technology Department 26

Head: E. Skytta

Livestock Department 27

Head: Dr G. Madata

Veterinary Investigation Centre 28

Address: Private Bag, Mpwapwa
Telephone: 81
Affiliation: Ministry of Livestock Development
Status: Official research centre.
Officer-in-charge: Dr J.I. Kitalyi
Graduate research staff: 2
Activities: Veterinary medicine, disease surveys and diagnosis.
Publications: Annual report, monthly reports.

Veterinary Investigation Centre, Arusha 29

Address: PO Box 1068, Arusha
Telephone: 3566
Affiliation: Tanzania Livestock Research Organization
Status: Official research centre
Officer-in-charge: Dr J.F.C. Nyange
Departments: Pathology and histology, Dr J.F.C. Nyange; bacteriology and serology, Dr M.M.M. Otaru; parasitology, Dr A.N. Mbise
Graduate research staff: 3
Activities: Animal disease, parasitology, serological survey of viral diseases, bacteriology.
Publications: Annual report, monthly reports.

THAILAND

Agricultural Research and Extension Institute 1

Address: Nakornsrithammaraj Agricultural Campus, Tungsong, Nakornsrithammaraj
Telephone: 411144
Telex: Nakornsrithammaraj Agricultural Campus
Affiliation: Institute of Technology and Vocational Education
Status: Official research centre
Director: Jim Nuuyimsai
Departments: Plant science, Sutin Tongnua-orn; animal science, Panya Netiwattanapong; farm mechanics, Ruengrit Pantong; related subjects, Sodsri Buchakorn; agricultural education, Pong Suwannarat
Graduate research staff: 12
Activities: Production and quality of tobacco species; utilization of local feedstuffs for dairy cattle.
Publications: Saiyai Journal, bimonthly.
Projects: Tobacco production (Preecha Aramuit); dairy husbandry (Shawlid Panichutra).

Department of Agriculture 2

Address: Phahonyothin Road, Bangkhen, Bangkok, 9
Telephone: 5790581; 5799151-8
Affiliation: Ministry of Agriculture and Cooperatives
Status: Official research centre
Director-General: Phaderm Titatarn
Graduate research staff: 1 390
Activities: Agricultural research on crop production, breeding, agronomy, horticulture, sericulture, rubber, plant pathology, entomology, seed technology, crop quality and standard control agricultural chemistry, post-harvest pest control, farm machinery development, seed multiplication, plant material and chemical import and export regulation.
Publications: Annual report.

See also entry for: Rubber Research Centre
Projects: Rice (Phairoth Sopanarath); rubber (Dr Slearmlarp Wasuwat); sericulture (Aree Keo-Ngarm); field crops (Dr Arwooth Na Lampang); horticulture (Pairoj Polprasid); national agriculture (Phaderm Titatarn); pesticide chemistry (Dr Prayoon Deema); agricultural engineering survey (Samnao Rugtrakul); agricultural chemistry analysis (Nongyow Thongton).

AGRICULTURAL CHEMISTRY 3
DIVISION

Address: Ngarnwongwarn, Bangkok
Telephone: 5790530
Director: Nongyow Thongtan
Graduate research staff: 116
Activities: Increased crop production, soil fertility, nutrition, soil chemistry; isotope use in agricultural research; alternative energy sources, agricultural engineering; crop quality and yield, agro-industrial production, food science, plant nutrition, post harvest; fertilizer and pesticide analysis, analytical chemistry.
Publications: Annual reports from the branches.
Projects: Soil chemistry, soil fertility and irrigation water quality for economic crops in Thailand (Wisit Cholitkul); fertilizer efficiency and root distribution of economic crops by nuclear technology (Patoom Snitwongse); production of vegetable oil and crop products (Vimolsri Devapalin); biogas production as alternative source of energy (Revadee Deemark); effect of hormones on yield and quality of economic crops (Dr Tongchai Kangvanvongse); plant nutrients and food quality of crop yield (Ladda Pakinnaka); characteristics of pesticides and analysis methods (Yobon Yingchol); characteristics of fertilizers and analysis methods (Nantapan Ratanatam).

Agricultural Chemistry Branch   4

Head: Dr Tongchai Kangvanvongse

Chemical Plants and Products   5
Research Branch

Head: Vimolsri Devapalin

Fertilizer Analysis Branch   6

Head: Nantapan Ratanatam

Fertilizer Research Branch   7

Head: Revadee Deemark

Local Laboratory Branch   8

Head: Prakong Chitasombati

Nuclear Research in Agriculture   9
Branch

Head: Patoom Snitwongse

Pesticide Analysis Branch   10

Head: Yobon Yingchol

Plant Analysis Branch   11

Head: Ladda Pakinnaka

Soil and Water Analysis Branch   12

Head: Thanom Dao-nyarm

Soil Chemistry and Fertility Branch   13

Head: Wisit Cholitkul

AGRICULTURAL ENGINEERING   14
DIVISION

Director: Samnao Rugtrakul
Graduate research staff: 40
Activities: Research and development of agricultural machines for tillage, planting, transplanting, harvesting and post-harvesting.
Publications: Annual reports.
Projects: Animal plough improvement, charcoal gas for running gasoline engines (Chanchai Rojanasaroj); field conditions for Chinese type transplanter, manual and machine types (Chalermchai Suksri); rice reaper development (Suraweth Krisanasreni).

Agricultural Extension Department   15

Head: Yookti Sarikaphuti

Agriculture Department   16

Head: Pradern Tititanon

Fishery Department   17

Head: Surang Charernpol

Forestry Department   18

Head: Pong Sono

Land Development Department   19

Head: Anant Komes

Livestock Development Department   20

Director-General: Dr Tim Bhannasiri
Sections: Disease control, Dr Piya Aranyakananda; veterinary service, Dr Prakan Samittanan; veterinary research, Dr Por Chindawanig; veterinary biology, Dr Pinit Suphavilai; animal husbandry, Prasoet Yutthawisut; animal nutrition, Phunudet Sutthat; artificial insemination, Dr Phat Sarikaphuti; livestock extension, Dr Wichian Khamnuanthong; feed quality control, Dr Yuanta Phruksarat; livestock development project, Dr Suntharaphon Rattanadilok Na Phuket
Graduate research staff: 15
Activities: Animal husbandry, veterinary medicine; increase of the livestock population for meat and dairy consumption and export; vaccine production for cattle, buffaloes, pigs, chickens and ducks; animal feed crop development.
Projects: Swamp buffalo semen characteristics (Dr Bunyawat Sanitwong Na Ayudhaya); deep frozen boar semen (Dr Aphon Bunkhum); dairy cattle progeny testing (Dr Prasoet Songsasen); pasteurella multocida serotype studies, swine erysipelas survey (Dr Saman Pipitkul); chronic respiratory disease (Dr Praphat Neramitmarnsuk); pathogenicity of fungi from animal feed (Dr Ladda Moolika); potency field trial of duck plague vaccine (Dr Urasi Tantaswasdi); egg drop syndrome virus in chickens, swine pseudorabies vaccine production (Dr Agree Sapcharoen); sarcocysts in cattle (Dr Tasanee Chompoochantra).

Royal Irrigation Department   21

Head: Sunthorn Runglek

AGRICULTURAL REGULATORY DIVISION 22

Head: Adul Worawisitthumrong

ENTOMOLOGY AND ZOOLOGY DIVISION 23

Head: Montri Rumakom

FIELD CROPS DIVISION 24

Director: Arwooth Na Lampang
Graduate research staff: 320
Activities: Breeding; cultural practice; soil fertility; seed technology research on major field crops, corn, sorghum, jute, kenaf, cotton, sugarcane, tobacco (native cultivar), soyabean, mung, peanut, castor, sunflower, cassava and some minor field crops.
Publications: Annual report.

Corn and Sorghum Branch 25

Head: Vichitr Benjasil

Cotton Branch 26

Head: Chookiat Ithrat

Fibre Crops Branch 27

Head: Navarat Sermsri

Oil Seed Crops Branch 28

Head: Amnuay Thong dee

Root Crops Branch 29

Head: Sophon Sindhuprama

Seed Production Branch 30

Head: Sripark Ves-Urai

Seed Technology Branch 31

Head: Sanit Kittikorn

Soil and Fertilizer Branch 32

Head: Samrit Chaiwanakupt

Sugarcane and Miscellaneous Crops Branch 33

Head: Prida Chatikavanij

HORTICULTURAL DIVISION 34

Telephone: 5790583
Director: Pairoj Polprasid
Graduate research staff: 99
Activities: Variety improvement; variety trials; seed production; cultural practice; seed technology; propagation and multiplication; post harvest soil and fertilizer usage.
Publications: Abstract of Horticulture Research, annual.
Projects: Cashew nut research and development (Pan Maliwan); cocoa development (Wit Suwanawut); coconut development (Maliwan Rattanapruk); viticulture for wine production (Prakit Duangpikool).

PLANT PATHOLOGY AND MICROBIOLOGY DIVISION 35

Director: Anong Chandrasrikul
Graduate research staff: 175
Activities: Plant protection; development of effective, economical, and nonhazardous control measures for diseases of major crops in Thailand; utilization and cultivation of beneficial microorganisms; phytopathology.
Publications: Annual reports, progress reports.
Projects: Rice diseases (Somkid Disthaporn); field crop diseases (Niyom Chew-Chin); fruit crop diseases (Kachonsak Bhavakul); post harvest crop diseases (Dara Buangsuwon); vegetables and ornamental crop diseases (Laksna Wannapee); fibre crops diseases (Sompark Siddhipongse); oil crop diseases (Preecha Surin); nematology (Charas Chunram); virology (Nuanchan Deema); mycology (Dr Piya Giatgong); applied microbiology (Phanthavee Puckdeedindan); bacteria and soil microbiology (Yenchai Vasuyat).

RICE DIVISION 36

Director: Phairote Sophanarat
Assistant director: Sermsak Awakul
Graduate research staff: 450
Activities: Collection and storage germplasm of rice cultivars; hybridization and selection to obtain new varieties: yield, disease and insect pest resistance; seed and cooking quality; maturity and plant types suitable for specific growing areas; soil problem and drought resistance; agronomy and cropping.
Publications: Annual report.
Projects: Lowland rice breeding (Sermsuk Awakul); genetics (Precha Khamphanonda); rice culture improvement, and seed storage (Darmkerng Chantarapanya; grain standard and quality (Sataporn Waitayakorn); disease and insect resistance (Prapas Weerapat); deep water rice (Chai Prechachart); rice fertilization (Chob Kanaresk); seed multiplication (Khean Kongjunteuk).

SERICULTURAL DIVISION   37

Director: Aree Kee-Ngarm
Graduate research staff: 38
Activities: Mulberry plant and silkworm rearing for silk production especially in the north east.

TECHNICAL DIVISION DEPARTMENT   38
OF AGRICULTURE

Director: Dr Winit Changsri
Graduate research staff: 34
Activities: Agronomy - cropping pattern, rice, field crops, soil science, plant physiology, horticulture; natural resources - remote sensing, agricultural resources, water resources; plant protection - weed control, rice, fieldcrops, horticulture crops.

Agronomy Branch   39

Head: Kleun Thongsan
Projects: Cropping patterns.

Botany Branch   40

Head: Umpai Yongboonkird
Projects: Plant introduction.

Climatology Branch   41

Head: Visut Chandharanysu
Projects: Crop environment.

Extension Branch   42

Head: Nopadol Kantitham

Irrigation Branch   43

Head: Dr Cherdchard Smitobol

Plant Pathology Branch   44

Head: Hansa Chakrabandu
Projects: Plant physiology.

Soil and Water Research Branch   45

Head: Pongpit Piyapongse
Projects: Remote sensing.

Soil Science Branch   46

Head: Dr Seri Sookakitch

Weed Science Branch   47

Head: Paitoon Kitipongse
Projects: Weed control.

Kasetsart University   48

Status: Educational establishment with r&d capability

RESEARCH AND DEVELOPMENT   49
INSTITUTE

– KURDI
Address: Paholyotin Road, Bangkok, 9
Telephone: 579 0032
Director: Dr Kamphol Adulavidhaya
Graduate research staff: 200
Activities: Agricultural research; plant and animal production; freshwater fisheries; food science; forestry and forest products; plant breeding, husbandry and protection of soyabean, tomato, cassava, sugarcane, cotton, rice, ornamental crops; poultry and buffalo breeding; husbandry and nutrition.
Publications: Annual report.
Projects: High-protein crop production; oil seed crop production; vegetable production; vegetable seed production; sugar cane production; root crops production; fibre crop production; cereal crop production; ornamental crop production. Animal beef production; poultry production; dairy cattle production; animal diseases and health. Forest management research and development; silviculture and wood properties; mangrove forests; aquaculture; agroindustry; agricultural business; agriculture and forest environment; fruit crops production and utilization; agricultural biology; applied engineering techniques for agriculture and industry development research.

Khon Kaen University *   50

Status: Educational establishment with r&d capability

FACULTY OF AGRICULTURE   51

Address: Khon Kaen
Telephone: 236199
Dean: Dr Kavi Chutikul
Departments: Agricultural economics, Dr Preeda
Prapruetchob; agricultural extension, Dr Adul Api-
nandra; agricultural products, Suwan Viratchkul; animal science, Dr Somchit Yodserani; entomology and
plant pathology, Dr Yongyoot Waikakul; plant science,
Dr Aran Patanothai; soil science, Dr Chaitat Pairin
Graduate research staff: 114
Activities: Agricultural technology and farming
systems for Northeast Thailand; plant production, field
crops (sorghum, peanut, grain, legumes), horticultural
crops (vegetables, tomatoes, tropical fruit crops); forage
crops, tropical forage legumes; animal production, water
buffaloes, beef and dairy cattle, swine, poultry, duck and
geese; fresh-water fisheries; food science, small scale
food processing; soil science, soil fertility, soil salinity.
Publications: Annual reports.
Projects: Pasture improvement (Dr Anake Top-
arkngarm); cropping system (Dr Terd Charoenwatana);
semi-arid crops (Dr Aran Patanothai); cassava nutrition
(Dr Sarote Kacharoen); home-processed legumes (Tip-
vanna Ngarmsak); shifting cultivation (Dr Chaitat
Pairin).

Ministry of Agriculture and   52
Cooperatives

Address: Rajadamnern Road, Bangkok, 2
Activities: While the Department of Agriculture is
mainly interested in crop production research, other
agricultural research activities are the responsibility of
different departments of the ministry.

Northeast Regional Office   53
of Agriculture and
Cooperatives

– NEROAC
Address: Tha Phra, Khon Kaen
Telephone: (043) 236704
Status: Official research centre
Director: Somchai Thamnoonragsa
Sections: Planning and evaluation; research and field
trial; operations; training and information; agricultural
data and information centre
Graduate research staff: 50
Activities: Planning, coordinating, implementing,
monitoring agricultural development programmes;
regional agricultural data and information is provided;

plant production includes crop husbandry, multiple
cropping, soil and water management, plant protection,
seed technology; farming system of field crops - rice,
cassava, kenaf, roselle, sugar cane, legumes and horticulture crops - vegetables, fruit trees; forestry; fresh-water
fisheries; livestock research, animal nutrition, genetic
(management), breeding, veterinary medicine, parasitology, disease diagnosis, farming systems of buffalo, cattle,
swine, layers, native chickens.
Publications: Annual report.
Projects: Integrated agricultural development in the
areas of middle and small scale irrigation ponds
(Tawinkarn Wangkahad); native chicken development
and improvement (Dr Sawat Thummabood); integrated
agricultural development in forestry - village projects
(Pittaya Namdang); integrated rural development in the
irrigated area of Nam Oon Dam (Chalermchai
Prasartsee); Tung Kula Rong Hai agricultural development (Kasem Chompoonutprapa); agricultural development in areas of electric water pumping stations (Dr
Waewchark Kongpolprom); dry-season cropping in irrigated areas (Prinya Srisawangwong); Tung Lui-Lai
development project (Withoon Wardthanabhuti);
agricultural development in areas around irrigable reservoirs (Nuttaporn Pensupa); northeast rainfed
agricultural development (Dr Utai Pisone); preliminary
crop development, (northeast region) (Somchai Thamnoonragsa); agricultural technology transfers on farming systems (Wisuthi Amaritsuthi); disease and pest
control of field crops, vegetables and fruit trees (Vilai
Prasartsee); vegetable development (Niyada Hornak);
fruit tree development (Sawai Surai); field crop development (Sitha Worajinda); livestock development (Dr
Worapong Suriyajantratong); soil analysis (Ladda Boon-
pakdee); soil and water conservation, soil improvement
(Paiboon Ratanaprateep); seed technology (Dr
Pornpimol Suriyajantratong); fresh-water fisheries
(Somdej Srikomut); forestry and forest products (Jesada
Leangcham); veterinary science (Dr Vichit Sukapes).

SUB STATION   54

Address: Huai-Si-Thon

Office of Agricultural Research and Extension   55

Address: PO Box Maejo, Prawn Street, Institute of Agricultural Technology Sansai, Chiangmai
Telephone: 234033; 234067
Director: Dr Boonrawd Supa-Udomlerk
Departments: Research, Channarong Doungsa-ard; extension, Sumeth Siriniran; training, Prapan Ostapan; animal and plant production, Thep Pongpanich
Graduate research staff: 13
Activities: Agricultural problems in rural areas, to contribute modern technology to farmers in cooperation with other government agencies, community education and training courses; soil science; plant breeding; plant protection; poultry breeding; livestock husbandry and nutrition; livestock physiology; horticulture; agricultural education.
Publications: Annual report, *Maejo Agricultural Journal*, three monthly.
Projects: Effect of Mimosa pigra contained in rations on growth and utilization of native chicken (Sakon Kaicom); detrimental assessment in soyabean caused by insect pests (Channarong Doungsa-ard); peanut breeding (Siriporn Loaterdpong); improvement of domestic chicken (Apichai Ratanavaraha); growth and flowering habit for flower seeds production (Songvut Phetpradap); Chinese geese feeding (Permsak Siriwan); effect of rapid fertility changes on school enrolment, labour supply and migration in Sansai community cooperative (Thep Pongpanich).

Phetchaburi Farm Demonstration Centre   56

Address: Amphur Thayang, Phetchaburi
Status: Official research centre
Director: Nareng Poolsilapa
Departments: Planning and follow-up, Chaluey Kongnoon; agriculture and cooperatives technique, Surapat Chiteng-art; project implementation and operation, Darm Tiensan; west region agricultural information, Prapa Tiensawai
Graduate research staff: 12
Activities: Agriculture of the west region; plant and animal production; fresh-water fisheries.
Publications: Annual report.
Projects: Agricultural integration planning (Chaluey Kongnoon); salt toleration of rice varieties (Surapat Chiteng-art); rice farm multiple cropping (Nareng Poolsilapa); dry season cotton growing after rice harvesting (Songkran Chaiyapant); soyabean production and cultural practices (Darm Tiensan); vegetable growing for increasing farm-income (Prapa Tiensawai); poultry production in the selected small farm (Sawasdee Tanawest); initiation of milk production in the west region (Traipant Kongkammert); farmer training programme in agriculture (Tular Teerachutimanant).

Pilot Farm Project, Huisithon   57

Address: Box 7, Kalasin
Telephone: (043) 811112
Status: Official research centre
Head: Prinya Srisavangwong
Graduate research staff: 9
Activities: Expansion of dry season agricultural production; analysis of economic, agronomic, environmental, social and other constraints affecting the various alternatives for dry season farming in the irrigated area; determination of comparative advantages of specific crops and farming systems.
Publications: Progress reports.
Projects: Irrigation agronomic research (Boripat Tanyaudom); irrigation agronomic trial and demonstration, training (Samer Jullavanich).

Royal Forest Department   58

Address: Phahonyothin Road, Bangkhen, Bangkok, 9
Telephone: 5791539
Affiliation: Ministry of Agriculture and Cooperatives
Status: Official research centre
Director-General: Pong Sono
Graduate research staff: 200
Activities: Seeking the most appropriate technique to manage the natural forest resource management; plantation technique improvement to obtain high returning yield; commercial tree species breeding for good quality seed source; lesser known species improvement and development; forest residue and industrial timber waste development; timber mechanical property improvement through chemical treatment research.
Publications: Occasional reports.
Projects: Tree improvement (Dr Aphichart Khaosa-ard); research and training in re-afforestation (Dr Thanit Yingvansiri); wood energy conversion research (Dr Aroon Chomchan).

FOREST MANAGEMENT DIVISION   59

Head: Udom Burana Kanonda

ROYAL FOREST PRODUCTS RESEARCH DIVISION   60

Head: Suthee Harnsongkram

SILVICULTURE DIVISION 61
Head: Swat Nicharat

Rubber Research Centre 62

– RRC
Address: Karnchanavanich, Hat Yai, Songkhla
Telephone: 244044; 244416
Affiliation: Ministry of Agriculture and Cooperatives, Department of Agriculture
Status: Official research centre
Director: Sribo Chaiprasit
Director: Sanit Samosorn
Training: Chamnong Kongsil
Graduate research staff: 67
Activities: Rubber production: breeding, exploitation, cultural practices; intercropping; pests and disease, soil science; soil survey; soil fertility; fertilizers; rubber technology; economics; development; training; audio production; project assessment and evaluation; planning and analysis.
Publications: Para Rubber Bulletin, quarterly.

AGRONOMY DIVISION 63
Head: Sompony Sookmark

DEVELOPMENT DIVISION 64
Head: Prationg Dolkich

ECONOMIC DIVISION 65
Head: Sucharit Promdej

RUBBER TECHNOLOGY DIVISION 66
Head: Kasem Intraskul

SOIL SCIENCE DIVISION 67
Head: Dr Likit Nualsri

TECHNICAL SERVICE DIVISION 68
Head: Muangchai Vimooktanant

Thailand Institute of Scientific and Technological Research 69

Status: Official research centre

AGRICULTURAL RESEARCH DIVISION 70
Address: 196 Phahonyothin Road, Bang Khen, Bangkok, 9
Telephone: 5791121 30
Director: Prapandh Boonklinikajorn
Graduate research staff: 21
Activities: Research and development of both indigenous and introduced crops/plants as raw materials for industries or export. Technical services available to government agencies or private sector on contract basis mainly in the area of plant production.
Projects: Essential oil crops for highlands (K. Wichapan); pigeon pea for the northeast (P. Burnasilpin); energy plants (S. Visuttipitakul); winged bean (K. Kovitvadhi); Dioscorea germplasm collection (W. Supatanakul); control of oriental fruit fly (J. Watanakul).

Crops Research Laboratory 71
Head: K. Wichapan

Agricultural Process Laboratory 72
Head: S. Visuttipitakul

TOGO

Institut de Recherche Agronomiques Tropicales et des Cultures Vivrières*   1

– IRAT
[Tropical Agriculture and Food Crops Research Institute]
Address: BP 1163, Lomé
Status: Official research centre
Director: M. Marquette

Université du Bénin*   2

[University of Bénin]
Address: BP 1515, Lomé
Telephone: 21-30-27
Telex: 52-58
Status: Educational establishment with r&d capability
Rector: Professor A.G. Johnson

HIGHER SCHOOL OF AGRICULTURE*   3
Dean: Dr K. Kpakote

TRINIDAD

Central Experiment Station   1

– CES
Address: Arima, Centeno
Telephone: 664-4335-7; 5158
Parent body: Ministry of Agriculture
Status: Official research centre
Technical Officer, crop research: Dr Ronald Barrow
Departments: Soils and chemistry, A.L. Bart; plant protection, Dr J.E. Pegus; agronomy, N.K. Pessad
Graduate research staff: 31
Activities: CES's main thrust is in agronomy with support coming from the plant protection and soil and chemistry sections. The soil science section deals with soil fertility and land use matters. In agronomy, vegetables, root crops, cereals and other grain crops, fruits, cocoa and coffee are under investigation. Plant protection covers the study of entomology, phytopathology, nematology, virology and plant quarantine.
Publications: Annual report, biannual report.
Projects: Integrated control programme for pests of cruciferous and solanceous crops (M. Jones); production of vegetables under controlled environmental conditions (Roy Griffith); selection and breeding for resistance to phytophthora palmivora, crinipellis perniciosus, cenatocystis fimbriata in cocoa (C. Gonzalves); investigation of high density planting and meadow orchard techniques for citrus and exotic fruit production (L. Andrews).

Forestry Division   2

Address: Port of Spain
Parent body: Ministry of Agriculture
Status: Official research centre
Conservator of forests: Dr Bal S. Ramdial
Publications: Annual report.

FOREST RESOURCE INVENTORY AND MANAGEMENT   3

Head: K. de Freitas

NORTHERN RANGE REAFFORESTATION PROJECT   4

Head: N.P. Lackhan
Projects: Watershed management techniques (B.P. Ramdial).

RESEARCH SECTION   5

Address: PO Bag 30, Port of Spain
Telephone: (62) 27476
Head: K.D. Musgrave
Graduate research staff: 1
Activities: Forestry and forestry products: collecting and processing data on all aspects of tropical forestry in order to produce results upon which sound management decisions can be made. Currently emphasis is being placed on areas such as species and provenance trials, tree improvement, nursery techniques, silviculture, forest protection, mensuration, inventory, wildlife and watershed management. Future areas of research are forest soils, timber-harvesting, forest engineering, utilization of forest products, and forest recreation.
Projects: Species trials (K.D. Musgrave); provenance trials of pinus caribaea and cordia alliodora, tree improvement of pinus caribaea and tectona grandis nursery techniques; silviculture and growth rates of pinus caribaea and tectona grandis, biological control of insect pest hypsipyla grandella which attacks species of the Meliaceae family (Dr M. Yaseen)

WILDLIFE SECTION   6

Head: B.P. Ramdial

University of the West Indies   7

– UWI
Address: St Augustine
Status: Educational establishment with r&d capability

FACULTY OF AGRICULTURE   8

Telephone: (662) 5511
Telex: STOMATA
Dean: Professor L.A. Wilson
Graduate research staff: 40
Activities: Improvement of crop and livestock production in the Caribbean through interdisciplinary research and development studies.
Publications: Annual research report.

Agricultural Economics and Farm Management Department   9

Head: Dr L.B. Rankine

Agricultural Extension Department   10

Head: Dr T.H. Henderson

Biological Sciences Department   11

Head: R.F. Barnes

Crop Science Department   12

Head: Professor L.A. Wilson
Projects: Root crops (Dr C. McDavid); grain legumes (Dr F.U. Ferguson); horticulture (Professor L.A. Wilson); cereal and grains (Dr R.A.I. Brathwaite).

Livestock Science Department   13

Head: Professor H. Williams

Soil Science Department   14

Head: Professor N. Ahmad
Projects: Soil chemistry (Professor N. Ahmad); soil physics and water management (Dr F. Gumba).

TUNISIA

Centre de Recherche du Génie Rural 1

[Rural Engineering Research Centre]
Address: BP 10, Ariana, Tunis
Telephone: (01) 231-624; 628; 630
Parent body: Ministry of Agriculture
Status: Official research centre
Director: El Amami Slaheddine
Departments: Fertility and salinity, T. Gallali; management, A. Bahri; agronomy, Mr Bouzaidi; physics, Z. Chaabouni; physiology, Mr Bouaziz; agricultural mechanics, K. Ben Khelil; horticulture, Mr Ben Othmane
Graduate research staff: 9
Activities: Water economy; irrigation methods; salt water irrigation; agricultural mechanics; water and soil conservation; solar energy; waste water.
Publications: Annual report.
Projects: Agricultural hydraulics; use of waste water; use of solar energy for pumping water and desalination (El Amami Slaheddine).

Institut National de la Recherche Agronomique de Tunisie 2

– INRAT
[National Institute for Agricultural Research of Tunisia]
Address: Avenue de l'Indépendance, Ariana
(01) 230 024
Parent body: Ministry of Agriculture
Status: Official research centre
Director: Mustapha Lasram
Graduate research staff: 45
Activities: Improvement of quantity and quality of agricultural products, including cereals, vegetables, fruits; sheep breeding.
Publications: Annual report.

AGRONOMY (PLANT SCIENCE) DEPARTMENT 3

Head: Hamouda Lakhoua
Projects: Development and growth of perennials.

ANIMAL SCIENCE DEPARTMENT 4

Head: Mongi Ben Dhia
Projects: Sheep breeding and nutrition; by-products for animal feeding (G. Khaldi).

ECONOMY AND BIOMETRY DEPARTMENT 5

Head: Mokhtar Essamet

FIELD CROPS DEPARTMENT 6

Head: Ali Maamouri
Projects: Cereal varieties improvement (A. Maamouri); cereal technology (M. Ben Salem).

HORTICULTURE DEPARTMENT 7

Head: Naceur Hamza
Projects: Vegetable breeding (N. Hamza); stone fruits improvement and nuts (M. Mlika, A. Gharbi); pome fruits improvement (S. El Aouini).

PLANT PROTECTION DEPARTMENT 8

Head: Ahmed Mlaiki
Projects: Biological control of insects (M. Cheikh); virus control on vegetables (C. Cherif).

Institut National de Recherches Forestières 9

– INRF
[National Institute for Forestry Research]
Address: Route de la Soukra, BP 2, Ariana, Tunis
Telephone: (01) 230420
Parent body: Ministry of Agriculture
Status: Official research centre
Director: Hechmi Hamza
Departments: Forest and pastoral ecology, Abdelmajid Hamrouni; forest produce and silviculture, Mohamed Dahman; forest economics and biometry, Hechmi Hamza; soil and water conservation, Hédi Berraies; environment and nature protection, Mohamed Charfi
Graduate research staff: 13
Activities: Forestry and forest products; ecology and protection of the environment.
Publications: Annual report; biannual bulletin.
Projects: Ecology (Abdelmajid Hamrouni); pedology (Mohson Selmi); plant physiology (M. Néjib Rjeb); forest genetics (M. Larbi Khoja); reafforestation (Mustapha Ksontini); pastoral techniques (Abdelmajid Hamrouni); silviculture (Moncef Hamdi); wood technology (Mohamed Dahman); forest protection and entomology (Hechmi Hamza).

Institute National de Nutrition et Technologie Alimentaire* 10

[National Institute of Nutrition and Food Science]
Address: 11 rue Aristide Briand Bab Saadoun, Tunis
Status: Official research centre
Director: Professor Zouhair Kallal
Publications: Nutrition Appliquée, two monthly.

Ministry of Agriculture 11

SOILS DIVISION 12

Address: Avenue de la République, Tunis
Telephone: (01) 246 232
Status: Official research centre
Head: Souissi Ahmed
Sections: Cartography and soil testing, Abderrahmane Mami; research and experimentation, Salah Khalfallah; geomorphology, Saída Selmi; laboratory, Ouahida Nanaa; teledetection, Ali Hamza
Graduate research staff: 20
Activities: Soil science; land use.
Publications: Annual bulletin.
Projects: Soil survey for irrigation and sugar beet (A. Mami); soil salinity - monitoring of irrigated areas (Habib Ben Hassine); ion selective electrodes (J. Susini); ions and water dynamics in irrigated soils (Mr Vieillefont); gypsum in soils of south Tunisia (Mr Vieillefont); crust-topped soils in south Tunisia (Amor Mtimet); erosion in watershed areas (Aida Selmi); Landsat imagery (A. Hamza).

TURKEY

Akdeniz Bölge Zirai Araştirma Enstitüsü 1

[Mediterranean Regional Agricultural Research Institute]
Address: PK 39, Antalya
Telephone: (3111) 1180
Parent body: Ziraat Isleri Genel Müdürlüğü
Status: Official research centre
Director: Şevket Özberk
Departments: Cotton, Süleyman Boyaci; second crop, Dr Atila Altunay; machinery, Süleyman Güleryüz; field and farm management, Haluk Daricioğlu, Murat Ahishali
Graduate research staff: 30
Activities: Plant production (cotton and second crops, maize, sorghum, soyabeans, rice, groundnuts, sesame) - plant breeding, crop husbandry, plant protection; animal production (chickens and cows) - livestock breeding, livestock husbandry and nutrition.
Publications: Annual report.
Projects: Maize - sorghum (Dr Mehmet Ali Tüsüz); rice (Erol Okumuşlar); soyabean (Salim Bozkurt); groundnuts (Cihangir Kayganaci); sesame (Cihangiv Kayganaci); forage crops (Nevin Fetullahoğlu); plant protection (Bülent Kaur); general economics (Mehmet Hacioğlu).

Alata Bahçe Kültürleri Araştirma ve Eğitim Merkezi Müdürlüğü 2

[Directorate of Alata Horticultural Research and Training Centre]
Address: Alata - Erdemli, Içel
Parent body: Ministry of Agriculture and Forestry
Status: Official research centre
Director: Enver Özyurt
Departments: Fruits, Caner Onur; viticulture, Dr Durmuş Ali Atalay; vegetables, Sami Kesici; plant nutrition, Nasbi Bayram; plant protection, Dr Şerif Ali Akteke; agricultural economics, Naşit Yakan; food technology, Orhan Yüncüler
Graduate research staff: 15
Activities: Plant production: tomato, pepper, cucumber, citrus, avocado, pecan, walnut, peach, plum, pomegranate, strawberry. The centre takes part in 9 national reserch projects.
Projects: Tomatoes (Sami Kesici); agricultural economics (Naşit Yakan); viticulture (Dr Durmuş Ali Atalay); sheltered vegetable production, (Sami Kesici); citrus (Dr Şerif Ali Aktekc); olives (Rehber Aydin); berries (Serap Onur); stone fruits; nut fruits (Caner Onur).

Ankara Nükleer Arastirma ve Egitim Merkezi* 3

[Ankara Nuclear Research and Training Centre]
Address: Besevler, Ankara
Telephone: 234439
Affiliation: Turkish Atomic Energy Commission
Status: Official research centre
Director: Dr Ilhan Ölmez
Sections: Agriculture, Professor Dr N. Özbek; physics, Associate Professor Dr A. Sinman; chemistry, Dr I. Ölmez; health physics, Dr G. Yülek; electronics, Z. Köksal
Graduate research staff: 60
Activities: Applied research on energy, agriculture, industry: topics included are environmental pollution, electronics, health physics.
Publications: Technical journal.

Ankara Üniversitesi* 4

[Ankara University]
Address: Tandoğen Meydani, Ankara
Telephone: 233245
Status: Educational establishment with r&d capability

FACULTY OF AGRICULTURE* 5

Dean: T. Güneş

FACULTY OF VETERINARY SCIENCE* 6

Dean: Professor Ismet Baran
Graduate research staff: 120
Activities: Animal production: veterinary medicine, livestock breeding, nutrition, and husbandry; food science; freshwater fisheries.
Publications: Journal.

Atatürk Bahçe Kültürleri Araştirma Enstitüsü 7

[Atatürk Horticultural Crops Research Institute]
Address: PK 15, Yalova - İstanbul
Telephone: (1931) 2520; 2521
Status: Official research centre
Director: Refet Ergin
Technical assistant to director: Dr S. Özelkök
Departments: Agricultural economics and marketing, Ayhan Yücel; agricultural mechanization, E. Semerci; field management, Hikmet Ergüler; food technology, Bumin Baykal; fruit crops, Sabri Demirören; greenhouse

ornamentals, Nurdal Ertan; landscape architecture, Ahmet Mengüç; mushroom growing, Dr S. Erol Işik; olive culture, Ali Ekber Fidan; plant nutrition, Dr Çağlar Gene; plant protection, Kemal Akar; postharvest physiology, Dr Ümit Ertan; training-extension, Orhan Adsan; vegetable crops, Dr Tamer Türkeş; viticulture, Ismet Uslu
Graduate research staff: 51
Activities: All aspects of horticultural crops (decidious fruits vegetables, viticulture, and ornamentals). Research interests include the introduction of new varieties (foreign and domestic), evaluation, breeding (annuals and biannuals), physiology, nutrition, protection, pre- and postharvest technology, and marketing. The mushroom research unit at Valova has an important role to play in training and demonstration. The institute holds seminars, symposiums, and training programmes as well as extension services, and is the coordimating centre of 6 nationwide projects.
Publications: Biannual journal.
Projects: Pome fruits (Mustafa Büyükyilmaz); stone fruits (Sabri Demirören); berry fruits (Dr Onur Konarli); nuts (Dr Gültekin Çelebioğlu); agricultural economics research and training (Ayhan Yücel); mushrooms (Dr S. Erol Işik).

Atatürk Universitesi 8

[Turkish University]
Address: Erzurum
Status: Educational establishment with r&d capacity

COLLEGE OF AGRICULTURE 9

Telephone: (0111) 13380; 13423
Dean: Professor Ali Özdengiz
Graduate research staff: 67
Activities: Soil science, drainage and irrigation, land use; plant production - plant breeding, crop husbandry; animal production (cattle, sheep, poultry) - livestock breeding, husbandry, and nutrition.
Publications: Ziraat Dergisi (Journal of Agriculture, quarterly).

Agricultural Engineering and Building Department 10

Head: Professor Ali Özdengiz
Projects: Irrigation problems of Iğdir valley (Dr Ali Özdengiz); changes in farm buildings in Erzurum during the last twenty years (Dr Nevzat Şişman); Malatya - Şahnahan valley soils: irrigation and limiting factors affecting productivity (Dr Mehmet Apan); relation between the variance of the in situ measured hydraulic conductivity values and the sampled soils volume (Dr

Ersan Gemalmaz); environmental conditions in the slotted floored open-front cold-confinement beef cattle buildings (Dr Ali Riza Uluata); locally available lining material for use in farm irrigation to prevent seepage loss (Dr Feridun Hakgören); present conditions, characteristics, and development possibilities of sheep barns in sheep farming enterprises in Ağri province (Dr Tahir Ekmekyapar); determination of mechanical properties of lightweight concretes made with natural aggregates, and the utilization of these concretes particularly in the construction of farm buildings in the Sarikamiş region (Dr Ümit Turgutalp); comparison of water advance estimation methods in field conditions (Dr Ünal Alici); effect of some soil properties on sedimentation in corrugated plastic drainpipes: establishment of criteria to determine limits of sedimentation danger in drainage areas (Dr Nevruz Yardimci); possible solutions of problems concerning new settlements and new farm buildings in villages affected by natural disasters in the Erzurum province (Dr Ismet Arici); commercial poultry houses in the Eastern Turkey and their improvement potentialities (Dr Mustafa Okuroğlu); environmental conditions (temperature, relative humidity, ventilation, lighting) in the 50 cattle stall barn in the university agricultural college: possibilities for improvement of this barn and planning of dairy barns in this region (A. Vahap Yavanoğlu); hay cutting, raking, field drying, baling, chopping, and handling mechanization in Erzurum valley - time, energy, and labour requirements, field capacity, cost operations; percentage of moisture loss in hay from moving to baling for field - cured hay (Dr Poyraz Ülger).

Animal Science Department   11

Head: Professor Ayhan Aksoy
Projects: Breeding Awassi sheep under farm conditions in Erzurum (Dr Yusuf Vanli, Dr Mustafa K. Özsoy); Merinos and Morkaraman crossbreeding (Dr Mustafa K. Özsoy, Dr Yusuf Vanli); feedlot performances and carcass characteristics of Brown Swiss and Eastern Anatolian red crosses (Dr Şakir Bayindir, Dr Oktay Yazgan); fur qualities of wild type marten - (mustela Martes) (Dr Erdoğan Selçuk); effect of stock water and amount of feed on growth rate and feed efficiency of Rainbow trout (Dr Sitki Aras); comparision of different calf rearing methods (Naci Tüzemen); effects of the type of utensil, daily feeding interval and feeding level on growth rate of rainbow trout (Dr Recep Bircan); comparison of open-shed and stall-based type of byre for dairy cows (Dr Ahmet Çakir, Dr Şakir Bayindir); fattening performance of Morkaraman, Merinos and Awassi purebred and their crosses, comparison of different fattening systems for male lambs Dr Ahmet Çakir, Dr Mustafa K. Özsoy; alleviation of flour toxicity of Mazi Daği row rock phosphate with aluminium sulphate (Dr Ahmet Çakir, Dr Nihat Özen); fattening performance and carcass characteristics of Zarut, Red Anatolian,

Red Anatolian and Simmental, and Red Anatolian and Brown Swiss F1 crosses, (Dr Ahmet Çakir, Cevat Geliyi); fattening and carcass characteristics of Eastern Anatolian red cattle of different ages; feedlot performances and carcass characteristics of Morkaraman, Merinos Breeds, and their crosses (Dr Şakir Bayindir); possibilities of increasing metabolic energy content of poultry feeds by the supplementation of oils and fats (Professor Ayhan Aksoy, Dr Ahmet Çakir); comparative evaluation of the protein quality of wheat, corn, barley, oats and sorghum in chicks (Dr Nihat Özen); comparative evaluation of various combinations of yellow corn and dehydrated alfalfa meal in broiler and laying hen diets (Dr Nihat Özen); possibilities of using aquatic plants as roughage covering a large protion of Iudir State Farm in steer and yearling lamb rations (Dr Sümer Haşimoğlu); estimation of the nitrogen and various mineral losses in alfalfa harvested at different times and stored with different moisture contents; nutritive value of cotton seed meal processed in different plants for medium and high yielding dairy cows (Dr Oktay Yazgan); physical properties of wool from different sheep breeds and their crosses (Dr Hakki Emsen).

Milk and Food Technology   12
Department

Head: Professor Ahmet Kurt
Projects: Some characteristics of yogurts which were made using different starter cultures and adding milk powder in different ratios (Professor A. Kurt, Dr M. Demirci, Dr H.H. Gündüz); qualitative properties of commercial flour used for bread making in Erzurum: some reasons for low quality bread (Dr Adem Elgün, Dr Zeki Ertugay); processing, composition, and bacteriological properties of ice cream consumed in Erzurum city (Dr Nurhan Akyüz); composition of and possibility of processing different food products from the rose hip grown naturally in Erzurum Province (Professor Bekir Cemeroğlu, Professor Ahmet Kurt).

Plant Production Department   13

Head: Professor İbrahim Manga
Projects: Improvement of dry-land ranges in Erzurum, Eastern Anatolia (Professor Fahrettin Tosun, Professor Ibrahim Manga, Dr Murat Altin, Dr Y. Serin); effect of time of haulm killing and digging time on the tuber yield and quality (Dr Erol Günel); edible seeds of legumes - their adaptation, breeding, and production in a cold climate (Dr A. Akçin, Dr Ali Gülümser); selection of apricots at Erzincan, Eastern Antatolia, 1980–83 (Dr Muharrem Güleryüz); chromosome doubling and improvement of lolium perenne and festuca pratenses (Dr Sevim Sağsöz); rotation of grasses and legumes in dryland conditions at Erzurum (Dr Murat Altin); yield, yield components and quality of foreign and native

winter wheat varieties (triticum aestivum L.) grown at Erzurum (Dr Coşkun Köycü); vines and ampelographic characteristics of some grape varieties at Igdir Valley in Eastern Anatolia (Dr Ferhat Odabas).

Soil Science Department   14

Head: Professor Dr Abdusselam Ergene
Projects: Usable methods in determining available nitrogen in the irrigated soils of Iğdir plain (Dr Turgut Sağlam, Dr Saim Karakaplan); effects of waste water of Erzurum city on soil pollution (Dr Sücaattin Kirimhan, Dr Turgut Sağlam); effects of soil compaction at various moisture contents on air and water permeability of the soils (Dr Saim Karakaplan); effect of phosphoric acid, triple superphospate and animal manure on aggregation, aggregate stability and crust strength of various soils; effect of molasses, ground beet pulp and vinasse on crust strength of a calcareous soil (Dr Koray Sönmez); effects of refuse compost on physical and chemical properties and productivity of soil (Dr Metin Bahtiyar); effects of refuse composts on the pollution of ground water in different soils (Dr Metin Bahtiyar); ion exchange resin (mixed bed) as a means of assessing plant available soil nutrients (Dr Orhan Aydemir); determination of index values concerning some parameters used in describing potassium supplying power of the soil, effects of applying N-Serve with urea on yield in rice production (Dr Yildirim Sezen); effect of environmental pollution on soil erosion in the vicinity of Murgul (Göktaş) province (Dr Hayati Çelebi); parameters and index values of wind erosion on Erzurum plain (Dr Hayati Çelebi); evaluation of the distribution of indigenons Vesicular-Arbuscular mycorrhizal fungi and determination of spore-types of the fungi in the soils of Eastern Anatolia; effect of Vesicular-arbuscular mycorrhiza on the growth and phosphourus-uptake of onion plants supplied with rock phosphate under greenhouse conditions; influence of soil inoculation with Vesicular-Arbuscular Mycorrhiza and a phosphate-dissolving bacterium on plant growth and phosphorus uptake (Dr Kemal Gür); distribution of residues of some chlorinated hydrocarbon insecticides in some soils in Çukurova Area and investigation of the biological decomposition of DDT and another insecticide, Lindane (Dr Hasan Hayri Tok).

Bağcilik Araştirma Enstitüsü   15
Tekirdağ

[Tekirdağ Viticultural Research Institute]
Address: PK 7, Tekirdağ
Telephone: (1861) 2042
Parent body: Ministry of Agriculture and Forestry
Status: Official research centre
Director: Cemal Bariş
Assistant director: Dr Erkal Gökçay
Departments: Vine breeding, Cemal Bariş; physiology, Ersan Kogamaz; plant propagation, Hilmi Eryldiz; technology, Ersan Kocamaz; extension, Dr Erkal Gökçay
Graduate research staff: 9
Activities: The institute is the coordination centre of the national viticultural experiment project. In this project basic subjects investigated are clonal selection, breeding, adaptation for varieties and rootstocks, training systems, fertilizer application in vineyards, and quality of standard varieties. The overall aim is to raise grape production and quality, and decrease costs through mechanization. Ten institutes located in different regions of Turkey take part in these studies.
Projects: Clonal selection on some economical table and wine grape varieties grown in Marmara and Thrace regions (Salih İnal); breeding for new early and late seedless table grape varieties (Cemal Bariş); national grape collection vineyard (Salih İnal); performance of Papazkarasi and Semillon grape cultivars on different rootskocks (Salih İnal); comparison of different trunk height of Papazkarasi and Semillon grape varieties in local conditions on cordon system (Seyfi Özişik); comparison of seven trellis systems for Hafizali and Semillon grape varieties (Salih İnal); vine spacing of Semillon on the Guyot and Goble training (Dr Erkal Gökçay); effect of application times of nitrogen fertilizers in vineyards (Ersan Kocamaz); determination of standard grape varieties for vineyard regions (Salih İnal).

Bağcilik Araştirma   16
İstasyonu Müdürlüğü

[Müdürlüğü Viticulture Research Station]
Address: PK 12, Nevşehir
Telephone: (4851) 1281
Parent body: Ministry of Agriculture and Forestry
Status: Official research centre
Director Yusuf Demirbüker
Departments: Viticulture, Erol Serttaş; technology, Erol Çaliş; agronomy, Hüsamettin Dabanli
Graduate research staff: 6

Activities: Viticulture: vine breeding, technology and techniques; propagation and distribution of rooted vine cuttings; agronomic experiments on potato production; extension.

Projects: Clone selection on grape cultivars; choice of new standard grape varieties for Nevşehir vineyards; effects of different training systems on some grape varieties of Nevşehir (Erol Serttas, Ismail Yüksel); quality and technological values of some local wine grape varieties (Hidir Sarioğlu, Erol Çaliş); potato production (Hüsamettin Dabanli).

Biyolojik Mücadele Araştirma Enstitüsü Müdürlüğü   17

[Biological Control Research Institute]
Address: PK 136, Antalya
Telephone: (3111) 4053; 3275
Parent body: General Directorate of Plant Protection and Plant Quarantine
Status: Official research centre
Director: Remzi Dikyar
Laboratories: Predators, Ergun Keçecioğlu; parasites, Üzeyir Genç; insect pathology, Kemal Çiftçi; weeds, Aynur Keleş; artificial diets, Naci Türkyilmaz
Graduate research staff: 18
Activities: Citrus, cotton, vegetables, apple trees: biological and integrated pest control; reduction of pesticide use.
Publications: Plant Protection Bulletin, annual.
Projects: Identification of natural enemies of citrus white fly, Dialeurodes citri R. on citrus in Antalya region and introduction of imported parasite Prospaltella lahorensis. how. into the region (Ergun Keçecioğlu); South Anatolia cotton pests: integrated control (Üzeyir Genş); effectiveness of Cryptolaemus montrouzieri as biological control agent against citrus mealybug Planococcus citri (Aynur Keleş); possible use of Chilocorus bipustulatus and Aphytis as biological control agents (Fahri Kumaş); population dynamics and natural enemies of Tetranydhus urticae (Ali Sdysal); population dynamics and natural enemies of Synanthedon myopaeformis.

Boğaziçi Üniversitesi*   18

[Boğaziçi University]
Address: PK2, Bebek, Istanbul
Telephone: 653400
Telex: 26411 BOUNTR
Status: Educational establishment with r&d capability

UNIVERSITY RESEARCH CENTRES*   19

Environmental Sciences Institute*   20
Director: Kriton Curi

Scientific Land Development Research Institute*   21
Director: Tancay Saydam

Bölge Zirai Mücadele Araştirma Enstitüsü, Erenköy-İstanbul   22

[Erenkoy-Istanbul Regional Plant Protection Research Institute]
Address: Bağdat caddessi 311, Erenköy-İstanbul
Telephone: 58 11 45; 58 11 46; 58 32 27
Parent body; General Directorate of Plant Protection and Plant Quarantine
Status: Official research centre
Director: Dr H. Hüseyin Karasu
Laboratories: Fruit and vine pests, Musa Altay; field crop pests, Turan Günaydin; subtropical plant pests, Ertan Seçkin; vegetable and food plant pests, Esen Atak; plant parasitic nematodes, Muzaffer Ağdaci; general pests, Emel Ilter; stored product pests, Güler Ilalan; industrial and ornamental plant pests, Erim Ünal; fruit diseases, Dr Erdoğan Erkam; vine diseases, Nevsat Özhendekçi; vegetable diseases, Erden Gülsoy; weeds, Cesarettin Özdemir; cereal diseases, Salih Bayezit; plant virus diseases, Abdullah Nogay; industrial and ornamental plant diseases Dr H. Hüseyin Karasu
Graduate research staff: 32
Activities: Control of plant pests and diseases.
Publications: Plant Protection Research Annual; Plant Protection Bulletin (quarterly).
Projects: Mostly in the Marmara region: apple pest control (Ali Gürses); Goat moth causing damage in fruit trees (Şafak Tüziin); sex pheromone and bait traps against grape berry moth (Suzan Gürkan); harmful and beneficial rice insects (Gürbüz Ersoy); harmful and beneficial maize insects (Turan Günaydin); sex pheromone traps against olive moth (Prays olea Bern.) (Ertan Seçkin); chemical control of cutworms (Agrotis spp.) in vegetable planting areas; chemical measures against colorado potato beetles (Esen Atak); damage by and control possibilities of potato root nematode (Ditylenchus destructor) and determination of other plant parasite nematodes in potato planting areas in Turkey (Muzaffer Ağdaci); crop losses due to Microtus spp. in cereal planting areas (Emel Ilter); crop losses due to Bruchus spp. in legumes in storehouses (Güler Ilalan); crop losses due to green peach aphid (Myzus persicae)

(Erim Ünal); common wheat bunt in Turkey; wheat resistance to powdery mildew (Erysiphe graminis) (Seçkin Finci); seed treatment against bacana of rice (Salih Bayezit); olive leaf spot (Cyclocanium oleaginum): effect of chemical control on crop yield (Dr Erdoğan Erkam); early warning system in chemical control of apple scab and powdery mildew of apple trees (Halil Ince); bioecology and control probabilities of agent (Phomopsis viticola) causing dead arm in vineyards (Nevzat Özhendekçi); weed control in onion and wheat fields, and among sunflowers (Cesarettin Özdemir); tip blight in tomatoes in Thrace; cucurbit viruses (Abdullah Nogay); primary inoculum sources of Phytophthora pepper blight and its control; transmission rates of causal agents of potato dry rot; effects of storage conditions on wastage and its control (Erden Gülsoy).

Bölge Zirai Mücadele Araştirma Enstitüsü, Erzincan
23

[Erzincan Regional Plant Protection Institute]
Address: PK 31, Erzincan
Telephone: 2748
Parent body: Generel Directorate of Plant Protection and Plant Quarantine
Status: Official research centre
Director: Aydoğan Ünal
Graduate research staff: 13
Activities: Plant protection.
Projects: Mostly in East Anatolia: crop losses and economic damage level of Citellus citellus (Orhan Benli); wheat fungal pathogens and their distribution (F. Yalçin Yilmazdemir); faunistic survey studies on cotton and cereal (Süleyman Uzunali); determination of integrated pest control on apricots (T. Mete Ergüden); faunistic survey studies on dry beans; chemical control of seed corn maggot (Delia platura) on dry beans in Erzincan (Mete Aydemir); fungal pathogens causing damage to apricot trees (M. İlhan Çataloğlu); determination of weeds in bean (Phaseolus vulgaris) fields; weeds in onion (Allium cepa) areas (Cihat Alsan); bioecology and natural enemies of Lepidosaphes ulmi in Erzincan (Selim Aydoğdu); pome fruits in Erzincan: insect pests (Fehmi Erden).

Bölge Zirai Mücadele Araştirma Enstititüsü, Kalaba
24

[Kalaba Regional Plant Protection Research Institute]
Address: Fatih Cadessi, Kalaba, Ankara
Telephone: (941) 164236
Parent body: General Directorate of Plant Protection and Plant Quarantine
Status: Official research centre
Director: A. Ulvi Kiliç
Laboratories: Fruit and grape pests, Dr Zekiye Iren; fruit and grape diseases, Avni Yürüt; vegetable and fodder pests, Sencer Çalişkaner; diseases of vegetables, industrial crops, fodder, and ornamental plants, Mübeccel Bariş; virus diseases, Seyhan Kurçman; nematology, Dr Sebahat Enneli; weed control, Metin Kurçman; taxonomy and plant protection museum, Dr Ayla Kalkandelen; general pests, Dr Erta',c Tutkun; pests of industrial crops and ornamental plants Yayla Öneş; storage pests, Nükhet Dörtbudak; wheat pests, YIldIrIm Dörtbudak; wheat diseases, Dr Çetin Çelik
Graduate research staff: 46
Activities: Plant protection: pests and diseases.
Publications: Plant Protection Research Annual; Plant Protection Bulletin.
Projects: Chemical control of weeds among chickpea at Konya Province and effects of herbicides available on nodulation (Suzan Çetinsoy); bioecology, in particular host plant relations, of the Colorado potato beetle in Central Anatolia (Alânur Has); distribution of sunpest species (Pentatomidae, Hemp.), and bionomics and control methods of E. maura in Ankara (Hatice Memişoğlu); determination of Trichogramma species on important Lepidopteran pests of fruit crops and their effectiveness on codling moth (Hüseyin Bulut); bioecology and control methods of the leaf scorch disease (Gnomonia erytrostoma) that damages sour cherry (Meral Erdem); factors influencing population density of cereal bug, (Aelia rostrata), crop losses caused by it, and its integrated control in Centrol Anatolia (Yildirim Dörtbudak); prime inoculum sources and control methods of pepper basal rot disease (Phytophthora capsici) in Central Anatolia (Mübeccel Bariş); control and bioecology of poppy root weevil in Central Anatolia (Yayla Öneş); inoculation of wheat varieties obtained by crossing to determine varieties resistant to wheat loose smut disease (Ustilago nuda tritici) and the pathogenic strains of the fungus (Dr Çetin Çelik); distribution, rate of damage, and possibility of control of potato rot nematode, and determination of other plant parasite nematodes, in potato planting areas in Turkey (Sabahat Enneli); biology and population fluctuations of oyster shell scale (Lepidosaphes ulmi), Europen fruit lecanium (Parthenolecanium corm), and Palaeolecanium bituberculatum on apple trees in Central Anatolia (Ali Okul).

Bölge Zirai Mücadele Araştirma Enstitüsü, Samsun

25

[Samsun Regional Plant Protection Research Institute]
Address: PK 3, Samsun
Telephone: (3611) 16020; 16021
Parent body: General Directorate of Plant Protection and Quarantine
Status: Official research centre
Director: Dr Fikret Dündar
Laboratories: Field pests, Nurettin Özdemir; fruit pests, M. Kemal Aykaç; subtropical crop pests, Ibrahim Bozan; hazelnut pests, Mehmet Işik; stored products pests, Ersan Yasan; industrial and ornamental plants pests, Metin Polat; vegetable and fodder crops pests, Nevzat Yilmaz; general pests and nematology, Ümit Tunçdemir; fruit diseases, Necati Altinyay; vegetable and fodder crops diseases, Mümin Şenyürek; grain diseases, Orhan Bilgin; industrial plants diseases, Metin Özkutlu; weed control, Ismail Korkut
Graduate research staff: 28
Activities: Plant protection: chemical tests on pests, diseases, and weeds. Short training courses on plant protection are given by the institute.
Projects: Apple pests in the Black Sea region: integrated control (M. Kemal Aykaç); ditylenchus dipsaci in the Black Sea region: control of hemp damage (U. Tunçdemir); population dynamics, natural enemies and control of scale insects causing damage to citrus in the Eastern Black Sea region (Ibrahim Bozan); hazelnut pests in the Eastern Black Sea region: integrated control (Mehmet Işik); primary inoculum sources and control possibilities of Phytophtora capsici in the peppers in the Black Sea region (Mümin Şenyürek); bioecology and control methods of Anthracnose in beans grown in Samsun Province (Yusuf Zavrak).

Çay Araştirma Enstitüsü

26

[Tea Research Institute]
Address: PK 23, Rize
Telephone: 1149; 2609
Telex: 83 137-RTKL-TR
Parent body: Ministry of Customs and Monopolies, General Directorate of Tea Organization
Status: Official research centre
Director: Mustafa Erden
Departments: Agrobotany, Turhan Tutgaç; soil science, Muammer Sarimehmet; biochemistry, Mustafa Bilsel; phytopathology, Muharrem Öksüz; entomology, Muharrem Öksüz; technology, vacant; research and coordination, vacant; experiments and statistics, vacant
Graduate research staff: 8
Activities: Tea cultivation: selection of tea plants for improved quality and quantity of production; dissemination of knowledge of best tea varieties among tea farmers.
Publications: Annual report.
Projects: Improved tea cultivation in the Eastern Black Sea region: plants produced by clonal selection (Mustafa Erden); determination of nutrient elements in the tea plant areas of Turkey (Muammer Sarimehmet); fixation of connections between plant and soil for the tea plant (Mustafa Bilsel, Muammer Sarimehmet); comparison of different pruning methods in tea plants (Hülya Mahmutoğlu, Mustafa Bilsel).

Çayir - Mer'a ve Zootekni Araştirma Enstitüsü

27

[Grassland and Animal Husbandry Research Institute]
Address: PO Box 453, Ankara
Telephone: (941) 231382; 231383; 236557
Parent body: Ziraat Isleri Genel Müdürlüğü
Status: Official research centre
Director: Dr Ali Karabulut
Departments: Range management, Dr Uğur Büyükburç, forage crops, production and breeding, Dı Geyhan Gençtan; sheep production and breeding, Dr Sabahat Cangir; cattle production and breeding, Mehmet Apaydin; feeding and animal nutriton, Bekir Ankarali; genetics and statistics, Dr Ahmet Gürbüz
Graduate research staff: 21
Activities: Land use; plant production (forage crops) - plant breeding, crop husbandry; animal production (cattle and sheep) - livestock breeding, livestock husbandry and nutrition. The institute takes part in 4 national research projects.
Projects: Range management (Dr Uğur Büyükburç); forage crops, production and breeding (Dr Geyhan Gençtan); sheep and goats (Dr Sezer Sabaz); cattle and buffalo (Dr Ahmet Gürbüz).

Çukurova Üniversitesi* 28

[Çukurova University]
Address: PK 444, Adana, Adana
Telephone: 19609
Status: Educational establishment with r&d capability
Rector: Professor Dr Mithat Özsan
Graduate research staff: 150
Activities: Horticulture, food science and technology, animal husbandry, agricultural machinery, irrigation, and landscape architecture in southern Turkey.
Publications: Quarterly report.

FACULTY OF AGRICULTURE 29

Address: PK 444, Adana
Telephone: 19609
Dean: Professor Dr Ercan Tezer
Graduate research staff: 150
Activities: Horticulture, food science and technology, animal husbandry, agricultural machinery, irrigation, and landscape architecture in Southern Turkey.
Publications: Quarterly report.

Agricultural Economics Department* 30

Head: Dr Onur Erkan

Agricultural Mechanization Department* 31

Head: Professor Dr Osman Tekinel
Senior staff: Professor Dr Vedat Oğuzei

Animal Science Department* 32

Head: Professor Dr Lüfti Özcan
Senior staff: Professor Dr Erdoğan Pekel, Professor Dr Naił Küçüker

Field Crops Department* 33

Head: Professor Dr Ibrahim K. Atakişi
Senior staff: Professor Dr Ibrahim Genç

Food Science and Technology Department* 34

Head: Professor Dr Kemâl Gökçe

Horticulture Department* 35

Senior staff: Professor Dr Nurettin Kaşka, Professor Dr Mithat Özsan

Landscape Architecture Department* 36

Head: Dr O. Türker Altan

Plant Protection Department* 37

Head: Professor Dr Ahmet Çinar
Senior staff: Professor Dr Özden Çinar

Soil Science Department* 38

Head: Professor Dr Hüseyin Özbek
Senior staff: Professor Dr M. Sefik Yeşilsoy, Professor Dr Nuri Güzel, Professor Dr Ural Dinç

Diyarbakir Bölge Zirai Mücadele Araştirma Enstitüsü Müdürlügü 39

[Regional Plant Protection Research Institute]
Address: Diyarbakir
Telephone: (8311) 13502
Parent body: General Directorate of Plant Protection and Plant Quarantine
Status: Official research centre
Director: Hasan Kıroğlu
Laboratories: Cereal pests, Ziya Şimşek; vegetable and fodder pests, Naşit Asena; industrial and ornamental plant pests, Şaban Karaat; fruit and vine pests, Sami Maçan; general pests, Ali Riza Akinci; stored products pests, Abuzer Yücel; plant parasitic nematodes, Nurdan Ertekin; cereal diseases, Ibrahim Aktuna; vegetable and fodder diseases, Ismail Ulukus; fruit and vine diseases, Ilhan Kural; weed science, Abdurrahman Uzun
Graduate research staff: 18
Activities: Plant protection: pest control.
Projects: Determination of and damage by insect species on Zea mays and Sorghum vulgare in east and south-east Anatolia (Ziya Şimşek): tobacco pests (Nicotiana tabacum) in eastern and south-eastern regions of Turkey: identification, distribution, and damage (Şaban Karaat); lesser brown bud-worm (Recurvaria nanella) in south-eastern Anatolia: bioecology and control (Sami Maçan); reactions of winter wheat varieties against Dwarf Bunt (Tilletia controversa) of wheat and chemical control measures in east Anatolia (İbrahim Aktuna); causal agents of bacterial diseases of tomatos and peppers in Elaziğ, Diyarbakir and Mardin provinces: damage, symptometology, identification, and control; primary inoculum sources and control methods of Pepper Crown Rot in south-eastern Anatolia (Ismail Ulukus); host ranges of some Phytophthora spp. and reactions of varieties (Abuzer Sağir).

Ege Bölge Zirai Araştirma Enstitüsü 40

[Aegean Regional Agricultural Research Institute]
Address: PK 9, Menemen, Izmir
Telephone: (51) 14 91 31; (5421) 1423
Parent body: Ziraat Isleri Genel Müdürlüğü
Status: Official research centre
Director: Dr Kâşif Temiz
Departments: Genetic resources of plants, Dr Kâşif Temiz; plant protection, Dr Cevdet Dutlu; cereals, Dr Polat Şölen; horticultural crops, Erol Uraz, Dr Nurten Gönülşen; industrial crops: food legumes, Dr Y. Ziya Kutlu; industrial crops: potatoes, Dr Belgin Gömeç; forage crops, Alper Ürem; animal husbandry, Alev Settar, Sencer Tümer.
Graduate research staff: 70
Activities: Crops, plant protection, soil fertility, agricultural economics, machinery, and technology - variety improvement of cereals, food legumes, corn, vegetables, fruits, forage crops, ornamental plants; livestock breeding and husbandry.
Publications: Quarterly journal.
Projects: Genetic resources of plants (Dr K. Temiz); wheat variety improvement: breeding for disease resistance (Dr C. Dutlu); wheat variety improvement: breeding (Dr P. Sölen); barley breeding (S. Güzel, F. Demirkan); sunflower breeding (E. Üçkardeşler); maize breeding (M. Buğdaycigil); sorghum breeding (Ş. Örmeci); potatoes (Dr B. Gömeç); food legunes (Dr Y.Z. Kutlu); aniseed breeding (M. Ekim); forage crops (A. Ürem); fruit (Dr N. Gönülşen, Dr E. Çetiner); viticulture (Dr N. Karabiyik); vegetables (E. Uraz); ornamentals (D. Özkahya); other economic plants (S. Çetiner); livestock husbandry (R.S. Tümer); apiculture (A. Setlar); agricultural statistics (Dr A. Kircalioğlu); agricultural economics (C. Balkan); seed certification (S. Telek); plant tissue culture (Dr N. Grönülşen).

Ege Üniversitesi* 41

[Aegean University]
Address: Bornova, Izmir
Telephone: 180110
Status: Educational establishment with r&d capability
Faculties: Medicine; agriculture; science; dentistry; textile engineering; pharmacy; food technology; earth sciences; civil engineering; mechanical engineering; chemistry

FACULTY OF AGRICULTURE 42

Dean: Professor Fevsi Sevgican
Departments: Agricultural economics; agricultural extension and communication centre; agronomy; animal husbandry; farm machinery; food technology; horticulture; plant protection; soil science
Graduate research staff: 80
Activities: Natural resources: soil science, drainage and irrigation, land use; plant production: plant breeding, crop husbandry, plant protection; animal production: livestock breeding, livestock husbandry and nutrition; agricultural engineering and building; freshwater fisheries; food science.
Publications: Quarterly review.

FACULTY OF FOOD SCIENCE* 43

Dean: Professor Dr Erdal Saygin

Fındik Araştirma ve Eğitim Merkezi Müdürlüğü 44

[Hazelnut Research Institute]
Address: PK 46, Giresun
Telephone: (0511) 1136
Status: Official research centre
Director: Bekir Çakir
Departments: Agricultural economics, Yüksel Yakut, Ali Kaya; fruit crop diseases, Ali Kaya; plant breeding, A. Nail Okay, Fehmi Baş; plant physiology, V. Yilmaz Küçük, Aynur Küçük, Cavidan Çakirmelikoğlu; soil science, Aysel Uzun, Gündüz Akoğlu, Talha Şenses; technology
Graduate research staff: 14
Activities: Soil science; plant breeding, crop husbandry, plant protection - filbert, hazelnuts; agricultural engineering and building.
Projects: Leaf and soil analysis with filbert, and fertilizer requirements (Dr Çağlar Genç); foliar fertilization and its effect on the productivity of filbert (Aynur Küçük); effect of hormonal treatment on the rooting of sprouts (Yilmaz Küçük); effect of soil exhaustion on hazelnut growing (Cavidan Cakirmelikoğlu); nitrogen fertilizer and fertilization methods - needs of filbert (Yüksel Yakut, Yilmaz Küçük); phosphorus fertilizer and fertilization methods - needs of filbert (Ali Kaya, Yilmaz Küçük); methods for improvement of filbert (Ahmet Nail Okay); fecundation of the round select filbert and new variety trials (Fehmi Baş).

Firat Üniversitesi*   45

[Firat University]
Address: Rektörlügü, Elazig
Telephone: 8826; 8827
Status: Educational establishment with r&d capability

FACULTY OF SCIENCE*   46

Dean: Professor Dr Sahabettin Elçi
Departments: Botany; Zoology; Mathematics; Geology Engineering; Physics Engineering; Chemical Engineering

FACULTY OF VETERINARY   47
MEDICINE*

Dean: Professor Dr Hümeyra Özgen

Animal Husbandry Division*   48

Sections: Animal nutrition and feeds; zootechnics; reproduction and artificial insemination; animal production, business economy and statistics; aquatic products, fisheries and wild life; history of veterinary medicine, deontology and veterinary service affairs

Clinical Division*   49

Sections: Internal medicine; surgery; obstetrics and gynaecology

Pre-Clinical Division*   50

Sections: Bacteriology and infectious diseases; parasitology; pathology; virology; food control and animal food product technology

Pre-Veterinary Science Division*   51

Sections: Anatomy, histology and embryology; physics and chemistry; botany and zoology; biochemistry; physiology; pharmacy and toxicology

Gübre Fabrikalari TAS*   52

[Fertilizer Company of Turkey Limited]
Address: Kasap Sokok 10, Esentepe, Istanbul
Telephone: 667265
Telex: 26235 gft tr
Status: Industrial company
Graduate research staff: 1
Activities: Superphosphates, normal and triple; chemistry and technology; phosphoric acid; sulphuric acid; corrosion; construction materials for fertilizer plants; phosphate rock; dicalcium phosphate, as produced as by-product of coke oven plant; compound fertilizers; MAP; utilization of gypsum; aluminium fluoride; cryolate.

Güneydoğu Anadolu Bölge   53
Zirai Araştirma Enstitüsü
Müdürlüğü

[South Eastern Anatolian Regional Agricultural Research Institute]
Address: PK 73, Díyarbakir
Telephone: 2561
Parent body: Ziraat Isteri Genel Müdürlüğü
Status: Official research centre
Regional director: A. Ertuǧ Firat
Departments: Plant breeding and crop husbandry, A. Ertuǧ Firat; livestock breeding and husbandry, Numan Kiliçalp
Graduate research staff: 15
Activities: Plant breeding: best varieties for the region; livestock breeding: best breeds for the region.
Publications: Annual project reports.
Projects: Wheat and barley breeding (A. Ertuǧ Firat); crop husbandry (Zülfü Keklikçi); lentil and chickpea breeding (Dr Mustafa Kaya); tomato breeding and production (Celâl Yaman); rice breeding (Sabri Karakaya); corn breeding (Nevzat Nergiz); sunflower breeding (Reslen Alagöz); plant production (Ali Kaygisiz); tobacco breeding (Vedat Uzunlu); livestock breeding and husbandry (Numan Kilicalp).

PLANNING AND INVESTMENTS   54
SECTION*

Head: Erdogan Durakbasa
Research engineer: Ismail Erdebil
Deputy manager: Ulus Koyas

Ipekböcekçiliği Araştirma Enstitüsü   55

[Sericulture Research Institute]
Address: PO Box 1, Bursa
Telephone: 31020; 31021
Parent body: Ziraat Isleri Genel Müdürlüğü
Status: Official research centre
Director: Sait Grüngör Ün
Departments: Mulberry cultivation, Orhan Sipahioğlu; genetics and breeding, A. Hamdi Özeler; silkworm rearing, Cahit Topmeşe; silkworm egg production, Yildirim Bağci; silkworm pathology, Betül Ömerbeyoğlu; cocoon and silk testing, Cemalattin Gülseren; marketing, Vedat Aksoy
Graduate research staff: 13
Activities: High yield silkworm breeding; rearing techniques to ensure good quality cocoons; mulberries - increased length of yield; cant reduction in silk products and silk reeling techniques; pebrine control.
Publications: Progress reports.
Projects: Selection of pure lines in batch system or individual system on the desired characters (Sait Gügör Ün); selection and preservation of pure lines (Mahmut Ayvaşik); comparison of biological, technological, and pathological characteristics of native and imported hybrids (Cahit Topmeşe); testing of technological characteristics of the cocoon from different regions of Turkey (Cemalettin Gülseren).

İstanbul Teknik Üniversitesi*   56

[Technical University of Istanbul]
Address: Taskisla Complex, Taksim Square, Istanbul
Telephone: 224200
Status: Educational establishment with r&d capability
Rector: Professor Dr Kemal Kafali

FACULTY OF MECHANICAL ENGINEERING*   57

Address: Gümüssuyu Complex, Taksim Square, Istanbul
Dean: Professor Dr Mustafa Gediktas

Agricultural Machinery Chair*   58

Senior staff: Professor Dr I. Hakki Öz, Professor Dr Irfan Saygili

Agricultural Machinery Test and Research Centre*   59

Istanbul Üniversitesi*   60

[Istanbul University]
Address: Horhor Caddesi 13, Fatih, Istanbul
Telephone: 211646
Status: Educational establishment with r&d capability
Rector: Professor Dr Cem'i Demiroglu
Vice-Rectors: Professor Dr Fikri Senocak, Professor Dr Akin Ilkin
Pro-Rector: Professor Dr Haluk Alp

FACULTY OF FORESTRY   61

Address: Büyükdere, Istanbul
Telephone: 62 40 50-56
Dean: Professor Selçuk Bayoğlu
Departments: Forest botany, Professor Muzaffer Selik; silviculture, Professor İbrahim Atay; soil science and ecology, Professor Necmettin Çepel; forest entomology and protection, Professor Hasan Çanakçioğlu; watershed management, Professor Selman Uslu; forest products chemistry, Professor Savni Huş; surveying and photogrammetry, Professor Tahsin Tokmanoğlu; forest policy, Professor Metin Özdönmez; forest products, Professor Yilmaz Bozkurt; forest yields and biometry, Professor Abdülkadir Kalipsiz; forest economy, Professor Muharrem Miraboğlu; forest law, Professor Hayri Bayraktaroğlu; forest engineering, Professor Orhan Uzunsoy; landscape architecture, Professor Ibrahim Atay; forest management, Professor Ismail Eraslan; Institute of Forest Products, Professor Yilmaz Bozkurt
Graduate research staff: 78
Activities: Basic and applied research in soil science, forest hydrology, forestry sciences, and forest products.
Publications: Orman Fakültes i Dergisi, biannual faculty journal.
Projects: Papermaking properties of tobacco stalks pulping characteristics of fast growing tree species: Pinus maritima, radiata pinea (Professor Dr Turan Tank); bleaching experiments on wood pulps (Dr Erol Göksal); rational use of manpower in roundwood transport (Professor Dr Selçuk Bayoğlu); rationalization of forest transportation in Turkey (Dr Ö. Bulend Seçkin); hydrological research in Belgrad Forest near Istanbul to study watershed management techniques in order to increase yield and quality of water for Greater Istanbul: nutrient cycling in mature oak-beech ecosystems within experimental watersheds; sediment discharge and precipitation-stream flow relationships in experimental watersheds instrumented with sharp crested V-notch weirs; water-balance studies in the same experimental watersheds (Professor Dr Nihat Balci); application of aerial photo interpretation to the watershed surveys (Ahmet Hizal); water-repellent soils affected by different types of vegetation, parent material, and forest fires in Armutlu peninsula, Turkey (Kamil Şengönül).

FACULTY OF VETERINARY MEDICINE* 62

Dean: Professor Dr Kemal Ozan

Karadeniz Bölge Zirai Araştirma Enstitüsü 63

[Black Sea Regional Agricultural Research Institute]
Address: PK 39, Samsun
Telephone: (361) 12087
Parent body: Ziraat Isleri Genel Müdürlüğü
Status: Official research centre
Departments: Maize breeding and production, Cengiz Güler; vegetable crops breeding and production, Aydin Uçak; wheat and barley breeding and production, Burhan Ergin; oil crops and tobacco breeding and production, Metin Kara; poultry husbandry Dr Ibrahim Okçu
Graduate research staff: 16
Activities: Agronomic studies on and improved varieties of maize, wheat, barley, sunflower, tobacco, and vegetable crops for the Black Sea region of Turkey. The institute conducts 8 national projects on these crops.
Publications: Annual report (in Turkish).
Projects: Improvement of: maize (Yusuf Ergün); second crop (Dr Atila Altinay); vegetable crops (Dr Ertekin Genç); wheat (Kamil Yakar); barley (Kamil Yakar); sunflower (Dr Erdoğan İndelen); tobacco (Dr Turhan Atay); poultry (Dr Avni Başdoğan).

Karadeniz Teknik Universitesi* 64

[Black Sea Technical University]
Address: Trabzon

FACULTY OF FORESTRY* 65

Dean: Professor H. Peker

Kavak ve Hizli Gelişen Orman Ağaçlari Araştirma Enstitüsü 66

[Poplar and Fast Growing Exotic Forest Trees Research Institute]
Address: PK 93, İzmit, Kocaeli
Telephone: (0211) 11878
Parent body: Ministry of Agriculture and Forestry
Status: Official research centre
Director: Orhan Acar
Departments: Silviculture, Ulvi Tolay; genetics, Yaşar Şimşek; entomology, Niyazi Yildiz; pathology, Metin Vural; technology, Acar Umaç; economics, Sencer Birler; biometrics, Orhan Acar
Graduate research staff: 40
Activities: Reafforestation; technical and economic data for man-made forests. The institute is conducting 86 research projects on poplar and encalyptus cultivation and reafforestation problems.
Publications: Annual bulletin.

Manisa Bağcilik Araştirma Enstitüsü Müdürlüğü 67

[Manisa Agricultural Research Institute]
Address: PK 12, Manisa
Telephone: (5511) 1293
Parent body: Ministry of Agriculture and Forestry
Status: Official research centre
Director: Ahmet Çalişkan
Departments: Vine growing and breeding, İsmail İlhan; vine physiology, Ahmet Çalişkan; plant protection: vine pathology, Atilâ Ertem; plant protection: vine damage, Niyazi Kacar; raisins technology, Engin Karagözoğlu; economy Mehmet Candemir; vine sapling growing, Mahmut Gerenli
Graduate research staff: 15
Activities: Vine growing, breeding, and training, vine affinity and adaptation, vine physiology and pruning, improved yield and quality of grapes and raisins, ampelography of grape varieties, vine pathology and damage, grape drying methods. The institute takes part in the national viticulture research project.
Projects: Ascertaining standard varieties of grape (Ahmet Çalişkan); clone selection and viticulture (İsmail İlhan); affinity and adaptation of rootstocks in vineyard regions (Nejat Yilmaz); training systems for vineyard regions (İsmail İlhan).

Marmara-Trakya Bölge **68** Zirai Araştirma Enstitüsü

[Marmara-Thrace Regional Agricultural Research Institute]
Address: PK 3, Halkali, Istanbul
Telephone: (11) 79 60 01
Parent body: Ziraat Isleri Genel Müdürlüğü
Status: Official research centre
Director: Enver Hüsemoğlu
Sections: Oil crops research department, Yüksel Yazici; cereal crops research department, Numan Ünman; forage crops research department, Fikret Acaroğlu; seed certification laboratory, Dr Nebahat Döşlüoğlu; technology laboratory, Nurhan Erol; chemical laboratory, Nezahat Baysal; vegetable crops research department, Esin Ünal
Graduate research staff: 20
Activities: Plant production - sunflowers, wheat, barley, oats, tomatoes, capsicam, carrots; plant breeding - sunflowers, wheat, barley.
Publications: Annual progress reports.
Projects: Research and education: sunflowers (Yüksel Yazici); wheat (Numan Ünman); barley (Muzaffer Kocaoğlu).

Ormancilik Araştirma **69** Enstitüsü *

[Forestry Research Institute]
Address: PK 24, Bahçelievler, Ankara
Telephone: (041) 13 17 34
Status: Official research centre
Director: Dr Osman Taşkin
Divisions: Afforestation and reafforestation, Ö. Sirri Erkulovlu; silviculture, Dr Ergün İlter; forest management, Dr Oktay Özkazanç; forest protection, Nejat Giray; forest economics, Celâl Çoban; land improvement and range management, forest products, Dr Yüksel Topçuoğlu; forestry public relations, Yalçin Anil; mathematical statistics, Dr Osman Sun
Graduate research staff: 689
Activities: Forest tree seed problems, nursery practice, afforestation problems, provenence trials and testing of fast-growing species, forest genetics, tending of stands, forest botany, plant sociology, mensuration, forest management, yield, forest fires, entomology, phytopathology, soil conservation, torrent control, forest influences, range management, work efficiency, logging and transportation, forest economics, standardization of forest products, wood anatomy, pathology, and conservation, physical, chemical, and mechanical properties of wood, wood in material, wood-based panels, minor forest products, application of mathematical statistics in forest research, forest inventory, forestry public relations.
Publications: Annual report (in Turkish); journal (in Turkish); technical bulletins (summaries in English, French, and German).

Orta Anadolu Bölge Zirai **70** Araştirma Enstitüsü Müdürlüğü

[Central Anatolian Regional Agricultural Research Institute]
Address: PK 226, Ankara
Telephone: (941) 15 32 44
Telex: 42 994 CIMY-TR
Parent body: Ziraat Isleri Genel Müdürlüğü
Status: Official research centre
Director: Dr Baydur Yilmaz
Departments: Agronomy, Dr Mengü Güler; breeding, Dr Kâmil Yakar; seed technology, Ayhan Atli; pathology, Lütfi Çetin; horticulture, Dr Emine Maden; field crops, Atakan Günay
Graduate research staff: 40
Activities: Cereal breeding, pathology, technology, and agronomy; food legume breeding and agronomy; soil preparation; weed control; horticulture (agronomy). The institute conducts research for the national cereals research project.
Publications: Annual report (in Turkish).
Projects: Cereals: pathology, breeding, seed technology, and agronomy (Dr Kâmil Yakar); food legumes: pathology, breeding, seed technology, and agronomy (Kader Meyveci); horticulture (Dr Emine Maden).

Pendik Veterinar **71** Mikrobiyoloji Enstitüsü *

[Pendik Veterinary Microbiological Institute]
Address: Istanbul
Telephone: 540100
Affiliation: Ministry of Agriculture and Forestry
Status: Official research centre
Director: Dr Mahmut Kurtkaya
Sections: Anaerobiology, Dr Sedat Güven; virology, Dr Hayri Ergin; mycoplasma, Dr Fahri Arisoy
Graduate research staff: 10
Activities: Improvement of vaccines against infectious diseases that affect domestic animals.
Publications: Biannual journal.

Sebzecilik Araştirma Enstitüsü * 72

[Vegetable Research Institute]
Address: PK 130, Antalya
Telephone: (3111) 1468
Affiliation: Directorate of Horticulture Affairs
Status: Official research centre
Director: Dr Erdoğan İbrişim
Departments: Plant production, vacant; plant breeding, Ertekin Genç; plant physiology, Cumhur Çetin; crop husbandry, Mevlüt Doğan; plant protection, Kamil Yelboğa
Graduate research staff: 20
Activities: Protective covering materials, such as glass, polyethylene, and ultraviolet transparent materials, for forcing vegetables in given climatic conditions; protective constructions for plants; mechanization: plant protective constructions and materials. Breeding of F, tomato hybrids suitable for growing in glass or polythene greenhouses, and resistant to soil-borne diseases such as verticillium and fusarium wilts.
Projects: Techniques and materials for growing plants under protective covering (Dr Erdoğan İbrişim); tomato breeding (Ertekin Genç).

Su Ürünleri Genel Müdürlüğü * 73

[General Directorate of Fisheries]
Address: Olgunlar Sokak 10, Bakanliklar, Ankara
Telephone: 258221
General director: Irfan Sahin
Departments: Marine Resources, Muzaffer Bumin; Inland Resources, Ismail Mert; Environmental Pollution Control, Ülkü Merter; Economics, Ibrahim Özbek
Graduate research staff: 30
Activities: Independent or joint research with international bodies on water pollution control.

Tavukculuk Araştirma Enstitüsü 74

[Poultry Research Institute]
Address: PK 47, Yeni Mahalle, Ankara
Telephone: (941) 15 60 35; 15 34 24
Parent body: Ziraat Isleri Genel Müdürlüğü
Status: Official research centre
Director: Ilhami Koca
Departments: Poultry research, Dr Mehmet Karazeybek
Graduate research staff: 16

Activities: Day-old hybrid chicks.
Publications: Teknik Tavukculuk Dergisi, quarterly.
Projects: Day-old white egg laying hybrid chicks - sex separation according to feather: parent and grandparent lines development (Ramazan Yetişir, Zekeriye Atik); brown egg layer parent lines development (Ibrahim Büyükbebebeci, A. Gazi Boga); white egg layer parent lines development (Bekir Kadioğlu, Halil Bilici).

Topraksu Genel Müdürlüğü 75

[General Directorate of Topraksu]
Address: Ulus, Ankara
Telephone: (941) 24 31 40
Parent body: Ministry of Village Affairs and Cooperatives
Status: Official research organization
General director: Erdoğan Bilgiç
Departments: Research, Somer Sarikatipoğlu
Graduate research staff: 187
Activities: Soil science; drainage and irrigation; plant production; agricultural engineering.
Publications: Annual report; research project reports.
Projects: Hydrology (Rifat Önal); soil and water conservation (Dr Orhan Doğan); agronomy (Nurettin Kayitmazbatur); drainage (Atilla Mavi); soil fertility; statistics and economics (Dr Necdet Yurtsever); irrigetion and soil physics (Dr Riza Kamber).

Türkiye Bilimsel ve Teknik Aratirma Kurumu * 76

[Scientific and Technical Research Council of Turkey]
Address: Atatürk Bulvari 221, Kavaklidere, Ankara
Telephone: 262770
Telex: 43186 BTAK TR
Status: Official research organization
President of the science board: Professor Dr Ratip Berker
Secretary general: Professor Dr Tevfik Karabag
Groups: Mathematical, physical and biological sciences research, Professor Dr Nihat Bozcuk; engineering research, Professor Dr Mehmet Ergin; medical research, Professor Dr Izzet Berker; veterinary medicine and animal husbandry research, Professor Dr Tayyip Calislar; agriculture and forestry research, Professor Burhan Kacar; environmental research, Professor Dr Ugur Büget
Activities: Development, organization and coordination of basic and applied research in natural, engineering, agricultural and veterinary sciences and in medicine.

Turunçgiller Araştirma Enstitüsü 77

[Citrus Research Institute]
Address: PK 35, Antalya
Telephone: (3111) 1465
Parent body: Ministry of Agriculture and Forestry
Status: Official research centre
Director: Kilinçarslan Morali
Assistant director: Dr A. Yilmaz Hizal
Departments: Horticulture, H. Yener Apaydin; fruit technology, E. Necdet Bağriyanik
Graduate research staff: 2
Activities: Plant production: plant breeding citrus, avocado, pecan nut, loquat, olive; crop husbandry - citrus, avocado, pecan nut, loquat. Food science - citrus and olive fruits. Agricultural and food economics - citrus, avocado, pecan nut, loquat, olive.
Projects: Citrus research and education (Türker Göral); pecan nut adaptation (Esma Faraçlar); loquat selection and adaptation (Şenes Demir); avocado adaptation (H. Ayhan Doğrular); olive selection and adaptation (Ahmet Salman).

Tütün Araştirma ve Eğitim Enstitüsü 78

[Tobacco Research and Training Institute]
Address: PK 9, Izmir
Telephone: (51) 14 91 31
Parent body: Ziraat Isleri Genel Müdürlüğü
Status: Official research centre
Director: Dr Maksut Selçuk
Departments: Tobacco genetics and breeding, Reşat Apti; plant protection, Ahmet Usturali; tobacco growing techniques, Yücel Müftüoğlu; tobacco technology, Attilâ Aykor; tobacco chemistry, Gülden Yazan; economy, Dr Ahmet Ağmaz
Graduate research staff: 13
Activities: Tobacco breeding: varieties resistant to disease and pests, superior in yield and quality to existing local varieties; fertilization, irrigation, manipulation, maintenance; oily tobacco seeds.
Projects: Tobacco research and training: main project (Dr Maksut Selçuk); breeding tobacco resistant to blue mould, analysis of tobacco populations in eastern and southeastern parts from the standpoint of morphology, yield, and quality; combination of features desired through hybridization method of breeding (Reşat Apti); hereditability of some morphologic and pathologic characters of tobacco, (Hacer Otan); resistance to economically important tobacco diseases (Ahmet Usturali); most suitable time and method of applying chemical fertilizers to Aegean tobaccos, effect of fertilizer levels on yield and quality of tobacco (Yücel Müftüoğlu); most convenient duration of time of natural fermentation process with Aegean tobaccos, effect of picking hards on yield and quality of seed tobacco, (Attilâ Aykor); breeding virus-resistant tobacco (Fatma Koca); effect of spacing on yield and quality of crop (Yücel Müftüoğlu); breeding and standardization of Turkish commercial types of tobaccos (Dr Maksut Selçuk); fertilizer experimentation with Bursa kind of tobaccos; blue mould resistant regional varieties: uniformity trials (Ülker Koyuncuoğlu).

Zeytincilik Araştirma Enstitüsü 79

[Olive Culture Research Institute]
Address: Bornova, Izmir
Telephone: (51) 16 10 35
Parent body: Ministry of Agriculture and Forestry
Status: Official research centre
Director: Ihsan Dikmen
Sections: Olive cultivation, Abdülgani Cavusoğlu; plant nutrition, Özgül Canözer; olive oil laboratory, Ayfer Pala; pickled olives laboratory, Nergiz Özgilmaz; economics and marketing, Birol Akman
Graduate research staff: 35
Activities: Olives: breeding, physiology, cultivation and harvesting techniques, plant health and processing, economics and marketing.
Publications: Annual research reports.
Projects: Clonal selection of olive trees and olive rootstock breeding; bed-type olive culture (Dogan Usanmaz); adaptation of olive trees (Hasan Özyilmaz); three types of fertilization application in intensive plantations (Mesut Çakir, O. Canözer); bioclimatological studies on olives (Ayhan Sayin); seasonal variations of macro and micro nutrients in Memecik cultivar (Gulter Püskülcü); methods to be used in determination of receivable phosphorous stocks in the olive groves of Aegean region soils (Mesut Çakir); determination of the amounts of nitrogen, potassium, and phosphorous removed by the seed and from the soil in different aged olive trees (Ulker Dimelik); effect of different treatments applied to the endocarp during germination of olive seeds; quantity of abscissic acid in olive seeds, and the variations occuring in storage and stratification processes (Bilman Yüce); effects of potassium on yield and some quality criteria of Gemlik olive cultivar (I. Moltay, C. Genc, H. Cetin); effects of fertilizers applied to leaves and soil on the yield and quality of Memecik olive cultivar (Özgül Canözer, Ayşe Çolakoğlu); cultivating short-trunked trees in intensive olive culture and the effect of periodic vegatative pruning on productivity of olive groves; rejuvenation and renovation of olive trees; harvesting effect of a multidirectional shaker on two major cultivars in Aegean region (Abdulgani Çavusoğlu); distances and space in intensive plantations (Ihsan Dikmen); three

different forms and two different spaces for some standard olive cultivars in olive groves planted in the Gaziantep area (A. Ulusaraç); comparisons of materials (grafted sapling and own-rooted sapling) used in establishment of olive groves (Ayten Uluskan); three irrigation methods and their effect on quality and quantity of olive trees; determination of wild and foreign olive varieties and establishment of collection grove (Özgül Canözer); determination of seasonal variations of soft wood cuttings rooted by the mist propagation method in some olive cultivars (Yahya Luma); natural maturation of olive fruit (Ayfer Pala, Abdülgani Çavusoğlu); physical and chemical properties of Turkish olive oils (Ayfer Pala, Aysun Oktar); quality of olive oils supplied from factories applying processing techniques in Gaziantep (N. Berk, R. Karaca, C. Kuru); reduction of the amount of oil that cannot be extracted from the olive residue with the addition of enzyme and the effects of the enzyme on the quality of the oil (Hasan Pala, Tamay Işikli); convenience of Gemlik variety for the production of Spanish type green olives (H. Çetin, F. Fidan); determination of sterol in Turkish olive oils (Ayfer Pala, Aysun Oktar); Turkish olive residue industry and the possibilities of olive residue as an edible oil (Hikmet Acar, R. Karaca, Tamay Işikli); preparation technology of marked olives (Birol Akman); technological investigation of products in factories applying dry and water systems (Hikmet Acar, Ayfer Pala); effects of various soil and fertilizer materials and their mixtures on vegetative growth in olive tree plantations (I. Moltay); changes in tocopherols, sterols and triterpenic alcohols in olive oil structure, during the storage of olives (Bahri Ersoy); preparation of Kalamata type (with vinegar) table olives (B. Erdemli).

Zeytincilik Araştirma Istasyonu, Edremit 80

[Edremit Olive Research Station]
Address: PK 40, Balikesir - Edremit
Telephone: (6640) 13781, 1557
Status: Official research centre
Director: Erhan Atalay
Departments: Olive growing, Hamdi Mustafa Dinçer; breeding and physiology, Yahya Luma technology of olives and olive oil, M. Bakir Erdemli
Graduate research staff: 8
Activities: Olive breeding: physiology and technology of olives and olive oil.
Projects: Clonal selection of olives; some olive varieties: seasonal variation in rooting semi-hardwood olive cuttings under mist (Yahya Luma); adaptation of some olive varieties on Edremit gulf region (Hamdi Mustafa Dinçer); preparing marked green table olives; black table olive technology; calamata-style with vinegar table olive preparation (M. Bakir Erdemli).

Ziraat Isleri Genel Müdürlüğü* 81

[General Directorate of Agricultural Research]
Address: Bakanliklar, Ankara
Telephone: 257560
Parent body: Ministry of Agriculture and Forestry
Status: Official research centre
Director: Dr Hayati Ölez
Graduate research staff: 180
See separate entries for: Akdeniz Bölge Zirai Araştirma Enstitüsü
Çayir-Mer'a ve Zootekni Araştirma Enstitüsü
Ege Bölge Zirai Araştirma Enstitüsü
Güneydoğu Anadolu Bölge Zirai Araştirma Enstitüsü Müdürlüğü
Ipekböcekçiliği Araştirma Enstitüsü
Karadeniz Bölge Zirai Araştirma Enstitüsü
Marmara-Trakya Bölge Zirai Araştirma Enstitüsü
Orta Anadolu Bölge Zirai Araştirma Enstitüsü Müdürlüğü
Tavukculuk Araştirma Enstitüsü
Tütün Araştirma ve Eğitim Enstitüsü
Zirai Araştirma Enstitüsü, Adana
Zirai Araştirma Enstitüsü, Edirne
Zirai Araştirma Enstitüsü, Eskişehir
Zirai Araştirma Enstitüsü, Gaziantep
Zirai Araştirma Istasyon Müdürlüğü, Isparta
Zirai Araştirma Istasyonu Müdürlüğü Igdir-Kars
Publications: Agricultural research journal.

ANIMAL HUSBANDRY RESEARCH DEPARTMENT* 82

Head: Dr Sezer Sabaz
Activities: Breeding and nutrition research.

FIELD CROPS RESEARCH DEPARTMENT* 83

Head: Yusuf Ergün
Activities: Crop science - breeding, agronomy, quality, economics, machinery development, plant protection, seed technology, range management.

RESEARCH CENTRES* 84

Agricultural Mechanization Research Institute* 85

Address: PO Box 226, Ankara

Agricultural Research Institute* 86

Address: PO Box 37, Afyon

Agricultural Research Institute*   87

Address: PO Box 25, Sakarya

Regional Agricultural Research   88
Institute*

Address: PO Box 257, Erzurum

Regional Variety Testing Directorate*   89

Address: PO Box 708, Ankara

Seed Testing and Certification   90
Institute*

Address: PO Box 415, Ankara

Seed Testing and Certification Station*   91

Address: Tarsus, Içel

Trial Station*   92

Address: PO Box 44, Kars

Zirai Araştirma Enstitüsü, Adana   93

[Adana Agricultural Research Institute]
Address: PK 300, Adana
Telephone: (711) 14558
Parent body: Ziraat Isleri Genel Müdürlüğü
Status: Official research centre
Director: Dr H. Bostancioğlu
Sections: Animals, Mehmet Güneyli; field crops, Isa Kafa; livestock, Cumali Saçmali
Graduate research staff: 2
Activities: Plant and animal production.
Projects: Wheat breeding (Isa Kafa); animal production (Mehmet Güneyli); livestock breeding (Cumali Saçmali); second crops (Dr H. Bostancioğlu)

Zirai Araştirma Enstitüsü, Edirne   94

[Edirne Agricultural Research Institute]
Address: PK 16, Edirne
Telephone: (1811) 1144; 1739
Parent body: Ziraat Isleri Genel Müdürlüğü
Status: Official research centre
Director: Erdoğan Indelen
Assistant director: Bülent Kiral
Departments: Cereals, Necati Hazar; oil seed crops, A. Özden Uludere; rice, Halil Sürek
Graduate research staff: 8
Activities: Plant yield, quality, and disease resistance: breeding, protection, agronomy, and seed production in wheat, barley, sunflowers, rice, and rapeseed; technological analysis of rice and sunflowers. The institute conducts national projects on sunflowers and rice.
Publications: Annual progres reports on projects.
Projects: Sunflowers (A. Özden Uludere, Erdoğan Indelen); rice (Halil Sürek, Bülent Kiral); rapeseed adaptation (Meliha Salihoğlu).

Zirai Araştirma Enstitüsü, Eskişehir   95

[Eskişehir Agricultural Research Institute]
Address: PK 17, Eskişehír
Telephone: (221) 11030
Parent body: Ziraat Isleri Genel Müdürlüğü
Status: Official research centre
Director: Dr Fahri Altay
Departments: Cereals, Turgut Çetinel; industrial crops, Bayram Bolat; food legumes, Muzaffer Işik; vegetable Tunçay Çetinel
Graduate research staff: 24
Activities: Plant breeding, crop husbandry, plant protection – wheat, barley, oats, food legumes, potatoes, maize, oil crops, vegetables.
Publications: Annual report.
Projects: Wheat (Bertan Süzen); barley (Bahattin Erginel); oats (Fikret Kaya); food legumes (Ziya Önceler); potatoes (Bayram Bolat); maize (Ural Soyer); oil crops (Derviş Engin); vegetables (Havva Kiraç).

Zirai Araştirma Enstitüsü 96
Gaziantep

[Gaziantep Agricultural Research Institute]
Address: PK 32, Gaziantep
Telephone: (851) 20363; 11057
Parent body: Ministry of Agriculture and Forestry
Status: Official research centre
Director: Ahmet M. Bilgen
Departments: Pistachio, Necip Uygur; viticulture, Temel Özen; olive, Azmi Ulusaraç; production, Gani Şekerden; technology, Nilüfer Berk; poultry, Özel Şekerden
Graduate research staff: 15
Activities: Plant production: pistachios - pistacia vere seedling; nursery stock. Animal production: poultry.
Projects: Grafting of Pistacia khinjuk seedlings in South-east Anatolia region (Ahmet M. Bilgen); rootstock selection for; quantity of Pistacia synthetic fertilizer for pistachios (Necip Uygur); artificial pollination methods in pistachios (Celal Kuru); selection of male types in pistachio (Necip Uygur, Celal Kuru); selection of female types in pistachio (Ahmet M. Bilgen); technologycal aspects of some pistachio varieties in the same caring and soil conditions (Nilüfer Berk); standard grape varieties (Temel Özen).

Zirai Araştirma Istasyon 97
Müdürlüğü, Isparta

[Isparta Agricultural Research Institute]
Address: Atatürk Bulvari, Isparta
Telephone: 3259
Parent body: Ziraat Isleri Genel Müdürlüğü
Status: Official research centre
Director: Ilyas Özmen
Departments: Essential oils, Erdal Güngör; aetheral and medical plants, M. Ümit Yağci, M. İlhan Tortopoğlu, G. Karataş
Graduate research staff: 6
Activities: Plants that supply ran materials for use in medicine, perfume, and cosmetics - agricultural and technological problems.
Projects: Cultivation and breeding of aetheral and medical plants (Ilyas Özmen); mint and thyme (Osman Nuri Aydin, Erdal Güngör); rose and lavender (M. Ilhan Tortopoğlu); astragalus (G. Karataş); etheral and medical plant genera (M. Ünit Yağci).

Zirai Araştirma İstasyonu 98
Müdürlüğü Iğidir-Kars

[Iğidir-Kars Agricultural Research Station]
Address: PK 12, IğdIr - Kars
Parent body: Ziraat Isleri Genel Müdürlüğü
Status: Official research centre
Director: Osman Can
Sections: Wheat, Ibrahim Sancar; barley, Ekrem Keski; sunflowers, Şükrü Kayak; grassland and forage crops, Ali Alagöz
Graduate research staff: 2
Activities: Plant production: wheat, barley, sunflowers, forage crops.

Zirai Mücadele ve Zirai 99
Karantina Genel
Müdürlügü*

[General Directorate of Plant Protection and Plant Quarantine]
Address: Necatibey Caddessi 98, Ankara
Telephone: 293575
Affiliation: Ministry of Agriculture and Forestry
Status: Official research organization
Director general: Vehbi Kesici
Graduate research staff: 267
Activities: The General Directorate is composed of 7 regional plant protection research institutes, 1 plant protection chemicals and equipment institute and 1 regional biological control research station. 228 projects are being carried out which include studies on the bioecology and control methods of plant diseases, pests and weeds of economic importance as well as registration of plant protection chemicals to be used in Turkey.
Publications: Plant Protection Research Annual; Plant Protection Bulletin.

UGANDA

Animal Health Research Centre 1

Address: Box 24, Entebbe
Telephone: (042) 20915/20192
Status: Official research centre
Director: G.L. Corry
Graduate research staff: 20
Activities: Veterinary diseases of importance to Uganda.
Publications: Annual report.

BACTERIOLOGY DEPARTMENT 2

Head: Dr Y.K. Ssentongo
Projects: Research into mycoplasma and brucellosis (Dr D.G. Ndyabahinduka).

ENTOMOLOGY DEPARTMENT 3

Head: J. Okello-Onen
Projects: Tick ecology and physiology (J. Okello-Onen).

HELMINTHOLOGY DEPARTMENT 4

Head: Dr D.W. Kakaire
Projects: Research into fasciolasis.

Kawanda Research Station 5

Address: Box 7065, Kampala
Telephone: (041) 67621-3
Status: Official research centre
Acting director: A.A. Mukasa-Kiggundu
Graduate research staff: 60

Activities: Natural resources; plant production; annual crops (maize, beans); perennial crops (coffee, cocoa, sugarcane).
Publications: Annual report.

AGRONOMY DEPARTMENT 6

Head: W.A. Sakira
Projects: Sugarcane agronomy (W.A. Sakira); legume research (T. Sengooba, D. Mulindwa).

BOTANY DEPARTMENT 7

Head: E.K. Rubaihayo
Projects: Cereals research.

COFFEE RESEARCH UNIT 8

Head: Dr Kibirige-Sebunya
Projects: Coffee, cocoa, oil palm research.

ENTOMOLOGY DEPARTMENT 9

Head: N.D. Bafokuzara
Projects: Crop protection.

HORTICULTURE DEPARTMENT 10

Head: Y.W. Mwaule
Projects: Pomology research (Y.W. Mwaule); vegetables research (J. Hakiza).

SOIL SCIENCE DEPARTMENT 11

Head: A.A. Mukasa-Kiggundu
Projects: Fertilizer trials (J.B. Kavuma); soil survey and soil classification (E.V. Sendiwanyo); soil chemistry research (A.A. Mukasa-Kiggundu).

Makerere University   12

Status: Educational establishment with r&d capability

FACULTY OF AGRICULTURE AND   13
FORESTRY

Address: PO Box 7062, Kampala
Telephone: (041) 56931-2; 42277
Dean: Professor J.S. Mugerwa
Graduate research staff: 52
Activities: The faculty carries out research within the university farm. The production-oriented research covers agricultural engineering, animal production (dairy cattle, pigs, poultry, goats - breeding, nutrition, husbandry), crop production (cereals, root crops, fruits, horticultural crops - breeding, husbandry (agronomy), protection, agro-forestry, soil chemistry, and farm management economics.
Publications: Annual research bulletin.

Agricultural Economics Department   14

Head: J.R. Bibangambah
Projects: Economics of small-scale farming (J. Nsereko); agricultural marketing (J.R. Bibangambah).

Agricultural Engineering Department   15

Head: E.W. Rugumayo
Projects: Solar energy/biogas products.

Animal Science Department   16

Head: J.S. Mugerwa
Projects: Use of by-products in animal feeding; forage production and utilization (J.S. Mugerwa); goat production (G.H. Kiwuwa).

Crop Science Department   17

Head: J.K. Mukiibi
Projects: Cropping systems (J.K. Mukiibi, D. Osiru); banana research (J.C. Ddungu); crop protection (J.K. Mukiibi).

Forestry Department   18

Head: A.J.W. Aluma
Projects: Agro-forestry.

Soil Science Department   19

Head: J.Y.K. Zake
Projects: Rhizobia research; soil fertility.

FACULTY OF VETERINARY SCIENCE*   20
Head: F.I.B. Kayanja

Namulonge Research   21
Station

Address: PO Box 7084, Kampala
Telephone: Namulonge 3
Telex: COTREST KAMPALA
Status: Official research centre
Director: E.K. Byaruhanga
Sections: Botany (breeding and pathology), C.A.D. Tadria; crop husbandry (agronomy, soil science), J.C.W. Odongo; entomology, Dr D. Maloba; animal husbandry and pasture, G.W. Napulu.
Graduate research staff: 28
Activities: Cotton breeding, pathology, pest control and agronomy. Soil science work is done on the use of fertilizers on beans, maize and cotton. There is also work on maize agronomy, use of herbicides, livestock nutrition, pasture agronomy and diseases of cassava and horticultural crops.
Publications: Annual report.
Projects: Cotton breeding (C.A.D. Tadria); cassava and cotton pathology (J. Kasirivu); agricultural meteorology (J.C.W. Odongo); crop fertilizer use (A.F. Kintukwonka); herbicides (M.P.E. Wetala); pasture agronomy (G.W. Napulu); cotton pest control (Dr D. Maloba); animal nutrition (Dr Y. Ajeabu).

Uganda Agriculture and   22
Forestry Research
Organization

– UAFRO
Publications: Annual record of research.

SORGHUM AND MILLETS UNIT,   23
SERERE RESEARCH STATION

Address: PO Soroti, Kampala, Soroti District/Eastern Region
Telephone: Serere 1434
Status: Official research centre
Principal research officer-in-charge: Vincent Makumbi Zake
Graduate research staff: 3

Activities: Plant production in sorghum finger millet and pearl millet through plant breeding procedures. Utilizing the existing considerable genetic variability, high grain yielding and high quality grain genotypes with desirable agronomic characteristics (resistant to lodging, insects, weathering and mould diseases, and tolerant to droughts) have been produced using traditional and population breeding methods.

Projects: Improve grain yield and grain quality of sorghum which is drought escaping and is resistant to leaf, weathering and moulding diseases; storage pests; to improve grain yield and grain quality of finger millet which is early maturing, resistant to blasts (Piricularia oryzea) lodging, and stalk borers (Chilo partelus and Sesemia species) (V. Makumbi Zake, B.W. Khizzah, J.R. Okello); improve grain yield in pearl millet which is early maturing, resistant to leaf and head diseases, birds and lodging (V. Makumbi Zake).

Uganda Forestry Department 24

Chief forest officer: B.K. Mwanga; *Deputy chief forest officer:* L. Kiwanuka
Utilization officer: Mr Kityo
Graduate research staff: 40
Publications: Woodsman (quarterly).

FORESTRY COLLEGE, UGANDA 25

Status: Educational establishment with r&d capability
Principal: Mr Naluswa

NAKAWA FORESTRY RESEARCH CENTRE 26

Address: PO Box 1752, Kampala
Telephone: (041) 56261-2; 59626
Telex: FORESTRY, KAMPALA
Status: Official research centre
Principal research officer: Peter K. Karani
Graduate research staff: 20
Activities: Forestry and the utilization of forest products; the environment; biology; silviculture; work science and studies; harvesting of wood: logging and transport; forest engineering; forest injuries and protection; forest mensuration, increment, development and structure of stands; survey and mapping; forest management; business economics of forestry; administration and organization of forest enterprises; marketing of forest products; economics of forest transport and wood industries; forest products and their utilization; social economics of forestry.

Projects: Power and energy (J. Calvaroh); charcoal (Mr Kawanguzi); timber preservation (Mr Kawoya); timber research (Mr Kasirye); entomology (Mr Oloya); forestry silviculture (Mr Kwikiriza); forestry protection (Mr Serwanga); forestry training (Mr Naluswa); senior silviculture (Mr Z. Dutchi); stand development of plantation grown trees (P.K. Karani); tree breeding (R. Musoke); wood utilization: logging wood properties (J. Calvaroh).

Uganda Freshwater Fisheries Research Organization 27

– UFFRO
Address: PO Box 343, Jinja
Telephone: (043) 20484
Status: Official research centre
Director: A.W. Kudhongania
Departments: Limnology, F.W. Bugenyi; population dynamics, D.L. Ocenodongo; fish biology, Dr T. Twongo; fish taxonomy, P. Basasibwaki; fishery economics, P. Karuhanga
Graduate research staff: 20
Activities: The aim of UFFRO is to supply adequate scientific information for the rational development, management and conservation of the fishery resources of Uganda with a view to fulfilling the nutritional, economic, and employment potential and providing scientific knowledge for the people and scientific community.
Publications: Annual report; *African Journal of Tropical Hydrobiology and Fisheries* (bi-annual).
Projects: Aquatic pollution on Lakes Victoria, George and Edward (F.W. Bugenyi); stock assessment studies on Lake Victoria (D.L. Ocenodongo); fishery of Lake Albert (T. Acere); fish population structure of Lake Kioga (Nile perch, Tilapia) (Dr T. Twongo); riverine investigations (J. S. Balirwa); economics of mechanized fishing on Lake Victoria (P. Karuhanga); investigations on minor lakes (J. Okaronon).

UNION OF SOVIET SOCIALIST REPUBLICS

All-Union VI Lenin Academy of Agricultural Sciences*   1

Address: Bolshoi Kharitonevsky Per 21, B78 Moskva, 107078
Status: Official research centre

All-Union Construction and Research Institute for Automation of Food Industry*   2

Address: ul Krasnova 6, Odessa
Status: Official research centre

All-Union Institute for Research in Karakul Sheep Breeding*   3

Address: ul K Marksa 47, Samarkand, 23
Status: Official research centre

All-Union Institute for Research in the Meat Industry*   4

Address: Industrial'naja 29, Semipalatinsk, 13
Status: Official research centre

All-Union Institute of Agricultural Microbiology*   5

Address: Gertsena 42, Leningrad
Status: Official research centre

All-Union Institute of Agricultural Power Engineering*   6

Address: Tomilinskaja ul 2, Moskva, 111 395

All-Union Institute of Fibre Crops Research*   7

Address: ul Lenina 45, Glerchov, Sumskaya Obl
Status: Official research centre

All-Union Legumes and Pulse Crops Research Institute*   8

Address: PO Streletskaye, Orel, 303112
Status: Official research centre

All-Union Plant Breeding and Genetics Institute* 9

Address: Ovidiopolskaja Dorogo 3, B36 Odessa, 270036
Status: Official research centre

All-Union Research and Construction Institute of Food Technology and Engineering* 10

Address: Novo-Chorosevskoe sosse 1, Moskva, 123308
Status: Official research centre

All-Union Research Institute for the Electrification of Agriculture* 11

Address: 2 I-St Veshnyakovsky pr, Zh-456 Moskva, 109456
Status: Official research centre

All-Union Research Institute of Animal Husbandry* 12

Address: Pos Dubrovitsky 14012, Paddsky rayon, Moskovskaya Oblast, 142012
Status: Official research centre

All-Union Research Institute of Dairy Industry* 13

Address: Ljusinovskaja ul 35, Moskva, 113093
Status: Official research centre

All-Union Research Institute of Experimental Veterinary Science* 14

Address: Kuzminki, Zh 472, Moskva, 109472
Status: Official research centre

All-Union Research Institute of Farm Animal Physiology and Biochemistry* 15

Address: Borovsk 249010, Kaluzhskaya Oblast
Status: Official research centre

All-Union Research Institute of Fermentation Products* 16

Address: Samokatnaja ul 46, Moskva, 109033
Status: Official research centre

All-Union Research Institute of Fertilizers and Soil Science* 17

Address: ul Prjanisnikova 31, Moskva, 127550
Status: Official research centre

All-Union Research Institute of Fish Farming* 18

Address: Pos Rybrios 141821, Moskovskaja Obl
Status: Official research centre

All-Union Research Institute of Forestry and Forestry Mechanization* 19

Address: Institutskaya 15, Puskino 141200, Moskovskaya Obl
Status: Official research centre

All-Union Research Institute of Grain Production* 20

Address: Shortandy, Celinogradskaya Obl
Status: Official research centre

All-Union Research Institute of Livestock Breeding and Genetics* 21

Address: Moskovskaya Shosse 550, Puskin, Leningrad, 188620
Status: Official research centre

All-Union Research Institute of Meat Industry* 22

Address: Ul Talalikhina 26, Moskva, 109029
Status: Official research centre

All-Union Research Institute of Medical Plants* 23

Address: PO Vilp 142790, Moskovskaya Obl
Status: Official research centre

All-Union Research Institute of Soil Erosion Control* 24

Address: ul Chelyuskintsev 28a, 4 Koursk, 305004
Status: Official research centre

All-Union Research Institute of Sugar Industry* 25

Address: ul Engelsa 20, Kiev, 24
Status: Official research centre

All-Union Research Institute of Tobacco and Markhova* 26

Address: PO Box 55, Krasnodar
Status: Official research centre

All-Union Research Institute of Winegrowing and Wine Production* 27

Address: R.F.F. S.R. Novocherkassk, 15
Status: Official research centre

All-Union Rice Research Institute* 28

Address: Pos Belozerny, Inskoi Royan, Krasnodarsky Krai, 353204
Status: Official research centre

All-Union Scientific Research Institute of Agricultural Economics* 29

Address: Orlikov Pereulok 1–11, Moskva
Status: Official research centre

All-Union Scientific Research Institute of Tea and Subtropical Plants* 30

Address: PO Anaseuli, Macharadze, Georgia
Status: Official research centre

All-Union Scientific Research Institute of Tea and Subtropical Plants* 31

Address: Ul Cacavadze 22, Suchumi
Status: Official research centre

All-Union Sugar Beet Research Institute*   32

Address: Kliniceskaja 25, Kiev
Status: Official research centre

Central Republic Botanical Garden   33

Address: Timiriazevskaya I, Kiev, 14
Telephone: 95 41 05
Affiliation: Ukranjan Academy of Sciences
Status: Official research centre
Director: Grodzinsky Andrei Mikhailovich
Publications: Introduction and acclimatization of plants.

Central Scientific Research Institute of the Silk Industry*   34

Address: Tepliy P 11, Moskva
Status: Official research centre

Moldavian Institute for Research in Irrigated Farming and Vegetable Growing   35

Address: Mira Str 50, 278013 Tiraspol, Moldavia
Telephone: 3 31 22
Affiliation: Ministry of Horticulture of Moldavia
Status: Official research centre
Director: V.P. Chichkin
Sections: Breeding V.C. Andryushcenco; industrial technology in vegetable production, V.G. Abacumov; irrigated farming, A.G. Scurtul

N.I Vavilov All-Union Research Institute of Plant Industry*   36

Address: ul Hertsena 44, Tsenti Leningrad, 190 000
Status: Official research centre

UNITED ARAB EMIRATES

Agricultural Experiment Station Digdaga 1

Address: PO Box 176, Ras Al Khaimah
Telephone: 28428
Telex: 99240 DIGDAG EM
Affiliation: Ministry of Agriculture and Fisheries
Status: Official research centre
Project manager: M. Hamad
Departments: Irrigation, A. Savva; plant protection, H.S. Abu Salih; horticulture, M. Hamad; soil management, Y.S. Puh; workshop, M. Pierconti
Activities: This centre is undertaking a project, on irrigation, crop water use, plant protection, soil fertility and plant nutrition, and fruit and vegetable production. The experimental results obtained are being transferred to the farmers of the United Arab Emirates in order to make them aware of modern agricultural technqiues.
Publications: Annual technical reports on irrigation, vegetables, fruits, plant protection and soil fertility.

Centre Expérimental Agricole 2

[Agricultural Experimental Centre]
Address: PO Box 1304, Al Ain, Abu Dhabi
Telephone: (03) 25125
Status: Research centre within an industrial company
Director: Jean-Pierre Moysan
Graduate research staff: 4
Activities: Vegetable production in arid zones, under greenhouses and sun shelters, and in open fields, using dipping irrigation.

UNITED KINGDOM

Agricultural Aviation Research Unit   1

– AARU
Address: Cranfield Institute of Technology, Cranfield, Bedfordshire MK43 OAL
Telephone: (0234) 75 08 51
Affiliation: Ciba Geigy, Basle
Status: Research centre within an industrial company
Director: R. J. Courshee
Departments: Chemistry, Dr S. Uk; engineering, Dr S. Parkin; entomology, Dr I. Outram; physics, Dr I. Lawson
Graduate research staff: 7
Activities: Reduction of crop losses caused by pests: pest bionomics monitoring and behaviour; nature, timing and degree of losses; selection of remedies including pesticides; pesticide application technology; interactions between pests and pesticides deposits; organization of resources for efficient operations; overall cost-benefit analysis.

Agricultural Research Council   2

– ARC
Address: 160 Great Portland Street, London, W1N 6DT
Telephone: (01) 580 6655
Telex: 291218 AGRECO
Affiliation: Department of Education and Science
Status: Official research organization
Chairman: Lord Porchester
Members: J.E. Cross, K. Dexter, Professor Sir Hugh Ford, J.S. Gibson, E.M.W. Griffith, R. Halstead, Professor J.L. Harper, Professor J. Heslop-Harrison, Professor D.L. Hughes, Professor J.L. Jinks, G. John, Professor Sir Hans Kornberg, C. Mackay, Professor J.

Mandelstam, Dr Anne McLaren, J. Maitland Mackie Jnr, W.H.G. Rees, The Earl of Selborne, E.J.G. Smith, Professor B.G.F. Weitz
Assessors to the Council: Dr W.O. Brown, Department of Agriculture, Northern Ireland; Professor J.M. Dodd, University of North Wales, Bangor
Assistant Secretaries, research: Animals and food, R.J. Harris; plants, soils and engineering, Dr B.G. Jamieson
Scientific Advisers to the Secretary: K.N. Burns, D.C.M. Corbett, Dr H. Fore, Dr J.K.R. Gasser, G. Jenkins, Dr J. Ingle, Dr J. Moorby, Dr J.S. Perry, Dr J.C. Tayler, Dr T.L.V. Ulbricht
Programmes Section: W.S. Wise
Clerk to the Council: E.S. Coltman
Information Officer: M.F. Goodwin
Graduate research staff: 3000
Activities: The Agricultural Research Council was established by Royal Charter in 1931, and its primary aims are to advance scientific knowledge relevant to agriculture, horticulture, and food supply; and to exploit this knowledge to increase the efficiency of the agricultural, horticultural, and food industries and to safeguard and improve the quality of food for the community
The Council has eight research institutes and six units under its direct control. In addition to these, there are other agricultural research institutes, fourteen in England and Wales and eight in Scotland, which, while retaining their own individuality, are financed wholly or in the main by grants made from Government funds. Most of these institutes have governing bodies of their own to which they are directly responsible. The support grants for institutes of this kind in England and Wales are administered by ARC; the Scottish institutes are borne on the vote of the Department of Agriculture and Fisheries for Scotland but the Department seeks the advice of the ARC in the consideration of research programmes and estimates.
The programmes of all institutes are coordinated and approved by the Council and are integrated with those of the independent research institutes in Scotland.
Grants for special investigations are made to universities

and other recognized research institutions and the Council annually awards research fellowships and training grants in the field of veterinary science.

Publications: Agricultural Research Service; Institutes and Units of the Agricultural Research Service; Opportunities in Agricultural Research; Index of Agricultural and Food Research; ARC Annual report.

Separate entries for:
ARC Institutes:
ARC Animal Breeding Research Organisation
ARC Food Research Institute
ARC Institute for Research on Animal Diseases
ARC Institute of Animal Physiology
ARC Letcombe Laboratory
ARC Meat Research Laboratory
ARC Poultry Research Centre
ARC Weed Research Organization
State-aided institutes, England and Wales:
Animal Virus Research Institute
East Malling Research Station
Glasshouse Crops Research Institute
Grassland Research Institute
Houghton Poultry Research Station
John Innes Institute
Long Ashton Research Station
National Institute for Research in Dairying
National Institute of Agricultural Engineering
National Vegetable Research Station
Plant Breeding Institute
Rothamsted Experiment Station
Soil Survey of England and Wales
Welsh Plant Breeding Station
Wye College, Department of Hop Research
State-aided institutes, Scotland:
Animal Diseases Research Association Moredun Institute
Hannah Research Institute
Hill Farming Research Organization
Macaulay Institute for Soil Research
Rowett Research Institute
Scottish Crop Research Institute
Scottish Institute of Agricultural Engineering

ARC INSECT CHEMISTRY AND PHYSIOLOGY GROUP   3

Address: University of Sussex, BN1 9RQ, Brighton
Telephone: (0273) 60 67 55 and 69 22 33
Head: Dr G.T. Brooks
Activities: Research to obtain basic information about the way insects function, with emphasis on the biological systems and enzymes involved in the regulation of insect development, in the production and metabolism of insect hormones, and in the metabolism within the insect of externally applied hormones, their synthetic analogues and antagonists, and insecticides. The long-term aim is to add to the basic information required for

the development of chemical agents for insect control that have improved biodegradability and are more selective toward insect pests in their action.

Projects: Physiology and biochemistry of insect juveniles, hormones and anti-hormones - insect endocrinology (Dr G.E Pratt); chemistry and isolation of insect hormones (Dr R.C. Jennings); biochemical pharmacology of insect hormones and insecticides (Dr G.T. Brooks).

ARC STATISTICS GROUP   4

Address: Department of Applied Biology, Pembroke Street, Cambridge, CB2 3DX
Telephone: (0223) 358381
Officer-in-charge: J.G. Rothwell
Graduate research staff: 3
Activities: The main function of the group is statistical consultancy and computing for ARC establishments in East Anglia. The group has ready access to the university computer and the powerful statistical packages associated with it. The main area of interest is in animal research, with a seconday interest in plants. Statistical research usually arises naturally out of the consultancy work, though its usefulness often extends well beyond its original area of application.

The group is accommodated within the university's Department of Applied Biology and provides limited assistance to this and other university departments through statistical consultancy, computational support, teaching, and examining.

ARC UNIT OF NITROGEN FIXATION   5

Address: The Chemical Laboratory, University of Sussex, Brighton, BN1 9QJ
Telephone: (0273) 606755 and 603446
Director: Professor John Postgate
Sections: Chemistry group, Dr G.H. Leigh
Graduate research staff: 30
Activities: The unit studies, at a fundamental level, the bio-inorganic chemistry, physiology, and molecular genetics of nitrogen fixation, a process of vital importance for plant growth and animal nutrition. The research may lead to the development of new biological systems for fixing nitrogen or to an understanding of the biological process such that it can be reproduced on a commercial scale and so contribute to the increasing world demand for food.

ARC UNIT OF STATISTICS 6

– ARCUS
Address: University of Edinburgh, James Clerk Maxwell Building, The King's Buildings, Mayfield Road, Edinburgh, EH9 3JZ
Telephone: (031) 667 1081
Telex: 727442 (UNIVED G)
Director: Professor D.J. Finney
Deputy director: Dr H.D. Patterson
Graduate research staff: 19
Activities: The unit acts as a collaborative and consultative body for agricultural research projects, primarily in Scotland, and also as a centre for research in statistical methodology having particular relevance to agriculture.
One major activity concerns statistical and computing methods for variety trials, development of appropriate designs, and responsibility for many aspects of the UK programme of testing new varieties of agricultural crops. This project is highly integrated in a computer package for its routine aspects, but also involves many research problems in experimental design and statistical analysis. Other activities of ARCUS include the methodology of quantitative genetics for animal breeding, collaboration in a large study of the management of suckler cows, theoretical analysis for animal growth curves, statistical method in a diverse range of problems in hill farming and in horticulture, and methods of biological assay. Numerous sample surveys for agricultural purposes are planned and analyzed with assistance from the unit. Techniques are sampled for estimating ecological resources and animal populations by aerial survey and by mark recapture.
Projects: Strategy, design, and analysis of crop variety trials (H.D. Patterson); design of genetic experiments and animal breeding (R. Thompson); sample estimation of mobile populations (G.M. Jolly); statistical programming (J. Muscott).

Agricultural Research Institute of Northern Ireland 7

Address: Large Park, Hillsborough, County Down BT26 6DR, Northern Ireland
Telephone: (0846) 682484
Affiliation: Department of Agriculture for Northern Ireland
Status: Official research centre
Director: Professor J.C. Murdoch
Graduate research staff: 9 (1981)
Activities: Crop husbandry; livestock breeding; livestock husbandry and nutrition; agricultural engineering and building.

Projects: Feeding and management of lactating cows (Dr F.J. Gordon); beef production (Dr R.J.W. Steen); management of the suckler cow and sheep production (Dr D.M.B. Chestnutt); management and feeding of sows and finishing pigs (Dr N. Walker); breeding and management of laying hens (Dr W.H. Foster); factors affecting cereal production (Dr D.L. Easson); farm mechanization (P.J. Frost).
Publications: Annual report.

CROP AND ANIMAL HUSBANDRY RESEARCH DIVISION 8

Head: Professor J.C. Murdoch
Activities: Research in crop and animal production (beef, sheep, pig and poultry); the administration of experimental facilities at the Institute, and specialist advice on husbandry matters.

Agriculture and Food Science Centre 9

Address: Newforge Lane, Belfast, BT9 5PX, Northern Ireland
Telephone: (0232) 661166
Affiliation: Department of Agriculture for Northern Ireland
Status: Official research centre

AGRICULTURAL AND FOOD BACTERIOLOGY RESEARCH DIVISION 10

Telex: 74487 QUB ADM
Head: Professor A.J. Holding
Graduate research staff: 10
Activities: In addition to the research activities of the division detailed below, the sections have specialist advisory functions which are mainly concerned with hygiene on farms and in poultry, meat, and fish processing plants, the incidence of mycotoxins in feeds, and the microbiological quality of vegetables and fruit. Microbiological analyses of milk and milk products are undertaken by the division for both statutory purposes and the Intervention Board for Agricultural Produce.
Publications: Annual report.

Meat and Fish Section   11

Head: Dr J.T. Patterson
Activities: Typing of staphylococci; incidence of salmonella; bacterial spoilage.
Projects: Influence of lipolytic and proteolytic bacteria on suitability of milk for processing (A. Gilmour, M.T. Rowe); pathogenic staphylococci in dairy products (A. Gilmour).

Milk and Milk Products Section   12

Head: Dr A. Gilmour
Activities: Bacterial enzyme production; heat treatment; incidence of staphylocci; bacterial attachment.
Projects: Typing of isolates of Staphylococcus aureus; bacteriology of packaged meat; development and application of fluorescent antibody technique; biochemical activities of bacteria leading to spoilage of meat; inhibition of salmonella by competing organisms.

Plant Conservation Section   13

Head: Dr A.P. Damoglou
Activities: Mycotoxin production in feed and fruit.
Projects: Interaction between bacteria and fungi in relation to mycotoxin production in feeds; environmental factors influencing mycotoxin production in feeds and fruit.

Plant Products Section   14

Head: Dr M.A. Collins
Activities: Yeast growth; activity of lipophilic compounds.
Projects: Studies on potential food spoilage yeasts; factors effecting the antimicrobial activity of lipophilic compunds in foods.

Soil and Water Section   15

Head: Dr J.E. Cooper
Activities: White clover and Lotus rhizobia.
Projects: Role of biotypes of rhizobia in the establishment of legumes on soils (J.E. Cooper); bacterial mineralization or organic phosphorus compounds (J.E. Cooper, A.J. Holding).

AGRICULTURAL AND FOOD   16
CHEMISTRY RESEARCH DIVISION

Head: Professor J.R. Todd
Sections: Animal nutrition, Dr N. Jackson; soil and plant nutrition, Dr J.S.V. McAllister
Graduate research staff: 20
Activities: Research into soils and plant nutrition, animal nutrition, plant and animal biochemistry and food chemistry; special analytical services.

AGRICULTURAL ENTOMOLOGY   17
RESEARCH DIVISION

Head: Dr Professor R.J. Marks
Graduate research staff: 9 (1982)
Activities: Epidemiology, pest status, and control of pests of grass and forage crops, cereals, potatoes, and horticultural crops.

Nematology Laboratory   18

Address: Feldon, Mill Road, Newtownabbey, Antrim, Northern Ireland
Telephone: (0231) 51361
Activities: Investigation of methods of minimizing damage by the more significant agricultural and horticultural pests.

BIOMETRICS DIVISION   19

Address: Castle Grounds, Stormont Estate, Belfast, BT4 3NR
Telephone: (0232) 63939
Head: Dr S.T.C. Weatherup
Graduate research staff: 6 (1981)
Activities: Computing; mathematical modelling; statistical methods; statistical consultancy service on experiments and surveys.
Projects: Discrimination between plant cultivars using multivariate statistical methods (Dr S.T.C. Weatherup); fish population prediction in Irish Sea (Dr D.A. Stewart); evaluation of sampling methods in agriculture (Dr T. McCallion); forest production prediction methods (Dr D.J. Kilpatrick); dairy herd management model (E.A. Goodall).

FIELD BOTANY RESEARCH DIVISION   20

Head: Professor C.E. Wright
Sections: Crop physiology, Dr B.R.M. Harvey; weed control, Dr A.D. Courtney; grassland, Dr A.S. Laidlaw
Graduate research staff: 12 (1982)
Activities: Fundamental and applied research on pasture and crop plants: physiology of growth of plants and some forest trees; inter-relationships of plants and cultivars in competition; physiological and ecological aspects of cultural and chemical weed control; assessment of yield and quality of crop plants; role of introduced and indigenous species in increasing farm productivity; breeding of new cultivars of certain crop species, especially of oats, ryegrasses, and potatoes.
Projects are being undertaken on the following topics: factors influencing grass and white clover associations in cut and grazed swards; genetic variation in grasses in magnesium uptake; mechanisms of herbicide tolerance in resistant grass varieties; micropropagation of coniferous forest trees through tissue culture; chemical pre-

harvest retting in flax; improvement of pasture production and utilization on less favoured areas.

Plant Breeding Station 21

Address: Manor House, Loughgall, Armagh BT61 8JB, Northern Ireland
Telephone: (076 289) 206
Officer-in-charge: Dr Brian E. Costelloe
Sections: Grasses, cereals, and flax, Dr J.S. Faulkner; potatoes, B.E. Costelloe; protectioning, Dr N. Evans
Graduate research staff: 4 (1982)
Activities: Breeding of new cultivars of certain crop species especially of oats, ryegrasses and potatoes; conservation of genetic material; trials on introduced and indigenous species for increasing farming productivity. Projects are being undertaken on the following topics: breeding grasses for tolerance to grass-killing herbicides; breeding perennial ryegrass and tall fescue varieties; breeding and testing of commercial potato varieties for the Mediterranean area; production of genetical variation in potatoes by protoplast culture; oat and barley breeding.

Plant Testing Station 22

Address: 50 Houston Road, Crossnacreevy, Castlereagh, Belfast BT6 9SH, Northern Ireland
Telephone: (023 123) 229
Officer-in-charge: Dr M.S. Camlin
Sections: Cereals, E. White; grasses, T. Gilliland
Graduate research staff: 3 (1982)
Activities: Determination of distinctness, uniformity, and stability (DUS) and value cultivation and use (VCU) of cereal, herbage, and potato cultivars as part of an integrated United Kingdom system for the award of Plant Breeders Rights and for entry on the United Kingdom National List.
Conduct of regional performance trials on cereals and herbage cultivars to identify cultivars best suited to Northern Ireland conditions and to issue Recommended Lists.
Projects: Relative competitive ability of grass species and cultivars; biochemical and physiological techniques to aid varietal identification; provenance and its effect on development and yield in cereals.

PLANT PATHOLOGY RESEARCH DIVISION 23

Head: Professor J. Peter Blakeman
Sections: Potato diseases, Dr C. Logan; potato certification, Dr R.B. Copeland; physiological plant pathology, Dr A.E. Brown; cereals and grasses, Dr P.C. Mercer; horticulture and forestry, D.A. Seaby; fungicides, Dr L.R. Cooke; virology, Dr P.R. Mills
Graduate research staff: 8 (1982)

Activities: Investigations into diseases in cereal and forage crops, potato and horticultural crops, forest trees, and top fruit; development of control measures: advent of pathogen strains resistant to specific fungicides.
Specialist technical services include tests for incidence of virus diseases in nuclear and commercial stocks of seed potatoes, examination of potatoes imported under quarantine and disease resistance assessments for National List Trials.
Current major projects are as follows: assessment of losses caused by fungi in grassland; control of seed-borne and foliar diseases in cereals; control of potato black leg and potato gangrene; air-borne apple diseases with particular reference to scab, canker and mildew; studies on the epidemiology of virus diseases of potatoes in Northern Ireland; physical and chemical factors at the surfaces of plants and their effect on fungal infection structures.

Allied Breweries Limited Technical Centre 24

Address: 107 Station Street, Burton-on-Trent, Staffordshire DE 15 9B
Telephone: (0283) 45 320
Product range: Ale, lager, wine, spirits, soft drinks.
Status: Research centre within an industrial company
Manager, central development services: Dr P.A. Brookes
Technical director: R.A. Young
Central technical services: P.A. Martin
Graduate research staff: 15 (1982)
Activities: Malting of cereals and wort production; fermentation; analytical methods; new product development. Major current projects are: rapid methods for detecting infection; yeast performance and fermentation, particularly the effect of lipids; origin and control of sulphur volatiles in beer.

Anglian Water Authority 25

Address: Ambury Road, Huntingdon, Cambridgeshire PE18 6NZ
Telephone: (0480) 56181
Status: Research centre within a public utility
Chairman: A. Morrison
Chief Executive: P.H. Bray
Scientific Directorate: Director: A.W. Davies
Assistant director, fisheries: A. Miller
Assistant director, water quality: J.A. Tetlow
Coordinator of research and laboratories: B.T. Croll

Activities: Water supply, water conservation, sewerage, sewage disposal, prevention of river pollution, fisheries, land drainage, and recreational use of waters. The authority undertakes research in areas directly concerned with present or probable future problems associated with its operations. To this end a coordinated research and development programme is operated consisting of projects carried out both in-house and by outside contract. Major areas of interest are as follows: nitrates and their removal from potable waters; eutrophication and its control in reservoirs; control of organic pollutants and their removal from potable waters; the disposal of sewage sludge to land.

Animal Diseases Research Association, Moredun Institute   26

– ADRA
Address: Moredun Institute, 408 Gilmerton Road, Edinburgh, EH17 7JH
Telephone: (031) 664 3262
Affiliation: State-aided institute, Agricultural Research Council
Status: Official research centre
Director: Dr W.B. Martin
Sections: Biochemistry, Dr A.C. Field; clinical studies, B. Mitchell; electron microscopy, E. Gray; microbiology, Dr I.D. Aitken; parasitology, M.G. Christie; pathology, Dr R.M. Barlow; physical chemistry, Dr D.L. Mould; physiology, Dr D.J. Mellor
Graduate research staff: 75
Activities: Research in the fields of livestock husbandry and nutrition, and veterinary medicine.
Projects: Research on non-respiratory microbial diseases (Dr I.D. Aitken); research on respiratory diseases (Dr I.D. Aitken, or Dr R.M. Barlow); nutritional deficiencies and excess in relation to animal health (Dr A.C. Field); physiology of pregnancy (Dr D.J. Mellor); parasitic diseases, (M.G. Christie), diseases of unknown aetiology, (Dr R.M. Barlow).

Animal Health Trust   27

Address: Lanwades Hall, Kennett, Newmarket, Suffolk CB8 7PN
Telephone: (0638) 751030
Status: Charity funding research
Director: W.B. Singleton
Graduate research staff: 45
Activities: Research into animal diseases, particularly equine and canine.

Projects: Hereditary eye diseases in the dog; infectious bovine keratoconjunctivitis (K.C. Barnett); investigation and management of chronic renal disease in the dog (D.F. Macdougall); fading puppy complex (A.S. Blunden); duck virus hepatitis and Chlamydia psittaci (P.R. Woolcock); hindlimb lameness; induced parturition (L.B. Jeffcott); forelimb lameness metabolic causes of lameness (C. Colles); epidemiology of equine infectious diseases (D.G. Powell); various research projects into equine haematology (B.V. Allen); various research projects into equine biochemistry (D.J. Blackmore); blood typing (A.M. Scott); viral diagnosis and research into equine respiratory diseases (Jennifer A. Mumford); reproductive and neonatal diseases (H. Platt); equine venereal diseases and anaerobic pathogens (Mary E. Mackintosh).
Publications: Annual report.

EQUINE RESEARCH STATION   28

Address: PO Box 5, Balaton Lodge, Snailwell Road, Newmarket, Suffolk CB8 7DW
Telephone: (0638) 61111
Sections: Pathology, Dr H. Platt; clinical, Dr L.B. Jeffcott; haematology, B.V. Allen; clinical chemistry, D.J. Blackmore; epidemiology, D.G. Powell; immunology, A.M. Scott; virology, Jennifer A. Mumford; microbiology

SMALL ANIMALS HEALTH CENTRE   29

Address: Lanwades Park, Kennett, Newmarket, Suffolk CB8 7PN
Telephone: (0638) 750543
Senior Scientist: Dr K.C. Barnett
Sections: Comparative ophthalmology, Dr K.C. Barnett; duck virus hepatitis, Dr P.R. Woolcock; renal, Dr D.F. Macdougall; fading puppy, A.S. Blunden; neurology; oncology; electroretinography; laboratory diagnostic service
Activities: Veterinary research on diseases of small domestic animals and poultry.

Animal Virus Research Institute   30

– AVRI
Address: Ash Road, Pirbright, Woking, Surrey GU24 0NF
Telephone: (0483) 232441
Telex: 859137
Affiliation: State-aided institute, Agricultural Research Council
Status: Official research centre
Director: Dr R.F. Sellers

Sections: Biochemistry, Dr F. Brown; bluetongue/ African swine fever, Dr W.P. Taylor; entomology, J. Boorman; epidemiology, Dr R.S. Hedger; experimental pathology, R. Burrows; genetics, Dr D. McCahon; vaccine research, Dr G.N. Mowatt; disease security, J.A. Mann

Graduate research staff: 50

Activities: The institute has traditionally concentrated on foot-and-mouth disease, but has more recently become involved in other virus research, notably swine vesicular disease, viruses of the rhabdovirus group (including rabies and vesicular stomatitis), African swine fever, African horse sickness, bluetongue and other arboviruses, and a number of herpes viruses. Specific aspects of research include: methods of preparation of vaccine; methods of increasing response to vaccine; biochemical studies on structure and function of viruses; production of polypeptides by genetic manipulation; genetic recombination studies; airborne transmission of foot-and-mouth; pharyngeal route of foot-and-mouth infection; routes of infection of pigs with swine vesicular disease; diagnostics; epidemiology and immunology of African swine fever and bluetongue.

In relation to foot-and-mouth disease, the institute is designated by the Food and Agriculture Organisation of the United Nations as the World Reference Laboratory for Foot-and-mouth disease and is responsible for the typing and classification of the virus from outbreaks of the disease in many countries overseas as well as in Britain.

ARC Animal Breeding Research Organisation   31

Address: The King's Buildings, West Main Road, Edinburgh, EH9 3JQ
Telephone: (031) 667 6901
Parent body: Agricultural Research Council
Status: Official research centre
Acting director: Dr R.B. Land
Sections: Applied genetics, Dr C. Smith; growth and efficiency, Dr St C.S. Taylor; computing and statistics, vacant; immunology, Dr R.L. Spooner; disease studies, Dr A.G. Dickinson; physiological genetics, Dr G. Wiener; experiments, Dr J.C. Alliston
Graduate research staff: 21
Activities: Research work in the form of breeding studies with livestock based on
Publications: Annual report.

DRYDEN FIELD LABORATORY*   32

Address: Roslin, Midlothian EH25 9PS
Telephone: (031) 440 2292 and 1016

ARC Food Research Institute   33

Address: Colney Lane, Norwich, NR4 7UA
Telephone: (0603) 56122
Telex: 975453
Parent body: Agricultural Research Council
Affiliation: University of East Anglia
Status: Official research centre
Director: Professor R.F. Curtis
Sections: Chemistry and biochemistry, Dr H.W-S. Chan; microbiology, Dr B.H. Kirsop; nutrition and food quality, Dr D.A.T. Southgate; scientific services and development, Dr A.G. Kitchell; process physics, Dr P. Richmond; national colection of yeast cultures, B.E. Kirsop
Graduate research staff: 80
Activities: The obectives of the institute are to carry out medium and long term research: to support the broad national interest of consumers in the quality, eg safety, nutritive value, and acceptability, of the food supply in the United Kingdom; in collaboration with the research associations, to assist the food manufacturing industry in maximizing its efficiency and effectiveness. The institute is concerned with the chemistry, physics, biochemistry, and microbiology of foods and food components. Special attention is paid to the behaviour of proteins, carbohydrates, and lipids during the manufacture of processed foods and there is a major interest in food composition and the nutritional quality of the national diet. There is a special interest in potatoes, soft fruit, vegetables, eggs, and poultry. The institute also houses the National Collection of Yeast Cultures.
Projects: The institute's research programme comprises about 40 projects with a similar number of project leaders. The projects are grouped into the following research areas: chemical toxicants in food; food-poisoning and toxigenic microorganisms; nutrition and food quality (excluding microbiology); microbial spoilage of foods and handling hygiene; basic chemical, biochemical, physical, and biological aspects of food; basic physiological, compositional, and taxonomic aspects of food microorganisms; raw materials, by-products and waste utilization; processing, storage, and distribution of food.
Publications: Biennial report; project list.

ARC Institute for Research on Animal Diseases   34

Address: Compton, Newbury, Berkshire RG16 0NN
Telephone: (063 522) 411
Parent body: Agricultural Research Council
Status: Official research centre
Director: Professor J.M. Payne
Sections: Biochemistry, Dr G.D. Hunter; pathology, Dr R.L. Chandler; functional pathology, Dr W.M. Allen; microbiology, Dr W. Plowright; immunology and parasitology, Dr K.G. Hibbit
Graduate research staff: 60
Activities: Research into economically important diseases of farm livestock, especially: atrophic rhinitis; enteritis; infertility; liver fluke; mastitis; production diseases; respiratory disease; salmonellosis; and swine dysentery.
Projects: Atrophic rhinitis (Dr J.M. Rutter); colostrum (Dr K.G. Hibbitt); cosreel (Dr A.M. Russell); enteritis (Dr D.J. Garwes); enteropathology (Dr G.A. Hall); infectious bovine keratoconjunctivitis (Dr R.L. Chandler); infertility (Dr I.M. Reid); lameness (Dr D.G. Baggott); liver fluke (Dr D.L. Hughes); mastitis (Dr A.W. Hill); production disease - hazards of new husbandry systems (Dr R.J. Heitzman); production disease - limiting factors (Dr B.F. Sansom); respiratory disease (Dr R.N. Gourlay); salmonellosis (Dr M.M. Aitken); swine dysentery (Dr R.J. Lysons).
Publications: Biennial report.

ARC Institute of Animal Physiology   35

Address: Babraham, Cambridge, CB2 4AT
Telephone: (0223) 832312
Parent body: Agricultural Research Council
Status: Official research centre
Director: Dr B.A. Cross
Sections: Physiology, Dr R.B. Heap; applied biology, Dr L.E. Mount; biochemistry, Dr R.M.C. Dawson; immunology, Dr A. Feinstein; cell biology, Dr M.W. Smith; biophysics, Dr A.D. Bangham; reproductive biology, Dr E.J.C. Polge
Graduate research staff: 66
Activities: The institute comprises two centres, at Babraham and at the Animal Research Station, Girton specially designed for work on large animals. Its main function is to extend knowledge of the basic physiology and biochemistry of farm livestock as a foundation for improved animal production. The work includes long-term strategic research specifically related to agricultural animals, basic studies on other species, and fundamental research on model systems. Among topics studied are reproduction, including ova transfer and the preservation of embryos and sperm, lactation, cardiovascular physiology, neuroendocrinology, environmental physiology and behavioural studies in farm animals, metabolism, ruminant digestion, and rumen biochemistry. Other research topics include growth factors, immunoglobulin receptors and transplantation antigens in cell membranes, transport processes in cells, culture of vascular tissue and biophysics of muscle contraction, cytogenetics, and animal blood groups.
The institute includes a Monoclonal Antibody Centre, the objective of which is to advance monoclonal antibody technology for use in agriculture and in veterinary fields.
Projects: Hypolthalmic control of the pituitary gland (B.A. Cross, R.G. Dyer); behavioural physiology of farm animals (B.A. Baldwin, D.B. Stephens, E.E. Walser); energy metabolism in pigs (L.E. Mount, D.L. Ingram, W.H. Close); neurochemistry (M. Holzbauer Sharman, D.F. Sharman); ruminant digestion and metabolism (F.A. Harrison, D.B. Lindsay, J.L. Mangan, J.Y.F. Paterson); lipids of ruminants (R.M.C. Dawson, P. Kemp, W.M.F. Leat); rumen microorganisms (G.S. Coleman, G.G. Orpin); cell biology and metabolism of sheep nematodes (E.A. Munn, P.V.F. Ward); intestinal transport physiology (M.W. Smith, F.V. Sepulveda); genetics and sheep blood groups, cytogenetics (E.M. Tucker, A.R. Dain); structure and function of immunoglobulins (A. Feinstein, D. Beale, C.P. Milstein); lymphocyte subpopulations and their functions (R.M. Binns, D.B.A. Symons); immunogenetics and mechanisms of immune responses and tolerance (J.C. Howard, B.J. Roser, M.J. Taussig); physiology, endocrinology, biochemistry, and electron microscopy of female reproductive processes (R.B. Heap, A.P.F. Flint, F.B.P. Wooding); physiology and endocrinology of the testis (B.P. Setchell).
Publications: Biennial report, *ARC Babraham.*

ANIMAL RESEARCH STATION   36

Address: 307 Huntingdon Road, Cambridge, CB3 OJQ
Telephone: (0223) 77222
Officer-in-Charge: Dr E.J.C. Polge
Projects: Hormonal aspects of reproduction in farm animals - artificial insemination and preservation of semen; early embryonic development - preservation and transplantation of embryos in farm and laboratory animals (C.E. Adams, S.M. Willadsen, M.A.H. Surani); physiology and biochemistry of spermatozoa and of the male tract (H.M. Dott, R. Jones, R.A.P. Harrison); egg physiology and ovarian endocrinology (R.M. Moor, D.G. Cran).

ARC Letcombe Laboratory 37

Address: Wantage, Oxfordshire OX12 9JT
Telephone: (02357) 3327
Parent body: Agricultural Research Council
Affiliation: University of Reading; close association with the University of Nottingham
Status: Official research centre
Director: Dr J.V. Lake
Sections: Chemistry and electronics, E.R. Mercer; physiology, Dr M.G.T. Shone; field studies, Dr R.Q. Cannell; statistics (jointly with Weed Research Organization), B.O. Bartlett
Graduate research staff: 50
Activities: Research at Letcombe is concerned mainly with the growth of crops in relation to soil conditions, cultivation systems, and drainage. The efficiency of fertilizer utilization and the microbial degradation of plant residues are also being studied. The programme comprises both field experiments, in which detailed observations are made on plants and the soil, and complementary laboratory studies of the physiological processes which control the growth and function of root systems. This dual approach aims not only to assist in answering specific practical problems directly but also to provide a fuller understanding of the underlying mechanisms involved in plant growth.
Projects:
Chemistry and Electronics Department: application and measurement of radioactive and stable isotopic tracers, tracer studies of the movement of water, nitrate, and other ions in soil and chalk (E.R. Mercer); chemical analysis of soil, plant, and water samples using manual and automated methods (W. Downs); isotope measurements by mass spectrometry and ultraviolet spectrophotometry in tracer studies with nitrogen-14, nitrogen-15, and deuterium (M.G. Johnson); development of apparatus for laboratory and field experiments, data logging and microprocessing, measurement of radioactivity (P.M. Lay)
Field Studies Department: effects of cultivation methods or direct drilling on soil conditions and crop growth (Dr R.Q. Cannell); effect of simplified cultivation on hydrology, drainage requirements, and leaching of major nutrients (R.Q. Cannell, Dr R.J. Dowdell, E.R. Mercer); lysimeter studies of effects of transient high water-tables on soil conditions and crop growth (R.Q. Cannell, Dr R.K. Belford); effects of controlled traffic on soil conditions for root and shoot growth in simplified tillage systems (Dr P.S. Blackwell); field studies of root growth of cereals, mathematical model of root growth of winter wheat (Dr P.L. Bragg); agronomy of simplified cultivation and direct drilling, effects of crop residues on the establishment of crops (D.G. Christian); soil aeration and denitrification in relation to soil management, direct measurement of gaseous nitrogen losses (Dr P. Colbourn); field and lysimeter studies on the fate of fertilizer nitrogen in grass/arable rotations (R.J. Dowdell); leaching of nitrate from shallow soils overlying chalk (R.J. Dowdell, E.R. Mercer); soil factors causing variation in yield of cereals (Dr K. Gales); effects of soil physical conditions on root growth with special reference to drainage and cultivation (Dr M.J. Goss); gas chromatography of permanent gases and volatile organic compunds (K.C. Hali).
Physiology Department: ion and water transport in plant roots - effects of drought (Dr M.G.T. Stone); processes affecting the regulation of division and development of cells in root apices - nuclear structure during cell development (Dr P.W. Barlow); physiology and structure of root systems in relation to ion and water uptake; physiological adaptation of cell development, membrane composition, and transport properties in roots at low temperatures (Dr D.T. Clarkson); inorganic nutrients - effect on the form and function of roots in cereals; waterlogging - factors contributing to damage to plant growth and plant adaptations to oxygen deficiency (Dr M.C. Drew); composting of straw - products of cellulolytic microbial populations (S.H.T. Harper); response of roots and shoots to waterlogged soil and to plant hormones (Dr M.B. Jackson); metabolic responses of roots to mineral nutrient deficiency, especially in relation to absorption of ions (Dr R.B. Lee); soil microorganisms - effects of their products on plants; microbial products from plant residues - their isolation and identification; root exudates in relation to plant growth and microbial activity (Dr J.M. Lynch); micro-autoradiography - development and application of techniques for soluble compunds (J. Sanderson).
Publications: Annual report.

ARC Meat Research Institute 38

Address: Langford, Bristol, BS18 7DY
Telephone: (0934) 852661
Telex: 449095 Metres G
Parent body: Agricultural Research Council
Affiliation: Close association with the University of Bristol
Status: Official research centre
Director: Professor Allen J. Bailey
Sections: Animal physiology, Dr David Lister; muscle biology, Dr Gerald Offer; food quality, Dr Terence Roberts; engineering and development, Colin Bailey; analytical chemistry, Dr Ronald Patterson
Graduate research staff: 98
Activities: The institute is devoted to identifying the basic scientific principles underlying all aspects of the production of fresh meat and meat products, from animal production through processing, storage, and retail to the consumer.
Current research topics include: neuroendocrinological control of growth and tissue distribution, use of male

animals for meat production, immunological methods of controlling growth, carcass dissection and evaluation, pre-slaughter handling and slaughter techniques; carcass utilization, new methods of post-slaughter handling, processing, packaging, meat refrigeration including basic thermophysical properties of meat and development of design data charts, methods of live body measurement using ultrasonics; carcass and meat hygiene, meat spoilage microbiology, pyrolysis-mass spectrometry to identify microorganisms, control of Clostridium botulinum; instrumental and sensory evaluation of meat quality; structure of muscle and connective tissue components, collagen cross-linking, role of myofibrils in meat texture and water holding; composition and texture of adipose tissue; analysis of meat flavour components, off-flavours and taints; immunological methods of identifying meat species in minces and meat products, metabolism of anabolic agents.

Publications: MRI Biennial Reports; MRI Biennial Report Appendix: Some Practical Research Applications .

ARC Poultry Research Centre   39

– PRC
Address: Roslin, Midlothian EH25 9PS
Telephone: (031) 440 2726
Parent body: Agricultural Research Council
Status: Official research centre
Director: Dr D.W.F. Shannon
Sections: Nutritional and environmental studies, Dr C. Fisher; ethology, Dr I.J.H. Duncan; reproductive physiology, Dr P.E. Lake; anatomy, Dr W.G. Siller; genetics, Dr G. Bulfield
Graduate research staff: 53
Activities: Research at the PRC is aimed at increasing the biological efficiency of poultry production whilst maintaining and improving poultry welfare. This is facilitated by basic and applied research in the fields of physiology, metabolism, nutrition, environment, behaviour, genetics, and pathology (metabolic). The remit excludes research on infectious disease, engineering aspects of poultry production, and most topics associated with food quality of poultry products which are the responsibility of other Agricultural Research Council institutes.
Specific topics of the centre's research programme are: productivity and management of breeding stock; regulation and manipulation of gene expression; productivity of laying stock and efficiency of commercial egg production; the physiological and genetic basis of the control of reproductin in poultry; growth and efficiency of meat production; poultry welfare behavioural and

physiological responses to perceived environmental variables; responses of poultry to the climatic environment; poultry feeds, ingestion, and nutrient requirements; chemistry of digestion, nutritional biochemistry, and metabolic pathology; neurobiological mechanisms controlling ingestion, behaviour, and reproductive processes; aspects of avian haematology and immunology.
Publications: Annual report.

ARC Weed Research Organization   40

Address: Bedbroke Hill, Yarnton, Oxford, OX5 1PF
Telephone: (08675) 3761
Parent body: Agricultural Research Centre
Affiliation: University of Reading
Status: Official research centre
Director: Professor John D. Fryer
Departments: Weed Control, J.G. Elliott; Weed Science, Dr K. Holly;
Groups: Annual crops, G.W. Cussans; grass and fodder crops, Dr R.J.Haggar; perennial crops, Dr J.G. Davison; herbicides, Dr R. Hance; weed biology, R.J. Chancellor; microbiology, M.P. Greaves; aquatic weeds and uncropped land, T.O. Robson; developmental botany, Dr D.J. Osborne; tropical weeds, C. Parker
Graduate research staff: 50
Activities: The Weed Control Department keeps abreast of current weed problems in British agriculture and horticulture and by field research on and off the station aims to improve the efficiency of existing control measures, develop new ones where necessary, and ultimately devise systems of control combining the best of the chemical and cultural methods. The Department of Weed Science research programme covers many aspects of herbicides including the interactions between herbicide and weed, crop, method of application, soil, and climate. The objectives are to provide a scientific background to the practice of weed control and a basis for continued improvement therein, and to avoid problems which might arise from extensive and repeated herbicide use.
Projects: Annual crops: herbicide treatments for the control of wild oat and blackgrass in cereals (Dr P.J. Lutman, M.E. Thomton); weed problems of minimum tillage (F. Pollard, S.R. Moss); long-term economic weed control in cereals (B.J. Wilson, P. Ayres); growth of cereals in reduced tillage systems (J.G. Elliott, F. Pollard); control of perennial grass weeds in cereal cropping systems (G.W. Cussans, P. Ayres); effect of high organic matter soils on use of herbicides (Dr P.J. Lutman, M.J. May); control of potato groundkeepers (Dr P.J. Lutman, G.W. Cussons); cereal to tolerance of herbicides (D.R. Totman, G.W. Cussans); success of weed beet in agricultural land (G.W. Cussans, C.J.

Bastian); effects of herbicides and weeds on oilseed rape (Dr P.J. Lutman, M.E. Thornton)

Grass and fodder crops: agro-ecology and control of important broad-leaved weeds (A.K. Oswald); role of herbicides in manipulating sward composition (Dr R.J. Haggar, F.W. Kirkham); minimum cultivation/herbicide systems for establishing grasses, legumes, and fodder crops in existing swards (N.R.W. Squires; Dr R.J. Haggar); agro-ecology and control of important grass weeds in leys and seed crops (A.K. Oswald, F.W. Kirkham)

Perennial crops: fruit crop tolerance of soil- and foliage-applied herbicides (D.V. Clay, Dr J.G. Davison); effect of important weeds on fruit production, response of newly planted fruit crops and nursery stock to weed competition and herbicides (Dr J.G. Davison, J.A. Bailey); evaluation of new herbicides for weed control in strawberries (D.V. Clay, Dr J.G. Davison)

Herbicide performance: evalution of new herbicides (W.G. Richardson, J. Holroyd); influence of formulation factors on the activity of herbicides (Dr D.J. Turner); methods of application of herbicides (W.A. Taylor, J. Holroyd); interaction of herbicides (Dr H.F. Taylor, M.P.C. Loader); herbicides for forestry (Dr D.J. Turner, W.G. Richardson)

Environmental studies: effect of environmental factors on the activity of herbicides and growth regulators (Dr J.C. Caseley, A.M. Blair, Dr D. Coupland, C.R. Merinlt, R.C. Simmons); techniques and equipment for monitoring the environment, environmental control (R.C. Simmons, Dr J.C. Caseley)

Chemistry: analysis of herbicides in soil, water, and plant environment, analytical methods for herbicides and their development (T.H. Byast, E.G. Cotterill); factors affecting performance of soil-applied herbicides (Dr R.J. Hance); influence of repeated applications of MCPA, tri-aliate, simazine, and linuron on the fertility of soil, persistence in soil of paraquat, effects of repeated applications of glyphosate (P.D. Smith)

Microbiology: effects of herbicides on natural microbial populations (J.A. Marsh, H.A. Davies); effects of herbicides on root microflora, interactions between herbicides and model microbial ecosystems (M.P. Greaves, G.I. Wingfield)

Weed biology: germination of weed seeds (R.J. Chancellor, Dr N.C.B. Peters) vegetative regeneration of weeds (R.J. Chancellor); grassland weed ecology (E.D. Williams, R.J. Chancellor); factors affecting weed/crop competition (Dr N.C.B. Peters); arable weed ecology (R.J. Chancellor, R.J. Froud-Williams); influence of light on seed germination and vegetative regeneration of weeds (R.J. Chancellor, J. Hilton)

Developmental botany: dormancy and viability of weed seeds, control of seed shedding in weed species (Dr D.J. Osborne, Dr J.A. Sargent, Dr R. Hooley); stress conditions in seed germination and seedling establishment, perennation and regeneration of plant parts (Dr D.J. Osborne, Dr J.A. Sargent, Dr M. Wright)

Aquatic weeds and uncropped land: chemical control methods for aquatic vascular plants and algae (T.O. Robson, P.R.F. Barrett); potential of grass carp for aquatic weed control (M.C. Fowler, T.O. Robson); herbicides and growth regulators in vegetation management on uncropped land (E.J.P. Marshali, T.O. Robson)

Tropical weeds: new herbicides for tropical crops, resistance of sorghum and millet varieties to a range of Striga species and strains (C. Parker).

Publications: Biennial report; technical reports; *Weed Abstracts; Plant Growth Regulator Abstracts.*

Associated Biscuits Limited*   41

Address: 121 Kings Road, Reading, RG1 3EF
Status: Industrial company
Research director: W.A. Palmer
Activities: Comprehensive research into the manufacture of biscuits and cakes.

Arthur Rickwood Experimental Husbandry Farm   42

Address: Mepal, Ely, Cambridgeshire CB6 2BA
Telephone: (03543) 2531
Affiliation: Ministry of Agriculture, Fisheries and Food, Agricultural Development and Advisory Service
Status: Official research centre
Activities: Subjects of interest are: potatoes - assessment of new varieties, herbicide evaluation, irrigation, cyst nematode control, growing systems; cereals - variety testing and agronomic trials; sugar beet - variety testing, weed control; and agronomic trials; carrots - production trials, agronomic trials, varietal assessment; celery - direct drilled and transplanted performance, irrigation; onions - mechanical planting, set size, population studies, ribbon systems, disease and weed control; soil mixing - treatments involving mixing subsoil and peat; weed control - investigations on soils having a high organic content are carried out by the Weed Research Organization.

Bass plc, Research Department 43

Address: 137 High Street, Burton upon Trent, Staffordshire DEl4 lJZ
Telephone: (0283) 45301
Telex: 341871 BASSHQC
Product range: Beers, soft drinks, wines and spirits.
Status: Research centre within an industrial company
Research director: Dr A.D. Portno
Brewing research managing: K.J. Fairbrother
Research manager: Dr S.W. Molzahn
Departments: Microbiology; genetics; and materials and process chemistry; flavour chemistry
Graduate research staff: 18
Activities: Commercial research of a short, medium and long-term nature aimed at improving product quality, reducing costs and developing new products. Main areas of research are concerned with the biochemistry and physiology of alcoholic fermentations, genetic manipulation of brewers yeast, materials and process chemistry of brewing and new product development.

Bedford College* 44

Address: Inner Circle, Regent's Park, London, NW1 4NS
Telephone: (01) 486 4400
Affiliation: University of London
Status: Educational establishment with r&d capability

DEPARTMENT OF BOTANY 45

Head: Professor W.G. Chaloner
Graduate research staff: 8
Projects: Control of flower longevity by pollination; interaction of auxin and ethylene in the control of higher plant growth (Dr A.D. Stead).

DEPARTMENT OF GEOGRAPHY 46

Head: Professor J.B. Thornes
Projects: Impact of forest clearance on soil-water chemistry in the Amazon (Professor J.B. Thornes); investigation of plant cover by remote sensing in relation to mineral resources (Professor M.M. Cole).

DEPARTMENT OF ZOOLOGY* 47

Head: Professor R.P. Dales

Beecham Group plc* 48

Address: Beecham House, Great West Road, Brentford, Middlesex TW8 9BD
Telephone: (01) 560 5151
Telex: 935986 BEECHM G
Status: Industrial company
Group Secretary: I.M.F. Balfour

BEECHAM PHARMACEUTICALS DIVISION 49

Address: Great Burgh, Epsom, Surrey KT18 5XQ
Telephone: (07373) 53344
Telex: 8814696
Director of Research: F.P. Doyle
Director of New Product Development: Dr M.J. Soulal
Director of Research Projects: Dr K.R.L. Mansford

BEECHAM PRODUCTS RESEARCH DEPARTMENT 50

Director of Research: Dr M. Brook
Vice President Research, Western Hemisphere: D.M. Huston
Medical Director, Medical Affairs and Safety Evaluation: Dr T.L.C. Dale
Director, Product Research, Food and Drink: D. Hicks
Director, Product Research, Proprietaries and Cosmetics: A.G. McGee
Director, Applied and General Research: Dr R. Swindells
Activities: Research into toiletries, proprietary medicines, health foods, health and soft drinks, cosmetics and adhesives.

Birkbeck College 51

Address: Malet Street, London, WC1E 7HX
Telephone: (01) 580 6622
Parent body: University of London
Status: Educational establishment with r&d capability
Master: Professor W.G. Overend
Vice-Master: Professor W.H. Barber
Graduate research staff: 513

FACULTY OF SCIENCE 52

Dean: Dr J.T. Temple
Vice-Dean: J.C.E. Jennings

Department of Botany   53

Head: Professor G.R. Stewart
Activities: Microbiology: soil bacteriology, plant pathology, fungal genetics and the metabolism of microalgae. Plant physiology and ecology: nitrogen metabolism, stress physiology and enzymology and the ecophysiology of plants from extreme environments such as salt marshes, rain forests and deserts. Taxonomy: study of small-scale evolutionary change. Molecular biology, with emphasis on DNA structure.

Department of Zoology   54

Head: Professor J.L. Cloudsley-Thompson
Activities: Animal physiology and biochemistry - excretion in insects, lipid metabolism in the mammalian nervous system, endocrine control of lipid metabolism in birds physiology of the avian embryo; acid-base balance, respiratory mechanisms and their control; biological rhythms - circadian rhythms in arthropods and reptiles, investigations of the mechanisms of biological clocks; photoperiodism and bird migration, feeding rhythms of birds; cell biology and genetics - problems of mitotic control and tissue organization; genetic and physiological aspects of ageing in Drosophila, mutagenesis in Drosophila, endopolyploidy and polyteny in insects; entomology - insect behaviour and neurophysiology, chemoreception, foraging behaviour of parasitoids, morphology and functioning of sense organs; palaeontology and systematics - vertebrate palaeontology.

Boots Company plc,   55
Research and Development

Address: Nottingham, NG2 3AA
Telephone: (0602) 56255
Telex: 37-128/9
Product range: Pharmaceutical and consumer products.
Status: Industrial company
Research Director: Dr B. Lessel
Departments: Pharmaceutical sciences (research), Professor S.S. Adams; pharmaceutical sciences (development), Dr C.J. Lewis; consumer products development, R.E. Collard; medical sciences, Dr J.W. Buckler; patents, T.S. Simpson; research services, Dr P.C. Risdall
Graduate research staff: 220
Activities: Chemical, biochemical, microbiological, pharmacological and biological research and development for medical, pharmaceutical, cosmetic, toiletry, household and food products.
See separate entry for: FBC Limited

Boxworth Experimental   56
Husbandry Farm

Address: Boxworth, Cambridgeshire CB3 8NN
Telephone: (09547) 372 and 391
Affiliation: Ministry of Agriculture, Fisheries and Food, Agricultural Development and Advisory Service
Status: Official research centre
Activities: Subjects of interest are: cereals - winter wheat including weed and disease control, cultivations, break crops, grain losses, and fertilizer effects, problem of wild oats; milk production - effects of heavier calving weights on milk production, nutritional and environmental aspects of calf rearing; beef production - performance of crosses with Fresian cows, feeding trials, maize silage; field beans - winter sown varieties as break crop; miscellaneous crops - winter rape, lucerne, maize, and linseed.

Brewing Research   57
Foundation

Address: Lyttel Hall, Coopers Hill Road, Nutfield, Redhill, Surrey RH1 4HY
Telephone: (073 782) 2272
Status: Independent research centre
Director: Professor B.A.D. Atkinson
Assistant director: Alan Clapp
Research director: Dr J.R. Hudson
Sections: Chemistry; brewing technology; biochemistry; analysis; microbiology; food safety
Graduate research staff: 30
Activities: Basic research and development for the brewing and related industries, including work on raw materials and processes concerned with barley, malt, hops and adjuncts: fermentation (including yeast genetics), malting and wort preparation, analysis and test brewing, process innovation and engineering.
Publications: Bulletin of Current Literature, monthly.

Bridget's Experimental   58
Husbandry Farm

Address: Martyr Worthy, Winchester, Hampshire S021 1AP
Telephone: (0962 78) 220 and 330
Affiliation: Ministry of Agriculture, Fisheries and Food, Agricultural Development and Advisory Service
Status: Official research centre
Activities: Investigations in the following subjects - cereals: winter wheat and spring barley cultivation, fertilizer applications, variety, cropping sequences, and use of fungicides; weed control experiments, wild oats

and perennial grass weeds; grain loss due to delayed harvesting. Oilseed rape: establishment of winter sown crop and effect of date sowing; new variety testing. Maize: population studies for grain and silage maize; overall assessment of potential. Grass: new methods of grassland management; Italian ryegrass seed production; effects of slurry applications. Milk production: problems associated with a large number of Friesian cows in a single unit, feeding regimes for cows and heifers; calf rearing methods, feeding with colostrum, whole milk, and milk substitutes. Sheep: intensive stocking; foot rot control.

British American Tobacco Company Limited   59

– BAT
Address: Westminster House, 7 Millbank, London, SW1P 3JE
Telephone: (01) 222 1222
Telex: 27 384 BAT TOB G
Parent body: BAT Industries plc
Product range: Manufactured tobacco products.
Status: Industrial company
Director, research and development: Dr L.C.F. Blackman

GROUP RESEARCH AND DEVELOPMENT CENTRE   60

Address: Regents Park Road, Millbrook, Southampton, SO9 1PE
Telephone: (0703) 782111
Telex: 477269
General manager: A.L. Heard
Deputy managers: Dr C.I. Ayres; Dr M.J. Hardwick
Information officer: Dr F. Marsh
Graduate research staff: 90
Activities: A range of scientific and engineering activities covering processing of tobacco, manufacturing of tobacco products, physical and chemical composition of tobacco and of tobacco smoke; product technology; behavioural studies; biological sciences.

British Food Manufacturing Industries Research Association   61

– BFMIRA
Status: Independent research centre

LEATHERHEAD FOOD RESEARCH ASSOCIATION   62

– LFRA
Address: Randalls Road, Leatherhead, Surrey KT22 7RY
Telephone: (037 23) 76761
Telex: 929846
Director: Dr A.W. Holmes
Deputy director (science): Dr B. Jarvis
Sections: Microscopy, Dr D.F. Lewis; biochemistry, Dr C.L. Walters; meat and fish, Dr J.M. Woods; microbiology, Dr P.A. Gibbs; process engineering, C.R. Elson; energy and productivity advisory service, W.E. Whitman; chocolate and confectionery, A.G. Dodson; analytical chemistry, Dr M.J. Saxby; mycotoxins, W.B. Chapman; oils and fats, Dr J.B. Rossell; information, R.J.D. Saunders; physics, Dr R.T. Roberts; process analysis and control, Dr N.D.P. Dang; fruit vegetables and textures, Dr J.C. Fry
Graduate research staff: 150 (1982)
Activities: The association derives its support mainly from private industry, but is also a major contractor to the Ministry of Agriculture, Fisheries, and Food and other government departments. Many companies and some other institutions are members of the association, in the United Kingdom and in 36 overseas countries, including the Republic of Ireland, the Netherlands, and the United States. Research is conducted on meat and fish products, chocolate and sugar confectionery, soft drinks, jams, pickles, sauces, oils and fats analysis, food engineering, and instrumentation.
Publications: Annual report; scientific and technical surveys; overseas food legislation manual; food legislation surveys; *UK Guide to Food Regulations; Food Additives in the UK.*

British Industrial Biological Research Association*   63

– BIBRA
Address: Woodmansterne Road, Carshalton, Surrey SM5 4DS
Telephone: (01) 643 4411
Telex: 25438
Status: Research association
Research director: Dr D.M. Conning
Departments: Biological chemistry, B.G. Lake; immunotoxicology, Dr K. Miller; pharmacology, D. Pelling; microbiology, I. Rowland; genetic toxicology, Dr D. Anderson; pathology, W.H. Butler
Graduate research staff: 40
Activities: To investigate the mechanisms underlying toxic effects and to develop new or improved toxicity tests; to give advice and information on toxicological problems; to provide a forum for communication bet-

ween industry and government; to extend the knowledge and understanding of toxicology through science of high quality.
Publications: Annual report; *BIBRA Bulletin.*

British Standards Institution*   64

– BSI
Address: 2 Park Street, London, WlA 2BS
Status: National standards authority
Director General: G.B.R. Feilden
Scope of interests: The recognized body for the preparation of national standards in the United Kingdom, and for UK participation in international standards activities.

British Sugar Corporation Research Laboratories   65

Address: Colney Lane, Colney, Norwich, Norfolk NR4 7UB
Telephone: (0603) 52576
Status: Research centre within an industrial company
Head of Research: M. Shore
Graduate research staff: 30
Activities: All aspects of sugar beet production and processing to sugar, molasses and animal feed. New by-products and food science.
Projects: Quality of sugar beet; storage of sugar beet; assessment of variety trials; monitoring of pesticides and herbicides used on sugar beet and measurement of residues; assessment of the effects of damage after harvesting; provision of methods for the removal of growth inhibitors from beet seed.
Publications: British Sugar Beet Review, quarterly.

Brunel University*   66

Address: Kingston Lane, Uxbridge, Middlesex UB8 3PH
Telephone: (0895) 37188
Telex: 261173
Status: Educational establishment with r&d capability
Vice-Chancellor and Principal: S.L. Bragg

SCHOOL OF BIOLOGICAL SCIENCES*   67

Head: Professor T.F. Slater

Buhler-Miag (England) Limited*   68

Address: 19 Station Road, New Barnet, Hertfordshire EN5 lNN
Telephone: (01) 440 6511
Telex: 21805
Managing Director: G. Murray
Activities: Research on flour mill machinery; feed mill machinery; brewing and malting machinery; conveying plants; chocolate machinery; refuse disposal plants; ink roller mills; plastic injection moulding machinery; diecasting machinery.

Building Research Establishment*   69

– BRE
Address: Building Research Station, Garston, Watford, WD2 7JR
Telephone: (092 73) 74040
Telex: 923220 BRSBRE G
Parent body: Department of the Environment and Department of Transport - Common Services
Official research centre
Director: Dr I. Dunstan
Deputy Director: M.E. Burt
Graduate research staff: 400 (in all departments)
Activities: The work of the Establishment is carried out by a number of Divisions, each dealing with a major building subject. Broad areas of research are as follows: planning; housing policy; building standards, design, construction, and materials; building performance and design; energy conservation and building services; components; foundations and substructures; structural design and performance; fire; construction management; materials utilization; maintenance and preservation; construction economics; general studies; environmental protection; overseas research.

COMPONENTS AND STRUCTURES DIVISION   70

Head.   J.F.S. Carruthers
Activities: Performance - based specifications for building components such as doors, windows, and their associated hardware; elements such as flat roofs and prefabricated wall units, including timber frame construction. The design and performance of structural timber components and elements are important aspects of the division work.

TIMBER AND PROTECTION DIVISION 71

Head: Dr J.W.W. Morgan
Activities: Properties, performance, losses due to biological or physical degrade, and utilization of timber and board materials. The division evaluates and provides advice on remedial and pre-treatment specifications for preservatives and protective coatings, and is also concerned with the assessment of the environmental impact of chemicals.

Cadbury Schweppes plc 72

Address: Leconfield House, Curzon Street, London, W1Y 7FB
Telephone: (01) 262 1212
Telex: 338011
Director: H. Lavery

CADBURY LIMITED 73

Address: PO Box 12, Bournville, Birmingham, B30 2LU
Telephone: (021) 458 2000
Telex: 338011
Research and Development Director: Dr W.A.W. Cummings
Senior staff: Dr R. Bradford
Graduate research staff: 38
Activities: Research into chocolate and confectionery products and processes, including control systems; quality control methods; plant and process commissioning and process optimization; legislation; nutrition standards; toxicology.

CADBURY TYPHOO LIMITED 74

Address: PO Box 171, Franklin House, Bournville Lane, Bournville, Birmingham, B30 2NA
Telephone: (021) 458 2000
Telex: 338011
Research and Development Director: Dr W.A.W. Cummings
Activities: Research into milk beverages, dairy products, biscuits, jams, jellies, canned fruit and vegetable products and processes; quality control.

LORD ZUCKERMAN RESEARCH CENTRE 75

Address: University of Reading, Whiteknights, Reading, RG6 2LA
Telephone: (0734) 868541
Telex: 849707
Product range: Confectionery, soft drinks, tea, coffee and a wide range of food and household products.
Director of Group Research: Dr J.R. Norris
Research executives, Reading: Dr G.G. Jewell, Dr D. McHale
Research executive, Development unit Somerdale/ Bristol: R.S. Ripper
Activities: The centre is concerned with long-range fundamental research into the science underlying the group's present and possible future products. Important areas of work include the understanding of the chemical and physical properties of fats in relation to the formulation and quality of the company's product range, flavour chemistry, the development and application of rapid methods in the study of microorganisms and the evaluation of eating quality using both instrumental and taste panel methods.

SCHWEPPES INTERNATIONAL RESEARCH AND DEVELOPMENT* 76

Address: 105 Brook Road, Dollis Hill Estate, London, NW2 7DS
Research and Development Director: Dr A.R. Grinham
Sections: Product development, P.A. Halsted; materials development, Dr F.B. Preston; services, A. Bruce; packaging development, J.A.W. Grey
Activities: The provision of comprehensive r&d, which currently markets products in 55 countries. Development covers all aspects from primary ingredient (flavour and fruit) processing, through beverage formulation to finished product packaging. Support is produced from microbiology, analytical chemistry, legal/library and international quality assurance.

Campden Food Preservation Research Association 77

Address: Chipping Campden, Gloucestershire GL55 6LD
Telephone: (0386) 840219
Telex: CFPRA-G 337017
Status: Research association
Director-general: K. Dudley
Sections: Chemistry and biochemistry, Dr D.J. Henshall; agriculture and quality, Dr V.D. Arthey;

engineering, D.A. Steele; food technology, D. Atherton; microbiology of heat processed foods, R.H. Thorpe; general microbiology, L.P. Hall
Graduate research staff: 34
Activities: Food preservation from raw material to final product.
Projects: Thermal process evaluation; alternative methods of food preservation; guidelines for continuous cookers (D. Atherton); microprocessor aplications: heat processing; pumping characteristics of foods (D.A. Steele); bacteriological quality control; hygienic design and post-process sanitation; spore studies; non-thermal methods of preservation (R.H. Thorpe); mathematical modelling (S.D. Holdsworth); assessment of varieties of fruit and vegetables for processing; potato quality for chipping (R. Billington); screening of agrochemicals for taint potential in processed foods (D. Lyon); microbiological assay methods for frozen foods (L.P. Hall); blanching of vegetables prior to processing; correlation of sensory and instrumental methods of quality appraisal (J.B. Adams); trace metals in foods; pesticide residues in foods (Dr J.D. Henshall).
Publications: Annual report; *Annual Statistics Review.*

AGRICULTURE AND QUALITY DIVISION 78

Head: Dr V.D. Arthey
Departments: Agriculture, R. Billington
Activities: Projects are undertaken on the following topics: assessment of fruit and vegetables for processing; development of quality standards on all food products.

FOOD SCIENCE DIVISION 79

Head: Dr J.D. Henshall
Departments: Chemistry and biochemistry, vacant
Activities: Projects are undertaken on pesticide residue analysis; enzymes - HTST; blanching; instrumental quality assessment; metal analysis; corrosion problems; headspace gas analysis; interaction of food components.

FOOD TECHNOLOGY DIVISION 80

Head: Dr C. Dennis
Departments: Food engineering, D.A. Steele; food technology, D. Atherton; general microbiology, L.P. Hall; heat processed foods microbiology, R.H. Thorpe
Activities: Projects are undertaken on process determination; new methods of preservation; container integrity; microprocessor control; frozen and heat processed foods; aseptic processing; heat resistance of micro-organisms and spores; methodology.

Cattle Breeding Centre 81

Address: Shinfield, Reading, Berkshire
Telephone: (0734) 883157
Affiliation: Ministry of Agriculture, Fisheries and Food, Agricultural Development and Advisory Service
Status: Official research centre
Head: Dr P.H. Lamont
Senior staff: Dr J.A. Foulkes; R.W. Sauders; Dr G.D.A. Wilson
Activities: Research into fertility of domestic animals with special reference to artificial insemination, breeding control and herd fertility.
Projects: Morphology and biochemistry of bull and boar semen, semen evaluation and processing, survival of disease agents during semen processing, freezing boar semen, cattle insemination techniques, heterospermic inseminations (Dr J. Foulkes); semen defects in bulls (Dr P.H. Lamont); oestrus detection in cattle and pigs (Dr J.A. Foulkes); oestrus synchronization and planned breeding, breeding efficiency in cattle, factors affecting herd fertility, fertility control programmes in dairy cattle (R.W. Saunders); periovulatory period (Dr G.D.A. Wilson).

Central Veterinary Laboratories 82

Address: New Haw, Weybridge, Surrey KT15 3NB
Telephone: (093 23) 41111
Affiliation: Ministry of Agriculture, Fisheries and Food, Agricultural Development and Advisory Service
Status: Official research centre
Director: A.J. Stevens
Deputy directors: Dr J.G. Ross, Dr P.H. Lamont
Departments: Bacteriology, Dr B.J. Shreeve; biochemistry, M.K. Lloyd; biological products and standards, I. Davidson; diseases of breeding, R. Bradley; epidemiology, G. Davies; medicines unit S.F.M. Davies; parasitology, Dr L.P. Joyner; pathology, Dr J.T. Done; poultry, Dr G.A. Cullen; virology, Dr D.H. Roberts; animal production, R.A. Huck
Graduate research staff: 200
Activities: Veterinary medicine, including: work initiated by the State veterinary service in relation to enabling the confirmation of diagnosis of some notifiable and scheduled diseases; the production and statutory testing of biological products; the licensing of veterinary medicines; the testing of samples for export certification; and the routine surveillance and monitoring of animal diseases.
The research programme supports these functions but also includes the study of those diseases of farm livestock which are of national economic importance, the study of emerging diseases, and of residues in meat.

The laboratory is also one of the three Biological Standard Laboratories of the World Health Organization, Food and Agriculture Organization reference laboratory for brucellosis and the world reference laboratory for Newcastle disease.

Projects:

Bacteriology - tuberculosis (T.W.A. Little); listeriosis (M. Gitter); mastitis (P.G. Francis); salmonellosis, colibacillosis (W.J. Sojka, C. Wray); mycoplasmosis (E. Boughton); pneumonia (D.G. Pritchard); leptospirosis, Johne's disease, Q Fever, anthrax (T.W.A. Little); mycoses (G.A. Pepin); monoclonal antibodies (J.A. Morris); determinative bacteriology, contagious equine metritis (J.E. Shreeve); swine dysentery (M. Cutler).

Biochemistry - endocrinology (N. Saba); general toxicology (G. Lewis); organophosphorus and other pesticides (A.F. Machin); chemical pathology (P.H. Anderson); toxic metals, metabolic disorders (G. Lewis); cobalt deficiency (R. St.J. Stebbings); copper deficiency and excess, iodine deficiency, cerebrocortical neurosis in sheep and cattle (G. Lewis); vitamin E and selenium (P.H. Anderson); fluorosis (M.K. Lloyd); Johne's disease, bracken poisoning (L.N. Ivins).

Biological products and standards - sterility testing of biological substances (Dr G.N. Frerichs); clostridial products (T.W. Lesslie); brucella abortus antigen (R. Martin); infectious bronchitis vaccine (Dr A.M.T. Lee); Merek's disease and, Newcastle disease vaccine, avian vaccines in general, infections bursal disease vaccine (Dr D.H. Thornton); tuberculins (D.W. Andrews); contagious pastular dermatitis (Dr G.N. Frerichs); ovine enzootic abortion vaccine (S.B. Woods); Johne's disease (I.W. Lesslie); viral arthritis, fowlpox vaccines, infections laryngotracheitis vaccines (Dr D.H. Thornton); escherichia coli salmonella, bordetella, pasteurella, swine erysipelas, and lungworm vaccines (Dr S.N. Hussaini); maliein (I.W. Lesslie); leptospira vaccines (S.B. Woods); infectious bovine rhinotracheitis vaccines (Dr G.N. Frerichs); egg drop syndrome 76 vaccines (Dr A.M.T. Lee); brucella abortus vaccine (J.C. Muskett); preservatives in vaccines (Dr G.N. Frerichs).

Diseases of breeding - Brucellosis: antigens (R.A. Beli, Dr M.J. Corbel); serological tests A.P. MacMillan, R.A. Bell, Dr M.J. Corbel, F.A. Stuart); automated Rose Bengal Plate Test (S.G.M. Gower); cross reactions, culture media, identification methods, culture preservation (Dr M.J. Corbel); investigation of anomalous reactions to brucella, antiobiotic sensitivity (F.A. Stuart); epidemiology (R.A. Bell). Campylobacteriosis: antigens (K.P.W. Gill); diagnostic and serological tests (K.P. Lander); culture media identification and classification, pathogenesis, zoonotic aspects, prevention (K.P. Lander); preservation and storage (K.P.W. Gill). Other bacteria (Dr M.J. Corbel); reproductive failure (R. Bradley); automated testing (R.A. Bell).

Epidemiology - observational studies (P.A. Williams); information collection and retrieval (J.C. Bell); epidemiological investigations (J. Wilesmith); surveys (M.S. Richards).

Parasitology - parasitic gastero-enteritis in sheep (Dr R.J.G. Cawthorne, Dr J.F. Michel); cattle nematodes, helminth control (Dr J.F. Michel); pig nematodes, serology (Dr I.J.B. Sinclair); trichomoniasis in cattle, coccidiosis in poultry, rabbits, sheep, and in pigs (Dr L.P. Joyner); fascioliasis incidence forecasts (Dr C.B. Ollerenshaw); diagnosis; warble fly, sheep blowfly, transmission of disease by cockroaches, sheep lice, flies affecting grazing livestock (Dr D.W. Tarry); parasitic rutes (A.C. Kirkwood); dourine (Dr L.P. Joyner); Babesia (J. Donnelly).

Pathology - cattle: general definitive pathology (L.M. Markson, M. Jeffrey); nervous diseases (L.M. Markson); developmental disorders of foetus (C. Richardson). Sheep: general definitive pathology (L.M. Markson, M. Jeffrey); nervous diseases (L.M. Markson); congenital disease of lambs, pathology of foetal wastage, wastage in adult sheep (C. Richardson). Pigs: general definitive pathology (G.A.H. Wells); reproductive failure, foetal pathology, cytomegalic inclusion body disease, virus infection in very early gestation (Dr A.E. Wrathall); atrophic rhinitis (Dr J.T. Done); disorders of muscle, nervous diseases, congenital tremor (G.A.H. Wells). Ultrastructural pathology.

Poultry - avian influenza paramyxo-viruses other than New castles disease (Dr D.J. Alexander); Newcastle disease (W.H. Allan); egg quality, diseases of ducks and geese (D. Spackman); infections bronchitis, psittacosis/ornithosis (Dr C.D. Bracewell); infections bursal disease (Dr G.A. Cullen); mycoplasmosis, salmonella (J.A. Thain).

Virology - rabies (A.A. King); swine fever (P.L. Roeder); respiratory disease in cattle (S. Edwards); respiratory disease in pigs, bovine leukosis and other retroviruses (Dr D.H. Roberts); reproductive loss in cattle (M.H. Lucas); bovine viral diarrhoea/mucosal disease, border disease in sheep (P.L. Roeder); viral enteritis in young pigs (Dr D. Chasey); reproductive failure in pigs, transmissible gasteroenteritis of pigs, epidemic diarrhoea, enteroviruses in pigs, Anjeszky's disease, porcine cytomegalovirus (S.F. Cartwright); viral diseases of sheep, chlamdial diseases of animals (M. Dawson); anti-viral agents (J.W. Harkness).

Animal production - parasite monitoring of minimal disease pig herd, dystokia in sheep, disease surveillance, pustular dermatitis of sheep (B.N.J. Parker); pietrain creeper syndrome, post-natal ataxia in piglets (Dr W.K.S. Wijertane).

Centre for Industrial Innovation 83

– CII
Address: 100 Montrose Street, Glasgow, G4 0LZ
Telephone: (041) 552 4400
Telex: 77472
Affiliation: University of Strathclyde
Status: Industrial research centre
Director: Edwin R. North
Project leaders: Dr C.A. Walker, Dr J. McKelvie
Graduate research staff: 7 (1983)
Activities: Providing a link between the university and industry, the centre is able to coordinate multi-disciplinary research and development contracts and is an official sub-contractor for SFTES, MAS and MAPCON schemes on behalf of the university. Development of university inventions to licensable level, including prototype building and liaison with industrial licensees, is an important service of the CII which is also available to industrial clients. Experience in this field has covered instruments, products and processes.
Current research projects within the CII relate to high temperature strain measurement, soya-based foodstuffs, and potential new methods of extracting energy from peat.

Centre for Overseas Pest Research 84

– COPR
Address: College House, Wright's Lane, London, W8 5SJ
Telephone: (01) 937 8191
Parent body: Overseas Development Administration, Foreign and Commonwealth Office
Status: Official research centre
Deputy director: T. Jones
Divisions: Biology I (ecology and taxonomy), Dr W.A. Sands; biology II (insect/host relationships), Dr R.F. Chapman; field (migrant pests and vectors), Dr J.C. Davies; chemical control (pesticide application), D. Lyon
Graduate research staff: 105 (1983)
Activities: COPR is one of the scientific units of the Overseas Development Administration. It exists specifically to help overseas governments, especially those of developing countries, to solve agricultural problems caused by crop-pests and health problems transmitted by disease carrying insects (vectors). This help includes field and laboratory research, the development and application of newer and more effective techniques of pest and vector control and the provision of scientific information, advice and training.

COPR has outstations at the Chemical Defence Establishment, Porton Down, near Salisbury (for chemical control), the British Museum (for natural history), the Royal Radar Establishment, Malvern, Worcester (for radar observations of insect flight), the Tropical Products Institute, (see separate entry) (for larvicides), and at Cromwell Road, London SW7 (for termite and some grasshopper taxonomy). The centre merged in 1983 with the Tropical Diseases Centre.
Projects: All projects are carried out in cooperation with the relevant international organizations and local governments. Examples are: control of rice planthoppers in the Far East (Dr R.F. Chapman); use of pheromones and viruses to control Egyptian cotton leafworm (Dr D.G. Champion); control of armyworm in East Africa (Dr D.J.W. Rose); ecology and control of termites as soil pests (Dr T.G. Wood); ecology and control of tsetse flies in Somalia (C.W. Lee); training courses in pest management in the United Kingdom and overseas (Dr I.H. Haines).
Publications: Tsetse and Trypanosomiasis Information Quarterly; Tropical Pest Management, quarterly; *Termite Abstracts*, quarterly; annual report.
See separate entry for:
Land Resources Development Centre
Tropical Products Institute

Chelsea College* 85

Address: Main Building, Manresa Road, Chelsea, London, SW3 6LX
Telephone: (01) 351 2488
Parent body: University of London
Status: Educational establishment with r&d capability
Principal: Dr C.F. Phelps

APPLIED BIOLOGY DEPARTMENT* 86

Head: T.R. Milburn
Graduate research staff: 9
Activities: Ecology and environmental monitoring, microbiology and mycology, physiology and biochemistry of invertebrates and plants, electron microscopy. Special attention is paid to ecology of natural waters; water conservation and pollution; fish and fisheries.

ZOOLOGY DEPARTMENT* 87

Head: Professor R.D. Purchon
Senior staff: G.E. Barnes, Dr W.P. Williams
Graduate research staff: 4
Activities: Ecology and environment.

Commercial Rabbit Association*   88

Address: Tyning House, Shurdington, Cheltenham, Gloucestershire GL51 5XF
Telephone: (0242) 862387
Status: Research association
Secretary: Peter Horne

Coypu Research Laboratory   89

Address: Jupiter Road, Norwich, NR6 6SP
Telephone: (0603) 405 990
Affiliation: Ministry of Agriculture, Fisheries and Food, Agricultural Development and Advisory Service
Status: Official research centre
Officer in Charge: Dr L.M. Gosling
Graduate research staff: 4
Activities: The aims of the laboratory are: to investigate fundamental aspects of coypu biology; to design effective control measures based on the information obtained. Techniques used include computer simulation of population behaviour.
Projects: Investigation of the population ecology and control of the UK, feral coypu population (L.M. Gosling).

Cranfield Institute of   90
Technology*

– CIT
Address: Cranfield, Bedford, MK43 0AL
Telephone: (0234) 750111
Telex: 825072
Status: Educational establishment with r&d capability
Activities: Cranfield Institute of Technology is a unique university institution offering mainly postgraduate degrees in advanced technology, applied science and management. Within the university sector, CIT is Britain's largest centre of applied research and development in industrial technology, ranging over most of the areas vital to both modern and developing societies. Cranfield is a centre for new ideas and concepts and its degree courses and programmes of continuing studies are concerned with the most up-to-date aspects of technology and management. The National College of Agricultural Engineering, which is a Faculty of CIT, is one of the largest undergraduate centres for agricultural engineering in the United Kingdom. Company and corporation sponsorship have important roles in supporting students while studying at CIT. There is a large research and development programme, mostly for private companies and public corporations. There is a strong emphasis on

international links especially with EEC and other advanced technological countries. CIT has a staff of 40 professors, 150 lecturers and nearly 300 research staff, forming a strong team for teaching and research in industrial technology and management.

FACULTY OF ENGINEERING*   91
Dean: Professor J.L. Stollery

College of Aeronautics*   92
Head and Professor of Aerodynamics: Professor J.L. Stollery
Senior staff: Aircraft design, Professor D. Howe; theoretical gas dynamics, Professor J.F. Clarke; air transport engineering, Professor D.G. Yeomans
Activities: Research into: aerodynamics of aircraft and missiles; hovercraft and road vehicles; gas dynamics including combustion; wind engineering, stability and control; agricultural aviation; overall design of manned and un-manned aircraft; structural design with emphasis on stability, reinforced plates, fracture and fatigue; aircraft systems, reliability and safety; design of offshore structures; management and appraisal of air transport systems, emphasizing economics and development; use of computer-based models; technical airline operations, maintenance and reliability; evaluation of transport aircraft; airline and air cargo marketing.

FACULTY OF SCIENCE AND   93
TECHNOLOGY*
Dean: Professor D.W. Saunders

Ecological Physics Research Group*   94
Research Professor: Professor G.W. Schaefer
Activities: Application of the philosophy and techniques of mathematics, physics and technology to ecological problems, both pure and applied; dispersion and management of the airborne pests of cattle, cereals, forests and cotton.

NATIONAL COLLEGE OF   95
AGRICULTURAL ENGINEERING
– NCAE
Address: Silsoe, Bedfordshire MK45 4DT
Telephone: (0525) 60428
Telex: 825 072 CITECH G
Head: Professor B.A. May
Departments: Building and processing, Professor B.A. May; soil and water engineering, Professor N.W. Hudson; design, machinery and mechanization, Professor R.W. Radley; market and product management, Professor R.W. Hill

Graduate research staff: 51
Activities: Applied research for the agricultural engineering and allied industries. Machinery design: tractor-implement control; tractor ride; harvesting and on-farm post-harvest treatment of crops; farm machinery for overseas conditions. Seed and grain technology: development of new machines for cleaning, handling and grinding based on studies of the properties of the materials involved and the principles of rheology. Crop drying: mathematical modelling of heat flow through stored crops; low-energy including natural drying processes. Drainage: modelling of soil response to drainage operations; improvements to mole and trenchless drainage procedures; use of drainage for land reclamation in arid environments. Irrigation and water management: crop water requirements; irrigation scheduling; techniques for accurate water placement including trickle and bubbler systems; aquifer recharge, crop yield - water use production functions. Soil erosion and conservation: erosion by rainsplash and overland flow; crop cover effects on erosion potential; surface roughness; modelling erosion and conservation systems. Tillage: soil-implement interactions and implications for seedbed preparation; powered tillage; subsoil loosening and deep fertilizer placement. Product management and marketing: decision-making processes; buying behaviour; innovation adoption. Crop quality assessment: assessment and prevention of mechanical damage to crops during handling and storage. Agricultural production systems planning: modelling and predicting optimum transport systems; locational analysis of agricultural processing plants; agricultural machinery and labour planning; assessment and utilization of agricultural wastes and residues.
Publications: Annual research and development booklet.

Dalgety Spillers Limited* 96

Status: Industrial company

PET NUTRITION UNIT* 97

Address: Bury Road, Kennett, Suffolk CB8 8QU
Research Controller, Spillers Agriculture: Dr L.G. Chubb
Information Scientist: H.B.C. Jones
Graduate research staff: 20
Activities: Nutritional research applied to pet foods.

RESEARCH AND TECHNOLOGY CENTRE 98

Address: Station Road, Cambridge, CB1 2JN
Telephone: (0223) 59181
Telex: 81671
Product range: Convenience foods and food ingredients; pet foods; flour; meat; malt; animal feeds and agricultural supplies; chemicals.
Director of research and technology: Dr P.W. Eggitt
Sections: Group research laboratory, research, Dr N.W.R. Daniels; group research laboratory, administration, J.C. Dickins; Spillers foods R&D, A.C.S. Gardiner; milling R&D, Dr P.S. Wood
Graduate research staff: 80
Activities: Applied research in food science and rheology; extrusion processing; the biochemistry of cereals and proteins; organic chemistry and advanced analysis. Product and process development in milling, baking, pet foods and convenience foods.

Dawson International plc* 99

Address: R & D Centre, Riverside Mills, Selkirk, TD7 5EF
Telephone: (0750) 20651
Telex: 727233 (FASRAN G)
Status: Research centre within an industrial company
Affiliation Dawson International plc Group
Group technical director; Geoffrey Allan Smith
Sections: Financial and physical processing, Rudi Kelhoffer; fibre analysis and identification, Justine Lawton; computer interfacing, William Laird; RF applications, Thomas McAulay; wet finishing and dyeing, Richard Young
Graduate research staff : 11
Activities: Preparation and upgrading of cashmere goats in Australia and application of the fibre to own company use; new yarns and methods of manufacture; computer use for production control, technology audits and interfacing within different manufacturing systems from raw material to final fabric - primarily, animal fibres; radio frequency heating technology as applied to wet processing of animal fibres; energy conversion as applied to textiles.

Department of Agriculture and Fisheries for Jersey 100

Address: 44 Esplanade, St Helier, Jersey, Channel Islands
Telephone: (0534) 23401
Chief executive officer: J.F. Abraham

Department of Agriculture and Fisheries for Scotland 101

Address: Chesser House, 500 Gorgie Road, Edinburgh, EH11 3AW
Telephone: (031) 443 4020
Telex: 72162
Affiliation: Scottish Office
Status: Government department
Sections: Scientific advisers unit, Dr A. Raven; research branch, A.I. Macdonald
Activities: The department funds seven research institutes.
Publications: Agriculture in Scotland, annual.
See separate entries for: Animal Diseases Research Association
Hannah Research Institute
Hill Farming Research Institute
Macaulay Institute for Soil Research
Rowett Research Institute
Royal Botanic Garden
Scottish Crop Research Institute
Scottish Institute of Agricultural Engineering

AGRICULTURAL SCIENTIFIC SERVICES 102

Address: East Craigs, Craigs Road, Corstorphine, Edinburgh, EH12 8NJ
Telephone: (031) 339 2355
Telex: 727 348 (DAFASS G)
Director: Dr D.C. Graham
Graduate research staff: 70 (1982)
Activities: East Craigs Agricultural Scientific Services was established in 1914 as the official seed testing and plant registration Station for Scotland. Its functions now are to provide scientific advice to the Department of Agriculture and Fisheries for Scotland, and to undertake statutory and regulatory work in connection with the department's national and international responsibilities relating to agricultural crops, especially certification, plant health, and pesticides, together with research and development necessary to support that work. The Station runs a 155 hectare farm which provides the land for testing and experimental work and for the maintenance of extensive reference collections of varieties of the crop plants for which it is responsible. There is a nuclear stock farm for seed potatoes near Biggar. Studies on pesticides and on pollution within an agricultural context.

Pest Control and Pesticides 103

Head: J.R. Cutler
Sections: Pesticide usage, J.R. Cutler; chemistry, G.A. Hamilton; crop entomology, L.A.D. Turl; wildlife investigation, J.H. Cuthbert; infestation control, M.P. Muttrie

Plant Varieties and Seeds Division 104

Telephone: (031) 339 2355
Head: R.D. Seaton
Sections: Official seed testing station, S.R. Cooper; cereals, J.L. Keppie; herbage and vegetable crops, W.G. Sutton
Graduate research staff: 15
Activities: The aim of the division, so far as research is concerned, is to carry out R&D projects which are necessary to maintain the efficiency of the routine scientific services provided in the spheres of seed quality control and registration of agricultural and horticultural plant varieties. Major current projects: seed pathology test methods and evaluation of the effect of seed-borne pathogens on seed quality, germination physiology studies, development and improvement of procedures, including data processing, and new descriptive characteristics related to variety registration and certification; study of electrophoretic techniques in relation to possible use in variety registration.

Potato and Plant Health Division 105

Head: M.J. Richardson
Sections: Potato, T.D. Hall; nematology, T.W. Mabbott; potato disease control, vacant; plant health, P.J.Howell. crop loss assessment, M.J. Richardson

FISHERIES RESEARCH SERVICES 106

Coordinator: Alan Preston
Activities: The coordinator of fisheries research and development for Great Britain is responsible jointly to the fisheries secretaries of the Ministry of Agriculture and Fisheries for Scotland and the Ministry of Agriculture, Fisheries, and Food. His duties are in addition to those of his directorship of fisheries research in the latter ministry.
See separate entries for: Freshwater Fisheries Laboratory
Marine Laboratory.

Department of Agriculture 107
for Northern Ireland*

Address: Dundonald House, Upper Newtownards Road, Belfast, BT4 3SB, Northern Ireland
Telephone: (0232) 650111
Status: Government department
Chief Scientific Officer: Dr W.O. Brown
Graduate research staff: 180
Activities: The Department finances practically all agricultural research work in Northern Ireland carried out by eight academic specialist Divisions (which also provide the staff of the Faculty of Agriculture and Food Science of Queen's University, Belfast), and by three agricultural colleges and four research centres run solely by the Department.
Publications: Annual report; *Agriculture in Northern Ireland* (series of project reports); scientific publications (including *Record of Agricultural Research*).

AGRICULTURAL ECONOMICS AND 108
STATISTICS DIVISION*

Head: Professor G.W. Furness
Activities: Projects on the economic prospects of the agricultural industry in Northern Ireland; changes in farm business structure and resource use; economic features of production and the marketing and processing of agricultural commodities.

Bush Fishery* 109

Address: Bushmills, Antrim, Northern Ireland
Manager: R.J.D. Anderson
Activities: Providing information on the management of natural stock of salmon; determining the economic feasibility of smolt rearing in connection with 'sea ranching'.

Fisheries Development* 110

Address: Movanagher Fish Farm, 152 Vow Road, Ballymoney, Antrim BT53 7NT, Northern Ireland
Chief Fisheries Officer: K.U. Vickers
Manager: S.D. Fidgeon
Activities: Research for information on hatching and rearing activities of trout; advising potential fish farmers; inspection for disease, pollution and other problems.

Drayton Experimental 111
Husbandry Farm

Address: Alcester Road, Stratford-upon-Avon, Warwickshire CY37 9RG
Telephone: (0789 29) 3057 and 2353
Affiliation: Ministry of Agriculture, Fisheries and Food, Agricultural Development and Advisory Service
Status: Official research centre
Activities: Research in the following areas: cereals (winter wheat, oats, and spring barley) - cultivation systems for heavy soil, wild oat control, cropping sequences, response to fertilizers and fungicides, grain losses due to delayed harvesting, variety testing; field beans - variety testing; grass - conservation as hay and as silage, additives, fertilizers, grazing management, conditioning and handling big bales of hay; sheep - effects of ewe nutrition on lambing performance; increasing prolificacy, effects of sheep integrated with beef cattle; beef production - four systems of beef production, performance testing of cross-bred calves.

East Malling Research 112
Station

– EMRS
Address: Maidstone, Kent MEl9 6BJ
Telephone: (0732) 843833
Affiliation: State-aided institute, Agricultural Research Council
Status: Official research centre
Director: Dr I.J. Graham-Bryce
Sections: Crop production, Dr J.E. Jackson - pomology, Dr J.E. Jackson, plant propagation, Dr B.H. Howard; crop protection, Dr T.R. Swinburne - plant pathology, Dr P.W. Talboys, plant protective chemistry, Dr D.J. Austin, zoology, Dr J.J.M. Flegg; fruit storage, Dr R.O. Sharples; plant physiology, Dr K.J. Treharne; fruit breeding, Dr R. Watkins; statistics, Dr A.A. Rutherford
Graduate research staff: 100
Activities: The general aims are to improve the productivity, reliability, quality, and marketability of the crops investigated and to advance relevant underlying sciences for possible application to any commodity. Specific aspects of research are:
Plants - crop production: temperate tree fruit (apple, pear, sweet cherry, plum); soft fruit (raspberry and other cane and bush fruit); includes research on crop environment, growth regulators, agronomy, plant anatomy. Plant propagation: vegetative propagation of fruit tree rootstocks and deciduous ornamental trees and shrubs; grafting and budding. Plant protection: fruit and hops; includes mycology, bacteriology, virology, production, and distribution of pathogenfree plants, entomology,

nematology, ornithology, chemistry. Plant physiology: investigation of carbon resources, stress physiology, plant hormones, tissue culture, and soil nutrient chemistry with particular reference to temperate fruit. Plant breeding: temperate fruit tree rootstocks, scion varieties, and self-rooted varieties; cane and bush fruit; ornamental fruit plants. Statistics: design of field experiments and data analysis; statistical advice on tropical and subtropical plantation crops.

Food - fruit storage: preservation of fresh apples, pears and, to a lesser extent, soft fruit.

Additional information: The station houses the Commonwealth Bureau of Horticulture and Plantation Crops. It also accommodates the National and Regional (South-east) Fruit Specialist Advisers of the Ministry of Agriculture.

Publications: Annual report.

East of Scotland College of Agriculture 113

See separate entry for: Edinburgh School of Agriculture

Edinburgh School of Agriculture 114

Address: West Mains Road, Edinburgh, EH9 3 JG
Telephone: (031) 667 1041
Telex: 727617 ESCA G
Affiliation: University of Edinburgh; East of Scotland College of Agriculture
Status: Educational establishment with r&d capability
Principal: Professor N.F. Robertson
Graduate research staff: 55 (1982)
Activities: Crop and animal production. Work on animals covers reproduction, physiology, nutrition, behaviour, health, and husbandry of dairy and beef cattle, sheep, pigs, and poultry. Crop research deals with variety testing, husbandry, nutrition, pathology and protection of cereals, potatoes, brassicas, legumes, flax, vegetable crops, soft fruit, and hardy nursery stock. Production and conservation of grass receive considerable attention. In the soils area the major aim is to identify and remove physical and chemical yield constraints associated with specific soil types. Research on nitrogen utilization is given some priority. Trace element work examines both deficiencies and excesses of trace elements in animals and plants. Research on farm buildings is mainly concerned with assessment of farm building design, and with the study of the environment inside buildings for both crops and livestock, but an active interest is taken also in energy conservation. Farm

machinery selection and systems evaluation are major components of the mechanization research and environmental control work is also important. Economics research includes the application of computer models to farm management extension work, commodity marketing and market forecasting, the economics of farming systems, and computerized management recording systems.

Increasing emphasis is being placed on the multi-disciplinary approach to computer modelling of production systems.

A considerable portion of the effort and resources of the school is devoted to development work which can be defined as the testing, modification, and synthesis of ideas shown to be of potential value to the agricultural industry, in order that problems of implementation under commercial conditions can be identified and resolved. The school fosters and maintains close links with the major agricultural research insitutions in order that research findings can be translated as quickly as possible through the applied stages and into farming practice.

The Edinburgh School of Agriculture is an amalgam of the University of Edinburgh Department of Agriculture and the East of Scotland College of Agriculture.

Projects: Computer software development for research, development, and extension purposes in animal production; growth of pigs: genetic improvement for high yields in dairy cows; management of genetically improved high yielding dairy cows at Langhill; evaluation of Hereford x Holstein and Hereford x Friesian cattle; identification of the causal forces of variation in level and character of lactation yield in dairy cows (Dr C.T. Whittemore); large dairy herd management, husbandry, and health (E.M. Bell); microbiology of milk (R.M. McLarty); beef cattle systems evaluation; suckler cow and calf nutrition and fertility; beef cattle feeding and finishing (B.G. Lowman); dairy bred calves: husbandry, health, and nutrition; lifetime performance of beef calves; silage: intake and utilization (C.E. Hinks); forage systems for livestock; beef cattle grazing studies (A.W. Illius); meat production potential of lamb genotypes; carcass and meat quality studies (J.H.D. Prescott); ewe nutrition and flock feeding strategy; early lambing of draft Blackface ewes (W.J.M. Black); group breeding; improved hill farm systems; hill sheep development; upland ewe flock management; grassland based sheep systems (House O'Muir); store lamb finishing systems; Suffolk sheep selection (M.D. Lloyd); epidemiological approach to the control of ovine toxoplasmosis (J.K. Miller); gastro-intestinal helminthiasis in sheep (G.B.B. Mitchell); experimental salmonella infection in sheep (K.A. Linklater); nutrition and reproduction in the pregnant and lactating gilt and sow (D. Wyllie); feeding, housing and systems management of sucking, weaned and young growing pigs; management of commercial sow herds (G.M. Hillyer); prediction of energy value of compound foods: evaluation of feed ingredients for pigs (C.A. Morgan); poultry management scheme evaluation (S. Brocklehurst); growth of chickens (G.C. Emmans); biochemical and food evaluation studies with poultry (J.P.F. D'Mello; microbiology associated with hatching, rearing and processing poultry; environmental bacteria and hygiene in animal prodution (S. Morgan-Jones); production responses of laying fowl following development of new practices for egg production (J.A. Anderson); reproductive physiology (R.H.F. Hunter); animal behaviour (D.G.M. Wood-Gush); investigations into energy utilization of farm animals by calorimetry (A.R. McLellan); physical and chemical processing of feeds (M. Lewis); environmental physiology (W.J. Guild); ration formulation and prediction of performance in farm animals involving the use of computer programs; silage-development studies (R.A. Edwards); exploring structures for co-operative honey handling and marketing; bees, disease, crop sprays, and pollination (R. Couston); lowland grass production (G. Swift, M.W. Morrison); hill pasture (J.B.D. Herriott); silage: biochemical and microbiological studies (A.R. Henderson, P. McDonald); silage: metabolism and balance studies (P. McDonald, R.A. Edwards); microbiology and chemical properties of hay (J. Robb); national list and college testing of herbage species and varieties (M.W. Morrison); cereal production (J.C. Holmes); incidence and control of and resistance of varieties to cereal diseases (J. Gilmour, J.H. Lennard); plant biochemistry including cereal grain development and germination (C.M. Duffus); grain quality; national list trials of pulse varieties; college and national list testing of cereal varieties (D.A.S. Lockhart); factors restricting cereal yield on fluvio-glacial sands and gravels; drainage of soils of low permeability (R.B. Speirs); cultivar assessment studies in cereals; genetic studies in brassica spp. (W. Spoor); potato production; college testing of potato varieties (R.W. Lang); control of potato diseases (J. Gilmour); brassica diseases (T. Brokenshire); rapid multiplication of stocks of potato varieties and horticultural stock by means of in vitro techniques (P.C. Harper); college and national list testing of root crop varieties (W.D. Gill); soft fruit production; vegetable crop production and storage; production and storage of bulb crops (W. Fordyce); weed control (general) (D.H.K. Davies); pest control (P. Osborne, R.G. McKinlay); crop systems simulation modelling as an aid to management decision-making; database management techniques in crop advisory work: design and production of computer software (D.G. Hughes); assessment of farm building design (A.D. Harper); climatic environment in farm buildings (B.D. Witney); mechanization systems selection (B.D. Witney, R.R. Morrison); financial management programs for extension work; farm systems studies; farm management manual (A. Blyth); mandatory work for Department of Agriculture and Fisheries for Scotland; Scottish seed potato marketing (J.L. Anderson); management accounting systems (J.D.

Rowbottom, A. Blyth); commodity costs, markets, and institutions (N.B. Lilwall); situation and outlook; sheep marketing (J.K.S. Volans); non-price barriers to inter-state trade in agricultural products (C. Blight, N.B. Lilwall); dairy enterprise studies (F.D. Mordaunt); marketing of beef from south east of Scotland (S. Scanlan); movement and disposal of barley harvest in the East College area (I.C. Beattie); economics of producing, harvesting, and marketing horticultural crops in south east Scotland (E. Wright); soil chemistry, fertility, and associated physical studies (J.C. Holmes, K.A. Smith, D.B. Naysmith); soil, plant and animal relationships of trace elements; (D. Purves, A.O. Mathieson); soil and water pollution by major and trace elements (D. Purves, R.M. McLarty); methodology of agricultural advisory effort (K.V. Runcie); basic microbiology (J.F. Wilkinson); computer organization, hardware, software, and resource requirements for advisory work in ESCA (A. Gibson); current and future needs of the market for scientific and technical information (K.V. Runcie, J.K.S. Volans).
Publications: Annual report.

ANIMAL DIVISION 115
Head: Professor J.H.D. Prescott
Head of advisory and development wing: Dr C.T. Whittemore
Sections: Animal nutrition, Dr R.A. Edwards; animal production, Dr C.T. Whitemore

CROPS DIVISION 116
Head of advisory and development wing: Dr J.C. Holmes
Sections: Crop protection, Dr J. Gilmour; soil science Dr K.A. Smith; trace elements and central analytical laboratory, Dr D. Purves; crop production, Dr J.C. Holmes; horticulture, W. Fordyce

EXTENSION DIVISION 117
Head: K.V. Runcie

FARMS DIVISION 118
Head: Dr W.J.M. Black

MICROBIOLOGY DIVISION 119
Head: Professor J.F. Wilkinson

VETERINARY DIVISION 120

Head: A.O. Mathieson
Senior staff: J.K. Miller, T.B. Nicholson, Dr K.A. Linklater, A.O. Mathieson

Efford Experimental Horticulture Station
121

Address: Lymington, Hampshire S04 OLZ
Telephone: (059 069) 73341
Affiliation: Ministry of Agriculture, Fisheries and Food, Agricultural Development and Advisory service
Status: Official research centre
Director: R.F. Clements
Departments: Vegetable crops, D.N. Antill; protected crops, T.M. Hinton-Mead; nursery stock, M.A. Scott; fruit, J.S. Wood
Graduate research staff: 8
Activities: Fruit - variety testing, soil management, and nutrition, pollination, virus diseases, and chemical control of growth in apples and pears; strawbery production under polythene tunnels, extension of fruiting season; varieties, pollination, mulching, container growing, pest and disease control in strawberries; variety trials of blackcurrants and raspberries; various aspects of viticulture. Protected cropping - variety trials and husbandry techniques for tomatoes, lettuce, celery, spray chysanthemums, and roses; work on pot plants, pot chrysanthemums, Rieger begonias, forcing azaleas, miniature cyclamen. Vegetables - variety trials, irrigation, pest and disease control, the main subjects being brussels sprouts, cauliflower, cabbage, runner beans, and autumn- and spring-sown onions, but also sweet corn, lettuce, early potatoes, leeks, and outdoor tomatoes. Nursery shock - nutrition, herbicides, growth regulation, protected cropping, and general husbandry of commercialy important conifers, deciduous, evergreen, and ericaceous species, as well as larger and more expensive subjects eg, magnolia, rhododendron, and camellia.
Publications: Efford EHS Annual Review.

Enniskillen Agricultural College
122

Address: Levaghy, Enniskillen, Fermanagh, Northern Ireland
Telephone: (0365) 3101
Affiliation: Department of Agriculture for Northern Ireland
Status: Educational establishment with r&d capability
Principal: R.B. Fulton
Graduate research staff: 1

Activities: Field drainage grassland improvement livestock production (beef). The overall aim of the college is to investigate and develop practical methods of increasing production from the marginal land areas of Northern Ireland, particularly the heavy wet soils of Fermanagh and Tyrone.

CASTLE ARCHDALE EXPERIMENTAL HUSBANDRY FARM 123

Address: Irvinestown, Enniskillen, Fermanagh, Northern Ireland
Telephone: (036 562) 245
Director: D.G. O'Neill
Projects: Phosphorus and potassium requirements of intensively managed silage swards; value of slurry applied during the non-growing season to grassland for silage; an investigation of the gravel tunnel drainage system.

Express Dairy UK Limited
124

Address: 430 Victoria Road, South Ruislip, Middlesex HA4 0HF
Telephone: (01) 845 2345
Telex: 934569 Express G
Parent body: Grand Metropolitan Limited
Product range: Food and dairy products for retail and manufacturing markets.
Status: Industrial company

RESEARCH AND DEVELOPMENT DEPARTMENT 125

Technical director: Dr J.F. Gordon
Graduate research staff: 35 (1982)

Farm-Electric Centre
126

Address: National Agricultural Centre, Stoneleigh, Kenilworth, Warwickshire CV8 2LS
Telephone: (0203) 58626
Status: Information and development centre
Section head, agriculture: P. Wakeford
Graduate research staff: 7 (1981)
Activities: Evaluation of economic uses of electrical energy in agriculture and horticulture, and establishment of their viability in the farm context.
Projects: Energy audit of dairy farms in association with PhD students at Seale-Hayne Agricultural College (Dr C.D. Mitchell); energy monitoring at National Agricultural Centre livestock units (C.M. Southall, Dr C.D. Mitchell); external monitoring of a number of

electro-agricultural projects concerned with livestock, crops, and horticulture (S.R. Pavey).

FBC Limited 127

Address: Chesterford Park Research Station, Saffron Walden, Essex CB10 1XL
Telephone: (0799) 30123
Telex: 817300 FISSAF G
Affiliation: 50 per cent owned by Boots Limited and 50 per cent owned by Fisons Limited
Status: Research centre within an industrial company
Director of Research and Development: Dr J.R. Corbett
Graduate research staff: 276
Activities: To discover new pesticides (including herbicides, fungicides, insecticides, and ectoparasiticides) and develop them for use internationally. The company will also undertake development of pesticides from third parties on a worldwide basis.
To discover new pesticides and develop them for use internationally.
Facilities: All that is necessary to achieve the above chemical synthesis, chemical analysis and biological development (growth rooms, specialist laboratories and overseas field units; laboratories for metabolism, residue analysis and toxicology; facilities for process and analytical development, formulations research and pesticide application research).
Contract work: No

BIOLOGY GROUP 128

Head: Dr S.B. Wakerley
Departments: Research screening, Dr D.T. Saggers; fungicide development, Dr G. Barnes; herbicide development, B.L. Rea; insecticide development, Dr Q.A. Geering

CHEMICAL DEVELOPMENT GROUP 129

Head: B.J. Needham
Departments: Repreparations, D.W.J. Lane; process development, P.D. McDonald; analytical development, P.L. Carter; formulations, W.T.C. Holden; application development, E.S.E. Southcombe

CHEMICAL SYNTHESIS GROUP 130

Head: Dr D.A. Evans

REGULATORY AFFAIRS GROUP 131

Head: Dr D.M. Foulkes
Departments: Metabolism, Dr L. Somerville; residue analysis, Dr R.J. Whiteoak; registration, Dr B. Thomas; toxicology, Dr G.J. Turnbull

Field Studies Council * 132

– FSC
Address: 62 Wilson Street, London, EC2A 2BU
Telephone: (01) 247 4651
Telex: 297371 GILCOM G (Attention Field Studies Council)
Status: Official research organization
Research director: Dr Jenifer M. Baker
Graduate research staff: 16
Activities: The Council exists to promote a better understanding of our environment and the way we use it; manages ten centres in England and Wales whose localities are chosen for the richness and diversity of their surroundings; provides working facilities and expert guidance; carries out research; runs short residential field courses. Areas of current research interest include: the effects of oil industry activities (through the Oil Pollution Research Unit - see below); hydrological studies (including movement of nutrients in small catchments); freshwater organisms with particular reference to eutrophication and pollution monitoring; surveys with conservation and management implications (including lowland rivers); invertebrate taxonomy.
Publications: Annual report, *Field Studies* (annual journal).

Field Centres: 133

Dale Fort Field Centre 134
Address: Haverfordwest, Dyfed SA62 3RD
Telephone: (064 65) 205
Warden and Director of Studies: D.C. Emerson

Drapers' Field Centre 135
Address: Betws-y-coed, Gwynedd LL24 0HB
Telephone: (069 02) 494
Warden and Director of Studies: A.J. Schärer

Epping Forest Conservation Centre 136
Address: High Beach, Loughton, Essex TG10 4AF
Telephone: (01) 508 7144
Warden and Director of Studies: P.A. Moxey

Flatford Mill Field Centre 137
Address: East Bergholt, Colchester, Essex CO7 6UL
Telephone: (0206) 298283
Warden and Director of Studies: Dr A. Hodges

Juniper Hall Field Centre   **138**
Address: Dorking, Surrey RE5 6DA
Telephone: (0306) 883849
Warden and Director of Studies: J.E. Bebbington

Leonard Wills Field Centre - Nettlecombe   **139**
Court
Address: Williton, Taunton, Somerset TA4 4HT
Telephone: (0984) 40320
Warden and Director of Studies: J.H. Crothers
Departments: Hydrology, Heather J. Howcroft; salt marsh ecology, J.H. Oldham; rocky shore ecology, J.H. Crothers; water table marshment in litoral sediments, D.L. Scales
Graduate research staff: 5
Activities: Primary activity of the Leonard Wills Field Centre is the conducting of one-week short courses in geography or ecology for school (A-level) and university staff and students. Research projects complement this activity and major current projects include monitoring the effects of oil and dispersant treatments on rocky shores and on salt marsh vegetation; fluctuation in the intertidal water table (influence on oil penetration in sediments); nitrate levels in stram water.

Malham Tarn Field Centre   **140**
Address: Settle, North Yorkshire BD24 9PU
Telephone: (072 93) 331
Warden and Director of Studies: Dr R.H.L. Disney
Graduate research staff: 4
Activities: Environmental studies and research. Projects include: taxonomic revision of world phoridae (diptera) - the scuttle flies; assessment of conservation values of sites in terms of invertebrate faunas; easy-to-use keys to British grasses; continuing documentation of the Malham Tarn region, across the entire range of the environmental sciences.

Orielton Field Centre   **141**
Address: Pembroke, Dyfed SA71 5EZ
Telephone: (064 681) 225
Warden and Director of Studies: Dr R.G. Crump

Preston Montford Field Centre   **142**
Address: Montford Bridge, nr Shrewsbury, SY4 1DX
Telephone: (0743) 850380
Warden and Director of Studies: J.A. Bayley
Graduate research staff: 6
Activities: Environmental impact assessment and monitoring particularly in freshwater, estuarine and marine habitats; investigations of macro invertebrate communities found in large rivers in the UK and their uses as indicator species; preparation of a manual of large river sampling procedures; determination of the extent and effect of refinery effluents on plant and animal communities in a South Wales bog; an investigation of the usefulness of ciliate communities in delimiting anoxic conditions in reservoirs and large lake basins.

Slapton Ley Field Centre*   **143**
Address: Slapton, Kingsbridge, Devon TQ7 2QP
Telephone: (0548) 580466
Warden and Director of Studies: A.D. Thomas

OIL POLLUTION RESEARCH UNIT*   144

– OPRU
Telephone: (064 681) 370
Head: Dr B.M. Dicks
Deputy Head: Dr K. Hiscock
Activities: OPRU forms a substantial part of the Field Studies Council's research effort, investigating primarily some of the biological effects of mineral oils, oil spills, spill clean-up, refinery effluents, refined products and offshore installations in the marine environment. Other marine research projects are undertaken within the Unit, some of which arise directly from oil-related work, and others of which are independent of such work. The Unit has a staff of 14.
Publications: Annual report.

Fish Diseases Laboratory*   145

Address: The Look-Out House, The Nothe, Weymouth, Dorset DT4 8UB
Telephone: (03057) 72137
Parent body: Ministry of Agriculture, Fisheries and Food, Directorate of Fisheries Research
Status: Official research centre
Activities: Research is carried out on diseases of fish, especially freshwater species. Granting of licences for the movement of live fish.

Fisheries Experiment   146
Station*

Address: Benarth Road, Conwy, Gwynedd LL32 8UB
Telephone: (0492 63) 3883
Parent body: Ministry of Agriculture, Fisheries and Food, Directorate of Fisheries Research
Status: Official research centre
Activities: Shellfish culture and the artificial breeding of certain species is carried out, and a mussel purification service is provided.

Fisheries Laboratory* 147

Address: Remembrance Avenue, Burnham on Crouch, Essex CM0 8HA
Telephone: (0621) 782658
Parent body: Ministry of Agriculture, Fisheries and Food, Directorate of Fisheries Research
Status: Official research centre
Activities: Research and advisory work on shellfish fisheries problems, non-radioactive marine and freshwater pollution.

Fisheries Laboratory 148

Address: Pakefield Road, Lowestoft, Suffolk NR33 0HT
Telephone: (0502) 62244
Telex: 97470
Parent body: Ministry of Agriculture, Fisheries and Food, Directorate of Fisheries Research
Status: Official research centre
Research director: A. Preston
Sections: Fish stock management, fish and shellfish culture, D.J. Garrod; aquatic environment protection) research support, H.W. Hill
Graduate research staff: 106
Activities: The aims of the laboratory area as folows: to provide scientific information and advice as a basis for the best management and exploitation of the stocks of fish and shellfish in and around the British Isles: this includes research into fish biology, population fluctuations and assessments, mathematical modelling, fisheries surveys, development of fisheries on under-exploited fish, ecology of fish and their eggs and larvae, fish physiology, and behaviour.
To provide scientific information and advice for developing artificial culture systems for fish and shellfish, and on diseases of cultivated fish and shellfish.
To provide scientific information and advice as to the distribution and effects of radionuclides, metals, pesticides, and other pollutants on the aquatic environment and aquatic organisms, and as to the effects of non-polluting man-made changes in the environment. These projects include research into radioactive waste disposal, dumping at sea, oil spillages, and safe levels of discharges.
Project: Coastal and inland fisheries (Dr F.R. Harden Jones); fish and shellfish culture (Dr C.E. Purdom); fish diseases (Dr B.J. Hill).
Publications: Fisheries Research Technical Reports; Aquatic Environment Monitoring Report, periodically; *Fishing Prospects,* annually.

Fisheries Research Laboratory 149

Address: 38 Castleroe Road, Coleraine, Londonderry BT51 3RL, Northern Ireland
Telephone: (0265) 4521
Affiliation: Department of Agriculture for Northern Ireland
Status: Official research centre
Head: Dr R.J. Boyd
Sections: Marine fisheries, Dr R.J. Boyd; freshwater fisheries, Dr G.J.A. Kennedy; marine environment, Dr J.G. Parker
Graduate research staff: 7 (1983)
Activities: The laboratory has a research and advisory role in the management of the freshwater and marine fish and shellfish stocks of Northern Ireland. Investigation of the major pelagic and demersal stocks of the North Irish Sea provides annual population parameters which are directly entered to the international stock assessment models for each stock. This biological advice results in the setting of Total Allowable Catches, minimum landing sizes, mesh regulations, open and closed fishing seasons and areas, and other legislation designed to maintain long-term yields to the fishing industry and to conserve these resources by optimal management.
Freshwater studies on salmonids of the River Bush and other stocks in the angling waters of the Department of Agriculture for Northern Ireland continue the development and understanding of the management of salmonid fisheries and provide advice on stocking of juvenile salmon and trout.
The laboratory's research and advisory role in the field of marine biological pollution provides data on marine pollutants in the inshore waters of the province and their occurrence and effects on shell fish and benthic communities.

Fisons plc 150

Address: Fison House, Princes Street, Ipswich, Suffolk, IP1 1OH
Telephone: (0473) 56721
Telex: 98240
Status: Chemical company
Chief executive: J.S. Kerridge
Chairman: Sir George Burton
Activities: The company's principal activities are to manufacture and supply agrochemicals, agricultural fertilizers, horticultural products, ethical and proprietary medicines, veterinary products, scientific apparatus, educational and laboratory equipment and supplies.

The turnover for the divisions in 1980 was as follows:
Fertilizer Division (£194)
Horticulture Division (£24m)
Pharmaceutical Division (£91m)
Scientific Equipment Division (£60m).

HORTICULTURAL DIVISION 151

Address: Paper Mill Lane, Bramford, Nr Ipswich
Telephone: (0473) 830492
Divisional chairman: J.S. Kerridge

SCIENTIFIC EQUIPMENT DIVISION 152

Address: Divisional Headquarters, Christopher Street,
Riverside Way, Uxbridge, Middlesex, UXB 24F
Telephone: (0895) 51100
Divisional chairman: C.A. Scroggs

Flour Milling and Baking 153
Research Association

Address: Chorleywood, Hertfordshire WD3 5SH
Telephone: (092 78) 4111
Telex: 8952883 FMBRA
Status: Research association
Director General: Professor Brian Spencer
Sections: Fundamental, Dr D.W.E. Axford; services,
R.A. Knight; baking technology, Dr N. Chamberlain;
milling, Dr D.J. Stevens; wheat and milling, Dr D.J.
Stevens; nutrition and toxicology, Dr N. Fisher
Graduate research staff: 50
Activities: FMBRA's main activity is to carry out
cooperative research in areas of interest to the milling
and baking industry, as well as offering advisory and
technical services. Subjects of research include: wheat
and flour testing; nutrition/toxicology; microbiology,
including identification of insects and possible sources of
contamination; test bakeries; problem ingredients or
products.
Publications: Annual report.

Food Science Laboratory 154

Address: Haldin House, Queen Street, Old Bank of
England Court, Norwich, NR2 4SX
Telephone: (0603) 611712
Telex: 97317
Affiliation: Ministry of Agriculture, Fisheries and
Food, Food Science Division
Status: Official research centre
Head: Dr M.E. Knowles

Departments: Statutory methods, Dr R. Wood;
strategic food studies, vacant; composition and
microbiology Dr G. Shearer; organic contaminant and
packaging, Dr J. Gilbert; additives and inorganic contaminants, Dr D.J. McWeeny
Graduate research staff: 22
Activities: Research and development in food science
topics arising directly from Ministerial activities and
responsibilities including the chemistry of food additives
and contaminants, the effect of food technology on food
composition, food surveillance and development of
analytical methodology and food microbiology, advice
on methods of analysis, and sampling for legislative
purposes.
Projects: Nitrozation reactions in foods (Dr R.
Massey); analysis of polycyclic aromatic hydrocarbons
and related compunds (Dr M. Dennis); trace inorganic
analysis in food and raw materials, studies of metal
speciation in food (J.A. Burreu); studies on the migration of packaging components, chemistry and analysis of
mycotoxins (Dr M. Shepherd); analysis of veterinary
drug residues in meat (Dr Warwick); chemistry and
analysis of colours, analysis of pesticides (Mr Reynolds);
applications of mass spectrometry to food analysis (Dr J.
Starting); microbiological methods of analysis, effect of
microorganisms on mycotoxin production (Dr J.E.L.
Corry).

FOOD LABORATORY 155

Address: Colney Lane, Norwich
Telephone: (0603) 56122

FOOD LABORATORY 156

Address: 65 Romney Street, London, SW1P 3RD
Telephone: (01) 212 7654

Great House Experimental 157
Husbandry Farm

Address: Helmshore, Rossendale, Lancashire BB4 4AJ
Telephone: (07062) 29663
Affiliation: Ministry of Agriculture, Fisheries and
Food, Agricultural Development and Advisory Service
Status: Official research centre
Activities: Research in the following areas: milk production - feeding trials involving conserved grass, milk
production from a spring calving herd, evaluation of
new sources of feed; dairy herd replacements - rearing
autumn- and spring-borne friesian heifers to calve at two
years; sheep - investigation of breeding potential and
encouragement of multiple births by feeding and
management, rearing of early weened lambs, prospects

of splitting twin lambs; pigs - long-term cross breeding trial, feeding trials, processing of piggery effluent; grassland - response to nitrogen, grazing systems, conservation as hay and as silage, sward composition under different treatments.

Forestry Commission   158

Status: Official research organization
Chairman: Sir David Montgomery
Director r&d: A.J. Grayson
Publications: Annual report.

FOREST RESEARCH STATION *   159

Address: Alice Holt Lodge, Wrecclesham, Farnham, Surrey GU10 4LH
Telephone: (042 04) 2255
Chief research officer (South): D.A. Burdekin
Sections: Engineering services, Dr W.H. Hinson; wildlife management, J.J. Rowe; silviculture, R.E. Crowther; site studies (South), Dr W.O. Binns; seeds, vacant; pathology, Dr J.N. Gibbs; entomology, D. Bevan
Graduate research staff: 80 (in both stations)
Activities: The aim of the Forestry Commission's research and development is to advance forest technology in order to improve the growth and quality of forests and increase labour productivity in ways consistent with sound land management. Work is concentrated mainly on applied research and development in both the private and public forestry sectors. A limited amount of basic research is done, mainly in the fields of soil physics, forest hydrology, genetics, physiology and biology of pests and diseases. The work covers all major practical aspects of the formation, protection, management and harvesting of forest crops.

NORTHERN RESEARCH STATION   160

Address: Roslin, Midlothian EH25 9SY
Telephone: (031) 445 2176
Chief research officer (North): D.T. Seal
Sections: Silviculture (North), J. Atterson; site studies (North), Dr D.G. Pyatt; genetics, R. Faulkner; physiology, Dr M.P. Coutts

Freshwater Biological   161
Association

– FBA
Address: The Ferry House, Ambleside, Cumbria LA22 OLP
Telephone: (096 62) 2468
Affiliation: Grant-aided institute, Natural Environment Research Council
Director: E.D. Le Cren
Assistant Director: Dr D.J.J. Kinsman
Status: Official research centre
Graduate research staff: 90
Activities: Basic and strategic research in freshwater biology (in the widest sense) and environmental physics and chemistry of freshwaters. While the general approach of the research is ecological, research in the taxonomy, physiology, and other aspects of the biology of freshwater organisms is also carried out. The main areas of research include: the hydrology, sedimentology, and substratum dynamics of rivers in relation to their ecology, physical limnology, analytical physical, and organic chemistry; microbiology, protozoology, mycology, algology and phytoplankton population and production dynamics; the biology of freshwater macrophytes; taxonomy, natural history, and physiology of freshwater invertebrates; population ecology of zooplankton stream invertebrates, physiology and population ecology of freshwater fish; and palaeolimnology. Most of the research is fundamental, but much has strategic relevance to the management of freshwaters and their resources, the water industry, and the impact on the freshwater environment of land-use, engineering, and pollution activities. Some applied and consultancy projects are undertaken.
Projects: Physical and chemical factors in the freshwater environment (Dr W. Davison); algal productivity, ecology, physiology (Dr J.F. Talling, Dr C.S. Reynolds, Dr A.F.H. Marker); microbiology of freshwaters (Dr J.G. Jones); ecology of macrophytes in rivers with some reference to land drainage (D.F. Westlake); palaeolimnology - the post-glacial history of lakes and their catchments (Dr E.Y. Haworth, Dr P.A. Cranwell); ecology, taxonomy, physiology of freshwater invertebrates (Dr G. Fryer, Dr J.M. Elliott, Dr M. Ladle, Dr L.C.V. Pinder, Dr J.F. Wright); ecology, population dynamics, and growth of freshwater fish (Dr T.B. Bagenal, Dr J.M. Elliott, Dr J.F. Craig, R.H.K. Mann); physiology of salmonid fish (Dr A.D. Pickering).
Publications: Annual report.

RIVER LABORATORY   162

Address: East Stoke, Wareham, Dorset BH20 6BB
Telephone: (0929) 462314
Officer-in-Charge: Dr A.D. Berrie
Sections: Analytical chemistry, H. Casey; physical
chemistry, Dr W.A. House; microbiology, Dr J.H.
Baker; algology, Dr A.F.H. Marker; macrophyte
botany, D.F. Westlake; invertebrate ecology, Dr M.
Ladle; invertebrate zoology, Dr L.C.V. Pinder; fish
ecology, R.H.K. Mann; analysis of river communities,
Dr J.F. Wright
Activities: Major current projects are: physical and
chemical environmental factors; biology, physiology,
growth, and production of submerged macrophytes;
development of an experimental river system; dissolved
and particulate organic carbon; freshwater invertebrates; ecology of Chironomidae; ecology of river
microorganisms; ecology of coarse fish; chalk river surveillance data; analysis of river communities; processes
controlling chemical fluxes; life-history tactics of fish;
invertebrate taxonomy.

TEESDALE UNIT   163

Address: c/o Northumbrian Water Authority, Lar-
tington Treatment Plant, Lartington, Barnard Castle,
County Durham
Telephone: (083 35) 600
Officer-in-Charge: Dr D.T. Crisp
Sections: Fish ecology, Dr D.T. Crisp; hydrology and
sediment transport, Dr P.A. Carling
Graduate research staff: 4
Activities: Major current projects are changes in fish
populations in the Upper Tees with the building of Cow
Green reservoir; effects of flow regime upon young
stages of salmonid fish; studies on effects of upland land
management.

WINDERMERE LABORATORY   164

Sections: Analytical chemistry, Dr J. Hilton; physical
chemistry,Dr W. Davison; phytoplankton ecology, Dr
C.S. Reynolds; phytoplankton physiology, J.F. Talling;
mycology, Dr G. Willoughby; microbiology, Dr J.G.
Jones; protozoology, Dr B.J. Finlay; crustacean ecology,
Dr G. Fryer; zoobenthos ecology, Dr J.M. Elliott; fish
physiology, Dr A.D. Pickering
Graduate research staff: 41
Activities: Major current projects are: physical and
chemical environmental factors; inorganic chemical
speciation; physical and chemical dynamics of summer
stratification in relation to phytoplankton production;
ecology of Saprolegniaceae; freshwater crustaceans;
effects of experimental and natural factors on population
dynamics of fish; microbial metabolic activity; microbial
population ecology; loss processes controlling

phytoplankton communities; quantitative samplers for
macroinvertebrates; iron and manganese transportation;
population dynamics of protozoa; denitrification; stress
response in teleost fish; remote sensing.

Freshwater Biological   165
Investigation Unit

Address: Greenmount Road, Muckamore, Antrim
BT41 4PX, Northern Ireland
Telephone: (084 94) 62660
Affiliation: Department of Agriculture for Northern
Ireland
Status: Official research centre
Director: Dr R.V. Smith
Sections: Biology, Dr C.E. Gribson; chemistry, Dr C.
Jordan
Graduate research staff: 4 (1983)
Activities: Continous measurement of parameters
reflecting eutrophication; factors leading to algal
blooms; examination of lakes in Northern Ireland.
Major projects include: impact of a phosphorus removal
programme on Lough Neagh; availability of different
forms of phosphorous for algal growth; physiological
ecology of cynobacteria; in situ phosphorus inactivation
in eutrophic lakes.

Freshwater Fisheries   166
Laboratory

Address: Faskally, Pitlochry, Perthshire PH16 5LB
Telephone: (0796) 2060
Parent body: Department of Agriculture and Fisheries
for Scotland, Fisheries Research Services
Status: Official research centre
Officer-in-charge: A.V. Holden
Sections: Salmonid stocks and exploitation, W.M.
Shearer; juvenile salmonid production, Dr H.J. Eg-
glishaw; salmonid rearing and behaviour, Dr J.E.
Thorpe; environmental chemistry, Dr D.E. Wells
Graduate research staff: 23
Activities: The laboratory is responsible for research in
Scotland into the status of salmon and sea trout fisheries;
levels of exploitation by sport and commercial fisheries;
factors affecting juvenile populations and their augmentation; studies of behaviour of juvenile salmonids and
optimum conditions for rearing; development of ocean
ranching of Atlantic salmon; effects of agricultural and
industrial pollutants on salmonid stocks; effects of acid
rain and afforestation on water quality and salmonid
stocks.

Projects: Electronic fish counting; salmon recruitment and exploitation (W.M. Shearer); salmon stock augmentation by fry planting (Dr H.J. Egglishaw); salmon rearing and ocean ranching (Dr J.E. Thorpe); pesticides and fisheries; afforestation, acidity, and salmonids (Dr D.E. Wells).
Publications: Triennial review of research.

Gallaher Limited R&D Division   167

Address: Henry Street, Belfast, BT15 1JE
Telephone: (0232) 744722
Telex: 74444
Parent body: American Brands Incorporated
Status: Research centre within an industrial company
Research director: vacant
Manager r&d, Northern Ireland: Dr P.W. Darby
Graduate research staff: 60
Activities: Research into physics and chemistry of tobacco and tobacco products.

Game Conservancy   168

Address: Fordingbridge, Hampshire SP6 1EF
Telephone: (0425) 52381
Status: Independent research centre
Chairman: C.V. Vandervell
Director: Richard van Oss
Research director: Dr G.R. Potts
Graduate research staff: 11
Activities: Research topics include: habits, pests and diseases of game birds and water fowl; their relationship to agricultural practices; the populations of predatory mammals and roe deer.
Publications: Annual review; game management booklets.
Projects: Population ecology of game, (Dr G.R. Potts); studies on mammalian predation, (Dr S.C. Tapper); ecology of the hare, (Dr R.W. Barnes); value of nesting cover to partridges, (M.W. Rands); factors affecting abundance of insects in cereal crops, (Dr G.P. Vickerman); predation on cereal aphids, conservation of polyhagous predators of cereal aphids, (Dr N.W. Sotherton); population dynamics of mallard, (D.A. Hill); ecology of gravel pits, (M. Street); strongylosis and red grouse, (Dr P.J. Hudson); ecology of sheep tick, (J.S. Duncan); ecology of woodcock, (Dr G.J. Hirons); ecology of redleg partridge, (Dr R.E. Green); pathology of game, (Dr J.V. Beer).

Glasshouse Crops Research Institute   169

– ACRI
Address: Worthing Road, Rustington, Littlehampton, West Sussex BN16 3PU
Telephone: (090 64) 6123
Affiliation: State-aided institute, Agricultural Research Council
Status: Official research centre
Director: Dr D. Rudd-Jones
Sections: Crop Protection and Microbiology Division, Dr N.W. Hussey - entomology, Dr N.W. Hussey; plant pathology and microbiology, Dr J.M. Lynch
Crop Science Division, Dr A.R. Rees - biomathematics, Dr D.P. Aikman; crop science, Dr A.R. Rees
Physiology and Chemistry Division, Dr D.V. Prue - biochemistry and plant nutrition, Dr G.W. Winsor; plant physiology, Dr D.V. Prue
Graduate research staff: 98
Projects: Protected crops: pest control using chemical and/or biological means (including microbial agents); disease control using fungicides or biological agents, or by manipulating the environment; damage by air pollutants; mathematical modelling of growth processes; nutrition in relation to environment, produce quality, and yield. Mushrooms: nutrition, physiology, and genetics; control of pests, bacterial, and fungal diseases; improved composting and cropping systems. Hardy nursery stock: pathogenicity of disease organisms and their control; nutrition; propagation. Protected and other crops (including mushrooms): study of viruses affecting them. Cereals: populations of aphids and their natural enemies, establishment of economic damage thresholds. Technology and crop husbandry, especially: nutrition using a range of composts; growth and breeding of midwinter lettuces; initiation of flowering in bulbs, chrysanthemums, roses, and tomatoes; growing crops in recirculating nutrient solution (nutrient film technique); computer regulation of glasshouse environment. Physiological, biochemical, and biophysical factors influencing: photosynthesis; plant senescence (including post-harvest behaviour); photobiology.
Publications: Annual report; growers' bulletins.

Glaxo Group Research Limited   170

Address: 6-12 Clarges Street, London, WLY 8DH
Telephone: (01) 493 4060
Telex: 25456
Parent body: Glaxo Holdings Limited
Product range: Ethical pharmaceuticals for human and veterinary use, nutritional products, OTC medicines. Subsidiaries include manufacturers of surgical equipment, operating tables, hospital furniture, and wholesaling chemists.
Status: Research centre within an industrial company
Chairman and Chief Executive: Dr D. Jack
Divisions: Greenford Research: Dr A.R. Williamson (Director);
Chemistry division: Dr B.J. Price (Head); chemical research, Dr S.M. Roberts; chemical development, Dr G.M. Wilson; physical chemistry, Dr G.I. Gregory
Microbiology division: Dr G.W. Ross (Head); fermentation research, Dr J. Herrman; chemotherapy, Dr P. Harper; microbial biochemistry, Dr J. Ward
Cell biology division: Dr M. Elves (Head); immunobiology, Dr A. MacKenzie; oncology, Dr C. Spilling; immunochemistry, Dr D.M. Turner
Pharmacology division: Dr D.W. Straughan (Head)
Medical: Dr R.N. Smith (Director); Greenford medical, Dr R.D. Foord; Ware medical, Dr D.M. Harris; clinical research services, Dr M. Mitchard; adverse reactions unit, Dr M.D.B. Stephens
Central Services: Dr W.G.E. Underwood (Director); computer science, V.S. Jacobson; science information, Dr S.E. Ward
Pharmacy: K.A. Less (Director)
Greenford: pharmaceutical analysis, Dr J.P. Jefferies; pharmaceutical formulation, Dr C. Walton
Ware: pharmaceutical sciences, Dr J. Padfield; analytical chemistry, Dr M. Martin-Smith
Ware Research: Dr R.T. Brittain (Director);
Pathology division: Dr D. Poynter (Head); histopathology, Dr D. Pratt; toxicology, Dr N. Spurling; animal services, J. Hastings
Chemistry division: Dr R.F. Newton (Head); chemistry research, (vacant); chemical development, Dr D. Hartley;
Pharmacology department, Dr P. Humphrey
Neuropharmacology department, Dr M. Tyers
Biochemical pharmacology division: Dr L.E. Martin (Head)
Ware: biochemical pharmacology, Dr I.F. Skidmore
Greenford: drug metabolism, Dr K. Childs
Regulatory Affairs and Planning: Dr E.W. Horton (Director); registration, Dr M.D.C. Scales (Coordinator); planning, Dr C.M. Towler (Coordinator)
Graduate research staff: 600

Activities: Glaxo Group Research Limited comprises Greenford and Ware Divisions in which the laboratory work needed for drug discovery and development is done. It also embraces a Medical and a Development Division which are responsible for the clinical evaluation of new products and, in concert with the Group's operating companies throughout the world, for communications with national drug authorities.
The principal research activities at Greenford are concerned with anti-bacterial agents, cancer, immunology, virology, the central nervous system and with animal health, whilst those at Ware are concerned with the cardiovascular and respiratory systems, with the alimentary tract and with metabolic diseases such as diabetes mellitus. Significant active substances have been identified in a number of these projects. An especially promising new entity, ranitidine, is now in course of development.

GREENFORD DIVISION*   171

Address: 891-995 Greenford Road, Greenford, Middlesex
Telephone: (01) 422 3434
Telex: 22134
Director: vacant
Sections: Animal health, P.G. Box, Dr I.D. Fleming; Medical Division, established products, Dr C.H. Dash, Dr J.C. Garnham, Dr I.M. Slessor; cell biology, Dr M.W. Elves; chemistry, Dr J. Elks; Medical Director, Greenford, Dr R.D. Foord; microbiology, Dr P.W. Muggleton; pharmacology, Dr D.W. Straughan; pharmacy, K.A. Lees; special projects, Dr W.F.J. Cuthbertson; computing and statistics, J.P.R. Tootill; science information, Dr S.E. Ward
Graduate research staff: 250

Gleadthorpe Experimental Husbandry Farm   172

Address: Meden Vale, Mansfield, Nottinghamshire NG20 9PF
Telephone: (0623 84) 4331
Affiliation: Ministry of Agriculture, Fisheries and Food, Agricultural Development and Advisory Service
Status: Official research centre
Director: R. Hart
Sections: General agriculture, M. Selman; poultry, Dr J. A. Hill
Graduate research staff: 7
Activities: Crop husbandry - cereals, sugar beet, potatoes; livestock husbandry and nutrition - beef cattle and grass, poultry.

Projects: Potatoes and sugar beet (Dr W.F. Cormack); cereals (H.G. McDonald); egg production (Dr A. Ballantyne); poultry meat (T.S. Bray); beef cattle (M. Selman).
Publications: Annual review; poultry booklet, biennialy.

Good Gardeners Association - International Association of Organic Gardens 173

Address: Arkley Manor, Arkley, Barret, Hertfordshire
Telephone: (01) 449 3031
Status: Independent research centre
Director: Dr W.E. Shewell-Cooper
Graduate research staff: 4
Activities: Also known as the International Association of Organic Gardeners, the Good Gardeners Association carry out research into soil science, no-soil inversion, no-chemical fertilizers or chemical sprays; crop husbandry; making compost.
Publications: Year book.

Grassland Research Institute 174

– GRI
Address: Hurley, Maidenhead, Berkshire SL6 5LR
Telephone: (062 882) 3631
Affiliation: State-aided institute, Agricultural Research Council
Status: Official research centre
Director: Professor A. Lazenby
Sections: Soils and plant nutrition, D.L.H.P. Jones; plant and crop physiology, Dr E.L. Leafe; agronomy, Dr R.J. Wilkins; animal production, Dr D.F. Osbourn; biomathematics, Dr J.H.M. Thornley
Graduate research staff: 80
Activities: GRI's research programme has two main objectives: to improve the efficiency of growth of the grass crop; and to improve the effectiveness with which it is utilized in animal production systems. Research topics include: grass growth; forage legumes; grass/clover mixtures; other forage crops; forage conservation; milk production; beef production; sheep production; permanent pasture; energy and the future.
Publications: Annual report; technical reports.

PERMANENT GRASSLAND DIVISION 175

Address: North Wyke, Okehampton, Devonshire

Greenmount Agricultural and Horticultural College 176

Address: 22 Greenmount Road, Antrim, BT41 4PU, Northern Ireland
Telephone: (084 94) 62114
Affiliation: Department of Agriculture for Northern Ireland
Status: Educational establishment with r&d capability
Principal: M. Boyd
Vice-Principal: G.M. Kennedy
Head of Research and Development: T.A. Stewart
Departments: Experimental Husbandry, I.I. McCullough; Hill Farm, S.J. McGaughey
Graduate research staff: 5 (1982)
Activities: Applied research programme includes work on grassland production and conservation; beef production and animal nutrition; potato, cereal, and flax production; soil fertility and plant nutrition; farm waste management; hill sheep and hill cattle production.
Projects: Grass and silage production (I.I. McCullough); clover production (T.A. Stewart); beef production (I.I. McCullough, T.A. Stewart); farm waste management (H.I. Gracey); potato production; cereal and arable crop production (A.R. Saunders); Hill Farm production (S.J. McCaughey).

Arthur Guinness Son and Company (Park Royal) Limited* 177

Address: Park Royal Brewery, London, NW10 7RR
Telephone: (01) 965 7700
Telex: 23498
Status: Industrial company
Technical Manager: F.O. Robson
Head, Research Laboratory: Dr R.E. Wheeler
Head, Production Research Department: J.B. Hedderick
Chief Chemist: Dr Margaret Jones
Graduate research staff: 10

Hannah Research Institute 178

– HRI
Address: Kirkhill, St Quivox, Ayr, KA6 5HL
Telephone: (0292) 76013 to 7
Affiliation: State-aided institute, Agricultural Research Council; Department of Agriculture and Fisheries for Scotland
Director: Professor M. Peaker

Sections: Animal nutrition and production, Dr P.C. Thomas; chemistry and physics of milk, Dr D.G. Dalgleish; lactational physiology and biochemistry, Dr R.G. Vernon; lipid biochemistry and enzymology, Dr W.W. Christie; milk utilization, Dr W. Banks; Director's group, Professor M. Peaker
Graduate research staff: 50
Activities: Studies relating to the production and utilization of milk - herbage production, conservation, and utilization by the dairy cow; feeding for milk production; milk composition in relation to market needs; milk analysis; milk going to manufacture; climate-related aspects of animal productivity.
Publications: Annual report; technical bulletins.

Hatfield Polytechnic*   179

Address: PO Box 109, College Lane, Hatfield, Hertfordshire
Telephone: (070 72) 68100
Telex: 262413
Status: Educational establishment with r&d capability
Director: Sir Norman Lindop

SCHOOL OF NATURAL SCIENCES*   180

Address: Bayfordbury, Hertford, Hertfordshire
Telephone: (0992) 53067
Dean: Dr W. Boardman

Biological Sciences*   181

Head: Dr K. Wilson
Activities: Research topics include: drug metabolism, pharmacokinetics, teratology and drug absorption; nerve transmission in the periphery, pharmacological receptors; assay of biocides, evaluation of the ecological effects of biocides; biodeterioration, plant pathology; metabolism in plant tissue culture, effects of growth inhibitors and anti-transpirants on plant tissue cultures; production of enzymes by continuous culture, fungal pelleting; the effect of micro meteorological factors on natural grassland communities.

Hazleton Laboratories   182
Europe Limited

Address: Otley Road, Harrogate, North Yorkshire HG3 1PY
Telephone: (0423) 67265
Telex: 57735
Affiliation: Hazleton Laboratories Corporation
Product range: Product safety evaluation services.
Status: Independent research centre
Managing director: P.S. Rogers
Research director: Dr A.K. Armitage
Scientific director: Dr P.J. Simons
Sections: Toxicology operations, W.E. Robinson; chemistry and metabolism, Dr T.H. Houseman
Graduate research staff: 31
Activities: Product safety evaluation for industry to fulfil the requirements of regulatory authorities worldwide the main services provided are toxicology, metabolism and pharmacokinetics and analytical chemistry. Specific activities include: conduct of acute, subchronic and chronic toxicity studies in the mouse, rat, dog and non-human primate using appropriate route of administration for test compound; teratology, fertility and carcinogenicity studies; identification and measurement of pesticide residues and environmental samples; biofluids and geochemical analysis; studies of the disposition and fate of compounds in man, animals, plants, and soil.

H.J. Heinz Company   183
Limited

Address: Hayes Park, Hayes, Middlesex UB4 8AL
Telephone: (01) 573 7757
Telex: 261477
Product range: Canned and bottled food products.
Status: Industrial company
General manager, research development and quality assurance: A.J. Skrimshire
Sections: Product development and packaging research, B.A. Harrison; new product development and process research, Dr W.N. Colby; quality assurance and laboratoires, G.R.W. Simpson
Activities: Development of new and improved products and packages within existing range; development of novel products and packages; development of new or improved processes; quality assurance; analytical and microbiological control procedures and new methods of analysis.

Heriot-Watt University* 184

Address: Edinburgh, EH1 1EX
Telephone: (031) 225 8432
Status: Educational establishment with r&d capability
Acting Principal: Professor T.D. Patten
Vice-Principal: Professor A.R. Rogers

FACULTY OF SCIENCE* 185

Dean: Professor J.R. Gray

Brewing and Biological Sciences Department* 186

Address: Chambers Street, Edinburgh, EH1 1HX
Telephone: (031) 449 5111
Head: Professor D.J. Manners
Professor of Microbiology: Professor C.M. Brown
Graduate research staff: 35
Activities: Brewing: physiology of malting; taxonomy of brewing yeasts; by-products of ethanol fermentation. Microbiology: ecological and applied aspects; sulphate-reducing bacteria; biodegradation of biodeterioration; the cell cycle in yeast.
Biochemistry: structure and metabolism of complex carbohydrates; fatty acid activating enzymes; biosynthesis of amino acids.
Marine biology: physiology, cytology and ecology of marine algae; reproduction of shellfish; benthic zoology; recirculation culture of marine flatfish.

High Mowlthorpe Experimental Husbandry Farm 187

Address: Duggleby, Malton, North Yorkshire YO17 8BW
Telephone: (09443) 646 and 647
Affiliation: Ministry of Agriculture, Fisheries and Food, Agricultural Development and Advisory Service
Status: Official research centre
Activities: Research in the following areas: cereals (winter wheat, spring barley, and some oats) - responses to cultivations, fertilizer applications, cropping sequences, fungicides to control leaf diseases, variety testing; potatoes - maintaining healthy stocks, effects of defoliation, performance of seed derived from virus tested stem cuttings; fodder roots - contribution as replacement for barley in diet of cattle and sheep; oilseed rape - prospects for autumn-sown rape; beef production - performances of crossbreeds; comparison of Friesian bulls with steers on a zero grazing system; sheep - time of lambing, effects of management levels on productivity, effects of blood copper levels on ewes and lambs.

Hill Farming Research Organisation 188

Address: Bush Estate, Penicuik, Midlothian EH26 OPY
Telephone: (031) 445 3401
Affiliation: State-aided institute, Agricultural Research Council; Department of Agriculture and Fisheries for Scotland
Status: Official research centre
Director: J. Eadie
Departments: Animal production, Dr T.J. Maxwell; animal nutrition, Dr J.A. Milne; grazing ecology, Dr J. Hodgson; plants and soils, Dr P. Newbould
Graduate research staff: 55
Activities: The organization was formed in 1954 as an independent grant-aided institute. Its present remit is to improve the viability of meat production from the hills and uplands. The Department of Agriculture and Fisheries for Scotland commissions research in five main areas: performance of sheep in hill and upland environments; synthesis of hill and upland farming systems: beef cattle in hill and upland environments; hill and upland pasture production; husbandry of red deer.
Projects: Environmental and genetic factors affecting reproductive rate in hill and upland sheep (Dr R.G. Gunn); early lamb growth in relation to ewe milk yield and intake of solid food, effectiveness of improved genotypes of hill sheep in utilizing better hill resources (Dr J.M. Doney); wool production from Blackface and crossbred ewes under improved management systems (W.F. Smith); voluntary intake of roughages by sheep (Dr J.M. Doney); interactions between nutrition and body composition in grazing sheep, nutritional physiology of the pregnant ewe (Dr A.J.F. Russel); nutritive value of heather to sheep (Dr J.A. Milne); supplementation of low quality roughage diets for sheep (Dr R.W. Mayes); mineral nutrition and animal performance (A. Whitelaw); metabolism of the grazing ewe (Dr J.A. Milne); nutritional and productivity consequences of land improvement (J. Eadie); utilization of hill and upland swards by grazing cattle and sheep (Dr J. Hodgson); improved systems of animal production from hill pastoral resources, year round grazing and in wintering (Dr R.H. Armstrong); methods for the economic evaluation of hill farming systems, improved systems of sheep production from upland pastoral resources (Dr T.J. Maxwell); simulation models of hill and upland sheep production systems (A.R. Sibbald); beef cattle - characterization of nutritional state, studies on reproduction, lactation, and calf growth (Dr A.J.F. Russel); nutrient requirements of white clover and sown grasses in hill soils (Dr A. Rangeley); seasonal patterns and different intensities of utilization on the growth of heather (S.A. Grant); growth and harvested production by continously grazed swards (Dr J. King); cycling of nutrients in grazed hill pasture (Dr M.J.S. Floate); fixation of nitrogen by white clover in hill pastures,

effect on nitrogen fixation on white clover/grass swards in the uplands (Dr A. Haystead); relationships amongst mycorrhiza, rhizobia, and the growth and performance of nodules on white clover roots (Dr A. Rangeley); effect of utilization by grazing hill sheep and beef cattle on growth and production of hill pastures (Miss S.A. Grant); interactions between acidity, aluminium, and phosphorus availability in hill soils (Dr M.J.S. Floate); practical problems associated with the application of animal husbandry methods to red deer kept under semi-intensive conditions (W.J. Hamilton).
Publications: Biennial report.

Home-Grown Cereals Authority 189

Address: Hamlyn House, Highgate Hill, London, N19 5PR
Telephone: (01) 263 3391
Status: Official research centre
General manager: C.J. Ames
Marketing and economics department: E.M. Low
Graduate research staff: 9
Activities: The authority provides a market intelligence service for the UK cereals industry.

Horticultural Centre 190

Address: Manor House, Loughall, Armagh BT61 8JB, Northern Ireland
Telephone: (076 289) 206
Affiliation: Department of Agriculture for Northern Ireland
Status: Official research centre
Director: G.H. McElroy
Departments: Fruit and nursery stock, J. Ward (nursery stock), B.S. Watters (fruit); protected crops and vegetables, W.M. Dawson, S. O'Neill (glasshouse), D. Moore (mushrooms), G. Maginnis (vegetables)
Graduate research staff: 8 (1982)
Activities: Applied research and development work on top and soft fruit, nursery stock, glasshouse crops, mushrooms, vegetables, novel sources of cellulose, and energy from biomass, to evaluate relevance for local growers of findings from fundamental research and new technological advances. This involves development of new and improved production techniques and the integration of these into production systems suitable for commercial growers.
Projects: Mushrooms: compost preparation, production systems (D. Moore); glasshouse crops: lettuce production, alternative low energy crops; vegetable crops: continuity of cropping, varietal evaluation, biomass pro-

duction (W.M. Dawson); fruit: disease control, intensification of production in Bramleys seedling, controlled atmosphere storage (B.S. Watters); nursery stock: rose rootstocks, production of containerised shrubs (J. Ward).
Publications: Annual report.

Houghton Poultry Research Station 191

– HPRS
Address: Houghton, Huntingdon, Cambridgeshire PE17 2DA
Telephone: (0480) 64101
Affiliation: State-aided institute, Agricultural Research Council
Status: Official research centre
Director: Professor P.M. Biggs
Sections: Microbiology, Dr H. Williams Smith; parasitology, Dr M.E. Rose physiology and biochemistry, Dr E.J. Butler; experimental pathology, Dr L.N. Payne
Graduate research staff: 31
Projects: Infectious bronchitis virus (IBV): selection and assessment of suitable strains for vaccine manufacture and immunological studies; genetic manipulation of IBV viruses for vaccine prodction. Marek's disease and lymphoid leukosis: vaccinal immunity, viral and tumour antigens, and genetic resistance; development and properties of Marek's disease-derived lymphoid cell lines; susceptibility of differing genotypes to Marek's disease; analysis of the genetic structure of the Marek's disease virus, genome. Bacteriology: study of immunizing capacity of E coli antigens; phage treatment of intestinal infections; antibiotic resistance of enterobacteria infecting domestic animals; factors influencing the excretion of salmonella organisms in chickens. Coccidiosis: effects of host infection by different species of Eimeria; the development of resistance to anticoccidial drugs; immune response of the host to coccidia; taxonomy, genetics, host-and site-specificity; immunological methods of control. Physiology and biochemistry: stress responses of the fowl; effects of stress and susceptibility to infection; mechanisms of thermoregulation and the control of metabolism by hormones; isolation of fowl immunoglobulins; liver damage caused by rapeseed meal heaptosis and fatty liver-haemorrhagic syndrome. Directors Department: new disease problems associated with intensive systems of management; turkey lentosis. Pathology: leg weakness problems of broilers.
Publications: Biennial report; *Current research in Poultry Science* , weekly.

HRS Limited* 192

Address: Howbery Park, Wallingford, Oxfordshire OX10 8RA
Telephone: (0491) 35381
Status: Independent research centre
Director: R.C.H. Russell
Sections: Overseas aid programme, C.L. Abernethy; offshore engineering, P.J. Rance; ports and coastal engineering, W.A. Price; estuaries and tidal processes, T.J. Weare; rivers, structures, drainage, A.J.M. Harrison
Graduate research staff: 120
Activities: Research in civil engineering and hydraulics. Specific topics include: wave prediction and analysis; offshore structures; submarine pipelines; seabed stability; offshore mooring; wave power conversion; coast and harbour protection; ship mooring; beach nourishment; land reclamation; effluent dispersal; offshore dredging; wave refraction; estuarial development; tidal propagation; sand and mud transport; tidal barriers; salinity intrusion; water supply and irrigation; land drainage; flood control; flow measurement; urban drainage; water management; soil erosion.

Hurst Crop Research and Development Unit 193

Address: Great Domsey Farm, Feering, Colchester, Essex CO5 9ES
Telephone: (0376) 71123
Telex: 99142
Status: Independent research centre
Director: David R.H. Dow
Departments: Agricultural crop breeding, Dr F.G.A. Bassi; oilseeds breeding, Dr P. Lapinskas; flower breeding, A. Hender; vegetable breeding, M. Thornton
Activities: Plant breeding with associated agronomy on species of agricultural flowers and vegetable crops grown from seed.

Imperial Chemical Industries Limited* 194

– ICI
Address: Imperial Chemical House, Millbank, London, SW1P 3JF
Telephone: (01) 834 4444
Telex: 21324
Status: Industrial company
ICI Research, Development and Technical Director: Dr C.H. Reece
General Manager, Research and Technology: Dr C.W. Suckling

Public Relations Officer for Research Matters: Dr D.M. Parker
Activities: Research and development programmes devoted to the improvement of existing products and processes and the invention and development of new ones, supporting the world-wide activities of the group, including systems analysis and mathematical modelling of chemical processes; chemical and biological topics and physics pertaining to the development and diversification of the chemical industry.

Central Toxicology Laboratory* 195

Address: Alderley Park, Near Macclesfield, Cheshire SK10 4TJ
Telephone: (0625) 582711
Activities: Investigation of the biological effects, metabolism and mode of action of industrial chemicals, pesticides, and food additives; research leading to new methods and techniques for assessing the carcinogenicity and other toxic effects of chemical compounds.

Corporate Laboratory* 196

Address: PO Box 11, The Heath, Runcorn, Cheshire WA7 4QE
Telephone: (092 85) 73456
Activities: Chemistry, physics and biological topics relevant to the development and diversification of the chemical industry; development of improved methods for the design and control of unit operations and chemical processes as a whole with special focusing of research interest in control on the design and control of complex business systems.

AGRICULTURAL DIVISION* 197

Address: PO Box 1, Billingham, Cleveland TS23 1LB
Telephone: (0642) 553601
Telex: 587443
Activities: Inorganic chemistry, including processes for the manufacture of inorganic compounds, fertilizer technology, catalyst behaviour, development of new catalytic processes; catalyst manufacture and building products.

PLANT PROTECTION DIVISION* 198

Address: Jealott's Hill Research Station, Bracknell, Berkshire RG12 6EY
Telephone: (0344) 24701
Telex: 847556 ICIPPJ-G
Activities: The discovery, characterization and development of novel crop protection chemicals and public health products (herbicides, fungicides, insecticides, aphicides, seed dressings and plant growth

regulators). Studies cover chemical synthesis, the assessment of biological efficacy, under controlled conditions and in the field, and formulation. Particular emphasis is placed on environmental aspects including the measurement of residues on the compounds themselves and the products of their decomposition, and of side effects in the environment.

Imperial College of Science and Technology* 199

Address: South Kensington, London, SW7 2AZ
Telephone: (01) 589 5111
Telex: 261503
Parent body: University of London
Status: Educational establishment with r&d capability
Rector: Lord Flowers
Graduate research staff: 800
Activities: The College consists of three closely linked colleges - the Royal College of Science, the Royal School of Mines and the City and Guilds College which between them carry out research in the physical sciences, life sciences, earth sciences, mining, metallurgy, engineering and computing.

DIVISION OF LIFE SCIENCES* 200

Department of Pure and Applied Biology 201

Address: Prince Consort Road, London, SW7 2BB
Head: Professor R.K.S. Wood
Senior staff: Applied zoology, Professor M.J. Way; plant physiology, Professor J. Barber; environmental technology, Professor G.R. Conway; insect ecology, Professor M.P. Hassell; insect physiology, Professor A.D. Lees; parasitology, Professor J.D. Smyth, pest management, Dr G. Matthews
Graduate research staff: 60
Activities: Field research work on the control of insect pests and nematodes is undertaken at Silwood Park. The Department's main research activities in soil science, land use, plant protection, agricultural engineering, forestry and forest products concern:
Morphology and taxonomy - insects, protozoa, and nematodes.
Ecology - population dynamics, energetics and genetics of insects and nematodes; computer models and systems analysis; insect predation, parasitism, distribution, competition, migration and intra-specific variability; nematodes and fungi.
Physiology and behaviour - insect endocrinology, invertebrate mechano- and chemo-receptors; arthropod photoreception, photoperiodism and vision, activity

rhythms, insect neurophysiology, neuropharmacology and behavioural studies of birds; insect ethology; protozoa; flatworms and nematodes.
Pest control and toxicology - pesticide residues, insecticide and nematicide action and resistance, insect pathology.
Physiology of plants - photosynthesis, respiration biochemistry, lignin degradation, germination biochemistry of cereal seeds and seedling carbohydrate metabolism.
Plant pathology - disease resistance, specificity in plant diseases; pathogenesis, leaf-infecting pathogens; epidemiology; chemical and biological disease control; viruses.
Ecology and physiology of plants in field conditions - plant communities; bryophyte ecology; productivity; pollution.
Physiology and development of fungi - cellular studies; host resistance.
Experimental taxonomy - seashore ecology; marine oil pollution.
Genetics - genetic recombination, mutation, genetic analysis techniques, natural populations and evolution.
Microbial biochemistry - cell fusion and somatic incompatibility; sensory responses of microorganisms.
Nitrogen fixation - lichens; environmental nitrogen.
Timber technology - wood decay and prevention; water movement through wood; electron microscope studies.
Palaeobotany - conifers; electron microscope studies; fossil charcoal; cuticle structure; taxonomy.
Projects: Atmospheric pollutants and the effect on crop plants, (Dr J.N. Bell); pest control and management, (Professor M. Waty, Dr G. Matthews); plant disease control: fungi, (Professor R.K.S. Wood, Dr B.E.J. Wheeler); plant disease control: nematodes, (Dr M. Evans); spraying equipment technology, (Dr G. Matthews); remote sensing of vegetation, (Professor J. Barber); insect pest behaviour and physiology, (Dr J. Brady, Professor Kennedy); preservation of wood and timber products, (Professor J. Levy).
Publications: Technical research report.

Silwood Centre for Pest Management 202
Address: Silwood Park, Sunninghill, near Ascot, Berkshire SL5 7PY
Telephone: (0990) 23911
Telex: 261503
Director: Dr G.A. Matthews
Chairman: Professor M.J. Way
Groups: Pesticide application; pesticide biology; ecology, decision-making; biological control; behaviour and physiology; plant pathology; weed science; nematology; parasitology.
Activities: Founded in 1981, the centre is sited in 100 hectares of parkland and is well equipped to study management problems of pests of almost every kind, whether temperate or tropical, plant or animal. Organized as a multidisplinary service it brings together

the Imperial College's wide-ranging expertise on insects, nematodes, parasites, plant pathogens and weeds, and the centre provides professional advice and undertakes contract research for the agriculture and chemical industries, for international organizations and for governments throughout the world. Current research covers: pesticides: performance of spray nozzles; equipment for small farms in developing countries; spraying for vector control; controlled droplet application; electrostatically charged sprays; biological effects of spray droplets; behaviour of pesticides in plants and soil, biological activity of pesticides on test surfaces; pesticides potentiation and controlled release; novel insecticide and nematicide modes of action; development of improved nematicide bioassays; herbicide uptake by plants. Ecology: insect-plant relations during succession; grazing effects on plant populations; plant quality and insect-herbivore communities; applied ecology and pest forecasting; predator-prey/parasitoid-host population dynamics; epidemiology and ecology of diseases; population biology of blood-sucking insects; mineral nutrients and plant competition. Decision making: sugar cane froghopper in Trinidad; cattle tick in Australia; sugar beet yellows in Britain; cocoa pod borer in Malaysia; savanna bush encroachment in southern Africa; pesticide resistance and world food production. Biological control: improved mass rearing systems for parasitoids; biology of protozoan/nematode insect diseases; chemistry of weed/insect-herbivore interactions, mathematical models for analysis of biocontrol. Behaviour and physiology: flight orientation control in odour plumes; plant and animal host-finding; computer modelling and analysis of behaviour; sensory physiology and neurothology; photo periodism, polymorphism and circadian rhythms. Plant pathology: epidemiology of temperate plant diseases; analysis of tropical plant pathogen populations; disease resistance and parasite specificity; microorganism damage mechanisms; viral genome organization and expression; disease control by fungicides and microorganisms. Weed science: herbicide activity and phloem mobility; membrane permeability and xenobiotic accumulation; adjuvant formulation effects on cuticle; herbicide choice for tropical crops; a weed index of growth, habit and IGR sensitivity; population growth of perennial weeds. Nematology: biology and survival of plant-parasitic nematodes; nematodes and nematicide physiology and biochemistry; taxonomy and bionomics of entomophagous nematodes; modes of action of anthelminthics. Parasitology: biochemistry and physiology of helminths; biology of malaria, coccidia, and trypanosomatids; epidemiology and immunology of leishmaniasis; cell biology and transmission of plasmodia; in vitro culture of Plasmodium falciparum; identification of insect bloodmeals; microsporidian and nematode pathogens of insects.

John Innes Institute 203

– JII
Address: Colney Lane, Norwich, NR4 7UH
Telephone: (0603) 52571
Affiliation: State-aided institute, Agricultural Research Council; University of East Anglia
Status: Official research centre
Director: Professor Harold W. Woolhouse
Departments: Applied genetics, Professor D.R. Davies; genetics, Professor D.A. Hopwood; virus research, Dr J.W. Davies
Graduate research staff: 41
Activities: The John Innes Institute, was founded at Merton, Surrey 1910. Over the years the institute has played a leading part in the development of the cytological and biochemical aspects of the science of genetics. The basic genetics has always been linked to applied work and at the present day this includes genetic manipulation of viruses, bacteria and higher plants for the production of gene vectors, bacteria with improved nitrogen fixing ability, and the breeding of new forms of ornamental higher plants and of peas for the dried pea industry.
In addition to the institute's complement of science group staff included in the staff there are approximately 50 visiting research workers, post-doctoral fellows, and research students registered for higher degrees at the University of East Anglia. The Trustees of the John Innes Charity provide post-doctoral fellowships and research studentships, and residential accommodation for visitors from overseas. Visiting workers from academic, government, and industrial research organizations in the United Kingdom and from overseas are accepted to work at the institute.
Projects: There are twelve main project areas: streptomyces genetics; genetics of bacterial plant pathogens, (agrobacterium and Xanthomonas); rhizobium genetics including the production of new strains of rhizobium; spiroplasmas and micoplasmas; structure and multiplication of plant RNA viruses; structure and multiplication of plant DNA viruses, and development of viruses as gene vectors; higher plant genetics, analysis and manipulation of plant genomes; regulation of gene expression in higher plants with special reference to the storage proteins of peas; plant breeding - breeding of dried peas as a field crop and of disease resistant carnations; protoplast tissue culture and plant regeneration studies; cytodifferentiation and biogenisis of plant cell walls, cell cell recognition mechanisms; high resolution electron microscopy and computer image reconstruction applied to biological specimens.
Publications: Biennial report.

Institute for Marine Environmental Research   204

– IMER
Address: Prospect Place, The Hoe, Plymouth, PL1 3DH
Telephone: (0752) 21371
Affiliation: Component institute, Natural Environment Research Council
Status: Official research centre
Director: R.S. Glover
Assistant to the Director: Dr P.N. Claridge
Graduate research staff: 75
Activities: Multidisciplinary research on processes which determine the performance of marine and estuarine ecosystems with particular emphasis on variability caused by natural factors, pollutants, and other technological changes. Principal programmes deal with estuaries and the pelagic system of the open seas and ocean. A project on ecotoxicology examines natural stress and pollution over a wide range from the cellular to ecosystem level and biological indices of water quality have been derived. A study of environmental radioactivity is designed to develop methods for detecting transuranic elements in marine systems and investigating the passage of radioactive materials through organisms and substrates. Simulation modelling and development of predictive techniques are a feature of the Institute's approach.
Publications: Annual report.

Institute of Marine Biochemistry   205

Address: St Fittick's Road, Aberdeen, AB1 3RA
Telephone: (0224) 875695
Affiliation: Component institute, Natural Enviromment Research Council
Status: Official research centre
Director: Dr P.T. Grant
Graduate research staff: 27
Activities: Research focuses on biochemical principles that underlie the biological function and adaptation of marine organisms. The main emphasis is on: the nature, metabolism, and function of lipids in marine biota; the nutritional requirements, nutritional imbalance, and pathology of marine and freshwater fishes; the specialised mechanisms of assimilation, metabolism, and detoxication of essential and other mineral elements by marine organisms. Research on nutrition has special reference to fish farming. Research on trace metals is relevant to marine pollution and its biological effects.

Institute of Oceanographic Sciences   206

– IOS
Address: Brook Road, Wormley, Godalming, Surrey, GU8 5UB
Telephone: (0428 79) 4141
Telex: 858833
Affiliation: Component institute, Natural Environment Research Council
Status: Official research centre
Director: Dr A.S. Laughton
Groups: Applied physics, Dr B.S. McCartney; biology, P.M. David; chemistry, Dr F. Culkin; geology, Dr A.H.B. Stride; geophysics, Dr T.G. Francis; marine information and advisory service, Dr N.C. Flemming; marine physics, J. Crease; ocean engineering, Dr J.S.M. Rusby
Graduate research staff: 235
Activities: The institute undertakes oceanographic research on a world-wide basis in the fields of physical oceanography (ocean circulation and fluid dynamics); marine geology, biology, and chemistry, and remote sensing. Coastal studies are carried out in sedementation and in applied wave research. Tidal computation is undertaken as well as studies of ocean tides and shelf-sea dynamics. IOS also operates the Marine Information and Advisory Service (MIAS) based on institute expertise and an extensive data bank which includes world-wide data.
Publications: Annual report.

BIDSTON OBSERVATORY   207

Address: Bidston, Birkenhead, Merseyside L43 7RA
Telephone: (051) 653 8633
Assistant director: Dr D.E. Cartwright
Sections: Tides and shelf sea dynamics group, Dr D.E. Cartwright; tidal instrumentation and engineering group, J.B. Rae; tidal computations group, vacant

TAUNTON LABORATORY   208

Address: Crossway, Taunton, Somerset TA1 2DW
Telephone: (0823) 86211
Telex: 46274
Assistant director: M.J. Tucker
Sections: Applied wave research group, E.G. Pitt; sedimentation group, Dr K.R. Dyer; instrument engineering group, Dr A.P. Salkield

Institute of Terrestrial Ecology 209

– ITE
Address: 68 Hills Road, Cambridge, CB2 1LA
Telephone: (0223) 69745 to 9
Telex: 817201 CAMITE
Affiliation: Component institute, Natural Environment Research Council
Status: Official research centre
Director: J.N.R. Jeffers
Sections:
Division of Animal Ecology, Dr J.P. Dempster - vertebrate ecology, Dr D. Jenkins; invertebrate ecology, Dr M.G. Morris; animal function, Dr I. Newton
Division of Plant Ecology, Professor F.T. Last - plant community ecology, Dr M.D. Hooper; soil science, Dr O.W. Heal; chemistry and instrumentation, S.E. Allen; data and information, Dr C. Milner; algal and protozoan culture, Dr J.R. Baker
Graduate research staff: 180
Activities: Research designed to contribute to the wise use and protection of the natural environment, as follows:
Survey, monitoring, and classification - establishing factual information on species/habitat distribution and changes; land use; monitoring population changes and pesticide levels; taxonomy; all using a wide range of methodology.
Ecological studies - understanding the ecological processes of species or habitats for conservation, pest/weed control, exploitation.
Assessing/modifying impact - changes in land use; effect of pollutant including SO_2, NO_x, heavy metals, organochlorines, radionuclides; reclamation of derelict land; modifying spoilation.
Projects: Theoretical and applied studies in land classification and land use (Dr O.W. Heal); effects of radionuclides on terrestrial ecosystems (S.E. Allen); flows and effects, direct and indirect, of sulphur, fluorine, and other airborne pollutants in terrestrial ecosystems - studies involving plants and animals and increasingly detailed investigation of soil processes (Professor F.T. Last); cycling of plant nutrients, with particular emphasis on woodland and grassland ecosystems (Dr O.W. Heal); synoptic limnology and environmental impact assessment of freshwater ecosystems (Dr P.S. Maitland); toxicology of organic pollutants and heavy metals in terrestrial and freshwater systems (Dr. I. Newton); plant demography and the dynamics of plant communities, including the effects of herbivores, with special reference to the management of woodland and grassland (Dr M.D. Hooper); genecology and the implications of variation, in plants and animals, in natural and man-made ecosystems (Dr M.D. Hooper); genetics and physiology of forest trees (Professor F.T. Last); physiology and population dynamics of invertebrates and vertebrates, including pest (rabbits, squirrels, deer, and bark beetles) and non-pest (butterflies, sparrowhawks, otters) species (Dr D. Jenkins, Dr M.G. Morris).
Publications: Annual report; research reports.

BANCHORY LABORATORIES 210

Address: Hill of Brathens, Glassel, Banchory, Kincardineshire AB3 4BY
Telephone: (033 02) 3434
Telex: 739396
Head of station: Dr D. Jenkins
Activities: The laboratory specialises in research on vertebrate ecology, moorlands and air pollution, and includes a small group concerned with research on upland game birds and some of their predators.

BANGOR RESEARCH STATION 211

Address: Penrhos Road, Bangor, Gwynedd LL57 2LQ
Telephone: (0248) 4001 to 5
Telex: 61224
Head of station: Dr C. Milner
Activities: Research on pedology, geochemical cycling, plant community dynamics, montane studies and mathematical modelling. The Terrestial Environment Information Services (TEIS), is being developed here which will provide advice and information based on research and surveys from many different sources.

CULTURE CENTRE OF ALGAE AND PROTOZOA 212

Address: 36 Storey's Way, Cambridge, CB3 0DT
Telephone: (0223) 61378
Head of station: Dr J.R. Baker
Activities: Maintenance and supply of cultures of algae and protozoa; for research, teaching and industry and carries out taxonomic and ecological research on these organisms. The Centre is an International Depository Authority under the Budapest Treaty.

EDINBURGH LABORATORIES 213

Address: Bush Estate, Penicuik, Midlothian EH26 0QB
or: 78 Craighall Road, Edinburgh EH6 4RQ
Telephone: (031) 445 4343 to 6; (031) 552 5596
Telex: 72579
Senior Officer: Dr M. Unsworth
Activities: Research on the physiology and ecology of trees, both as individuals and in forests and plantations. The laboratory is also involved in research on air pollution, particularly its effect on trees and bryophytes.

FURZEBROOK RESEARCH STATION   214

Address: Wareham, Dorset BH20 5AS
Telephone: (092 93) 361 to 2
Telex: 418326
Head of station: Dr M.G. Morris
Activities: The main areas of research are: heathland ecology - plant demography, nutrient cycling, effects of fragmentation and isolation on invertebrates, spider ecology; comparative ecology of ants, genetic variation, and social structure; population dynamics of butterfly spp; population dynamics of oystercatchers and their prey, mussels; genetic structure and demography of grasses; wetland ecology - variation in Sphagnum; rotational management of scrub plots and its effects; effects of grassland management on invertebrate populations; maintenance of a data bank on plant/insect linkages.

MERLEWOOD RESEARCH STATION   215

Address: Grange over Sands, Cumbria LA11 6JU
Telephone: (044 84) 2264 to 6
Head of station: Dr O.W. Heal
Activities: Research covers, primarily, woodlands, uplands and soil microbiology. Computer-based ecological studies are being developed at the laboratory, which also houses the main chemical and library services and a radionuclide research facility.

MONKS WOOD EXPERIMENTAL   216
STATION

Address: Abbots   Ripton,   Huntingdon, Cambridgeshire PE17 2LS
Telephone: (048 73) 381 to 8
Senior Officer: Dr M.D. Hooper
Activities: The Station is a base for research in invertebrate, plant community and bird ecology; pollution by heavy metals and pesticides. The Biological Records Centre and part of the chemical and library services are also housed at the station.

Institute of Virology   217

– IOV
Address: Mansfield Road, Oxford, OX1 3SR
Telephone: (0865) 512361
Telex: VIROX 83147
Affiliation: Component institute, Natural Environment Research Council
Status: Official research centre
Acting director: J.S. Robertson
Graduate research staff: 36

Activities: Basic and strategic laboratory and field research on the characterization, biochemistry, and taxonomy of insect viruses, insect cell lines, and the ecology of viruses and their use in insect pest control. Also, parallel research in comparative virology and on viruses of wild birds, trees, and grasses. Main field sites for applied research on viruses control of forest pests currently in Wales and Scotland. IOV, sometimes in collaboration with other bodies, also works on pest control problems overseas including East and West Africa, the Mediterranean, the Seychelles, and Papua New Guinea.
Projects: Viruses of insects - baculoviruses (Dr D.C. Kelly); viruses of insects - RNA viruses (Dr N.F. Moore); viruses birds (Dr K.A. Harrap); epizootiology of insect viruses (P.F. Entwistle); viruses of trees and wild grasses (Dr J.I. Cooper); comparative virology (Dr T.W. Tinsley); resistance of insects to viruses (J.S. Robertson).

Institute of Zoology   218

Address: The Zoological Society of London, Regent's Park, London, NW1 4RY
Telephone: (01) 722 3333
Affiliation: The Zoological Society of London
Status: Independent research centre
Director of Science: Professor J.P. Hearn
Graduate research staff: 30
Activities: Major areas of research are - basic science, conservation, comparative medicine, education and collaboration, in an effort to improve animal and human welfare, and include: reproductive physiology of exotic species, primates and man; biochemical indicators of inbreeding and mechanisms of inheritance; infectious diseases in animals and man. ELISA and luminescence assays; essential fatty acids and brain development; energetics of ruminant nutrition; non-invasive methods for monitoring reproductive events and for diagnosing disease; neonatal nutrition and artificial rearing; comparative pathology; haematological diagnosis of diseases; comparative anaesthesia; in vitro fertilization; embryo storage and transfer; radioimmunoassay of plasma and urinary hormones; artificial insemination; cryopreservation. Improvement of animal management and breeding. Control of embryonic diapause and implantation.
Publications: Journal of Zoology (monthly), *International Zoo Yearbook*

CURATORS' RESEARCH   219
DEPARTMENT

Units: Mammals and Aquarium, Dr B.C.R. Bertram; birds and reptiles, P.J.S. Olney; Whipsnade, V.J.A. Manton; improvement of animal management and breeding; keepers' research projects

GENETICS DEPARTMENT 220

Head: Dr D.B. Whitehouse
Activities: Biochemical indicators of inbreeding; mechanisms of inheritance; sexing of monomorphic species; gene assignments in equids.

INFECTIOUS DISEASES DEPARTMENT 221

Head: Dr G.R. Smith
Units: Immunoassay Dr A. Voller
Activities: Botulism in birds; mycoplasma and anaerobic bacterial infections in animals and man; diagnostic methods for human and animal diseases; ELISA and luminescence assays; adjuvants.

NUTRITIONAL BIOCHEMISTRY DEPARTMENT 222

Head: Professor M.A. Crawford
Activities: Essential fatty acids and brain development; nutritional requirements of neonatal animals and man; placental function; metabolism of fat soluble vitamins; energetics of ruminant nutrition.

RADIOLOGY DEPARTMENT 223

Head: Professor G.H. du Boulay
Activities: Arterial spasm; nutrition and stroke; comparative bone development; non-invasive methods for monitoring reproductive events and for diagnosing disease.

REPRODUCTION DEPARTMENT 224

Head: Professor J.P. Hearn
Units: Behaviour, Dr A.F. Dixson; developmental biology, Professor J.P. Hearn; endocrine, Dr J.K. Hodges; gamete biology, Dr H.D.M. Moore
Activities: Reproductive physiology of exotic species, laboratory primates and man; neuroendocrine control of pituitary function, aggressive and sexual behaviour; puberty; control of embryonic diapause and implantation; superovulation. In vitro fertilization; embryo storage and transfer; neonatal development; radioimmunoassay of plasma and urinary hormones; corpus luteum function; pregnancy; steroid metabolism; sperm maturation and epididymal function; fertilization; artificial insemination; cryopreservation.

VETERINARY SCIENCE DEPARTMENT 225

Head: D.M. Jones
Units: Haematology Unit, Dr Christine M. Hawkey
Activities: Comparative anaesthesia; laparoscopic diagnosis; neonatal nutrition and artificial rearing; field projects in the Sudan and Niger; computerised diagnostics; haematological diagnosis of diseases; red cell sickling; coagulation; hypertrophic cardiomyopathy.

International Distillers and Vintners Limited* 226

Address: 1 York Gate, London, NW1 4PU
Telephone: (01) 935 4446
Parent body: Grand Metropolitan Hotels Limited
Status: Industrial company

GROUP TECHNICAL SERVICES* 227

Address: Vintner House, River Way, Harlow, Essex
Telephone: (0279) 26801
Telex: 817968 IDVGTS
Head: Allan C. Simpson
Activities: Flavour and stability of wines and spirits in relation to methods of production, storage, and packaging.

Kellogg Company of Great Britain Limited* 228

Address: Park Road, Stretford, Manchester, M32 8RA
Telephone: (061) 865 4411
Telex: 667031
Status: Industrial company
R & D Manager: K.C. Yates
Graduate research staff: 6
Activities: Product and process development in areas of breakfast cereals, dry mix products, frozen foods and other convenience food products. Packaging research relevant to the above is carried out to a limited extent, as is relevant specific applied food research.

Kings College London* 229

Address: Strand, London, WC2R 2LS
Telephone: (01) 836 5454
Parent body: University of London
Status: Educational establishment with r&d capability
Dean: vacant
Graduate research staff: 300

FACULTY OF NATURAL SCIENCE *   230

Dean: Professor E.A. Bell
Vice-Dean: Professor G.R. Wilkinson

Department of Plant Sciences   231

Address: 68 Half Moon Lane, London, SE24 9JF
Telephone: (01) 733 5666
Head: Professor J.W. Bradbeer
Graduate research staff: 24

Department of Zoology and Animal   232
Biology

Acting: Professor F.W. Darwin
Senior staff: Professor F.E.G. Cox, Professor C.B. Cox
Activities: Current research projects include: physiology and ecology of marine and estuarine organisms; ectoparasite-host relationships; immunological responses of rodents and primates to malaria and piroplasmosis; investigations of the ultrastructure; behavioural, physiological and anatomical aspects of ultrasonic communication in small mammals; Mesozoic vertebrates; behavioural and physiological aspects of the ecology of mammals; investigations of nucleo-cytoplasmic relations in amoebae; physiology and ecology of marine and estuarine organisms; ecology of arboreal communities; theoretical and experimental investigations of the population dynamics of host-helminth parasite systems.

Kirton Experimental   233
Horticulture Station

Address: Kirton, Boston, Lincolnshire PE20 1EJ
Telephone: (0205 722) 391
Affiliation: Ministry of Agriculture, Fisheries and Food, Agricultural Development and Advisory Service
Status: Official research centre
Activities: Research in the following areas - vegetable crops: variety testing; assessment of cultural systems; control of pests, diseases, and weeds. Vegetable storage: systems and methods of harvesting, barn storage, controlled temperature storage, and ice-bank cooled storage; shelf-life of Arored products. Plant establishment: techniques and ideas leading towards precise, uniform, and rapid establishment of vegetable crops. Silt soil management: maintenance and improvement of silt soil structure. Bulb flower crops (narcissms, tulips, and some minor bulb crops): storage, physiological treatments, and planting of bulbs; flower forcing methods; control of disease, pests, and weeds; mechanization of harvesting bulbs and flowers. Strawberries: evaluation of varieties for silt soil; mechanical harvesting and handling; weed, pest and disease control.

Laboratory of the   234
Government Chemist

– LGC
Address: Cornwall House, Stamford Street, London, SE1 9NQ
Telephone: (01) 928 7900
Parent body: Department of Industry
Status: Official research centre
Government Chemist: R.F. Coleman
Deputy Director (Customer Services): Dr D.C. Abbott
Deputy Director (Resources): Dr P.G. Jeffery
Sections: Research and Special Services Division, Dr R.J. Meoley
Activities: The laboratory is responsible for sponsoring the development of biotechnology in the UK. It evaluates proposals received from industry for DI Support to assist in new products, processes and demonstration plants. The laboratory also provides a comprehensive service of analysis, advice and studies based on chemistry to government departments and other official bodies, public institutions, local authorities and official international organizations, and, where appropriate, the private sector (provided that the statutory functions are not impaired for industry). The Government Chemist has statutory functions as official analyst or referee under various Acts of Parliament.
The Laboratory undertakes research on the development and improvement of methods of analysis, as appropriate to its terms of reference, including the study and application of physical methods of analysis such as electron microscopy, inductively-coupled plasma optical emission spectroscopy, mass spectrometry, nuclear magnetic resonance, infrared and X-ray fluorescence spectroscopy. The Laboratory is active in the application of computers to chemical analysis and in the development of automatic methods. Chromatographic techniques are applied in many studies.
Special investigations include: work on pesticide residues in agricultural crops and food; contaminants arising from processing and storage of foodstuffs; antibiotics and hormones in animal tissue and milk; radioactive nuclides in water supplies; the development of glass-ionomer cements for dental and other applications. Work on public health and environmental protection is concerned with foods, drugs, agriculture, pesticides, toxic metals, exposure to toxic gases and vapours, radiochemistry and oil pollution.
The Laboratory maintains six advisory and information services dealing with Agricultural Materials (AMAIS), Chemical Nomenclature (CNAS), Consumer Hazards (CHAIS), Dangerous Goods (DAGAS), General Analytical Methods (GAMIS) and Oil Spillage (OSAIS).
The organization has branch laboratories, at the following addresses: Cornhill, Liverpool, L18 JJ; Royal Clarence Yard, Gosport, Hants PO12 1AY;

LGC Unit, National Physical Laboratory, Teddington, Middlesex TW11 0LW.
Publications: The Report of the Government Chemist, annually.

Land Resources Development Centre   235

– LRDC
Address: Tolworth Tower, Surbiton, Surrey KT6 7DY
Telephone: (01) 399 5281
Telex: 263907
Parent body: Overseas Development Administration, Foreign and Commonwealth Office
Director: A.J. Smyth
Deputy Director: M.A. Brunt
Graduate research staff: 40
Activities: LRDC assesses and develops land resources in Third World countries. Studies range from evaluation of land and water resources, forest inventories, social and economic appraisals, feasibility studies and project identification, to pilot-scale developments, project supervision and monitoring, training in land use, erosion control and extension, culminating in detailed recommendations for development. Current major projects include: development of new settlement areas in Sumatra; integrated rural development in Nepal (agriculture, forestry, communications health and cottage industry); assessment of water requirements in eastern Cyprus and feasibility of conveyance from the west; three land capability and development projects in Tanzania and Kenya; land use survey/tsetse control in Somalia; coordination of agricultural research centre in Yemen Arab Republic; assessment of groundwater resources and agricultural potential in north-west Sri Lanka.
Publications: Progress reports, technical bulletins, land resource studies and bibliographies.

Lee Valley Experimental Horticulture Station   236

Address: Ware Road, Hoddesdon, Hertfordshire EN11 9AQ
Telephone: (61) 63623
Affiliation: Ministry of Agriculture, Fisheries and Food, Agricultural Development and Advisory Service
Status: Official research centre
Activities: Subjects of research include: vegetables - celery, cucumber, lettuce, mushrooms, tomatoes, aubergine, Chinese cabbage, sweet peppers; flowers - carnation, chrysanthemum, rose, peony, bedding plants, pot plants; fuel saving; frost protection; waste heat simulation studies; film and rigid plastic greenhouses, prototype insulated film plastic mushroom unit; variety trials for higher yield, better quality, and pest and disease resistance; peat substrates, straw beds, and hydroponics; short-term storage and shelf-life tests; irrigation, ventilation, and heating systems.

Leicester Polytechnic*   237

Address: PO Box 143, Leicester, LE1 9BH
Telephone: (0533) 551551
Status: Educational establishment with r&d capability
Director: David Bethel
Assistant Director: Dr J.W.L. Warren

FACULTY OF SCIENCE*   238
Dean: Dr Jeanne M. Thompson

School of Life Sciences*   239
Head: Dr J.M. Thompson
Activities: Plant developmental biology, including regulation of development of the storage of sugar beet, hormone regulation of root growth, control of root growth in root crop species, and introduction of a nitrogen-fixing bacterium into plant cells; pharmacology, including interactions of hoglycaemic agents with steroid hormones; ecology, including utilization by birds of a disused gravel working.

Life Science Research*   240

Address: Stock, Essex CM4 9PE
Telephone: (0277) 840101
Telex: 995126 LIFSCI G
Parent Body: IMS International Incorporated
Status: Independent research centre
Director of research: Dr Kenneth H. Harper
Sections: General toxicology, Roger Ashby; short-term toxicology, Simon A. Buch; inhalation toxicology, Simon A. Buch; oncogenicity, Trevor W. McSheehy; reproductive toxicology, Dr John M. Tesh; pathology, J. Peter Finn; genetic toxicology, James Bootman; metabolism, Dr John W. Daniel; analytical chemistry, Peter M. Brown; clinical services, Dr Neville W. Shephard; consumer studies, Dr Philip C. Rofe
Graduate research staff: 90
Activities: Research into toxicology: acute, long-term and carcinogenicity studies, reproductive studies, inhalation studies, irritance and sensitization studies; pharmacokinetics; genetic toxicology; metabolic and analytical chemistry; pathology; veterinary studies; clinical trials and other volunteer studies; consumer studies.

ELM FARM LABORATORIES*   241

Address: Occold, Suffolk IP23 7PX
Telephone: (037971) 491
Telex: 975389 LIFESCI G
Activities: Provides laboratory facilities for both Life
Science Research and Medical Science Research.

Liscombe Experimental   242
Husbandry Farm

Address: Liscombe, Dulverton, Somerset TA22 9P2
Telephone: (064 385) 291
Affiliation: Ministry of Agriculture, Fisheries and
Food, Agricultural Development and Advisory Service
Status: Official research centre
Director: L.R. Curnett
Deputy director: M. Appleton
Sections: Sheep, M. Appleton, D. Done; cattle, D.
Chapple; silage/hay, D. Holness, I. Rigby; grassland
production/utilization, L. Curnett
Activities: Research in animal production from grass:
beef cattle - sucklers and dairy calves; sheep - lamb
production from housed prolific ewes, and from hill
ewes; silage - techniques, additives, chop length; grass-
land - utilization and management.
Publications: Annual review; cattle, sheep, and grass
bulletins.

Liverpool School of   243
Tropical Medicine*

Address: Pembroke Place, Liverpool, L3 5QA
Telephone: (051) 708 9393
Telex: 627095
Affiliation: University of Liverpool
Status: Educational establishment with r&d capability
Dean: Professor H.M. Gilles

MEDICAL ENTOMOLOGY   244
DEPARTMENT*

Head: Professor W.W. Macdonald
Staff: Dr M.W. Service, Dr H.Townson, Dr M.H.
Birley, Dr M.J. Roberts, Dr R.D. Ward
Graduate research staff: 7
Activities: Vector population dynamics; mosquito ecol-
ogy.

VETERINARY PARASITOLOGY   245
DEPARTMENT*

Head: Dr W.N. Beesley
Lecturer: Dr A.J. Trees
Graduate research staff: 3
Activities: Parasitic diseases of the horse; the warble
and tsetse flies; biology of cattle ticks.

Long Ashton Research   246
Station

Address: Long Ashton, Bristol, BSl8 9AF
Telephone: (027 580) 2181
Affiliation: State-aided institute, Agricultural
Research Council; University of Bristol
Status: Official research centre
Director: Professor J.M. Hirst
Graduate research staff: 90 (in all departments)
Activities: The station serves agriculture, horticulture
and the food and drink industries through an unusual
diversity of research directed, where possible, through
fruit and cereal crops. Topics include the biochemistry of
nitrogen metabolism; of ripening and senescence; the
activities of plant growth regulators. The chemical,
cultural, and biological protection of crops against pest
and diseases and the influence of weather on these
phenomena. The breeding of improved strawberries,
plums and apples and the improvement and stabilization
of yields. The production of fruit juices and fermented
products together with studies of their chemistry,
microbiology and sensory assessment.

POMOLOGY DIVISION   247

Head: Dr A.I. Campbell
Activities: Improvement of fruit crops by clonal selection, breeding, and mutation breeding; improvement of
woody ornamentals; orchard management, including
high density systems, pollination, and ground cover
control; windbreaks (the station also houses the national
willows collection); cider pomology.
Projects: Production of healthy and improved fruit
trees (Dr A.I. Campbell); pollination and fruit set in
apples, cider and perry orcharding (R.R. Williams);
micropropagation of tree fruits and strawberry, mutation and regeneration in vitro (Dr A.J. Abbott); apples -
factors influencing components of yield - optimum cropping (D.L. Abbott); apples - factors influencing components of yield - optimum cropping (D.L. Abbott);
strawberry - factors affecting cropping (C.G. Gutteridge, H.M. Anderson); growth regulators in fruit
production, new planting systems for fruit trees (R.D.
Child); micropropagation of woody ornamentals in
vitro, regeneration from tissues, cells, and protoplasts

(D.R. Constantine); improvement of trees and shrubs (C.S. Gundry); strawberry - factors limiting productivity and selection method, - calcium nutrition (C.G. Gutteridge); plum breeding (R.P. Jones); induction of mutations in plants (Dr C.N.D. Lacey); trees for windbreaks, biomass, basket making, and amenity, herbicides for sward control in orchards (K.G. Sott); strawberry breeding (Dr D. Wilson); water stress and waterlogging in relation to changes in plant growth substances levels, transpiration and its control by endogenous hormones including abscisic acid, winter dormancy - the role of endogenous hormones such as abscisic acid and its metabolites (Dr S.T.C. Wright).

CROP PROTECTION DIVISION 248

Head: Dr K.J. Brent
Sections: Organic chemistry, Dr N.H. Anderson; plant pathology, Dr R.J.W. Byre; zoology, Dr B.D. Smith; spray application, Dr N.G. Morgan
Activities: Pests and diseases of fruit and arable crops; chemical and biological control methods; synthesis, application, performance, and fate of fungicides; basic studies of mode of action of fungicides, resistance problems, leaf surfaces, and the plant's response to infection.

FOOD AND BEVERAGE DIVISION 249

Head: Dr F.W. Beech
Activities: Chemistry, microbiology, and technology of cider and wine production; quality assessment of cider, wine, and fruit; use of preservatives - spoilage and taint problems; preservation and use of food in the home.
Projects: Technology of ciders, wines and fruit juices, preservatives (Dr F.W. Beech); lactic, acetic and other bacteria of ciders, fruit juices, and wines, sulphur dioxide and other antimicrobial substances - effect on lactic and acetic acid bacteria of cider; malo-lactic fermentation in ciders, microbial spoilage and hazards from foods cooked and preserved in the home (Dr J.G. Carr); factors affecting chemical composition of fruit juices, cider, and wines, yeast nutrition - study of nitrogenous compounds of apple juices and ciders, sulphur dioxide - reactions with constituents of juices and ciders (Dr L.F. Burroughs); microecology of yeasts and yeast-like organisms in natural and industrial environments, identification and classification of yeasts and yeast-like organisms (Dr R.R. Davenport); post-harvest biochemistry and physiology of vegetables and fruit, compositional analyses of plant constituents (P.W. Goodenough); production of ciders, wines, and juices under controlled conditions (J.G. Williams); phenolics - isolation and identification in apples, cider, and wines, anthocyanins - study for non-toxic food colours, identification and reactions of red wine pigments (Dr C.F. Timberlake); fermented beverages - isolation and identification of taints, chemistry of flavour compunds of fruits, vegetables, and fermented products (Dr O.G. Tucknott); yeasts - organic acid metabolism (R.A. Coggins); fruit and beverage quality, volatile aroma components, sensory evaluation of fruits and beverages (Dr A.A. Williams).

PLANT SCIENCE DIVISION 250

Head: Dr K. Treharne
Sections: Plant nutrition, Dr D.G. Hill-Cottingham; plant physiology and biochemistry, Dr E.J. Hewitt; environmental physiology, vacant
Activities: Plant nutrition, physiology, and biochemistry, especially nitrogen metabolism; environmental physiology.
Projects: Inorganic nitrogen metabolism (including nitrogen oxides) and functions of micronutrients, especially molybdenum in plants (Dr E.J. Hewitt); nitrogen metabolism of Vicia faba, inter cultivar differences in seed protein (Dr D.H.P. Barratt); plants - distribution, form, and mobility of mineral elements, especially calcium (Dr E.G. Bradfield); cytokinins in plants - relationship with nitrate reductase (Dr J.S. Chalice); uptake, movement, and metabolism of nitrogen constituents of grain legumes (Dr D.G. Hill-Cottingham); plant hormones and their role in control of growth and fruiting (Dr G.V. Hoad); nitrate reductase - structure and mechanism and electron donors in non-chlorophyllous tissues, nitrate reduction - regulation and relation to photosynthesis (Dr D.P. Huckleby); nitrate assimilation in cereals - genetic, physiological, environmental, and pathogenic effects (D.M. Jaes); nitrogen uptake and carbon utilization in field beans (C.P. Lloyd-Jones); molybdenum and tungsten complexes in plants, nitrate reductase-formation, structure, and function (Dr B.A. Notton); cereals - interrelation between disease and water relations (D.B.B. Powell); soil fertility and crop nutrition problems, tropical crop agronomy (Dr J.B.D. Robinson); metabolism of amines and ployamines in plants, especially cereals - biosynthesis, oxidation, concentrations, and derivatives (Dr T.A. Smith).

Loughborough University of Technology* 251

Address: Ashby Road, Loughborough, Leicestershire LE11 3TU
Telephone: (0509) 63171
Telex: 34319
Status: Educational establishment with r&d capability
Chancellor: Sir Arnold Hall

SCHOOL OF PURE AND APPLIED SCIENCE* 252

Dean: Professor K.W. Bentley

Department of Chemical Engineering 253

Head: Professor D.C. Freshwater
Activities: Current research topics include: optimum plant capacity; electrical discharges; aeration with a plunging jet; development of a microwave discharge reactor; catalytic two-phase oxidation processes; polymer manufacture; effect of polymeric additives on the precipitation of solids; design techniques - economic aspects of design; particle packing and characterization. Computer simulation of the chemical industries; stability and oscillations in biochemical reactors; models for fixed bed catalytic reactors; modelling of porous media; rising damp; assessment of scientific excellence using the citation index; a novel density measuring apparatus; hydrodynamic and model-independent chromatography theory; trickle flow in packed beds; reactor theory; rheology of pastes; polyelectrolyte flocculation; development of tape recorder and operator decision interface to quantimet image analysing computer; aerosol migration and size analysis; behaviour of fibrous filters under load; lunar dust; instability of a liquid jet; hydraulic conveying; thermal and electrical conductivity of a powder; coincidence in stream scanning; variations in continuous mixing; flow through packed beds; water absorption and the physical properties of powders; particle size analysis; classification and flow properties of powders; development of laboratory testing techniques for cake filtration equipment design; the performance of hydrocyclones - viscosity effects; liquid flow patterns in centrifuges; mechanism of particle capture in gas filtration; design of mass transfer equipment; heat transfer; flow fluctuations in natural circulation reboilers; pool boiling; effect of hydrodynamics and physical properties of fluids, mass diffusivity; properties of binary liquid mixtures; condensation; the design of deaerators and water chillers; jet equipment; prediction of thermodynamic properties of mixtures; reduced equations of state for pure fluids and fluid mixtures; plant; reliability engineering and loss prevention in the process industries; control of major hazard plants; the propagation of faults; and the use of reliability information in decision making on maintenance of process plants; alarm systems in process control and alarm analysis using a process computer; food processing technology; ultrafiltration of waste blood plasma proteins; physical and biochemical properties of blood plasma protein fibres; modelling in the process industries; edible oils; processing of fish waste and vegetables.

Loughry College of Agriculture and Food Technology 254

Address: Cookstown, Tyrone, BT80 9AA, Northern Ireland
Telephone: 62491
Parent body: Department of Agriculture for Northern Ireland
Status: Educational establishment with r&d capability
Principal: H.R. Kirkpatrick
Vice-Principal: G.I.A. Lang
Departments: Agriculture, Robert E. Haycock; food technology, George I.A. Lang; communication and Rural Development, Robin C. Stevenson
Activities: The college provides education, training and in research and development in food production, food processing and agricultural Communication, and gives specialist advice and help to the Agriculture and Food Industry. Applied research and development is particularly aimed at solving local problems and assessing likely useful new ideas in some aspects of food production, food processing and in education and rural development. The emphasis at the college is very much on development rather than applied research.

Luddington Experimental Horticulture Station 255

Address: Luddington, Stratford-upon-Avon, Warwickshire CV37 9SJ
Telephone: (0789) 75 06 01
Affiliation: Ministry of Agriculture, Fisheries and Food, Agricultural Development and Advisory Service
Status: Official research centre
Director: D.J. Harrison
Departments: Fruit, H.A. Pudwell; glasshouse, J.R. Chrimes; vegetables, A.C.W. Davies; ornamental nursery stock, P.D.E. Cooper
Graduate research staff: 8
Activities: Research into: drainage and irrigation; crop husbandry; plant production. Specific areas include: top fruit - variety trials of apples and plums, pollination studies, pest and disease control experiments, fruit nutrition, herbicide treatments, irrigation, storage, fruit mechanization; soft fruit - culture experiments with blackcurrants, raspberries, and strawberries, variety trialling, mechanical handling; vegetables - variety trials, storage, etc of cauliflower, brussels sprouts, cabbages, spring greens, onions, and leeks, crop establishment and salad crops, soil sterilization, herbicides, nutrition; glasshouse crops - winter lettuce and tomatoes, wind loading problems of film plastic clad structures; soil warming, nutrient film work; nursery stock - quality of ornamental

trees and shrubs, home production of seedling root stocks, propagation trials.

The National Bee Unit which has a major national advisory function, is now located at Luddington, and research is undertaken into pollination of fruit, potential of some agricultural crops for honey production, and ways of reducing risks from pesticides.

Projects: Vegetables - plant establishment, soil management (A.C.W. Davies); variety trials, propagation (D.J. Harrison); fruit - rootstocks and varieties, raspberry (L. Andrews); training systems, soft fruit variety trials (H.A. Pudwell); glass - NFT culture, alternative crops. (J.R. Chrimes); nursery stock - tree and shrubs produced from seed, herbicides for field grown nursery stock, propagation for cuttings (P.D.E. Cooper); irrigation investigations (all sections); shelter investigation (D.J. Harrison); storage of fruit and vegetables; shelf life and marketing studies.

Publications: Annual review.

Macaulay Institute for Soil Research   256

Address: Craigiebuckler, Aberdeen, AB9 2QJ
Telephone: (0224) 38611
Affiliation: State-aided institute, Agricultural Research Council; Department of Agriculture and Fisheries for Scotland
Status: Official research centre
Director: Dr T.S. West
Sections: Mineral soils, Dr R.C. Mackenzie; peat and forest soils, R.A. Robertson; spectrochemistry, Dr R.O. Scott; soil organic chemistry, Dr G. Anderson; plant physiology, Dr P.C. DeKock; microbiology, Dr J.F. Darbyshire; soil fertility, Dr B.W. Bache; statistics, R.H.E. Inkson; soil survey, R. Grant
Graduate research staff: 100
Activities: Surveying and mapping of soil types in Scotland, including peat. Assessment of their fertility in pot and field experiments, particularly their potential with regard to crop production and forestry. Chemical studies to measure their contents of major nutrients and trace elements, their mineralogical composition, the nature and properties of their organic components. Plant physiological studies and soil-root relationships. Studies of soil microorganisms and their effects. Devising and improving analytical techniques for measuring soil components.
Projects: Study of the development and composition of mineral soils and their size fractions - characterization of minerals, organo mineral complexes, and major constituents (Dr R.C. Mackenzie); characterization of trace elements (Dr R.O. Scott); survey and classification of the mineral soils of Scotland (R. Grant); nature and properties of soil organic matter (Dr G. Anderson);

investigation of the role of soil microorganisms in soils and in soil-plant relationships (Dr J.F. Darbyshire); nature and distribution of organic soils and peat in Scotland (R.A. Robertson); investigations on the fertility of soils and the yield of agricultural crops (Dr B.W. Bache); factors affecting crop composition (Dr P.C. DeKock); fertility of forest soils (Dr H.G. Miller).

Publications: Annual report.

Malaysian Rubber Producers' Research Association   257

– MPPRA
Address: Tun Abdul Razak Laboratory, Brickendonbury, Hertford, SG13 8NL
Telephone: (0992) 54966
Telex: 817449 MRRDBL G
Parent body: Malaysian Rubber Research and Development Board
Status: Research centre within and industrial company
Director of Research: Dr L. Mullins
Deputy Director: Dr D. Barnard
Assistant to the Director: Dr P.W.Allen
Graduate research staff: 75
Activities: Research, development, and promotional activities relating to the consumption of Malaysian rubber in the western hemisphere and in assisting towards the development of rubber-based industries in Malaysia. Current activities include: fundamental research on the abrasion, adhesion, friction, and crystallization of natural rubber; modification of natural rubber to produce thermoplastic forms; development of winter tyres; design of bearings to isolate buildings from earthquakes; improvements in vulcanization and latex technology; mechanism of rubber production; effect of yield stimulants.
Publications: Rubber Developments, NR Technology, both quarterly; annual report.

Marine Biological Association of the United Kingdom 258

– MBAUK
Address: The Laboratory, Citadel Hill, Plymouth, PL1 2PB
Telephone: (0752) 21761
Affiliation: Grant-aided institute, Natural Environment Research Council
Status: Official research centre
Director: Professor Eric James Denton
Graduate research staff: 60
Publications: Journal of the Marine Biological Association of the UK.

Marine Laboratory 259

Address: PO Box 101, Victoria Road, Torry, Aberdeen, AB9 8DB
Telephone: (0224) 876544
Telex: 73587
Parent body: Department of Agriculture and Fisheries for Scotland, Fisheries Research Services
Status: Official research centre
Director: A.D. McIntyre
Sections: Fishery resources research, A. Saville; fish cultivation, disease, and biochemistry, Dr A.L.S. Munro; fish capture D.N. Maclennan; environmental research, R. Jones; statistical and computer services, J.A. Pope
Activities: Marine research, relevant to commercial fisheries; various aspects of aquaculture. The prime function of the laboratory is to provide information for the Department of Agriculture and Fisheries for Scotland. Most of the work is biological in character, involving studies of the behaviour of the principal food fishes, including shellfish, and catching techniques. Interest mainly centres on the distribution, reproduction, growth, population dynamics and parasites of these animals. As his research requires background knowledge of the conditions under which the animals live so that investigation of the water conditions (temperature, salinity, and nutrient content, etc, and currents), the free-floating life (plankton), and bottom-living animals, and an assessment of marine productivity form an essential part of the work of the laboratory. Contact is maintained with those responsible for similar research activities in other countries, mainly through the International Council for the Exploration of the Sea, the Food and Agriculture Organization of the United Nations, and the International Commission for the Northwest Atlantic Fisheries.

Publications: Fisheries of Scotland Report; Marine Laboratory Triennial Review of Research; Scottish Fisheries Bulletin; Scottish Fisheries Information Pamphlets; Scottish Fisheries Research Reports.

Mars Limited* 260

Address: Dundee Road, Slough, Berkshire
Telephone: (0753) 23932
Telex: 851-848253 (MARSSL G)
Status: Industrial company
Research Director: Dr Graham Harris
Sections: Dundee Road development, P.T. Turner; Liverpool Road development: Dr A.R. Gibbons; research project group, R.G. Windred
Graduate research staff: 45
Activities: Investigation of raw materials, processing and finished goods in confectionery, snack foods and ancillary areas. Quality assurance techniques, microbiology of products.

May & Baker Limited 261

– M&B
Address: Rainham Road South, Dagenham, Essex RM10 7XS
Telephone: (01) 592 3060
Telex: London 28691
Parent body: Rhône-Poulenc SA
Product range: Medical and veterinary specialities; agricultural, horticultural and garden chemicals; photographic, laboratory and fine chemicals.
Status: Industrial company

ONGAR RESEARCH STATION 262

Address: Fyfield Road, Ongar, CM5 0HW
Telephone: (0277) 362127
Telex: London 28691
Director: Dr J.A. McFadzean
Sections: Agrochemicals research, Dr B. Savory; animal health product development, A.J. Colegrave; animal health research, Dr M. Griffin

Meat and Livestock Commission 263

– MLC
Address: PO Box 44, Queensway House, Bletchley, Milton Keynes, MK2 2EF
Telephone: (0908) 74941
Telex: 82227
Status: Independent development authority
Director of Research and Veterinary Services: Dr D.R. Melrose
Scientific Secretary: Dr J.E. Duckworth
Activities: No fundamental research laboratories but facilities to permit cattle and pig testing and on-farm testing services for cattle, sheep and pigs for genetic improvement programmes, recording services covering commercial production, carcass appraisal facilities. Economics reporting and forecasting services. Marketing, advisory and promotion services and a design and operational advisory service for meat plants and abbatoirs and also the retail trade. Research is commissioned at universities, research institutes, etc. Development work and services operate in animal production, marketing, economics and veterinary service.
Publications: Handbooks; leaflets; annual series - newsletters; veterinary publications; AI publications; *United Kingdom Meat and Livestock Statistics; Marketing and Meat Trade Technical Bulletins; MIDAS Bulletins.*

Medical Research Council* 264

Address: 20 Park Crescent, London, W1N 4AL
Telephone: (01) 636 5422
Telex: 24897
Status: Official research organization
Chairman: The Rt Hon the Lord Shepherd
Deputy Chairman and Secretary: J.L. Gowans
Second Secretary: S.G. Owen
Activities: To promote the balanced development of medical and related biological research in the United Kingdom, receiving and administering funds for general or specific disease research. There are also units in Jamaica and the Gambia.

MRC LABORATORY ANIMALS CENTRE* 265

Address: Medical Research Council Laboratories, Woodmansterne Road, Carshalton, Surrey SM5 4EF
Telephone: (01) 643 8000
Director: G.H. Townsend
Activities: To make available for biomedical research throughout the United Kingdom animals of the appropriate species, strain, and quality by supplying breeding nuclei (conventional, specified-pathogen-free, and germ-free) to those wanting to establish their own colonies; administering the voluntary accreditation and recognition schemes for commercial breeders and suppliers; research into breeding, nutrition, environment, disease, animal husbandry, and laboratory animal science generally; training technical and scientific staff; collecting and disseminating information, and maintaining national and international contacts (the Centre is a WHO Collaborating Centre for Defined Laboratory Animals and an International Committee on Laboratory Animals Virus Reference Centre).

MRC LABORATORY OF MOLECULAR BIOLOGY* 266

Address: MRC Centre, University Medical School, Hills Road, Cambridge, CB2 2QH
Telephone: (0223) 48011
Telex: 81532
Director: S. Brenner
Deputy director: H.E. Huxley
Sections: Cell biology, J.B. Gurdon
Protein and nucleic acid chemistry, Dr F. Sanger
Structural studies, H.E. Huxley, Dr A. Klug
Activities: Interpreting biological phenomena at the molecular level by a wide range of studies of the structure of proteins, nucleic acids and macromolecular assemblies, and by research on the mechanisms and control of gene expression especially as related to cell differentiation and other developmental processes.

MRC MAMMALIAN DEVELOPMENT UNIT* 267

Address: University College London, Wolfson House, 4 Stephenson Way, London, NW1 2HE
Telephone: (01) 387 9521
Director: Anne McLaren
Activities: Studies on early mammalian embryos and on the interactions of cell populations in genetically composite embryos.

MRC MAMMALIAN GENOME UNIT*   268

Address: Department of Zoology, West Mains Road, Edinburgh, EH9 3JT
Telephone: (031) 667 1081
Director: Professor P.M.B. Walker
Activities: Study of the structure of higher organism DNA using biochemical methods, and also the functions of chromosomes by manipulating mammalian cells in culture, with a view to strengthening and encouraging links between fundamental and applied studies in the fields of cancer and human genetic disorders.

MRC REPRODUCTIVE BIOLOGY   269
UNIT*

Address: Centre for Reproductive Biology, 37 Chalmers Street, Edinburgh, EH3 9EW
Telephone: (031) 229 2575
Director: Professor R.V. Short
Honorary clinical advisor: Professor D.T. Baird
Activities: Study of the fundamental aspects of the reproductive processes in experimental animals (laboratory animals, monkeys and sheep) and development of a number of novel approaches for controlling fertility in man and animals by immunological and pharmacological means. Clinical studies are aimed at a greater understanding of the way in which lactation regulates fertility in women in developed and developing countries.

MRC TOXICOLOGY UNIT*   270

Address: Medical Research Council Laboratories, Woodmansterne Road, Carshalton, Surrey SM5 4EF
Telephone: (01) 643 8000
Director: Dr T.A. Connors
Deputy director: Dr W.N. Aldridge
Sections: Biological chemistry, Dr T.A. Connors
Molecular toxicology, Dr W.N. Aldridge
Biochemical pharmacology, Dr F. De Matteis
Clinical section (at St Bartholomew's Hospital Medical College), Professor P.J. Lawther
Activities: Clinical and scientific research on the mechanism of action of toxic chemicals. (The Unit is designated the WHO International Reference Centre on Clinical and Epidemiological Aspects of Air Pollution and a WHO Collaboratory Centre for the assessment of mammalian toxicity of new pesticides to laboratory animals.)

MRC VIROLOGY UNIT*   271

Address: Institute of Virology, Church Street, Glasgow, G11 5JR
Telephone: (041) 339 8855
Honorary director: Professor J.H. Subak-Sharpe
Activities: Research on the structure, function and information content of viruses with particular reference to changes in animal cells infected by lytic and tumour viruses.

Metal Box plc Research   272
and Development
Department

Address: Denchworth Road, Wantage, OX12 9BP
Telephone: (02357) 2929
Telex: 837929
Product range: Metal containers, plastics containers, paper and board containers, packaging machinery, kitchen ware, central heating radiators and boilers, bank cheque books.
Status: Research centre within an industrial company
Managing Director, r&d division: Dr John McIntosh
Graduate research staff: 400
Departments: Business development, F. Fidler; divisional services, Dr D.J. Bland; food science and customer services, M.G. Alderson; operations, A.C. Mapp; engineering, M. Newman
Activities: Development and evaluation of new packaging containers; biological safety of products and processes; food microbiology and preservation methods; electrochemical corrosion of metal containers; compatibility of food with container materials; analysis of trace metals in foods; properties and processing of packaging materials (metals, plastics, paper and board); analytical techniques for these materials; curing and performance of protective and decorative coatings on all types of container; printing technology on all container materials; security printing of cheques, stamps, labels, tickets; design of machinery for manufacture and closing of containers; CAD/CAM; electronics and computing; intrumentation and control engineering; noise control; energy conservation; materials recycling; sheet metal forming.

Ministry of Agriculture, Fisheries and Food 273

Address: Whitehall Place, London, SW1A 2HH
Telephone: (01) 217 3000
Telex: 889351
Status: Government department
Permanent Secretary: Sir Brian Hayes
Chief Scientific Adviser (Food): Dr G.A.H. Elton

AGRICULTURAL DEVELOPMENT AND ADVISORY SERVICE 274

– ADAS
Director-general: Dr K. Dexter
Deputy director-general: E.S. Carter
Chief agricultural officer: J.J. North
Head of Agricultural Science Service: W. Dermott
Chief surveyor: D.B.S. Fitch
Chief engineer: G. Cole
Chief veterinary officer: W.H.G. Rees
Controller, experimental horticulture stations: E.R. Bullen
Activities: The Ministry's professional, scientific and technical services are mainly incorporated in the Agricultural Development and Advisory Service - ADAS. Specialist advice is available on agriculture generally, including arable farming, horticulture, and horticultural and farm management; disposal of agricultural effluents; land drainage; and the control and eradication of notifiable diseases. Architects and chartered surveyors can provide guidance on the design, layout, construction, or improvement of all types of farm constructions. In addition, microbiologists, entomologists, plant pathologists, soil scientists, etc are available to provide, at a charge, services for analyses of various kinds. A leaflet describing these services, and the charges, is obtainable from the Ministry's local offices. ADAS also operates a number of experimental farms and horticultural stations in England and Wales.
Publications: A guide to the work of experimental stations is published annually; completed experiments are described in *Experimental Husbandry and Experimental Horticulture*
See separate entries for:
Agricultural Science Service:
Cattle Breeding Centre
Central Veterinary Laboratory
Coypu Research Laboratory
Harpenden Laboratory
Royal Botanic Gardens, Kew
Slough Laboratory
Tolworth Laboratory, Rodent Pests Department
Veterinary Laboratory, Lasswade
Worplesdon Laboratory
Experimental Horticulture Stations:

Efford Experimental Horticulture Station
Kirton Experimental Horticulture Station
Lee Valley Experimental Horticulture Station
Luddington Experimental Horticulture Station
National Fruit Trials
Rosewarne Experimental Horticulture Station
Stockbridge House Experimental Horticulture Station
Experimental Husbandry Farms:
Arthur Rickwood Experimental Husbandry Farm
Boxworth Experimental Husbandry Farm
Gleadthorpe Experimental Husbandry Farm
Weat House Experimental Husbandry Farm
High Mowthorpe Experimental Husbandry Farm
Liscombe Experimental Husbandry Farm
Pwllpeiran Experimental Husbandry Farm
Redesdale Experimental Husbandry Farm
Rosemaund Experimental Husbandry Farm
Terrington Experimental Husbandry Farm
Trawsgoed Experimental Husbandry Farm

Field Drainage Experimental Unit 275

Address: Anstey Hall, Maris Lane, Trumpington, Cambridge, Cambridgeshire
Telephone: (0223) 840011
Telex: 81131
Head: S. Le Grice
Sections: Regional developments, J.G. Rands; soils and engineering, C.W. Dennis; computer and statistics, Dr A.C. Armstrong; hydrology, G.L. Harris
Graduate research staff: 12
Activities: All work is concerned with evaluation field drainage in commercial farming practice, priority being given to problems which are widespread in England and Wales. Other topics studied are: economics, filter wraps; catchment hydrology, drainage of saline soils.
Projects: Drainage/cultivations interactions (G.L. Harris); soil strength parameter and mole drainage success (A.A. Thorburn); soil strength parameters and poaching problems (P.A. Waters); drainage of heavy clay soils (Drayton EHF) (C.W. Dennis); drainage of restored opencast coal land (N.C. Bragg); modelling of drainage needs (Dr A.C. Armstrong); alleviation of iron ochre deposition (C.W. Dennis).
Publications: Progress report, triennial, technical reports.

DIRECTORATE OF FISHERIES RESEARCH 276

Director of Fisheries Research: A. Preston
Deputy Director of Fisheries Research: H.W. Hill
Sections: Fish Stock Management Division: North sea section, A.C. Burd; western section, Dr J.G. Shepherd; inshore and coastal section, Dr F.R. Harden Jones
Fish and Shellfish Cultivation Section, Dr C.E. Purdom: fish cultivation group, Dr V.J. Bye; shellfish cultivation

group, B.T. Hepper; fish diseases laboratory, Dr B.J. Hill

Aquatic Environment Protection Division: radiobiology section, Dr N.T. Mitchell; non-radioactivity section, P.C. Wood; physical environment section, Dr R.R. Dickson

Research Support Group, J.W. Ramster: information section, A.R. Margetts; electronics and instrumentation section, R.B. Mitson; computing and statistics section, C.A. Goody

Activities: The Directorate conducts research into marine and freshwater stocks of fish and shellfish (to assess potential yields to fisheries), and the effects of fishing (so as to formulate scientific advice on conservation measures); the effects on fish stocks of water movements and physical and chemical changes in the water; use of electro-acoustic apparatus for detection of fish, fish counting and for other research activities; acoustic and radio fish tracking tags; biology of fish and shellfish, including their migration and behaviour; biology of planktonic and benthic animal and plant communities; fish and shellfish culture and diseases; effects of radioactive waste disposal on both the environment and man; effects of non-radioactive waste and effluent disposals on the environment and its flora and fauna; permissible levels and rates of waste disposal; climatic effects on the marine environment.

Publications: Report of the Directorate of Fisheries Research, triennially or biennially.

See separate entries for:
Fish Diseases Laboratory
Fisheries Experiment Station
Fisheries Laboratory (Burnham on Crouch)
Fisheries Laboratory (Lowestoft)
Torry Research Station.

FOOD SCIENCE DIVISION *   277

Address: Great Westminster House, Horseferry Road, London, SW1P 2AE
Telephone: (01) 216 6225
Telex: 21271 MAFWSL
Head of Division: P.J. Bunyan
Activities: The division provides the scientific and technological background on which Government food policy and legislation are based. Advice is given on the chemical and biological aspects of the composition of food, food additives and contaminants, and on various aspects of human nutrition, including the interpretation of dietary statistics and the initiation of studies on the treatment and composition of foods.

See separate entry for: Food Science Laboratory

VETERINARY INVESTIGATION   278
SERVICE

– VIS
Address: Hook Rise South, Tolworth, Surbiton, KT6 7NF, Surrey
Telephone: (01) 337 6611
Telex: 22203/4
Divisional veterinary officer: Dr W.A. Watson
Activities: The VIS comprises 23 separate centres distributed throughout England and Wales. Its major functions are to provide: a diagnostic, advisory, and consultative service; support for statutory and health schemes; educational facilities for veterinary surgeons and others connected with agriculture; research and development all centres undertake applied research into local disease problems in farm animals.

See separate entry for: Veterinary Investigation Centres

Murphy Chemical Limited   279

Address: Wheathampstead, St Albans, Herfordshire AL4 8QU
Telephone: (058 283) 2001
Telex: 82414
Status: Research centre within an industrial company
Technical director: R.J. Roscoe
Departments: Technical, Dr J.R. Dowsett; chemistry, V.P. Lynch
Activities: Research into plant protection.

National Fruit Trials,   280
Brogdale Experimental
Horticulture Station

Address: Brogdale Road, Faversham, Kent ME13 8XZ
Telephone: (079 582) 5462/3/4
Telex: 96107
Affiliation: Ministry of Agriculture, Fisheries and Food, Agricultural Development and Advisory Service
Status: Official research centre
Director: J. Ingram
Sections: Apples and strawberries, L.H. Clark; pears and raspberries, M.V. Helliar; cherries and plums, D. Pennell; blackcurrants and gooseberries, R.F.V. Williams; genetic banks, H.F. Ermen
Graduate research staff: 6
Projects: Variety trials - apples and strawberries (L.H. Clark); variety trials - pears and raspberries, SARD, rootstock and irrigation trial (M.V. Helliar); variety and cultural trials on cherries and plums, micropropagation (D. Pennell); variety trials on blackcurrants and gooseberries, apple planting systems, cultural trials

(R.F.V. Williams); variety collections (genetic banks) pollination and propagation of fruit crops (H.F. Ermen).
Activities: Evaluation of new varieties of fruit crops and cultural experiments covering rootstocks, crop spacing, pollination, fruit thinning, irrigation, specific apple replant disease, soil management, nutrition, nutrient film culture, and biennial cropping and raspberries and rubus. Micropropagation and propagation from hardwood cuttings. Maintenance of genetic banks of most hardy fruits.
Publications: Annual review.

National Institute of Agricultural Botany 281

– NIAB
Address: Huntingdon Road, Cambridge, CB3 OLE
Telephone: (0223) 276381
Telex: 817455
Affiliation: Ministry of Agriculture, Fisheries and Food
Status: Official research centre
Director: Dr G. Milbourn
Branches: Cereals, W.E.H. Fiddian; grasses and herbage legumes, D.T.A. Aldrich; potato, sugarbeet, oil seeds and fodder, D.E. Richardson; vegetables, T. Webster; plant pathology, Dr R.H. Priestley; chemistry and quality assessment, Dr S.R. Draper; regional trials, W.M. French; statistics and data processing, V. Silvey
Graduate research staff: 85
Activities: The institute is the technical organization in England and Wales which gives impartial advice on all aspects of seed quality - of which one of the most important is choice of variety. The institute's lists of recommended varieties are widely used by farmers.
NIAB tests all new varieties of most agricultural and vegetable crops. Field characters, disease resistance, quality, and yield are assessed by specialists at Cambridge and at 13 regional centres with a range of soil types and weather conditions in the main crop production areas of England and Wales. The new varieties are also tested for distinctness, uniformity, and stability.
Projects: Detection of seed weakness (D.B. Mackay).
Publications: Journal of the National Institute of Agricultural Botany, annually.

National Institute of Agricultural Engineering 282

Address: Wrest Park – NIAE, Silsoe, Bedfordshire MK45 4HS
Telephone: (0525) 60000
Telex: 825808
Affiliation: State-aided institute, Agricultural Research Council; British Society for Research in Agricultural Engineering
Status: Official research centre
Director: Professor Ronald L. Bell
Divisions: Crop engineering, Dr D.S. Boyce; engineering design and construction, L.E. Maher; farm buildings, H.J.M. Messer; glasshouse and scientific information, S.W.R. Cox; machine, T.C.D. Manby; instrumentation and control, M.E. Moncaster; overseas, R.D. Bell; tractor and cultivation, J. Matthews
Graduate research staff: 150
Activities: Research is concerned with agricultural engineering. Seven divisions are engaged on work for British agriculture, the eighth is concerned with problems of agricultural engineering and mechanization for the developing countries. Close liaison is maintained with many organizations in the sphere of research, education and industry both in the United Kingdom and abroad.
Projects:
Crop engineering cultivations - soil compaction (E. Audsley). Crop protection - spray physics, electrostatics (J.A. Marchant); field sprayer design, spray boom stability (H.J. Nation); air-carried sprays (R.P. Sharp). Forage and grain crops - simulation of forage conservation (E. Audsley); agricultural crop drying (Dr M.E. Nellist); simulation of cereal grain drying and harvesting costs (Dr D.S. Boyce, E. Audsley). Operational research - optimun labour and machinery for arable farms (E. Audsley); annual cost of machinery (E. Audsley)
Farm building: livestock - automated animal feeding (H.J.M. Messer); environmental engineering, ventilation and air treatment systems, temperature control, dust filtering (G.A. Carpenter, Dr J.M. Randall); waste engineering, mechanized slurry handling, reduced atmospheric pollution and odour intensity (R Q Hepherd, T.R. Cumby). Buildings - wind loading (R.P. Hoxey); retaining walls for agricultural materials (P. Moran)
Glasshouses: protected crops - efficient use of energy in greenhouses (Dr B.J. Bailey); greenhouse heating (Dr B.J. Bailey); monitoring and control of the greenhouse environment (G.S. Weaving)
Instrumentation and control: forage and grain crops - moisture measurement (G.E. Bowman). Livestock - performance monitoring of livestock (M.J.B. Turner); dairy parlour engineering (Dr A.D. Burgess); statistical time series analysis (Dr P.F. Davies)
Machines: tractors and vehicles - farm materials transport (J.B. Holt); machine dynamic performance

and overload protection (Dr C.J. Chisholm). Cultivations - wear (Dr C.J. Chisholm). Forage and grain crops - performance of crop mowing and conditioning equipment (O.D. Hale, W.E. Klinner); forage chopping (A.C. Knight, W.E. Klinner); application of preservatives during hay harvesting (M.R. Holden, R.E. Arnold). Forage and group crops - methods and machines for compacting and transporting straw (Dr M.J. O'Dogherty, W.E. Klinner); separation of grain from straw and other materials at harvest (R.E. Arnold). Rowcrops - establishment, drilling seeds (F.R. Brown); establishment, transplanting (W. Boa); sugarbeet topping mechanisms (W.P. Billington); harvesting and preparation of vegetables in the packing shed (W.P. Billington, W.Boa). Protected crops - materials handling and allied operations in glasshouses (J.R. Sharp)

Overseas division: agricultural mechanization development programmes including studies of alternative methods of cultivation, harvesting, and threshing, in Paraguay, Sudan, Yemen Arab Republic, India, Mexico, and Sri Lanka; development and introduction of animal-drawn machinery to remove cotton stalks, in Costa Rica and Peru; improved efficiency of animal-powered cultivation in Swaziland; adaptation of small thresher for sorghum; harvesting equipment for rice; collection and harvesting of grass seeds and guinoa.

Tractors and cultivation: tractors and vehicles - tyres, wheels, and traction (Dr D. Gee-Clough); ride vibration (R.M. Stayner); tractor driver monitoring and control, steering (D.J. Bottoms); tractor cab climatic environment (D.H. O'Neill); tractor and machinery noise (J.D.C. Talano); tractor monitoring and control (G.O. Harries); design of tractors and vehicles (Dr M.J. Dwyer). Cultivations - soil dynamics and design of cultivation implements (Dr J.V. Stafford); cultivation implements and seed drills for cereals (D.E. Patterson).

Publications: Programme of Research, annual; biennial reports.

National Vegetable Research Station 283

– NVRS
Address: Wellesbourne, Warwickshire CV35 9EF
Telephone: (0789) 840382
Affiliation: State-aided institute, Agricultural Research Council
Status: Official research centre
Director: Professor J.K.A. Bleasdale
Departments: Biochemistry, Dr R.S.S. Fraser; entomology, G.A. Wheatley; plant breeding, Professor N.L. Innes, Dr A.G. Johnson; Vegetable Gene Bank, Dr D. Astley; plant pathology, Dr R.T. Burchill, Dr J.A. Tomlinson; plant physiology, Dr P.J. Salter, Dr T.H. Thomas; scientific liaison, Dr G.F. Forster, F.K. Nien-

dorf; soil science, Dr D.J. Greenwood; statistics, Dr G.H. Freeman; weeds, H.A. Roberts
Graduate research staff: 90
Activities: Research at NVRS is concerned with improving the efficiency and profitability of vegetable growing, thus ensuring a plentiful supply for consumers. Some of this research is designed to resolve particular problems such as control of pests, diseases, and weeds, or the improvement of agronomic techniques. Longer-term projects seek ways of growing produce which is at present imported.
A Vegetable Gene Bank, funded by OXFAM, is situated at the NVRS.
Projects: Biochemistry - resistance to tobacco mosaic virus in tomato; proteins associated with altered susceptibility to tobacco mosaic virus; biochemical basis of resistance to bean common mosaic virus; radioassay of abscisic acid; gas chromatography of cytokinins.

Entomology - control of carrot fly, cabbage root fly, cabbage white fly, and cabbage aphid; insecticide residues in carrots; incorporation of insecticides into fluid drilling gels; precision equipment for applying granular insecticides; thermal requirements for development of cabbage root fly; monitoring cabbage root fly populations; influence of wind on cabbage root fly dispersal; chemical constituents of crucifers; microbial influence on the incidence of root flies; attractants for onion fly; cabbage white fly - effects on plant growth; lettuce aphids; biochemistry of lettuce in relation to aphid attack; carrot fly distribution; resistance of carrots to carrot fly; resistance of radish and cauliflowers to cabbage root fly; resistance of onions to onion fly.

Plant breeding - breeding systems and interspecific hybrids in brassica, onions, and beans; brussels sprouts; carrots; cauliflower; broccoli; lettuce; onions; phaseolus beans; experimental seed production; multiplication and testing of NVRS cultivars; genetic conservation of crucifers.

Plant pathology - neck rot of bulb onions; botrytis diseases of salad onions; leek rust; canker of brassicas; seed treatment studies; fungal diseases of ornamentals; white rot disease of onions; clubroot of brassicas; brassica seeding establishment; downy mildew of lettuce; halo-blight of beans; rhizobium inoculation of dwarf beans; virus diseases of lettuce; virus diseases of brassicas; watercress chlorotic leaf spot disease; isolation and characterization of a new soil-borne virus; virus disease of ornamentals; virus inactivation; viruses in navy beans; viruses in rhubarb; electron microscopy.

Plant physiology - growth and development studies - peas, navy beans, and onions; flowering in brussels sprouts; distribution of assimilates in root crops; hormonal control of soyabean and wheat growth; hormonal control of celery seed germination; ribosomal RNA integrity in seeds; effects of osmotic seed treatments on germination; effects of calcium peroxide seed treatment on emergence; effect of thermotherapy on seed germination; carrot seed development; carrot seed production;

seed quality in overwintered onions; plant size variation in crops; production of crisphead lettuce; transplant establishment; fluid drilling; commercial assessment of growth regulators; onion storage.

Soil science - computer simulation of crop water extraction; instrumentation for soil compaction studies; cultivation systems; deep fertilizer incorporation; seed and seedling environment; movement of water through seedlings; measurements of root resistance to water flow; measurements of leaf water potential; stern water potential; stern diameter; effect of cellulose xanthate on seedling emergence; breakdown of cellulose xanthate in soil; effects of cellulose xanthate on soil fauna, and on water and nutrient uptake; sap nitrate concentrations in vegetables; effects of early nitrogen stress on yields; potassium requirements of vegetable seedlings; nutrient placement; internal browning in brussels sprouts; cavity spot of carrots; sap analysis for phosphorus, potassium, and sodium; hydrochloric acid extraction of cations from plants; analysis of plant nitrogen and phosphorus; filtration technique for soil extracts; determination of bromide in plants; site-to-site variation in yield; nitrate leaching.

Statistics - hydraulic conductivity and soil-water diffusivity; experimental designs with some balance for neighbours; studies of variable estimates; methods for analysing experiments on soil pests.

Weeds - weed seed populations; population dynamics of annual weeds; inhibition of photosynthesis by methazole; prediction of crop responses to herbicide residues in soil; variability in herbicide application and degradation rate; cellulose xanthate - implications for weed control; herbicide evaluation.

Publications: Annual report; practical guide leaflets for gardeners. Ministry of Agriculture, Fisheries and Food.

Natural Environment Research Council 284

- NERC
Address: Polaris House, North Star Avenue, Swindon, Wiltshire SN2 lEU
Telephone: (0793) 40101
Telex: 444293 ENVRE G
Parent body: Department of Education and Science
Status: Official research organization
Chairman: Sir Hermann Bondi
Secretary to the Council: Dr J.C. Bowman
Second Secretary: Dr P.F.G. Twinn
Director, Science Division: J.D.D. Smith
Director, Scientific Services Division: B.F. Rule; Jonathan Lawson

Members: P. Ackers, Consultant, Binnie and Partners; Professor J.A. Allen, University Marine Biological Station Millport; Professor M.G. Audley-Charles, Queen Mary College, University of London; Professor Sir James Beament, University of Cambridge; Professor R.J. Berry, University College, London; Professor J.C. Briden, University of Leeds; Professor E.H. Brown, University College, London; Dr A.A.L. Challis, Department of Energy; Dr D.S. Davies, Department of Industry; Dr G.A.H. Elton, Ministry of Agriculture, Fisheries and Food; H. Fish, Thames Water Authority; Dr M.W. Holdgate, Departments of the Environment and Transport; Professor B.E. Leake, University of Glasgow; Professor J.L. Monteith, University of Nottingham; Professor W.D.P. Stewart, University of Dundee; G. Williams, United Kingdom Offshore Operators Association Limited; Earl of Cranbrook, Royal Commission on Environmental Pollution; Professor R. Edwards, University College of Wales Institute of Science and Technology; Professor J. Simpson, University College of North Wales, Bangor.

Sections: Science Division - earth sciences, Dr B. Kelk; terrestrial and freshwater life sciences, Dr P.J.W. Saunders; marine life sciences, P. Foxton; oceanography, hydrology and atmospheric sciences, D.B. Smith

Activities: NERC was established in 1965 under the Science and Technology Act with responsibility to encourage, plan, and execute research in physical and biological sciences relating to man's natural environment and its resources. The fields of research can be defined broadly as: the solid Earth - its physical properties and mineral resources; the seas - their characteristics and living and mineral resources; inland waters - their characteristics and living resouces; the terrestrial environment - structure, interactions, and productivity of plant and animal populations and communities; atmosphere - its structure and interactions; and interdisciplinary studies of the physical and biological properties of the Antarctic environment.

The council does research and training through its own institutes and grant-aided associations and by grants, fellowships, and other post-graduate awards to universities and other institutes of higher education.

NERC is uniquely able to provide government, industry, and commerce with highly trained personnel backed by the most modern equipment, data banks, and other facilities for contract work across the spectrum environmental science. Its institutes undertake work on an individual and collaborative basis and can offer either projects or integrated programmes in areas such as resource exploration and evaluation, environmental management, and impact assessment.

Publications: Annual report; research reports
See separate entries for: Component institutes:
British Antarctic Survey
Institute for Marine Environmental Research
Institute of Geological Sciences
Institute of Hydrology

Institute of Marine Biochemistry
Institute of Oceanographic Sciences
Institute of Terrestrial Ecology
Institute of Virology
Sea Mammal Research Unit
Grant-aided institutes:
Freshwater Biological Association
Marine Biological Association of the United Kingdom
Scottish Marine Biological Association
Unit of Comparative Plant Ecology
Unit of Marine Invertebrate Biology

EXPERIMENTAL CARTOGRAPHY UNIT   285

Address: Wingate House, 58 Prospect Place, Swindon, SN1 3LN
Telephone: (0793) 40101
Head: Dr M.J. Jackson
Activities: The unit provides a service to NERC institutes on the application of digital cartography techniques and cooperates with appropriate bodies working in this field. It also carries out research and development in experimental cartography in relation to environmental sciences.

New University of Ulster*   286

Address: Coleraine, County Londonderry BT52 1SA
Telephone: (0265) 4141
Status: Educational establishment with r&d capability
Vice-Chancellor: Dr W.H. Cockcroft

SCHOOL OF BIOLOGICAL AND ENVIRONMENTAL STUDIES   287

Dean: Dr K.L. Wallwork

Department of Biology and Ecology   288

Head: Professor A. Macfadyen
Director of Freshwater Biology Field Station: Professor R.B. Wood
Graduate research staff: 18
Activities: Ecological and physiological research on freshwater and terrestrial systems and organisms with an emphasis towards management and pollution control, systems analysis, and microbial conversion of waste into single cell protein; microstructure, entomology, parasitology, numerical taxonomy, genetics, plankton, fish biology and molluscan behaviour.

Department of Environmental Science   289

Head: Dr J.C. Roberts
Graduate research staff: 7
Activities: The department is conducting a research project into the effect of arterial drainage on the hydrology of the River Main, County Antrim, to establish quantitatively the impact of a major arterial drainage scheme on the stream flow, ground water, soil moisture and water quality characteristics of the upper River Main basin.

Department of Geography*   290

Head: Professor K.M. Barbour
Graduate research staff: 16
Activities: Land use; agriculture; the perception of the environment and the concept of relative deprivation.

Nickerson RPB Limited*   291

Address: Joseph Nickerson Research Centre, Rothwell, Lincoln, LN7 6DT
Telephone: (0472 89) 471
Telex: 527072 AGSERV G
Affiliation: Nickerson Seed Company Limited
Status: Research centre within an industrial company
Research Director: D. Mason
Activities: Cereal plant breeding and evaluation; forage grass breeding and evaluation; oilseed rape and forage legume evaluation; investigations into seed vigour; cereal husbandry, hybrid wheat, hybrid barley and disease resistance breeding and evaluation.

Norfolk Agricultural Station   292

Address: Morley, Wymondham, Norfolk NR18 9DB
Telephone: (0953) 605511
Status: Independent research centre
Director: S.P. McClean
Senior technical officers: Sugar beet, oilseed rape, W.E. Bray; cereals, beef cattle, D.B. Stevens
Graduate research staff: 9
Activities: Sugar beet: weed control, cultivations, methods of improving crop establishment. Cereals: disease control, integration of all appropriate husbandry practices for high input/high output production. Oilseed rape: cultivations. Beef cattle: feeding of arable by-products.
Projects: Weed control in sugar beet, cultivation systems for sugar beet, other aspects of beef husbandry cultivation systems for oilseed rape (W.E. Bray),disease control in cereals, overall management of new cereal varieties feeding, arable by-products to beef cattle (D.B. Stevens).

Publications: Annual reports; *Farm Guide*; *Morley Bulletin*, (magazine for members of the Norfolk Agricultural Station).

North Atlantic Fisheries Commission   293

Address: Room 224, Great Westminster House, Horseferry Road, London, SW1
President: J.C.E. Cardoso

North East London Polytechnic*   294

– NELP
Address: Romford Road, London, E15 4LZ
Telephone: (01) 590 7722
Status: Educational establishment with r&d capability

FACULTY OF ENVIRONMENTAL STUDIES*   295

Address: Forest Road, Waltham Forest, London, E17 4JB
Telephone: (01) 527 2272
Dean: Dr J.L. Taylor

Department of Land Surveying*   296

Head: J. Holloway
Graduate research staff: 6
Activities: Design and survey networks; post processing of inertial survey data; determination by remote sensing of phenomena associated with derelict and reclaimed land.

FACULTY OF SCIENCE*   297

Telephone: (01) 555 0811
Dean: G. Wright

Department of Biology*   298

Head: Dr G. Beedham
Senior staff: Dr S.J. Ball
Graduate research staff: 18
Activities: Areas of research are as follows: animal physiology; algal ecology; biometry; pollution; marine biology; limnology; insect ecology; parasitology; palynology; plant physiology; cryobiology; micropropagation of crop plants; mutation breeding in vitro and in vivo.

Department of Mathematics*   299

Head: Dr Jon V. Pepper
Activities: Research in the following areas: time series; applications of microprocessors; database and accounting; sail design; biometry; economic modelling; integral equations; group representations.

Ordnance Survey*   300

– OS
Address: Romsey Road, Maybush, Southampton, SO9 4DH
Telephone: (0703) 775555
Deputy Director, Planning and Development: Colonel C.N. Thompson
Principal Survey Officer (Development): I.T. Logan
Activities: The Ordnance Survey carries out research and development mainly in connection with its own production requirements. Current development activity is concentrated mainly on enhancement of the digital mapping and on the proposed creation of a national cartographic database. Research is also carried out in geodetic levelling, automation in field surveying, and other related fields.

Paisley College of Technology*   301

Address: High Street, Paisley, Renfrewshire PA1 2BE
Telephone: (041) 887 1241
Telex: 778951 PCT LIB
Status: Educational establishment with r&d capability
Principal: T.M. Howie
Vice Principal: Dr T.C. Downie
Director of Liaison with Industry and Commerce: Dr J.A. Wylie

SCHOOL OF SCIENCE*   302

Dean: Dr R.R. Burnside

Department of Biology*   303

Head: Professor J.C. Smyth
Activities: Interests include: faunal, floral and biochemical aspects of marine coastal and freshwater pollution; thermophilic microorganisms; the biology of arachnids; meiobenthos; freshwater algae; lignin degradation; plant lipids; enzyme inhibition; ecology of non-human primates; arable weed vegetation; genetics of nitrate metabolism in algae.

Department of Chemistry* 304

Head: Professor T.G. Truscott
Activities: Photochemistry of systems of commercial, biological and medical importance; chromatography and lipids in green plants; electroanalytical chemistry; polarography; colloid science; oxide chemistry; antitumour complexes; archaeological textiles; colour; reflectance; general analysis; environmental hazards in industry; industrial hygiene; properties and formulation of plastics; tribology.

Penarth Research Centre 305

Address: Otterbourne Hill, Winchester, Hampshire SO21 3HJ
Telephone: (042 15) 2465
Telex: 477437 Penres G
Status: Independent research centre
Director: Barry A. Richardson
Laboratory manager: T.R.C. Cox
Graduate research staff: 2
Activities: Research in forest products: use of wood and wood products; preservation of wood against biodeterioration due to bacteria, fungi, insects, marine borers; investigation of defects, failures, etc including chemical analysis; development of preservative and biocide formulations for use in foreots, mills, buildings, etc; development of water repellants, decorative systems, fire retardants; assesment of treatment processes.

Pfizer Limited* 306

Address: Sandwich, Kent CT13 9NJ
Telephone: (030 46) 3511
Telex: 96114
Status: Industrial company

AGRICULTURAL PRODUCTS RESEARCH* 307

Director: Dr C.A.E. Briggs
Activities: Veterinary and agricultural research directed towards animal health products for disease control and productivity.

CHEMICAL PRODUCTS RESEARCH* 308

Director: Dr P.G. James
Activities: Fine chemicals for the food industry and for general industrial use. Fermentation products.

Plant Breeding Institute 309

Address: Maris Lane, Trumpington, Cambridge, CB2 2LQ
Telephone: (0223) 840411
Affiliation: State-aided institute, Agricultural Research Council
Status: Official research centre
Director: Dr Peter R. Day
Sections: Cereals, Dr F.G.H. Lupton; forage, oil and potatoes, Dr A.J. Thomson; sugar beet, Dr M.H. Arnold; cytogenetics, Dr C.N. Law; pathology and entomology, Dr M.S. Wolfe; chemistry, D.B. Smith; statistics, R.A.Kempton; physiology, Dr R.B. Austin
Graduate research staff: 85
Activities: The Plant Breeding Institute carries out research related to improvement of agricultural crops by breeding. Special attention is paid to the development of breeding techniques and to the study of factors controlling crop productivity and quality in relation to environment and utilisation. The crops include the cereals; winter and spring wheat, winter and spring barley, and triticale: and the broad-leafed crops; sugar beet, potatoes, oilseed rape, field beans and red clover. Research in cytogenetics, on wheat and its relatives, includes work on chromosome structure and behaviour, qualitative inheritance, developmental genetics, and genetic manipulation by recombinant DNA techniques. Research in plant pathology and entomology is directed to better exploitation of genetic resistance to the major cereal diseases and aphids and includes comparative studies of variety mixtures and pure crops. Research in physiology is concerned with maximising cereal plant adaptation and devising selection procedures for attributes associated with high yield and quality. Breeding is also supported by chemists and statisticians.
Publications: Annual report.

Plymouth Polytechnic* 310

Address: Drake Circus, Plymouth, Devon PL4 8AA
Telephone: (0752) 21312
Telex: 45423
Status: Educational establishment with r&d capability
Director: Dr R.F. Robbins
Deputy Director (Academic): Dr I.C. Cannon
Deputy Director (Resources): Captain G.R. Hughes

FACULTY OF SCIENCE* 311

Dean: Dr K.C.C. Bancroft

Biological Sciences Department* 312

Head: Dr L.A.F. Heath
Reader: Dr M.J. Manning
Senior staff: Dr E.S. Martin, Dr E.B. Dunn, Dr P. Milton, Dr C.B. Munn, Dr R.A. Matthews
Activities: Diseases of fish under fish farming conditions, chemotherapy and the genetics of disease resistance; effects of water quality on fish respiration; population structure of estuarine and marine fish; marine pollution, marine fouling; screening new fungicides and insecticides; crop-water relations and anti-transpirants; influence of industrial effluents, anti-transpirants and herbicide sprays on plant function; forestry management and ecological impact of forestry practices; carboxylating enzymes; biological and integrated pest control; rotation of crops in the control of cabbage root fly; integrated control of brassica pests; honeybee diseases; germination requirements of buried seeds; pod carbon utilization and yield in peas; effects of pollution on stream and estuarine organisms.

Geographical Sciences Department* 313

Head: Dr J.C. Goodridge
Senior staff: Dr B.S. Chalkley
Graduate research staff: 2
Activities: Industrial geography of small firms; urbanization, transport services, the housing market; displacement of non-conforming industry; mining communities; lowland reservoir construction; effects of upland forestry, soil treatments, etc on hydrology, nutrient status and erosion; effects of acid rain on soil weathering; soil properties/hydrological relationships on agricultural land; rainfall variations in an upland area; groundwater investigations in weathered granite; fisheries policy.

Polytechnic of Central London* 314

Address: 309 Regent Street, London, W1R 8AL
Telephone: (01) 580 2020
Telex: 25964
Status: Educational establishment with r&d capability
Rector: Professor Colin Adamson
Pro-Rector: Professor Terence E. Burlin

SCHOOL OF ENGINEERING AND SCIENCE* 315

Address: 115 New Cavendish Street, London, W1M 8JS
Dean: Professor G. Holt
Heads: Engineering, Dr J.H. Gridley; mathematics and computing, J. Ettinger; science, Dr J. Mellerio
Graduate research staff: Engineering, 34; mathematics and computing, 13; science, 70

Research Groups 316

Applied Energy Research Group* 317
Sub-units: Aquatic; Biometric and botanical; microbial ecology; pollution
Activities: Microbial ecology: microbial growth and activity in industrial condenser tubes; effects of oils on the growth and development of freshwater phytoplankton.
Pollution: biological assessment of pollution; role of indicator organisms in fresh waters; effects of run-off and pollutants from roads and vehicles on aquatic ecosystems.
Aquatic: zooplankton; ecology of fish parasites.

Bio-organic Research Group* 318
Activities: Antibiotics; microbial enzymology and biotechnology including the conversion of waste organic matter into useful raw materials; mutagenesis.

SCHOOL OF THE ENVIRONMENT* 319

Address: 35 Marylebone Road, London, NW1 5LB
Dean: F.C. Garrison
Heads: Architecture, A. Cunningham; building, Dr M.S. Romans; civil engineering, J.A. Percival; planning, M. Roberts; surveying, C.W. Parfitt
Graduate research staff: Architecture, 1; building, 10; civil engineering, 13

Research Groups 320

Soil Mechanics* 321
Activities: Soil mechanics; geotechnics.

Polytechnic of the South Bank* 322

Address: Borough Road, London, SE1 0AA
Telephone: (01) 928 8989
Status: Educational establishment with r&d capability
Director: Dr R.J. Beishon

FACULTY OF SCIENCE AND ENGINEERING* 323

Department of Applied Biology and Food Science* 324

Head: Dr P. Wix
Activities: Subjects of research include: mechanism of action of mixed preservatives on food spoilage by yeasts; low water activity problems in respect of yeasts in foods; effect of low water activity of xerophylic fungal growth; vitamin B enhancement of certain Ghanaian foods by yeasts; factors influencing the deterioration of black tea during storage; factors affecting the stability of Betamin when used as a food colour; deodorization of oils and fats; effect on properties of cocoa butter; oxidation studies on sesame seed oil; biology and physiology of some aquatic invertebrates associated with public water supplies.

Polytechnic of Wales* 325

Address: Treforest, Pontypridd, Mid Glamorgan CF37 1DL
Telephone: (0443) 405133
Director: Dr J.D. Davies

FACULTY OF ENVIRONMENTAL STUDIES* 326

Department of Estate Management and Quantity Surveying* 327

Head: A.J. Biker

Department of Science* 328

Head: Dr W.O. George
Graduate research staff: 29
Activities: Research into correlation of heart disease and fat diet; synthetic organic chemistry; spectroscopic studies of hydrogen bonding; variable temperature and matrix isolation of chlorinated hydrocarbons by infrared; low pressure measurements of gases; solar energy conversion; material science of ceramics; nuclear magnetic resonance studies of phospholipid membranes as biological models; anaerobic digestors and methane production; heavy metal and organic pollutants in the environment; coastal processes; ecology of water and forestry systems.

The Polytechnic, Wolverhampton* 329

Address: Molineux Street, Wolverhampton, WV1 1SB
Telephone: (0902) 710654
Status: Educational establishment with r&d capability
Director: G.A. Seabrooke
Deputy Director: C.J.M. Lee

FACULTY OF SCIENCE* 330

Dean: J.A. Sandbach

Department of Biological Sciences 331

Head: Dr F.C. Webber
Senior staff: Biomedical research, Dr N.J. Birch; biochemistry research, Dr M.J. Connock; plant biochemistry, Dr T.J. Hocking; ecology/animal behaviour, Dr J.R. Packham; tropical products, Dr S. New
Activities: Research in the following areas: biomedical - pharmacology and physiology of lithium, trace metals in human disease and following surgery, metal transport in cells in hypertension, rapid microanalysis of serum enzymes and constituents; animal biochemistry - metabolism of, and significance in disease states of, peroxisomes particularly with respect to obesity; development of automated microanalysis for glucose in neonates; plant biochemisty/crop physiology and pathology - increase in yield of wheat, use of agrochemical in treatment of diseases of wheat, carbon dioxide exchanges in crop canopy; ecology - Shropshire flora project, genecology of birh species.

Potato Marketing Board 332

Address: 50 Hans Crescent, Knightsbridge, London, SW1X 0NB
Telephone: (01) 589 4874
Telex: 912192 PMBLDNG
Status: Statutory regulatory body
General Manager: Brigadier E.B. Forster
Department: Research and development, C.P. Hampson; marketing development, J.A. Markham; marketing - overseas relations, W. Sharpe; publicity, R.M. Meredith
Publications: Commercial Assessment of Recently Introduced Potato Varieties, Potato Processing in Great Britain, Potato Statistics Bulletin, all annually; annual report and accounts; *Maincrop Production Techniques in Great Britain,* triennially.

SUTTON BRIDGE EXPERIMENTAL STATION 333

Address: Sutton Bridge, Spalding, Lincolnshire PE12 94B
Telephone: (0406) 350528
Manager: N.H. Twell
Sections: Experimental, T.J. Dent; commercial, N.H. Twell
Activities: The main concern is with performance of stored potatoes under experimental and commercial conditions. Investigations are aimed at reducing losses through disease control by means of chemical treatment and/or environmental control.

Preston Polytechnic* 334

Address: Corporation Street, Preston, Lancashire PR1 2TQ
Telephone: (0772) 51831
Director: Dr H.D. Law

FACULTY OF SCIENCE AND TECHNOLOGY* 335

Dean: J.J. Betts

Sciences School* 336

Head: A.M. Short

Biology Division* 337
Graduate research staff: 7
Activities: Effects of biocides on biodeteriogens, effects of environmental factors on biocidal activity, colonization of inservice timber joinery by microfungi; analysis of foods, effects of diet and exercise on metabolism; measurement of pigment absorption and fluorescence, growth and development of blue-green algae, effects of and tolerance to heavy metals in freshwater organisms.

Processors and Growers Research Organisation* 338

Address: The Research Station, Great North Road, Thornhaugh, Peterborough, PE8 6HJ
Telephone: (0780) 782585
Status: Research association
Director: A.J. Gane
Sections: Agronomy, J.M. King; biology, A.J. Biddle; botany, vacant
Graduate research staff: 6

Pwllpeiran Experimental Husbandry Farm 339

Address: Cwmystwyth, Aberystwyth, Dyfed SY23 4AB
Telephone: (0947 22) 229 and 261
Affiliation: Ministry of Agriculture, Fisheries and Food, Agricultural Development and Advisory Service
Status: Official research centre
Activities: Grassland research for sheep and beef: pasture improvement; effects of improved nutrition on ewe and lamb performance; management of ewes and lambs on improved pasture; wintering systems for ewes; effects on beef production of improved grazing and good quality silage for overwintering.

J.A. Pye Research Centre* 340

Address: Walnut Tree Manor, Haughley, Stowmarket, Suffolk IP14 3RS
Telephone: (044 970) 444
Status: Independent research centre
Agricultural Ecologist, and Research Director: Dr R. Kowalski
Graduate research staff: 4
Research activities: To investigate the ecological, economic and nutritional viability of conventional and alternative agricultural systems.

Queen Elizabeth College* 341

Address: Campden Hill Road, London, W8 7AH
Telephone: (01) 937 5411
Parent body: University of London
Status: Educational establishment with r&d capability
Director: Dr G. Ayrey

BIOLOGY DEPARTMENT 342

Senior staff: Professor P.B. Gahan, Professor G. Chapman
Graduate research staff: 18
Activities: Biochemistry of cereals and vegetable crops; nutrition and growth studies related to fish-farming; plant cytochemistry; diseases of cereal, vegetable, and forage crops.
Projects: Physiology and biochemistry of wheat grains (M. Black, J. Chapman); nutrition, growth, and energetics of grass carp and tilapia (A.E. Brafield); taxonomy of brassica crops using biochemical methods (J.G. Vaughan); effects of growth rate and photosynthesis on cambial activity and wood density, cytodifferentiation of roots and shoots (Professor P.B. Gahan); identification of plant diseases (J.B. Heale).

FOOD SCIENCE AND NUTRITION DEPARTMENT   343

Head: Professor I.D. Morton
Graduate research staff: 8
Projects: Emulsion stability, rheological, and texture problems (Dr P. Sherman); trace constituents in oils, legume based products (Professor I.D. Morton); sensory analysis, food flavours, correlation of chemical and sensory data (Dr G. MacLeod); post-harvest metabolism and deterioration of fruits and vegetables (Dr. S.N. Thorne); formulation and storage of freeze-thaw stable foods (Miss M.A. Hill); interdisciplinary studies on world food production (Dr J. Jones).

Queen Mary College*   344

Address: Mile End Road, London, E1 4NS
Telephone: (01) 980 4811
Parent body: University of London
Status: Educational establishment with r&d capability
Principal: Sir James Menter

FACULTY OF SCIENCE*   345

Dean: Professor V. Moses

Plant Biology and Microbiology Department*   346

Head: Professor E.A. Bevan
Senior staff: Genetics, Professor E.A. Bevan; botany, Professor J.G. Duckett; microbiology, Professor V. Moses
Graduate research staff: 28

Zoology and Comparative Physiology Department*   347

Head: Professor J.D. Pye
Graduate research staff: 24

Queen's University of Belfast*   348

Address: Belfast, BT7 1NN
Telephone: (0232) 45133
Telex: QUBADM 74487
Status: Educational establishment with r&d capability

FACULTY OF AGRICULTURE AND FOOD SCIENCE   349

Address: Newforge Lane, Belfast, BT9 5PX
Dean: Professor C. Dow
Graduate research staff: 69

Agricultural and Food Bacteriology Department   350

Head: Professor A.J. Holding
Activities: Research relating to the quality and safety of human foods and animal feedstuffs; nutrient cycling in soil and freshwaters.

Agricultural and Food Chemistry Department   351

Head: Professor J.R. Todd
Activities: Soil science, plant nutrition; plant and animal biochemistry, animal nutrition; food chemistry.

Agricultural Biometrics Department   352

Head: S.T.C. Weatherup
Activities: Design and analysis of experiments and surveys.

Agricultural Botany Department   353

Head: Professor C.E. Wright
Activities: Plant physiology; grassland agronomy; weed control; plant and seed testing; breeding of new varieties of cereals, grasses, and potatoes.

Agricultural Economics Department   354

Head: Professor G.W. Furness
Activities: Agricultural economics and farm management; marketing and agricultural statistics.

Agricultural Zoology Department   355

Head: R.J. Marks
Activities: Study of insects and nematode pests.

Crop and Animal Production Department   356

Head: Professor J.C. Murdoch
Activities: Cereal and grass production; feeding and management of dairy and suckler cows; finishing beef animals, sheep, pigs, and poultry.

Mycology and Plant Pathology Department   357

Head: vacant
Activities: Fungal and virus diseases of plants.

Veterinary Science Department 358

Head: Professor C. Dow
Activities: Animal diseases including enteritis in young animals, brucellosis, mastitis, hypomagnesaemia, selenium metabolism, Newcastle disease.

FACULTY OF APPLIED SCIENCE AND TECHNOLOGY* 359

Dean: Professor P.P. Benham

Town and Country Planning Department* 360

Head: Professor R.L.G. McKie
Graduate research staff: 5

FACULTY OF SCIENCE* 361

Dean: Professor R.S. Asquith

Botany Department* 362

Address: David Keir Building, Stranmillis Road, Belfast, BT9 5AG
Telephone: (0232) 661111
Head: Professor Eric W. Simon
Sections: Mycology; biochemistry; cell biology; Palaeoecology Centre, B.C.S. Wilson; microbiology, Professor E.W. Simon
Graduate research staff: 19

Zoology Department* 363

Head: Professor Lawrence T. Threadgold
Graduate research staff: 16

Marine Biology Station* 364
Director: Dr Patrick J. Boaden
Graduate research staff: 3

Redesdale Experimental Husbandry Farm 365

Address: Rochester, Otterburn, Northumberland NE19 1SB
Telephone: (0830 20) 608 and 672
Affiliation: Ministry of Agriculture, Fisheries and Food, Agricultural Development and Advisory Service
Status: Official research centre
Activities: Research in the following areas: sheep-management and feeding of hill ewes, hill pasture improvement, early weaning twin lambs, headfly control, scrapie in Swaledale sheep, effect of birthweight on subsequent weaning weight; beef production - nutritional requirements of hill suckler cow and calf, over-wintering and summering of weaned spring-born calves; direct drilled turnips - quick growing modern varieties on rough hill ground.

RHM Research Limited 366

Address: Lord Rank Research Centre, Lincoln Road, High Wycombe, Buckinghamshire HP12 3QR
Telephone: (0494) 26191
Telex: 837445
Parent body: Rank Hovis MacDougall Limited
Status: Research centre within an industrial company
Research director: Professor J. Edelman
Sections: Crop science, Dr J.R.S. Ellis; animal nutrition and toxicology, G. Edwards; aquaculture, Alan Walker
Graduate research staff: 100
Activities: Provision of a comprehensive service for the design and installation of process control systems in food production; design, development, and commissioning of prototype production plant and machinery for food and feed processes; studies for the utilization of natural materials using fermentation pilot plant engineering and related technologies; assessment and understanding of the compositional, sensory, and textural properties of food and food ingredients which affect their quality and storage stability; investigation of the structure, composition, and physical properties of food materials and the manner in which the various components interact to give observed properties and functional behaviour in products; biology of plants, animals, and man in the context of farming, food, and feed production and quality; identification and investigation of new areas of technology and their applicability to RHM commercial operations; analysis and microbiology of foods.
Projects: Disease resistance evaluation of cereal varieties; variety identification by electrophoresis; rat diet for toxicity studies; nutritional requirements of pigs and sheep; dietary fibre; cultivation of eels and other species in industrial waste warm water.
Publications: Research and development report, biennially.

Robert Gordon's Institute of Technology* 367

Address: Schoolhill, Aberdeen, AB9 1FR
Telephone: (0224) 574511
Status: Educational establishment with r&d capability
Principal: Dr P. Clarke
Vice Principal: B.L. Gomes da Costa
Graduate research staff: 32

FACULTY OF ARTS* 368

Dean: Professor R.T. Hart

School of Business Management 369 Studies*

Address: 352 King Street, Aberdeen, AB9 2TQ
Head of School: Professor R.T. Hart
Activities: Impact of oil on local manufacturing and service companies; employment and traditional industries; fish marketing and Aberdeen's position in the distribution network; West German retailing; shopping centre planning applications.

Rosemaund Experimental 370 Husbandry Farm

Address: Preston Wynne, Hereford HR1 3PG
Telephone: (0432 78) 444
Affiliation: Ministry of Agriculture, Fisheries and Food, Agricultural Development and Advisory Service
Status: Official research centre
Director: S.C. Meadowcroft
Deputy director: R.W. Clare
Sections: Hops, F.J. Dickens; grassland and cattle, R. Hardy; cereals and root crops, D. Boothroyd; sheep, D.C. Brown
Graduate research staff: 6
Activities: The farm aims to provide information on all aspects of agricultural production: hops - work on cultivations, use of insecticides and fungicides, and variety trials; grass - manurial trials, control of pests, and silage; cattle - feeding and housing systems; cereals - time of sowing and nitrogen timing trials, applications of fungicides and herbicides; roots - fodder beet, length of growing season, seed priming; sheep - comparison of performance of different breed ewes, feeding and housing trials, and work on prevention of disease in lambs.
Publications: Farm review; hop review, both annually.

Rosewarne Experimental 371 Horticulture Station

Address: Camborne, Cornwall TR14 0AB
Telephone: (0209) 716673
Affiliation: Ministry of Agriculture, Fisheries and Food, Agricultural Development and Advisory Service
Status: Official research centre
Director: M.R. Pollock
Graduate research staff: 6

Activities: Field experiments and/or varietal improvement principally in bulb and other outdoor ornamental crops and in winter maturing cauliflower and early potatoes.
Projects: Bulb crops experiments, (A.A. Tomsett); anemone breeding, (L.M. Gill); Roscoff cauliflower breeding and vegetable crops experiments, (B.H. Houghton); narcissus breeding, (B.M. Fry).
Publications: Rosewarne and Isles of Scilly Annual Review.

Rothamsted Experimental 372 Station

Address: Harpenden, Hertfordshire AL5 2JQ
Telephone: (05827) 63133
Affiliation: State-aided institute, Agricultural Research Council
Status: Official research centre
Director: Sir Leslie Fowden
Departments: Biochemistry, Dr B.J. Miflin; biomathematics, Dr J.A. Nelder; entomology, Dr T. Lewis; farms, R. Moffitt; field experiments, J. McEwen; insecticides and fungicides, Dr M. Elliott; molecular structures, Dr Mary R. Truter; nematology, Dr A.R. Stone; physiology and environmental physics, Dr R.K. Scott; plant pathology, E. Lester; soil microbiology, Dr J.E. Beringer; soils and plant nutrition, Dr P.B. Tinker
Graduate research staff: 325
Activities: Soils and crop nutrition: study of the physical, chemical, and biological factors influencing the genesis of soils and their mineralogical composition; structure and fertility of soils; release and exchange of nutrient ions between soil particles and the soil solution; movement of nutrients, gases, and water both within the rhizosphere and through the soil profile; rhizobia and mycorrhiza in relation to biological fertilization of crops, with emphasis on effective strain recognition, selection, culture, and establishment in field soils.
Plant science: effects of environment on plant growth (plant physics and whole crop physiology); nature of the physiological and biochemical processes governing the assimilation of deposition of photosynthate, including protein, in the different tissues and organs of crops in relation to growth, yield (harvest index) and quality; genetic engineering research, including research on the confirmation of organic molecules able to coordinate cations, ion movement across plant cell membranes, and identification of chemicals useful for the control of plant processes such as transpiration.
Crop protection: basic biology of relevant pest and disease organisms, and of the ecological/environmental factors governing crop infestation or infection, including the monitoring and prediction of situations in which crop damage will result. These biological studies complement a chemically-based programme of research.

Agricultural practices are explored to seek methods of pest and disease control that in the longer-term may reduce reliance on conventional pesticides, whilst immediately effort is concentrated on the more accurate and timely application of safer pesticides to provide adequate control yet lowered environmental pollution. Agronomy and multi-component studies: research on sugar beet physiology and agronomy, and multidisciplinary field experiments on winter wheat, winter barley, and field beans to study the effects of factors, and their interactions, on variations of yields. Biomathematical sciences: support in the design and analysis of laboratory and field experiments: development of computer-based statistical programs; research on statistical theory.
Publications: Rothamsted Report, annually.

BROOMS BARN EXPERIMENTAL STATION   373

Address: Higham, Bury St Edmunds, Suffolk
Telephone: (0284) 810363
Head: Dr R.K. Scott
Study groups: Crop establishment, R.A. Dunning, Dr M.J. Durrant; crop productivity, Dr P.V. Biscoe, Dr K.W. Jaggard; pests and diseases, Dr R.A. Dunning, Dr W.J. Byford
Graduate research staff: 22
Activities: The objective of the station is to improve the yield and productivity of the sugarbeet crop in England through research in the fields of crop establishment, crop productivity, and pest and disease control, together with extensive education and advisory work.
Projects: Beet cyst nematode (Dr D.A. Cooke); weed beet (Dr P.C. Longden); soil structure (K.W. Jaggard); seed and seedling pests and diseases and their control (Dr R.A. Dunning, Dr W.J. Byford); seed quality studies (Dr P.C. Longden); physical and chemical conditions of seedbeds (Dr K.W. Jaggard); virus yellows (Dr G.D. Heathcote); aphid control (Dr G.D. Heathcote); fungal leaf diseases (Dr W.J. Byford); crop growth and productivity (Dr A.P. Diaycott); relationship between bolting and the environment (D.J. Webb); herbicides and weed competition (Dr P.V. Biscoe).

Rowett Research Institute   374

Address: Greenburn Road, Bucksburn, Aberdeen, AB2 9SB
Telephone: (0224) 712751
Affiliation: State-aided institute; Agricultural Research Council; Department of Agriculture and Fisheries for Scotland
Status: Official research centre
Director: Dr W.P.T. James

Sections: Protein biochemistry/lipid biochemistry, Dr G.A. Garton; microbial biochemistry, Dr P.N. Hobson; nutritional biochemistry, Dr C.F. Mills; applied sciences division, Dr A.S. Jones; physiology/surgery/radiography, Dr R.N.B. Kay; energy metabolism/feed evaluation, Dr J.C. MacRae; experimental pathology, Dr B.F. Fell
Graduate research staff: 82
Activities: The institute's principal activities are concerned with the improvement of the productivity of farm livestock (mostly cattle, sheep, and pigs) as it is affected by nutrition. The work involves formulation of diets and their evaluation in practice, as well as fundamental studies related to the physiology, biochemistry, microbiology and nutritional pathology of ruminant and non-ruminant animals.
The Duthie Experimental Farm manages the institute's land, supplies animals, materials, and facilities for activities involving large farm stock, and conducts research projects under practical conditions.
Projects: Structure and biosynthesis of lipids and proteins of animal tissues; nutritive value of proteins for non-ruminants; nutritive value of proteins and other nitrogenous compunds for ruminants; nutritive value of feeds for pigs; nutritive value of energy-yielding feeds for ruminants; minerals as essential and toxic dietary constituents; physiology, biochemistry, and nutrition of the new-born animal; physiology of pregnancy and parturition; pathological conditions associated with diet and animal management; deer and rabbits as meat-producing animals; microbial degradation of feeds and faeces.
Publications: Annual report of studies in Animal Nutrition and Allied Sciences.

Rowntree Mackintosh Limited*   375

Address: York, YO1 1XY
Telephone: (0904) 53071
Telex: 57846
Status: Industrial company
Research Director: J.W. Colquhoun
Activities: Research into problems that affect, or may affect, the operational efficiency, safety and survival of flying personnel; physiological and psychological effects of the flight environment.

Royal Botanic Garden   376

Address: Inverleith Row, Edinburgh, EH3 5LR
Telephone: (031) 552 7171
Parent body: Department of Agriculture and Fisheries
for Scotland
Status: Official research centre
Regius Keeper: D.M. Henderson
Garden Outstations at: Younger Botanic Garden, Benmore; Logan Botanic Garden; Dawyck Arboretum
Graduate research staff: 20
Activities: Taxonomic research including studies on
Ericaceae, Crucifera, Labiatae, Musaceae, Fungi
(agaricales and rusts), and lichens; morphology including ultrastructural studies using scanning and transmission electron microscopy; flora-writing - Bhutan, S W
Asia including Arabia, Brazil, China; general taxonomic
work on ferns, conifers, and cultivated plants.

Royal Botanic Gardens   377

Address: Kew, Richmond, Surrey
Telephone: (01) 940 1171
Affiliation: Ministry of Agriculture, Fisheries and
Food, Agricultural Development and Advisory Service
Status: Official research centre
Director: Professor J.P.M. Brenan
Departments: Herbarium, P.S. Green; library, S.M.D.
FitzGerald; Jodrell Laboratory, Professor K. Jones;
museums division, R.C.R Angel; living collections, divison, J.B. Simmons
Graduate research staff: 105
Activities: The main function of the gardens is to serve
as a research institute for the accurate identification of
plants and for the provision of information in the field of
pure and applied botany. The Herbarium is primarily
engaged on taxonomic research and systematic accounts
of the flora of various parts of the world. Palynological
research is also carried out here, and much work on
plant conservation. Anatomical, cytological, biochemical, and physiological research is undertaken in the
Jodrell Laboratory. (Staff are also working at the
Wakefield Place Gardens, Ardingly, Haywards Heath,
Sussex).
Publications: Annual reports.

Royal Holloway College *   378

Address: Egham Hill, Egham, Surrey TW20 0EX
Telephone: (87) 34455
Parent body: University of London
Status: Educational establishment with r&d capability

BOTANY DEPARTMENT *   379

Head: Professor J.D. Dodge
Graduate research staff: 12

ZOOLOGY DEPARTMENT *   380

Head: Professor C.T. Lewis
Graduate research staff: 11

Royal Horticultural Society   381

– RHS
Address: 80 Vincent Square, London, SW1P 2PE
Telephone: (01) 834 4333
Sections: RHS Garden, Wisley, C.D. Brickell
Graduate research staff: 8
Activities: The aim of the society is the furtherance of
horticultural knowledge in all its forms. Plant protection, propagation, etc are carried out, as is horticultural
research.
Publications: The Garden, monthly.

Royal Institution of Great   382
Britain *

Address: 21 Albemarle Street, London, W1X 4BS
Status: Learned society conducting research
President: His Royal Highness The Duke of Kent
Director: Professor Sir George Porter
Secretary: H.J.V. Tyrrell
Sections: Astronomy, Professor Antony Hewish
Chemistry, Professor Sir George Porter
Experimental medicine, Sir Peter Medawar
Experimental physics, Professor C.A. Taylor
Natural philosophy, Professor David Phillips
Physics, Professor Ronald King
Physiology, Sir David Phillips
History of science, Dr Frank Greenaway
Graduate research staff: 26-30
Activities: Scientific research and the presentation of
science to non-specialist audiences; study of the history
of science.
Publications: Proceedings of the Royal Institution, annually.

Royal Society* 383

Address: 6 Carlton House Terrace, London, SW1Y 5AG
Status: Learned society
President: Lord Todd
Biological Secretary: Sir David Phillips
Physical Secretary: Dr T.M. Sugden
Executive Secretary: Dr R.W.J. Keay
Activities: Natural sciences including mathematics; meetings, publications, international scientific relations, grants and fellowships for research, etc.

Royal Veterinary College 384

Address: Royal College Street, London, NW1 0TU
Telephone: (01) 387 2898
Parent body: University of London
Status: Educational establishment with r&d capability
Dean: Dr A.O. Betts
Graduate research staff: 60
Activities: The College has an extensive research programme including basic and applied studies into problems of veterinary and medical importance.
Publications: Annual report.

CLINICAL STUDIES DIVISION 385

Chairman: Professor J.A. Laing

Animal Husbandry and Hygiene Department 386

Head and Courtauld Professor: Professor J.A. Laing
Projects: Progesterone assay and pregnancy diagnosis, fertility in thoroughbred mare (Professor J.A. Laing); rapeseed meal in pigs diet, influence of trace elements on bacterial disease, determination of selemium by atomic absorption spectrophotometry, influence of nutrition on sheep production (R. Hill); rapeseed meals in ruminants' diets (J.A. Stedman); mechanisms of bacterial pathogenicity, streptococcal infection, cystitis and pyelonephritis in sows, perinatal mortality in sheep, animal health schemes - economics (J.E.T. Jones); reproduction in Dorset Horn sheep, genotype-environment interactions in adult sheep, puberty in sheep (H.W. Williams); mastitis in ewes (A.J. Madel); reproductive disorders in sows, pregnancy detection in the bitch by ultrasound, bacterial endometritis in pigs (M.J. Meredith); genetic control of Perthes disease, chromosome analysis (Heather G. Pidduck).

Medicine Department 387

Head: Professor F.R. Bell
Section: Clinical veterinary medicine, Professor R.H.C. Penny

Surgery and Obstetrics Department 388

Head: Professor L.C. Vaughan
Sections: Veterinary obstetrics and diseases of reproduction, Professor D.E. Noakes
Projects: Ovarian and endocrine activity during induced pseudo-pregnancy in mares, bacterial flora of the canine genital tract, pregnancy diagnosis in bitches, endocrine changes during induced parturition in mares (W.E. Allen); glaucoma in dogs, retinopathies in all species, electroretinography (P.G.C. Bedford); equine anaesthesia and premedication, neuromuscular blocking agents in dogs (Kathleen W. Clarke); arthroscopy, cardiovascular surgery, compression fixation of fractures, hip joint replacement in dogs (D.G. Clayton-Jones); equine colic, carbon fibre in tendon repair and replacement (G.B. Edwards); control of wound infection, pharmokinetics and toxicity of phenylbutazone in horses, ischaemia in equine colic (E.E.L. Geering); reproduction in the male dog, placental transfer in the ewe, reproductive physiology, renal function, and mineral metabolism, uterine mobility and parturition in the sow, parturition and dystocia in sheep and cows, foetal and embryotic death in sheep and cows (D.E. Noakes); genetic control of Perthes disease (P.M. Weddon); muscle and tendon injuries, and growth plate disturbance, in dogs (L.C. Vaughan); facial sinuses of cats (Arlene Coulson); canine blood bank (N.J.H. Sharp).

PARACLINICAL STUDIES DIVISION 389

Chairman: Professor E. Cotchin

Microbiology and Parasitology Department 390

Head: Dr I.M. Smith
Sections: Bacteriology; parasitology; virology; gnotobiote
Activities: Respiratory bacterial infections of pigs, horses and birds; helminth infections of cattle and other domesticated animals.
Projects: Viral and bacterial infections of respiratory tract in horses (N. Edington); atrophic rhinitis in pigs, Bordetella bronchiseptica in pigs, influence of trace elements on systemic infections in mammals and birds (I.M. Smith); porcine cytomegalovirus, malignant catarrhal fever, equine herpesvirus 1 (N. Edington); helminth parasites in pigs and dogs, hookworm infection in dogs (D.E. Jacobs); nematode parasites in cattle (M.T. Fox); sewage sludge and parasite transimission (Ann Dean); epidemiology of Toxocaris canis infections (Elizabeth

Pegg); Newcastles disease virus (P.H. Russell).
Publications: Annual report.

Pathology Department   391

Head: Professor E. Cotchin
Projects: Tumours of captive wild animals (E.C. Appleby); pathology of adrenal glands of pigs (J.B.A. Smyth); neoplasms in domesticated mammals, equine pathology (E. Cotchin); liver damage in the rat (J.G. Evans); biochemical monitoring of tumours in dogs, membrane bound enzymes as indicators of toxic and pathological change (P.W. Gould); pathology of respiratory disease, arthritis in small animals (A.H.S. Hayward); renal disease in cats and dogs (M. Sabri); Salmonella in horses (P.L. Ingram); Staphyllococcus hyicus and 'greasy pig disease' (R.D. Hooper); immunopathology of African trypanosomiasis, canine nasal lesions, paraquat poisoning in dogs and cats (J.A. Longstaffe); experimental toxicology (R. Schoental); porcine atrophic rhinitis (D. Silveira).

PRECLINICAL STUDIES DIVISION   392

Chairman: vacant

Anatomy Department   393

Acting Head: Dr R.F.S. Creed
Projects: Impotency in the bull, structure and function of penes in domesticated mammals (R.R. Ashdown); reproductive patterns in wild mammals (R.F.S. Creed); clinical infertility of dogs, radiological study of quail eggs (S.P. Dean); environmental factors affecting structure of respiratory tract (S.H. Done); spermatogenesis in bull and boar (J.L. Hancock); grass sickness in horses (Norma P. Hodgson).

Physiology and Chemistry Department 394

Head: Professor M.G.M. Jukes
Projects: Respiration, acid base regulation, and hatching in birds (C.M. Dawes); cytochrome oxidase (F.A. Holton); toxicology of natural and man-made materials in domestic and other animals (D.J. Humphreys); processing of nociceptive information in the spinal chord (P.M. Headley); physiology of diving animals, neural substrates of sexual behaviour in male rats (M.C.M. Jukes); chemical restraint in the pig, pharmacokinetics and toxicology of phenylbutazone in the horse (P. Lees); anaesthetics and neurotransmitters (D. Lodge); hotophosphorylation and light induced enzyme activities in chloroplasts (R.H. Marchant); hepatic biochemistry and function using the perfused liver (P.A. Mayes); salt appetite, electrolyte physiology, Na-k adenosine triphospathease, and hypertension (A.R. Michell); nervous and muscular systems of molluscs

(Jennifer M. Plummer); metabolism of adenosine triphosphate (D.D. Tyler); semen preservation and reproductive physiology (P.F. Watson).

ROYAL VETERINARY COLLEGE FIELD   395
STATION*

Address: Hawkshead House, Hawkshead Lane, North Mymms, Hatfield, Hertfordshire AL9 7TA
Telephone: (070 72) 55486

Royal Zoological Society of   396
Scotland*

Address: Scottish National Zoological Park, Murrayfield, Edinburgh EH12 6TS
Telephone: (031) 334 9171
Status: Learned society
Director: R.J. Wheater
Activities: Promotion of zoological sciences; the conservation of Scottish wild animal life; development and maintenance of the National Zoological Gardens of Scotland; all aspects of zoological education research.

School of Agriculture,   397
Aberdeen

Address: 581 King Street, Aberdeen, AB9 IUD
Telephone: (0224) 40291 488291
Telex: 73538
Note: The School of Agriculture, Aberdeen is made up of the Department of Agriculture, University of Aberdeen, in association with the North of Scotland College of Agriculture.
Status: Educational establishment with r&d capability
Principal: Professor G.A. Lodge
Graduate research staff: 150
Activities: Drainage; crop husbandry; plant protection; livestock husbandry and nutrition; veterinary medicine; agricultural engineering and building; agricultural economics and marketing.
Publications: Annual report.

AGRICULTURAL ECONOMICS DIVISION 398

Head: Professor J.S. Marsh
Activities: Profitability of farming; marketing situation and outbook studies; college farm accounts studies; effects of capital taxation on agriculture in Scotland; design and use of a series of modal farms representing typical situations in the north of Scotland; evaluation of the farm and horticulture development scheme within the north of Scotland; market analysis and feasibility studies; farmers retirement decisions; computerized diet formulation; evaluation of viability of the scottish pig sector; whole crop cereal harvesting; evaluation of heifer improvement scheme in Western Isles; dairy cow control programme.

ANIMAL PRODUCTION AND HEALTH GROUP 399

Chairman: Professor J.F.D. Greenhalgh

Animal Husbandry Division* 400

Head: Dr M. Kay
Activities: Effect of anabolic steroids on the growth of beef cattle; response of suckler cow genotypes to winter environment and nutrition; dairy cattle; sheep.

Poultry Husbandry Divsion 401

Head: W. Michie
Graduate research staff: 4
Activities: Broiler, turkey and guinea fowl dietary studies; egg production.

Veterinary Hygiene and Physiology Division 402

Head: D.C. Macdonald
Graduate research staff: 4
Activities: Response of suckler cow genotypes to winter environment and nutrition.

Veterinary Investigation Division 403

Head: L.G. Donald
Graduate research staff: 30
Activities: Maintaining a low incidence of sub-clinical mastitis in the absence of dry cow therapy; biochemical examination of milk from suckler cows with clotting abnormalities; trial of a vaccine designed to control necrotic enteritis in pigs; veterinary diagnostic service; veterinary investigation and advisory service; brucellosis eradication; radiographic studies of pathological changes in the foetus and neonate of farm animals.

CHEMISTRY AND MICROBIOLOGY GROUP 404

Chairman: Dr J.H. Topps

Bacteriology Division 405

Head: Dr A.M. Paton
Graduate research staff: 4
Activities: Maintaining a low incidence of subclinical mastitis in the absence of dry cow therapy; survey of udder washing techniques and water supplies; rapid methods for bacterical identification; response of suckler cow genotypes to winter environment and nutrition; agricultural waste treatment processes; incidence of Salmonella on agricultural land; infection of potatoes with blackleg - its source, extension and control; use of polyacrylic resins for the incorporation of seed dressings; direct epi-fluorescent technique for counting bacteria; control of seed-borne bacterial diseases; non-pathogenic infection of plant tissue by L-phase bacteria.

Chemistry and Biochemistry Divsion 406

Head: Dr J.H. Topps
Activities: Food intake, milk yield and weight changes of dairy cows as affected by condition at calving; anabolic compounds in cattle and sheep; performance of calves on different suckling regimes; land drainage studies.

CROP PRODUCTION AND PROTECTION GROUP 407

Chairman: D. Morrison

Agricultural Botany Division 408

Head: Dr S. Matthews
Graduate research staff: 15
Activities: Biology and control of foot rotting diseases of cereals; appraisal of fungicides for control of cereal diseases; foliar diseases of barley and their effect on the quality of grain for malting; monitoring crop diseases; weed control in arable crops; weed control in grassland.

Agricultural Zoology Division 409

Head: G.F. Burnett
Graduate research staff: 9
Activities: Broad initiation within the winter cluster of the honeybee; insecticide treatment on yield of grassland (leys); sward deterioration and renovation; monitoring crop pests; potato aphids; plant parasitic nematodes; selective assessment of control measures for crop pests.

Crop Husbandry Division   410

Head: D. Morrison
Activities: Cereal variety testing; response of cereals to fertilizer; cereal husbandry; appraisal of fungicides for control of diseases in cereals; grain quality monitoring; screening of crop protection chemicals; forage crop variety testing; forage crop husbandry; potato variety testing and husbandry; national list trials to evaluate new varieties for cultivation and use in the United Kingdom.

Grassland Husbandry Divsion   411

Head: G.J.F. Copeman
Activities: Screening and testing herbage species, varieties and mixtures including National List Trials; salt spray damage; turf cultivar testing; effect of silage making techniques on silage quality and on subsequent performance of beef cattle; sward deterioration and renovation; relationships between herbage yield and quality; grazing systems for beef and sheep production; establishment of grass and assessment of grass production assessment on machair; development of tetraploid red clover; agronomy of white clover.

Horticultural Division   412

Head: Dr G.R. Dixon
Activities: Vegetable cultivar trials; vegetable husbandry; soft fruit cultivar trials; soft fruit husbandry; rose rootstocks; bulb husbandry; nursery stock production; propogation of virus free narcissus; protected crop production; shelter belts.

ENGINEERING AND BUILDINGS   413
GROUP

Chairman: J.G. Shiach

Engineering Division   414

Head: J.G. Shiach
Graduate research staff: 4
Activities: Anaerobic digestion; energy saving equipment for farm dairies; forsinard, reclamation of blanket peat and grass establishment for cutting and drying.

Farm Buildings Division   415

Head: W.A.G. Gerrie
Graduate research staff: 2
Activities: Mesh floor development for pig pens.

SCOTTISH FARM BUILDINGS   416
INVESTIGATION UNIT

Head: S.H. Baxter
Graduate research staff: 7
Activities: Housing requirements for nutritionally deprived piglets; design and layout of farrowing pen components; farm building costs and indices; beef housing handbook; parturition in the sow, the effect of environmental constraints imposed by housing; feeding environment of the pig; reduction of piglet mortality using a farrowing box; effect of stocking density and group size on the economic performance and behaviour of growing/finishing pigs; injury to livestock in buildings; safety of man in livestock housing; automatically controlled natural ventilation; glass reinforced concrete as a flooring material in livestock buildings; thermal simulation of a suckler cow; estimating accomodation requirements for a sow herd; relationship of voluntary energy intake of newly weaned piglets to their thermal environment; weaned rabbit production; sudden deaths in 'sweat-box' piggeries; biological basis for the design of animal space in farm buildings.

WOLFSON UNIT OF SEED   417
TECHNOLOGY

Telephone: (0224) 40291
Telex: 73538
Director: Dr S. Matthews
Graduate research staff: 6
Activities: Fundamental and applied research of relevance to seed technology, and where appropriate, the transfer of information to the seed industry. Topics include - seed physiology, with particular reference to differences in seed quality which cause problems in field emergence and seed storage; physiology of seed ageing, in particular membrane deterioration; reversal of ageing in seeds at high moisture contents, physiological background and the application of hydration treatments to improve seed vigour; non destructive tests of seed viability; development of non-pesticidal chemical treatments to improve the performance of crop seeds, particularly soya beans.

Scottish Agricultural Industries, Process and Product Development Department

418

Address: 124 Salamander Street, Edinburgh, EH6 7LA
Telephone: (031) 554 3156
Status: Research centre within an industrial company
Manager: Dr E. Davidson
Graduate research staff: 8
Activities: Research and development into fertilizer products and processes.

Scottish Crop Research Institute

419

– SCRI
Address: Mylnefield, Invergowrie, Dundee, DD2 5DA
or: Pentlandfield, Roslin, Midlothian EH25 9RF
Telephone: Invergowrie - (082 67) 731; Pentlandfield - (031) 445 2171
Affiliation: State-aided institute, Agricultural Research Council; Department of Agriculture and Fisheries for Scotland
Status: Official research centre
Director: Professor C.E. Taylor
Sections: Crops research, Dr P.D. Waister. Plant breeding - cereals, A.M. Hayter; chemistry, M.J. Allison; forage brassicas, A.B. Wills; potatoes, G.R. Mackay; soft fruit, Dr D.L. Jennings. Mycology, R.A. Fox. Virology, Dr B.D. Harrison. Zoology, Dr D.L. Trudgill
Graduate research staff: 70
Activities: The work of the SCRI is to improve the productivity and quality of crops, particularly those grown in northern Britain, by studying their breeding, culture, and protection from diseases and pests; fundamental research is also done which contributes to the establishment of scientific principles. Potatoes, spring barley (especially for malting), forage brassicas (especially swedes, rapes, and kales), raspberries, and blackcurrants from the major crops in the research programme.
The institute was formed by the merging of the Scottish Plant Breeding Station at Pentlandfield and the Scottish Horticultural Research Institute. The former organization was managed by the Scottish Society for Research in Plant Breeding. That society will continue in existence as a support organization for the new institute. It is intended that the work carried on at the Pentlandfield station will be transferred to Invergowrie over the next 2 to 4 years.
Publications: Annual report.

Scottish Institute of Agricultural Engineering

420

– SIAE
Address: Bush Estate, Penicuik, Midlothian EH26 OPH
Telephone: (031) 445 2147
Affiliation: State-aided institute, Agricultural Research Council; British Society for Research in Agricultural Engineering
Status: Official research centre
Director: Dr D.P. Blight
Sections: Agriculture, Dr B.D. Soane; engineering, Dr D.P. Haughey; instrumentation, Dr R. Parks
Graduate research staff: 50
Activities: The institute is concerned with the application of engineering to agriculture, the effect of machinery on soil or crop, new techniques for the use of existing machinery, and the development of new machinery.
The current programme is particularly concerned with: Mechanization of the growing and handling of the potato crop - development of a new planter; development of better ridging equipment; haulm pulling; design and construction of improved harvesting machinery; equipment and packaging to reduce tuber damage; methods of tuber sizing and inspection. Behaviour of tractors and machinery on sloping ground - reasons for control and stability loss; instrumentation to predict the limits of safe working; mathematical modelling of the behaviour of tractors and associated equipment on hill slopes; design criteria for safe and effective working on slopes. Design of equipment for the mechanical harvesting of raspberries. Equipment and techniques for the measurement of soil compaction under tractors and heavy implements; techniques for avoiding or reducing soil compaction, the design of equipment, and development of techniques for direct-drilling cereals and the classification of soils and localities according to their suitability for direct drilling. Drying of hay and grain with air at near-ambient temperatures, including the use of solar collectors, and the mathematical modelling of the grain drying process. Application of automatic controls to farm machinery, including survey and telemetry equipment to establish working loads on machine components and frequency of operation of manual controls.
Projects: Development of coulters for introducing seeds, etc, into cultivated and uncultivated ground (J.A. Pascal); effect of compaction and traffic intensity on crop yields (D.J. Campbell); factors affecting safety of tractors and machines on sloping land (H.B. Spencer); experimental work on hay and grain drying and the use of solar energy for drying (P.H. Bailey); development of machinery for all aspects of potato crop mechanisation (D.C. McRae); automatic survey of machine performance and survey of tractor use (J. Palmer); identification of disease and damage in produce by optical, spectral and acoustical methods (R.L. Porteous);

mechanical harvesting of raspberries (A.M. Ramsay).
Publications: SIAC Biennial report; technical reports.

Scottish Marine Biological Association   421

– SMBA
Address: Dunstaffnage Marine Research Laboratory,
PO Box 3, Oban, Argyll PA34 4AD
Telephone: (0631) 62244
Telex: 776216
Affiliation: Grant-aided institute, National Environment Research Council
Status: Official research centre
Director: Professor R.I. Currie
Assistant Director: Dr S.O. Stanley
Graduate research staff: 60
Activities: The association is mainly concerned with understanding processes which control the marine ecosystem, particularly in Scottish coastal waters and the adjacent part of the North Atlantic Ocean. The emphasis is on an experimental approach but a certain amount of survey work is conducted. Main activities include: physical oceanography of the Rockall Channel, continental shelf and coastal waters west of Scotland; ecology of the deep water of the Rockall Channel and continental slope; phytoplankton and primary production; ecology of marine sediments; fish physiology and behaviour; aquaculture; marine pollution and environmental impact of industrial developments. Contract work is undertaken in the UK and overseas.
Publications: Annual report.

Sea Fish Industry Authority   422

Address: Sea Fisheries House, 10 Young Street, Edinburgh, EH2 4JQ
Telephone: (031) 225 2515
Telex: 727225 WFA G
Status: Official research centre
Chief technical officer: Dr N.M. Kerr
Marine farming units: Hunterston, K.T. Howard, S.T. Hull; Ardloe, M.J.S. Gillespie
Graduate research staff: 15 (1981)
Activities: Dover sale and turbot farming: refining of techniques, improving performence, and proving commercial viability by running a pilot scale production module; scallop farming: development of systems suitable for United Kingdom conditions - prediction and collection of scallop spat outfalls, assessment of various ongrowing methods; preliminery trials to assess halibut farming's potential.
Publications: Annual report.

SEA FISH INDUSTRY AUTHORITY   423
INDUSTRIAL DEVELOPMENT UNIT

Address: Saint Andrew's Dock, Hull, North Humberside
Telephone: (0482) 27837
Telex: 527261
Research director: R. Bennett
Activities: Research into new or alternative fish stocks; development of new and improved instruments, fishing methods and gear, mechanization on board and performance trials on vessels and machinery; application of operation techniques in fisheries; processing and distribution.

Sea Mammal Research Unit   424

– SMRU
Address: c/o British Antarctic Survey, High Cross, Madingley Road, Cambridge, CB3 0ET
Telephone: (0223) 311354
Affiliation: Component institute, Natural Environment Research Council
Status: Official research centre
Director: Dr R.M. Laws
Graduate research staff: 12
Activities: The Unit has a major responsibility for providing scientific advice to government departments, international agencies, and others on conservation and management issues relating to seals and whales. Research topics include; assessment of size and status of British seal populations and certain whale stocks; evaluation of the role of seals and whales in the marine ecosystem; effects of present and future management policies.

Shell Research Limited*   425

Address: Shell Centre, London, SE1 7PG
Telephone: (01) 438 2389
Telex: 22585
Affiliation: Royal Dutch/Shell
Status: Research centre within an industrial company
Chairman: Dr Ir H.L. Beckers
Publications: Shell Agriculture.

AGROCHEMISTRY DIVISION   426

Head: Duncan Colville
Activities: Pesticide research and development - insecticides, herbicides, fungicides, plant growth regulators; forest protection.

SITTINGBOURNE RESEARCH CENTRE* 427

Address: Sittingbourne, Kent ME9 8AL
Telephone: (0795) 24444
Telex: 96181 Shell G
Director: Dr J.A. Abbott
Laboratories: Biosciences Laboratory; Toxicology Laboratory
Activities: Agricultural chemicals; oil production; toxicology; physiology and pathology in relation to agricultural chemicals.

Shipowners Refrigerated Cargo Research Association* 428

Address: 140 Newmarket Road, Cambridge, CB5 8HE
Telephone: (0223) 65101
Telex: 81604 SRCRA
Status: Research association
Director: G.R. Scrine
Activities: Transport of perishable cargoes in containers and refrigerated spaces in cargo ships.

Slough Laboratory 429

Address: London Road, Slough, Berkshire SL3 7HJ
Telephone: (0753) 34626
Affiliation: Ministry of Agriculture, Fisheries and Food, Agricultural Development and Advisory Service
Status: Official research centre
Head: Dr G.H.O. Burgess
Divisions: Biology, Dr G.H.O. Burgess; storage pests, Dr D.A. Griffiths; pest control chemistry, S.G.B. Heuser; microbiology, J.J. Panes; analytical chemistry, P.J. Ballinger
Activities: Biology and control of insects, mites and fungi associated with stored products and their control by means of contact insecticides and fumigants; problems associated with the storage of grain and other commodities.
Projects: Pests of cereals and foodstuffs (C.W. Coombs); pest control, physiology, and behaviour (D.B. Pinniger); pest resistance and toxicology (C.E. Dyte); pest control biology and identification (Dr D.G.H. Halstead); storage mycology (Dr J.H. Clarke); storage chemistry (D.G. Rowlands).
Publications: Pesticide Science; Storage Pests, both annually.

Soil Survey of England and Wales 430

Address: Rothampsted Experimental Station, Harpenden, Hertfordshire AL5 2JQ
Telephone: (05827) 63133
Affiliation: State-aided institute, Agricultural Research Council
Status: Official research centre
Head: D. Mackney
Sections: Field surveys, J.M. Hodgson; supporting research and services, P. Bullock
Graduate research staff: 50
Activities: The overall aim of the survey is to evaluate the soil resources of England and Wales and contribute to their efficient use. In more direct terms its role is to collect and organize information about the soil types of the country in the form of classification, maps, and explanatory texts that can be used in transferring results of research and experience in land use from one area to another, and in planning further research. Specific areas of interest are: soil classification; soil water retention; soil structure; soil information system; soil erosion; applications of soil survey.
Projects: Soil mapping of England and Wales at 1:250 000, the maps to accompany six regional bulletins on soils and their use: northern England (R.A. Jarvis); midland and western England (J.M. Ragg); eastern England (C.A.H. Hodge); south east England (M.G. Jarvis); south west England (D.C. Findlay); Wales (C.C. Rudeforth).
Publications: Annual report; bulletins; special surveys; records; maps.

South Western Industrial Research Limited* 431

– SWIRL
Address: University of Bath, Claverton Down, Bath, BA2 7AY
Telephone: (0225) 63637
Telex: 449097 UOBATH
Parent body: University of Bath
Status: University-based contract research company
Chairman: Professor P.T. Matthews
Directors: Professor J. Black, Professor L. Broadbent, Dr S. Butler, W.G. Carter, P.T. Clother, Dr B. McEnaney, Dr K.G. Major, Professor D.A. Norton, Professor R.E. Thomas
Director and Secretary: M.R. Forsey
Activities: SWIRL has been set up at Bath University to carry out collaborative research and development work for industry. The resources of the technology-based university are at its disposal, and short and long-term project work on a contractual basis can be carried

out, the main fields being: management - organization, marketing, accounting, and business efficiency; engineering - mechanical, electrical, control, chemical, production, and fluid power; architecture and building technology - materials, metals and alloys, ceramics and glass, polymers, and composites; biology - horticulture, biochemistry, bacteriology, and microbiology; pharmacy - pharmaceutical products, quality control; mathematics - numerical analysis, statistics; physics - solid-state, geophysics, optics; chemistry - analytical, physical, organic, and inorganic.

Sports Turf Research Institute*   432

Address: St Ives Research Station, Bingley, West Yorkshire BD16 1AU
Telephone: (0274) 565131
Status: Independent research centre
Research director: J.R. Escritt
Graduate research staff: 7
Activities: Self-financed and contract research on: turfgrass species and cultivars; wear tolerance of turf; fertilizers, pesticides and herbicides for turf; soils for turf; drainage of turf areas; various other aspects of turfgrass science and management.
Publications: Annual journal; *Sports Turf Bulletin*, quarterly.

States of Guernsey Horticultural Advisory Service*   433

Address: Experimental Station, Burnt Lane, St Martin's, Guernsey, Channel Islands
Telephone: (0481) 35741
Status: Official research centre
Director: R.D. Pollock
Graduate research staff: 3
Publications: Annual report.

Stockbridge House Experimental Horticulture Station   434

Address: Cawood, Selby, North Yorkshire
Telex: (075 786) 275 and 276
Affiliation: Ministry of Agriculture, Fisheries and Food, Agricultural Development and Advisions Service
Status: Official research centre
Director: J.D. Whitwell
Deputy director: R.E. Butters
Sections: Glasshouse and protected crops; outdoor vegetable crops; soft fruit crops; rhubarb, outdoor and forced
Graduate research staff: 6
Activities: Experimental and development work in horticulture for the North of England, to serve commercial growers of glasshouse crops, outdoor vegetables, soft fruit, and rhubarb.
Projects: Glasshouse crops - vegetables only: energy saving techniques including use of waste heat; culture in rockwool; comparison of glass and plastic structures; environmental studies on tomatoes cucumbers and lettuce; disease control and soil sterilization; alternative crops, cultural techniques. Outdoor vegetables - cultural techniques on vegetable crops for the fresh market and for processing; weed control; pest and disease control; transplanting techniques and use of multiseeded blocks; study of carrot cavity spot disorder; variety trials in conjunction with the National Institute of Agricultural Botany. Soft Fruit - varieties for the North of England. Rhubarb - testing of varieties bred at Stockbridge House; virus-tested stocks; rapid propagation; forcing techniques.
Publications: Stockbridge House Experimental Horticulure Station Annual Review.

Sunderland Polytechnic*   435

Address: Langham Tower, Ryhope Road, Sunderland, Tyne and Wear SR2 7EE
Telephone: (0783) 76231 and 76191
Status: Educational establishment with r&d capability
Rector: Dr E.P. Hart

FACULTY OF SCIENCE*   436

Address: Chester Road, Sunderland, Tyne and Wear SR1 3SD
Dean: J.F. Reed

Department of Biology 437

Telephone: (0783) 76191
Head: A. Peat
Activities: Animal Parasitology - ecophysiology of parasitic helminths, effects of drugs on parasitic protozoa in blood of mammals; environmental toxicology - studies of polluted rivers, fluoride in vegetation, food chains in fluorosis in mammals; environmental and clinical virology/mycology; insect ecophysiology - biology of parasitic wasps; microbial physiology; plant cell culture; population genetics; ultrastructure of blue-green algae.
Publications: Annual report.

Tate and Lyle Limited, Group Research and Development, Philip Lyle Memorial Research Laboratory 438

Address: Whiteknights, PO Box 68, Reading, Berkshire, RG6 2BX
Telephone: (0734) 861361
Telex: 847915 Talres G
Parent body: Tate and Lyle, Limited
Status: Research centre within an industrial company
Research director: Dr R.C. Righelato
Programmes: Fine chemicals, Dr R.A. Khan; process engineering, Dr M. Donovan; food and biotechnology, Dr C. Bucke; agricultural products, Dr S.G. Lisansky; speciality chemicals, Dr C.J. Lawson
Graduate research staff: 80
Activities: Research into sweeteners and related food products, chemicals and biochemicals derived from carbohydrates, and agricultural and industrial technology for sugar cane and other crops.
Publications: Tate and Lyle's Sugar Industry Abstracts, bimonthly.

TALRES DEVELOPMENT* 439

Address: Knowsley Industrial Estate, Penrhyn Road, Prescot, Merseyside L34 9HY
Telephone: (051) 548 8840
Telex: 627829
Activities: Research into and production of polysaccharides (in cooperation with Hercules Powder Company Limited) and surfactants. The surfactants have applications in detergents, skin cure products, cosmetics, human food, animal feed, and polymers. Sweeteners are being produced from the Thoumatococcus plant.

Terrington Experimental Husbandry Farm 440

Address: Terrington St Clement, Kings Lynn, Norfolk PE34 4PW
Telephone: (0553 828) 621 and 622
Affiliation: Ministry of Agriculture, Fisheries and Food, Agricultural Development and Advisory Service
Status: Official research centre
Activities: Research in the following areas: cereals (mainly winter wheat but also spring barley and spring wheat) - foliar disease control, varietal performance, pest control (especially slugs), cropping sequences, maximizing yields; potatoes (maincrop production) - seed production, varietal performance, fertilizer requirements, sprouting and handling seed, maximizing yield, reducing labour and machinery inputs; sugarbeet - obtaining improved establishment of crops drilled to a stand, pest control; oilseed rape - dates of sowing and varieties; pigs - nutrition of sow and growing pig, rearing systems, long-term cross-breeding programme, disposal and use of slurry.

Thames Polytechnic* 441

Address: Wellington Street, Woolwich, London, SE18 6PF
Telephone: (01) 854 2030
Status: Educational establishment with r&d capability
Director: Dr N. Singer

FACULTY OF ARCHITECTURE AND SURVEYING* 442

Dean: J.D.A. McWilliam

School of Architecture and Landscape* 443

Head: Dr J. Paul
Deputy Head: P. Arvanitakis
Senior staff: Architectural design, J. Lowman; landscape architecture, M.L. Lancaster

School of Surveying* 444

Head: J.D.A. McWilliam
Senior staff: Building economics and costs, vacant; construction technology and building management, S.A. Smith; land use and planning, D.G. Constable
Activities: Current research interests include: urban regeneration with particular reference to London Dockland; construction technology, including processes, management, building services, building maintenance and maintenance management; building economics and costs.

FACULTY OF SCIENCE AND MATHEMATICS*   445

Dean: M.D. Morisetti

School of Biological Sciences*   446

Head: M.D. Morisetti
Deputy Head: Dr B.N. Kliger
Senior staff: Environmental biology, Dr A.R.W. Smith; physiology and cell biology, Dr D.J. Beadle
Activities: Topics currently being investigated include: ultrastructural studies of animal cell systems, eg of the arthropod nervous system; the mode of action of parathyroid hormone; cancer studies using the Erhlich's ascites tumour as a model system; cell cycle studies in developing callus systems; biochemical studies of embryogenesis in plant cell culture; mitochondrial developments in Neurospora crassa; microbial dissimilation of herbicides; the biochemistry and genetics of Acinetobacter species; the phytogeography of Scandinavian montane heath; the development of environmental sensors.

Tolworth Laboratory, Rodent Pests Department   447

Address: Government Buildings (Toby Jug Site), Hook Rise South, Tolworth, Surbiton, Surrey KT6 7NF
Telephone: (01) 337 6611
Telex: 22203
Affiliation: Ministry of Agriculture, Fisheries and Food, Agricultural Development and Advisory Service
Status: Official research centre
Head: D.C. Drummond
Departments: Mammals and birds, E.N. Wright; pest control chemistry, Dr P.I. Stanley
Activities: Investigation of the behaviour, ecology and control of rodents; chemical and biochemical aspects of vertebrate pest control; methods of detecting the presence of pesticide residues in mammals and birds, and of interpreting their significance; the effect of agricultural chemicals on wildlife; biochemical toxicology and field evaluation; chemical analysis of odours from intensive animal houses.

Torry Research Station   448

Address: PO Box 31, 135 Abbey Road, Aberdeen, AB9 8DG
Telephone: (0224) 877071
Telex: 73587 Prefix Message TRS
Parent body: Ministry of Agriculture, Fisheries and Food, Directorate of Fisheries Research
Status: Official research centre
Director: Dr J.J. Connell
Assistant directors: Dr G. Hobbs; Dr R. Hardy
Sections: Fish bacteriology, Mr D.C. Cann; microbial biochemistry, Dr D.M. Gibson; shellfish and quality assessment, Dr A. Aitken; physics, Dr M. Kent; computing and statistics, Mr W.R. Sanders; flavours and pollution, Dr. A. McGill; protein chemistry, Dr I. Mackie; resource development, Dr K. Whittle; engineering, Mr J. Graham; intrinsic quality factors, Dr R.M. Love
Graduate research staff: 21
Activities: The research and development programme is directed towards all aspects of handling, processing, preservation, wholesomeness, and transport of fishery products. Its aim is to fulfill the requirements of both government departments and industry in this area
The major areas of research are concerned with: handling and processing of fishery products, including the design of equipment, investigation of the changes in the properties resulting from handling and processing, and optimizing the use of fishery products; fish qualtiy, including studies of the physical, chemical, and microbiological changes occurring during storage, the effects of preservation techniques on spoilage processes and on food poisoning organisms, and physical, chemical, microbiological, and sensory methods of assessing quality; new fish products including the development of new consumer products and optimizing the use of under-utilized species; fishery byproducts including studies into improving traditional processes such as fish meal production and the development of novel byproducts for either animal or human consumption.
Publications: Annual report.

HUMBER LABORATORY   449

Address: Wassand Street, Hull, HU3 4AR
Telephone: (0482) 27879
Officer-in-charge: Dr J.R. Burt
Sections: Fishery byproducts, J. Wignall; processing technology and product quality, R.M. Storey
Graduate research staff: 12
Activities: The Humber Laboratory provides a base for investigation the particular problems of the Humber fish handling and processing industries and helps maintain contact with areas of the British Isles less easily accessible from Aberdeen. Its research and development programme is an integral part of that of the Torry Research Station, Aberdeen.

Transport and Road Research Laboratory* 450

– TRRL
Address: Crowthorne, Berkshire RG11 6AU
Telephone: (034 46) 3131
Telex: 848272
Parent body: Department of the Environment and Department of Transport - Common Services
Status: Official research centre
Director: R. Biddle
Deputy Director: L.B. Mullett
Assistant Director: W.A. Lewis
Activities: The Laboratory provides technical and scientific advice and information to help in formulating, developing and implementing government policies relating to roads and transport, including their interaction with urban and regional planning. Research is carried out with related activities, in highway engineering, traffic engineering and safety, and in more general transport subjects.

TRANSPORT GROUP* 451

Transport Operations Department* 452

Head: Dr A. Hitchcock

Access and Mobility Division* 453
Head: Dr C.E.B. Mitchell
Activities: The Division's research covers: improvements in processes and practice of planning for personal public transport impact studies; forecasting; social studies of relations between people, activities, land uses and transport provision.

Trawsgoed Experimental Husbandry Farm 454

Address: Trawsgoed, Aberystwyth, Dyfed SY23 4HT
Telephone: (097 43) 307 and 308
Affiliation: Ministry of Agriculture, Fisheries and Food, Agricultural Development and Advisory Service
Status: Official research centre
Activities: Research in the following areas: milk production - herd management, nutrition, grazing systems, and grassland management; beef production - feeding and management of cows and their calves; sheep husbandry; cereals (spring oats and spring barley for stock feed) - variety testing, disease control; pigs - long-term breeding programme, nutrition of sow and growing pig, effluent disposal.

Trent Polytechnic* 455

Address: Burton Street, Nottingham, NG1 4BU
Telephone: (0602) 48248
Telex: 377534 POLNOT G
Status: Educational establishment with r&d capability
Director: R. Hedley
Deputy Directors: J.B. Neilson, J. O'Neill
Academic Registrar: A.E. Foster
Graduate research staff: 65

SCHOOL OF ENGINEERING AND SCIENCE* 456

Dean: R. Stock

Life Sciences Department* 457

Acting Head: C.K. Mercer
Graduate research staff: 17
Activities: Plant pathology, effects of herbicides on plant physiology, farm effluent treatment; microbial metabolism of detergents and cyclohexane, role of transglutaminases in fibrosis, anti-ulcerogenic and anti-inflammatory action of calcitonin, metabolic disorders in childhood, effects of pesticides on insect neurophysiology, fish farming, fish physiology and pathology, mycotoxins, environmental pollution.

Tropical Products Institute 458

– TPI
Address: 56-62 Gray's Inn Road, London, WC1X 8LU
or: 127 Clerkenwell Road, London EC1R 5DB
Telephone: (01) 242 5412; (01) 405 7943
Parent body: Overseas Development Administration, Foreign and Commonwealth Office
Status: Official research centre
Director: Dr E.M. Thain
Deputy directors: Dr N.R. Jones; A.M. Morgan Rees
Departments: Animal products and feeds, J.G. Disney; central operations, Dr J. Nabney; industrial development, D. Adair; marketing and industrial economics, E. Orr; non-food commodities, J.H.S. Green; plant food, D.G. Coursey; storage, D.J.B. Calverley
Graduate research staff: 200
Activities: TPI's research is mainly concerned with the following: storage, processing, and marketing of food grains; preservation, processing, and marketing of horticultural products; processing, preservation, and marketing of fish, including the utilization of by-catch and other waste fish; meat technology, including processing and utilization of abattoir by-products; development of animal feed resources of developing countries; use of insect sex pheromones for pest control; processing,

quality control, and marketing of spices and essential oils; sources of raw material for pulp and paper production in developing countries; incidence and control of mycotoxin contamination of food; processing, quality control, and marketing of oilseeds and edible nuts; beverage crops and plant enzymes.

Publications: Tropical Science, (TPI quarterly journal); biennial report; *Tropical Stored Products Information*, twice yearly; *Tropical Storage Abstracts*, six times a year; *Oil Palm News*, annually.

TROPICAL STORED PRODUCTS CENTRE   459

Address: London Road, Slough, SL3 7HL

UKF Fertilisers Limited   460

Address: Ince, Chester, CH2 4LB
Telephone: (09282) 2777
Telex: 627407
Status: Research centre within an industrial company
Marketing director: S.J. Cooper
Departments: Marketing development, P.A. Squire; field research and development, G.H. Mackenzie
Graduate research staff: 4
Activities: Plant production; yield and quality of grass, cereal, potatoes, oil seed rape; animal production from grass, using sheep and dairy cows.
Projects: Fertilizer use on grass (M. Daly); fertilizer use on arable crops (D. Mitchell).
Publications: Annual report, occasional trial progress reports.

Ulster Polytechnic*   461

Address: Shore Road, Newtownabbey, County Antrim BT32 0QB
Telephone: (0231) 65131
Telex: 747493
Status: Educational establishment with r&d capability
Pro-Rector (Industrial Liaison): R.I. Houston
Pro-Rector (Development): Dr R.H. McGuigan

FACULTY OF SCIENCE*   462

Dean: Professor J.A. Magowan

School of Computer Science   463

Director of studies: Professor R.W. Ewart
Activities: Artificial intelligence; data allocation in a database environment; microprocessors and microcomputers in agriculture; automated programming assessment; security against programming errors; theoretical and practical aspects of systems design.
Projects: Data storage and access strategies, multi-site linking of homogeneous and heterogeneous database systems via osi networks (D.A. Bell); nursing resource management systems (W. Blackburn); security is modular multiprogramming (C.J. Copeland); potential for computerization within the northern health board, application of distributed databases to medical informatics, microcomputer system for administration of TEC and BEC courses (Professor R.W. Ewart); development of distributed concurrent systems and sequence of events recorder (Dr M.E.C. Hull); introduction of computer systems to small business (R.D. McLarnon); authorship of literary texts (Dr M.W.A. Smith).

School of Environmental Sciences   464

Director of studies: Professor B.W. Langlands
Activities: (with the School of Life Sciences) Pike population studies; cystic fibrosis, a biochemical study, serum electrofocussing, isolation and characterization of CF factor protein(s), development and assessment of bioassay systems; physiology of bovine lymphatic smooth muscle; purification and characterization of a glycoprotein which stimulates the proliferation of white blood cells; biochemical determinants of phytoplankton growth; plankton analysis; structure and physiology of parasitic flat worms; analysis of genetic control systems in Aspergillus nidulans; action of inert dusts on plant materials; environmental study on the larger fungi occurring in an area of coniferous forest; polyphosphates in the food industry; microbial polysaccharides; microbiology of liver pâté preparations; lecithiases from milk psychrophiles; frozen food regeneration; depositional environments in the carboniferous limestone and their relation to mineralization.
Projects: Ecology of plant communities in County Antrim (Dr A. Cooper); comparative volcanological studies (Dr B.P. Kokelaar); petrology of basaltic rocks in County Antrim, Iceland and the zeolite zones of Mull (Dr P. Lyle); interpretation of quarternary ice-movements in Ireland (Dr A.M. McCabe); heavy metal pollution of Belfast and Carlingford loughs and the river Lagan, biological assessment of water quality in rivers of northern Ireland (Dr N. Manga); assessment of variations in coal quality (G.H. Nevin); autecology of stellaria media (chickweed) (Dr D.H. Sobey).

School of Life Sciences 465

Director of studies: Dr S.M. Brown
Projects: The physiology of smooth muscle (Dr J.M. Allen); photosynthetic control of growth (Dr R.H. Bishop); pike population studies in Lough Erne (Dr S.M. Brown); a long term environmental study of the large fungi occuring in an area of coniferous forest, effect of pesticides and additives on leaf physiology (Dr D.W. Eveling); biochemical study of colony-stimulating factor (W.S. Gilmore); polyphosphates in meat processing (D.A. Halliday); investigation of the structure and physiology of parasitic flat worms (Dr S.W.B. Irwin); field studies in Capanagh Forest, County Antrim (G.H. McCourt); analysis of genetic control systems in Aspergillus nidulans (Dr W. McCullough); study of the cytolytic staphylococcal ß haemolysin (D.A. McDowell); ultrastructure and physiology of tapeworms from elasmobranch fish (G. McKerr); plankton of Belfast Lough (Dr T.H. Maxwell); cystic fibrosis, a biochemical study (Dr G.B. Nevin); characterization of phospatidylcholine hydrolyzing enzymes from psychrotrophic bacteria, characterization of lactobacilli in vacuum packed bacon, psychrotrophic bacteria in milk and their effects on cheese production (Dr J.J. Owens); experimental validation of Bloom's taxonomy (J. Whiting).

School of Physical Science 466

Director of studies: Professor D.W. Grant
Activities: Photolysis of sulphur-containing amino acids; electrochemistry of organic compounds; investigation of starch degrading enzymes; synthesis and testing of fungicides; some exploratory calculations of the electronic structure and properties of hydrogen-bonded complexes; colorimetric measurement of colour films; synthesis and characterization of inherently coloured and photoresponsive polymers; physics of the upper atmosphere.
Projects: Investigation of microbial starch degrading enzymes (Dr K.R. Adams); interactions of polypropylene additives (Dr M.J. Baillie); nitrification studies in the river Lagan (Dr. W. Byers); formation of acetals of D-galactitol and its derivatives (Dr R.F.J. Cole); electrochemical fixation of carbon dioxide (Dr B.R. Eggins); development of an electrical impedance plethysmograph (G.W.A. Fogarty, A.T. Gardner); some explanatory calculations of the electronic structure and properties of hydrogen-bonded complexes (Dr S.G.W. Ginn); integrated drive circuits for electroluminescent displays (Dr T.W. Hall); colorimetric measurement of colour films (Dr M.B. Meyers); synthesis and characterization of inherently coloured and photoresponsive polymers (J.E. Riordan); measurement of colour(J.E. Riordan); investigation of winds and temperatures in the polar thermosphere (Dr R.W. Smith); photochemistry of penicillamine disulphide and related compounds (J.H. Stewart); stabilization of polypropylene (E. Tyrrall).

Unigate Limited* 467

Address: Unigate House, Western Avenue, London, W3 05H
Status: Industrial company

COW & GATE LIMITED* 468

Address: Cow & Gate House, Manvers Street, Trowbridge, Wiltshire BA14 8HZ
Telephone: (02214) 3611
Telex: 449328 COWGAT G
Chief Scientific Adviser: R.A. Hendey
Chief Chemist: Dr J.V. Stevens
Activities: Paediatric nutrition and adult nutrition; development of nutritional projects and dietetic products.

Unilever PLC* 469

Address: Unilever House, Blackfriars, London, EC4P 4BQ
Telephone: (01) 822 5252
Telex: 28395
Status: Industrial company
Head of Unilever research: Sir Geoffrey Allen
Activities: In 1980 Unilever, incorporating Unilever PLC and Unilever NV, spent £144m on research and development. This sum was divided almost equally between the Central Research Division, based largely in the United Kingdom and the Netherlands, and the development departments of Unilever's operating companies, with laboratories in over 40 countries. The Research Division has a total staff of approximately 4 000, well over 1 000 of whom are graduates. These are mainly concentrated in three large laboratories - Port Sunlight and Colworth in the United Kingdom, and Vlaardingen in the Netherlands - each of which has a scientific staff of approximately 1 000. Smaller research laboratories are also located in Germany (Unilever Forschungs-Gesellschaft mbH, Hamburg), India and the United States.

UNILEVER RESEARCH COLWORTH 470
LABORATORY*

Address: Colworth House, Sharnbrook, Bedfordshire
Telephone: (0234) 781781
Telex: 82229 UNILAB G
Head: Dr D.L. Georgala
Activities: A wide range of research in the food, agriculture, biotechnology, and engineering sciences, particularily devoted to the development of foods (frozen, meat, edible fats, etc), animal feeds and chemicals.

Unit of Comparative Plant 471
Ecology

Address: Department of Botany, The University, Sheffield, S10 2TN
Telephone: (0742) 78555
Telex: 54348 HLSH EFG
Affiliation: Grant-aided institute, National Environment Research Council; University of Sheffield
Status: Official research centre
Director: Dr I.H. Rorison
Sections: Plant growth analysis, Dr R. Hunt; field ecology, Dr J.G. Hodgson; chemical analysis, Dr P.L. Gupta; electronics and computing, Mr F. Sutton
Graduate research staff: 12
Activities: The unit is making a fundamental study of the interaction of plants with their environment and in particular of the mechanisms controlling plant distribution and the structure of vegetation. In doing so, it is increasing understanding of the responses and tolerance of species and of individuals to a range of environments, and formulating predictive approaches to plant competition and distribution. It is also concerned with the transmission of information based on its research to applied scientists and non-biologists dealing with vegetation management and habitat reconstruction. The unit's objectives include recognition of the main pathways of evolutionary specialization in autotrophic plants and description of the role of plant strategies in vegetation processes and in the determination of the structure and species compostion of plant communities. During the current quinquennium (1979-84) our efforts have been concentrated so far on these major projects.
Publications: Annual report.

Unit of Marine Invertebrate 472
Biology

Address: Marine Sciences Laboratories, Menai Bridge, Gwynedd LL59 5EH
Telephone: (0248) 712641
Affiliation: Grant-aided institute, Natural Environment Research Council; University College of North Wales
Status: Official research centre
Honorary Director: Professor D.J. Crisp
Graduate research staff: 16
Activities: The Unit's research is concerned with the biology of marine invertebrate larvae and is basic to problems affecting a number of economically significant areas of marine biology, for example, shellfish culture, marine fouling, and the effects of marine pollution.

University College, 473
Cardiff*

Address: PO Box 78, Cardiff, CF1 1XL
Telephone: (0222) 44211
Affiliation: University of Wales
Status: Educational establishment with r&d capability
Principal: Dr C.W.L. Bevan
Deputy Principals: Professor C. Hooley, Professor J.L. Evans, Professor K.D. George

CARDIFF UNIVERSITY INDUSTRY 474
CENTRE

– CUIC
Chairman: Dr C.W.L. Bevan
Activities: The Centre assists college departments to develop research projects to an industrial scale and helps industry in finding research and consultancy services from within the college. Currently six projects are being evaluated on an industrial scale: a new textile insulating material; roof insulation; anaerobic digestion; high strength steels; building projects made from waste glass; horticultural systems.

FACULTY OF SCIENCE* 475

Dean: Dr W.A.L. Evans
Assistant dean: Dr R. Potter

Department of Chemistry   476

Head: Professor A.H. Jackson
Sections: Organic chemistry, Professor A.H. Jackson; inorganic chemistry, Professor R.D. Gillard; physical chemistry, Professor M.W. Roberts, Professor E. Whittle
Activities: Coordination complexes of the transition elements; optical activity; theoretical chemistry; spectroscopy; kinetics of nucleophilic reactions; biological action of inorganic ions and coordination compounds; synthesis of molecules of pharmacological interest; biosynthesis of haem and chlorophyll; mass spectroscopy of natural products: medical and biological aspects; gas chromatography and mass spectrometry of organic compounds; X-ray crystallography; heterocyclic chemistry; chemistry of brewing; microanalysis; nuclear magnetic resonance spectroscopy; very fast reactions; electron spin resonance; organometallic compounds; polymerization; organic fluorine compounds; photochemistry of vapours and solutions; reactions of nitrogen dioxide; heavy-metal analysis in the environment; gas phase decomposition reactions; enzyme systems; adsorption and heterogeneous catalysis; dental research; gamma-irradiation studies; surface chemistry and catalysts, photoelectron spectroscopy and low energy electron diffraction.

Department of Genetics   477

Chairman: Professor A.G. Smith
Activities: Ecological genetics of plants and animals; behavioural genetics; adaptation of plant populations to contrasting environments.

Department of Physical Chemistry*   478

Head: Professor M.W. Roberts
Senior staff: Dr R.J. Breakspere, Dr J.C. Evans; Dr E.D. Owen
Activities: Coordination complexes of the transition elements; optical activity; theoretical chemistry; spectroscopy; kinetics of nucleophilic reactions; biological action of inorganic ions and coordination compounds; synthesis of molecules of pharmacological interest; biosynthesis of haem and chlorophyll; mass spectroscopy of natural products: medical and biological aspects; gas chromatography and mass spectrometry of organic compounds; X-ray crystallography; heterocyclic chemistry; chemistry of brewing; microanalysis; nuclear magnetic resonance spectroscopy; very fast reactions; electron spin resonance, organometallic compounds; polymerization; organic fluorine compounds; photochemistry of vapours and solutions; reactions of nitrogen dioxide; heavy-metal analysis in the environment; gas phase decomposition reactions; enzyme systems, adsorption and heterogeneous catalysis; dental research; gamma-irradiation studies; surface chemistry and catalysts,

photoelectron spectroscopy and low energy electron diffraction.

Department of Plant Science   479

Head: Professor A.G. Smith
Assistant director of laboratories: R. Harvey
Graduate research staff: 32
Activities: Ecology; plant physiology; genecology; mycology and aerobiology; palaeobotany; molecular biology (including genetic engineering). Current projects: Control of duplication of genetic material (DNA); genetic engineering of plant cells; plant virus replication; gravity sensing and geotrophic control mechanisms in plants; phototrophically triggered responses in near weightlessness (experiment to be flown in Skylab late 1984); relationship between DNA replication and DNA amount in a range of agriculturally important species; waterlogging tolerance, root formation and function.

Department of Zoology   480

Head: Professor D. Bellamy
Sections: Parasitology, Professor D.A. Erasmus
Activities: Aquaculture; population dynamics; community structure; energy flow in ecosystems; ageing and development; comparative endocrinology; entomology; parasitology; genetics; behaviour; biospeleology; taxonomy; digestion; locomotion; heat tolerance; defence; communication.

University College London*   481

Address: Gower Street, London, WC1E 6BT
Telephone: (01) 387 7050
Parent body: University of London
Status: Educational establishment with r&d capability
Chairman: Sir Bernard Waley-Cohen
Graduate research staff: 800 (in all departments)

FACULTY OF SCIENCE   482

Botany and Microbiology Department   483

Head: Professor P.R. Bell
Activities: Research concentrates on higher plants and microorganisms; electron microscope studies; tissue culture; cytochemistry; taxonomy; autecology; genetics; morphology; photosynthesis.

Geography Department   484

Head: Professor W.R. Mead
Senior staff: Professor E.H. Brown, Professor D. Lowenthal
Activities: Economic and social geography: resources and planning; spatial organization of agriculture in developed countries; rural development in Europe and the Third World; peri-urban planning. Urban studies: labour migration; urbanization. Physical geography: geomorphology in semi-arid and arid regions; quantitative study of soils and plant communities; applied climatology and hydrology; management of environmental resources; quaternary technics and palaeogeography.

Zoology and Comparative Anatomy   485
Department

Head: Professor N.A. Mitchison
Senior staff: Biology, Professor M.C. Raff
Activities: Anatomy; morphology; immunology; cell biology; land ecology; parasitology; vertebrate palaeontology.

University College of North   486
Wales, Bangor

Address: Bangor, Gwynedd LL57 2DG
Telephone: (0248) 51151
Telex: 61100
Affiliation: University of Wales
Status: Educational establishment with r&d capability

FACULTY OF SCIENCE   487

Department of Agriculture   488

Address: Memorial Buildings, Deiniol Road, Bangor, Gwynedd LL57 2UW
Telephone: (0286) 51151
Telex: 61100 UCNWSL G
Head: Professor J.B. Owen
Sections: Agronomy, Dr M.B. Alcock; animal health, R.F.E. Axford; animal production, G.L. Williams; farm mechanization and buildings, G. Farley; farm management, T.H. Thomas; agricultural economics, R.W. Howarth; animal nutrition, A.G. Chamberlain
Graduate research staff: 30 (1982-3)
Activities: The aim of the university department is to provide a link with other university departments - forestry, biochemistry/soil science, animal biology and plant biology in the process of tackling agricultural problems of local and world importance. Overseas collaboration through the agricultural development section includes studies of sheep breeding, dairy cattle improvement and cattle nutrition with Syria, Saudi Arabia, Kenya, Egypt and Israel for example. Bangor is a major centre for animal breeding, particularly the application of group breeding schemes in the improvement of hill sheep, lowland sheep, pigs and beef cattle. Other activities include: evaluation of grass varieties and fertilizing procedures through the animal; animal nutrition, ruminent degradation properties of feedstuffs and the computer formulation of diets for cattle and sheep; agronomic studies of forage crops for animal consumption; soil science, drainage and irrigation, land use; agricultural engineering and building; forestry forest products, agricultural forestry.
Projects: Evaluation of animal production from perennial ryegrass varieties (Dr M.B. Alcock); rumen degradability of forages and diet formation (Dr A.G. Chamberlain); group breeding schemes for the improvement of beef cattle and sheep (G.L.I. Williams); complete diets for feeding dairy and beef catle and sheep, use of milk records for improvement of dairy animals (Professor J.B. Owen).
Publications: Research reports, occasional.

Department of Applied Zoology   489

Address: Deiniol Road, Bangor, Gwynedd LL57 2UW
Head: J. Hobart
Graduate research staff: 9
Activities: Plant protection, aspects of veterinary medicine and livestock husbandry related directly to agriculture, forestry, the farm stock.

Department of Biochemistry and Soil   490
Science

Head: Professor D.W. Ribbons
Graduate research staff: 37
Activities: In both teaching and research, the department attempts to relate biochemistry to a wide range of life forms, including microbes, plants and animals, and to show how these interact with their environment, terrestrial and marine. Research activities can be divided into four groups dealing with microbial biochemistry, plant biochemistry, animal biochemistry and soil science. Collaboration is a useful feature of our research programmes, not only between groups within the department, but also with individuals in other departments. There are also several research associations with other institutions in the community. Specific research interests of academic staff are: microbial biochemistry and genetics; plant and animal biochemistry and nutrition; soil science.

Department of Forestry and Wood Science 491

Address: Deiniol Road, Bangor, Gwynedd LL57 2UW
Head: Professor L. Roche
Sections: Forestry operations, genetics, economics, Professor L. Roche; properties and utilization of wood, Dr G.K. Elliott
Graduate research staff: 15
Activities: The relationship between the anatomy and properties of wood; environmental aspects of forestry; forestry management and operations; degradation and properties of wood; technology of wood based composite material manufacture.
Projects: Variation, selection and breeding of tree species (Professor L. Roche); forest mensuration and inventory (Dr J.C. Hetherington); forest economics, recreation and landscape economics (Dr C. Price); watershed management, land use conflicts (Dr D.M. Harding); industrial economics and wood based product marketing (R.J. Cooper); materials science of wood (Dr G.K. Elliott); technology of wood processing (Dr A.J. Bolton, Dr W.B. Banks); physiology of wood development (Dr M.P. Denne).
Facilities: Comprehensive facilities for mechanical testing of wood-based products; pilot plant for wood preservation, drying, and manufacture of composites.
Consultancy: Yes

Department of Zoology 492

Head: Professor J.M. Dodd
Graduate research staff: 39
Activities: Freshwater fisheries are a principal field of study.

School of Animal Biology 493

Graduate research staff: 57
Projects: Relationship of insect pests to forest trees with special reference to Adelges cooleyi (J. Hobart); biological control of red spider mites (Dr J.B. Ford); immune responses of animals to infection with animal parasites especially tapeworms and ticks (Dr I.V. Herbert); leaf cutting ants with regard to attractants and arrestants (Dr J.M. Cherrett); physiology of helminth parasites of livestock (Dr A.J. Probert); flight behaviour of African Armyworm (Dr A.G. Gatehouse); age determination of insects (Dr M.J. Lehane); reproduction and endocrine control of slugs (Dr N.W. Runham); fish ecology, modelling of commercial fisheries (Dr T.J. Pitcher); ecology of running waters (Dr M.A. Lock).

School of Plant Biology 494

Address: Memorial Buildings, Bangor, Gwynedd LL57 2UW
Telephone: (0248) 51151
Telex: 61100 UCNWSL
Head: Professor G.R. Sagar
Senior staff: Professor G.R. Sagar, Professor P. Greig Smith, Professor W.S. Lacey
Graduate research staff: 50
Activities: Herbicides; whole plant physiology, experimental horticulture; weeds biology; plant architecture - modular construction; branching patterns and the invasion of space (particularly related to rhizomatous plants); population in Penicillium claviforme; colony formation and the concept of the individual in various microfungi; resource development in selected organisms of different reproductive type; carbon in plant growth - modelling of carbon allocation and the control of dark respiration rate; physiology of plants infected with biotrophic fungi; physiological ecology; effects of breeding system on the levels, distribution of genetic variation and neighbourhood size in natural populations - surveys of isoenzyme variation by gel electrophoresis and mathematical modelling; co-evolution of plant species. Physiology of grasses and cereals - tiller and stolon interrelationships, photosynthesis, translocation of assimilates and minerals; plant growth regulators; grassland ecology; computer simulation of plant systems; plant competition; population dynamics of biotrophic plant pathogenic fungi, especially rusts and Phytophthora infestans; grassland diseases; genetical analysis of Phytophthora spp; fluorimetric DNA analysis of fungal nuclei; experimental and orthodox taxonomy of bryophytes, bryophyte cytology, distribution in Britain; preparation of a liverwort flora of Great Britain and Ireland; physiology of host-parasite relations; bacterial plant pathogens, especially toxigenic species of Pseudomonas and Xanthomonas; plant growth-promoting rhizosphere bacteria; acclimatization of plants to chilling and freezing temperatures, lipid and protein changes in plant cell membranes at low temperature; mitochondrial respiration; algal ecology; physiological ecology of micro-algae, particularly chlamydomonads; size in diatoms; algal ecology of Australian ricefields. Pattern in vegetation; classification and ordination of vegetation; plant demography; role of predation in population dynamics of plants; demography of leaves - birth and death rates; resource allocation and reproductive strategies; ecological significance of sex; consequences of interactions between plants. Upper Devonian/Lower Carboniferous transition floras in Eire; Lower Carboniferous floras in North Wales; characteristics of silica deposits in grasses and cereals.

University College of Swansea
495

Address: Singleton Park, Swansea, SA2 8PP
Telephone: (0792) 25678
Affiliation: University of Wales
Status: Educational establishment with r&d capability
Principal: Professor R.W. Steel
Vice-Principals: Professor J. Dutton, Professor R.B. Gravenor, Professor D.G. Pritchard

FACULTY OF SCIENCE
496

Dean: Professor D.V. Ager

Department of Biochemistry
497

Head: Professor E.G. Brown
Activities: Biochemistry of plant constituents especially N-heterocycles and lipids; nucleotide analysis by high pressure liquid chromatography; cyclic nucleotides; nitrogen-fixation by blue-green algae; metabolism by marine invertebrates and effects of pollutants.

Department of Botany and Microbiology
498

Head: Professor P.J. Syrgett
Graduate research staff: 10
Projects: Mechanisms of resistance of crop plants to Verticillium wilt disease (Dr J.M. Milton); salt tolerance of herbaceous plants (Dr S.J. Warnwright).
Activities: Plant physiology and biochemistry - metabolic physiology of algae and fungi, seed germination, salt tolerance; mycology and plant pathology - taxonomy; cultural morphology and physiology of vascular parasites.

Department of Genetics
499

Address: Natural Sciences Building West, Singleton Park, Swansea, SA2 8PP
Head: Professor J.A. Beardmore
Activities: Biochemical genetics - flower pigmentation; cytogenetics - population and in vitro studies of mammalian chromosomes; evolutionary genetics - genetic polymorphism and ecological variety, marine ecosystem pollution, enzyme polymorphism, genetic effects of parental age and stabilising selection, population genetics; human genetics - mortality and morbidity in the new born, social and behavioural studies; molecular genetics - conditional lethal mutations, physiological and genetic influences on mutation, recombination and cell death, nuclear DNA content, environmental chemical studies, cancer and DNA metabolism; organelle genetics - plastid inheritance and development, embryo development.

University College of Wales, Aberystwyth
500

Address: Old College, King Street, Aberystwyth, Dyfed SY23 3DD
or: Penglais Campus, Aberystwyth, Dyfed SY23 3BZ
Telephone: (0970) 3177; (0970) 3111
Telex: 35181 ABYUCW-G
Affiliation: University of Wales
Principal: Dr Gareth Owen

FACULTY OF SCIENCE
501

Dean: Professor H. Rees

School of Agricultural Sciences
502

Address: Penglais, Aberystwyth, Dyfed SY23 3DD
Principal: Dr Gareth Owen
Graduate research staff: 50
Publications: Annual report.

Agricultural Botany Department
503

Head: Professor H. Rees
Senior staff: Professor J.P. Cooper
Graduate research staff: 30 (1982)
Activities: Control of chromosome behaviour with particular emphasis upon species hybrids and with the view to improving fertility in hybrids important in plant breeding; epidemiology and control of crop plant diseases; symbiosis of legumes and nitrogen fixing bacteria, constrution of bacterial strains appropriate to particular grassland systems; grassland physiology and management; in collaboration with the Welsh Plant Breeding Station, the regeneration and manipulation of cell and tissue cultures.
Projects: Chromosome evolution and behaviour (H. Rees); fungal diseases of cereals, rhizobium/clover symbiosis (E. Griffiths); heritable changes in linum (A. Durrant); crop physiology (D.B. James).

Agricultural Economics Department
504

Head: Professor D.I. Bateman
Activities: Economics of farming industry (special interests in hills and uplands, in Wales, and in organic farming); marketing of food and agricultural products (special interests in meat processing and distribution, European legislation on competition); economics of land use and the environment; cooperative organizations in agriculture and industry; less developed countries.
Projects: Farm management survey covering 500 farms in Wales (D.A.G. Green); economics of low input farming systems (D.I. Bateman); hills and uplands research (W. Dyfri Jones); agricultural marketing (Michael Haines).

Agriculture Department 505

Head: Professor J.D. Hayes

Activities: Current research in crop husbandry is concerned with grassland, including the effects of nitrogen application and cutting and grazing management, on different grasses and legumes, and with early potatoes and other arable crops in Pembrokeshire, where the effects of seed-crop, husbandry, variety and other agronomic factors are under investigation. Animal research is mainly concerned with the production and health of ruminants, with emphasis on the physiological and environmental factors which affect reproductive performance and output of livestock under farm conditions. A new field of study concerns the breeding and management of horses. Land management and the productivity of agricultural systems in pre-industrial Britain is also under investigation.

Projects: Cultural factors influencing production of potatoes, oil seed rape and cereals (J.D. Hayes); growth and nutritive value of grasses and herbage legumes (D. Wilman); factors controlling reproduction in sheep, selenium deficiency in sheep diets, horse exertion physiology (J.L. Lees).

Biochemistry and Agricultural Biochemistry Department 506

Head: Professor R.B. Beechey (acting)

Graduate research staff: 25

Activities: The composition, structure and function of biological membranes; bacterial metabolism of aliphatic and aromatic compounds, including role of plasmids; biotechnological aspects of microbe metabolism; biochemistry of pigments, including structure elucidation and stereochemistry of carotenoids and formation and function of carotenoids in tetrapyrroles in animals, plants and microbes; photosynthesis in algae and cyanobacteria, particularly structure, function and evolution of electron transfer proteins; mode of action of fungicides and herbicides; ruminant nutrition, particularly nitrogen utilization; aspects of plant nutrition and soil chemistry, with a special interest in sports turf management, sportsfield construction and land rehabilitation.

Soil Science Unit 507

Activities: Chemical, physical and biological factors affecting agricultural productivity.

Projects: Restoration of farmland after opencast mining, sportsfield contruction, influence of earthworms on soil development (Dr V.I. Stewart); mineralogy of soils on lower palaeozoic sediments, hill soils, sports turf management (Dr W.A. Adams).

Botany and Microbiology Department 508

Address: Dyfed SY23 3DA

Sections: Microbiology, Professor J.G. Morris; botany, Professor M.A. Hall

Graduate research staff: 62

Activities: Botany - research into the structure, occurrence and mode of action of plant growth regulators; metabolism and mode of action of ethylene, metabolism of abscisic acid and cytokinins, control of tuberization in potatoes, plant tissue culture.

Microbiology - studies of anaerobic fermentation processes, solvent production by Clostridium acetobutylicum; application of electrochemical methods to microbial systems; development of novel sensors; gene transfer/manipulation in species of Bacillus and Clostridium; applied mycology - biodeterioration, plant pathology; applied phycology, fresh-water algae.

Welsh Plant Breeding Station 509

Address: Plas Gogerddan, Aberystwyth, Dyfed SY23 3EB

Telephone: (0970) 87255

Director: Professor J.P. Cooper

Sections: Developmental genetics, Dr D. Wilson; herbage breeding, Dr E.L. Breese; arable crop breeding, Dr D.A. Lawes; grassland agronomy, Roy Hughes; seed multiplication and herbage seed, D.H. Hides; plant pathology, Dr A.J.H. Carr; cytology, Dr H. Thomas; chemistry, Dr D.I.H. Jones; cell physiology, Professor J. Heslop-Harrison; plant biochemistry, J.L. Stoddart; legume breeding, W.E. Davies

Activities: Improvement by plant breeding of agricultural crops, particularly forage grasses and legumes, and supporting research in genetics and cytology, plant physiology and biochemistry and plant pathology.

Zoology Department 510

Head: Professor Bryn M. Jones

Activities: Animal behaviour, cell biology, development biology, immunology, parasitology, animal ecology, animal physiology, freshwater biology, marine biology.

Projects: Comparative biochemistry of parasitic helminths, particularly energy metabolism pathways and their control (J. Barrett); physiological ecology of marine invertebrates; littoral mollusc studies (J.D. Fish).

University of Aberdeen* 511

Address: Regent Walk, Aberdeen, AB9 1FX

Telephone: (0224) 40241

Telex: 73458 UNIABN G

Status: Educational establishment with r&d capability

DEPARTMENT OF AGRICULTURE   512

Address: School of Agriculture, 581 King Street, Aberdeen, AB9 1UD
Note: The School of Agriculture, Aberdeen is made up of the Department of Agriculture, in association with the North of Scotland College of Agriculture.
Head: Professor G.A. Lodge
Senior Staff: Professor of Agricultural Economics, Professor J.S. Marsh; Professor of Animal Production, Professor J.F.D. Greenhalgh; Professor of Agriculture, Professor A. Martin
Activities: See entry for School of Agriculture, Aberdeen.

DEPARTMENT OF BOTANY   513

Head: Professor Charles H. Gimingham
Graduate research staff: 28
Activities: This is a plant science department, with research actitivites spanning all levels of organization from the molecular to the ecological. The main thrusts are in the fields of plant physiology and ecology. The former includes investigations on the uptake and transport systems in plants (electrophysical aspects of mineral uptake by roots, the scanning and transmission electron microscopy of sieve tubes, also Laser-Doppler microscopy of flow in living cells). Related work includes the preparation of plant protoplasts for studying membrane proteins and lectins by immunochemistry, tissue culture and micropropagation, investigations on the influence of light on plant growth and development, and the mechanism of stomata. The latter (ecology) concerns production and nutrient cycling, including the effects of pollutants such as lead and fluoride; the ecology of mycorrhizas and their influence on the nutrition of forest trees and of orchids; and aspects of vegetation management (eg burning and grazing of heathlands) and conservation. Members of staff participate in an Institute of South-east Asian biology. Research is in progress on the cytology and reproductive biology of economically important tropical plants and on the recovery of tropical forest after disturbance. Other research areas include experimental taxonomy and chemotaxonomy, vegetative and floral morphogenesis, and environmental physiology of conifers.

DEPARTMENT OF FORESTRY   514

Address: St Machar Drive, Aberdeen, AB9 2UU
Telephone: (0224) 40241
Telex: 73458 UNIABN G
Head: Professor J.D. Matthews
Sections: Forest management, M.S. Philip; silviculture and policy, D.G. Cumming; forest pathology and ecology, C.S. Millar; forest harvesting and utilization, J. Henderson

Graduate research staff: 19
Activities: Cytology, genetics and breeding of scottish birches; physiology and genetics of scots pine; tissue culture and cell culture of birch and alder; fuels from forest biomass; forest mycology and ecology; biology of sap sucking insects attacking forest trees; movement of fluids and gases in wood; physical effects of wind and snow on trees; integration of forestry with agriculture in the highland region; design of shelterbelts; economic surveys of private forestry; techniques for forest business management; forest harvesting systems.
Publications: Economic Surveys of Private Forestry, annual.

DEPARTMENT OF MICROBIOLOGY*   515

Address: Marischal College, Aberdeen, AB9 1AS
Head: Dr W.A. Hamilton
Activities: Sulphate reducing bacteria and oil industry, presence and activity in injection waters, corrosion; mechanisms of nitrification in soil; wall synthesis and mechanism of anti-fungal antibiotics; acid tolerance of food spoilage organisms.

DEPARTMENT OF SOIL SCIENCE*   516

Head: Professor Joseph Tinsley

DEPARTMENT OF ZOOLOGY   517

Address: Tillydrone Avenue, Aberdeen, AB9 2TN
Telephone: (0224) 40241
Telex: 73458
Head: Professor W. Mordue
Zootelemetry Research Laboratory Limited: Dr I.G. Priede
Graduate research staff: 80
Activities: Manufacture of radio and acoustic telemetry equipment for use in ecological and physiological research.

Aberdeen University Marine Studies   518
Limited

Telephone: (0224) 491723
Telex: 73458 UNIABN G
Director: Dr R. Ralph
Senior staff: Dr Peter Boyle
Graduate research staff: 7
Activities: Marine fouling of North Sea oil and gas platforms; biology of key fouling species; effects of fouling on corrosion; corrosion protection systems; field trials of coatings and paints in the marine environment; rocky shore survey and monitoring; analysis of tissue hydrocarbon burden; marine biology consultancy services.

**Zootelemetry Research Laboratory 519
Limited***

University of Aston in 520
Birmingham

Address: Gosta Green, Birmingham, B4 7ET
Telephone: (021) 359 3611
Telex: 336997
Status: Educational establishment with r&d capability

FACULTY OF SCIENCE 521

Dean: Professor M.R.W. Brown

Biodeterioration Centre 522

Address: St Peter's College, College Road, Saltley, Birmingham, B8 3TE
Telephone: (021) 328 5950
Director: Dr H.O.W. Eggins
Contract Research Manager: Dr K.J. Seal
Information and Services Manager: Dr D. Allsopp
Graduate research staff: 6
Activities: The centre is associated with the Department of Biological Sciences and deals with all aspects of damage to products, premises and processes caused by living organisms including bacteria, fungi, algae, insects, rodents, birds and plants and also aspects of the upgrading of wastes to produce useful products and control environmental pollution; pest control and food storage.
Publications: Annual report.
Projects: Fungal upgrading of waste straw for animal feed (Dr K.J. Seal).
Additional information: The centre specializes in wood, plastics, engineering fluids, and control of biodeterioration.
Facilities: Unique literature collection on biodeterioration of materials; wide ranging capability to test materials for biological susceptibility.
Consultancy: Yes
Publications: International Biodeterioration Bulletin; biodeterioration research titles; waste materials biodegradation research titles.

Department of Biological Sciences 523

Head: Professor A.J. Matty
Assistant head: Dr N. Bromage
Graduate research staff: 3
Activities: Research is mainly of an applied nature and is carried out in the following areas: aquaculture and freshwater biology (fish endocrinology, physiology and nutrition, ecology of sewage, eutrophication, biotic ind-

ices); general animal physiology (fish and amphibian immunology, cell proliferation, intestinal absorption, carbohydrate metabolism and relationship with diabetes and obesity); microbiology (fungal ecology and physiology, effects of pollution on soil mycology, biology and biochemistry, lichen ecology, edible fungi, mushroom farming, microbiological breakdown of fabricated materials, biodeterioration).

Fish Culture Unit 524

Head: Professor A.J. Matty
Graduate research staff: 20
Activities: Fish culture, freshwater fisheries and production methods for aquaculture including nutritional, endocrinological and other physiological studies of the important farmed species of finfish and also investigations of methods of water reuse, oxygenation and filtration in aquaculture, mushroom science.
Projects: Tilapia nutrition, control of fish reproduction, nutritional studies in salmonid fish, maturation in male salmonids, waste materials and fish nutrition (Professor A.J. Matty, Dr N. Bromage).

University of Bath 525

Address: Claverton Down, Bath, BA2 7AY
Telephone: (0225) 61244
Telex: 449097
Status: Educational establishment with r&d capability

SCHOOL OF BIOLOGICAL SCIENCE 526

Head: Professor A.H. Rose, Biology and horticulture: Professor L. Broadbent
Senior staff: Microbiology, Professor A.H. Rose; biochemistry, Professor P.D.J. Weizman; plant biology, Professor G.G. Henshaw
Graduate research staff: 30
Activities: Food and pesticide microbiology; tree pathology; weed ecology and mechanisms of herbicide action; insect physiology, ecology and control of behaviour; seed physiology and production; biochemistry of plant storage tissues, plant cell and tissue culture; nutrient film production of horticultural crops; agricultural and horticultural cooperation; heavy metal pollution of water; fish hormones and physiology; lipid metabolism and heart disease; mammalian cell membranes; enzymology and regulation of metabolism.
Projects: Tissue culture in relation to crop improvement (Professor G.G. Henshaw, A.W. Flegmann, H. Wainwright); biochemical changes in plant storage (Dr P.P. Rutherford); plant molecular biology (Dr Janet A. Pryke); physiological, biochemical and ultrastructural studies of seeds and seedlings (Dr K.G. Moore); mechanisms of herbicide action, chlorophyll formation and destruction (Dr A.D. Dodge); whole plant physiology of

watercress (Dr L.W. Robinson); seed production and technology (R.A.T. George); nutrient film production and utilization of wastes (D.C. Cull); weed ecology and pesticide application (R.J. Stephens); agricultural and horticultural cooperation, labour use in horticulture (Dr M.J. Sargent); hormonal control of behaviour in insects (Dr S.E. Reynolds); biology of insect guts (Dr A.K. Charnley); water balance of adult locusts (Dr A.K. Charnley); mushroom initiation and development (Dr L. Jacobs); cell wall degrading enzymes of plant pathogens, vascular wilt diseases (Dr R.M. Cooper); viruses of woody ornamentals (Dr R.G.T. Hicks); tree pathology, population and community structure of fungi in decaying wood (Dr A.D.M. Rayner); pesticide effects on soil and aquatic micro-algae and protozoa (Dr S.J.L. Wright); microbiology of eggs and meat products, shelf-life studies (Dr R.G. Board).
Publications: Annual report.

University of Birmingham*   527

Address: PO Box 363, Birmingham, B15 2TT
Telephone: (021) 472 1301
Status: Educational establishment with r&d capability
Vice-Chancellor and Principal: Lord Hunter of Newington

FACULTY OF SCIENCE AND ENGINEERING*   528

Dean: P.A. Garrett

School of Biological Sciences*   529

Chairman: Dr D.A. Wilkins

Genetics Department*   530
Head: Professor J.L. Jinks
Senior staff: Honorary Professor N.L. Innes, Honorary Professor Sir Kenneth Mather
Graduate research staff: 23

Microbiology Department*   531
Head: Professor H. Smith
Graduate research staff: 13

Plant Biology Department   532
Telex: SPAPHYS 338938
Head: Professor J.G. Hawkes
Senior staff: Honorary Professor J.K.A. Bleasdale
Graduate research staff: 13
Activities: Research on plant production; plant breeding; plant protection; crop plant genetic resources; crop physiology agroecology.
Projects: Origins and evolutionary relationships of wild and cultivated potatoes (Professor J.G. Hawkes); investigation of the suitability of True Potato Seed (TPS)

in third world agriculture (Dr M.T. Jackson); cytogenetics of wild Beta species and their use in plant breeding (Dr B.V. Ford-Lloyd); evolutionary relationships of the eggplant (Solanum melongena) (Dr R.N. Lester); long-term storage of seeds for genetic resources purposes (Dr P.M. Mumford); long-term storage of meristems for genetic resources purposes (Dr J.H. Dodds); control of plant enzyme activities with special reference to phytochrome and its influence on mustard seedlings (Dr H.J. Newbury); salt tolerance and aluminium toxicity studies in rice (Dr D.A. Wilkins); ion uptake, with reference to pH and carbon dioxide supply; soil aeration (Dr L.W. Poel).
Publications: Annual report.

Zoology and Comparative Physiology   533
Department*
Head: Professor L.H. Finlayson
Senior staff: Parasitology, Professor J. Llewellyn
Graduate research staff: 17

UNIVERSITY EXTRA-FACULTY   534
ORGANIZATIONS

Biochemistry Department*   535

Head: Professor S.V. Perry
Senior staff: Professor G.A. Gilbert, Professor D.G. Walker
Graduate research staff: 42
Activities: Microbial chemistry; plant biochemistry and photobiology; metabolism of drugs and related compounds; growth and ageing; nucleic acid metabolism.

British School of Malting and Brewing   536
Telephone: (021) 472 1301
Head: Professor J.S. Hough
Senior staff: Cereals biochemistry, Dr D.E. Briggs; microbiology and fermentation, Dr T.W. Young
Graduate research staff: 8
Activities: The school carries out fundamental and applied research (including research training) in malting and brewing along with other aspects of industrial biochemistry. The present areas of research are - cereals biochemistry, especially germination studies; microbiology and fermentation, particularly genetic manipulation of yeasts, zymocide secretion and brewery bacteria; beer proteins involved in beer foam and beer haze; plastic pipes for beer dispensers in public houses.

University of Bradford*   537

Address: Bradford, West Yorkshire BD7 lDP
Telephone: (0274) 33466
Telex: 51309 UNIBFD G
Status: Educational establishment with r&d capability
Vice Chancellor: Professor J.C. West

BOARD OF STUDIES IN   538
ENGINEERING

Postgraduate School of Studies in   539
Control Engineering

Chairman: Professor M.G. Mylroi
Graduate research staff: 12
Activities: The main research effort of the school falls into two areas: computer control of industrial processes, and correlation of noise techniques in measurement, with particular application to flow measurement of difficult fluids, including pollution measurement. Other individual research projects include: computer control of medical problems; simulation studies to investigate long term effects of industrial investment in a mainly agricultural economy; and measurements relevant to the understanding of volcanic mechanisms.
Projects: Time domain and transformed domain techniques for the analysis and design of nonlinear systems (Dr B. Kouvaritakis).

BOARD OF STUDIES IN LIFE   540
SCIENCES

Postgraduate School of Studies in   541
Biological Science

Chairman: Dr M.J. Merrett
Senior staff: Professor E. Lees
Graduate research staff: 10
Activities: Research is currently being pursued in the following areas: algal and plant biochemistry; biochemistry of the cell-cycle; entomology; fish biochemistry; molluscan physiology; actinomycete ecology and taxonomy; parasitology; plant cell organelles; plant ecology; protein biochemistry.
Projects: Synthesis and segregation of organelle proteins in castor bean endosperm (Dr J.M. Lord).

Postgraduate School of Studies in   542
Environmental Science

Chairman: Dr M.R.D. Seaward
Senior staff: Professor M.J. Delaney
Graduate research staff: 10
Activities: Research in the following areas: physical and chemical hazards in the working environment; nature and toxicity of metallurgical fumes; spectroscopic techniques for atmospheric pollution monitoring; odour and microbiological aspects of sludge disposal to land; industrial waste water treatment; biological monitoring of urban and industrial complexes; heavy metal uptake by organisms; recreation ecology of heathlands near conurbations.

BOARD OF STUDIES IN SOCIAL   543
SCIENCES*

Postgraduate School of Studies in   544
Planning*

Project Planning Centre for Developing   545
Countries*
Address: 26 Pemberton Drive, Bradford, West Yorkshire BD7 lRA
Graduate research staff: 20
Activities: Topics of research include: industrial development in developing countries; methods of appraising industrial projects in developing countries; foreign investment projects for industrial development; sources and methods for financing industrial development; planning methodology for capital goods industries and agro-industries in developingcountries; small-scale industrial developments.

University of Bristol   546

Address: Senate House, Tyndall Avenue, Bristol, BS8 ITH
Telephone: (0272) 24161
Status: Educational establishment with r&d capability

FACULTY OF MEDICINE*   547
Dean: Professor D.C. Berry

Animal Husbandry Department   548

Address: Langford House, Langford, Bristol, BSl8 7DU
Telephone: (0934) 852581
Head: Professor A.J.F. Webster
Graduate research staff: 20
Activities: Research on animal production - livestock husbandry and nutrition, veterinary medicine; agricultural building design.
Projects: Protein quality of feeds for dairy cattle; rearing systems and calf welfare (Professor A.J.F. Webster); aerobiology and respiratory disease in calves (Dr C. Wathes, Professor A.J.F. Webster); oestrus detection in ruminants; feed selection in poultry (Dr G.C. Perry); immune responses in newborn pigs and calves (Dr T.J. Newby); cytogenetics and disorders of reproduction (Dr S.E. Long).

Bacteriology Department*   549

Senior staff: Professor D.C.E. Speller

Biochemistry Department*   550

Senior staff: Professor J.B. Chappel, Professor H. Gutfreund

Veterinary Medicine Department   551

Address: Langford House, Langford, Bristol, BSl8 7DU
Telephone: (0934) 852581
Head: Professor F.J. Bourne
Graduate research staff: 14
Activities: Research in the following areas: livestock husbandry and nutrition; veterinary medicine; drug therapy; vaccine development.
Projects: Enzyme development in gut of calf and pig; influence of diet on enzyme development (Dr D.E. Kidder); infections of the bovine eye (K. Bazeley); drug resistance in calves and chickens (Dr A. Linton); plasmid transfer in the chicken and calf (M. Hinton, Dr A. Linton); Salmonella in the calf; uterine immunology in the horse (Professor J. Bourne); gut immune response to infectious disease; gut immune response to dietary antigens (Dr C. Stokes, Professor Bourne); respiratory disease in the calf (Dr F. Taylor, Professor J. Bourne); herpes virus infection in the bovine (Dr R. Gaskell).

Veterinary Surgery Department*   552

Address: Langford House, Langford, Bristol, BS18 7DU
Head: Professor H. Pearson

FACULTY OF SCIENCE*   553

Dean: Professor R.N. Dixon

Agriculture and Horticulture Department*   554

Address: Long Ashton Research Station, Long Ashton, Bristol, BSl8 9AF
Telephone: (027 580) 2181
Head: Professor J.M. Hirst
Activities: Plant breeding; crop protection; plant sciences; foods and beverages studies.

Botany Department   555

Address: Woodland Road, Bristol, BS8 1UG
Telephone: (0272) 24161
Head: Professor A.E. Walsby
Senior staff: Professor M. Wills
Graduate research staff: 36
Projects: Effects of heavy metals on specific components and processes within a contaminated ecosystem (Dr M.H. Martin); physiology and development of Nematophtora gynophila, a fungus parasite of cyst eelworms (Dr M.F. Madelin).

Zoology Department   556

Address: Woodland Road, Bristol, BS8 1UG
Telephone: (0272) 24161
Head: Professor B.K. Follett
Graduate research staff: 49
Activities: Research on animal production (birds and mammals), and livestock breeding.
The department houses the Agricultural Research Council's research group on photoperiodism and reproduction (Director, B.K. Follett), investigating aspects of endocrinology, reproduction, light, and photoperiodism.
Projects: Seasonal reproduction in mammals and birds - photoperiodism, neuroendocrinology of gonadotrophin secretion, radioimmunoassay; mechanics of flight in birds and bats; physiology of vision - diel changes in behaviour and retinal morphology of fishes; organization and physiology of simple nervous systems in relation to behaviour - giant axon systems, and integrative mechanisms in nervous system of insects; biology of free-living protozoa - ultrastructure, physiology, systematics, ultra structure of invertebrates; reproduction in insects - hormonal and pheromonal studies; assessment of freshwater pollution - biology of chironomids; mollusc biology - ecology, morphology, physiology, embryology; ecology, physiology, behaviour of reptiles; and nematode ecology and physiology - parasites and free-living; ecology of foxes and badgers in urban areas; extracellular developmental biology - morphogenesis of arthropod cuticle and plant cell walls,

biophysics; ecology of marine and brackish water animals; systematics of fishes - fish phenology; differentiation of cells.

Comparative Animal Respiration Unit   557
Head: Professor G.M. Hughes

University of Cambridge   558

Address: Board of Graduate Studies, 4 Mill Lane, Cambridge, CB2 1RZ
Telephone: Central Administrative Offices (0223) 358933
Status: Educational establishment with r&d capability

FACULTY OF BIOLOGY 'A'   559

Address: 19 Trumpington Street, Cambridge
Telephone: (0223) 358381

Department of Applied Biology   560

Address: Pembroke Street, Cambridge, CB2 3DX
Telephone: (0223) 358381
Head: Professor Sir James Beament
Graduate research staff: 70
Activities: Applied research in the broadest sense in biological fields, and such basic research as may be needed to underpin the science of applications, but especially multi-disciplinary problems.
Main areas of expertise and current projects include: animal nutrition, especially ruminants; small mammal ecology and pests; conservation of large tropical mammals; weeds and land reclamation and restoration; pollination and floral biology, especially by bees; integrated insect pest control and management; crop plant hormones; agronomy and physiological age, especially in cereals and potatoes; plant breeding, especially beans; epidemiology of plant diseases; post-harvest biology, especially surface phenomena and coating fruit; design of experiments, statistics and biometry.

Department of Botany   561

Address: Downing Street, Cambridge, CB2 3EA
Telephone: (0223) 61414
Head: Professor R.G. West
Graduate research staff: 32
Activities: The department can offer research facilities and supervision in a wide range of the plant sciences. In the field of plant physiology special interest is taken in studies of uptake and movement of ions, in regulation of plant metabolism in relation to differentiation, in developmental botany and control of plant development, in the synthesis of chloroplast proteins, and in some aspects of whole plant physiology. Mycological interests include fungal ecology and plant pathology. There is a major interest in quaternary research, catered for by a sub-department linked with earth science and archaeology departments and supported by radiocarbon and stable isotope laboratories. Experimental ecology is also a major interest. The department has an important herbarium, supporting active taxonomic research. Other research interests include genetic recombination of fungi, experimental and numerical taxonomy and phytosociology. In several of these fields important projects are at present supported by research councils. The university Botanic Garden provides a wealth of plant material and experimental glasshouse and plot facilities.

Botanic Garden   562
Address: 1 Brookside, Cambridge
Telephone: (0223) 350101
Director: Dr S.M. Walters

Sub-department of Quaternary Research   563
Address: Godwin Laboratory, Cambridge
Telephone: (0223) 358381
Head: Professor R.G. West

Department of Genetics   564

Address: Downing Street, Cambridge
Telephone: (0223) 69551
Head: Dr K.J.R. Edwards
Graduate research staff: 20
Activities: Research into most branches of genetics, including: population, ecological and quantitative genetics; evolution; developmental genetics; cytogenetics; the analysis of the genome in both procaryotes and eucaryotes. The organisms studied are bacteria, Aspergillus, yeast, Drosophila and other insects, mice, barley and other plants. The approaches used are those of classical genetics combined with modern techniques utilizing the techniques of recombinant DNA.

Department of Parasitology   565

Head and Director of Molteno Institute: Dr B.A. Newton
Graduate research staff: 9
Activities: Current research concerns the nutrition, metabolism and developmental biology of trypanosomes, amoebae and helminths parasitic in the alimentary tract of vertebrates. Studies of the mechanism of action of trypanocidal drugs are also in progress. The institute provides facilities for most biochemical techniques.

Department of Zoology   566

Address: Downing Street, Cambridge, CB2 3EJ
Telephone: (0223) 358717
Head: Professor G. Horn
Sections: Agricultural Research Council Unit, Dr J.E. Treherne
CRC mammalian DNA repair laboratory, Dr R.T. Johnson
Unit on development and integration of behaviour, Professor R.A. Hinde
Graduate research staff: 37 full-time, 22 part-time
Activities: Research activities are largely directed to the solution of fundamental problems in the biological - including biotechnological - and the biomedical sciences. Current programmes include reseach on: mechanisms of DNA repair in normal and cancer-prone patients and in tumour cells; movement of electrolytes across cell membranes; physiological and pharmacological studies of transmission between nerve cells; regulation of cell function through studies of 'second messengers'; neural bases of learning; the physiology of flight in insects and birds; endocrinology of reproductive behaviour; development of social behaviour in humans; marine and terrestrial ecology; evolution at the molecular and morphological levels.

FACULTY OF BIOLOGY 'B'   567

Address: 19 Trumpington Street, Cambridge
Telephone: (0223) 358381

Department of Anatomy   568

Address: Downing Site, Cambridge
Telephone: (0223) 68665
Head: Professor R.J. Harrison
Graduate research staff: 18
Activities: The research interests of the department encompass cell biology and histochemistry at light and electron microscopic levels; functional and structural adaptions to different environments; computer studies; experimental embryology and teratology; neuroendocrinology of behaviour and reproduction in primates and other species; circulatory studies; comparative anatomy and reproduction.

Sub-department of Veterinary Anatomy   569
Director: Professor R.J. Harrison

Department of Biochemistry   570

Address: Tennis Court Road, Cambridge
Telephone: (0223) 51781
Head: Professor Sir Hans Kornberg
Senior staff: Professor D.H. Northcote
Graduate research staff: 63
Activities: The major research projects of the department include: structure and function of membrane proteins in bacterial and mammalian systems; structure and synthesis of microbial surface layers; biochemistry of opportunistic fungi; regulation of plant cell growth and differentiation; the mechanisms of photosynthesis and of mitochondrial energy conservation; regulation of metabolism of fatty acids; the structure, function and assembly of proteins including enzymes, viruses, and chromatin; the control of nucleic acid and protein synthesis in normal and virus-infected eukaryotic cells. In addition to the normal equipment of a biochemistry department there are several. electron microscopes, analytical and preparative ultracentrifuges, amino acid analysers, extensive facilities for handling and counting of radioactive materials, electrophoresis rooms, facilities for plant and animal tissue culture and an animal house for small mammals. The department has a P1-level containment laboratory for work on recombinant DNA and shares a P2 containment laboratory with the Department of Genetics.

Sub-department of Chemical Microbiology   571
Director: Professor E.F. Gale

OTHER DEPARTMENTS,   572
INDEPENDENT OF ANY FACULTY

Department of Clinical Veterinary   573
Medicine

Address: Madingley Road, Cambridge, CB3 0ES
Telephone: (0223) 355641
Telex: 81240
Head: Professor E.L. Soulsby
Divisions: Animal pathology, Dr W.F. Blakemore; medicine, Dr R.F.W. Goodwin; surgery, J.E.F. Houlton; animal health, Dr D.W.B. Sainsbury
Graduate research staff: 54
Activities: The general scope of research is in veterinary medicine with particular reference to: bacterial and virus diseases of pigs (eg Streptococcal meningitis of pigs, enzootic pneumonia); immunology of parasitic infections, with particular reference to immunosuppression in pregnancy and lactation and immune unresponsiveness in neonatal animals; epidemiology of parasitic infections of sheep; characterization of basic immune phenomena of domesticated animals, with particular reference to the characters and functions of lymphoid cells and non-specific effector cells. Other studies in-

clude neuropathology, studies of myelination and demyelination, the hormonal control of parturition and the environmental aspects of housing on animals and poultry with reference to optimal production.

Projects: Enzootic pneumonia in pigs (Dr P. Whittlestone); streptococcal infections in pigs (Dr T.J.L. Alexander); neuro-pathological disorders (Dr A.C. Palmer); myelination (Dr W.F. Blakemore); respiratory physiology and anaesthesia (Dr L.W. Hall); immune response of neonatal animals to parasitism (Professor E.J.L. Soulsby); immune response to Fasciola hepatica (Dr W.P.H. Duffus); epidemiology of parasitic infections (Dr R.M. Connan); salmonellosis in poultry (Dr D.R. Wise); pig health control measures (Dr R.F.W. Goodwin); avian viral infections (Dr H.P. Chu); housing environment and health (Dr D.W.B. Sainsbury); immunological/haemological disorders of animals (E.T. Rees Evans).

Department of Land Economy 574

Address: 16-21 Silver Street, Cambridge, CB3 9EP
Telephone: (0223) 355262
Head: Professor G.C. Cameron
Graduate research staff: 37
Activities: The department is concerned with the analysis of processes of regional, urban and rural change primarily, though not exclusively, in economically advanced nations. It lays particular stress upon the interactions, within a legal framework, of private and public agents which shape local economic, political and environmental welfare. It also seeks to analyse and evaluate the range of public policy mechanisms which attempt to modify processes of local change.

AGRICULTURAL ECONOMICS UNIT 575
Director: I.M. Sturgess
Graduate research staff: 7
Activities: Continuous work, performed under contract to the Ministry of Agriculture, Fisheries and Food, involves collection and analysis of economic data on farming in the eastern counties and on selected agricultural enterprises in UK. Other areas of research are in part extensions of these contractual commitments (for example a study of economics of size and the marginal productivity of capital in eastern counties farming and the supply of and demand for malting barley) but also reflect particular broader interests of staff and graduate students. Areas of research recently completed or in progress include: food aid, planning models for West Africa, concessional trading in oil-seed products, and land prices.
Projects: Economics of farming in the eastern counties of England (M.C. Murphy); economics of pig production (R.F. Ridgeon); economics of cereal production and storage (J.G. Davidson); economics of horticulture: production and marketing (W.L. Hinton).

University of Dundee* 576

Address: Dundee, DD1 4HN
Telephone: (0382) 23181
Telex: 76293
Status: Educational establishment with r&d capability
Principal and Vice-Chancellor: Dr Adam Neville

FACULTY OF SCIENCE* 577
Dean: Professor K.J. Standley

Biological Sciences Department 578
Telephone: (0382) 23181 extn 324
Head: Professor W.D.P. Stewart
Senior staff: Professor P.S. Corbet, Professor J.A. Raven
Graduate research staff: 80
Activities: The department has a strong research interest in all aspects of the nitrogen cycle, how this affects crop productivity, and how agricultural practise affects the natural environment, particularly agricultural soils and freshwaters. Research is aimed at improving crop production with particular emphasis on nitrogen and phosphorus fertilizer usage. There is a strong group working on insect pests, and another on nitrogen cycling in tropical agriculture. Recently the department has obtained substantial funds to support plant biotechnology research.
Projects: Biological nitrogen fixation particularly in relation to rice production; nitrogen fixation by algae and bacteria; nitrogen cycling (Professor W.D.P. Stewart); nitrogen fixation by legumes (Dr J. Sprent); mycorriza and crop production (Dr M.J. Daft); insect pests of crop plants (Professor P. Corbet); plant genetics and biotechnology (Professor W.D.P. Stewart, Dr A.C. Hastie, Dr G.J. Warren).

University of Durham 579

Address: Old Shire Hall, Durham, DH1 3HP
Telephone: (0385) 64466
Status: Educational establishment with r&d capability

FACULTY OF SCIENCE 580

Botany Department   581

Address: South Road, Durham, DH1 3LE
Telephone: (0385) 64971
Telex: DURLIB G 537351
Head: Professor D. Boulter
Sections: Genetic engineering for crop improvement, Professor D. Boulter; rational design of drugs, Dr J.W. Payne; heavy metal tolerance of microorganisms in relation to pollution and mining, Dr B.A. Whitton
Graduate research staff: 50
Activities: Multi-interdisciplinary research into genetic engineering for crop improvement of rice, cotton and beans; primary structure determinations of plant proteins in relation to functional properties for potential use in the food industry; enzyme technology; nitrogen fixation in the field; the role of microorganisms in pollution control, pollution rescue and mining.
Projects: Physiological constraints to yield in rice (Dr P. Gates); yield stability in field beans; improving nutritional quality of legumes, screening methods and the relationship between protein and functional properties (Professor D. Boulter); model breeding in Pisum (Dr J. Gatehouse); basis of biochemical resistance to Bruchid (Dr A. Gatehouse); crop improvement by genetic engineering (Dr R. Croy).

Zoology Department   582

Head: Professor D. Barker
Readers: Dr K. Bowler, Dr J.D. Horton, Dr J.C. Coulson
Graduate research staff: 28
Activities: Research is mainly of an ecological and physiological nature. Ecological projects are concerned with the biology and population dynamics of moorland invertebrates; colonial breeding in vertebrates; and the ecology of sea birds and small mammals. The physiological research is largely concerned with the nervous system, immunology, and with the effect of temperature at the cellular level. Neurological studies include work on muscle receptors, insect muscle, vertebrate muscle innervation, the vertebrate retina, and the biochemical properties of the plasma membranes of nerve and muscle cells. Research is also carried out on social behaviour in vertebrates; bird navigation and migration; the endocrine control of insect metabolism; the role of the thymus in the development of immunity in amphibians; and the control of the appearance of lipogenic enzymes in differentiating tissues.

University of East Anglia*   583

Address: Norwich, Norfolk NR4 7TJ
Telephone: (0603) 56161
Status: Educational establishment with r&d capability
Vice-Chancellor: Professor M. Thompson

SCHOOL OF BIOLOGICAL   584
SCIENCES*

Dean: Dr Michael John Selwyn
Senior staff: Professor D.D. Davies, Professor A.F.G. Dixon, Professor B.B. Folkes, Professor E.E. Rojas
Activities: Yeast metabolism and drug resistance; plant metabolism; mitochondrial energetics and ion transport; biochemistry of organometallic compounds; electron transfer using rapid reaction techniques and magnetic circular dichroism; nitrogen uptake in plants, amino acid metabolism and protein synthesis and turnover; biophysics of lens and cataractogenesis; electrophysiology of secretion in ß-cells of islets of Langerhans; population dynamics of aphids; ecology of freshwater invertebrates and fish; bacterial mutational systems in relation to environmental mutagen testing; penetration of plants by fungal pathogens; toxigenicity of microorganisms in relation to crop plant disease.

SCHOOL OF ENVIRONMENTAL   585
SCIENCES

Telephone: (0603) 56161
Dean: Dr J.R. Tarrant
Senior staff: Professor K.M. Clayton, Professor B.M. Funnell, Professor T. O'Riordan, Professor F. J. Vine
Graduate research staff: 60
Activities: Research and teaching in the environmental sciences, including research in agricultural development, climatology, mineral resources, water resources, geophysics, geochemistry, oceanography, sedimentology, glaciology, terrestrial and freshwater ecology, atmospheric and water chemistry, rural and urban planning, environmental economics and environmental policy.
Major current projects include: palaeoenvironmental reconstruction of global mesonzoic sedimentation; experimental determination of electrical conductivities and seismic velocities of rocks; the chemistry of crustal fluids; methods of soil survey and land evaluation; environmental impact assessment for the River Yare Flood Alleviation Barrier; river regulation and channel stability; microelectrophoresis studies of natural waters; controls on ecosystem structures in the Norfolk Broads and related drainage dykes; involvement in the Cyprus Deep Drilling Project; groundwater movement and chemistry in limestone aquifers; materials recycling; rural services; national and international food and agricultural policies.

University of Edinburgh* 586

Address: Old College, South Bridge, Edinburgh, EH8 9YL
Telephone: (031) 667 1011
Status: Educational establishment with r&d capability
Vice-Chancellor: Dr J.H. Burnett

FACULTY OF MEDICINE* 587

Address: Teviot Place, Edinburgh, EH8 9AG
Dean: Professor G.J. Romanes

Bacteriology Department* 588
Head: Professor J.G. Collee

FACULTY OF SCIENCE* 589

Dean: Professor W. Cochran

Forestry and Natural Resources Department 590
Address: Darwin Building, King's Buildings, Mayfield Road, Edinburgh, EH9 3JU
Telephone: (031 667) 1081
Telex: 727442
Head: W.E.S. Mutch
Sections: Plant science, Dr J. Grace, Professor P.G. Jarvis; hydrology,Dr D. Ledger; forestry, Dr D.C. Malcolm; fisheries, Dr D.H. Mills; computer modelling, Dr R. Muetzelfeldt; wildlife ecology, Dr I.R. Taylor
Graduate research staff: 30
Activities: Interdisciplinary research in biological and management subjects, especially: plant science-photosynthesis, CO_2 and water vapour exchange, stomatal responses and environmental control, plant responses to wind, boundary layer characteristics of leaves; forestry - fertilizer effects on conifers, nutrition of trees on peat and peaty grey soil, nitrogen availability in mixed tree stands, forest entomology and pathology; wildlife ecology - population ecology, feeding ecology and management problems associated with roe deer, red deer, squirrels, pine martens, barn owls, small birds in forest, tropical and other environments; fisheries - growth and performance of migratory and non-migratory fish, especially salmonids, in relation to land management, afforestation , and acid rain problems; hydrology land use, recreational land management, industrial land reclamation.

School of Biology* 591

Departments: The School consists of the following departments in two faculties: Botany, Genetics, Microbiology, Molecular Biology, Zoology (Faculty of Science); Bacteriology, Biochemistry, Pharmacology, Physiology (Faculty of Medicine).

Botany Department* 592
Head: Professor M.M. Yeoman

Genetics Department* 593
Head: Professor J.R.S. Fincham

Microbiology Department* 594
Head: Professor J.F. Wilkinson
Activities: Biology and exploitation of methane-oxidizing bacteria; biological resistance to radiation damage; biosynthesis and structure of extracellular polysaccharides in various bacteria; microbiology of silage and other conserved crops.

Molecular Biology Department* 595
Address: King's Building, Mayfield Road, Edinburgh, EH9 3JR
Head: Professor K. Murray
Activities: Bacterial and bacteriophage genetics; cell division and replication; genetic engineering; biochemistry of nucleic acids particularly enzymology related to DNA and DNA sequence determination; protein chemistry; enzymology; electron microscopy; molecular developmental biology.

Zoology Department* 596
Address: West Mains Road, Edinburgh, EH9 3JT
Heads: Professor J.M.Mitchison, Professor A.W.G. Manning

FACULTY OF VETERINARY MEDICINE 597

Address: Summerhall, Edinburgh, EH9 1QH
Telephone: (031) 667 1011
Telex: 727442 (UNIVED G)
Dean: Professor K.M. Dyce

Animal Health Department 598

Head: Professor G.S. Ferguson
Activities: Research on animal production and health; animal husbandry and nutrition.
Projects: Linkage studies in Landrace pigs in relation to halothane response; detection of carriers of the halothane gene by combining halothane and succinyl chloride; sex effect on survival in hereditary pig lymphosarcoma (Dr P. Imlah); nutritional requirements of goats (Dr D. Cuddeford); epidemiology of Orf (J.B. Lawson); immunological castration in male cattle (Dr I.S. Robertson); evolution of horse breeds, particularly native ponies (R. Beck); dairy herd health and productivity service; equine nutrition (J.M. Kelly); reservoirs of Trichinella spiralis infection (G. Owen); epidemiology of Type A flu virus infection; small animal database (M.V. Thrusfield).

Centre for Tropical Veterinary Medicine   599

Head: Professor D.W. Brocklesby
Activities: Research on livestock husbandry and nutrition in the tropics; tropical veterinary medicine.
Projects: Nutritional requirements of draught animals; reproductive capacity of cattle maintained at high ambient temperatures; nutrition of goats; comparison of different methods for measured sweat rate of cattle (Dr A.J. Smith); bovine cysticercosis; fascioliasis (Dr M.M.H. Sewell); Rickettsial diseases of domestic animals; epidemiology of orf; streptothricosis (Dr G.R. Scott); trypanosomiasis; theileriosis (C.G.D. Brown).

Veterinary Anatomy Department   600

Head: Professor K.M. Dyce
Activities: Research on mammalian, avian and piscine anatomy with particular reference to its application to veterinary medicine and animal production.
Projects: The canine cornea in health and disease, with particular reference to dystrophies, degeneration and infiltrations (S.M. Crispin); studies of the mucosa of the equine large intestine (Dr S.A. Kempson); a scanning and transmission electron microscope study of prenatal muscle development in the mouse (Dr N.C. Stickland); morphology of the alimentary canal of the rainbow trout (Dr W.M. Stokoe).

Veterinary Medicine Department   601

Head: Professor J.T. Baxter
Activities: Advancement of animal health through investigations into the aetiology, epidemiology, immunology, biochemistry and clinical features of endocrine, metabolic, allergic and infectious diseases of domestic animals.
Projects: Immunology and cytology of normal horses and horses affected with chronic obstructive pulmonary disease (COPD); effects of drugs on the respiratory system in normal and COPD-affected horses (E.A. McPherson); immunology and chemotherapy of pigs affected with lymphosarcome (H.S. McTaggart); isoenzyme analysis in ruminants; serum protein fractionation by electrophoretic and chemical methods in domestic animals (Dr D.L. Doxey); thyroid function in normal and hypothyroid dogs (Professor J.T. Baxter); thyroid function in health and disease in cats (K.L. Thoday); endocrine disorders of the pancreas in the dog (C.P. Mackenzie); aetiology and management of maldigestion and malabsorption in the dog (J.W. Simpson).

Veterinary Pathology Department   602

Acting Head: I.S. Beattie
Activities: Studies of various aspects of disease in domestic animals and poultry with particular reference to aetiological agents, pathogenesis, pathology and immunity.
Projects: Lymphosarcoma in pigs; fibropapilloma and adenocarcinoma in sheep - small intestine (K.W. Head); possible antigenic markers of malignancy in canine and feline mammary tumour cells (Dr R.W. Else); adenomatosis complex in pigs (A.C. Rowland); wild type and temperature sensitive mutants of Newcastle disease virus (Dr G. Fraser); the taxonomy of Campylobacter sputorum ss mucosalis and its relationship to tissue culture cells (Dr G.H.K. Lawson); bacterial L-forms in animal and bird tissues (Dr J.E. Phillips).

Veterinary Pharmacology Department   603

Head: Professor F. Alexander
Activities: Research in the following areas: electrolyte metabolism in equines; drug metabolising enzymes in poultry; calcium metabolism in the ewe and foetus; the mode of action of anthelmintics.
Projects: Studies on electrolyte metabolism in the horse (Professor F. Alexander); pharmacology of nitrofurans and carbon tetrachloride (Dr A.L. Bartlet); calcium metabolism in the sheep (Dr S.S. Carlyle); morphology of neuromuscular junctions of Ascaris suum; pharmacology of inhibitory receptors on Ascaris suum muscle; neuropharmacology of reticulo-spinal neurons (Dr R.J. Martin).

Veterinary Physiology Department   604

Head: Professor A. Iggo
Activities: The research work of the department seeks to achieve the following: to understand the mechanisms of sensation at a peripheral and central level; to understand principles of neuronal organisation; to find the central nervous sites of action of anti-hypertensive drugs; to identify the factors that control the normal oestrous cycle in sheep; to determine the mechanism of extrinsic nervous control of gastrointestinal motility; to understand the mechanisms of pain; to elucidate the role of Merkel cells.
Projects: Studies of somatosensation (Professor A. Iggo); electrophysiology, ultrastructure and quantitative morphology of identified axons and neurones in spinal cord (Dr A.G. Brown); neurophysiological and autoradiographic study of the action of clonidine (Dr A.L. Haigh); uterine control of ovarian function in sheep (Dr K.P. Bland); gastrointestinal sensory and motor systems (Dr D.F. Cottrell); electrophysiological studies of the central mechanisms involved in somatosensation and in particular the phenomenon of pain (Dr V. Molony); electrophysiological and immunocytochemi-

cal study of cutaneous mechanoreception (Dr E.J. Cooksey).

Veterinary Surgery Department   605

Head: Professor J.R. Campbell
Sections: Large animal unit, J.A. Fraser; reproduction unit, A.C. Wilson
Activities: Veterinary research; tendon prosthesis; ruminant anaesthesia; orthopaedic research; navicular disease; electrical stimulation of healing; dental disease.
Projects: Navicular disease in the horse (J.A. Fraser); equine laryngeal paralysis (Dr P.M. Dixon); regurgitation and ruminal reflex during anaesthesia (M.A. Camburn); immunocastration (J.C. Wilson); repeat breeder cows (Dr C.D. Munro).

University of Essex*   606

Address: Wivenhoe Park, Colchester, CO4 3SQ
Telephone: (0206) 862286
Telex: 98440 UNILIB G
Status: Educational establishment with r&d capability

SCHOOL OF SCIENCE AND   607
ENGINEERING*

Department of Biology   608

Chairman: Professor T.R.G. Gray
Senior staff: Developmental genetics, Professor J.G.M. Shire
Graduate research staff: 23
Activities: The department has been awarded several grants for research projects on the subjects of: photosynthesis and molecular genetics; nuclear and mitochondrial genes in eukaryotes; cell and molecular biology. Collaborative research with industry deals with: photosynthesis (marine bacteria that corrode off-shore oil-rigs); ultracentrifugation; the use of monoclonal antibodies to be used as reagents in clinical tests; monitoring the environmental effects of waste disposal by local industrial firms and local autohorities.

University of Exeter*   609

Address: Northcote House, The Queen's Drive, Exeter, EX4 4QJ
Telephone: (0392) 77911
Telex: 42894
Status: Educational establishment with r&d capability

FACULTY OF SCIENCE*   610

Dean: Professor D.R. Davies

Biological Sciences Department   611

Address: Hatherly Laboratories, Prince of Wales Road, Exeter, EX4 4PS
Head: Professor J. Webster
Senior staff: Professor D. Nichols; plant pathology, Dr D. Pitt; plant ecology, Dr M.C.F. Proctor; zoology, Dr C.R. Kennedy; plant physiology, Dr I.D.J. Phillips
Activities: Biodegradation of synthetic molecules by microorganisms in pure culture.
Projects: Ecology and taxonomy of aquatic and aero-aquatic conidial fungi; biology of entomophthora infections on aquatic insects (Professor J. Webster); reproductive biology of echinoderms; population structure and migrational behaviour in sea-urchins; coastal biology in south west England (Professor D. Nichols); animal/microbial interactions: feeding and nutrition of soil animals, grazing effects on bacterial and fungal populations; decomposition in temperate and tropical forests; nutrient flux pathways in soils, particularly nitrogen (J.M. Anderson); bacteriology of natural waters, especially estuaries, in relation to sewage pollution; antibiotic-resistant bacteria and factors affecting bacterial conjugation in water environments (A.E. Anson); taxonomy and cytology of Polychaeta; littoral ecology and distribution of annelids; ecology of estuarine communities, especially those involving polychaetes (T. Harris); taxonomy and ecology of flowering plants; analysis of taxonomic and ecological data (R.B. Ivimey-Cook); ecology of animal parasites, especially population dynamics and community structure of parasites of freshwater and migratory fish; biology and management of freshwater fish; ball clay pollution in freshwater (C.R. Kennedy); distribution, ecology and behaviour of European and African mammals; colour change in fishes (I.J. Linn); experimental population genetics of plants; theoretical modelling of plant and animal populations (M.R. Macnair); physiology of development and responses to the environment in higher plants (I.D.J. Phillips); physiology of growth and reproductive development in fungi; plant host/parasite physiology; development of the lytic compartment in plants (D. Pitt); vegetation ecology and plant autecology; physiological ecology of bryophytes (M.C.F. Proctor); physiology and biochemistry of plant steroids, particularly the solanaceous glycoalkaloids (J.G. Roddick); osmotic and ionic regulation in aquatic animals, particularly branchial mechanisms of ionic regulation in teleosts; physiology of the elasmobranch rectal gland; osmoconformity in Arenicola marina (T.J. Shuttleworth); cell biology; cellular and intracellular movement (H. Stebbings); ecology of ants; insect/fungus interaction of leaf-cutting ants; biology of theraphosid spiders; analysis of light-trap data (D.J. Stradling); gene-

tics: mutagenesis; incompatibility in fungi (N.K. Todd); biochemistry of methylotrophic fungi; biochemistry of aliphatic glycol metabolism by micro-organisms; biodegraduation of xenobiotic compounds by micro-organisms (A.J. Willets); functional morphology and evolution of insect flight systems, particularly of wings; palaeontology, palaeoecology and evolution of insects, especially Hemiptera and lower Pterygota (R.J. Wotton).

Physics Department   612

Address: Stocker Road, Exeter, EX4 4QL
Head and Professor of Theoretical Physics: Professor G.N. Fowler
Senior staff: Physics, Professor A.F.G. Wyatt; solid-state physics, Dr J.R. Drabble; electromagnetism, Dr W.G.V. Rosser; medical physics, Dr F.C. Flack; nuclear spectroscopy, Dr R.E. Meads
Activities: Research topics include: conduction electrons in metal; phonons in liquid helium; velocity of sound in solids; internal elastic strains liquid crystals; Mössbauer studies; deep traps in semiconductors; critical point theory; molecular scattering; long range radio transmission; ionospheric studies; geomagnetism; carbon dioxide monitoring; thermal response of buildings; environmental monitoring; large scale Schlieren photography; design of equipment for diagnosis and treatment of velopharyngeal incompetence; design of flow instrumentation for use in haemodialysis systems; design of aids for the deaf; design of equipment for diagnosis and treatment of urinary incontinence; design of intravenous infusion apparatus; design of equipment for an interpretation of data from respiratory sounds in horses.

INTER-FACULTY DEPARTMENT   613

Agricultural Economics Unit   614

Address: Lafrowda House, St German's Road, Exeter, EX4 6TL
Telephone: (0392) 73025/6
Head: R.C. Rickard
Graduate research staff: 10
Projects: Farm organization and incomes in south-west England (R.C. Rickard); an input/output model of the UK livestock and meat industry (K.S. Howe); land use, farm recreation and tourism (E.T. Davies); farm finance and taxation (W.J. Dunford); horticultural production and marketing (P.M.K. Leat); economics of livestock and milk production in South-West England.

University of Glasgow *   615

Address: Glasgow, G12 8QQ
Telephone: (041) 339 8855
Telex: 778421
Status: Educational establishment with r&d capability
Principal: Alwyn Williams

FACULTY OF MEDICINE *   616

Pathological Biochemistry   617
Department *

Head: Professor H.G. Morgan

FACULTY OF SCIENCE *   618

Agriculture Department   619

Affiliation: West of Scotland Agriculture College
Head: Professor J.M.M. Cunningham
Activities: Research in the following areas: science and technology of milk production; treatment of milk and the production of milk products; nutrition of ruminant livestock; poultry production with particular emphasis on systems and nutritional aspects; treatment and utilization of animal wastes; economics of milk, beef and sheep production; grassland production and utilization; weeds, pests and diseases of grassland and arable crops; soil management.

Botany Department   620

Head: Professor M.B. Wilkins
Sections: Agricultural botany, Dr A.M.M. Berrie
Graduate research staff: 22
Activities: Research in the following areas: hormonal and chemical control of plant growth; environmental studies on algae; seed dormancy; photosynthesis in lower and higher plants; plant cell walls; biological nitrogen fixation; steroids in fungi; fungal disease in stored crops; historial ecology; plant viruses; taxonomy of bryophytes.

Cell Biology Department   621

Head: Professor A.S.G. Curtis
Graduate research staff: 10
Activities: Research in the following areas: tumour cells; inflammation; automated enzyme and chemical reaction monitoring; clinical tests in connection with inflammation and diagnosis of various diseases; histocompatibility; sponge cells.

Chemistry Department*   622

Agricultural Chemistry   623
Head: Dr A.M.M. Berrie
Graduate research staff: 13
Activities: Research in the following areas: soil studies and analysis; industrial land reclamation; effects and application of herbicides and pesticides; food composition and crop storage.

Dairy Science Department*   624

Affiliation: Hannah Research Institute
Head: Professor Malcolm Peaker
Activities: Research of industrial interest concerns the following: use of conserved forages, especially silage, for milk products; mechanisms of milk secretion and their regulation; lipid metabolism in ruminants; use of milk for manufacture.

Marine Biological Station - Millport   625

Address: Millport, Isle of Cumbrae, Strathclyde
Telephone: (047 553) 581/2
Head: Professor J.A. Allen
Activities: Supply of biological specimens to schools, colleges, universities and to commercial and industrial users; tertiary teaching primarily to undergraduates; research: marine invertebrate biology, benthic ecology and behaviour (marine microbiology and immunology, marine pollution research, shellfish culture).
Publications: Annual report.

Microbiology Department   626

Address: Anderson College, 56 Dumbarton Road, Glasgow, G11 6NU
Head: Professor A.C. Wardlaw
Graduate research staff: 21
Activities: Research in the following areas: adhesion of microorganisms to surfaces; fouling; bacterial pollution of rivers and estuaries; bacterial vaccines and toxins; slime moulds; dental microbiology; spore-forming bacteria; genetic engineering; microbiology of shellfish.

FACULTY OF VETERINARY MEDICINE   627

Address: Bearsden Road, Bearsden, Glasgow
Telephone: (041) 942 2301
Dean: Professor D.D. Lawson
Graduate research staff: 42
Activities: Research in the following areas: respiratory diseases of animals; neonatal diseases of animals; immunological and chemical control of parasitic diseases; cancers in animals; applied animal nutrition with particular reference to dietary supplementation; metabolism of drugs in animals; naturally occurring neuromuscular diseases; infectious and immunological disease of the dog and cat.
Projects: Canine parvovirus infection (Dr I.A.P. McCandlish, Dr H. Thompson); naturally occurring models of neuromuscular disease (Dr I. Griffiths); renal diseases in the dog and cat (Professor N.G. Wright, Professor E.W. Fisher, Mr A. Nash, Dr H. Thompson); liquid nutrient composition for ruminant nutrition (Professor R.G. Hemingway, Dr J.J. Parkins, Dr N.S. Ritchie); wart viruses and cancer; leukaemia vaccine (Professor W.F.H. Jarrett); colostral immunity in the calf (Professor I.E. Selman, Dr L. Petrie); a radiation-attenuated vaccine against lungworm disease of cattle (Dictol) (Professor W. Mulligan, Professor I. McIntyre, Professor W.F.H. Jarrett, Professor G.M. Urquhart, Dr F.W. Jennings).

Animal Husbandry Department   628
Head: Professor R.G. Hemingway

Veterinary Anatomy Department   629
Head: Professor N.G. Wright

Veterinary Medicine Department   630
Head: Professor W.I.M. McIntyre

Veterinary Parasitology Department   631
Head: Professor G.M. Urquhart
Graduate research staff: 9
Activities: Helminth infection of farm animals; trypanosomiasis; immunology of parasitic infections.

Veterinary Pathology Department   632
Head: Professor W.F.H. Jarrett

Veterinary Pharmacology Department   633
Head: Professor J.A. Bogan

Veterinary Physiology Department   634
Head: Professor P.H. Holmes
Graduate research staff: 10
Activities: Research in the following areas: pathogenesis and immunology of African bovine trypanosomiasis - factors influencing the virulence of trypanosome infections in susceptible and trypanotolerant animals; interactions between host nutrition and gastrointestinal sheep infected with Haemonchus contortus; immune responses to gastrointestinal helminths - studies on rats infected with Nippostrongylus brasiliensis, examining the roles of local intestinal immunity and gut motility in parasite expulsion.

Veterinary Surgery Department   635

Head: Professor D.D. Lawson

University of Hull*   636

Address: Cottingham Road, Hull, HU6 7RX
Telephone: (0482) 46311

FACULTY OF SCIENCE*   637

Department of Biochemistry   638

Head: Professor E.A. Dawes
Senior staff: Professor J. Paul
Graduate research staff: 23
Activities: The department concentrates on the following research areas: biotechnology - continuous culture of microorganisms, fundamental metabolism studies, pilot scale metabolite production, bacterial exopolysaccharides and microbial production of oils and fats; molecular biology - genetic engineering, characterization of nucleic acids, macromolecules and biopolymers; microbial physiology - metabolic pathways in bacteria, metabolism of methylated amines and methylated sulphur compounds for application in the paper and brewing industries; membrane studies - cell membranes of prokaryotic and eukaryotic microorganisms and phospholipid models, behaviour of membrane components such as guinone coenzymes and microbial iron-binding compounds, acid resistance of sulphur-oxidizing bacteria; assay methods for creative kinase, aldehyde determination.
Projects: Fundamental biochemistry and physiology of nitrogen-fixing bacteria (Azotobacters) eg poly-ß-hydroxybutyrate and alginic acid production (Professor E.A. Dawes); oils and fats chemistry, biochemistry and analysis (Dr C. Ratledge); metabolism of amines and methylated sulphur compunds by bacteria and yeasts (Dr P.J. Large); acceptability of beeswax to bees (Dr B. Chipperfield).

Department of Plant Biology   639

Head: Professor J. Friend
Senior staff: Genetics, Professor D.A. Jones; plant protection, Dr J.R. Coley-Smith; soil science, Dr D.J. Boatman; plant physiology, Dr A.J. Peel; plant biochemistry, Dr D.R. Threlfall
Graduate research staff: 11
Activities: Research in the following areas: plant pathology and disease physiology - control of pests and diseases, development of pesticides, herbicides and fungicides, including development of phenolic and terpenoid antifungal compounds (terpenoid quimones, carotenoids, and plant sterols); plant breeding - identification of seed lipids and constituent fatty acids; plant physiology - root identification of timber species, age estimation of established stands; soil science - pollution measurements in soils, rivers and estuaries, estimation of soil aeration status, uptake of organic compounds into bacteria and genetic control of active transport systems.
Projects: The biology and control of white rot of Allium species (Sclerotium cepivorum); biology and control of bottom rot of protected lettuce (Rhizoctonia solani); biology and control of leaf and pod spot of field beans (Ascochyta fabae); biology and control of black root rot of cucumbers (Phomopsis sclerotioides) (Dr J.R. Coley-Smith); macromolecular basis of resistance of potatoes to Phytophthora infestans; the role of cell wall degrading enzymes in the establishment of the onion diseases, white rot and neck rot, caused by Sclerotium cepivorum and Botrytis allii (Professor J. Friend); biosynthesis and metabolism of antifungal terpenoids in diseased potatoes (Dr D.R. Threlfall); genetic control of cyanogenesis in potential food and fodder crops (Professor D.A. Jones); physiological basis of flood-resistance in tolerant species; root aeration in agriculturally important plants (Dr W. Armstrong); transport of pesticides, hormones and nutrients in plants (Dr A.J. Peel); development of peat bogs (Dr D.J. Boatman).

Department of Zoology*   640

Head: Professor I.M.L. Donaldson
Senior staff: Professor G. Goldspink, Professor R.F. Chapman
Graduate research staff: 16

University of Keele*   641

Address: Keele, Staffordshire ST5 5BG
Telephone: (0782) 621111
Telex: 36113 UNKLIBG
Status: Educational establishment with r&d capability
Vice-Chancellor: Dr D. Harrison
Director of Information Services: B.G. Rawlins

BOARD OF NATURAL SCIENCES*   642

Department of Biological Sciences 643

Head: Professor C. Arme
Graduate research staff: 22
Activities: Areas of research include: reclamation of coal spoil heaps; utilization of sewage sludges; liposomes as oral therapeutic agents; fish parasitology; ecology and taxonomy of freshwater invertebrates, especially insects; effects of heavy metal pollution on soft-bodied invertebrates.

Department of Chemistry 644

Head: Professor I.T. Millar
Senior staff: Inorganic chemistry, Professor C.T. Mortimer; physical chemistry, Professor P.H. Plesch

University of Kent at Canterbury* 645

Address: Canterbury, Kent CT2 7NZ
Telephone: (0227) 66822
Telex: 965449
Status: Educational establishment with r&d capability
Vice-Chancellor: Dr D.J.E. Ingram

FACULTY OF NATURAL SCIENCES 646

Dean: Professor K.A. Stacey
Graduate research staff: 194

Biological Laboratory 647

Director: Professor K.A. Stacey
Senior staff: Molecular biology, Professor K.A. Stacey; microbiology, Professor A.T. Bull; biology, Dr R.B. Cain; biochemical parasitology, Dr W.E. Gutteridge; biochemistry, Dr C.J. Knowles
Graduate research staff: 25 (1981)
Activities: Biochemistry, microbiology, and microbial genetics: anaerobic degradation and methane production; microbial technology; environmental microbiology; microbial evolution; dissociated hypothalamic cells in primary culture; effect of alcohol consumption on steroid metabolism by liver microsomes; investigation of the mechanism of steroid directed differentiation; microbial ecology of soils and sediments; degradation of aromatic compounds and detergents by micro-organisms and its regulation; biodegradation of pyridine and pyridinium compounds with special reference to herbicide inactivation; mechanism of protein disulphide bond formation and characterization of enzyme catalysing protein disulphide interchange; organization and function in microsomal membranes; metabolism of foreing compounds by microsomal monooxygenases, the effects of toxic products and the induction of specific enzyme species; cytoskeletal proteins of physarum polycephalum; characterization of the microtubule organizing centres of eukaryotic cells; microtubules as target organelles for pesticides; host-parasite interactions in mycoparasitism; ultrastructural studies of zygomycete fungi; utilization of nitriles by bacteria; effect of haem on streptococci; production of cyanide as a model microbial secondary metabolite; effects of xenobiotics on the hepatocyte cytoskeleton; control of haem metabolism in isolated rat and avian hepatocytes, distribution of capacities for xenobiotic metabolism in subpopulations of isolated hepatocytes; regulation of protein synthesis in mammalian tissues; genetics and molecular biology of bacterial cell division; genetics of recombination and mutation; regulation of folate biosynthesis.

Chemical Laboratory 648

Director: Professor J.A. Connor
Senior staff: Organic chemistry, Professor R.F. Hudson, Dr A. Williams; physical chemistry, Professor E.F. Caldin
Honorary staff: Physical chemistry, Professor Sir George Porter; chemistry, Professor R.L. Wain; medicinal chemistry, Professor C.R. Ganellin
Graduate research staff: 30 (1981)
Activities: Subjects of research include: organophosphorus chemistry and related topics; fast reactions in solution; agricultural chemistry; nuclear and radiochemistry; analytical chemistry; colloids; photo-induced reactions; organometallic chemistry; coordination and organometallic chemistry; thermochemistry of transition metal compounds; spectroscopy; X-ray structure analysis; solid-state chemistry; mass spectrometry; theoretical chemistry; synthetic and gas-phase organic reactions; nuclear magnetic resonance; biological organic chemistry; synthetic organic chemistry; environmental chemistry, trace element distributions.

SCHOOL OF MATHEMATICAL STUDIES* 649

Chairman and Professor of Statistics: G.B. Wetherill
Senior staff: Applied mathematics, Professor J.S.R. Chisholm; pure mathematics, Professor M.E. Noble; biometry, Professor S.C. Pearce
Graduate research staff: 14
Activities: Research topics include: biometrical methods for the study of intercropping different crops on the same land; statistical principles of quality control procedures for saturation analysis radioimmunoassay kits; pesticide evaluation; toxicity trials; computer-aided design of response surfaces; design of computer algorithms for parallel processors.

University of Lancaster*   650

Address: Lancaster, LA1 4YW
Telephone: (0524) 65201
Telex: 651 11 Univlib. Lancstr
Status: Educational establishment with r&d capability
Vice-Chancellor: Professor P.A. Reynolds

DEPARTMENT OF BIOLOGICAL   651
SCIENCES

Head: Professor W.T.W. Potts
Senior staff: Plant physiology, Professor T.A. Mansfield; plant ecology, Professor C.D. Pigott, plant biochemistry, Dr A.R. Wellburn; plant pathology, Dr P.G. Ayres; biophysics, Dr S. Hunt; animal ecology, Dr J.B. Whittaker; nutritional biochemistry, Dr F.W. Heaton; microbial genetics, Dr A. Upshall; neuromuscular physiology, Dr H. Huddart
Graduate research staff: 74 (1982)
Activities: Research groups are involved in 47 projects covering diverse topics. The areas receiving major financial support from external sources are air pollution effects on plants, plant-water relations, fungal diseases of plants, chloroplast biochemistry, ecology of herbivorous insects, nitrogen fixation by bacteria, mitotic recombination in fungi, molecular biology of structural proteins and polysaccharides, physical biochemistry of connective tissue, muscle structure and function, and osmotic regulation in animals.
Publications: Annual report.

DEPARTMENT OF ENVIRONMENTAL   652
SCIENCES*

Head and Professor of Hydrology: T. O'Donnell
Senior staff: Environmental physics, Professor A.N. Hunter
Graduate research staff: 30
Activities: Subjects of research programmes include:
Solid earth geophysics and geology - geomagnetism to determine the electrical conductivity of the earth's crust and upper mantle; crustal structure and isotasy; volcanism and tectonics in the Midland Valley of Scotland, the rift valley in Kenya, Saudi Arabia, USA, and Greenland.
Surface processes - sedimentary geochemistry; hydrogeology in the Lake District and carboniferous limestone and millstone grit areas around Lancaster; hydrology as part of a real time radar-based flood forecasting procedure; water resources; soil formation on carboniferous limestone; Quaternary geology in Greenland.
Atmospheric science - disturbances, irregularities and anomalies of the ionized upper atmosphere; boundary layer studies; atmospheric pollution, including planetary studies - centred on the comparison of like features on different planets which have suffered different environmental conditions; a terrestrial application of planetary remote sensing techniques is explored in a study in which NASA satellite data are used to detect sea surface pollutants, particularly in connection with the North Sea oil industry, and several novel image processing techniques have been developed to support this programme.

University of Leeds*   653

Address: Leeds, LS2 9JT
Telephone: (0532) 31751
Status: Educational establishment with r&d capability

FACULTY OF APPLIED SCIENCE*   654

Dean: Dr P.T. Speakman

Animal Physiology and Nutrition Department   655
Telephone: (0623) 431751
Head: Professor A.D. Care
Graduate research staff: 20 (1982)
Activities: Physiology and nutrition of farm animals; animal production. Particular attention is given to mineral requirements and their supply, and the control of voluntary food intake.

Procter Department of Food Science*   656
Head: Professor David S. Robinson
Activities: Emphasis on chemistry and biochemistry of foodstuffs; food enzymology; food analysis; food additives and colours; flavour identification; food texture and rheology; food colloids; food processing and engineering; nutritional value of convenience foods.

University Field Station*   657
Director: J. Dalley

FACULTY OF MEDICINE*   658

Dean: Professor D.R. Wood

Biochemistry Department*   659
Address: 9 Hyde Terrace, Leeds, LS9 9LS
Head: Professor Donald S. Robinson

Microbiology Department*   660
Address: Old Medical School, Thoresby Place, Leeds, LS2 9NL
Senior staff: Professor Mary E. Cooke, Professor D.H. Watson

Pharmacology Department*   661
Senior staff: Professor A.M. Barrett, Professor D.R. Wood

FACULTY OF SCIENCE* 662

Dean: Dr T.D. Talintyre

School of Biological Sciences* 663

Chairman: Professor A.C.T. North
Departments: The School consists of the following departments in three faculties: Biophysics (Astbury Department of), Genetics, Plant Sciences, Pure and Applied Zoology (Faculty of Science); Animal Physiology and Nutrition (Faculty of Applied Science); Biochemistry, Microbiology, Pharmacology, Physiology (Faculty of Medicine).

Astbury Department of Biophysics* 664
Head: Professor A.C.T. North
Activities: Three-dimensional structure of biological macromolecules, especially proteins, polysaccharides and drugs; relationship between structure and function of enzymes and the structural basis of drug/receptor interactions.

Genetics Department* 665
Head: Professor D.T. Cove

Plant Sciences Department 666
Telephone: (0532) 431751
Head: Professor J. Elston
Senior staff: Professor G.F.Leedale
Graduate research staff: 34 (1982)
Activities: Agricultural chemistry: pesticide chemistry and use, plant nutrition, soil biology; crop science: agricultural meteorology, plant pathology; plant biology: fresh water and marine algae, electron microscopy, genetics and plant breeding, marine biology, plant ecology, plant physiology.

Pure and Applied Zoology Department 667
Head: Professor D.L. Lee
Senior staff: Professor R.McN. Alexander
Activities: Parasitic diseases of farm animals; insect and nematode pests of crops.

Wellcome Marine Laboratory* 668
Address: Robin Hood's Bay, Whitby, North Yorkshire YO22 4SL
Director: J.R. Lewis

Earth Sciences Department* 669
Chairman: Professor E.H. Francis
Senior staff: Professor J.C. Briden, Professor E.H. Francis

University of Leicester* 670

Address: University Road, Leicester, LE1 7RH
Telephone: (0533) 554455
Status: Educational establishment with r&d capability
Vice-Chancellor: M. Shock

FACULTY OF SCIENCE* 671
Dean: Professor R. Whittam

School of Biology* 672
Chairman: Professor H.C. Macgregor
Graduate research staff: 99

Botany Department 673
Head: Professor H. Smith
Projects: Identification of sources of microfossils etc (Professor H. Smith); the breeding relationships and ancestry Festuca rubra aggregate (Dr C.A. Stace).

Zoology Department 674
Head: Professor H.C. Macgregor
Projects: Sequence analysis of transcribed satellites of Triturus Christatus; chromosome organization and evolution in amphibia (Professor H.C. Macgregor).

LEICESTER BIOCENTRE 675
Status: University-based contract research institute
Head, research programme: Professor Barry Holland
Graduate research staff: 12 (1982)
Activities: The centre, established in 1982, is housed on the site of the University of Leicester, and the running costs are shared between several industrial sponsors including Dalgety Spillers Limited, Gallaher Limited, John Brown Engineers and Constructors Limited, and Whitbread and Company plc. Representatives from these bodies form the management committee of the biocentre.
The centre undertakes a core programme of fundamental scientific research, complemented by other, more applied research, sponsored on a contract basis, with particular emphasis on the use of genetic manipulation technology, initial emphases being on the gene structure of yeasts and higher plants. The core research programme involves three themes: molecular biology of yeast plasmid replication; studies of plant gene expression in yeast and in bacteria; and mechanistic studies of protein secretion in yeast.

University of Liverpool   676

Address: Senate House, Abercromby Square, PO Box 147, Liverpool, L69 3BX
Telephone: (051) 709 6022
Telex: 627095
Status: Educational establishment with r&d capability
Vice-Chancellor: Professor R.F. Whelan
Research committee, 1981-82: Professor A.M. Breckenridge, chairman; Professor R.F. Whelan, Vice-Chancellor; D.H. Jennings, Pro-Vice-Chancellor; Professor D. Hull, chairman of grants committee; Professor P.E.H. Hair, chairman of committee on university awards; Professor P. Holmes, chairman of PhD committee; Professor J.M. Cassels, chairman of university research fellowships selection committee; Professor J.L. Alty; Professor A.D. Bradshaw; Professor C.A. Finn; Dr K.A. Kitchen; Professor A.P.L. Minford; Professor R. Shields
Graduate research staff: 3 000 (1982)
Activities: 1980-81 research funds were received from the following sources: research councils, £2.8m; government departments, £0.1m; other bodies, £1.5m.
Publications: Annual research report.

ANIMAL HUSBANDRY DEPARTMENT   677

Research correspondent: Professor J.O.L. King
Activities: Joint work with the Department of Veterinary Pathology on the role of nutrition and husbandry in the pathogenesis of Escherichia coli infections, and on the occurrence of gastric abnormalities in the pig; selection programme in the pure bred sheep flock: development of a two trait index figure for each ewe, measurement of testis size as indicator of prolificacy in the male; artificial rearing of lambs: formulation of diet giving optimum growth rate and efficient food utilization; neonatal calf health and growth; collaborative work with the Department of Veterinary Pathology on the role of husbandry factors in the pathogenesis of calf enteric and respiratory disease.

BIOCHEMISTRY DEPARTMENT   678

Research correspondent: Professor T.W. Goodwin
Senior staff: Professor J. Glover
Activities: DNA biochemistry; genetic engineering; Fast Atom Bombardment mass spectometry; structural elucidation of biological compunds; moulting hormones in ticks and nematodes; control of development of the free-living nematode, Caenorhabditis elegans; hexokinase studies; interactions of new insecticides with mixed function oxidases in insects.
Projects: Nature of the hormonal systems controlling filarial nematode development (Dr Rees); structure and regulatory characteristics of hexokinase isoenzymes; theory of the time-response of enzyme sequences, including metabolic pathways; regulation of glycolytic activity in normal and ischaemic heart through intracellular location of enzymes (Dr Easterby); isolation and structural elucidation of transfer ribonucleic acids from the chloroplast of the seaweed, Codium fragile, and location of their genes on the chloroplast, genome; effects of iron restriction on modification to transfer ribonucleic acid and transcriptional control of aromatic amino acid biosynthesis in Pseudomonas aerquinosa (Dr Jones); plant sterol side ·chain biosynthesis on a steroid hormone metabolism during amphibian oocyte maturation; interconversions of steroids in birds during egg development (Dr Goad); intracellular proteinases (Dr Beynon); function of membranes and control of metabolism in yeast (Dr Haslam); molecular biology of plant viruses (Dr Wilson); industrial application of affinity chromatography with special reference to dye ligand biochemistry (Dr Peter Dean); photosynthesis: means by which the activity of a number of Calvin Cycle enzymes is controlled (Dr Powls); biosynthetic· problems; carotenoprotein (Dr Britton); blue-green bacteria (Dr Carr).

BOTANY DEPARTMENT   679

Research correspondent: Professor A.D. Bradshaw
Senior staff: Professor D.H. Jennings
Activities: The population biology of plants is a major interest of the department, which conducts a mixture of pure and applied research. Applied research ranges from heavy metal sensitivity of algae in tropical sewage treatment systems to hybridization processes in potato blight.
Projects: Possible role of the introduced grass carp as an alternative to herbicides (Dr Eaton); population dynamics of common agricultural weeds (Dr Mortimer); evalution of herbicide resistance in weeds (Dr Putwain); heavy metals: mechanism of metal tolerance (Dr Thurman); genetic exchange in streptomycetes (Dr Williams); development of engineering designs to optimize biological processes in sewage. treatment systems in warm European countries (Dr Pearson).

CENTRE FOR MARINE AND COASTAL   680 STUDIES

Director: Dr D.F. Shaw
Activities: This Centre, created in 1980, is a focal print for marine and coastal research work being carried out in a wide range of departments within the university. It provides a single point to which outside enquiries can be directed and acts as a forum for the encouragement of inter-disciplinary work.
Associated departments cover biochemistry, botany, civic design, civil engineering, computing, economics, geography, geology, geophysics, marine biology, marine transport, mathematics, mechanical engineering, oceanography, parasitology, veterinary pathology, and zoology.

ENVIRONMENTAL ADVISORY UNIT   681

Director: Dr G.D.R. Parry
Graduate research staff: 7 (1982)
Activities: This self-funding advisory and research unit of the university is situated in the Botany Department but incorporates a large range of environmental research skills. Major areas of interest are environmental survey, pollution, and land reclamation. Current research contracts include the following: site assessment techniques for contaminated land surveys; the movement of contaminants in reclaimed sites; evaluation of cover materials for waste disposal; environmental and ecological surveys of the Mersey estuary; environmental impact of new mineral developments.

GENETICS DEPARTMENT   682

Research correspondent: Professor D.A. Ritchie
Activities: Repair of damaged DNA; molecular genetics; population studies; development of new plasmid vectors for genetic engineering; recombination processes in bacteriophage TI; nitrate metabolism in Aspergillus nidulans; taxonomic genetics of streptomyces species. Grants to support this work have been obtained from the North West Cancer Research Fund, the Medical Research Council, and the Science and Engineering Research Council.
Projects: DNA repair in bacterial cells: genes carried by two unrelated bacterial plasmids which both affect the accuracy with which repair processes operate (Dr Strike); ultrastructural changes in eukaryotic chromosomes and chromatin during the activation of ribonucleic acid transcription, in the oocytes of amphibians and insects (Dr R.S. Hill); hybrich dysgenesis in Drosophila melanogaster: induction of male recombination events (Dr P. Eggleston); new quantitive approach to the measurement and analysis of competitive interactions (Dr P. Eggleston); bacteriophage TI (Professor D.a. Ritchie); comparative gentics of wild isolates of streptomyces (Professor D.A. Ritchie); the fungus Aspergillus nidulans: regulation of nitrate assimilation (Dr A.B. Tomsett); maintenance of a palatability spectrum (polymorphism) in the butterfly Danaus chrysippus (Dr D.O. Gibson); development of cereal variety mixtures (Dr J. Barrett); occurence and chromosomal characterization of free-martinism in cattle (Dr J.J.B. Gill).

MARINE BIOLOGY DEPARTMENT   683

Address: Port Erin, Isle of Man
Telephone: (0624) 832027
Acting director: Dr D.I. Williamson
Activities: Irish Sea fisheries; mariculture; algal biomass; pollutants; nutrients; ecology, physiology, development, growth, behaviour, and systematics of marine organisms.
Publications: Collected reprints (annual).

MEDICAL ENTOMOLOGY DEPARTMENT   684

Research correspondent: Professor W.W. MacDonald
Activities: Filariasis and its vectors; dynamics of mosquito and housefly populations.
Projects: Linkage relationships of three genes governing susceptibility of Aedes aegypti to three species of Dirofilaria and Brugia (H.A. Matthews); the Simulium damnosum complex in West Africa (Dr R.J. Post); taxonomic study of the non-onspheline mosquitoes of the Afrotropical region (Dr Service).

VETERINARY ANATOMY DEPARTMENT   685

Head: Professor A.S. King
Activities: Morphometry of avian lung and of the lung of the bat; relationship between the structure of avian bones and the forces acting upon them: in collaboration with members of the Departments of Dental Sciences and Civil Engineering; localization and surgical removal of cysts of Coenurus cerebralis in sheep: in collaboration with members of the Department of Veterinary Preventive Medicine.

VETERINARY CLINICAL STUDIES DEPARTMENT   686

Head : Professor R.J. Fitzpatrick
Senior staff: Professor E.J.H. Ford, Dr J.E. Cox
Activities: Farm animals: medicine, surgery, radiology, reproduction. Topics investigated include the following. mechanism of the toxicity of organic preparations of copper; factors affecting the toxic response of calves to organophosphorus insecticides; horses and ruminants: diagnostic importance of the measurement of enzyme activity in the blood after release from liver cells, together with the measurement of liver blood flow; effect of various types of tissue damage on the release of S-nucleotidase into plasma; disturbances of energy metabolism; reproductive endocrinology: control of ovulation in sheep; transfer of early embryos between high grade, expensive, donor cows and low grade, cheap, foster mothers; adrenal influence on reproductive function in bulls; tumours of the orbit and eyelid of horses: treatment by radiotherapy and immunotherapy.

VETERINARY FIELD STATION 687

Address: Neston, Wirral, Cheshire L64 7TE
Telephone: (051) 336 3921
Sections: Animal husbandry, Dr T.L.J. Lawrence
Graduate research staff: 6 (1982)
Activities: Farm animals: nutrition, breeding, and environmental studies. The departments of animal husbandry, veterinary clinical studies, and veterinary preventive medicine are situated at the station.

VETERINARY PARASITOLOGY 688 DEPARTMENT

Head: Dr W.N. Beesley
Activities: Internal parasites of horses; immunity to coccidiosis in poultry; Onchocerca worms of British cattle; sheep: strain of gut nematodes which demonstrates resistance to benzimidazole compounds. Outside the university area studies are undertaken on parasite ecology and control in the flock of North Ronaldsay sheep maintained by the Rare Breeds Survival Trust on Linga Holm in Orkney.

VETERINARY PATHOLOGY 689 DEPARTMENT

Head: Professor D.F. Kelly
Senior staff: Dr R.M. Batt
Activities: Inherited chondrodystrophy in the Pointer dog; naturally-occuring small intestinal disease in the dog; diseases of wild grey seals; role of husbandry and diet in the pathogenesis of Escherichia coli infections of pigs and ulceration of the non-glandular zone of the porcine stomach: joint project with the Department of Animal Husbandry.

VETERINARY PHYSIOLOGY AND 690 PHARMACOLOGY DEPARTMENT

Head: Professor C.A. Finn
Senior staff: Dr A. Knifton, T. Nicholson
Activities: Nature and characteristics of synaptic transmission in the cerebellum and the mechanisms by which prolonged transmitter action leads to the death of brain cells; factors affecting ruminant stomach motility; effects of thyroid hormones upon the mechanical and biochemical properties for mammalian skeletal muscle; control of interine motility in the goat and the role of oxytocin in cyclic ovarian activity; pharmacokinetics of anticonvulsant drugs in dogs; effect of different routes of oxytetracycline administration on the attainment of therapeutic blood levels in dogs; decidual cell reaction in mice: breakdown of the decidua on cessation of hormone treatment.

VETERINARY PREVENTIVE MEDICINE 691 DEPARTMENT

Head: Professor M.J. Clarkson
Senior staff: Dr J.R. Walton, G.S. Walton; avian medicine, Dr F.T.W. Jordan
Activities: Conditions affecting food producing animals (including cattle, sheep, and pigs), with the emphasis on herd or flock problems. Research topics include the following: effect of growth promoting agents on the intestinal villi and underlying lamina propria of poultry and pigs; cutaneous responses to staphylococci and streptococci; avian medicine, including a study of leg weakness among turkeys and chickens; role of plant pollens in contact allergies in domestic animals; hydatid disease; chlamydial abortion in sheep; hypocuprosis, muscular dystrophy, coccidiosis, and toxoplasmosis in sheep.

ZOOLOGY DEPARTMENT 692

Address: Brownlow Street, PO Box 147, Liverpool, L69 3BX
Head: Dr R.G. Pearson
Groups: Cell physiology, Professor C.J. Duncan; endocrinology, Dr D.M. Ensor; freshwater biology, Dr J.O. Young; parasitology, Dr J.C. Chubb; freshwater fisheries, Dr K. O'Hara; population ecology, Professor A.J. Cain
Graduate research staff: 29 (1982)
Activities: Physiology: effects of calcium on various cell systems; effects of temperature on membranes. Freshwater fisheries, freshwater biology, endocrinology, parasitology: ecology of British freshwaters, particularly the biological effects of pollutants in lacustrine, riverine, and estuarine sites. Endocrinology: controls and feedback systems in which prolactin is involved.

University of Manchester 693

Address: Oxford Road, Manchester, M13 9PL
Telephone: (061) 273 3333
Status: Educational establishment with r&d capability
Vice-Chancellor: Professor M.H. Richmond

FACULTY OF ECONOMIC AND 694 SOCIAL STUDIES

Dean: S.A. Moore

Agricultural Economics Department 695

Address: Dover Street, Manchester, M13 9PL
Telephone: (061) 273 7121
Head: Professor D.R. Colman
Graduate research staff: 8
Activities: The department was originally established by the university in association with the Ministry of Agriculture, Fisheries, and Food. It undertakes investigations in the fields of farm incomes and farm costs, with some advisory work on farm and horticultural management and related aspects, in the counties of Lancashire, Cheshire, Shropshire, and Staffordshire, largely on contract to the ministry.
Data sources for research are mainly within the United Kingdom and the European Economic Community, but also extend to less developed countries and North America. Most of the department's research has marked quantitative and theoretical elements, covering the application of economic and econometric analysis to policy problems in agriculture, fisheries, and natural resources.
Projects: Analysis of efficiency of transmission of agricultural policy prices; measurement of aggregate agricultural supply response (Professor D.R. Colman); analysis of impact of generic milk and meat advertising (Dr J. Strak); simulation of impact of agricultural price policies on agricultural input markets (Dr W.B. Traill); economic study of world turkey industry (D.I.S. Richardson); measurement of fishing effort and effect of fishery licensing arrangements (J.A. Butlin); estimating productive efficiency in Italian agriculture (Dr N. Russell); measurement of economic efficiency in agriculture (Dr T. Young).
Publications: Annual report.

FACULTY OF MEDICINE* 696

Dean: Professor J.M. Evanson

Bacteriology and Virology Department 697

Address: Stopford Building, Oxford Road, Manchester, M13 9PT
Telephone: (061) 273 8241
Telex: Manumed, Manchester
Head: Professor A. Percival
Sections: Bacteriology, Professor A. Percival, Dr N.W. Preston; virology, Professor Maurice Longson; molecular microbiology, Professor M.H. Richmond; oral microbiology, Dr Conrad Russell
Activities: Immunology of trauma; development of immunodiagnostic tests for fish diseases, development of vaccines for fish diseases and studies of fish vaccination; occupational bladder cancer; carcinogenesis and testing of compounds for anti-carcinogenic properties.

FACULTY OF SCIENCE* 698

Dean: Professor J. Zussmann

Botany Department 699

Address: Williamson Building, Oxford Road, Manchester, M13 9PL
Head: Professor Elizabeth G. Cutter
Sections: Cryptogamic botany, Professor A.P.J. Trinci; cytology, Dr R.D. Butler; genetics, Dr R.J. Wood
Activities: Plant pathology: biology of diseases of agricultural and horticultural crops caused by fungi, bacteria and viruses in temperate and tropical regions; control of crop diseases; methods of testing the efficiency of fungicides.
Applied crop physiology: hydroponics; commercial application of artificial illumination.

Pollution Research Unit 700

Telephone: (061) 273 7121
Director: Professor M. Gibbons
Assistant Directors: Dr E.G. Bellinger, Dr R.F. Griffiths
Activities: Investigations into waste water treatment and the treatment of industrial pollutants; monitoring of air pollutants with special reference to heavy metals; monitoring of pollutants within the working environment; fouling of marine structures, including ships and buoys; use of organisms as monitors of pollution; chemical monitoring of freshwater pollutants; removal of nitrogen from waters by biological methods.
The unit is a joint venture by the University of Manchester and the University of Manchester Institute of Science and Technology. The subject is taught and researched on an interdisciplinary basis, with research laboratories on both campuses. The unit is active in work for industry, using its own resources and, where appropriate, the facilities, equipment, and expertise available throughout both the university and UMIST. Such work can range from sampling and analysis of site problems with staff consultancy to the setting up of longer-term research and development studies.

Zoology Department 701

Head: Professor D.M. Guthrie
Sections: Zoology, Professor D.M. Guthrie, Dr L.M. Cook, Dr R.R. Baker; cytology, Dr R.D. Butler; entomology, Dr R.R. Askew; genetics, Dr R.J. Wood; Honorary Professor of Marine Biology, Professor D.J. Crisp

UNIVERSITY INSTITUTIONS 702

Manchester Museum 703

Director: A. Warhurst
Deputy director: Dr R.M.C. Eagar
Keepers: Archaeology, Dr A.J.N.W. Prag; botany, Dr J.W. Franks; conservation, C.V. Horie; entomology, C. Johnson; geology, Dr R.M.C. Eagar; zoology, Dr M.V. Hounsome
Activities: Promotion of life and earth sciences, including botany, entomology, geology, and zoology, together with archaeology, Egyptology, ethnology and numismatics.

University of Newcastle Upon Tyne 704

Address: Newcastle Upon Tyne, NE1 7RU
Telephone: (0632) 328511
Telex: 53654
Status: Educational establishment with r&d capability
Vice-Chancellor: Professor Laurence Martin
Registrar: W.R. Andrew

FACULTY OF AGRICULTURE 705

Address: 6 Kensington Terrace, Newcastle upon Tyne
Telephone: (0632) 32 85 11
Dean: Dr Robert J. Thomas
Graduate research staff: 81 (in all departments)
Activities: The University has farms at Cockle Park and Nafferton and a field station at Close House, which provide a wide range of experimental and commercial conditions for the work of all departments in the Faculty.
Projects: Efficacy of land drainage systems (Dr S. Wilcockson); reclamation of opencasted land; clover nitrogen contribution to grassland productivity (Dr A. Younger); agronomy of cereals and oil seed rape (Dr E.J. Evans); selection, breeding, and nutrition of pigs (M. Ellis); silage feeding and protein supplementation of lactating dairy cows (Dr P. Rowlinson); gastro-intestinal parasitology of grazing livestock and intensively managed pigs (Dr R.J. Thomas); control and environment in crop stores (T.T. McCarthy); low cost heat exchangers for agricultural applications; thermophilic oxidation of animal wastes (Dr J.L. Woods); underwater earthmoving and anchoring (Dr A.R. Reece); tractors for developing countries (Dr B.M.D. Wills); mechanics of root growth in compacted soils (Dr D. Hettiaratchi); studies in digestion in the ruminant animal with particular reference to amino acid supply and the fate of carbohydrates; factors affecting magnesium absorption and utilization in the ruminent (Professor D.G. Armstrong); nitrogen recycling in the digestive tract of the ruminant and rabbit (Dr D. Parker); lipid contents of oil seeds and their fate in the rumen (R. Smithard);

relationship between nutrition and endocrine response in ruminants; Selection indices for pig breeding based on biochemical parameters (Dr T.E.C. Weekes); rumen microbiology; the use of continuous fermentors alternatives for United Kingdom agricultural policy (K.J. Thomson, A.E. Buckwell, Dr D.R. Harvey); rural depopulation in England and Wales (M.C. Whitby).

Department of Agricultural Biochemistry and Nutrition 706

Head: Professor David G. Armstrong
Activities: Areas of research include: ruminant digestion; metabolic pathways of end products of digestion; the metabolism of magnesium in ruminants, and of calcium and phosphorus in pigs and rats; protein quality evaluation using amino acid profiles and rat assay techniques; endocrinological studies in pigs and ruminants.

Department of Agricultural Biology 707

Head: Professor Gordon E. Russell
Activities: Areas of research include: biology, taxonomy, distribution, population dynamics and behaviour of invertebrates; insect pathology; biological control of pests; environmental effects on animals in soil and other habitats, including reclaimed industrial sites; diseases of cereals, grasses and sugar beet; chemical methods of crop protection; natural products controlling pest behaviour; controlled release pesticides; seasonal variation in growth of crop plants; biochemistry of plant responses to low temperatures of seed viability; agronomy of oil and protein producing crop plants; cytogenetics of polyploid crop plants; inheritance of yield components in the potato; microbial transformations and fixation of nitrogen.

Department of Agricultural Economics 708

Head: Professor John Ashton
Activities: Research into: economic analysis of the agricultural industry and the rural economy; farm management and enterprise analysis; agricultural input and output markets; land use studies; rural industrialization; agricultural trade and development.

Agricultural Adjustment Unit 709

Director: Professor John Ashton
Activities: Areas of research include: agricultural policy, national and international trade and commodity studies; computer models of agricultural systems; problems of rural land use and resource development including employment and recreational studies; farm management including long-term surveys of farm accounts and enterprises on North of England farms.

Department of Agricultural Engineering 710

Head: Derek J. Greig
Activities: Areas of research include: agricultural machinery design; soil-vehicle mechanics; critical state soil mechanics; mechanism of root growth in compacted soils; direct drilling machinery; sea-bed soil mechanics; crop processing, drying and storage; soil water and nitrogen movement; drainage of layered soils; pollution and utilization of waste; slurry handling; thermophilic fermentation of wastes; simulation of agricultural enterprises and mechanization; electronics, instrumentation and control applications.

Department of Agricultural Marketing 711

Head: Professor Christopher Ritson
Activities: Areas of research include: structure and performance of the UK livestock and meat markets; comparative analysis of objectives and behaviour in agricultural marketing; exploration of consumer attitudes to foods; consumer taste discrimination; scaling techniques and image measurement; consumer and farmer decision-making; agricultural cooperation, and direct farm marketing; European Economic Community marketing policies.

Department of Agriculture 712

Head: Professor Gordon R. Dickson
Activities: Areas of research include: animal breeding and genetics - selection studies in pig populations; animal production - parasitology in cattle, sheep and pigs, host-parasite relationships, climatic effects and pasture infestations; nutritional studies in dairy cattle, beef production and pigs; crop production - agronomy of oilseed crops, forage peas and forage roots, grassland production and hill sward improvement, opencasted site restoration problems.

Department of Soil Science 713

Head: Professor Peter W. Arnold
Activities: Areas of research include: fundamental physical, chemical and mineralogical studies; response of soil to applied mechanical stress, surface charge characteristics of highly weathered soils, cation adsorption by soils; soil reclamation problems on industrial sites; aspects of slurry utilization; nutrient balance in upland soils; pedogenesis and soil-vegetation relationships in Malaysia, Nigeria and central Brazil; land classification and nature conservancy.

FACULTY OF MEDICINE* 714

Graduate research staff: 89 (in medicine and dentistry)

Medical School* 715

Dean: Professor Sir John Walton
Postgraduate Dean: Professor James Parkhouse
Associate Dean: Professor David A. Shaw

Department of Anatomy* 716
Head: : Professor J.S.G. Miller
Activities: Areas of interest include: the neural control of movement in animals and man and its relationship to the rehabilitation of neurological patients; light microscopical and ultrastructural studies of nerve cells; investigation of the morphological basis of learning in the chick; neurophysical and neuroanatomical studies of cells in cerebral cortex in relation to consciousness; basic immunology; developmental biology; stereological approaches to organ growth.

FACULTY OF SCIENCE* 717

Dean: Professor D.G. Murchison
Graduate research staff: 241 (in all departments)

Department of Genetics* 718
Head: Professor S.W. Glover
Activities: Areas of research include: genetic control of restriction and modification of DNA in bacteria; genetic and physical properties of bacterial plasmids; tissue culture techniques for research in higher plant genetics, mutant selection in plant cell cultures; genetic control of cyanoglucoside metabolism in Trifolium repens; cytogenetics and development of a Drosophila melanogaster translocation heterozygote; control of gene expression using the fungus Aspergillus nidulans; protein and glycoprotein synthesis in tissue culture animal cells infected with strains and mutants of the paramyxovirus, Newcastle Disease Virus.

Department of Geography* 719
Head and Professor of Regional Development Studies: Professor J.B. Goddard
Senior staff: Professor E.S. Simpson
Activities: Areas of research include: urban and regional studies - new towns, telecommunications and regional development, office location, labour mobility, inner city development, marketing, migration and urban development in southern Africa; Caribbean economic development; transport studies; spatial organization of elections; agrarian revolution in East Anglia and Northumberland; classification of land; geomorphology; climatology; biogeography-environmental studies of West Africa, the Sahelian zone of the Niger, South East Mauritania.

Department of Plant Biology   **720**
Head: Dr A.W. Davison
Graduate research staff: 12 (1982)
Activities: Fouling of marine structures: basic research on settling and growth of organisms and effects on paint and metal; control of commercial mushroom diseases; environmental monitoring and problems associated with monitoring of fluoride; application of electron microscopy to applied problems such as causes of air filter failure, structure of paper etc.

Department of Zoology *   **721**
Head and Professor of Comparative Physiology: Professor John Shaw
Senior staff: Zoology, Professor R.B. Clark

Dove Marine Laboratory *   **722**
Activities: Subjects of research include: osmoregulation and ion transport in crustaceans and insects; neurophysiology of insect behaviour; biochemistry and physiology of parasitic organisms, with special reference to plant behaviour of polychaetes; mate selection and flock behaviour in birds; locomotion of polychaetes; vertebrate palaeontology; mimicry in tropical butterflies; ecology of lakes and streams with emphasis on the energetics of benthic organisms; production and stability of marine benthic communities, marine zooplankton production, reproductive biology of marine invertebrates, consequences of effluent disposal including heavy metals into the sea; tropical marine biology.

University of Nottingham   723

Address: University Park, Nottingham, NG7 2RD
Telephone: (0602) 56101
Telex: 37346 NUSCLB G
Status: Educational establishment with r&d capability

FACULTY OF AGRICULTURAL   724
SCIENCE

Address: School of Agriculture, Sutton Bonington, Loughborough, Leicestershire LE12 5RD
Telephone: (05097) 2386
Dean: Professor W.J. Whittington
Graduate research staff: 79 (1981)
Activities: Soil science; plant production (all temperate crops); crop husbandry; plant protection; plant physiology; plant pathology; animal production (cattle, sheep, pigs, poultry); livestock breeding; livestock husbandry and nutrition; agricultural engineering; food science; food economics; farm management and economics; environmental science; microclimatology of tropical crops; microbiology and bacteriology. The Biometry Unit deals with the mathematical and statistical aspects.
Publications: Triennial report.

Department of Agriculture and   **725**
Horticulture

Head: Professor J.D. Ivins
Sections: Agronomy, Dr P.D. Hebblethwaite; animal production, Dr D.J.A. Cole; farm management, H.W.T. Kerr; horticulture, Dr P.G. Alderson
Graduate research staff: 24 (1982)
Activities: Horticultural crops: vegetative propagation; plant tissue culture; diseases caused by viruses; effects of within-plant competition on vegetative and reproductive growth; control of flowering and shoot elongation.
Nutrition, production, and factors affecting reproductive performance of pigs; energy and protein nutrition of pigs and poultry; nutrition and production of dairy and beef cattle; control of reproduction in sheep.
Physiological aspects of grass seed production, peas, and other arable crops; agronomy and crop physiology; mechanization of raising and transplanting seedlings; drying, processing, and utilization of whole crop cereals.
Accounting and budgeting techniques for agriculture; studies of production economics of arable and livestock enterprises and influencing farmers' incomes; capital investment; marketing of meat, milk, and milk products; direct marketing methods; parameters of efficiency in milk production; economic and social factors associated with family-run dairy farms in Derbyshire; application and effectiveness of advisory services; development of personnel management techniques for agricultural work situations and the training of individual workers.
Publications: Farming in the East Midlands.

Department of Applied Biochemistry   **726**
and Food Science

Head: Professor D. Lewis
Graduate research staff: 50 (1982)
Activities: Protein metabolism in animals with special reference to growth and the control of protein deposition; protein synthesis and degradation in muscle; purine and pyrimidine metabolism; nitrogen transactors in the ruminant digestive tract; amino acid needs of poultry, pigs, and ruminants; regulation of feeding in poultry; egg shell quality; fats in ruminant and poultry diets; endogenous amino acid oxidation; iron availability from food; myofibril assembly, disassembly, phosphorylation, and function; lipid and protein biosynthesis in developing rapeseed; softening associated with food ripening; cell wall degradation; ethylene synthesis and respiration; abscission; meat science; meat pigments; recovery and texturization of proteins from wasted and under-utilized animal sources; utilization of vegetable proteins in meat products; storage-life of meat products; unequivocal assay of meat proteins in processed foodstuffs; post-slaughter electrical stimulation and accelerated tenderizing; calorimetry of meat proteins; intermediate moisture technology; iron availability from protein foods; natural colours; pectin biochemistry; computer control and simulation; starch granule structure and

function; polysaccharide structure in food systems as studied by X-ray, light, and neutron scattering, electron spin resonance and nuclear magnetic resonance; physical chemistry of starch-based snacks and frozen confectionery; food rheology; functional properties of proteins and polysaccharides; food extrusion.

Department of Physiology and Environmental Science 727

Head: Professor W.J. Whittington
Sections: Animal physiology, Professor G.E. Lamming; environmental physics, Professor J.L. Monteith; plant science, Professor W.J. Whittington
Graduate research staff: 36 (1982)
Activities: Genotype-environment interactions and crop growth; environmental factors affecting the incidence of brown rust of barley; characterization of cauliflower mosaic virus gene expression; synthesis and function of polygalacturonase in ripening tomatoes; effects of atmospheric saturation deficit on crop establishment; soil compaction and plant growth; relation between net radiative exchange and solar radiation; modelling the direct effects of temperature, humidity, wind, and radiation on animals; control of ovarian development following weaning in the sow; Agricultural Research Council research group on hormones and farm animal reproduction; control of gonadotrophin secretion and ovarian function in domestic animals; use of Gn-RH preparations for improving reproductive performance in cattle and sheep; transmembrane transport of amino acids in the mammary gland: in vitro studies.
Projects: Photosynthesis and transpiration of crops: heat balance of animals; instrumentation for micrometeorology and crop ecology (J.L. Monteith); uptake and distribution of inorganic plant nutrients and heavy metals; earthworm colonization of opencast coal sites (Dr M.J. Armstrong); aerodynamic transfer in and above crop canopies; greenhouse climate; climatic physiology of poultry and livestock (Dr J.A. Clark); cycling of nutrients and pollutants between soil and plant; uptake of air-borne pollutants; effects of heavy metals on biochemical processes in soils (Dr D.V. Crawford); heat exchange between farm animals and their environment (Dr A.J. McArthur); storage and movement of water in soils; profile drainage; extraction of water by crop roots: soil structure, shrinkage, and swelling (Dr M. McGowan); biology and ecology of soil micro-arthropods and soil nematodes; soil-opulation studies of open-cast coal-mined sites; arboreal and grass-herbage mites; predator-prey relationships of mesostigmatic mites; light-trap studies of Macrolepidoptera (Dr P.W. Murphy); populations of delphacid planthoppers in hedgeside vegetation and cereal crops; diapause and rate of developments in planthoppers (J.Y. Ritchie); radiation climatology; uptake and effects of sulphur plants; atmospheric transport of spray droplets to crops (Dr M.H. Unsworth).

FACULTY OF PURE SCIENCE 728
Dean: Professor Leslie Crombie

Department of Botany 729
Head: Professor E.C.D. Cocking
Activities: Plant genetic manipulations (including fusion and transformation of protoplasts); culture of plant cells; aspects of plant pathology; cell wall synthesis; fine structure studies of plant cells; physiological ecology of plants growing in water-logged habitats; botanical aspects of water pollution; cytology, physiology, and biochemistry of fungi; ecology of microorganisms; interrelationships of the soil microflora.

Department of Zoology 730
Head: Professor P.N.R. Usherwood
Activities: Facilities are available for postgraduate research in a variety of fields, including developmental zoology, cell biology, ecology, animal behaviour, ecophysiology, fish, insect, and reptile physiology, osmoregulation and excretion, parasitology, neurobiology (including neuropharmacology), and digestive physiology.
The department encourages interdisciplinary research with other departments in the faculty, with departments in the Faculties of Medicine and Agricultural Science, and with industrial organizations through the Science Research Council CASE Award Scheme.

University of Oxford* 731

Address: University Offices, Wellington Square, Oxford, OX1 2JD
Telephone: (0865) 56747
Status: Educational establishment with r&d capability

FACULTY OF BIOLOGICAL AND AGRICULTURAL SCIENCES* 732

Department of Agricultural Science 733
Address: Parks Road, Oxford, OX1 3PF
Telephone: (0865) 57245
Telex: 83147 Agrox
Head: Professor D.C. Smith
Graduate research staff: 63 (1981)
Activities: Agricultural plant production; farm management; soil science.
Projects: Control and regulation of intracellular chlorella symbionts by host digestive cells in green hydra; biology of the establishment of the green hydra symbiosis; taxonomic studies on Solanum L. Section

Solanum (Professor D.C. Smith); nature of sludge: soil interaction; long-term effects of potentially harmful elements present in sewage sludge (Dr P.H.T. Beckett); isolation and identification of phosphate-solubilizing compounds in the rhizosphere of rape: Brassica napus (P.H. Nye, Dr R.E. White); biology of mast cells in the uterus; pre-natal mortality and the endocrine control of implantation and pregnancy in the vole, bank vole, and wood mouse (Dr J.R. Clarke); effect of root and shoot temperature on plant growth and solute uptake (P.H. Nye); physiological and biochemical basis of the selective action of phenoxyacetic and herbicides on plants; herbicide action (Dr B.C. Loughman); factors involved in vitamin E/Se deficiency (S.D. Baird).
Publications: Annual report.

Department of Botany   734

Address: South Parks Road, Oxford, OX1 3RA
Telephone: (0865) 54211
Head: Professor F.R. Whatley
Activities: Inter-departmental research involving this department covers the following: biochemical aspects of chloroplasts, peroxisomes, biosynthesis of phenolics, tissue cultures, fungal growth, seed germination and biological energetics; plant physiology; meristematic cell cycles and differentiation; cytogenetics, cytotaxonomy and chromosomal evolution; plant economy and taxonomy; mycology; the production of industrially-important biochemicals by tissue culture.

Department of Forestry,   735
Commonwealth Forestry Institute

Address: South Parks Road, Oxford, OX1 3RB
Telephone: (0865) 511431
Head: Professor M.E.D. Poore
Activities: Economic aspects of agriculture and forestry are dealt with over a wide field. Studies range from forest utilization and mechanization, forest management and working plans, to the planning of farming enterprises; aspects of land-use planning; social aspects; agricultural policy; supply, demand, and prices of agricultural products; econometric techniques applied to agricultural problems; agricultural planning and development in underdeveloped countries and the control of their growth-rates; population and fertility problems and techniques.

Unit of Invertebrate Virology   736
Address: 5 South Parks Road, Oxford, OX1 3UB
Telephone: (0865) 52081
Director: Professor T.W. Tinsley
Activities: Research topics include: viral diseases of insects, their structure and replication; immunological and biophysical characterization, and epidemiology; viruses of forest trees.

Unit of Tropical Silviculture*   737
Address: Commonwealth Forestry Institute, South Parks Road, Oxford, OX1 3RB
Telephone: (0865) 51 1431
Telex: 83147 FOROX
Head: Peter J. Wood
Graduate research staff: 18 (1983)
Activities: Tropical forest research, especially on social forestry, plantations, agroforestry, and rain forest management and mensuration. Other topics include forest botany, computing, and wood quality.
Projects: Tropical hardwoods for dry and arid areas; central American pines and hardwoods (C.E. Hughes); mycorrhizas of tropical pines (Dr M.H. Ivory); wood quality in provenance trials of Pinus caribaea and Pinus oocarpa (R.A. Plumptre); secondary tropical species data-base (Dr C.I. Goodwin-Bailey); research systems for agroforestry (P.J. Robinson); intensive study of tropical pine gene resources; genetics of tropical forest trees (Dr R.D. Barnes).
Publications: Annual report.

Department of Zoology   738

Address: South Parks Road, Oxford, OX1 3RB
Telephone: (0865) 56789
Head: Professor T.R.E. Southwood
Section: Entomology, Professor D. Spencer Smith
Activities: Inter-departmental work on: animal physiology; evolution (paleaentological and experimental); animal behaviour; the entomology group in Zoology covers areas such as insect population ecology, plant/insect relationships, the behaviour and epidemiology of parasitic insects, and insect-borne diseases as well as taxonomic studies. The animal ecology group in Zoology is concerned with: energy flow in ecosystems, particularly woodlands; population studies of wild mice, voles, owls, and tropical ecology in East Africa. The Edward Grey Institute is particularly concerned with the ecology and physiology of breeding birds and is making a series of studies of sea birds at its field station in Wales.

Institute of Agricultural Economics*   739

Address: Dartington House, Little Clarendon Street, Oxford, OX1 2HP
Telephone: (0865) 52921
Director: G.H. Peters
Graduate research staff: (1981)
Activities: The institute conducts research into farm organization; supply, demand and price relationships; labour economics and income distribution, land use policy, agricultural policies, and rural social organization in both the United Kingdom and the European Economic Community; international trade, agribusiness, and aspects of agricultural economics of developing countries.

Projects: Economic implications of changes in United Kingdom food and drink supply and demand (S.T. Parsons); methodology of farm economic studies in underdeveloped countries (Dr R.W. Palmer-Jones); socio-structural policy within the framework of the European Economic Community (Dr R. Fennell); land tenure in Western Europe (A.H. Maunder); agricultural policy, food prices, and land use policy in the United Kingdom and Europe (G.H. Peters); world food situation 1980-1990 (G.R. Allen); effect of technological change on employment and income distribution in agriculture (Dr G.J. Tyler).

University of Reading   740

Address: Whiteknights, Reading, Berkshire RG6 2AH
Telephone: (0734) 85123
Telex: 847813
Status: Educational establishment with r&d capability

FACULTY OF AGRICULTURE AND   741
FOOD

Dean: Professor G.M. Waites
Sub-dean: I.T. Cawte

Department of Agricultural Botany   742

Head: Dr R.W. Snaydon
Graduate research staff: 33
Activities: Research into biological systems which inhibit the maximization of usable yield and inter-specific hybridization in crop genera, especially Fragaria, Arachis, and Vicia; the evolutionary history, taxonomy and genetics of capsicum; the genetics, ecological genetics and breeding of Lupinus spp; physiological properties of vacuoles isolated from crop plants; factors affecting legume-Rhizobium symbiosis; the developmental physiology of wheat and barley; interactions between crop species, pasture species, and between crops and weeds; environmental control of crop growth, development and yield; the physiological and ecological basis of weed control methods; agricultural climatology, especially in tropical regions.

Department of Agricultural   743
Development Overseas *

Head: Professor A.H. Bunting

Department of Agricultural Economics   744

Address: 4 Earley Gate, Whiteknights Road, Reading, Berkshire
Head: Professor R.H. Tuck
Senior staff: Professor J.P. McInerney
Graduate research staff: 26
Activities: Research in agricultural and food economics. Environmental, land use and investment studies; historical studies; food and commodity studies, including the economics and structure of the food industry; management and production studies; marketing studies; overseas development studies; trade policies and policy studies.
Publications: Development studies; *Farm Business OATA* (annual).

Department of Agriculture and   745
Horticulture

Address: Earley Gate, Reading, Berkshire RG6 2AT
Head and Professor of Agricultural Systems: Professor C.R.W. Spedding
Senior staff: Crop production, Professor E.H. Roberts; applied entomology, Dr H.F. van Emden; horticulture, Professor G.F. Pegg
Graduate research staff: 42
Activities: Subjects of research in all major areas of agriculture and horticulture in the UK and overseas include: use and production of energy in agriculture; productivity of small-scale production units in agriculture; production by carp and grass carp; long term seed storage for genetic conservation; agronomy of potato, lupin, and mixed-cropping systems; green crop fractionation; control of growth and development in glasshouse crops; artificial light in horticultural crop production; ultra-intensive systems for apple production; interrelations of insects and plants and crop protection; physiology of tropical grain legumes. Farm based research units include: crop research unit (carrying out research on potatoes, seed storage, tropical legumes and mixed cropping). Animal production research unit (carrying out research on pigs, poultry, sheep and beef animals). Agricultural eugineering research workshop. Fish research unit. Horticultural research laboratory.
Publications: Research studies series.

Department of Food Science   746

Address: London Road, Reading, Berkshire RGl 5AQ
Telephone: (0734) 85234
Head: Professor H.E. Nursten
Graduate research staff: 40
Activities: Areas of research centred on 3 main themes are better use of natural resources (home and abroad) as human food; the improvement of human nutrition; and sensory and acceptability studies. Within these themes, particular attention is paid to proteins, lipids, and

vitamins, to plant and dairy produce, and to fundamental research on sensory evaluation.
Publications: Annual report.

Department of Soil Science   747

Address: London Road, Reading, Berkshire RG1 5AQ
Telephone: (0734) 875234
Head: Professor A. Wild
Departments: Soil physics, D. Payne; soil chemistry, D.L. Rowell; soil microbiology, Dr P.J. Harris; pedology and soil survey, Dr J.B. Dalrymple
Graduate research staff: 50
Activities: Areas of research include: soil survey; hydrology of catchments; paleosols; soil formation; physics of field soils and effect of cultivations; drainage and irrigation; land use; kinetics of soil enzymes; chemistry of nitrogen, phosphorus, potassium and other elements in soil and crop uptake; soil salinity; oxygen effects on seedlings. The general aim is to improve the knowledge of soils and their use throughout the world, with particular reference to Britain, the humid tropics and the seasonally arid tropics and laboratory, greenhouse and field work is undertaken on soils in all these regions. Major projects are being carried out on water use in relation to crop root development in the seasonally arid tropics (Middle East) with a view to increasing yields of barley and chick pea; phosphate in soils of the humid and seasonally arid tropics, especially on the potential use of ground rock phosphate; nitrogen leaching and cycling, including the use of ^{15}N, in UK ecosystems; proposed for further work in other countries; other nutrients - trace element deficiencies and metal toxicities in UK and Middle East soils; physical properties of soils subject to different cultivations, especially clay soils in southern England; microbiology of Rhizobium and V.A. mycorrhiza, and their interaction; pedogenesis using field observations, thin sectioning and other techniques, (especially in relation to slope); variability of soils as a problem in describing and mapping them; water movement in relation to catchments (humid tropics) and water use by crops.

National College of Food Technology*   748

Address: St George's Avenue, Weybridge, Surrey KT13 0DE
Telephone: (0932) 43991
Principal: Professor E.J. Rolfe
Senior staff: Professor N. Blakebrough
Activities: Subjects of research include: instrumental and sensory studies of food texture; hygienic design of food equipment; physical properties of foodstuffs; food dehydration; mathematical modelling of heat sterilization processes for foodstuffs; chemical and physiological studies on sugars and starches; extraction of vegetable proteins and their utilization; utilization of agricultural residues as carbon source in fermentation; thermal resis-

tance of microorganisms; microbiology of frozen foods and intermediate moisture foods; application of high pressure liquid chromatography to food analysis.

FACULTY OF SCIENCE   749

Dean: Professor A. Wild
Sub-dean: T.C.F. Sibly

Department of Applied Statistics   750

Head: Professor R.N. Curnow
Graduate research staff: 18
Activities: Research and advice on statistical methods in agriculture (particularly in developing countries), including the following 60 pics: nuclear accountacy and safety of nuclear reactors; experiments for intercropping; population and quantitative genetics; analysis of climatic data, particularly in relation to tropical agriculture; statistical computing; application of statistics to audiology, ecology, cell biology and epidemiology.
Projects: Agricultural epidemiology, (Dr A.J. Woods); agro-meteorological research, (Dr R.D. Stern); intercropping experimentation, spatial pattern and competition, (R. Mead); models of cell growth and division, quantitative genetics, (Professor R.N. Curnow); response surfaces and experimental design, (Dr D.J. Pike).

Department of Botany   751

Head: Professor V.H. Heywood
Senior staff: Professor J.R. Harborne, Professor D.M. Moore
Activities: Research includes the following topics: taxonomy and systematics of flowering plants; phytochemistry; mycology; cytogenetics; ecology; bryology; plant physiology; control of germination; hormonal control of plant development; light and plant development; inorganic nitrogen metabolism; cytophysical basis of plant reproduction; ultrastructural studies.

Department of Microbiology   752

Address: London Road, Reading, Berkshire RG1 5AQ
Head: Professor C. Kaplan
Activities: Research interests in microbial genetics, physiology and biochemistry include: microbiological aspects of the disposal of agricultural, domestic and industrial wastes, including production of liquid fuels; porphyrin biosynthesis and permeability in Escherichia coli; genetical analysis of biochemical variation; biochemistry of microorganisms using methanol, etc; biodegradation of alkyl ethers; problems with microbial degradation of liquid hydrocarbon fuels and lubricants; granulosis viruses of insect pests; immunology and virulence of viruses; epidemiology of Campylobacter from man and animals; microorganisms in the ecology of rivers.

Department of Physiology and Biochemistry 753

Address: Whiteknights, Reading, Berkshire
Telephone: (0734) 875123
Head: Dr N.K. Jenkins
Sections: Biochemistry, Professor R.R. Dils; physiology, Professor G.M.H. Waites
Graduate research staff: 24
Activities: Research into physiology and biochemistry in animal production, breeding and nutrition: molecular biology (control of nucleic acid synthesis, RNA polymerase), structure of chromatin; metabolism of foreign compounds, hepatic microsomal monoxygenases; reproductive physiology and biochemistry in male and female animals using sheep and chickens; biochemistry of mammary gland and lactation, mammary cell differentiation; mammalian metabolism and nutrition, digestion, liver metabolism and metabolic turnover; electrophysiology of ovulation and of secretion in the testis; respiratory physiology and its control through chemoreception; homeostasis of thirst and of mineral balance.

Department of Zoology 754

Senior staff: Professor K. Simkiss, Professor G. Williams
Activities: Research is concerned with a wide range of interests in both pure and applied aspects of zoology, including whole organism studies; physiology and biomaterials; heavy metal detoxification; ecology, freshwater biology, marine biology; behaviour and ecology of pest species of slugs; applied zoology linked to agriculture, ecology and control of pests.

FACULTY OF URBAN AND REGIONAL STUDIES 755

Dean: Professor W.D. Biggs
Sub-dean: P.J. Must

Department of Land Management and Development 756

Head and Professor of Land Management: Professor H.W.E. Davies
Senior staff: Land appraisal, Professor D. Cadman
Activities: Research covers problems of land use in town and country including land and development; the planning process and its application; aspects of valuation and appraisal; recreational land management.

UNIVERSITY ASSOCIATED INSTITUTES AND DEPARTMENTS 757

National Institute for Research in Dairying 758

Address: Shinfield, Reading, Berkshire RG2 9AT
Telephone: (0734) 883103
Affiliation: State-aided institute, Agricultural Research Council
Director and Head of Biological Sciences Division: Professor J.W.G. Porter
Deputy Director and Head of Production Division: Dr C.C. Balch
Head of Processing Division: Dr G.C. Cheeseman
Head of Statistics Department: D.R. Westgarth
Sections:
Biological Sciences Division - basic ruminant nutrition, Dr R.H. Smith; microbiology, Dr L.A. Mabbitt; nutrition, Dr M.I. Gurr; physiology, Dr Isabel A. Forsyth (acting); small animals unit, A. Mowlem
Production Division - dairy husbandry, Dr F.H. Dodd; feeding and metabolism, Dr J.H.B. Roy; applied pig nutrition, Dr K.G. Mitchell
Processing Division - analytical services, E. Florence; physical sciences, Dr E.W. Evans; process technology, Dr H. Burton
Graduate research staff: 110
Activities: The Institute's research programmes are organized on a multidisciplinary basis in the two broad areas of the production of milk and of its processing and utilization. The titles of the research programmes are as follows:
Production - dairy cow feeding; mastitis; reproduction; nutrition and metabolism of the young animal; nutritional processes in the ruminating bovine; metabolism in the lactating cow; physiology and biochemistry of lactation; applied pig nutrition.
Processing and utilization - quality of raw milk; processed liquid milk; cheese and fermented products; whey and new products; butter and cream; nutritional value of dairy products and related foods;
In each area a Coordinating Committee is responsible for the planning and detailed review of the programmes. A Research Advisory Committee, under the Chairmanship of the Director and which includes individuals from outside organizations who are actively engaged in r&d in both areas of the Institute's research, reviews the immediate and long-term objectives of the programmes and advises on research policy.
Projects: Dairy cow feeding (Dr C.C. Balch); mastitis (Dr F.H. Dodd); reproduction (Dr M.J. Ducker); nutrition and metabolism of the young animal (Dr J.H.B. Roy); nutritional processes in the ruminating bovine (Dr R.H. Smith); metabolism in the lactating cow (Dr J.D. Sutton); physiology and biochemistry of lactation (Dr A.T. Cowie); applied pig nutrition (Dr K.G. Mitchell); quality of raw milk (Dr L.A. Mabbitt); processed liquid milk (Dr H. Burton); Cheese and fermented products (Dr M.L. Green); butter and cream (Dr L.W. Phipps); nutritional value of dairy products and related foods (Dr M.I. Gurr).
Publications: Annual report.

University of St Andrews*   759

Address: College Gate, North Street, St Andrews, Fife KY16 9AJ
Telephone: (0334) 76161
Telex: 76213
Status: Educational establishment with r&d capability
Principal and Vice Chairman: J.S. Watson
Vice Principal: Professor F.D. Gunstone

FACULTY OF SCIENCE*   760

Dean: Dr D.M. Finlayson

Department of Biochemistry and   761
Microbiology

Chairman: Dr S. Bayne
Graduate research staff: 19
Activities: Research into marine and river pollution from sewage outfalls, biochemical energetics; connective tissue; tumour biochemistry; human genetics disorders; structure and function of membranes.

Department of Botany   762

Acting Chairman: Professor R.M.M. Crawford
Graduate research staff: 9
Activities: Botanical research (including field studies) in all aspects of living plants.

Department of Zoology   763

Chairman: Dr J.B. Tucker
Graduate research staff: 5
Activities: Anatomy, biochemistry, physiology, behaviour, ecology, heredity, development and evolution throughout the animal kingdom, including man.

University of Salford*   764

Address: Salford, M5 4WT
Telephone: (061) 736 5843
Telex: 668680 (Sulib)
Status: Educational establishment with r&d capability

FACULTY OF SCIENCE*   765

Dean: Professor L.S. Bark

Department of Biology   766

Chairman: Professor T.A. Villiers
Senior staff: Dr A.V. Silver
Activities: Areas of research include: effects of environmental factors upon the microfauna of the crop/soil interface; feeding habits of triatomid bugs; behaviour of tsetse flies; biology of green-veined white butterfly; biology of dragonflies and of chironomids; ecological surveys of estuaries; coarse fisheries; control of Simulium fly by chemical and biological means in West Africa; effects of recreational activities on reservoirs, sand dunes and woodlands; soil fauna of agricultural and forestry land; Dutch elm disease; feeding habits and diseases of freshwater fish; ecology and developmental biology of gliding bacteria and of Herpetesiphon species; development cycle of myxobacterium Myxococcus xanthus; neuroglia in the nervous system of molluscs and insects; host-parasite relationship in organisms responsible for the tropical diseases trypanosomiasis and filariasis; ecology of parasites and host-parasite relationship; fungal diseases of wheat; plant cytology; seed physiology; locomotory organs of fish; respiration of birds; amphibian and reptilian tooth development.

University of Sheffield*   767

Address: Western Bank, Sheffield, S10 2TN
Telephone: (0742) 78555
Telex: 54348 ULSHEF G
Status: Educational establishment with r&d capability
Vice-Chancellor: Professor G.D. Sims

FACULTY OF MEDICINE*   768

Address: Beech Hill Road, Sheffield, S10 2RX

Medical School*   769

Dean: Professor D.S. Munro
Graduate research staff: 70 (in all departments)

Department of Biochemistry*   770

Department of Physiology*   771

FACULTY OF PURE SCIENCE*   772

Dean: Professor W. Galbraith

Department of Botany 773

Head: Professor A.J. Willis
Units: Comparative plant ecology (NERC), Dr I.H. Rorison; ARC research group on photosynthesis, Professor D.A. Walker
Graduate research staff: 26
Activities: Research, both academic and applied, is on a wide front in the botanical field. The major areas are plant physiology and biochemistry, plant pathology and mycology, plant ecology, biosystematics and pollution. The two Research Council units are involved mainly with internal factors limiting photosynthesis and with mechanisms controlling plant distribution and the structure of vegetation. The chief aims include the elucidation of photosynthetic phenomena with special regard to yield, and of the ecology of plants with respect to an understanding and the management of vegetation.
Other research involves the ultrastructure and physiology of host-symbiotic interactions and includes studies on infection of leaves by fungi, on mycorrhizas, lichens and mycoparasites. In development physiology, research concerns hormonal control of growth and differentiation, the development of adventitious roots and physiological aspects of seed dormancy. In mycology the biology of Pythium is studied in relation to the control of soil-borne pathogens; other lines of research are the physiology and ecology of psychrophilic fungi and the morphogenesis of fungal sclerotia.
In plant ecology, factors controlling the composition of mire communities are being examined, with special regard to rich fen systems and management of peatlands. Pollution studies concern the effects of heavy metals on plants, and the mechanisms of heavy metal tolerance. Other applied work includes the effects of sulphur dioxide on plants and the influence of herbicides on soil fungi. In biosystematics, a detailed investigation is being made of the important genus Allium.
Publications: Annual reports from each unit.

Department of Geography 774

Head: Professor R.J. Johnston
Graduate research staff: 25
Activities: Areas of research include: site selection and locational change by industrial firms; land evaluation and landform-soil relationships; environmental management; microclimatology; solutes in natural waters and water quality; remote sensing of ecosystems; geomorphology, biogeographical studies.

Department of Microbiology 775

Head: Professor J.R. Quayle
Senior staff: Professor J.R. Guest
Graduate research staff: 13
Activities: Biochemistry and growth physiology of C_1-utilizing microorganisms; biochemistry and molecular genetics of the tricarboxylic acid cycle; biochemistry and molecular genetics of sporulation; biochemistry of cell wall structure and function; effects of herbicides, fungicides and pollutants on microbial flora of the soil.

Department of Zoology 776

Head: Professor J.N. Ball
Senior staff: Professor F.J.G. Ebling
Activities: Research is carried out in the following five main areas: Behavioural ecology of birds and amphipods. Comparative endocrinology: (steroid hormones and receptors, and evolution throughout the vertebrates; hypothalamic control and ultrastructure and physiology of the teleost pituitary; pituitary hormone physiology in chemical carcinogenesis. Neurosecretion and endocrine relationships in insects). Marine biology: (factors controlling water and ion transport across biological mambranes). Neurobiology: (neural basis of behaviour in cephalopod molluscs, including the functional organisation of the higher parts of the motor-control system and the central neural control of the unique chromatophore system). Renal physiology: (fluid homeostasis in teleosts, elasmobranchs, amphibians and mammals, including the Brattleboro strain of rat, particularly with regard to the actions of renin, neurohypophysial and adrenocortical hormones; interrelationships between sodium intake, glomerular filtration and the cardio-vascular system, using techniques of free-flow renal tubular micropuncture and perfusion of nephrons in vitro).

University of 777
Southampton*

Address: Highfield, Southampton, SO9 5NH
Telephone: (0703) 559122
Telex: 47661
Status: Educational establishment with r&d capability

FACULTY OF ENGINEERING AND 778
APPLIED SCIENCE*

Dean: Professor R.C. Smith

Department of Civil Engineering*   779

Head and Professor ·of Soil Mechanics: Professor R. Butterfield
Senior staff: Professor P.B. Morice; Professor T.E.H. Williams
Graduate research staff: 4

Institute of Irrigation Studies*   780
Director: Dr J.R. Rydzewski

FACULTY OF SCIENCE*   781

Dean: Professor G.A. Kerkut

Department of Biology   782

Address: Medical and Biological Sciences Building, Bassett Crescent East, Southampton, S09 3TU
Chairman: Professor Michael A. Sleigh
Departments: Chemical entomology, Dr P.E. Howse; insect ecology, Dr S.D. Wratten; chemical entomology, Dr P.E. Howse; plant breeding, Dr J. Smartt; ecology of domestic animals, Dr R.J. Putman; parasitology (veterinary and fisheries), Dr C.E. Bennett; plant pathology, Dr J.G. Manners; plant improvement, Dr P.K. Evans; plant (crop) physiology, Dr D.A. Morris; animal genetics, Dr N. Maclean; plant taxonomy, Dr F.A. Bisby; animal cell and development biology, toxicology, Professor M.A. Sleigh; ecology and conservation, Dr I.F. Spellerberg
Graduate research staff: 68
Activities: Plant Production: Plant breeding; development of strategies for exploitation of germplasm resources in grain legumes; production of high-yielding, disease and pest-free cultivars with a wider environmental adaptation; development of new disease and pest resistant strains of Vicia; information service on Vicieae; limits of yield of winged bean; somatic cell hybridization leading towards the production of novel legume hybrids and gene transfer across incompatibility barriers. Plant development/crop physiology; transport and action of growth regulators in plants. Better understanding, and possible exploitation of mechanisms regulating assimilate partitioning. Plant protection; cereal diseases, especially rusts and mildews; applied ecology of cereal insects; damage and the role of natural enemies; use of pheromones in monitoring and control systems; effects of pesticides on beneficial insects; provision of a service for research and development work on the use of pheromones in integrated control.
Animal production:
Livestock - ecology of domestic animals; ecology and management for expoitation of conventional and unconventional farm animals.
Veterinary medicine; the use of monoclonal antibodies in dissecting the immune response of hosts to veterinary parasites to determine protective antigens. Freshwater

Fisheries: the investigation of interactive factors on parasite populations in trout farms; improvement of rainbow trout stocks by direct introduction of desirable genes from other species; use of trout as a model system for work on gene manipulation.

Department of Oceanography   783

Chairman and Professor of Physical Oceanography: Professor H. Charnock
Senior staff: marine biology, Professor A.P.M. Lockwood; marine chemistry, Dr J.D. Burton; marine geology/geophysics, Dr E.A. Hailwood; marine physics, Professor H. Charnock.
Graduate research staff: 30
Activities: Current research projects include oceanic circulation in relation to climate modelling; bioturbation studies; plume distribution of geothermal water release; trace metal distribution and speciation in oceanic and estuarine waters; use of satellite imagery in determining sediment particle distribution and movement in coastal waters, geomagnetism of sediments and rocks, magnetic fabric studies; offshore UK geological and geophysical investigations; ecological studies on the benthos of the Fleet, Southampton Water and salt marshes, circadian and tidal rhythms of physiology and behaviour in invertebrates; physiological and biochemical responses to salinity and temperature change and the influence of pollutants; effects of light quality on algal metabolism; reproduction of hyperiid amphipods and ion levels in mesopelagic decapods; precise oxygen measurements in relation to primary productivity.

Wolfson Unit of Chemical Entomology   784
Sections: Chemistry, Professor R. Baker; biology, Dr P.E. Howse
Graduate research staff: 10
Activities: The unit was established jointly by the Departments of Biology and Chemistry, Southampton University, to provide a service for industry and others interested in novel methods of insect pest control. The unit is interdisciplinary and uses a range of advanced and integrated biological and chemical techniques. It can give advice on the use of pheromones, attractants and repellents in insect control, as well as synthesise and supply these chemicals in kilogram quantities. The unit designs and tests traps with which the attractants are often used both in the laboratory and in the field, and specializes in the formulation of these chemicals for use in monitoring and control of insect pests of agriculture and public health. In addition, it is able to test the toxicity of insecticides on a whole range of laboratory reared insects. Research and development work on new chemicals controlling insect behaviour is also undertaken, as is non-routine types of trace chemical analysis. Major current projects include kilogram-scale synthesis of pheromones for mating disruption experiments; analysis and characterization of pheromone components

of several important insect species; development of new aerodynamically designed insect traps.

University of Stirling   785

Address: Stirling, FK9 4LA
Telephone: (0786) 3171
Telex: 778874
Status: Educational establishment with r&d capability
Principal and Vice-Chancellor: Kenneth Alexander

DEPARTMENT OF BIOLOGICAL   786
SCIENCE

Head: Professor W.R.A. Muntz
Graduate research staff: 29
Activities: Research topics include: Physiology: energetics of fish locomotion; physiology of reproduction and osmotic regulation in teleosts, marine and estuarine invertebrates; physiology of molluscs, of hibernation, of mammalian hormones; and comparative studies on vision. Ultrastructure and developmental biology of plant cells; functional plant anatomy; plant water relations; physiology of stomata, photosynthesis and respiration; physiology of seed development and germination.
Ecology: ecology of birds, foraging, breeding and energetics; animal population dynamics; plant ecology; ecology of metalliferous soils; tropical rain forests; plant insect interactions; numerical techniques in ecology; co-evolution of species in natural and model systems; littoral, estuarine and freshwater ecology, including pollution studies.
Biochemistry: mechanisms of enzyme action; enzyme sub-unit interactions; protein folding, biochemical aspects of development in micro-organisms, plants and mammals; proteinases and development; biochemistry of spore and seed germination; enzymes under conditions of water-limitation; structure and formation of biological membranes; metabolism of seed storage carbohydrates; structure and function of complex plant polysaccharides.
Microbiology: ecology of fungi, litter decomposition, wood rots, bioconversions; evolution and ecology of bacterial viruses; histopathology and microbiology of disease of teleost fishes; microrespiration and energy balance of protozoa, crustaceae and fresh water nematodes; cellular slime mould development. Fungal spore germination.

UNIVERSITY INSTITUTES   787

Institute of Aquaculture   788

Telex: 777738
Director: Professor R.J. Roberts; *Deputy director:* Dr R.H. Richards
Graduate research staff: 30
Activities: Research and training in all aspects of aquaculture, disease, nutrition and husbandry technology of cultured aqualic organisms, including temperate and tropical, freshwater and marine, finfish and shellfish. Major research projects include: nutrition, genetics, sex-reversal and husbandry of tilapias; hybridization and genetic improvement of tilapias and hatchery technology; development of fish meat substitutes in salmonid diets; studies of environmental impact of intensive cage culture in UK and tropics; renal pathology of farmed fishes and study of proliferative kidney disease; development of fish vaccines for vibriosis, etc; improvement of strains of brown trout; diseases of crayfish and cephalopods.
Publications: Annual report.

Wolfson Unit of Seed Technology*   789

Director: Dr S. Matthews
Activities: Research and development in collaboration with the UK seed industry.

University of Strathclyde*   790

Address: 204 George Street, Glasgow, G1 1XW
Telephone: (041) 552 4400
Telex: 77472 UNSLIB G
Status: Educational establishment with r&d capability
Principal and Vice-Chancellor: Dr Graham J. Hills

FACULTY OF SCIENCE*   791

SCHOOL OF MATHEMATICS,   792
PHYSICS AND COMPUTER SCIENCE*

Dean: Professor George Eason

Department of Bioscience and Biotechnology   793

Address: 131 Albion Street, Glasgow, G1 1SD
Telephone: (041) 552 2071
Telex: 77472
Head: Professor John Hawthorn
Divisions: Food science; Professor W.R. Morrison; applied microbiology; biology (including horticulture); biochemistry
Activities: Academic and applied research in the food science division comprises: cereal starches, cereal lipids (composition and technological properties); water relations of foods, intermediate moisture foods; automated enumeration of microorganisms by means of fluorescence microscopy; characterization of mixed populations of particles by microscopy and automated image analysis; food proteins; descriptive analysis of whisky flavour, and flavour of other beverages and foods.

University of Surrey*   794

Address: Guildford, Surrey GU2 5XH
Telephone: (0483) 71282
Telex: 859331
Status: Educational establishment with r&d capability
Vice-Chancellor: A. Kelly

FACULTY OF BIOLOGICAL AND CHEMICAL SCIENCES*   795

Dean: Professor P.R. Davis

Department of Biochemistry   796

Head: Professor D.V.W. Parke
Sections: Clinical biochemistry, Professor V. Marks; human nutrition, Professor J.W.T. Dickerson; toxicology, Professor J.W. Bridges; food science, Dr R. Walker
Graduate research staff: 29
Activities: Clinical and medical biochemistry; experimental pathology; enzymology; food science and technology; human nutrition; entero-pancreatic research; microbial biochemistry; antibody production; toxicology.
Projects: Development and use of radioimmunoassays for cytotoxic drugs in the treatment of cancer; diagnosis and monitoring of diabetes; effect of a high fat and low fat basal diet on the entero-pancreatic axis (Professor V. Marks); role of the pineal and its secretions in ovine reproduction (Dr Josephine Arendt); short-term tests for the detection of potential carcinogens/mutagens (Dr C. Ioannides); yeast cytochrome P-450: drug and carcinogen metabolism (Dr A. Wiseman); lymphocyte hybridoma production of monoclonal antibodies (Dr R. Hubbard); vitamins and cancer (Professor D.V. Parke);

mechanisms of activation of effects of dietary bran; effect of carcinogens and drugs of lysosomal stability; drug toxicity (Professor D.V. Parke); mechanisms of activation of drugs and carcinogens to toxic intermediates (Dr g.g. Gribson); mechanism of azo reductase in memmalian liver (Dr r. Walker); investigation of the entero-insular axis in health and disease (Dr Linda Morgan); production of antisera for passive immunization of sheep and cattle to improve fecundity (B.A. Morris; Professor V. Marks); metabolism and toxicology of DDT and its degradation products in birds and mammals (Dr L.J. King); development of rapid methods of food analysis; fate of woodsmoke constituents that are incorporated in foods (Dr M.N. Clifford).

Department of Microbiology   797

Head: Professor J.E. Smith
Sections: Plant microbiology, Dr R.M. Jackson; virology, Dr M. Butler
Graduate research staff: 13
Activities: Electron microscopy; environmental microbiology; food microbiology; genetics and molecular biology; immunology; industrial microbiology; medical microbiology; microbial biochemistry and physiology; microbial taxonomy; plant microbiology.
Projects: Campylobacter pathogenesis, epidemiology, and laboratory identification (O.M. Murphy); pathogenesis of scrapie (Dr M. Butler); influence of biocides, fertilizers, and other environmental factors on the production of fusarium toxins in cereals; temporal and spatial distribution of diatoms in the River Wey; use of microorganisms to assay mycotoxins, elucidation of mechanisms of action with a view to developing biochemical assay methods; effects of penicillium purpurogenum and rubratoxin on aflatoxin production by aspergillus flavus (Dr M.O. Moss); host-parasite relationship of crop plants infected with broomrape - Orobanche Sp. (Dr P.J. Whitney); mycorrhizas of Sitka Spruce (Dr R.M. Jackson); antibiotic production in Streptomyces cattleya; analysis of the secondary metabolites of Streptomyces cattleya (Dr M.E. Bushell).

WOLFSON BIOANALYTICAL CENTRE*   798

Director: Dr E. Reid
Activities: Analytical methodology and instrumentation; applying techniques of 'wet' analysis to meet needs of pharmaceutical and food industries; cell biochemistry; clinical enzymology; trace organic assay techniques.

UNITED KINGDOM

University of Wales Institute of Science and Technology* 799

– UWIST
Address: King Edward VII Avenue, Cardiff, CF1 3NU
Telephone: (0222) 42522
Affiliation: University of Wales
Status: Educational establishment with r&d capability
Principal: Dr A.F. Trotman-Dickenson
Vice-Principal: Professor D. Wallis

SCHOOL OF APPLIED SCIENCES* 800

Dean: Professor J. King

Department of Applied Biology 801

Address: Redwood Building, King Edward VII Avenue, Cathays Park, Cardiff, CF1 3NU
Telephone: (0222) 42522
Head: Professor R.W. Edwards
Senior staff: Dr R.E. Hughes
Graduate research staff: 20 (1982)
Activities Biotechnology; pollution studies; aquatic surveillance.
Research projects include the following topics: Ecology - feeding of aquatic invertebrates; biology and taxonomy of the aquatic stages of chironomid midges; pollution of inland waters; ecology of freshwater benthic algas. Biochemistry and physiology comparative endocrinology and mechanisms of hormone action; effects of air pollution on vegetation. Microbiology - antiseptic and antibiotic resistance in bacteria; bacterial indicators of sewage pollution; genetic studies on bacterial plasmids; microbial ecology of freshwater habitats; effects of pesticides on freshwater bacteria; microbial degradation of synthetic compounds; genetics of microbial populations and communities.
The department has a field station in mid-Wales.
Publications Annual report.

University of York* 802

Address: Heslington, York, YO1 5DD
Telephone: (0904) 59861
Status: Educational establishment with r&d capability
Vice-Chancellor: Professor S.B. Saul

Department of Biology and Biochemistry 803

Head: Professor M.H. Williamson
Sections: Biology, Professor J.D. Carrey, Professor R. Leech; biochemistry, Professor J.R. Bronk
Graduate research staff: 54 (1981)
Activities: Research into ecological energetics; optimal harvesting strategies; mineral nutritional studies with particular reference to plant growth on colliery spoil; renovation of waste water; analysis of activated sludge plants; biochemistry and microbiology of activated sludge and anaerobic sludge digestion; biophysical characteristics and sedimentation of activated sludge; biomechanics, development of short-term tests to detect environmental and occupational carcinogens.
Projects: Reclamation of colliery spoil (Dr M.D. Chadwick); biological control: population dynamics of insects (Dr J.H. Lawton); chloroplasts in wheat: agronomics significant of variation in number (R.M. Leech).

Veterinary Investigation Centres 804

Address: Hook Rise South, Tolworth, Surbiton, Surrey KT6 7NF
Telephone: (01) 337 6611
Telex: 22203/4
Affiliation: Ministry of Agriculture, Fisheries and Food, Veterinary Investigation Service
Assistant chief veterinary officer: Dr W.A. Watson
Regional veterinary officer: Dr B.M. Williams
Deputy regional veterinary officer: A.R.M. Kidd
Divisional veterinary officers: Dr G. Jackson; Dr I.A. Marsh; Dr J.C. Bell
Deputy regional veterinary officers: Wales - E.T. Davies; south west region - Dr A.H. Pill; eastern region - Dr P.S. Dawson; northern region - K.B. Baker; south east region - J.A. Benson; west midland region - D.C. Ostler
Graduate research staff: 90 (at all centres)
Activities: Research into husbandry methods, diseases, pathology, etc. of cattle, sheep, pigs, poultry, and fish.
Projects:
Cattle - bedding, Enkamat K with minimal bedding, effect on cleanliness, comfort, and mastitis incidence (Bristol); bovine respiratory syncytial virus, indirect haemagglutination test (Reading); bovine viral diarrhoea, foetal pathogenicity (Worcester); exit race automatic teat sprayer, efficiency (Bristol); fertility control pilot exercise in dairy herds (most centres); foetal pathology (Reading); foetopathogenic effect of unknown agents, determination by intrafoetal inoculation (Worcester); infertility (Starcross); Johne's disease, clinical disease investigation (most centres); leptospirosis, survey of bovine abortions (Carmarthen); mastitis sur-

veillance scheme (most centres); milk fever incidence, effect of pre-partum Ca intake (Bangor); Nematodirus battus, monitoring of field hatching with reference to meteorological conditions (Starcross); permanent cubicle beds without litter retaining lips, study of various falls (most centres); respiratory diseases of housed cattle, epidemiology, and pathogenesis (Carmarthen); selenium deficiency in a dairy herd (Reading); teat orifice condition, effect of milk flow rate (Bristol), gross pathology and histopathology (Shrewsbury); tickborne fever, causal organism (Leeds).

Sheep - Brucella abortus, transmission in naturally infected pregnant ewes (Thirsk); campylobacters, assessment of a more critical method of identification (Worcester); chlamydia, foetal pathogenicity (Newcastle); coccidiosis control (Reading); ensiled broiler litter, feeding to fattening lambs (Thirsk); fascioliasis forecasting (most centres); maedi/visna, attempted eradication (Lincoln), serological study during control in a flock (Wye); nematodes, relationship of serum pepsinogen and herbage larval levels (Northampton); pasteurellosis, epidemiology, and microbiology (Carmarthen and Thirsk), haemolytica, casualties in field cicumstances (Aberystwyth); salmonella arizonae, experimental infection (Thirsk); tick borne fever/tick pyaemia, oxytetracycline as a prophylactic (Carmarthen); toxoplasma, abortion, diagnosis (Reading), infection, role of the ram (Lincoln), infection in foetal tissue (Leeds); toxoplasma gondii, examination of foetal fluids for antibodies (Leeds).

Pigs - Aujeszky's disease, surveillance of restocked premises after depopulation following outbreak (Leeds); corynebacterium suis, isolation from carcases (Thirsk); haemophilus pleuropneumonia (Leeds); isospora in piglets, field study (Wye); negatively charged air ions, effect on environment of pig house and on growth rate and feed conversion (Lincoln); spirochaetes in faeces and slurry, (Leeds); streptococcus suis type II, development of ELISA test for antibodies, (Cambridge); swine vesicular disease (Leeds); transmissible gastroenteritis (Norwich); treponema hyodysenteriae, identification by enzymatic methods (Leeds); Yersinia enterocolitica, survey (Cambridge and Leeds).

Poultry - ensiled poultry litter, effect of fermentation on bacterial content and nutritive value (most centres); hygiene in a poultry processing plant (Bangor); infectious bronchitis serology, (Cambridge, Reading, Shrewsbury and Sutton Bonington); lameness in broilers, tenosynovitis (Shrewsbury); leg weakness in broilers, (most centres); mortality survey, commercial laying flock kept in an aviary (Gloucester); salmonella hadar, monitoring and epidemiology in a turkey enterprise (Norwich); turkey hatching eggs, laying system to produce bacteriologically cleaner eggs (Liverpool).

Fish - bacterial kidney disease in salmonids (Penrith).

Other species - tuberculosis in badgers (Gloucester and most other centres); survey of mortality among swans (most centres).

Non-specific and multi-specific - antibiotic sensitivity testing (Norwich); campylobacter, isolation from faeces and post-mortem specimens (Wye), preservation of cultures (Shrewsbury), relationship between infections in man, animals and poultry (Gloucester); intra-uterine growth retardation (Aberystwyth); lead poisoning, blood porphyrin determination as a rapid test (Shrewsbury); farm control of parasites (most centres); parasitic gastroenteritis (most centres); quality control working party, pilot survey (most centres); salmonella, monitoring of imported animal feeds for (Liverpool), serology (Thirsk); salmonella dublin infection, epidemiology (Carmarthen); selenium content of soil and grassland herbage (most centres); toxoplasmosis, assessment of the microtitre latex aglutination test (Leeds); trace elements, study on Bodmin Moor (Truro); tuberculosis in wild life (Truro); virus isolations, use of chicken kidney tissue culture (Reading).

Veterinary Laboratory   805

Address: Eskgrove, Lasswade, Midlothian
Telephone: (031 663) 6525
Affiliation: Ministry of Agriculture, Fisheries and Food, Agricultural Development and Advisory Service
Status: Official research centre
Head: Dr J.G. Ross
Senior staff: A.R. Hunter; R.D. Lapraik; C.J. Randall
Activities: Research in diseases of poultry, cattle, sheep, and red deer.
Projects: Bovine abortions other than brucella (R.D. Lapraik); ovine prenmonias (A.R. Hunter); chlamydia (J.W. Macdonald); infections bronchitis in poultry, avian artritis, mycoplasma infections, avian blindness, leg weakness in poultry, electron microscopy (C.J. Randall); poultry virus infections, virus diseases of turkeys (J.W. Macdonald); avian clinical pathology, (Dr G.A. MacKenzie); fasciola hepatica, helminth immunity (Dr J.G. Ross); lungworm in red deer (R. Munro).

Veterinary Research   806
Laboratories

Address: The Farm, Stoney Road, Stormont, Belfast BT4 3SD, Northern Ireland
Telephone: (0232) 760011
Affiliation: Department of Agriculture for Northern Ireland
Status: Official research centre
Director: Professor C. Dow
Deputy Director: Dr J.B. McFerran

Departments: Virology, Dr J.B. McFerran; bacteriology, Dr J.J. O'Brien; biochemistry, Dr C.H. McMurray; pathology, Dr R.M. McCracken; parasitology, Dr S.M. Taylor immunology, Dr E.F. Logan; physiology, Dr W.J. McCaughey
Graduate research staff: 24
Activities: Diagnosis and control of diseases of farm animals.
Projects: Arthropod-borne haematozoal diseases of cattle in Northern Ireland (S.M. Taylor); leptospiral and mycoplasmal mastitis (D.P. Mackie); mechanisms of immunity in the bovine udder (G.R. Pearson); respiratory disease in young cattle (D. Bryson); mycoplasma species in respiratory disease (H. Ball); studies on the causes of bovine abortion (W.A. Ellis); control of reproduction in ruminants; whole-herd investigation into infertility in dairy cattle; pregnancy diagnosis in farm animals (W.J. McCaughey); neonatal immunity (E.F. Logan); purification and biochemical characterization of viruses (M.S. McNulty); development of an Aujeszky's disease vaccine (J.B. McFerran); role of vitamin E and selenium; copper deficiency in ruminants; selenium in ruminant nutrition in Northern Ireland (C.H. McMurray).

VETERINARY INVESTIGATION CENTRE 807

Address: 43 Beltany Road, Coneywarren, Omagh, Tyrone, Northern Ireland
Telephone: (0662) 3337
Director: Professor C. Dow
Activities: As a satellite laboratory to that at Stormont, providing research into and diagnosis of farm animals, poultry and fish.

Welsh Plant Breeding Station 808

Address: Plas Gogerddan, Aberystwyth, Dyfed SY23 3EB
Telephone: (0970) 828255
Telex: 35181 ABYUCW
Affiliation: State-aided institute, Agricultural Research Council; University College of Wales, Aberystwyth
Status: Official research centre
Director: Professor J.P. Cooper
Sections: Herbage breeding, Dr E.L. Breese; clover breeding, W. Ellis Davies; arable crop breeding, Dr D.A. Lawes; developmental genetics, Dr D. Wilson; seed production, D.H. Hides; grassland agronomy, J.M.M. Munro; plant pathology, Dr A.J.H. Carr; plant biochemistry, Dr J.L. Stoddart; chemistry, Dr D.I.H. Jones; cytology, Dr Hugh Thomas

Graduate research staff: 75
Activities: Current research on herbage plants includes a genetic appraisal of a wide range of agronomic and nutritive characteristics from material collected in different ecological environments in Europe and elsewhere. Particular emphasis is given to features of grasses and legumes that are important for increasing animal production from grassland in British agriculture. Recent developments include the synthesis by chromosome manipulation, of new species-hybrids and several tetraploid hybrid ryegrasses are already being marketed. In arable crops the main objective is to increase harvestable yield under both intensive and less favourable conditions. Research includes work on breeding methodology, disease, and pest resistance. Improved varieties of barley and winter and spring oats are being developed.
Plant breeding is supported by well coordinated research programmes in genetics, cytology, plant physiology, biochemistry, chemistry, and plant pathology. These disciplines provide basic information on those biological features of the crop which determine its potential production and nutritional quality, and hence assist the breeder in developing suitable screening techniques for crop improvement
Facilities are provided at the station for overseas scientists to carry out research work on plant breeding and allied problems. As a university department, the station accepts postgraduate students who, if suitable qualified, can register for higher degrees of the University of Wales.
Projects: Grass breeding: genetic and cytogenetic manipulation including tissue culture techniques; improved breeding methods and improved grass varieties; genetic resources (Dr E.L. Breese). Herbage legume breeding: competition and cooperation between grasses and clovers; improved nitrogen fixation from clover/Rhizobium associations; improved varieties of clover and lucerne (W.E. Davies). Arable crops breeding: improved breeding techniques and methods including the use of barley F1 hybrids and double haploids; cereal resistance to fungal parasites and nematode; pests developing high yielding, disease resistant varieties of oats and barley (Dr D.A. Lewes). Grassland agronomy: environmental and management factors limiting grassland production; evaluation of potential new grass and legume cultivars (J.M.M. Munro); seed production: husbandry methods to maximize seed yield of herbage crops (D.H. Hides). Chemistry: chemical and physical properties of grasses, clovers, and grains in relation to the nutritive value of the crop; selection criteria for breeding programmes (Dr D.I.H. Jones). Developmental genetics: physiological, morphological, and developmental bases of genetic variation in utilizable yield of forage grasses and legumes; selection criteria for breeding programmes (Dr D. Wilson). Plant biochemistry: potential limitations to crop production; genetic variability; screening procedures for breeding programmes (Dr J.L.

Stoddart). Plant pathology: crop/pathogen cycles, managements to minimize yield loss; nature and inheritance of resistance in herbage and arable crops to pathogens and pests (Dr A.J.H. Carr). Cytology: cytogenetic relationships between cultivated species and their wild relatives; manipulative methods for genetic exchange through species hybrids and their derivatives (Dr H. Thomas).
Publications: Annual report.

Welsh Water Authority* 809

Address: Cambrian Way, Brecon, Powys LD3 7HP
Status: Public utility conducting research
Director of Scientific Services: W. Roscoe Howells
Sections: Water quality, D.H. Myers; sewage treatment, Dr P. Hulme; water treatment/health, Dr L. Clark; tidal waters, W. Halcrow; fisheries r&d, Dr G.S. Harris
Activities: The Authority has formed 7 multi-purpose divisions responsible for water supplies, sewage treatment and disposal, water conservation, land drainage, pollution control, fisheries and recreation.

West of Scotland Agricultural College 810

Address: Auchincruive, Ayr, KA6 5HW
Telephone: (0292) 520331
Telex: 777400 WSAC G
Affiliation: Department of Agriculture and Fisheries for Scotland
Status: Educational establishment with r&d capability
Principal: Professor J.M.M. Cunningham
Deputy Principal: Dr D.J. Martin
Graduate research staff: 100 (1982)
Activities: The college has three main spheres of activity: teaching a wide range of courses, from diploma to postgraduate level, carrying out applied research and development work, and providing a general and specialist advisory service for farmers and growers. The research and development programme aims to obtain results which have potential for wide application in agriculture and horticulture, but with particular attention focussed on meeting the needs of the region served by the College. This necessitates covering a range of subjects concerned with crop and animal production, marketing and milk processing. Areas being concentrated on are the science, husbandry and economics of milk, beef and sheep production, milk utilization, farm waste treatment and its use, grass production and its utilization, and glasshouse crops production.

Projects: Distribution of concentrates to cows offered silage ad libitum (J.D. Leaver, W. Taylor; buffer feeding for dairy cows under winter feeding and summer grazing conditions (J.D. Leaver, C.J.C. Phillips); in-place cleaning of milking installations and hygiene measures at milking on dairy farms (J. Bruce, Elizabeth G. Cruickshank); quality of milk and milk products; cheese production methods (R.J.M. Crawford, Janet H. Galloway); treatment and utilization of animal excreta; laboratory studies (M.R. Evans); effects of grazing and cutting management on sward productivity (R.D. Harkess, J. Frame); yield and quality responses of grasses and clovers to fertilizer nitrogen (J. Frame, A.G. Boyd); hill land improvement in Argyll (J.R. McCreath, A. MacLeod); college and National List testing of cereal varieties; manurial requirements of tillage crops (cereals) (W.G.W. Paterson); studies of cereal leaf diseases and methods of control (A.G. Channon, K. Mawson); barley root and stem base diseases (Marjorie R.M. Clark); effect of different temperatures on the growth of tomatoes (R.E. Johnston, G.M. Hitchon); effect of nutritional factors on the production and quality of glasshouse crops (D.A. Hall, G.C.S. Wilson); glasshouse production of lettuce (R.E. Johnston, J.H.F. Smith); glasshouse production of sweet peppers (G.M. Hitchon, R.E. Johnston); evaluation of substrates other than soil for glasshouse production (A.D.N. Hutchison, R.E. Johnston); linear programming of glasshouse crops production (A.S. Horsburgh, D.D. Mainland); methods of selling beef cattle; catering demand for meat into the 1980s (J. Clark, R.G. Aitken); investigation of consumer preference for cheese and butter (J. Clark, C.R. Groves); milk net margins investigation (J. Clark, J.M. Tweddle); farm capital investigations (P.G. Smith, S.W. Ashworth).
Publications: Annual report.

ADVISORY AND DEVELOPMENT DIVISION 811

Director: H.A. Waterson
Sections: Agronomy, Dr J. Frame
Crichton Royal Farm, Dr J.D. Leaver
Farm buildings, Dr M. Kelly
Glasshouse Investigational Unit for Scotland, R.E. Johnston

Kirkton Farm 812

Address: Crianlarich, Perthshire
Telephone: (08384) 210
Head: T.H. McClelland

AGRICULTURAL CHEMISTRY DIVISION 813

Head: Dr A.J. McGregor

AGRICULTURAL ECONOMICS DIVISION 814

Head: J. Clark

BIOLOGICAL SCIENCES DIVISION 815

Head: Dr S. Baines
Sections: Botany, Dr R.K.M. Hay; microbiology, Dr S. Baines; zoology, J.W. Newbold; plant pathology, Dr A.G. Channon

EDUCATION DIVISION 816

Head: Professor J.M.M. Cunningham
Sections: Agriculture, M. Buckett
Agricultural engineering, G.C. Mouat
Dairy technology, Dr R.J.M. Crawford
Horticulture and beekeeping, Dr H.J. Gooding
Poultry husbandry, P. Dun
Brickrow Farm Unit, R. Laird

VETERINARY MEDICINE DIVISION 817

Address: Veterinary Investigation Centre, Auchincruive, Ayr, KA6 5AE
Telephone: (0292) 520318
Head: C.L. Wright
Section heads: N.S.M. Macleod, M.A. Bonniwell

Dumfries V.I. Centre 818

Address: St Mary's Industrial Estate, Dumfries, DG1 1DX
Veterinary Investigation Officer: N.S.M. Macleod

Oban V.I. Centre 819

Address: Glencruitten Road, Oban, Argyll PA34 4DW
Telephone: (0631) 3093
Veterinary Investigation Officer: C.C. Bannatyne

Weston Research Laboratories Limited 820

Address: 644 Bath Road, Taplow, Maidenhead, Berkshire SL6 0PA
Telephone: (06286) 4741
Telex: 849135 WESLAB
Parent body: Associated British Foods Limited
Product range: Flour, bread, biscuits, crispbread, canned fruit and vegetables, margarine, jam, ice cream, starch, gluten.

Status: Research centre within an industrial company
Director: Dr William Elstow
Sections: Research and development, Dr E. Filmore; cereals, D.J. Wallington; analytical, R. Watkinson
Graduate research staff: 11 (1983)
Activities: New product and process development; uses for wheat starch; new analytical methods.

Worplesdon Laboratory* 821

Address: Tangley Place, Worplesdon, Surrey GU8 3LQ
Telephone: (0483) 232581
Affiliation: Ministry of Agriculture, Fisheries and Food, Agricultural Development and Advisory Service
Status: Official research centre
Sections: Mammals, Dr Wendy A. Rees; birds, E.N. Wright
Graduate research staff: 28
Activities: The laboratory undertakes research on the damage, control, and biology of birds and mammals which are considered to cause significant economic damage. The most important mammals currently being studied are rabbits, foxes, badgers, and moles. The principal birds whose biology is being investigated are bullfinches, starlings, rooks, and gulls. In addition, damage by wood-pigeons, feral pigeons, collared doves, and by three species of goose is being monitored and methods of alleviating this damage are being investigated. The problem of birds and their impact on aircraft is studied, particularly in respect to gulls, as are problems of wildlife conservation in relation to agriculture.

Wye College 822

Address: Wye, Ashford, Kent TN25 5AH
Telephone: (0233) 812401
Telex: 96118 ANZEEC G
Parent body: University of London
Status: Educational establishment with r&d capability
Principal: I.A.M. Lucas
Graduate research staff: 106 (1983)
Activities: Agriculture, horticulture, rural environment.

AGRICULTURE DEPARTMENT   823

Head: Professor W. Holmes
Section: Animal production, Dr R.C. Campling
Activities: Crop production: influence of various agro-chemicals on the development, seed yield, and quality of oil seed rape (B. napus); agricultural engineering and mechanization; improved works rates and reduced power requirements, mechanization and efficient use of support energy in protected cropping; pig production: utilization of animal effluents as slurries to use the plant nutrients they contain and to control odour; animal breeding: on-farm testing of pigs, performance testing of beef cattle, selection to increase prolificacy in Romney, sheep, pig selection; numinant production: dairy and beef cattle - factors influencing their voluntary intake on conserved forages and on pasture, interaction of animal behaviour, management, and production.
Projects: Nutrition of ruminants; land use in food production (Professor W. Holmes); ruminant studies (Dr R.C. Campling); animal breeding and production (Dr M.K. Curran); crop physiology (Dr R.W. Daniels); pig production and behaviour (Dr I.J. Lean); crop physiology and agronomy (Dr D.H. Scarisbrick); tillage studies; crop processing (I.B. Warboys); tractor, implement performance; tillage studies (J.M. Wilkes).

BIOCHEMISTRY, PHYSIOLOGY AND   824 SOIL SCIENCE DEPARTMENT

Head: Professor J.H. Moore
Sections: Agricultural biochemistry, Dr A.E. Flood; agricultural chemistry, Professor J.H. Moore; animal physiology, Dr A.H. Sykes
Activities: Animal biochemistry - detoxication mechanisms, lipid biochemistry; plant biochemistry - carbohydrate metabolism, membrane transport; crop protection - chemical structure-activity relationships, mode of toxic action and selectivity of pesticides; animal physiology - ultrastructure, intestinal and climatic physiology; soil science - effects of soil structure and of the distribution of nutrients in the soil profile on plant growth and crop yield.

BIOLOGICAL SCIENCES   825 DEPARTMENT

Head: Professor D.A. Baker
Sections: Agricultural botany, Professor D.A. Baker; zoology, Dr D.S. Madge; plant breeding and genetics, Dr G.P. Chapman, Dr W.E. Peat; weed biology, Dr T.A. Hill; plant virology, Dr K.W. Bailiss; entomology, Dr M.J.W. Copland, Dr C.J. Hodgson; microbiology, Dr J.M. Lopez-Real
Activities: Interrelation between biological processes and agriculture. Plant physiology: influence of growth substances on the partitioning and compartmentation of assimilates with specific reference to phloem loading and long-distance transport of solutes within various crop plants. Applied zoology: biology of resistant and susceptible forms of the damson-hop aphid to insecticides, including comparative morphology, behaviour, and biochemistry; effects of agricultural practice and land management on the biology of soil invertebrate animals. Plant breeding and genetics: physiology and genetics of Cicia faba (field bean) - improved adaptability to a range of environments; chromosomes ultrastructure and the application of new deoxyribonucleic acid technology to crop improvement. Weed biology: nutritional aspects of the competition between annual weeds and crops; growth and physiology of rhizomatous plants. Plant virology: physiology of virus-induced symptoms; effect of host water status and the establishment of infection; legume viruses. Physiological plant pathology: mechanisms of disease resistance in plants; biochemical basis for specificity in plant/parasite interactions. Entomology: parasitic Hymenoptera; biological and integrated control of insect pests in field and glasshouse crops. Microbiology: recycling of agro-industrial organic waste materials - high temperature composting; microbiology of organic compost production and the subsequent rhizosphere development of organic compost grown crops.

ENVIRONMENTAL STUDIES AND   826 COUNTRYSIDE PLANNING DEPARTMENT

Head: Professor R.H. Best
Section: Countryside planning, Professor G.P. Wibberley
Activities: Land use, rural conservation; land resource studies: land capability and quality assessment, restoration of worked-out sand and gravel quarries, deterioration of soil in storage heaps; social and economic studies; international and comparative studies.
Projects: National and international land use and land competition (Professor R.H. Best); environmental policy and planning (Dr M.A. Anderson); pedology and land evaluation (Dr C.P. Burnham); ecology and conservation (Dr B.H. Green); land resources and reclamation (Dr S.G. McRae); agrarian social development in Latin America (Dr M.R. Redclift); rural planning and housing (A.W. Rogers).

HOP RESEARCH DEPARTMENT 827

Head: Dr R.A. Neve
Activities: Hop breeding: disease and pest resistance studies, high brewing value; agronomic studies: effects of irrigation, use of herbicides in relation to no-cultivation techniques, replacement of traditional and hand work by chemical or mechanical methods; engineering investigations: hop drying and the design of improved wirework systems; chemistry of hop resins and essential oils.
The department is grant aided by the Agricultural Research Council, the Hops Marketing Board, and the Brewers' Society.
Projects: Plant breeding (Dr R.A. Neve); herbicides and field trials: viticulture (R.F. Farrar); organic chemistry (Dr C.P. Green); engineering (M.W. Shea); plant physiology (Dr G.G. Thomas).

HORTICULTURE DEPARTMENT 828

Head: Professor W.W. Schwabe
Graduate research staff: 27 (1982)
Activities: Applied and fundamental aspects of whole plant physiology and related areas of research. Some specific topics are as follows: reproduction (flowering), propagation and cloning by a variety of techniques including tissue culture, juvenility in plants, hormone relations in morphogenetic control, induction of parthenocarpy in top fruit and regulation of tillering and reproductive growth in cereals, environmental control of morphogenesis in plants, measurement of physical parameters of the environment, computer control of glasshouse environments, landscape ecology, and reclamation studies.

SCHOOL OF RURAL ECONOMICS 829

Head: Dr E.S. Clayton
Units:
Agricultural economics, D.K. Britton; agrarian development overseas, E.S. Clayton; farm business, Professor J.S. Nix; marketing, D.H. Pickard
Activities: Agricultural and horticultural economics; agrarian development overseas: rural employment and income distribution, rural water supplies and irrigated agriculture, and the supply and use of credit in developing countries are among topics studied; farm business; marketing.
Projects in the Agrarian Development Overseas Unit: Agrarian problems in economic development; rural employment and development (Dr E.S. Clayton); development of livestock industries in the United Kingdom and overseas (G. Allanson); planning rural water supplies and irrigation (Dr I.D. Carruthers); FME in East Africa (Dr M.J. Dorling); irrigation economics in West Africa (Dr R.W. Palmer-Jones).

Projects in the Agricultural Economics Unit: Econometrics applied to agriculture (A. Burrell); wealth and income in agriculture; economics of size and tenure (Dr N.W.F.B. Hill); comparative agricultural efficiency in European countries; agricultural policy (J.R. Medland); economics of horticulture (D. Ray).
Projects in the Farm Business Unit: Farm management techniques, research, practice (Professor J.S. Nix); farm finance and accounting (G.P. Hill); management systems: development of new technology (J.A.H. Nicholson); decision models in agriculture: micro-computers (Dr J.P.G. Webster); production economics, micro-computers (N.T. Williams).
Projects in the Marketing Unit: Evolution and operation of input and food marketing systems; agricultural input marketing (P. Newbound); marketing of flowers; import trade in fresh fruit, vegetables, and flowers into north west European markets; marketing of horticultural produce from the Channel Islands (A.R. Hunt).

Centre for European Agricultural Studies Centre 830

– CEAS
Address: Wye College, Wye, Ashford, Kent, TN25 5AH
Telephone: (0233) 81 21 81
Telex: 96118 ANZEEC G
Status: Research centre within a university teaching establishment.
Director: I.G. Reid
Graduate research staff: 7 (1983)
Activities: Research and consultancy activities are aimed at assisting companies and other organizations whether governmental, quasi-governmental or trade in their planning exercises at both strategic and tactical levels. Current research is focused on marketing systems and trade flows of agricultural and food products, European Community food aid and agricultural finance.
Projects: European agricultural finance and taxation, land ownership and tenure; West German and United Kingdom agriculture (I.G. Reid); European meat/livestock marketing systems, fruit, and vegetables (Anne McLean Bullen); systems simulation, European Economic Community food aid evaluation, agricultural trade policy (M.K. Mitchall); wine policy, economics of wine/Mediterranean products (Dr S. Shea); European potato marketing systems, agricultural centrally sponsored organizations (N.A. Young).
Publications: Annual report.

Yorkshire Water Authority* 831

Address: West Riding House, 67 Albion Street, Leeds, LSl 5AA
Telephone: (0532) 448201
Status: Public utility conducting research
Chairman: D.B. Matthews
Research director: Dr A. J. Shuttleworth
Sections: Chemistry I (analytical), Dr J.M. Carter; chemistry II (treatment), D.A. Bailey; biology, Dr C. Urquhart; microbiology, H. Fennell
Graduate research staff: 20
Activities: Research into drinking water quality, water treatment, resource development and management, sewage treatment, sludge treatment, river management, instrumental analysis.

UNITED STATES OF AMERICA

Acarology Laboratory 1

Address: 484 West 12th Avenue, Columbus, OH 43210
Telephone: (614) 422 7180
Affiliation: Ohio State University
Status: Educational establishment with r&d capability
Research director: Professor Donald E. Johnston
Graduate research staff: 6
Activities: Research and dissemination of acarological knowledge on a world basis. Research areas of agricultural importance are identification and ecology of soil mites, genetics and population biology of spider mites, and physiology of ticks. The laboratory maintains one of the largest collections of specimens and literature in the world and provides identification services on a world basis.
Publications: Directory of Acarologists, quarterly.
Projects: Physiology of ticks (Dr Glen R. Needham); physiology of mites (Dr G.W. Wharton); genetics and population biology of spider mites (Dr Dana L. Wrensch); anatomy and classification of mites (Dr Donald E. Johnston).

Agricultural Research and Education Center, Lake Alfred 2

Address: 700 Experiment Station Road, Lake Alfred, FL 33850
Telephone: (813) 956 1151
Affiliation: University of Florida
Status: Official research centre

Agrigenetics Research Park 3

Address: 5649 East Buckeye Road, Madison, WI 53716
Telephone: (608) 221 5000
Parent body: Agrigenetics Corporation
Status: Research centre within an industrial company
Vice President, Director, Advanced Research: Timothy C. Hall
Sections: Biochemistry; cell biology; genetics; microbiology
Graduate research staff: 30
Activities: The laboratory conducts fundamental research to improve crop quality and yield. Attemps are being made to understand plant breeding at the molecular level and make crosses which cannot be made by classical methods.
Projects: Recombinant DNA research in major crop plants (M. Murray, L. Hoffman); protoplast and cell culture technology (D. Hansen, V. Tilton, S. Loesch-Fries, J. Ranch); nitrogen metabolism and microbial genetics (J. Cramer, C. Sengupta-Gopalan); protein biochemistry and quality improvement of seed (J. Brown, Y. Ma).

Agronomic Services Laboratory 4

Address: PO Box 639, 1087 Jamison Road NW, Washington CH, OH 43160
Telephone: (614) 335 1562
Parent body: Agrico Chemical Company
Status: Research centre within an industrial company
Management: R.B. Lockman
Graduate research staff: 5
Activities: Soil science - maintenance of a series of plots that are used for: calibrating soil and plant analysis; determining effects of continued fertilizer practices upon soil test, plant analysis, yield, quality, and economics.

The laboratory has greenhouse and field plot capability to evaluate various products as required. The primary responsibility however is soil and plant analysis services for the customers.
Projects: Maintaining long-term plots (1964 to date) (M.G. Molloy).

Agway Incorporated, Research and Development Division* 5

Address: 56 Roland Street, Charlestown, MA 02129
Telephone: (617) 776 2734
Status: Research centre within an industrial company
Activities: Dairy and food research.

Alabama A&M University 6

Address: PO Box 183, Normal, AL 35762
Status: Educational establishment with r&d capability

SCHOOL OF AGRICULTURE 7

Telephone: (205) 859 7319
Dean: Dr Winfred Thomas
Graduate research staff: 60
Activities: The school operates within the total mission of the university's land-grant function of research, service and instruction. The main thrust of the school's research programme is in the areas of food science, animal nutrition, soil science, land use, forestry and forest products, plant breeding and production (soyabean and triticale) and rural development. More specifically, major research activities involve: developing new food items with improved nutritional qualities; breeding higher yielding soyabean varieties with resistance against cyst nematodes; breeding triticale with improved yield and protein qualities; evaluating factors related to levels and patterns of living in the rural South; applying remote sensing technology to the management of agricultural and natural resources.
Publications: School of Agriculture, Environmental Science and Home Economics Annual Report.
Projects: Soyabean breeding and production; triticale breeding and production (V.T. Sapra); application of remote sensing technology (O.L. Montgomery); developing Christmas trees for commercial production in Alabama (G.F. Brown); evaluation of R. Japonicum strains and soyabean germplasm (M. Floyd); broadening the food preferences of preschool children (B. Auclair); factors related to levels and patterns of living (G. Wheelock); rabbit production, management and utilization (C.B. Chawan); nutritional and healthful aspects of

fermented milk (D.R. Rao); utilization of agricultural wastes for production of ethanol (B. Singh).

Agribusiness Education Department 8
Head: Dr P. Preyer

Community Planning Department 9
Head: D. Outland

Division of Home Economics 10
Head: Dr V. Caples

Food Science and Animal Industries Department 11
Head: Dr G. Sunki

Natural Resource and Environmental Studies Department 12
Head: Dr G. Sharma

American Cocoa Research Institute 13

– ACRI
Address: 7900 Westpark Drive, Suite 514, McLean, VA 22102
Telephone: (703) 790 5011
Telex: 710-833-0898
Status: Independent research funding organization
President: Richard T. O'Connell
Science Advisor: Dr Russell E. Larson
Activities: The institute does not carry out research itself, but is a non-profitmaking association supporting cocoa production research in the fields of plant protection, plant breeding, and plant pathology.
Publications: CMA/ACRI Annual Report.

American Institute of Baking* 14

Address: 1213 Bakers Way, Manhattan, KS 66502
Telephone: (913) 537 4750
Status: Trade association
Activities: Research and education into nutrition, baking processes, and food composition.
Publications: Technical bulletin, monthly.

American Plywood Association   15

– APA
Status: Trade association

TECHNICAL SERVICES DIVISION   16

Address: PO Box 11700, Tacoma, WA 98411
Telephone: (206) 565 6600
Telex: 32 7430
Director: Thomas R. Flint
Departments: Panel technology, Dr Michael R. O'Halloran; engineering technology, Daniel H. Brown
Graduate research staff: 15
Activities: The research centre provides technical background for the correct usage of plywood and other structural wood panel products. Basic engineering properties are derived for construction and industrial grades of plywood. Maintenance of Product Standard PS-1 for plywood is carried out as well as development of performance standards for major categories of use of structural wood panel products. Research relating these panel products to building construction uses and industrial uses is an ongoing function.
Projects: Mechanical properties of panels (Paul W. Post); performance standards for siding (Harry D. Jorgensen); product standards (Thomas E. Batey); plywood for industrial uses (Raymond C. Mitzner); construction uses of panel products (John D. Rose).

American Tobacco Company   17

Parent body: American Brands Incorporated
Status: Industrial company

DEPARTMENT OF RESEARCH AND DEVELOPMENT   18

Address: PO Box 899, Hopewell, VA 23860
Telephone: (804) 748 4561
Vice President, Research and Development: R.S. Sprinkle III
Graduate research staff: 50
Activities: Research and development related to tobacco and tobacco products.

Applied Agricultural Research Incorporated*   19

Address: 1305 E Main Street, Lakeland, FL 33801
Status: Industrial company
Activities: Research on management of citrus fruits and improvements in forage production.

Archer Daniels Midland Company*   20

Address: PO Box 1470, 4666 Faries Parkway, Decatur, IL 62525
Status: Industrial company
Research director: Dr Kenneth E. Beery
Activities: Developments of processes and products from agricultural raw materials, including soyabeans, sunflowers, cotton.

Armour Research Center   21

Address: 15101 North Scottsdale Road, Scottsdale, AZ 85260
Telephone: (602) 991 3000
Parent body: Armour Food Company
Status: Research centre within an industrial company
Vice President, Director of Food Research: R.B. Sleeth
Sections: Food chemistry, K. Sato; cured and processed meats, R.B. Rendek; sausage development, G.J. Jedlicka; frozen convenience foods, R.F. Grotts; food packaging, B.R. Lundquist; poultry and dairy development, G.S. Grauman
Graduate research staff: 30
Activities: Research and development on foods, including fresh and processed meats, fresh and processed poultry products, dairy products and frozen foods.

Auburn University*   22

Address: Auburn, AL 36830
Status: Educational establishment with r&d capability

ALABAMA AGRICULTURAL EXPERIMENT STATION   23

Address: 107 Comer Hall, Auburn University, Auburn, AL 36849
Telephone: (205) 826 4840
Director and Dean for Research: Dr Gale A. Buchannan
Graduate research staff: 322
Activities: To promote efficiency in increased production and marketing of crops and livestock, to protect and preserve the natural resources and environment.
Publications: Annual report; *Highlights of Agricultural Research*, quarterly.
Projects: Farm management, livestock and grain marketing, recreation, and rural sociology (Dr J.H. Yeager); soil and water, power and machinery, farm structures, animal environmental, and animal waste management (Dr Paul K. Turnquist); soil chemistry and fertility, forage crops, plant breeding, weed control varietal work and turf management (Dr Coleman Y. Ward); animal breeding, animal nutrition, biochemistry, and meats (Dr David G. Topel); physiology, microbiology, animal health, and parasitology (Dr P.C. Smith); plant nematology, forest pathology, and bacteriology (Dr Paul A. Lemke); fish nutrition, fish diseases, management, and aquaculture (Dr E.W. Shell); wood technology and anatomy, silviculture, forest management, economics, harvesting and processing (Dr Emmett F. Thompson); housing, nutrition, textiles (Dr Ruth Galbraith); vegetable breeding, fruit and nut crops, ornamentals, and food science (Dr Donald Y. Perkins); nutrition, housing, breeding, diseases and management (Dr Claude H. Moore); insect control, genetics, biological control, forest-entomology, and wildlife (Dr Kirby L. Hays).

Department of Agricultural Economics and Rural Sociology   24

Head: Dr J.H. Yeager

Department of Agricultural Engineering   25

Head: Dr Paul K. Turnquist

Department of Agronomy and Soils   26

Head: Dr Coleman Y. Ward

Department of Animal and Dairy Sciences   27

Head: Dr David G. Topel

Department of Animal Health Research   28

Head: Dr Paul C. Smith

Department of Botany, Plant Pathology, and Microbiology   29

Head: Dr Paul A. Lemke

Department of Fisheries and Allied Aquacultures   30

Head: Dr E.W. Shell

Department of Forestry   31

Head: Dr Emmett F. Thompson

Department of Home Economics Research   32

Head: Dr Ruther L. Galbraith

Department of Horticulture   33

Head: Dr Donald Y. Perkins

Department of Poultry Science   34

Head: Dr Claude H. Moore

Department of Research Data Analysis   35

Head: Dr R.M. Patterson

Department of Research Information   36

Head: E.L. McGraw

Department of Research Operations   37

Head: Lavern Brown

Department of Zoology and Entomology   38

Head: Dr Kirby L. Hayes

SCHOOL OF VETERINARY MEDICINE   39

Dean: J.T. Vaughan
Associate Dean for Research and Graduate Studies: S.D. Beckett
Graduate research staff: 50
Activities: Veterinary medicine - development of methods for diagnosis, treatment, and prevention of disease conditions in food and companion animals.

Projects: Thermography in horses (R.C. Purohit); pathogenesis and therapy of intestinal parasites in calves (L.A. Hanrahan); persistence of natural infection in calves born to brucellosis-infected dams (P.C. Smith); immune response of new born pigs; immune response associated with bovine leukosis (R.D. Schultz); virological aspects of bovine respiratory disease (P.C. Smith); congenital heart defects in dogs (C.E. Branch); central nervous system conditions in dogs (R.W. Redding); stereotaxic anatomy of animals (D.F. Buxton); neurological diseases of companion animals (B.F. Hoerlein); reconstructive surgery (S.F. Swaim).

Department of Anatomy and Histology 40

Head: C.L. Holloway

Department of Large Animal Surgery and Medicine 41

Head: D.F. Walker

Department of Microbiology 42

Head: P.C. Smith

Department of Pathology and Parasitology 43

Head: L.G. Wolfe

Department of Physiology and Pharmacology 44

Head: C.H. Clark

Department of Radiology 45

Head: J.E. Bartels

Department of Small Animal Surgery and Medicine 46

Head: C.D. Knecht

Scott-Ritchey Research Department 47

Head: B.F. Hoerlain

Baltimore Spice Company* 48

Address: PO Box 5858, Baltimore, MD 21208
Telephone: (301) 363 1700
Status: Industrial company
Laboratory director: Simon G. Statter
Activities: Research on spices, essential oils and oleoresins.

Beatrice Foods Research Center* 49

Address: 1526 S. State Street, Chicago, IL 60605
Telephone: (312) 791 8200
Parent body: Beatrice Foods Company
Status: Research centre within an industrial company
Research director: Dr R. Tjepkema
Activities: Developments and improvements to dairy and grocery products; nutritional studies.

Bee Management Research Station* 50

Address: 436 Russell Laboratories, University of Wisconsin, Madison, WI 53706
Telephone: (608) 262 1732
Affiliation: University of Wisconsin, Department of Entomology
Status: Official research centre
Activities: Bee management; methods of handling bees and honey; selection and breeding of bees; pollination of specific crops.

Beet Sugar Development Foundation 51

– BSDF
Address: PO Box 1546, Fort Collins, CO 80522
Telephone: (303) 482 8250
Status: Independent research centre
Manager: James H. Fischer
Graduate research staff: 5
Activities: The foundation supports basic research in plant breeding, seed improvement, nutrition studies and integrated pest management.
Projects: Many projects are supported at each department of agriculture station conducting research on sugar beets. Additional projects are supported at state universities.

Bio-Technical Resources 52

– BTR
Address: 7th and Marshall Streets, Manitowoc, WI 54220
Telephone: (414) 684 5518
Status: Independent research centre
President: Michael R. Sfat
Departments: Engineering, James A. Doncheck; biochemistry, Thomas J. Skatrud; analytical chemistry, Bruce J. Morton
Graduate research staff: 12
Activities: development of processes and products employing biotechnology, soil science, food science, food economics.
Projects; Microbial leaching; carbohydrate bioconversion (T.J. Skatrud); malting and brewing (J.A. Doncheck).

Brigham Young University 53

– BYU
Address: Provo, UT 84602
Affiliation: Church of Jesus Christ of Latter Day Saints
Status: Educational establishment with r&d capability

COLLEGE OF BIOLOGICAL AND AGRICULTURAL SCIENCES 54

Address: 301 WIDB, Brigham Young University, Provo, UT 84602
Telephone: (801) 378 2007
Telex: 389-452
Dean: A. Lester Allen
Graduate research staff: 100
Activities: Research is done to train students to improve standards of living in developed and in developing nations. Subjects covered are natural resources (botany and range science), plant production (plant breeding and crop husbandry), animal production (dairy, beef, poultry, hogs, rabbits and goats - breeding, husbandry, nutrition and management), food science (quality control, new product development), agricultural economics (farm, food and business economics and management), and range science (ecology and management).

Agricultural Economics Department 55

Head: William L. Park

Agronomy and Horticulture Department 56

Head: Laren R. Robison
Projects: Small scale agriculture.

Animal Science Department 57

Head: Leon E. Orme

Bean Museum 58

Head: Richard Bauman

Benson Institute 59

Head: D. Delos Ellsworth

Botany and Range Science Department 60

Head: Jerran T. Flinders

Food Science and Nutrition Department 61

Head: Clayton S. Huber

Microbiology Department 62

Head: Richard D. Sagers

Zoology Department 63

Head: Ferron L. Anderson

California Department of Water Resources 64

Status: Government agency

OFFICE OF WATER CONSERVATION 65

Address: PO Box 388, Sacramento, CA 95802
Telephone: (916) 445 9958
Chief, Agricultural Branch: Edward Craddock
Graduate research staff: 5
Activities: Research in the following areas: drainage and irrigation water management; drainage and irrigation technology; plant breeding; distribution system improvements; brackish water irrigation; evaporation-transpiration reduction.
Publications: Water Conservation News, bimonthly; annual project reports.

Projects: Cropping practices alternatives for water conservation (Dr Robert Hagan); evaporation-transpiration relationships (Dr Tsiao, Dr Hatfield); brackish water irrigation (Dr John Letey); brackish water irrigation of cotton (Dr James Rhoades); drainage water re-use on crops (Dr James Rhoades); California irrigation management information system (Dr Elias Ferreres); plant breeding for water conservation (Dr Iver Johnson).

California Fig Institute* 66

– CFI
Address: PO Box 709, Fresno, CA 93712
Status: Independent research centre
Activities: Improvement of cultural practices and control of nitidulids (fruit beetle) by the use of insecticides, attractants, repellants and biological control.

California State University, Chico 67

Address: Chico, CA 95929
Status: Educational establishment with r&d capability

SCHOOL OF AGRICULTURE AND HOME ECONOMICS 68

Telephone: (916) 895 5131
Dean: Lucas Calpouzos
Departments: Animal science, agricultural business, education and mechanics, Dr Lal Singh; plant and soil science, Dr Michael Maynard; home economics, Dr Marilyn Ambrose
Graduate research staff: 5
Activities: Research in the following areas: applied aspects of crop production - edible legumes, cotton, vegetables, tree fruit and vine crops; applied aspects of animal production - breeding, feeding, and carcase evaluation of sheep, swine, and beef; natural resources - soil fertility, drainage and irrigation, land use.
Projects; Plant protection (Dr J. Burleigh); range management (Dr Henricus Jansen); crop irrigation and drainage (Dr Herb Paul); tree fruit trials (Dr R. Baldy); field crop trials (Buel Mouser); soil fertility (Dr John Hart); alternative energy applications (Ron Borge); bovine nutrition (Dr Ron Hutchings); economics of natural resources (Dr Tom Dickinson); human nutrition (Dr Doug Buck).

Campbell Institute for Research and Technology 69

Address: Campbell Place, Camden, NJ 08101
Telephone: (609) 964 4000
Parent body: Campbell Soup Company
Status: Research centre within an industrial company
Chief research officer: A.E. Denton
Sections: Food science and technology (Dr C.L. Duncan); poultry research (Dr R.H. Forsythe); mushroom research (Dr R.E. Miller); vegetable research (Dr M.A. Stevens).

DAVIS LABORATORY 70

Address: Route 1, Box 1314, Davis, CA 95616
Telephone: (916) 753 2116
Telex: 377421
Vice President, Vegetable Research: Dr M. Allen Stevens
Graduate research staff: 6
Activities: Plant production, breeding, and crop husbandry - tomatoes, carrots, cucumbers, water-chestnuts and peas.
Projects: Tomato breeding (S.J. Warnock); tomato quality, heat and cold tolerance (M.A. Stevens); pea breeding (S.J. Warnock); tomato and cucumber production practices, and water-chestnut production (H.R. Hikida); carrot breeding (G.S. McWalter); disease resistance in tomatoes (K. Fisher).

NAPOLEON LABORATORY 71

Address: P-152 R12, Napoleon, OH 43545
Telex: (419) 592 8015
Telex: 286451
Director of Pest Management Research: Richard C. Henne
Graduate research staff: 4
Activities: Plant breeding - to improve yield and quality of tomatoes and cucumbers; plant protection - to develop control systems to reduce crop losses due to pests in vegetable crops.
Projects: Tomato breeding (W.S. Taylor); cucumber breeding (Dr G.E. Tolla); variety evaluations (Dr R.D. Peel); pest management (R.C. Henne).

Ciba Geigy Agricultural Division* 72

Address: 410 Swing Road, PO Box 11422, Greensboro, NC 27409
Telephone: (919) 292 7100
Parent body: Ciba Geigy Corporation
Status: Research centre within an industrial company
Research and Development Manager: H.B. Camp
Activities: Research into herbicides, insecticides, plant growth regulators, lawn and garden products.

Clemson University* 73

Address: Clemson, SC 29631
Status: Educational establishment with r&d capability

EDISTO EXPERIMENT STATION* 74

Address: Box 247, Blackville, SC 29817
Activities: Soyabeans, watermelon, sweet potatoes and studies on crops suitable for the coastal plain area of South Carolina.

PEE DEE EXPERIMENT STATION* 75

Address: PO Box 271, Florence, SC 29503
Telephone: (803) 662 3526
Activities: Development of new varieties of cotton, corn and tobacco.

Coca Cola Research Laboratory* 76

Address: Hammondsport, NY 14840
Telephone: (607) 569 2111
Parent body: Coca Cola Company
Activities: Research and development into wine chemistry, soft drinks, tea and coffee blends, and citrus fruit flavourings.

Colorado Serum Company* 77

Address: 4950 York Street, Denver, CO 80216
Status: Industrial company
Director: L.E. Drehle
Activities: Tissue culture; veterinary medicine; agronomy; chick embryo virus propagation.

Colorado State University 78

Status: Educational establishment with r&d capability

COLLEGE OF AGRICULTURAL SCIENCES 79

Address: Fort Collins, CO 80523
Dean: Dr Donal D. Johnson

Agricultural Experiment Station 80

Telephone: (303) 491 6272
Head: Dr J.P. Jordan
Graduate research staff: 300
Activities: The emphasis of research is placed on the following: soil and water, conservation and management; basic crop production in arid and semi-arid climates; irrigation; beef cattle breeding; embryo transplant; animal reproduction; wheat, potatoes, sugar beets, corn, barley, tree fruit.

Department of Agricultural Economics 81

Head: K.C. Nobe

Department of Agricultural Engineering 82

Head: J.M. Harper

Department of Agronomy 83

Head: W.F. Keim

Department of Animal Science 84

Head: B.M. Jones

Department of Botany and Plant Pathology 85

Head: G.M. McIntyre

Department of Horticulture 86

Head: K.M. Brink

Extension Service 87

Head: L.H. Watts

Connecticut Agricultural Experiment Station* 88

Address: 123 Huntington Street, PO Box 1106, New Haven, CT 06504
Telephone: (203) 789 7214
Director: Dr Paul E. Waggoner
Activities: Plant genetics, plant pathology, botany, entomology, soils.

Cornell University* 89

Status: Educational establishment with r&d capability

CORNELL UNIVERSITY AGRICULTURAL EXPERIMENT STATION 90

Address: 292 Roberts Hall, Ithaca, NY 14850
Telephone: (607) 256 5429
Acting director: T.L. Hullar
Activities: The station coordinates and conducts the agricultural research work of the university; an outline of major areas of study is given under the various departments.

NEW YORK COOPERATIVE FISHERY RESEARCH UNIT* 91

Address: 118 Fernow Hall, Cornell University, Ithaca, NY 14853
Telephone: (607) 256 2151
Activities: Ecology of stream fishes; management of warm-water fisheries; aquatic pollution; toxic interaction of metals and natural organic compounds.

NEW YORK COOPERATIVE WILDLIFE RESEARCH UNIT* 92

Address: 118 Fernow Hall, Ithaca, NY 14853
Leader: Dr M. Richmond
Activities: Wildlife conservation and management.

NEW YORK STATE COLLEGE OF AGRICULTURE AND LIFE SCIENCES 93

Address: Cornell University, Ithaca, NY 14853
Dean: David L. Call
Publications: Annual report.

Agricultural Economics Department 94

Chairman: O.D. Forker
Activities: Research in the following areas: economics of dairy produce, beef, fruit and vegetables, ornamental plants, cereals and oilseed, and coastal fishing industries; labour management; local, national, and international marketing practices; agricultural cooperatives; population migration; computerized farm management; food demand and consumption; socioeconomic factors and rural land use; agricultural economics of the American tropics and Latin America; political economy of food in developing countries, particularly the Asian rice and grain economy.

Agricultural Engineering Department 95

Chairman: N.R. Scott
Activities: Research in the following areas: mechanization in cultivation, harvesting, and handling of agricultural products; aspects of animal housing and other farm buildings, including energy conservation, environmental stress and animal behaviour, and numerical air-flow prediction; mechanical properties of stabilized soils; mathematical modelling of biological problems; alternative energy, including production of ethanol and alcohol from agricultural products, biomass, and wastes, solar heating, wind energy, and waste-heat greenhouses; water and sludge management; irrigation; environmental science, including waste, water, and energy systems, pollution monitoring and control, and environmental impacts on agricultural efficiency.

Agronomy Department 96

Chairman: R.F. Lucey
Activities: Research in the following areas: origin, transformation, and management of soil nitrogen; energy analysis of the production and application of fertilizers; soil studies, including sorption of heavy metals, chemistry of soil humus, and mechanical properties of freezing soils; weed and crop competition; forage systems; plant breeding for cold and drought resistance; phenology, weather, and crop yields; dynamics and energetics of the soil-plant-atmosphere continuum.

Animal Science Department   97

Chairman: R.J. Young .
Activities: Research in the following areas: all aspects of dairy herd management, including nutrition and metabolism, physiology, reproduction, disease prevention and control, and milk production; sheep and beef cattle production; evaluation of selection techniques in animal breeding; environmental stress and animal behaviour; animal monitoring techniques, including hormonal, ultrasonic, and other measurements; forage, silage, animal manure, and agricultural by-products as animal feeds.

Biological Sciences Division   98

Sections: L.H. Bailey Hortorium, D.M. Bates; biochemistry, molecular and cell biology, R.E. McCarty; botany and genetics, P.J. Bruns; ecology and systematics, B. Chabot; neurobiology and behaviour, T.R. Podleski; physiology, W. Hansel
Activities: Research in the following areas: plant taxonomic and classification studies; cell biology, including DNA replication, protein synthesis, genetic mutation, photosynthesis, enzyme function, cancer and virus studies; botany, genetics and plant development, including hormonal control of senescence, cellular ageing, investigation of the shoot apex at various stages of plant growth, palaeobiochemistry, ion transport, and photosynthetic energy conversion; ecology and systematics of aquatic, forest, and temperate agricultural environments; neurobiological studies, including principally orientation and navigation mechanisms of birds, amphibians and reptiles, honeybees, and cockroaches, and also pharmacologically active secretions in insects and arthropods; behavioural studies and primate sociobiology; physiology of cattle and other vertebrates.

Entomology Department   99

Chairman: M. Tauber
Activities: Research in all areas of insect ecology, including biochemistry and reproductive physiology; population dynamics and methods of control; insect damage to crops, and plant resistance, particularly through the development of glandular trichomes in the potato plant; integrated pest management, including computer-aided strategies, pesticide development and assessment, and non-chemical pest control.

Floriculture and Ornamental Horticulture Department   100

Chairman: C.F. Gortzig
Activities: Research in the following areas: mineral nutrition of greenhouse and nursery crops; weed control; development of turfgrasses; seed production; and tissue culture as a means of micropropagation of ornamental plants.

Food Science Department   101

Chairman: J.E. Kinsella
Activities: Research in the following areas: the development of new food products and processing technologies, including milk and fermented milk-based products, convenience foods from eggs, and utilization of fish proteins and flavours; microbiology, including the study of lactic acid fermentation, the synthesis of cocoa fat, proteolytic enzymes in food fermentation, and lipid oxidation in fish-based foods; isolation of natural anti-oxidants; availability of iron and folic acid in selected foods; vegetable storage.

Microbiology Department   102

Chairman: R.P. Mortlock
Activities: Research in the following areas: microbiology of waters, waste waters, and wastes; microbial nitrate reduction; utilization of xylitol by soil bacteria; investigation of microbial populations by electron microscopy; the evolution of enzymatic activities.

Natural Resources Department   103

Chairman: W.H. Everhart
Activities: Research in the following areas: the ecology and wildlife of New York State, and the environmental stresses resulting from recreational uses, hunting, and fishing, and from oil and sulphur dioxide pollution; plant and wildlife surveys; population dynamics of birds and control of bird damage to crops; limnology of New York lakes and rivers; fish culture; forestry and forest management.

Nutritional Sciences Department   104

Activities: Research in the following areas: nutrition and intestinal parasitism; effects of fat-altered diets; nutritional aspects of multiple sclerosis; amino acid and polyamino metabolism; availability and utilization of dietary magnesium; cysteine metabolism in animals; sodium intake and essential hypertension; vitamin A deficiency in the Philippines; lipoprotein transport and arteriosclerosis; effect of nutritional variations on brain and behaviour; nutrition and chemical carcinogenesis; energy use and conservation in the food system.

Plant Breeding and Biometry Department   105

Chairman: W.D. Pardee
Sections: Biometrics unit, D.L. Solomon
Activities: Research in the following areas: statistics and biometry; microcomputers for data analysis; breeding improved cultivars of the following crops - corn (Zea mays), potatoes resistant to the potato cyst nematode, cool-season forage species, alfalfa for increased nitrogen fixation, wheat, oats, small grain cereals, beans (Phaseolus vulgaris), sweet corn, Brassica campestris, and cold-resistant rice.

Plant Pathology Department   106

Chairman: B. Fry
Activities: Research in the following areas: epidemiology and control of plant diseases, including fire blight of apples and pears, barley yellow dwarf virus, vascular wilt pathogens, canker diseases of trees, fusarium blight on turfgrass, and other viral, bacterial and fungal diseases; the attacking mechanisms of plant pathogenic fungi; ecology and control of plant parasitic nematodes and their effect on crop yields; development of new disease control methods, and disease-resistant strains of plant crops.

Pomology Department   107

Chairman: W.J. Kender
Activities: Research in the following areas: postharvest physiology of fruits; biochemistry of flavonoid compounds; photo-periodic control of apple ripening on the tree; mineral nutrition of fruits; shoot growth in the apple tree; metabolism of pesticides, and pesticide residues in agricultural commodities.

Poultry and Avian Sciences   108

Chairman: R.C. Baker
Activities: Research in the following areas: protein and energy requirements for poultry production; environmental and hormonal influences on ovulation in chickens; selenium and vitamin E in the nutrition of poultry and other animals; immune response and poultry breeding; eggshell quality; composition, nutritive value and stability of poultry meat and egg products; White Peking duck production; mutagenicity of environmental chemicals in poultry.

Vegetable Crops Department   109

Chairman: R.D. Sweet
Activities: Research in the following areas: health, physiology, breeding, handling, harvesting, storage, and processing of vegetable crops, including lettuce, potatoes, red beets, dry beans, and tomatoes; weed control, herbicides, and weed-crop competition; nitrogen fertilization; cultural methods for commercial crops.

NEW YORK STATE COLLEGE OF VETERINARY MEDICINE*   110

Address: Cornell University, Ithaca, NY 14853

James A. Baker Institute for Animal Health   111

Telephone: (607) 277 3044
Director: Dr Douglas D. McGregor
Graduate research staff: 9
Activities: The institute conducts research on animal diseases in order to increase knowledge about their nature and to develop methods for disease control. Simultaneously, it provides opportunities for advanced training of scientists in comparative medicine.
Publications: Annual report.
Projects: Canine infectious diseases (Dr L.E. Carmichael); canine nutrition (Dr B.E. Sheffy); canine viral diseases (Dr M.J.G. Appel); osteoarthritis (Dr G. Lust); immunogenetics of domestic animals (Dr D.F. Antczak); immunoparasitology (Dr R.G. Bell).

Cotton Incorporated Research Center*   112

Address: 4505 Creedmoor Road, PO Box 30067, Raleigh, NC 27612
Telephone: (919) 782 6330
Status: Independent research centre
Sections: Textiles, H.E. Brockmann; economics, David W. Ox; agriculture, J.K. Jones

Dairy Research Incorporated 113

Address: 6300 North River Road, Rosemount, IL 60018
Telephone: (312) 696 1870
Affiliation: Dairy Research Foundation
Status: Trade association
Executive Vice President: Ray Mykleby
Sections: Dairy Research Foundation, William W. Menz; Commercial Development Division, Jasper W. Reaves
Graduate research staff: 3
Activities: The organization sponsors research projects in food and dairy science and technology which are carried out at various university and research centres. The main areas of research concern fluid milk, milk protein, milk fat, whey and whey products, and cultured dairy products.
Publications: Dairy Research Digest.

DCA Food Industries Incorporated* 114

Address: 919 Third Avenue, New York, NY 10022
Telephone: (305) 727 0660
Parent body: J. Lyons & Co, UK
Status: Industrial company
Director, central research laboratory: H. Roth
Activities: Prepared food mixes; food coatings; research into frying fats and oils, and into cereal and flour chemistry.

Deere Technical Center 115

Address: 3300 River Drive, Moline, IL 61265
Telephone: (309) 757 5378
Parent body: Deere and Company
Status: Research centre within an industrial company
Director: Robert Wismer
Sections: Product technology, James Toal; engineering science, Richard Strunk; materials technology, Duane Olberts
Graduate research staff: 200

Delta Branch Experiment Station* 116

Address: PO Box 197, Stoneville, MS 38776
Telephone: (601) 689 9311
Parent body: Mississippi Agricultural and Forestry Experiment Station
Status: Official research centre
Superintendent: Dr Charles G. Shepherd
Activities: Farm management and costs, fertilizers and soils, horticulture, livestock, soyabeans, and cotton genetics.

Dow Chemical Company 117

Status: Industrial comapny

AGRICULTURAL PRODUCTS DEPARTMENT 118

Address: 9008 Building, Midland, MI 48640
Telephone: (517) 636 0875
Telex: 227455
Vice President, Agricultural Research and Development: Dr P.J. Gehring
Activities: Research in the following areas: plant protection - herbicides, insecticides, fungicides, nematicides; plant growth regulators; fertilizer additives; animal growth promoters.

Duke University* 119

Address: Durham, NC 27706
Status: Educational establishment with r&d capability

DEPARTMENT OF BOTANY* 120

Duke University Phytotron 121

Telephone: (919) 684 6523
Director: Boyd R. Strain
Graduate research staff: 12
Activities: Research in the following areas: physiological plant ecology; plant productivity; plant water relations; global carbon dioxide enrichment effects on plants.
Projects: Carbon dioxide effects on plant growth (B.R. Strain); C-11 technique for carbon allocation research (John D. Goeschl); tundra response to global carbon dioxide enrichment (W.D. Billings); root chilling effects on crops (Paul J. Kramer); weed ecology (David T. Patterson); population ecology (Janis Antonovics).

Elm Research Institute   122

Address: Main Street, Harrisville, NH 03450
Telephone: (603) 827 3048
Status: Independent research centre
Executive Director: John P. Hansel
Resident Arborist: Paul Ayers
Activities: Research into methods of combatting Dutch elm disease.
Publications: Newsletter, quarterly.
Projects: Developing an American elm resistant to Dutch elm disease (Dr Eugene Smalley); various control methods of combatting Dutch elm disease (Dr Ganga Nair).

Florida A&M University   123

Status: Educational establishment with r&d capability

DIVISION OF AGRICULTURAL SCIENCES   124

Address: Tallahassee, FL 32307
Telephone: (904) 599 3429
Director: W.L. Johnson
Sections: Entomology, Dr W.L. Peters; agri-business, N. Saylor; animal sciences, Dr L.E. Evans; horticulture
Graduate research staff: 12
Activities: Research is geared towards meeting the needs of small-acreage farm operators. The subject areas presently being researched are: soil science; beef husbandry and nutrition; food science; and plant protection.
Projects: Water quality studies and aquatic insects (Dr W.L. Peters); peanut protein characterization (Dr S.K. Pancholy); Florida marsh ecology (Dr C.B. Subramanyam); cultural food habits (Dr C.M. Marquess); beef cattle/cow-calf operations (Wilbur Bate); rural poverty (Dr J.S. Dhillon)

Florida Agricultural Experiment Stations*   125

Address: 1022 McCarty Hall, Gainesville, FL 32611
Telephone: (904) 392 1784
Status: Indpendent research centre
Dean, research: Dr F.A. Wood
Graduate research staff: 390
Activities: All fields of research.
Publications: Sunshine State Agricultural Research report, quarterly.

Foremost-McKesson Research and Development Center   126

Address: PO Box 2277, Dublin, CA 94566
Telephone: (415) 828 1440
Status: Research centre within an industrial company
Vice-President, Research and Development: Dr Theodore W. Craig
Graduate research staff: 41
Activities: Food technology; product research and development.

CONTRACT RESEARCH DEPARTMENT   127

Manager: William A. Hoskins
Sections: Industrial accounts, Sandra L. Donatoni, Peter Q. Little; water environmental services, Dr Warren C. Steele

PRODUCT/PROCESS DEVELOPMENT DEPARTMENT   128

Manager: Dr David G. Holmes
Sections: Consumer products, Dr Lynn V. Ogden; food ingredients, Dr Alan G. Hugunin; pilot processing, Gaylord Palmer

TECHNICAL SERVICES DEPARTMENT   129

Manager: Richard G. Henika
Sections: Analytical services, Dr Rod P. Kwok; sensory, Marianne E. Lane

Forest Service   130

Address: PO Box 2417, Washington, DC 20013
Affiliation: United States Department of Agriculture
Status: Official research centre
Chief: R. Max Peterson
Deputy Chief, Research: Robert E. Buckman

FOREST ENVIRONMENTAL RESEARCH   131

Director: vacant
Senior staff: Watershed and aquatic habitat, J.F. Corliss; forest recreation, G.H. Moeller; range, wildlife and fish habitat, W.R. Goforth

FOREST FIRE AND ATMOSPHERIC SCIENCES (ROSSLYN)   132

Director: C.C. Chandler
Senior staff: Meteorology, W.T. Sommers; fire control technology, C.B. Lyon; fire planning and economics, J.B. Davis

FOREST INSECT AND DISEASE   133

Director: G.W. Anderson
Senior staff: Forest insects, M.W. McFadden; forest diseases, E.F. Wicker

FOREST PRODUCTS AND ENGINEERING   134

Director: J.W. Erickson
Senior staff: Wood technology, R.C. Koeppen; structural and materials engineering, R.L. Tuomi; energy, D.B. Johnson; engineering, G.W. Brown; forest engineering, F.E. Biltonen

FOREST RESOURCES ECONOMICS   135

Director: R.S. Whaley
Senior staff: Resources evaluation, C.C. Van Sickle; renewable resources economics, E.F. Bell; trade analysis, D. Hair; evaluation methods; C. Row

INTERNATIONAL FORESTRY   136

Director: R.W. Brandt
Senior staff: Resources information, L.M. LaMois

REGIONAL HEADQUARTERS   137

Alaska Region   138

Address: Federal Office Building, Box 1628, Juneau, AK 99802
Regional Forester: John A. Sandor

Eastern Region   139

Address: 633 W Wisconsin Avenue, Milwaukee, WI 53203
Regional Forester: Steve Yurich

Intermountain Region   140

Address: Federal Building, 324 25th Street, Ogden, UT 84401
Telephone: (804) 626 3011
Regional Forester: Jeff M. Sirmon

Northern Region   141

Address: Federal Building, Missoula, MT 59807
Telephone: (406) 329 3011
Regional Forester: Charles T. Coston

Pacific Northwest Region   142

Address: 319 SW Pine Street, Box 3623, Portland, OR 97208
Regional Forester: R.E. Worthington

Pacific Southwest Region   143

Address: 630 Sansome Street, San Francisco, CA 94111
Regional Forester: Zane G. Smith

Rocky Mountain Region   144

Address: 11177 W 8th Avenue, Box 25127, Lakewood, CO 80225
Regional Forester: Craig W. Rupp

Southern Region   145

Address: 1720 Peachtree Road NW, Atlanta, GA 30367
Regional Forester: Lawrence M. Whitfield

Southwestern Region   146

Address: 517 Gold Avenue, SW, Alburquerque, NM 87102
Regional Forester: Milo Jean Hassell

RESEARCH CENTRES   147

Forest Products Laboratory   148

Address: PO Box 5130, Madison, WI 53705
Telephone: (608) 264 5600
Director: Robert L. Youngs
Deputy director: George G. Marra
Sections: Process and protection research, George A. McSwain; engineering and economics research, Billy Bohannan; chemistry and paper research, John W. Koning; planning and applications, Gary R. Lindell; research support services, Michiel J. Noordewier
Graduate research staff: 105
Activities: The mission of the laboratory is to find ways to extend timber supply by using the available timber supply most effectively. This can be done by increasing the yield from timber being harvested through improved concepts of processing and products; by making it possible to use economically timber that is not now being harvested; by increasing the efficiency of use through improved grading, engineering, and structural use; and

by extending the useful life of wood products through effective protection.

Publications: Prospectus, Age of Wood, both every 2 years.

Projects: Forest mycology research (John G. Palmer); biodegradation of wood (Wallace E. Eslyn); wood anatomy research (Regis B. Miller); improved adhesive systems (Robert H. Gillespie); protection of wood used in adverse environments (Rodney C. DeGroot); structural composite products (John A. Youngquist); quality and yield improvement in wood processing (Robert R. Maeglin); improvements in drying technology (William T. Simpson); engineering properties of wood (B. Alan Bendtsen); engineering design criteria (Jen Y. Liu); engineered wood structures (Russell C. Moody); fire design engineering (Erwin L. Schaffer); load duration design criteria (Joseph F. Murphy); national timber and wood products requirements and utilization economics (Robert N. Stone); criteria for fibre product design, (Vance C. Setterhoma); high-yield nonpolluting pulping (Necmi Sanyer); wood fibre product and process development (Donald J. Fahey); improved chemical utilization of wood (John W. Rowe); microbial technology in wood utilization (T. Kent Kirk); corrugated package engineering (James F. Laundrie).

Intermountain Forest and Range Experiment Station   149

Address: 507 25th Street, Ogden, UT 84401
Telephone: (801) 626 3361
Director: Roger R. Bay
Projects: Population dynamics of Mountain Pine Beetle (Walter E. Cole); resources evaluation (Dwane D. Van Mooser).

Forestry Sciences Laboratory, Boise   150
Address: 316 East Myrtle Street, Boise, ID 83706
Projects: Minimizing nonpoint source pollution in the northern Rocky Mountains (Walter F. Megahan); ecology and silviculture of Rocky Mountain Douglas-fir and Ponderosa Pine ecosystems (Russell A. Ryker).

Forestry Sciences Laboratory, Bozeman   151
Address: PO Box 1376, Bozeman, MT 59715
Projects: Silviculture of northern Rocky Mountain subalpine forest ecosystems (Wyman C. Schmidt); engineering technology for improved forest resource management and protection (Edward R. Burroughs Jr).

Forestry Sciences Laboratory, Logan   152
Address: 860 North 12th East, Logan, UT 84321
Projects: Ecology and management of aspen lands in the west (Walter F. Mueggler); mined-land reclamation in the intermountain and northern Rocky Mountain regions (Eugene E. Farmer).

Forestry Sciences Laboratory, Missoula   153
Address: Drawer G, Missoula, MT 59806
Projects: Wilderness management research (Robert C. Lucas); ecology and the management of forest wildlife habitats in the northern Rocky Mountains (L. Jack Lyon); forest ecosystems (Robert D. Pfister); improving wood resources utilization in the Intermountain West (Michael J. Gonsior); economics of multiple use management on public forest lands (Ervin G. Schuster).

Forestry Sciences Laboratory, Moscow   154
Address: 1221 South Main Street, Moscow, ID 83843
Projects: Insects of northern Rocky Mountain forest trees and associated wildland shrubs (Malcolm M. Furniss); diseases of natural and nursery-grown seedlings in the central and northern Rocky Mountains (Alan E. Harvey); genetics and pest resistance of Rocky Mountain conifers (Raymond J. Hoff); quantitative analysis of forest management practices and resources for planning and control (Albert R. Stage); silviculture of cedar, hemlock and grand fir ecosystems.

Northern Forest Fire Laboratory   155
Address: Drawer G, Missoula, MT 59806
Projects: Fire occurrence (Donald M. Fuquay); fire behaviour (Richard C. Rothermel); fire technology (Charles W. George); fire effects and use in the interior west (James K. Brown); synthesizing fire management techniques.

Renewable Resources Center   156
Address: 920 Valley Road, Reno, NV 89502
Projects: Ecology and management of pinyon-juniper woodlands in the Great Basin (Richard O. Meeuwig).

Shrub Sciences Laboratory   157
Address: 735 North 500 East, Provo, UT 84601
Projects: Ecology and management of shrub-herb rangelands of the Great Basin (Warren P. Clary); shrub improvement and revegetation (Arthur R. Tiedemann).

North Central Forest Experiment Station   158

Address: 1992 Folwell Avenue, St Paul, MN 55108
Telephone: (612) 642 5207
Director: Robert A. Hann
Deputy director: Denver P. Burns
Sections: Research, northwest, Arne K. Kemp; research, southeast, Richard V. Smythe; planning and application, Roger W. Leonard; research support services, John F. Prokop
Graduate research staff: 48
Projects: Resources evaluation in the north central region (Burton L. Essex); multiple use evaluation and modelling of forest ecosystems in the north central region (Lewis F. Ohmann); wildlife habitat management (Richard R. Buech); plant resistance, climate, and genetics as related to forest insect outbreaks (Harold O. Batzer); canker, foliar, and root diseases of forests,

plantations, and Christmas tree plantings (Thomas H. Nicholls); backcountry river recreation management (David W. Lime).

Field Station, Chicago   159
Address: 5601 North Pulaski Road, Chicago, IL 60646
Telephone: (312) 588-7650
Head: John F. Dwyer Jr
Graduate research staff: 2
Projects: Enhancing urban forest recreation opportunities (John F. Dwyer Jr).

Field Station, Columbia   160
Address: 1-26 Agriculture Building, University of Missouri, Columbia, MO 65201
Telephone: (314) 882 2667
Head: Ivan L. Sander
Graduate research staff: 8
Projects: Silviculture and ecology of the oak-hickory forest ecosystem (Ivan L. Sander); land use impacts on wildlife habitats of the central oak-hickory ecosystem (James M. Sweeney).

Field Station, Duluth   161
Address: 118 Old Main Building, University of Minnesota, 23rd Avenue East and Fifth Street, Duluth, MN 55812
Telephone: (218) 724 1046
Head: Edwin Kallio
Graduate research staff: 4
Projects: Regional economics of forest resources.

Forestry Sciences Laboratory, Carbondale,   162
Address: Southern Illinois University, Carbondale, IL 62901
Telephone: (618) 453 2318
Head: Howard N. Rosen
Graduate research staff: 12
Projects: Culture, genetics, and protection of black walnut, white ash, and white oak (Richard C. Schlesinger); new and improved systems, methods, and techniques for processing hardwoods (Howard N. Rosen).

Forestry Sciences Laboratory, Grand Rapids   163
Address: Route 3, Grand Rapids, MN 55744
Telephone: (218) 326 8571
Head: John W. Benzie
Graduate research staff: 7
Projects: Aspen, birch, and conifer research and development (Reid W. Goforth); ecology and culture of aspen, birch, and conifer forests (John W. Benzie); water quality management in northern Lake States forests (Elon S. Verry).

Forestry Sciences Laboratory, Houghton   164
Address: Forest Hill Road, Houghton, MI 49931
Telephone: (906) 482 6303
Project leader: Rodger A. Arola
Graduate research staff: 8
Activities: Research in forestry engineering.
Projects: Engineering technology for managing northern forest stands (Rodger A. Arola).

Forestry Sciences Laboratory, Marquette   165
Address: 806 Wright Street, Marquette, MI 49855
Telephone: (906) 225 1323
Head: Thomas R. Crow
Projects: Silviculture and ecology of northern hardwoods in the Lake States (Thomas R. Crow).

Forestry Sciences Laboratory, Rhinelander   166
Address: Star Route 2, Rhinelander, WI 54501
Telephone: (715) 362 7474
Head: David H. Dawson
Graduate research staff: 17
Projects: Maximum yield of wood and energy from intensively cultured plantations (David H. Dawson); intensively cultured plantations for fibre and energy production (Edward A. Hansen); physiology and raw material evaluation of intensively cultured plantations (Judson G. Isebrands); genetics of northern forest trees (Hans Nienstaedt); pioneering research in physiology of wood formation (Philip R. Larson).

Northeastern Forest Experiment Station   167

Address: 370 Reed Road, Broomall, PA 19008
Telephone: (215) 461 3104
Director: vacant
Graduate research staff: 168
Activities: Research performed by the station encompasses the following major areas of forest research: timber management, forest watershed management, wildlife habitat, forest recreation· and related human amenities, forest insects and disease, forest products and engineering, forest resource evaluation, and forest economics and utilization.
Projects: Resources evaluation (Joseph Barnard); urban forestry (Richard M. Degraaf); physiology of mycorrhizae (Edward Hacskaylo); physiology of growth, wood formation, and ageing (John A. Romberger); reclamation of surface-mined areas (Willie R. Curtis); relationship of air pollution to forests (Leon S. Dochinger); ecology and management of northern hardwood forests (Carl H. Tubbs); discolouration and decay in forest trees (Alex L. Shigo); ecology and management of forest insect pests (William E. Wallner); insect pathology and microbial control (Franklin B. Lewis); diebacks and declines of forest trees (David R. Houston); hardwood timber harvesting (Penn A. Peters); culture of northeastern spruce-fir forests (Barton M. Blum); culture of central Appalachian hardwoods (Henry C.

Smith); low-grade hardwoods utilization (Charles J. Gatchell); silviculture of Allegheny hardwoods (David A. Marquis).

Forestry Sciences Laboratory, Delaware   168
Address: Box 365, Delaware, OH 45013
Assistant station director: Albert N. Foulger

Forestry Sciences Laboratory, Durham   169
Address: Concord-Mast Road, Box 640, Durham, NH 03824
Assistant station director: David T. Funk

Forestry Sciences Laboratory, Morgantown   170
Address: 180 Canfield Street, Morgantown, WV 26505
Assistant station director: Robert E. Phares

Pacific Northwest Forest and Range   171
Experiment Station

Address: 809 NE 6th Avenue, Portland, OR 97232
Telephone: (503) 231 2094
Director: Robert L. Ethington
Assistant director: E.M. Estep
Sections: Continuing research North, Kenneth H. Wright; continuing research Central, Donald F. Flora; continuing research South, Don H. Boelter; planning and applications, Eldon M. Estep
Graduate research staff: 124
Activities: Forestry and forest products research to provide the scientific basis for the management and use of regional forest and range and related resources of Oregon, Washington, and Alaska. Research encompasses the broad range from forest establishment through protection and management to harvesting, processing, and wood-product use.
Projects: Forest residues and energy (Richard O. Woodfin Jr); Canada/US spruce budworms (Ron Stark); renewable resources evaluation (John H. Poppino); wildlife habitat and range (Jack W. Thomas); intensive culture of Douglas-fir (Dean S. DeBell); forest engineering (Charles N. Mann); reafforestation/vegetation management (Peyton W. Owston); forest genetics (Roy R. Silen); forest insects (Gary E. Daterman); forest diseases (Earl E. Nelson); watershed management/introduced chemicals (Logan A. Norris); ecology of SE Alaska forests (Donald C. Schmiege).

Forestry Sciences Laboratory, Anchorage   172
Address: 2221 E Northern Lights Boulevard, Suite 106, Anchorage, AK 99504

Forestry Sciences Laboratory, Corvalis   173
Address: 3200 Jefferson Way, Corvalis, OR 97331

Forestry Sciences Laboratory, Juneau   174
Address: Box 909, Juneau, AK 99802

Forestry Sciences Laboratory, Olympia   175
Address: 3625 93rd Avenue SW, Olympia, WA 98502

Forestry Sciences Laboratory, Portland   176
Address: 809 NE 6th Avenue, Portland, OR 97232

Forestry Sciences Laboratory, Seattle   177
Address: 4507 University Way NE, Seattle, WA 98105

Forestry Sciences Laboratory, Wenatchee   178
Address: 1133 N Western Avenue, Wenatchee, WA 98801

Institute of Northern Forestry   179
Address: Fairbanks, AK 99701

Range and Wildlife Habitat Laboratory   180
Address: C Avenue and Gekeler Lane, Route 2, Box 2315, LaGrande, OR 97850

Silviculture Laboratory   181
Address: 1027 NW Trenton Avenue, Bend, OR 97701

Pacific Southwest Forest and Range   182
Experiment Station

– PSW
Address: PO Box 245, Berkeley, CA 94701
Telephone: (415) 486 3292
Director: Robert Z. Callaham
Sections: Assistant director for continuing research, Northern California, Benjamin Spada; Assistant director for continuing research, Southern California, Charles W. Philpot; Assistant director for planning and applications, Richard L. Hubbard
Graduate research staff: 120
Activities: Natural resources - soil science, drainage and irrigation; forestry and forest products.
Projects: Pioneering research - hybridization and evolution of forest trees (William B. Critchfield); pioneering research - integrated management systems for forest insect pests and diseases (Carroll B. Williams); vegetation management alternatives for Chaparral and related ecosystems research and development programme (George Roby); management of Chaparral and related ecosystems in Southern California (C.E. Conrad); environmental hydrology of the snow zone of the Sierra Nevada and the coast ranges of California (Neil Berg); processes affecting management of Pacific coastal forests on unstable lands (Raymond Rice); genetics of western forest trees (Thomas Ledig); intensive timber culture of northwestern California conifers (David F. Olson); urban forestry (J. Alan Wagar); biology and control of diseases in forests of California (Robert V. Bega); field evaluation of chemical insecticides (Patrick J. Shea); insecticide evaluation (Michael J. Haverty); biology and control of forest insects in California (Richard H. Smith); silviculture of Sierra Nevada conifer types (Douglass F. Roy); fire-management planning and economics (Thomas J. Mills); meteorology for forest and brushland management (Michael A. Fosberg); range management research in California (John Kie); protection and management of sensitive species in

California (Jared Verner); timber and watershed management research in Hawaii (Roger G. Skolmen); Pacific Islands (Territories) forestry research (Craig D. Whitesell); forest protection research in Hawaii (Charles S. Hodges); maintenance of native Hawaiian forest ecosystems (C.J. Ralph).

Forest Fire Laboratory   183
Address: 4955 Canyon Crest Drive, Riverside, CA 92507

Institute of Pacific Islands Forestry   184
Address: 1151 Punchbowl Street, Honolulu, HI 96813

Redwood Sciences Laboratory   185
Address: 1700 Bayview Drive, Arcata, CA 95521

Rocky Mountain Forest Experiment   186
Station

Address: 240 W Prospect, Fort Collins, CO 80526
Director: Charles M. Loveless

Forest Research Laboratory   187
Address: South Dakota School of Mines and Technology, Rapid City, SD 57701
Telephone: (605) 343 0811
Telex: FTS 782-1451
Project Leader: Ardell J. Bjugstad
Graduate research staff: 15
Activities: Developing means to increase production of wood, water, and animal products in the Black Hills; to maintain and reestablish the prairie woodlands; and promoting the reestablishment of plants, animals, and hydrologic stability on surface mine spoils and water impoundments in the northern Great Plains.
Projects: Improvement of forest, range, and water productivity in the Black Hills; improvement of wildlife habitat in wooded areas in the plains; revegetating surface-mined land in the plains.

Forestry Sciences Laboratory, Flagstaff   188
Address: Northern Arizona University, Flagstaff, AZ 86001

Forestry Sciences Laboratory, Lincoln   189
Address: East Campus, University of Nebraska, Lincoln, NE 68583

Forestry Sciences Laboratory, Tempe   190
Address: Arizona State University, Tempe, AZ 85281

Great Plains Wildlife Research Laboratory   191
Address: Box 4249, Texas Tech University, Lubbock, TX 79409

Shelterbelt Laboratory   192
Address: First and Brander, Bottineau, ND 58318

Southeastern Forest Experiment   193
Station

Address: PO Box 2570, Asheville, NC 28802
Telephone: (704) 258 2850
Telex: FTS 672-0758
Director: Dr Eldon W. Ross
Deputy Director: Robert L. Scheer
Assistant Director, Research in Virginia and Carolina: John C. Hendee
Assistant Director, Research in Georgia and Florida: Thomas H. Ellis
Activites: The aims of research are as follows: to increase efficiency of timber production by increasing growth, reducing mortality, improving utilization and marketing of important timber species, and reducing economic and environmental costs; to increase the efficiency of the integrated multiresource production system, including production of environmental amenities, by improving technology of management, production, and service delivery while maintaining a quality environment; to maintain or enhance the diversity and abundance of native faunal and floral communities associated with forest ecosystems and the urban environment.
Projects: Population genetics of forest trees (Gene Namkoong); Loblolly pine management research and development (Samuel F. Gingrich) forest soil productivity in the southeast (Carol G. Wells); Loblolly pine stand management (William R. Harms); stand establishment of Loblolly pine (David L. Bramlett); Loblolly pine tree improvement (John F. Kraus); stand development composition, and growth of southern Appalachian hardwoods (Donald E. Beck); silvicultural guidelines for managing Piedmont hardwoods (Douglas R. Phillips); biological potential for timber production in eastern forests (Stephen G. Boyce); water, soil, and aquatic responses to management of southern Appalachian-Piedmont forests (James E. Douglass); endangered and threatened wildlife in southern forests (Michael R. Lennartz); urban forestry research in the south (Harold K. Cordell); detection, evaluation and control of damaging southeastern forest insects (H.A. Thomas); insecticides for control of bark beetles in southern pines (Felton L. Hastings); processing hardwood trees, logs, and timber (James G. Schroeder); forest resources in the southeast (Joe P. McClure); economic returns from forestry investments in the southeast (George F. Dutrow); silviculture and tree improvements of eucalyptus in Florida (Jack Stubbs); management alternatives for pine species in the SE coastal plain (Raymond H. Brendmudhl); intensive management practices assessment (B.F. Swindel); mensurational concepts for forest management decisions (Kenneth D. Ware); silviculture and genetics of Slash and Longleaf pine in the southeastern coastal plain (S.V. Kossuth); combustion processes in wildland fuels (Charles K. McMahon); fire science (Von Johnson); forestry weather data systems (James T.

Paul); biology, ecology, and control of cone and seed insects of southern forests (Harry O. Yates); diseases of southern pine plantations and seed orchards (Harry R. Powers); integrated pest management (Thomas Miller); mycorrhizal research and development (Donald H. Marx); utilization of southern timber (Joseph R. Saucier); wood products research (Gerald A. Koenigshof).

Forest Resources Laboratory   **194**
Address: Box 938, Lehigh Acres, FL 33936

Forestry Sciences Laboratory, Athens   **195**
Address: Carlton Street, Athens, GA 30602

Forestry Sciences Laboratory, Research   **196**
Triangle Park
Address: Box 12254, Research Triangle Park, NC 27709

Naval Stores and Timber Production   **197**
Laboratory
Address: Box 70, Olustee, FL 32072

School of Forest Resources and Conservation   **198**
Address: University of Florida, Gainesville, FL 32611

Southern Forest Fire Laboratory   **199**
Address: Georgia Forestry Center, Riggins Mill Road, Box 5106, Macon, GA 31208

Southern Forest Experiment Station   200

Address: T-10210 Postal Services Building, 701 Loyola Avenue, New Orleans, LA 70113
Telephone: (504) 589 6787
Director: Laurence E. Lassen
Graduate research staff: 100
Activities: Research into forestry and forest products in midsouth USA and Puerto Rico.
Publications: Research report, three times a year.
Projects: Silviculture research; biometrics research; forest tree genetics research; tree seed technology and research; forest watershed research; forest fire prevention research; forest insect and disease research; forest products utilization research; forest engineering research; forest range management research; wildlife habitat research; forest economics research, renewable resources evaluation.

Alexandria Forestry Centre   **201**
Address: 2500 Shreveport Highway, Pineville, LA 71360
Projects: Silviculture of southern pines (J.P. Barnett); intensive culture of southern pines (E. Shoulders); range management for Longleaf-Slash pine blue stem and Loblolly-Shortleaf pine ecosystems (H.A. Pearson); pest management for southern pine beetle (P.L. Lorio); processing southern woods (P. Koch).

Forest Hydrology Laboratory   **202**
Address: PO Box 947, Oxford, MI 38655
Telephone: (601) 234 2744
Project Leader: S.J. Ursic
Graduate research staff: 5
Activities: The aim of research is to evaluate and develop the capability to predict the hydrologic responses of southern pine forests to harvesting and cultural measures.
Projects: Evaluation of impacts of pine plantation harvesting on hydrology and water quality; evaluation of impacts of site preparation for pine regeneration on water quality and hydrology on small catchments; determination of nutrient and sediment yields from undisturbed pine plantations; laboratory techniques for water quality derterminations.

Forestry Sciences Laboratory, Gulfport   **203**
Address: Box 2008, GMF, Gulfport, MS 39503
Projects: Genetics of southern pines (C.F. Bey); control of wood biodeterioration (J.K. Mauldin); diseases of southern pine (G.A. Snow).

Forestry Sciences Laboratory, Starkville   **204**
Address: Box 906, Starkville, MS 39759
Projects: Technology of eastern forest tree seed (F.T. Bonner); genetics and breeding of southern hardwoods (J.A. Pitcher); fire prevention technology (M.A. Doolittle).

G.W. Andrews Forestry Sciences Laboratory   **205**
Address: Devale Street, Auburn, AL 36849
Projects: Control of undesirable vegetation (W.D. Boyer); engineering systems for forest management (D.L. Sirois).

Southern Hardwoods Laboratory   **206**
Address: Box 227, Stoneville, MS 38776
Projects: Management of southern hardwoods (R.L. Johnson); southern hardwoods insects and diseases (T.H. Filer).

Silviculture Laboratory   **207**
Address: SPO Box 1290, Sewanne, TN 37375
Projects: Integrated resource management in the Cumberland Plateau region (C.E. McGee).

University of Puerto Rico Agricultural   **208**
Experiment Station
Address: Box AQ, Rio Piedras, PR 00928

TIMBER MANAGEMENT   209

Director: S.L. Krugman
Senior staff: Conifer ecology and management, R.M. Burns; hardwood ecology and management, N.S. Loftus Jr; botany, B.H. Honkala; bioresources for energy, H.E. Wahlgren

General Mills Incorporated 210

Status: Industrial company

JAMES FORD BELL TECHNICAL 211
CENTER

Address: 9000 Plymouth Avenue North, Minneapolis, MN 55427
Telephone: (612) 540-3564
Telex: 29-0326
Senior Vice President and Technical Director: Dr John V. Luck
Corporate Research and Development Directors: G.V. Daravingas, E.H. Borochoff, S.R. Sapakie
Division Research and Development Directors: V.E. Weiss, D.E. Weinauer, W.L. McKown, R.B. Ward
Sections: Technical and quality control services, D.F. Emery; engineering and administrative services, G.W. LaLone
Graduate research staff: 325
Activities: Food science - product development of dry mixes, desserts, and cereals; process development - cooking, extrusion, and drying.
Publications: Annual progress report.

Gerber Research 212
Laboratory*

Address: 445 State Street, Fremont, MI 49412
Telephone: (616) 928 2000
Parent body: Gerber Products Company
Status: Research centre within an industrial company
Sections: Nutrition science, G.A. Purvis; chemical science, V.J. Kelly
Activities: Product development and applied research in food technology and products.

Great Western Sugar 213
Company

Status: Industrial company

AGRICULTURAL RESEARCH CENTRE 214

Address: 11939 Sugarmill Road, Longmont, CO 80501
Telephone: (303) 776 1802
Telex: 910 913 2692
Director: James F. Gonyou
Sections: Variety development, Dr R.K. Oldemeyer; plant protection, Dr E.F. Sullivan
Graduate research staff: 8

Activities: Sugarbeet variety improvement projects - breeding for improved quality, disease, insect, and nematode resistance; vegetable improvement projects - includes onion, dry edible bean, carrot, melon and several others; contract research in selective herbicides and plant growth regulants - crops included are malting barley, sunflower, dry edible beans, and sugarbeet; sugarbeet seed emergence and post-harvest pile storage improvement methods, chemical and physiological; environmental impact of various pesticides, herbicides and other crop protection chemicals.
Projects: Sugarbeet variety development (Dr R.K. Oldemeyer); selective sugarbeet herbicides (Dr E.F. Sullivan); plant growth regulators, sugarbeets and other crops (Dr E.F. Sullivan); sugarbeet pile storage studies (Dr W.R. Akeson).

Hawaiian Sugar Planters' 215
Association

– HSPA
Status: Research centre supported by industry

EXPERIMENT STATION 216

Address: PO Box 1057, 99-193 Aiea Heights Drive, Aiea, HI 96701
Telephone: (808) 487 5561
Director: Dr Don J. Heinz
Departments: Crop science, H.W. Hilton; engineering, C.M. Kinoshita; entomology, A.K. Ota; genetics and pathology, T.L. Tew; sugar technology, G.E. Sloane
Graduate research staff: 37
Activities: The aims of the association are to maintain, advance, improve and protect the sugar industry in Hawaii by supplying information and services to member companies to enable them to improve the quality and productivity of their operations, with the goal of increasing their profits and to support the development of agriculture in general. The main areas of research are: breeding and propagation of sugarcane; selection of commercial sugarcane cultivars; drainage and irrigation; fertilization, soil science and nutrition; growth and development of sugarcane; control of weeds, rats, insects and sugarcane diseases; factory processing of sugarcane; agricultural engineering.
Publications: Annual report.
Projects: Interaction of water and nutrition on growth and yield of sugarcane (R.P. Bosshart, K. Ingram, L. Santo); physiology of stress - salinity, water deficiency, remote sensing (R. McKenzie, R.P. Bosshart, P.H. Moore); movement of photosynthate into storage cells and the influence of growth regulators on growth and storage in various cultivars (A. Maretzki, R.V. Osgood); high performance liquid chromatography of sugars and

non-sugar components in factory streams (N. Nomura); use of tissue culture techniques for study of genetic improvement, transport of sugars, growth regulators and other substances (A. Maretzki, P.H. Moore); feasibility and economics of subsurface irrigation (W. Bui); centralized seed processing (L. Jakeway); reassessing the economics of alcohol from final molasses and assessing the economic feasibility of producing alcohol from B and A molasses (W.O. Gibson); evaluation of bagasse drying and handling processes (C.M. Kinoshita); control of ant damage to drip irrigation tubes (A.K. Ota); pest management of the New Guinea sugarcane weevil, Rhabdoscelus obscurus (A.K. Ota); control of a ground nesting wasp, the western yello-jacket, Vespula pennsylvanica, on Hawaiian sugarcame plantations (V.C.S. Chang); cross commercial varieties to produce seedlings for selection of improved varieties (T.L. Tew); develop tissue culture techniques as a plant breeding tool (K.K. Wu); screen breeding materials and potential commercial varieties for resistance to smut disease, tolerance to stress factors such as drought, salinity and herbicides, and resistance to disease such as leaf scald, eye spot and red rot (J. Comstock, S. Ferreira); conduct basic studies of sugarcane smut to identify the genetics of host resistance, pathogenicity and virulence and develop and understanding of pathogen race dynamics (S. Ferreira); lotus roll evaluation (K. Onna); effect of cane ripeners on molasses exhaustibility (T. Moritsugu); removal of potash and other salts from stillage and cane juice (D. Hsu).

H.J. Heinz Company Research and Development Laboratory* 217

Address: 1062 Progress Street, Pittsburgh, PA 15212
Status: Research centre within an industrial company
Director: Dr R.E. Heinz
Activities: Food processing and packaging research; development of improved tomato strains.

Horticultural Research Institute 218

– HRI
Address: 230 Southern Building, Washington, DC 20005
Telephone: (202) 737 4060
Parent body: American Association of Nurserymen
Status: Research organization of trade association
Administrator: Duane F. Jelinek

Activities: The Horticultural Research Institute is a non-profit organization devoted exclusively to the support of research necessary for the advancement of the nursery industry. Its main function is the collection of research information and communicating that information to the nursery industry, nursery researcher and interested public. To help fulfill that goal, HRI offers limited support to horticultural research projects throughout the United States and Canada which are of concern to the nursery industry. This research is conducted by outside institutions, since HRI does not have research facilities of its own.
Projects: HRI sponsors research projects on the following subjects: bionomics, behaviour, and control of bronze birch borer (David G. Nielsen); cytogenetics, breeding, and evaluation of Malus (Donald R. Egolf); increasing the new University of Minnesota hardy azaleas by tissue culture (Paul E. Read); pest management of clearwing borers infesting woody ornamentals (Daniel A. Potter); protecting urban trees from environmental stress (Bruce R. Roberts); accelerating the growth of taxus cuspidata capitata with selected mycorrhizae (Larry Kuhns); a screening programme to evaluate juniper cultivars to innoculations with cedar-apple, cedar-hawthorne, and cedar-quince rusts, and the phomopsis blight fungus (Frederick J. Crowe); capillary irrigation for nursery containers, dormancy and requirements for growth in rhododendron rooted cuttings (John R. Havis); cold hardiness of southeastern native and ornamental taxa (M.A. Dirr, H. Pellett); chemical basis of root weevil resistance in rhododendron (Robert P. Doss); critical effects of fertility on root growth of landscape plants; critical effects of light intensity on plant growth (David F. Hamilton); effect of acid or base on root initiation in woody ornamentals; seed dormancy of trifoliate maples (Dennis P. Stimart); importance of some mycorrhizal fungi in establishment and performance of tissue culture explants (Dale M. Maronek); nursery management with micro-computers (P. James Rathwell); propagation of ericaceous plants by regeneration of adventitious shoots from leaf discs (William R. Krul); selection, propagation and culture of native and exotic perennial plants for use in the Chuhuahuan Desert area (Jimmy L. Tipton); tissue culture propagation of perennial plants (Chiko Haramaki).

Humbolt State University 219

Address: Arcata, CA 95521
Status: Educational establishment with r&d capability

SCHOOL OF NATURAL RESOURCES 220

Telephone: (707) 826 3561
Dean: Dr Richard L. Ridenhour
Graduate research staff: 45
Activities: A wide variety of research projects are conducted in the discipline areas of the academic departments. Most research is short-term and is conducted by graduate students directed by individual faculty members. There are no specific areas of research to which the primary efforts of the School of Natural Resources are directed.

Department of Fisheries 221

Head: Dr George H. Allen
Projects: Utilization of wastewater for production of salmonids.

Department of Forestry 222

Head: Dr Gerald L. Partain

Department of Natural Resources Planning and Interpretation 223

Head: Dr Mark B. Rhea

Department of Oceanography 224

Head: Dr John E. Pequegnat

Department of Range Management 225

Head: Dr Norman E. Green

Department of Watershed Management 226

Head: Dr F. Dean Freeland

Department of Wildlife Management 227

Head: Dr David W. Kitchen

International Flavors and Fragrances Research and Development Laboratory* 228

Address: Foot of Rose Lane, Union Beach, NJ 07735
Telephone: (201) 765 5500
Status: Independent research centre
Director: I.D. Hill
Graduate research staff: 100
Activities: Biochemistry, chemical engineering, food technology, and toxicology.

International Harvester 229

Status: Industrial company

SCIENCE AND TECHNOLOGY LABORATORY 230

Address: 16 W 260 83rd Street, Winsdale, IL 60521
Telephone: (312) 887 2343
Research director: Steven J. Gage
Sections: Advanced concepts, Andrew G. Watson; biomass energy, Tom F. O'Connell; advanced harvesting, Bob N. Alverson; automation technology, R.N. Nagel; engines and power trains, Paul N. Blumberg
Graduate research staff: 110
Activities: Research in the following areas: plant production efficiency; advanced mechanization; farming system research; aquaculture; agricultural engineering and building.
Projects: Precision agriculture (Raul Baumweckel); biomass farming (Gerry Murphy); international agricultural development (Andrew Watson); biomass fuels (Tom O'Connell); alternative fuel engines (Paul Blumberg); advanced crop harvesting (Bob Alverson); biological desalination (Andrew Watson); remote sensing (Gerry Murphy).

Iowa State University 231

Address: Ames, IA 50011
Status: Educational establishment with r&d capability

IOWA AGRICULTURE AND HOME ECONOMICS EXPERIMENT STATION 232

Telephone: (515) 294 4763
Director: Dr Lee R. Kolmer
Graduate research staff: 126
Activities: The aims of agricultural research are as follows: developing methods of adjusting production and marketing of agricultural production to meet the demand and provide for adequate reserves; conserving, developing and using natural resources essential to the public in Iowa; protection of plants and animals from pests, pollution, deficiencies; production of an adequate supply of farm products at decreasing real production costs; improvement of marketing efficiency; improvement of quality and variety of farm products available to the consumer; protection of human health through food quality; expansion of export markets for Iowa farm products; development of technical assistance programmes to developing nations; assistance to rural Americans to adjust and achieve a comparative standard of living to urban peoples; facilitating adjustments brought about by new technologies; improving capacity to develop and

disseminate new knowledge; conservation and use of energy - substitution by renewable or non-critical energy sources and forms, and development of technology for use of alternative sources and forms of energy.
Publications: Annual report.

Agricultural Economics Department 233

Head: R.R. Beneke
Projects: Evaluation of alternative rural freight transport systems, storage, and distribution systems (Phillip Baumel); factors affecting marketing performance of the dairy industry (George W. Ladd); marketing and delivery of quality cereals and oil-seeds in domestic and foreign markets (Carl Bern); organization and control of US food production systems (Marvin Hayenga).

Agricultural Education Department 234

Head: Harold R. Crawford
Projects: Evaluation of the effectiveness of strategies for structuring agriculture and agribusiness education (Alan A. Kahler).

Agricultural Engineering Department 235

Head: Howard Johnson
Projects: Energy and by-products from animal manure (R.J. Smith); development and evaluation of conservation tillage systems; continuous on-farm corn-alcohol production - energy and management needs (W.F. Buchele).

Agronomy Department 236

Head: John T. Pesek
Projects: Environmental accumulation of nutrients as affected by soil and crop management (M.A. Tabatabai); relating soil and landscape characteristics to land use (T.E. Fenton); forage production and utilization systems as a base for beef; crops, soil and animal management systems for upper Midwest driftless soils production (W. Wedin); characterization of climate and assessment of impact on agriculture and other renewable resources (R.H. Shaw).

Animal Ecology Department 237

Head: R.C. Summerfelt
Projects: Agricultural stream ecosystems, structure and function for understanding water quality benefits (R. Bachmann).

Animal Science Department 238

Head: S.A. Ewing
Projects: Improvement of dairy cattle through breeding (A.E. Freeman); nature and utilization of genetic variation influencing economic traits in the fowl (A.W. Nordskog); genetic improvement of efficiency in the production of quality pork (L.L. Christian); methods for improvement of fertility in cows postpartum (L.L. Anderson).

Biochemistry Department 239

Head: James A. Olson
Projects: Steady state approach to vitamin A status.

Centre for Agricultural and Rural 240 Development

Telephone: (515) 294-3133
Director: Dr Earl O. Heady
Graduate research staff: 35
Activities: The aim of research is to evaluate the economic impact of agricultural, community, and developmental policies on resource use, farm income, consumer price, income distribution, agricultural productivity.
Projects: Modelling agricultural development and policy; evaluation of land and water resource policies; evaluation of changes in agricultural structure and policies; price and production impacts of energy production in agriculture; development of analytical and policy models in agriculture; agricultural policy and development analysis; world food production and demand; income distribution in agriculture.

Entomology Department 241

Head: Paul Dahm
Projects: Environmental implications and interactions of pesticide usage (P. Dahm); arthropod management and economic losses of insects on livestock (E.S. Krafsur); phenology and genetics as ecological bases for the management of the European corn borer (W.D. Guthrie).

Food Technology Department 242

Head: William W. Marion
Projects: Improvement of thermal processes for proteinaceous foods (L.A. Wilson); occurrence of mycotoxins in feeds and foods and their effects on animal and human health (A.A. Kraft).

Forestry Department 243

Head: George W. Thomson
Projects: Establishment and early growth of intensively cultured trees (H.S. McNabb).

Genetics Department   244

Head: Alan Atherly
Projects: Soyabean genetics and cytogenetics (K. Sadanaga); genetics of the genus glycine (R. Palmer, K. Sadanaga).

Home Economics Department   245

Head: Ruth Deacon
Projects: Quality of life as influenced by area of residence (E. Morris); communication strategies to improve nutrition practices of families (R. Deacon).

Horticulture Department   246

Head: Charles V. Hall
Projects: Scion/rootstock and interstem effects on apple tree growth and fruiting (G. Buck); quality and nutritive value of processed potatoes (W.L. Summers).

Microbiology Department   247

Head: Fred Williams

Plant Pathology, Seed and Weed   248
Science Department

Head: Abe Epstein
Projects: Detection, survival, and control of plant pathogenic bacteria on seed and plant propagative materials (J.M. Dunleavy); seed production of breeding lines of insect pollinated legumes (I. Carlson); reduction of corn losses due to nematodes in the North Central region (D.C. Norton).

Special Research and Development   249
Department

Head: H.L. Self
Projects: Sunflower production, oil extraction, and oil utilization as a tractor fuel (W. Lovely).

Veterinary Medicine Department   250

Head: P.T. Pearson
Projects: Bovine respiratory diseases (D.E. Reed); avian respiratory disease control for consumer protection and market quality (M.S. Hofstad); role of effector cells in the resistance of cattle to bovine respiratory disease (M.L. Kaeberle).

University Attached Institute   251

World Food Institute   252
Address: 102 E.O. Building, Ames, IA 50011
Telephone: (515) 294 7699
Telex: ISU Intl Ames 9105201157
Director: Charlotte E. Roderuck
Activities: Research in world food and food-related areas within existing structure of Iowa State University, including: natural resources; plant production; animal production (including wildlife, veterinary medicine); agricultural engineering; fisheries and aquaculture; food science; forestry; nutrition. The institute also sponsors foreign scholars and research fellows for world food-related activities.
Projects: Annual analysis of international trade flows of feed and foodgrains (Robert Wisner); the rational utilization of South American guanaco male groups for meat and wool production on arid lands (William Franklin); development advisory team training programme (D.M. Warren).

Jenkins Research   253

Address: PO Box 801, Salinas, CA 93902
Telephone: (408) 757 4228
Status: Research centre within an industrial company
Senior Plant Breeder: Dr B. Charles Jenkins
Graduate research staff: 4
Activities: To develop and further improve the new grain triticale (a combination of wheat and rye). Research in all aspects related to plant breeding concerning wheat, rye and triticale, including botany, genetics, cytogenetics, plant physiology and the new techniques in genetic engineering.
Projects: Winter, forage and hybrid triticale breeding (Dr B. Charles Jenkins); spring and hybrid triticale breeding (Dr Stan Nalepa).

Kansas State University   254

Status: Educational establishment with r&d capability

BRANCH STATIONS   255

Colby Branch Experiment Station   256

Address: Colby, KS 67701
Telephone: (913) 462 7575
Head: L.D. Robertson
Activities:

Fort Hays Branch Experiment Station   257

Address: Hays, KS 67601
Telephone: (913) 625 3425
Head: W.M. Phillips
Activities: Plant breeding and beef cattle raising, including studies on cereals, sorghum, forage crops, and dry land farming.

Garden City Branch Experiment Station   258

Address: Eminence Route, Garden City, KS 67846
Telephone: (913) 276 8286
Head: G.L. Greene
Activities: Agricultural engineering, animal husbandry and entomological problems.

Southeast Kansas Branch Station   259

Address: Parsons, KS 67357
Telephone: (316) 412 4826
Head: R.J. Johnson

Tribune Branch Experiment Station   260

Address: Tribune, KS 67879
Head: R.E. Gwin
Activities: Plant science and irrigation.

COLLEGE OF VETERINARY MEDICINE   261

Dean: D.M. Trotter

Anatomy and Physiology Department   262

Head: R.A. Frey

Diagnostics Laboratory   263

Head: A.D. Anthony

Laboratory Medicine Department   264

Head: E.H. Coles

Pathology Department   265

Head: S.M. Dennis

Surgery and Medicine Department   266

Head: J.R. Coffman

KANSAS AGRICULTURAL EXPERIMENT STATION   267

Address: Waters Hall 113, Manhattan, KS 66506
Telephone: (913) 532 6147
Director: John O. Dunbar
Graduate research staff: 650
Activities: The mission of the station can be subdivided into the following objectives: to achieve maximum efficiency in crop production to ensure a viable economic base for Kansas agriculture; to integrate animal production enterprises with crop production and range use to enhance Kansas agricultural income; to protect plants and animals from insects, diseases, and other pests; to achieve effective marketing of Kansas agricultural products to ensure that commodities generate maximum income; to develop conservation practices that ensure efficient and effective use of water, soil, and energy resources; to develop processing, handling, and packaging technology for food and fibre that will economically benefit producers and consumers in Kansas; to improve the quality of life for citizens of Kansas, especially those in rural environments.
Publications: Biennial report.

Agricultural Economics Department   268

Head: Milton Manuel

Agricultural Engineering Department   269

Head: G.E. Fairbanks

Agronomy Department   270

Head: George E. Ham

Animal Sciences and Industry Department   271

Head: D.L. Good

Biochemistry Department   272

Head: D.J. Cox

Biology Department   273

Head: T.C. Johnson

Chemistry Department   274

Head: K.J. Klabunde

Entomology Department   275

Head: R.G. Helgesen

Forestry Department	**276**

Head: J.K. Strickler

Grain Science and Industry   **277**
Department

Head: C.W. Deyoe

Horticulture Department   **278**

Head: C.W. Marr

Plant Pathology Department   **279**

Head: L.E. Claflin

KANSAS WATER RESOURCES   **280**
RESEARCH INSTITUTE

Head: F.W. Smith

Kraft Incorporated   281

Status: Industrial company

RESEARCH AND DEVELOPMENT   **282**

Address: 801 Waukegan Road, Glenview, IL 60025
Telephone: (312) 998 3702
Telex: 72-4320
Vice-president and Director of research and development: John F. White

Krause Research and   283
Development Laboratory*

Address: 4222 W Burnham Street, Milwaukee, WI 53215
Telephone: (414) 355 7500
Parent body: Krause Milling Company
Status: Research centre within an industrial company
Activities: Research in cereal chemistry, fermentations and food processing.

Lancaster Laboratories   284
Incorporated

Address: 2425 New Holland Pike, Lancaster, PA 17601
Telephone: (717) 656 2301
Status: Independent research centre
President: Dr Earl H. Hess
Sections: Foods and feeds chemistry, David V. Schumacher; microbiology, C. Robert Graham; water quality, Robert F. Beisel; instrumentation, J. Wilson Hershey; air quality/industrial chemistry, Barbara J. Felty; Franklin Division, Howard E. Holzman
Graduate research staff: 27
Activities: Development of analytical methods for composition and additives in foods and animal feeds; by-products and waste utilization for the food industry; nutritional studies; food product and process development.

Land O'Lakes Incorporated,   285
Corporate Research and
Product Development
Laboratories*

Address: PO Box 116, Minneapolis, MN 55440
Telephone: (612) 481 2222
Status: Research centre within an industrial company
Director: Dr S. Henig
Activities: Product development for the food and the food processing industries.

Lilly Research Laboratories   286

Address: PO Box 708, Greenfield, IN 46140
Telephone: (317) 467 4528
Telex: 272208
Parent body: Eli Lilly and Company
Status: Research centre within an industrial company
Vice-president: Dr Edwin F. Alder
Divisions: Animal science, Dr C.E. Jordan; plant science, Dr R.L. Mann; agricultural chemistry, Dr E.W. Shuey
Activities: Research in the following areas: plant protection; weed control; veterinary medicine; livestock nutrition; livestock parasite control.

Lincoln University 287

Address: 303 Damel Hall, Jefferson City, MO 65101
Telephone: (314) 636 5511
Status: Educational establishment with r&d capability
Dean of research: Dr Edward M. Wilson
Senior staff: Agriculture, natural resources, and home economics, Dr E.M. Wilson; human nutrition, Dr I.G. Molnar
Graduate research staff: 16
Activities: Cooperative research in the fields of animal science, plant and soil science, agribusiness, natural resource management, and food and nutrition.
Projects: Ecological study of a unique management technique to restore a remnant prairie (Fred Hassien); relationship of adrenal function to reproduction in swine (Dr Diana Killian); nutritional parameters in the aetiology of diabetes in the experimental rat (Dr I.G. Molnar); seed pre-treatment, planting methods on germination, emergence, growth, and yield of vegetable crops (Dr K.B. Paul); behavioural oestrus in ewes treated with vaginal progestogen and pregnant male serum; gonadotropin at three periods of anoestrous (Dr John Warren); an integrated hormonal and foliar mineral nutritional approach to soyabean production (Dr Ikbal Rashid Chowdhury); use of light stimulation to reduce the lambing interval; the induction of twinning in beef cows; specialized feeding of lactating dairy cows (Dr David Snyder).

Arthur D. Little Incorporated 288

Address: Acorn Park, Cambridge, MA 02140
Telephone: (617) 864 5770
Telex: 921436
Status: Industrial company
President: John F. Magee
Graduate research staff: 1500
Activities: Consultants and researchers in science and engineering, including agricultural problems.

Louisiana State University 289

Status: Educational establishment with r&d capability

LOUISIANA AGRICULTURAL EXPERIMENT STATION 290

Address: Post Office Drawer E, University Station, Baton Rouge, LA 70893
Telephone: (504) 388 4181
Director: Doyle Chambers
Graduate research staff: 320
Publications: Louisiana Agriculture, quarterly.

Agricultural Economics Department 291
Acting head: Dr Les J. Guedry

Agricultural Engineering Department 292
Head: Dr W.H. Brown

Agronomy Department 293
Head: Dr J. Preston Jones

Animal Science Department 294
Head: Dr James W. Turner

Biochemistry Department 295
Head: Dr R.S. Allen

Cotton Fibre Department 296
Head: Wilbur Aguillard

Dairy Science Department 297
Head: Dr J.B. Frye Jr

Entomology Department 298
Head: Dr J.B. Graves

Feed and Fertilizer Department 299
Head: Hershell F. Morris Jr

Food Science Department 300
Head: Dr Auttis M. Mullins

Forestry Department 301
Head: Dr T. Hansburgh

Horticulture Department 302
Head: Dr D.W. Newsom

Microbiology Department 303

Head: Dr M.D. Socolofsky

Plant Pathology Department 304

Head: Dr W.J. Martin

Poultry Science Department 305

Head: Dr W.A. Johnson

Rural Sociology Department 306

Head: Dr Hart C. Nelsen

Sugar Station 307

Head: Dr M.J. Giamalva

Veterinary Science Department 308

Head: Dr Kirklyn M. Kerr

Louisiana Tech University 309

Status: Educational establishment with r&d capability

LIFE SCIENCES RESEARCH DIVISION 310

Address: PO Box 10198, Ruston, LA 71272
Telephone: (318) 257 4331
Director: Dr John L. Murad
Departments: Agriculture education/business, Dr Larry Allen; animal industry, Dr C. Reid McLellan; agronomy/horticulture, Dr John A. Wright; botany/bacteriology, Dr Dallas D. Lutes; forestry, Dr J. Lamar Teate; zoology, Dr Margaret H. Peaslee
Graduate research staff: 22
Activities: Research in the following areas: natural resources - soil science, land use; plant production - plant breeding, crop husbandry, vegetable crops, grains, plant protection; animal production - livestock breeding, livestock husbandry and nutrition, animal diseases, agricultural engineering and building; fresh-water fisheries; forestry and forest products - biomass and energy conservation; water quality for domestic consumption.
Publications: Annual research report.
Projects: Herbicides in pine release (Dr Richard Neill); longleaf, loblolly, slash pines (Dr Fred Jewell); above ground biomass (Dr George Woodson); herbicide projects (Dr Richard Neill); desulphurization of coal (Dr Harold Hedrick); hardwood flakeboard project (Dr George Woodson); lignin-bonded dry-formed fibreboard (Dr George Woodson); lignite media for plant growth and production (Dr Bob Wright); beetle-killed southern pine (Dr George Woodson); soil residue in cotton (Dr Charles Winstead); unused forest biomass in Louisiana (Dr Richard Neill); bluegill project (Dr John Wakeman); furbearer project (Dr Paul Ramsey); itchgrass project (Dr James White); pine-site hardwood project (Dr Clyde Vidrine); forestry flooding; forestry wildlife (Dr James Dyer); water quality project (Dr Bill Davis); conservation of drinking water for North Louisiana (Dr James Michael).

Maine Forest Service 311

Address: State House Station 22, Augusta, ME 04333
Telephone: (207) 289 2791
Status: Official research centre
Director: Kenneth G. Stratton
Sections: Spruce budworm management, A. Temple Bowen; entomology, Clark A. Granger
Activities: The service implements its research programme primarily through contractors. Research on forestry and forestry products, including the development of management strategies and biological and chemical insecticides to reduce spruce budworm populations, and the development of improved survey and detection systems and spray application technology; actual research work is conducted principally at the University of Maine at Orono.
Projects: Various projects in applied research on spruce budworm (Choristoneura fumiferana) control and management, funded by the Forest Service (Thomas Rumpf); limited in-house applied research conducted on other forest pests (Clark Granger).

Merck Sharp & Dohme Research Laboratories 312

– MSDRL
Address: PO Box 2000, Rahway, NJ 07065
Telephone: (201) 574 6775
Parent body: Merck & Company Incorporated
Status: Research centre within an industrial company
President: Dr P. Roy Vagelos
Scientific information director: Dr Horace D. Brown
Sections: Basic research: Dr Ralph Hirschmann; biological research, Dr C. Stone
Activities: Discovery and development of new antiparasitic agents for poultry, livestock and domestic animals; development of new genetic strains of poultry for improved quality and production.

POULTRY GENETICS 313

Address: Hubbard Farms Incorporated, Bellows Falls Road, Walpole, NH 03608

Michigan State University* 314

Status: Educational establishment with r&d capability

COLLEGE OF AGRICULTURE AND NATURAL RESOURCES* 315

Dean: James H. Anderson

COLLEGE OF VETERINARY MEDICINE* 316

Dean: John R. Welser

MICHIGAN AGRICULTURAL EXPERIMENT STATION* 317

– MAES

Address: 109 Agriculture Hall, Michigan State University, East Lansing, MI 48824
Telephone: (517) 355 0123
Director: Dr Sylvan H. Wittwer
Sections: Centre for Remote Sensing, Dr Jon F. Bartholic; Agricultural and Natural Resources Education Institute, Dr Carroll H. Wamhoff; agricultural economics, Dr Larry J. Connor; agricultural engineering, Dr D.M. Edwards; animal sciences, Dr Ronald H. Nelson; anthropology, Dr Bernard Gallin; biochemistry, Dr Charles C. Sweeley; botany and plant pathology, Dr Edward Klos; crop and soil sciences, Dr Dale D. Harpstead; electrical engineering and systems science, Dr John B. Kreer; entomology, Dr James E. Bath; family and child ecology, Dr Eileen Earhart; fisheries and wildlife, Dr N.R. Kevern; food science and human nutrition, Dr Lawrence Dawson; forestry, Dr Larry W. Tombaugh; horticulture, Dr John F. Kelly; human environment and design, Dr Gertrude Nygren; Institute of Water Research, Dr Howard Johnson; Kellogg Biological Station, Dr George H. Lauff; large animal surgery and medicine, Dr Edward C. Mather; microbiology and public health, Dr Paul T. Magee; School of Packaging, Dr Chester J. Mackson; park and recreation resources, Professor Louis F. Twardzik; pathology, Dr Janver D. Krehbiel; Pesticide Research Centre, Dr Fumio Matsumura; pharmacology and toxicology, Dr Theodore M. Brody; physiology, Dr Harvey L. Sparks; Plant Research Laboratory, Dr Charles J. Arntzen; resource development, Dr Paul E. Nickel; sociology, Dr Jay W. Artis; COMNET/MIS, Dr Stephen B. Harsh
Graduate research staff: 300

Activities: The station is supported by Michigan State University and the College of Agriculture and Natural Resources; state support amounts to $12 million out of a total annual budget exceeding $20 million (1979-80), and approximately 300 faculties in 30 departments and 8 colleges receive support for a total of almost 500 research projects.
Research is conducted in forestry and other natural and renewable resources, fisheries, wildlife, packaging, parks and recreation resources, human nutrition, human health, human ecology, improvement of human welfare in rural Michigan, home beautification, animal disease control, engineering, biochemistry, microbiology, plant protection, food technology, safety and handling, and computer and systems sciences.
Major interdisciplinary programmes include: food contamination and toxicology research, a pesticide research centre, integrated pest management, energy resources in the home, remote sensing of land and water resources, human nutrition, biological nitrogen fixation, animal disease control, and utilization of by-products in agriculture.
Frontiers in research are being explored in cellular approaches to plant breeding; field bean variety development; hybrid carrots, cucumbers, peaches, cherries, blueberries and grapes; soil microbiology; toxic chemicals in the environment; labour saving technologies; energy conservation, generation, and management; greater photosynthetic efficiency; nitrogen fixation; crop and livestock resistance to competing biological systems and environmental stresses; satellite and remote sensing of land, water, crop resources, and wildlife; controlled environments for flowers, vegetables, ornamentals, forest tree seedlings, and dairy cattle; improved reproduction efficiencies in livestock and poultry; underwater parks; unique approaches to packaging; and the utilization of marginal lands for food and biomass resources.
Publications: Science in Action (quarterly).

Michigan Technological University* 318

Status: Educational establishment with r&d capability

SCHOOL OF FORESTRY AND WOOD PRODUCTS 319

Address: L'Anse, MI 49931
Telephone: (906) 524 6181
Dean: Eric R. Bourdo
Graduate research staff: 20
Activities: Research in the following areas: soil science; land use; forestry and forest products.
Projects: Soil-site relationships; management of northern forest types; logging and primary utilization; sawmilling.

Ford Forestry Centre   320

Mississippi State University* 321

Status: Educational establishment with r&d capability

MISSISSIPPI AGRICULTURAL AND FORESTRY EXPERIMENT STATION 322

– MAFES
Address: PO Box ES, Mississippi State, MS 39762
Telephone: (601) 325 3005
Director: Dr R. Rodney Foil
Associate directors: Dr A.D. Seale, Dr J. Charles Lee, Dr W.K. Porter
Departments: Agricultural and biological engineering, Dr W.R. Fox; agricultural economics, Dr V.G. Hurt; agronomy, Dr R.G. Creech; biochemistry, Dr R.P. Wilson; dairy science, Dr H.J. Bearden; entomology, Dr T.J. Helms; agricultural and experimental statistics, Dr W.J. Drapala; animal science, Dr Bryan Baker; forestry, Dr Douglas Richards; home economics, Dr Jean Snyder; horticulture, Dr C.C. Singletary; plant pathology and weed science, Dr Charles Laughlin; poultry science, James E. Hill; sociology and rural life, Dr Arthur Cosby; wildlife and fisheries, Dr Dale H. Arner; foundation seed, Bennie Keith; services, Mitchell Roberts
Graduate research staff: 150
Activities: The objectives of the Mississippi Agricultural and Forestry Experiment Station include the following: the stabilization of agricultural production through efficient resources management; protection of forests, crops, and livestock from hazards and disease; the improvement of marketing efficiency and export markets; the improvement of nutrition and living standards in the USA; environmental improvement and natural habitat conservation.
The station maintains ten branch stations in different areas of the state of Mississipi.
Publications: MAFES Highlights (monthly); *CRIS Progress Reports* (annual).

A.B. McKay Food and Oenology Laboratory 323

Head: Dr B.J. Stojanovic

Animal Research Centre 324

Head: Frank T. Withers

College of Veterinary Medicine   325
Dean: Dr J.G. Miller

Monsanto Agricultural Products Company 326

Address: 800 North Lindbergh Boulevard, St Louis, MO 63166
Telephone: (314) 694 1000
Status: Research centre within an industrial company

RESEARCH DEPARTMENT 327

Research director: Dr John T. Marvel
Departments: Biological evaluation, Dr Reuven Sacher; chemical research, Dr K. Wayne Ratts; environmental science Dr Dexter B. Sharp; international reseach, Dr John A. Stephens, new project identification, Dr Robert J. Kaufman; operations, James C. Barnett
Graduate research staff: 300
Activities: Research in the following areas: exploratory biology; plant science, pesticide chemicals - synthesis, evaluation, processes, residues, metabolism, toxicology; plant growth regulators.
Publications: Annual report.

Montana State University* 328

Status: Educational establishment with r&d capability

COLLEGE OF AGRICULTURE* 329

Central Agricultural Research Centre 330

Address: Moccasin, MO 59462
Telephone: (406) 423 5227
Head: Art Dubbs
Graduate research staff: 1
Publications: Annual progress report.
Projects: Dryland cereal investigations; dryland forage crops; cropping systems (Art Dubbs); weed control on drylands; soil fertility (Pat Rardon).

Eastern Agricultural Research Centre   331

Address: PO Box 393, Sidney, MO 59270
Telephone: (406) 482 2208
Head: Jerald W. Bergman
Publications: Annual progress report.
Projects: Improvement of soil fertility; oil seed crop investigations; irrigated and dryland crop variety (Jerald Bergman); weed control in cereal grains and row crops; cultural practices for sugar beet production (Ronald Anderson).

Northern Agricultural Research Centre   332

Address: Star Route 36, Havre, MO 59501
Telephone: (406) 265 6115
Head: Donald Anderson
Graduate research staff: 1
Publications: Annual progress report.
Projects: Beef cattle recurrent selection (Daniel Doornbos); livestock crossbreeding (Donald Anderson); cropping systems; weed control; crop variety testing (Greg Carlson); range renovation; soil fertility; tillage practices (Harald Houlton).

Northern Plains Soil and Water   333
Research Centre

Address: PO Box 393, Sidney, MO 59270
Head: J. Kristian Aase

Northwestern Agricultural Research   334
Centre

Address: 1570 Montana 35, Kalispell, MO 59901
Telephone: (406) 755 4303
Head: Vern Stewart
Graduate research staff: 1
Publications: Annual progress report.
Projects: Pasture renovation; forage crop investigations (Leon Welty); crop variety testing (Leon Welty, Vern Stewart); weed control in small grains; potato improvement; pulse crop investigations; cropping practices (Vern Stewart).

Southern Agricultural Research Station 335

Address: Huntley, MO 59037
Telephone: (406) 348 3400
Head: Gil Stallknecht
Graduate research staff: 1
Publications: Annual progress report.
Projects: Dryland and irrigated field crop investigations; weed control in field crops (Gil Stallknecht); forage crops investigations; livestock nutrition (John H. Williams); irrigated soils; soil fertility, irrigated and dryland (Vincent A. Haby).

Western Research Centre   336

Address: 531 NE Quast, Corvallis, MO 59828
Telephone: (406) 961 3332
Head: Don Graham
Graduate research staff: 1
Publications: Annual progress report.
Projects: Biological weed control (Jim Story); soil nutrient deficiencies; irrigation systems (Don Graham); small fruit production; winter hardy fruit trees (Nancy Callan).

Western Triangle Research Centre   337

Address: Box 1471, Conrad, MO 59425
Telephone: (406) 278 7707
Head: Greg Kushnak
Graduate research staff: 1
Activities: Research in the following areas: land use; cropping systems; variety testing.
Publications: Western Triangle Annual Progress Report.
Projects: Dryland and irrigated cereal grain investigations; cropping systems; weed control.

MONTANA AGRICULTURAL   338
EXPERIMENT STATION

Address: 202 Linfield Hall, Bozeman, MO 59717
Telephone: (406) 994 3681
Director: James R. Welsh
Associate director: Arne Hovin
Activities: Basic and applied research is aimed at improving productivity of Montana agriculture: livestock and cereal grain are the major commodities. Transportation of agricultural products to markets is becoming a key factor in the state's competitive position. Efficient utilization of natural resources, improved cropping systems and reduced losses from pests and adverse climate are the primary research areas in field crops, horticultural crops and livestock. Research is cooperative with United States Department of Agriculture - Agricultural Research Service in some areas.

Agricultural Economics Department   339

Head: Bruce Beattie
Projects: Agricultural resources (O. Burt, R. Stroup); marketing (G. Cramer, C. Greer); economics (R. Taylor).

Agricultural Education Department   340

Head: Max Amberson

Agricultural Engineering Department   341

Head: William E. Larsen
Projects: Water and irrigation science (T. Hanson, G.L. Westesen).

Animal and Range Sciences Department 342

Head: Arthur C. Linton
Projects: Animal husbandry and nutrition; beef breeding (D. Kress); sheep breeding (R. Blackwell).

Biology Department 343

Head: James Pickett
Projects: Ornithology (R. Eng); large game (H. Picton).

Chemistry Department 344

Head: Edwin H. Abbott
Projects: Chemical reactions (R. Geer); chemistry of agricultural wastes (J. Robbins).

Home Economics Department 345

Head: Margaret A. Briggs
Projects: Food science (R. Johnson).

Plant and Soil Science Department 346

Head: Dwane G. Miller
Projects: Alfalfa breeding (R. Ditterline); barley breeding (R. Eslick, E. Hockett); wheat breeding (A. Taylor); quantitive genetics (J. Martin); weed science in field and horticultural crops (P. Fay, R. Lockerman); plant/soil systems (H. Fergusen); water stress (E. Skogley); physiology of crop yield (J. Brown); climatic influences on crop husbandry (J. Caprio).

Plant Pathology Department 347

Head: Eugene L. Sharp
Projects: Viral plant diseases (T. Carroll); soil-borne pathogens (D. Mathre); bacteriology (D. Sands); fungal diseases (E. Sharp); plant physiology (G. Strobel); grain insect pests (W. Morrill); plant hormones and plant insect populations (S. Visscher).

Veterinary Research Laboratory 348

Head: David M. Young
Projects: Respiratory diseases (G.P. Epling); reproduction (B. Firehamner, L. Myers).

Montana Forest and Conservation Experiment Station 349

– MFCES
Address: School of Forestry, University of Montana, Missoula, MO 59812
Telephone: (406) 243 5521
Director: Dr Benjamin B. Stout
Wilderness Institute: Dr R.R. Ream
Graduate research staff: 26
Activities: Forestry research; range, wildlife, and wilderness management research.
Publications: Annual progress report.
Projects: Mission oriented research (Dr Robert Pfister); elk logging study (Dr C.L. Marcum); border grizzly project (Dr Charles Jonkel).

SCHOOL OF FORESTRY* 350

Dean: Dr Benjamin B. Stout

National Dairy Council 351

– NDC
Address: 6300 North River Road, Rosemont, IL 60018
Telephone: (312) 696 1020
Status: Industrial organization funding research
President: Dr M.F. Brink
Departments: Nutrition research, Dr Elwood W. Speckmann; nutrition education, Dr Gloria G. Kinney
Activities: The National Dairy Council conducts research through the annual Nutrition Research Grant-in-Aid Programme, awarding grants to scientists across the United States. The objective of this programme is to provide support for nutrition research which will answer critical questions regarding the specific role of dairy foods in attaining and maintaining optimal health, and to further the understanding of the þiological value of the nutrients in milk, their interactions, and factors which may affect their availability to the human body. The scope of the programme extends from the most fundamental aspects of the nutritional importance of the components of dairy foods, as determined with experimental animals and/or human subjects, to the testing of the nutritional value of specific dairy foods or fractions thereof. In addition, the programme is designed to consider those human diseases or disturbances which may be related to, or alleviated by, the consumption of dairy foods.
Projects: Nutritional evaluation of milk's minerals (Dr David A. McCarron, Dr Robert R. Recker, Dr Gary S. Rogoff); nutritional evaluation of milkfat (Dr David Kritchevsky, Dr Jon C. Lewis, Dr Donald J. McNamara); relationship of dietary cultures to gut ecology (Dr Barry R. Goldin, Dr Sherwood L. Gorbach, Dr Khem M. Shahani); availability of nutrients from dairy

foods (Dr Richard S. Rivlin, Dr Vernon R. Young); dietary dairy food intake and plasma concentrations of cholesterol, fats and other nutrients (Dr R.B. Alfin-Slater); interactions between cow milk proteins, other food proteins and the immune system (Dr S.A. Back, Dr C.D. May); dairy fats in modern day diets (Dr D.L. Costill); bioavailability of milk iron and copper (Dr J. Hegenauer, Dr P. Saltman); small intestinal injury and milk consumption and tolerance in infants and children (Dr E. Lebenthal); dairy product consumption and lipoproteins in young adults (Dr L.K. Massey); lactase deficiency in indeterminate abdominal pain of childhood (Dr A.D. Newcomer); effects of milk protein on cell-mediated immunity and cancer incidence (Dr R.L. Nutter); orotic acid metabolism in man (Dr J.L. Robinson); effects of exercise on riboflavin requirement in young adult women (Dr D.A. Roe); tolerance to cow's milk with normal or reduced lactose content in children with protein-energy malnutrition (Dr N.W. Solomons, Dr B. Torun); interaction of milk's nutrients, particularly protein, calcium, and phosphorus (Dr George M. Briggs, Dr Janet L. Greger, Dr Victor Herbert, Dr Hellen M. Linkswiler); special nutrient and micro-constituent properties of dairy foods (Dr E.L. Barrett, Dr E.B. Collins, Dr Richard A. Bernhard, Dr John C. Bruhn, Dr Dennis Hsieh, Dr R.L. Merson, Dr Charles Shoemaker).

National Food Processors Association*   352

Status: Industrial company

NORTHWEST RESEARCH LABORATORY*   353

Address: 1600 S Jackson Street, Seattle, WA 98144
Telephone: (206) 323 3540
Status: Research centre within an industrial company
Director: R. DeCamp

New Mexico State University*   354

Status: Educational establishment with r&d capability

AGRICULTURAL EXPERIMENT STATION   355

Address: PO Box 3-BF, Las Cruces, NM 88003
Telephone: (505) 646 3125
Associate Dean and Director: Dr Koert J. Lessman
Graduate research staff: 89
Activities: The station is the land-grant university for New Mexico, and consequently has the responsibility to conduct research in the areas of agriculture and home economics which will be beneficial to the people of the state and nation. Livestock (beef), alfalfa, cotton, chilli, small grains, and horticultural crops are the major research areas to receive emphasis.
Publications: Agri-search (quarterly).

Agricultural Economics/Agricultural Business Department   356

Head: Dr G.R. Dawson
Projects: Impacts of relative price changes of feeds and cattle on the marketing of US beef (Dr B. Gorman).

Agricultural Engineering Department   357

Head: Dr G. Abernathy
Projects: Agronomics of recycling sewage solids (Dr G. O'Connor).

Agricultural Information Department   358

Head: N. Newcomer

Agronomy Department   359

Head: Dr M. Niehaus
Projects: Onion improvement and physiology (Dr J. Corgan).

Animal and Range Science Department   360

Head: Dr A. Nelson
Projects: Clayton - cattle health and care (Dr G. Lofgren).

Branch Stations*   361

Española Valley Branch Experiment Station*   362
Address: PO Box 86, Alcalde, NM 87511
Superintendent: Professor Frank B. Matta

Middle Rio Grande Branch Experiment Station*   363
Address: 1036 Miller Street SW, Los Lunas, NM 87031
Superintendent: R.F. Hooks

Northeastern Branch Experiment Station*   **364**
Address: PO Box 689, Tucumceri, NM 88401
Coordinator: R.E. Kirksey

Plains Branch Experiment Station*   **365**
Address: Star Route, Clovis, NM 88101
Superintendent: Dr R.E. Finkner

San Juan Branch Experiment Station*   **366**
Address: PO Box 1018, Farmington, NM 87401
Superintendent: E.J. Gregory

Southeastern Branch Experiment Station,*   **367**
Address: Route 1, Box 121, Artesia, NM 88210
Superintendent: C.E. Barnes

**Entomology and Plant Pathology   368
Department**

Head: Dr E. Huddleston
Projects: Breeding of disease and insect resistant alfalfa
(Dr B. Melton); rearing and biology of the range cater-
pillar (Dr E. Huddleston).

Experimental Statistics Department   369

Head: Dr M. Finkner

**Fishery and Wildlife Sciences   370
Department**

Head: Dr C. Davis
Projects: Antelope habitat study (Dr R. Pieper).

Home Economics Department   371

Head: Dr M. Hoskins

Horticulture Department   372

Head: Dr F. Widmoyer

North Carolina State   373
University*

Status: Educational establishment with r&d capability

**NORTH CAROLINA AGRICULTURAL   374
RESEARCH SERVICE**

– NCARS
Address: PO Box 5847, North Carolina State University, Raleigh, NC 27650
Telephone: (919) 737 2718
Director: D.F. Bateman
Sections: Agricultural communications, D.M. Jenkins;
animal science, C.A. Lassiter; biochemistry, S.B. Tove;
biology and agricultural engineering, F.J. Hassler;
botany, J.P. Miksche; crop science, B.E. Caldwell;
economics and business, W.D. Toussaint; entomology,
R.J. Kuhr; food science, D.R. Lineback; forest
resources, E.L. Ellwood; veterinary medicine, T.M. Curtin; genetics, J.G. Scandalios; home economics, Naomi
Albanese; horticultural science, A.A. De Hertogh;
microbiology, J.B. Evans; plant pathology, Robert
Aycock; poultry science, R.E. Cook; sociology and
anthropology, R.C. Wimberley; soil science, C.B. Mc-
Cants; statistics, D.L. Solomon; zoology, J.G. Vanden-
bergh
Graduate research staff: 400
Activities: The Agricultural Research Service serves as
the statewide agency with responsibility for developing
and conducting a comprehensive research programme in
support of agriculture and rural development. It works
closely with United States Department of Agriculture
research services in areas of mutual interest.
Publications: Research Perspectives (quarterly); annual
report.
Projects: Approximately 400 project leaders supervise
550 authorized projects in all areas of agriculture, forestry, home economics, veterinary medicine, and rural
development.

North Dakota State   375
University of Agriculture
and Applied Sciences*

Status: Educational establishment with r&d capability

**NORTH DAKOTA AGRICULTURAL   376
EXPERIMENT STATION**

Address: Box 5435, State University Station, North
Dakota State University, Fargo, ND 58105
Telephone: (701) 237 7654
Director: Dr H. Roald Lund
Associate Director: D.E. Anderson
Departments: Agricultural economics, Dr D.F. Scott;
agricultural engineering, Dr G.L. Pratt; agronomy, Dr
J.F. Carter; animal science, C.N. Haugse; bacteriology,
Dr K.J. McMahon; biochemistry, Dr H.J. Klosterman;
botany, Dr Harold Goetz; cereal chemistry and technol-

ogy, O.J. Banasik; entomology, Dr J.T. Schulz; horticulture and forestry, Dr E.P. Lana; plant pathology, Dr R.L. Kiesling; soils, Dr J.E. Foss; veterinary science, Dr M.H. Smith.
Graduate research staff: 180
Activities: The station conducts 265 projects in all areas of agricultural research.
Publications: Progress report.

Norton Simon Incorporated, Research & Development Department*

377

Address: 1645 West Valencia Drive, Fullerton, CA 92634
Telephone: (714) 871 2100
Status: Research centre within an industrial company
Corporate director, science and nutrition: W.L. Clark
Activities: Product-oriented research on food, food processing, food packaging and agriculture.

Nutrition International Incorporated*

378

Address: 725 Cranbury Road, East Brunswick, NJ 08816
Status: Industrial company
Corporate research officer: Dr C. Wo
Activities: Applied research and development of foods, animal health products, agricultural chemicals and cosmetics.

Oakite Products Incorporated*

379

Address: 50 Valley Road, Berkeley Heights, NJ 07922
Telephone: (201) 464 6900
Status: Research centre within an industrial company
Vice-President, Research and Product Development: Edward Heinzelman
Activities: Chemical cleaners and surface treatments; food hygiene.

Oklahoma State University*

380

Address: Stillwater, OK 74078
Status: Educational establishment with r&d capability
President: Dr L.L. Boger

DIVISION OF AGRICULTURE*

381

Dean: Professor Charles B. Browning

Agricultural Experiment Station

382

Director: Professor Charles B. Browning
Associate director: Professor J.A. Whatley
Assistant director: Professor J.C. Murray
Special stations: Stillwater agronomy research station, H.R. Myers; Stratford agronomy research station, W.N. Stokes; Perkins agronomy research station, D.W. Hooper; Caddo research station, Q.O. Opitz; Eastern research station, Haskell, J.W. Walker; irrigation research station, Altus, P.D. Kruska; Kiamichi forestry research station, Idabel, B.J. Smith; North Central research station, Lahoma, Dr R.J. Sidwell; Oklahoma pecan research station, H.L. Davis; Oklahoma vegetable research station, Bixby, D.B. Bostian; Panhandle research station, Goodwell, Professor J.P. Alexander; Sandyland research station, Magnum, Professor R.W. Foraker; Sarkeys research station, Lamar, J.L. Luna; South Central research station, Chickasha, J.D. Avis; Southwest agronomy research station, Tipton, G.L. Strickland; Southwestern livestock and forage research station, El Reno, Professor R.N. Sprowls; United States Southern Great Plains field station, Woodward, Professor Phillip L. Sims; veterinary research laboratory, Pawhuska, Professor E.J. Richey
Activities: The station was created in 1890 as the research arm of the State's land-grant education institution. The main station and most of the researchers are at Stillwater, and research is aimed at improved crops and livestock, greater understanding of human nutritional needs, control of plant and animal diseases and insect pests, better farm and ranch management practices, basic research towards combatting human diseases, energy-saving technology, and new methods and facilities for marketing crops and livestock. In 1980 there were 300 projects being conducted in the fields of agricultural economics, education and engineering; animal sciences; biochemistry; biological sciences; entomology; forestry; horticulture; nutrition; plant pathology, and veterinary sciences.

Agricultural Economics Department
383
Head: Professor J.E. Osborn

Agricultural Engineering Department
384
Head: Professor C.T. Haan

Agronomy Department 385
Head: Professor P.W. Santelmann

Animal Science Department 386
Head: Professor Robert Totusek

Biochemistry Department 387
Head: Professor R.E. Koeppe

Biological Sciences Department 388
Head: Professor G.W. Todd

Entomology Department 389
Head: Professor D.C. Peters

Forestry Department 390
Head: Professor J.E. Langwig

Home Economics Research Department 391
Head: Professor M.M. Scruggs

Horticulture Department 392
Head: Professor H.G. Vest

Plant Pathology Department 393
Head: Professor D.F. Wadsworth

Statistical Laboratory 394
Director: Professor J.L. Folks

VETERINARY MEDICINE COLLEGE 395

Dean: Professor P.M. Morgan

Oregon State University* 396

Address: Corvallis, OR 97331
Status: Educational establishment with r&d capability

AGRICULTURAL EXPERIMENT STATION 397

Director: John R. Davis
Graduate research staff: 268
Activities: Protection of crops and animals from insects, diseases and other hazards; efficiency of agricultural production; environment, nutrition and quality of living; consumer protection; new agricultural products and processes; enhanced product quality; marketing; community development, economic and public services for Oregon's rural and urban people.
Publications: Oregon's Agricultural Progress (quarterly).

Agricultural Chemistry Department 398

Head: Virgil Freed

Agricultural Economics Department 399

Head: Gene Nelson

Agricultural Engineering Department 400

Head: Ron Miner

Animal Sciences Department 401

Head: Jim Oldfield

Botany and Plant Pathology Department 402

Head: Tom Moore

Crop Science Department 403

Head: Dale Moss

Entomology Department 404

Head: Bruce Eldridge

Fisheries Department 405

Head: Richard Tubb

Food Science Department 406

Head: Paul Kifer

Home Economics Department 407

Head: Marge Woodburn

Horticulture Department 408

Head: C.J. Weiser

Microbiology Department 409

Head: John Fryer

Poultry Department 410

Head: George Arscott

Statistics Department 411

Head: Lyle Calvin

NUTRITION RESEARCH INSTITUTE 412

Address: 101 Milam Hall, Oregon State University, Corvallis, OR 97331
Telephone: (503) 754 3561
Status: Educational establishment with r&d capability
Director: Dr Suk Y.Oh
Activities: The institute has as its broad objectives the stimulation, encouragement, facilitation, and coordination of research efforts in the varied fields of nutrition as practised in the departments and schools of Oregon State University. These objectives are served through the sponsorship of interdepartmental and institutional seminars and symposia, methodology workshops, support of promising preliminary investigations and acquisition of laboratory equipment, coordination of nutrition course offerings, and through encouraging entry of qualified scientists and graduate students into nutrition research through their particular disciplines.
Publications: Annual report; quarterly newsletter.
Projects: Effect of diets on high density and other lipoproteins; dietary eggs, plasma lipids and lipoproteins.

SCHOOL OF VETERINARY MEDICINE 413

Dean: E.E. Wedman

Pennsylvania State University* 414

Address: University Park, PA 16802
Status: Educational establishment with r&d capability

Pet Incorporated* 415

Parent body: IC Industries
Status: Industrial company

CONTECH LABORATORIES* 416

Address: Louis Latzer Drive, Greenville, IL 62246
Telephone: (618) 664 1554
Vice-president, r&d: Dr V.L. Stromberg
Sections: Food science, R.E. Arends; contract research, Dr J.J. Betscher
Activities: Contract research for food science and technology, dairy technology, and nutrition.

Pfizer Incorporated* 417

Status: Industrial company

PFIZER CENTRAL RESEARCH* 418

Address: Groton, CT 06340
Telephone: (203) 445 5611
Graduate research staff: 300
Activities: Applied research towards human pharmaceutical, agricultural, veterinary, and fine chemical products.

Plum Island Animal Disease Center 419

– PIADC
Address: PO Box 848, Greenport, NY 11944
Telephone: (516) 323 2500
Telex: 510 6312
Affiliation: United States Department of Agriculture, Agricultural Research Services, Northeast Region
Status: Official research centre
Director: J.J. Callis
Associate director: J.H. Graves
Research sections: Biochemistry and biophysics, J. Polatnick (acting); cytology, C.H. Campbell; diagnostics, A.H. Dardiri; immunology, P.D. McKercher; pathology, C.A. Mebus
Graduate research staff: 42
Activities: The centre was established by an Act of Congress with the purpose of research and development of diagnostic competency for diseases that do not exist in the United States.
Projects: Recombinant viral DNA cloning (D. Moore); African swine fever research (C.A. Mebus); Rift-valley fever research (R.J. Yedloutschnig); foreign animal diseases diagnosis (A.H. Dardiri); foot-and-mouth disease vaccine development (P.D. McKercher); exotic bluetongue research (C.H. Campbell).

Purdue University 420

Address: West Lafayette, IN 47907
Status: Educational establishment with r&d capability

INDIANA AGRICULTURAL EXPERIMENT STATION 421

– AES
Telephone: (317) 494 8360
Director: Dr Bill R. Baumgardt
Graduate research staff: 270
Activities: The station is the research arm of Purdue University's agricultural complex, providing the facility, funding and administrative framework for the conduct of scientific investigation and experimentation. Personnel of the experiment station are housed in the various academic departments of agriculture, home economics, and veterinary science and medicine. Their research deals with production, protection, marketing, natural resources, product development and quality of living. It takes place not only in the campus laboratory, but also on various university farms and woodland, and often at the site of clientele need - in the business establishment, community, farmers field or on public or private land.

SCHOOL OF AGRICULTURE 422

Departments: Agricultural economics, Dr P.L. Farris; agricultural engineering, Dr G.W. Isaacs; agronomy, Dr H.W. Phillips; animal science, Dr W.R. Woods; biochemistry, M.A. Hermodson; botany and plant pathology, Dr T. Hodges; entomology Dr E.E. Ortman; food science, Dr P. Nelson; forestry and natural resources, Dr M.C. Carter; horticulture, Dr B.C. Moser

SCHOOL OF CONSUMER AND FAMILY SCIENCES 423

Dean: N. Compton
Departments: Foods and nutrition, Dr P. Abernathy

Animal Disease Diagnostic Laboratory 424

Head: Dr F.R. Robinson

SCHOOL OF VETERINARY MEDICINE 425

Dean: J. Stockton
Departments: Veterinary anatomy, Dr M.W. Stromberg; veterinary microbiology, Dr R.M. Claflin; veterinary physiology, Dr G. Coppoc

Quaker Oats Company* 426

Status: Industrial company

JOHN STUART RESEARCH LABORATORIES* 427

Address: 617 West Main Street, Barrington, IL 60010
Telephone: (312) 381 1980
Vice-President, chemical r&d: A.P. Dunlop
Activities: Research on food products, pet foods, and appropriate processes; some basic research in biochemistry and nutrition.

Ralston Purina Company 428

Address: Checkerboard Square, St Louis, MO 63188
Publications: Annual report.

R.J. Reynolds Industries Incorporated* 429

R.J. REYNOLDS RESEARCH DEPARTMENT* 430

Address: 115 Chestnut Street, Winston-Salem, NC 27102
Telephone: (919) 748 2136
Status: Research centre within an industrial company
Director of research: Dr Alan Rodgman
Sections: Analytical division, J.A. Giles III; electrical division, Dr D.H. Piehl; agricultural research division, J.D. Shiffert
Activities: Chemical, physical and analytical studies of tobacco products and smoke.

Richard B. Russell Agricultural Research Center 431

Address: PO Box 5677, Athens, GA 30613
Telephone: (404) 546 3311
Parent body: United States Department of Agriculture, Agricultural Research Service
Status: Official research centre
Area Director: David E. Zimmer
Research units: Animal physiology, Robert R. Kraeling; field crops, James A. Robertson; food protection and processing, James E. Thomson; meat quality, Leroy C. Blankenship; plant physiology, Horace G. Cutler quality evaluation, Gerald G. Dull; tobacco safety, Orestes T. Chortyk; toxicology and biological constituents, William P. Norred
Graduate research staff: 75

Activities: The programme of the centre is guided by major national goals for increased agricultural productivity and improved nutrition, environmental quality, rural area development, and food safety. New uses and more efficient processing methods are being sought for important agricultural commodities. Expansion of both domestic and foreign markets is a major objective.

Projects: Endocrinological and physiological mechanics regulating reproduction and body growth and development in livestock (Dr R.R. Kraeling); improved technology for processing forages, feed grains, oilseeds, and other agricultural products (J.A. Robertson); procedures for improving processing efficiency, quality, shelf-life, and safety of meat products, with emphasis on poultry (J.E. Thomson); composition and physical, chemical, and organoleptic properties of poultry tissues and products derived from them (L.C. Blankenship); improvement of biological conversion of solar energy for increased crop production by increasing efficiency of photosynthesis, translocation and associated metabolism (H.G. Cutler); basic and applied research on fruits, vegetables, and tree nuts (Dr G.G. Dull); development of safer cigarette (Dr O.T. Chortyk); toxicological and pharmacological properties of natural toxicants; chemical, pharmacological, and toxicological properties of fungal metabolites (W.P. Norred).

Robey Wentworth Harned 432 Research Laboratory

Address: PO Box 5367, Mississippi State, MI 39762
Telephone: (601) 323 2230
Parent body: United States Department of Agriculture, Agricultural Research Service
Status: Official research centre
Director: T.B. Davich
Research sections: Boll weevil, T.B. Davich; crop science and engineering, J.N. Jenkins
Graduate research staff: 40
Activities: The mission of boll weevil research is: to eliminate, or greatly reduce, the economic impact of the boll weevil on cotton production; and to reduce the amount of insecticides used for its control. The mission of crop science and engineering research is to provide increased crop production with greater efficiency in the southeastern United States by developing equipment, cropping systems, pest resistant strains with improved agronomic traits, and decision-making models to reduce costs and conserve natural resources.

Projects: Boll weevil - biology and behaviour (W.H. Cross); chemistry (P.A. Hedin); field evaluation (T.B. Davich); mass bearing engineering (J.G. Griffin); nutrition, sterility and quality control (J.E. Wright); crop science and engineering - corn host plant resistance (G.E. Scott); cotton host plant resistance (J.N. Jenkins); crop simulations (D.N. Baker); forage research (W.E. Knight).

Rocky Mountain Forest and 433 Range Experiment Station

Address: 240 West Prospect Street, Fort Collins, CO 80526
Telephone: (303) 221 4390
Affiliation: United States Department of Agriculture, Forest Service
Status: Official research centre
Director: Charles M. Loveless
Deputy director: Dixie R. Smith
Assistant director, research planning and applications: J.S. Krammes
Assistant director, research support services: Jacqueline Cables
Assistant director for research - north: Clyde A. Fasick
Graduate research staff: 98
Activities: The station is one of eight regional experiment stations, plus the Forest Products Laboratory and Washington Office staff, that make up the Forest Service research organization, which aims to develop knowledge and technology needed to balance economic and environmental demands for resources on forests and related rangelands. Research is focused on water, wildlife and fish habitat, range, timber, recreation, human and community development, multiresource evaluation, and protection against fire, diseases and insects.
Projects: Resource planning techniques (Thomas Hoekstra); multiresource management of montane and subalpine zones (Robert Alexander); mountain snow and avalanche research (M. Martinelli, Jr); resources evaluation techniques (Richard Driscoll); national site classification system (Daniel Merkel); national resource inventory techniques (H. Gyde Lund); activity level fire planning techniques (Peter Roussopoulos); forest meteorology and air quality (Douglas Fox); forest diseases in Rocky Mountains and southwest (Frank Hawksworth).

FOREST, RANGE, AND WATERSHED 434 LABORATORY

Address: 222 South 22nd Street, Laramie, WY 82070
Telephone: (307) 742 6621
Projects: Hydrology of sagebrush lands and high plains snow management (Ronald Tabler); effects of multiple-use land management on wildlife in the Central Rocky Mountains (A. Loren Ward).

FOREST RESEARCH LABORATORY 435

Address: South Dakota School of Mines and Technology, Rapid City, SD 57701
Telephone: (605) 343 0811
Projects: Environmental improvement and multiple use management in the northern Great Plains (Ardell Bjugstad).

FORESTRY AND RANGE SCIENCES 436 LABORATORY

Address: 2205 Columbia S.E., Albuquerque, NM 87106
Telephone: (505) 766 2384
Projects: Disturbed site reclamation in the Southwest (Earl Aldon).

FORESTRY SCIENCES LABORATORY 437

Address: Arizona State University, Tempe, AZ 85281
Telephone: (602) 261 4365
Assistant director for research - south: Richard G. Krebill
Projects: Multiresource management of mixed conifer and chaparral watersheds (Leonard De Bano); wildlife habitat research on southwestern forests and rangelands (David Patton); fuel management in the Central Rockies and southwest (John Dieterich).

FORESTRY SCIENCES LABORATORY 438

Address: Northern Arizona University, Flagstaff, AZ 86001
Telephone: (602) 779 3311
Projects: Culture of southwestern conifers (Frank Ronco, Jr); multiresource management analysis and evaluation (Dave Garrett).

FORESTRY SCIENCES LABORATORY 439

Address: University of Nebraska East Campus, Lincoln, NE 68583
Telephone: (402) 471 5178
Projects: Protection and improvement of trees in the Great Plains (Glenn Peterson).

GREAT PLAINS WILDLIFE RESEARCH 440 LABORATORY

Address: Texas Technical University, PO Box 4249, Lubbock, TX 79409
Telephone: (806) 762 7671
Projects: Management of wildlife habitat in southern Great Plains (Fred Stormer).

SHELTERBELT LABORATORY 441

Address: 1st and Brander Street, Bottineau, ND 58318
Telephone: (701) 228 2259
Projects: Forestry practices in the northern Great Plains (Richard Tinus).

Rutgers - The State 442 University of New Jersey*

Status: Educational establishment with r&d capability

NEW JERSEY AGRICULTURAL 443 EXPERIMENT STATION

Address: Cook College, Rutgers University, PO Box 231, New Brunswick, NJ 08903
Telephone: (201) 932 9447
Director: Grant F. Walton
Assistant director: David R. Burns
Graduate research staff: 100
Activities: The aim of the Agricultural Experiment Station is to conduct basic and applied research in order to discover, apply and disseminate knowledge in the public interest for the following purposes: to ensure that high quality food will be readily available at lower cost to the people of New Jersey; to improve the physical environment of the state and to improve living standards of the people of the state. The primary area of research within the station concerns vegetable crops, corn, soya beans and forages: animal production ranks second, with dairy cattle the principal livestock type.
Publications: Research Update (biannual).

Agricultural Economics and Marketing 444 Department

Head: vacant
Projects: Identification of market forces which determine milk prices (R. Stammer); impact of in and out migration and population redistribution in the northeast (D. Thatch); socioeconomic factors related to land use (D. Derr); change in rural communities (R. Beck); contract production and marketing of broilers and eggs (A. Meredith); impact of tourism on rural economic development (R. Koch).

Animal Sciences Department   445

Head: Dr Colin Scanes
Projects: Mineral interrelationships - calcium and copper nutrition (J. Evans); genetic bases for resistance to avian diseases (I. Kujdych); nitrogen utilization and interactions with fibre digestibility, energy source and exercise in horses (M. Glade); effects of nitrogen, energy, minerals and their interrelationships in metabolic diseases in ruminant animals (J. Wohlt).

Biochemistry and Microbiology   446
Department

Head: Dr Stanley Katz
Projects: Nematode-trapping fungi and biological control of agricultural pests (D. Pramer); pyridine nucleotide-dependent dehydrogenases in metabolism of aldehydes and alcohols (T. Chase); enzymatic degradation of polymers produced by plant and fungal cells (J. MacMillan); relationship of non-nutritive additives in feeds to bacterial resistance problems in man (S. Katz); metabolism and fate of pesticides and their residues in agricultural commodities (R. Bartha); functional properties of protein (D. Strumeyer); assembly, structure and energy transfer in multicomponent protein complexes (P. Kahn).

Biological and Agricultural Engineering 447
Department

Head: William Roberts
Projects: Engineering greenhouse systems and environments (D. Mears); mechanical harvesting and handling of fruits and vegetables (R. Wolfe); hydrology of suburban areas (K. Nathan).

Entomology and Economic Zoology   448
Department

Head: Dr Herbert Streu
Projects: Investigations of insects and other animals attacking tree fruits (F. Swift); investigation of persistence of insecticides on granules in the environment (W. Carey); generation of pesticide residue data for label clearances (K. Helrich); blueberry insect investigations (P. Marucci); insecticides and synergist action and insect resistance gypsy moth behaviour, the enemies and insecticides (A. Forgash); population dynamics, feeding physiology, and toxicology of resistant and susceptible Colorado potato beetles (J. Lashomb); metabolism and fate of pesticides and their residues in agricultural commodities (G. Winnett); population dynamics, habitat utilization and foraging strategies of roosting blackbirds (D. Caccamise); cranberry insect investigations (P. Marucci).

Environmental Physiology Department   449

Head: vacant
Projects: Glucose and lactate metabolism following E. coli. endotoxin challenge in avian and mammalian species (G. Merrill); role of hormonal modification of adrenergic stimuli in swine, neural control of renal tubular sodium reabsorption (E. Zambraski); biophysical factors affecting energy requirements for poultry production, renal function in salt loaded and fluid retaining swine (H. Frankel).

Environmental Resources Department   450

Head: vacant
Projects: Development and implementation of environmental law (W. Goldfarb).

Environmental Sciences Department   451

Head: vacant

Food Science Department   452

Head: Dr Stephen Chang
Projects: Protein quality in processed food (M. Solberg); biophysical factors affecting energy requirements for poultry production (R. Squibb); quality maintenance and control in the marketing and storage of vegetables (G. Carman); human nutrition improvement; vitamin A research (P. Lachance); new food products from underutilized fish species (R. Morse); texture of viscoelastic and slurry type food (J. Kokini); migration of indirect additives from package to food (S. Gilbert); basic lipid chemistry for parenteral nutrition (S. Chang).

Horticulture and Forestry Department   453

Head: Dr Bernard Pollack
Projects: Management of the Cream Ridge Farm, display gardens, ornamental and vegetable farms, ornamental and vegetable greenhouses (B. Pollack); apple breeding (L. Hough); genetic improvement of cultivated strawberry (G. Jelenkovic); blueberry breeding (G. Jelenkovic), asparagus breeding (J. Ellison); holly breeding (E. Orton); role of oxidative metabolism in the regulation of fruit ripening (C. Frenkel); dessication stresses and resulting injury of overwintering containerized nursery stock (S. Wiest); growth and physiology of plants grown under conditions of high root temperature (H. Janes); mineral nutrition of apple grown on dwarfing rootstock I. calcium; phosphorus, potassium, and trace element requirements of the American cranberry, Vaccinium macrocarpon (P. Eck); genetics and physiology of sweet corn quality, pest resistance and yield; studies of multiple forms of enzymes of the starch biosynthetic pathway of higher plants (C. Boyer); dynamics of the New Jersey hunting

population (J. Applegate); light quality and species alteration (B. Zeide); genetics of precocious fruit pigmentation and the evolution of new cultivars in Cucurbita (O. Shifriss); production of hybrid plants by fusion of somatic protoplasts; production of haploid plants by tissue culture (C. Chin); flowering response of strawberry plants to varying environmental interactions (C. Smith).

Meteorology and Physical Oceanography Department   454

Projects: Impact of climatic variability on agriculture (M. Shulman).

Nutrition Department   455

Projects: Amino acid and protein metabolism (H. Fisher); metabolism and biochemical actions of cadmium in animals; zinc nutrition and the synthesis of zinc binding protein (R. Cousins); early nutritional experiences in the development of obesity (M. Kaplan); human nutrition improvement (H. Fisher); role of amino acids and diet in liver disease (R. Steele).

Oyster Culture Department   456

Projects: Natural seed oyster beds of Delaware Bay (H. Haskin).

Plant Pathology Department   457

Head: Dr Raymond Capellini
Projects: Bacterial spot of peach: ecology and control (M. Davis); pathogenicity and control of fungi associated with turfgrass diseases (P. Halisky); injury to vegetation by air pollutants (I. Leone); physiological and biochemical response of vegetation to air pollutants (E. Brennan); spiroplasmas and the yellow diseases of plants; impact of nematode population dynamics on crop growth and yield (T. Chen); control of postharvest decays of fruits and vegetables (R. Cappellini); reducing the influence of air pollution on plant productivity in the northeast (I. Leone); rhizosphere ecology as related to plant health and vigor (R. Myers); tree fruit diseases and their control (J. Springer); disease control in ornamental plants (J. Peterson); vegetable disease control (G. Lewis).

Soils and Crops Department   458

Head: Dr Donald Riemer
Projects: Correlation of soil tests and plant analysis results with plant growth response (R. Flannery); mechanisms of phosphate retention and fixation by aluminium and iron in acidic soils; characterization of clay minerals in soils (P. Hsu); origin, development, characteristics and classification of New Jersey soils (J.

Tedrow); turgrass breeding and evaluation (C. Funk); fundamental factors of weed science (R. Ilnicki); biology and control of aquatic weeds (D. Riemer); soil microbiology (L. Douglas); low maintenance grasses for roadsides, rights of way, fields, lawns, and areas needing erosion control (R. Duell); factors affecting turfgrass herbicide efficiency (R. Engel); evaluation of small grains and forage crop varieties and production methods for use in New Jersey (J. Justin); forage - livestock systems for land with soil and site limitations (M. Sprague).

Statistics and Computer Science Department   459

Projects: Management information services research (R. Mather).

South Dakota State University*   460

Address: Brookings, SD 57007
Status: Educational establishment with r&d capability

Agricultural Research and Extension Centre*   461

Address: Beresford, SD 57004

College of Agriculture and Biological Sciences*   462

Address: Agricultural Hall, Box 2207, Brookings, SD 57007
Dean: Dr Delwyn Dearborn

South Dakota Agricultural Experiment Station   463

Telephone: (605) 688 4149
Director: Dr Raymond Moore
Departments: Agricultural information, John Pates; agriculture engineering, Dennis Moe; animal science, John Romans; biology, Ernest Hugghins; dairy science, John Parsons; economics, John Thompson; horticulture/forestry, Ron Peterson; microbiology, (acting) Robert Pengra; plant science, Dr Maurice Horton; rural sociology, James Satterlee; station biochemistry, David Hilderbrand; veterinary science, Mahlon Vorhies; wildlife and fisheries sciences, Charles Scalet; home economics, Ardyce Gilbert
Graduate research staff: 90
Activities: Crops production; plant breeding; plant physiology; wheat science; plant protection; animal production; dairy science; beef, pork and sheep production; veterinary medicine; animal nutrition and physiology;

soil science; drainage and irrigation; land use; wildlife and fisheries; forestry; horticulture; food science; agricultural economics; rural sociology; poultry production; farm power machinery; alternative energy sources. *Publications: SD Farm and Home Research* (quarterly); field station annual reports.

Southeast South Dakota Experiment Farm 464

Address: RR 3 Box 93, Beresford, SD 57004
Telephone: (605) 563 2989
Director: Raymond Moore
Projects: Plant nutrition (Paul Carson); livestock nutrition (Dan H. Gee); soil testing (Paul Carson); soil management and cultural practices (Fred E. Shubeck); variety testing (Joe Bonneman); chemical weed control (Leon Wrage).

Southern Illinois University – Carbondale* 465

Address: Carbondale, IL 62901
Status: Educational establishment with r&d capability

SCHOOL OF AGRICULTURE 466

Telephone: (618) 453 2469
Dean: Gilbert H. Kroening
Activities: Plant production (corn, soya beans, small grains): plant breeding, soil science, crop husbandry, weed research; animal production (beef and dairy cattle, hogs, sheep, poultry, horses): breeding, husbandry, nutrition, physiology; agricultural mechanization; forestry and forestry products: forest resource management, outdoor recreation resource management, wood science and technology; agribusiness economics. In general, the research programme aims to benefit industry, farming and forestry in southern Illinois and throughout the midwestern region of the United States.

Agribusiness Economics Department 467

Head: William McD. Herr
Projects: Agricultural marketing problems in Illinois (W. Willis); small scale farming systems (L. Solverson).

Agricultural Education and Mechanization Department 468

Head: Thomas Stitt
Projects: Soya bean oil as diesel fuel additive (R. Wolff).

Animal Industries Department 469

Head: Anthony Young
Projects: Swine growth and composition (S. Powell); soya bean utilization in cattle (H.D. Woody).

Forestry Department 470

Head: Howard Spalt
Projects: Remotely sensed data use for forest and coal mining planning (C. Budelsky).

Plant and Soil Science Department 471

Head: Gerald Coorts
Projects: Abiotic factors affecting nodulation in soya beans (B. Klubek); land-use effects on infiltration (S. Chong); various organic materials for topsoil replacement on mined prime farmland (J. Jones); roadside vegetative effects on particulate levels (G. Aubertin); breeding for cyst nematode resistance in soya beans (O. Meyers).

Southern University* 472

Address: Post Office Box 9614, Baton Rouge, LA 70813
Telephone: (504) 771 2262
Status: Educational establishment with r&d capability
President, University System: Dr Jesse N. Stone, Jr
Chancellor, Baton Rouge Campus: Dr Roosevelt Steptoe
Research Director: Dr George E. Robinson, Jr
Activities: Cooperative research at Southern University is involved in four basic research programme areas: agricultural economics and rural sociology; animal science; human nutrition and food sciences; plant and soil sciences. The aims and general scope of the programme is to concentrate on research findings that can be applied by small farmers to enhance their capabilities of utilizing their resources to the fullest extent and to increase their economic standards.
Publications: Annual progress report; annual research bulletin.
Projects: Analysis of credit for small farm growth (Dr Leroy Davis); social and economic characteristics of farm residents and adoption to new technology (Dr Christopher Hunte); economic analysis of the earning potential of small farm families in Louisiana (Dr Dewitt Jones); digestibility and nutritive value of ryegrass forage and sorghum silage (Dr Alonzo Chappell); use of milk progesterone assays in monitoring reproduction activities of dairy cattle (Dr Kirkland E. Mellad); role of steroid hormones in cholesterol and total fat disposition in egg and tissues of domestic fowl (Dr M.T. Malek); effects of feeding grain dust on growth, carcass traits, and reproductive performance of swine (Dr George E.

Robinson, Jr); role and nutrient management and soil factors in increasing soya beans yields (Dr P.S.C. Reddy); distribution and control of soya bean parasitic nematode in Louisiana (Dr V.R. Bachireddy); adaptability and physiology of flower abortion of soya beans (Dr Kit L. Chin); dietary practices and nutritional health of selected low income families in Louisiana (Dr Bernestine McGee); quality profiles of commercially blended soya protein and ground beef (Freddie L. Johnson).

COLLEGE OF AGRICULTURE*   473

Dean: Dr Leroy Davis

COLLEGE OF HOME ECONOMICS*   474

Dean: Dr Eula Masingale

Stanford University   475

Address: Stanford, CA 94305
Telephone: (415) 497 2534
Status: Educational establishment with r&d capability
Activities: Research on materials and related problems is being conducted within eleven different academic departments of Stanford University as well as in the Hansen Laboratory. The research programmes are organized into two groupings: (a) the research programme of members of the Center for Materials Research who have participated in one of the five major Thrust group programmes supported by the National Science Foundation - MRL block grant; and (b) individual research programmes. Most research projects received support from one or more of the following agencies of the United States government: Defense Advanced Research Projects Agency; Department of the Air Force; Department of the Army; Department of Commerce; Department of Energy; Department of Health, Education and Welfare; Department of the Interior; Department of the Navy; National Aeronautics and Space Administration; National Bureau of Standards; National Science Foundation; Veterans Administration.
Individual Research Programmes (name of principal investigator in parentheses): Acoustic and magnetic waves and devices (B.A. Auld); materials synthesis using vapour deposition techniques (T.W. Barbee, Jr); deformation in solids (D.M. Barnett); photoelectronic materials physics (C.W. Bates, Jr); structures of amorphous materials and synchrotron radiation (A.I. Bienerstock); properties of adsorbents and catalysts (M. Boudart); primary photochemistry of photosynthesis (S.G. Boxer); intrinsic chemical reactivity in the gas phase and solvation effects from condensed phases (John I. Brauman); geochemical studies of amorphous and crystalline silicates and silicate melts (G.E. Brown); photoelectronic properties of solids (R.H. Bube); improved nonlinear, electro-optic and laser materials (R.L. Byer); catalysis studies to model metalloenzymes, to develop homogeneous catalysts and invert electrode catalysts (J.P. Collman); theoretical physics of condensed matter, theoretical and experimental biophysics (S. Doniach); coal slag phenomena in open cycle magnetohydrodynamic generators (R.H. Eustis); experiments at liquid helium temperatures on macroscopic quantum effects, material properties, general relativity, and superconducting accelerators (W.M. Fairbank, C.W.F. Everitt, R.P. Giffard, H.A. Schwettman); molecular excited-state dynamics in the condensed phase (M.D. Fayer); crystal growth studies to enable the preparation of materials with closely defined properties (R.S. Feigelson, D. Elwell, R.K. Route); macromolecular research on the theory of rubber elasticity, liquid crystals, and anisotropy (R.J. Flory); photophysics of amorphous solid-state polymer blends - compatibility and relaxation behaviour (C.W. Frank); photophysics of polymers in solution - energy transfer and intramolecular rotational diffusion (C.W. Frank); theory of metal fatigue (H.O. Fuchs, A.K. Miller, D.V. Nelson); ion implantation and laser annealing of semiconductors (J.F. Gibbons); Mössbauer effect and related phenomena (S.S. Hanna, D.L. Clark, H.T. King); pseudopotential methods in physics (W.A. Harrison); high-resolution and in-situ investigations of small particles and reactions on surfaces (K. Heinemann); electron-density approach to binding energies of molecules and condensed systems (C. Herring); structural studies of metalloenzymes and metalloproteins (K.O. Hodgson); solid-state electrochemistry (R.A. Huggins); acoustic interactions with solids (G.S. Kino); elastic-plastic stress analysis with particular emphasis on metal-forming applications (E.H. Lee); superconductivity and molecular physics (W.A. Littel); surface reactivity (R.J. Madix); the role of F- and V-centres present in oxygen-ion type electrolytes in electrocatalyzing gaseous reactions (D.M. Mason); physical chemistry of lipid, lipid-protein or lipid-detergent bilayer membranes (H.M. McConnell); computer-aided engineering of semiconductor integrated circuits (J.D. Meindl); deformation and fracture in structural materials (A.K. Miller); multiaxial fatigue of structural metals (D.V. Nelson); relations between crystalline imperfections and the physical properties of crystals (W.D. Nix); wave propagation in solids with fluid filled pores (A.M. Nur); photon production from high energy electron beams (R.H. Pantell); physical chemistry of oxides and oxide surfaces (G.A. Parks); gas-liquid metal reactions and interactions (N.A.D. Parlee); electrical, optical, and metallurgical properties of semiconducting material (G.L. Pearson); studies of the dynamics of molecules and macromolecules in liquids (R. Pecora); continuum characterization of beam structures applied to cancellous bone (R.L. Piziali);

micromechanical models for the expansive process in expansive cement concrete (C.W. Richards); protein adsorption to polymer films (C.R. Robertson); spectroscopy and quantum electronics (A.L. Schawlow, T.W. Hansch); interactions of optical and acoustic radiation with solids (H.J. Shaw); mechanical behaviour of solids (O.D. Sherby); ultra-short optical pulse generation and applications (A.E. Siegman); resonant Raman spectroscopy of the visual pigment rhodopsin (A.E. Siegman, L. Stryer); studies of surfaces and interfaces in silicon using transmission electron microscopy and Auger electron spectroscopy (R. Sinclair, C.R. Helms); high resolution electron microscopy of phase transformations (R. Sinclair); surface, interface, and electronic studies (W.E. Spicer, I. Lindau, C.R. Helms); solid-state physical chemistry (D.A. Stevenson); process induced defects in silicon (R.M. Swanson); thermophotovoltaic solar energy conversion (R.M. Swanson); fundamental aspects of reactivity in inorganic systems (H. Taube); semiconductor processing, gas discharges, growth-dissolution and surface behaviour of crystals, biomaterials (W.A. Tiller); state-to-state reaction dynamics (R.N. Zare)

FOOD RESEARCH INSTITUTE 476

Telephone: (415) 497 3941
Director: Walter P. Falcon
Graduate research staff: 57
Activities: Agricultural economics.
Publications: Food Research Institute Studies, 3 times a year; *Food Research*, biennial.
Projects: Cassava in Indonesia (Walter P. Falcon, Scott R. Pearson); US Mexico relations (Clark W. Reynolds).

State University of New York* 477

Status: Educational establishment with r&d capability

COLLEGE OF ENVIRONMENTAL SCIENCE AND FORESTRY 478

Address: Syracuse, NY 13210
Telephone: (315) 470 6500
Telex: 7105410555
President: Dr Edward E. Palmer
Vice-president for Programme Affairs: Dr Donald F. Behrend
Assistant vice-president for research programmes: Dr James W. Geis
Assistant vice-president for academic programmes: Dr Robert H. Frey
Graduate research staff: 250

Activities: The college is an upper division/graduate centre offering educational training and performing research in the environmental sciences. Expertise areas include biochemistry, biology, botany, building construction, cellular ultrastructure, chemical ecology, chemistry, ecology, engineering, entomology, environmental studies, fibre physics, forest technology, land use, landscape architecture, limnology, materials marketing, meteorology, microscopy, mycology, natural products engineering, organic materials science, outdoor recreation, paper engineering, paper science pathology; photogrammetry, physiology, regional planning, remote sensing, resource management, resource policy, silviculture, soils, thermodynamics, urban analysis, water resources, wildlife, wood products engineering, wood science, world forestry, and zoology. Recent areas of research: limestone quarry reclamation; polymeric materials for artificial human organs; nonchemical control measures for insect pests; ecology of Antarctic birds; new wood pulping processes leading to pollution-free water and air effluents; ecological effects of winter navigation in the Great Lakes and the St Lawrence River.

Graduate Programme in Environmental Science 479

Director: Dr Robert D. Hennigan

Institute for Environmental Programme Affairs 480

Executive Director: Dr James W. Geis
Activities: The institute coordinates the research effort of the College with the efforts of other academic institutions, public agencies and private industries for a concerted attack on environmental problems.
Projects: Effects of acid precipitation on terrestrial ecosystems; reclamation study of mines and quarried lands; biomass potentials for energy production; stream channel response to land use changes; social and political factors affecting disposal of sewage on land areas.

School of Biology, Chemistry and Ecology 481

Dean: Dr Stuart W. Tanenbaum

DEPARTMENT OF CHEMISTRY 482
Chairman: Dr Kenneth J. Smith

DEPARTMENT OF ENVIRONMENTAL AND FOREST BIOLOGY 483
Chairman: Dr Robert L. Burgess

POLYMER RESEARCH INSTITUTE 484

Director: Dr Israel Cabasso
Activities: Pure and applied polymer chemistry: development of living polymers; study of anionic polymerization and electron-transfer initiation; permeation of gases through polymeric films.

School of Continuing Education 485

Dean: Dr John M. Yavorsky

School of Environmental and Resource 486 Engineering

Dean: Dr William P. Tully

DEPARTMENT OF FOREST 487 ENGINEERING

Chairman: Dr Robert H. Brock

DEPARTMENT OF PAPER SCIENCE 488 AND ENGINEERING

Chairman: Dr Bengt Leopold

DEPARTMENT OF WOOD PRODUCTS 489 ENGINEERING

Chairman: Dr George H. Kyanka

EMPIRE STATE PAPER RESEARCH 490 INSTITUTE

Head: Dr Bengt Leopold
Activities: The institute is a basic research organization in the pulp and paper field. It performs investigations in cooperation with the Empire State Paper Research Association, which is comprised of 78 pulp and paper companies in 14 countries. Almost all aspects of pulping and papermaking are covered, including additive retention, oxygen pulping and bleaching, effluent control, sheet drying, printability, and energy efficiencies.

NELSON COURTLANDT BROWN 491 LABORATORY FOR ULTRASTUCTURE STUDIES

Director: Dr Wilfred A. Côté
Activities: The laboratory is a teaching, research and service facility of the College. It is equipped to handle virtually every type of modern microscopy, including light, scanning electron, and transmission electron microscopy.

School of Forest Technology 492

Director: James E. Coufal

School of Forestry 493

Dean: Dr John V. Berglund

School of Landscape Architecture 494

Dean: Robert G. Reimann

Stephen F. Austin State University* 495

Status: Educational establishment with r&d capability

SCHOOL OF FORESTRY 496

Address: PO Box 6109, SFASU, Nacogdoches, TX 75962
Telephone: (713) 569 3304
Dean: Dr Kent T. Adair
Graduate research staff: 23
Activities: The overall aim is to provide the geo-data base required for scientific management of forestlands. The scope of the reseach programme includes tree physiology, forest hydrology, wildlife management, biometrics and forest recreation. Both basic and applied research are included; however, most studies are data-based, field oriented studies of the potential or real impact of forest practices on natural resources.
Projects: Population sampling of the pine tip moth; annotated bibliography and FAMULUS file for the southern pine beetle (Dr David L. Kulhavy); Little Oil Company wood chemical research; quality and yield of plywood and lumber produced from fast growing, short rotation sulphur pine trees (Dr Leonard F. Burkart); radiotelemetric study on movements of whitetail deer in east Texas forests (Dr James C. Kroll); small animal population inventories on a mixed pine-hardwood forest of east Texas (Dr R.M. Whiting); ecological assessment of 2,4-D treatment on pine-hardwood ecosystem (Dr James C. Kroll, Dr David L. Kulhavy); nonpoint sources of soil and water losses from harvested forests in east Texas (Dr M. Chang).

Stokley-Van Camp Incorporated* 497

Status: Industrial company

CENTRAL LABORATORIES* 498

Address: 6815 East 34th Street, Indianapolis, IN 46226
Telephone: (317) 542 9291
Director: Dr T. Crawford
Activities: Applied research on canned, frozen and processed foods.

Strasburger & Siegal Incorporated* 499

Address: 1403 Eutaw Place, Baltimore, MD 21217
Telephone: (301) 523 5518
Activities: Food technology and analysis including canned, frozen, fresh, and dehydrated foods, meat analysis, and nutritional analysis.

Swift & Company* 500

Parent body: Esmark Incorporated
Status: Industrial company

RESEARCH AND DEVELOPMENT CENTRE* 501

Address: 1919 Swift Drive, Oak Brook, IL 60521
Telephone: (312) 325 9320
Research directors: H.F. Bernholdt; L.E. Klinger; T.V. Kueper
Activities: Nutrition; food; development of new products and new processes and equipment for processing.

Tennessee Valley Authority 502

– TVA
Address: Muscle Shoals, AL 35660

NATIONAL FERTILIZER DEVELOPMENT CENTER 503

– NFDC
Address: Muscle Shoals, AL 35660
Telephone: (205) 386 2601
Status: Official research centre
Manager: B.J. Bond
Sections: Chemical development, C.H. Davis; agricultural development, J.T. Shields; chemical operations, E.A. Lindsay
Graduate research staff: 250
Activities: New fertilizer technology: improved products and more efficient manufacturing processes; agricultural development in the Tennessee River Valley region.
Publications: Annual report.
Projects: Energy-efficient fertilizer manufacturing processes (J.G. Getsinger); improved phosphate processing methods (Z.T. Wakefield); coal gasification to make hydrogen for ammonia production (D.A. Waitzman); alcohol fuels from agricultural crops and wood (J.M. Stinson); new fertilizer granulation technology (J.G. Getsinger); improved fluid fertilizer technology (J.G. Getsinger); beneficial uses of waste heat from industrial plants (C.M. Madewell); integrated systems for recovering and using nutrients in animal waste (J.J. Maddox); increasing crop efficiency of nutrient use (E.C. Sample).

Texas A and I University* 504

Address: Kingsville, TX 78363
Status: Educational establishment with r&d capability

COLLEGE OF AGRICULTURE 505

Address: PO Box 156, Station 1, Kingsville, TX 78363
Telephone: (512) 595 3711
Dean: Charles A. DeYoung
Graduate research staff: 8
Activities: Agricultural production problems of the South Texas region: natural resources - drainage and irrigation; plant production: forages, crop husbandry; animal production - sheep, goats, cattle, livestock husbandry.
Publications: Annual progress reports for individual projects.
Projects: Productivity of selected pasture grasses in South Texas (James D. Arnold); fluctuation of the water table under soils of Kleberg County as they relate to developing surface soil salinity problems in cultivated fields and uncultivated wildlife habitats (David D. Neher); electric deer proof fence (Darroll L. Grant).

Caesar Kleberg Wildlife Research Institute 506

Address: Campus Box 218, Kingsville, TX 78363
Telephone: (512) 595 3922
Director: Dr Charles DeYoung
Graduate research staff: 40
Activities: Rangeland in South Texas: wildlife diseases, native plants, commercial utilization of wildlife, basic ecology of important native plants and animal species.
Publications: Annual report.
Projects: Seasonal feed intake and digestibility of white-tailed deer; digestive efficiency of nilgai and goats; effects of maternal nutrition on fawn growth; biomedical implications of antlerogenesis in deer (Dr Robert Brown); productivity and vegetation response of mesquite/mixed brush community utilized by five herbivores (Dr Terry McLendon); development of uses of mesquite (Dr Peter Felker); avian use of South Texas aquatic habitat; nesting and brood rearing of bobwhite quail in South Texas, quail production (Daniel Everett); current status of white-fronted doves in South Texas (Dr Charles DeYoung); dredged material islands and nesting birds (Dr Allan Chaney).

Texas A and M University Systems   507

Address: Texas
Status: Educational establishment with r&d capability
Chancellor: Dr Frank W.R. Hubert
Deputy chancellor for agriculture: Dr Perry L. Adkisson

PRAIRIE VIEW A & M UNIVERSITY*   508

Address: Prairie View, TX 77445

College of Agriculture*   509

Agricultural Research Center   510
Address: PO Box 2854, Prairie View, TX 77445
Telephone: (713) 857 3311
Director: Dr Ocleris Simpson
Assistant research director: Harold L. Hauser
Graduate research staff: 34
Activities: Small farm technology, dairy goats, food safety, rural development, urban agriculture.
Publications: Dairy Goat Management: Physiological Phases of Production.
Projects: Nitrogen fixation (Dr Ronald Humphrey); animal tissue (Dr Eugene Brams); dairy goat management (Dr Frank Pinkerton); boron and sulphur in soils (Dr Arthur Mangaroo); Zea Mays (development control) (Dr George Brown); pork quality feeding regime analysis (Lindsey Weatherspoon).

International Dairy Goat Research Center   511
Telephone: (713) 857 3927
Director: Dr Frank Pinkerton
Graduate research staff: 8
Activities: Generation of technical information concerning dairy goat management.
Publications: Technical bulletins; scientific journals.
Projects: Veterinary health care (Dr Aubrey Watkins); nutrition/ forages (Dr Frank Pinkerton); reproduction; environmental physiology; food technology; marketing.

TARLETON STATE UNIVERSITY*   512

Address: Stephenville, TX 76402
Telephone: (817) 968 9000

School of Agriculture and Business*   513

Dean: Dr J.L. Tackett

TEXAS A AND M UNIVERSITY   514

Address: PO Box 2854, Prairie View, TX 77445
Interim president: Charles H. Samson

Texas Agricultural Experiment Station   515
– TAES
Telephone: (713) 845 8484
Director: Dr Neville P. Clarke
Departments: Agricultural communications, Dr William E. Tedrick; agricultural economics, Professor John A. Hopkin; agricultural education, Professor Earl H. Knebel; agricultural engineering, Professor Edward A. Hiler; animal science, Professor Z.L. Carpenter; biochemistry and biophysics, Professor Eugene K. Sander; entomology, Professor Fowden G. Maxwell; forest science, Professor Wayne K. Murphey; horticultural sciences, Professor Creighton Miller; plant sciences, Professor C.R. Benedict; poultry science, Professor Willie F. Krueger; range science, Professor Joseph L. Schuster; recreation and parks, Professor Leslie Reid; rural sociology, Professor Steven Murdock; soil and crop sciences, Professor Edward C.A. Runge; wildlife and fisheries sciences, Professor Wallace G. Klussmann
Graduate research staff: 1 200
Activities: Improved production methods in agriculture aimed at combating short-and long-term crises related to energy, inflation, and limitations in land, water, and other natural resources: new crop varieties, improved water efficiency, better livestock production, disease and pest control; economic and marketing research.
Publications: Research progress reports.

Amarillo Research and Extension Center   516
Address: 6500 Amarillo Boulevard West, Amarillo, TX 79106
Telephone: (806) 359 5401
Resident research director: Dr G.B. Thompson
Graduate research staff: 20
Activities: Feeder cattle; irrigation; wheat and feedgrains; pest control.

Beaumont Research and Extension Center   517
Address: Route 7, Box 999, Beaumont, TX 77706
Telephone: (713) 752 2741
Resident director: Dr Julian P. Craigmiles
Graduate research staff: 3
Activities: Basic, field, greenhouse and laboratory research on rice and soya beans: varietal improvement, management, pest control.
Projects: Soils and plant nutrition (F.T. Turner); soya bean assessment (J.W. Sij); soya bean breeding (G.R. Bowers); disease control (N.G. Whitney); insect control (C.C. Bowling); weed control (E.F. Eastin); water management (G.N. McCauley).

Blackland Research Center   518
Address: Box 748, Temple, TX 76501
Telephone: (817) 774 1201
Director: Dr Earl Burnett
Graduate research staff: 5
Activities: Crop adaptation, soil and water research.

Chillicothe-Vernon Research and Extension   **519**
Center
Address: PO Box 1658, Vernon, TX 76384
Telephone: (817) 552 9941
Resident director: Professor Earl Gilmore
Graduate research staff: 20
Activities: Field and vegetable crops, livestock and range management.

Corpus Christi Research and Extension   **520**
Center
Address: Highway 44, PO Box 10607, Corpus Christi, TX 78410
Telephone: (512) 265 9201
Resident director: Professor George Slater
Graduate research staff: 14
Activities: Field crops; agriculture and commerce; mariculture and aquatic foods.

Dallas Research and Extension Center   **521**
Address: 17360 Coit Road, Dallas, TX 75252
Telephone: (214) 231 5362
Resident director: Professor Al Turgeon
Graduate research staff: 14
Activities: Urban agriculture; turf and landscape crops; forage and field crops.

El Paso Research and Extension Center   **522**
Address: 1380 A and M Circle, El Paso, TX 79927
Telephone: (915) 859 9111
Resident director: Professor J. Dan Hanna
Graduate research staff: 15

Lubbock Research and Extension Center   **523**
Address: Route 3, Lubbock, TX 79401
Telephone: (806) 746 6101
Resident research director: Professor Bill Ott
Graduate research staff: 24
Activities: Cotton and grain sorghum, vegetables, oilseeds and water conservation.

McGregor Research and Extension Center   **524**
Address: Box 447, McGregor, TX 76657
Telephone: (817) 840 2878
Farms manager: Samuel S. Pegues
Graduate research staff: 3
Activities: Cattle breeding.

Overton Research and Extension Center   **525**
Address: Drawer E, Overton, TX 75684
Telephone: (214) 834 6191
Resident research director: Professor William H. Smith
Graduate research staff: 16
Activities: Forage and livestock; fruits and vegetables.

San Angelo Research and Extension Center   **526**
Address: Route 2, Box 950, San Angelo, TX 76901
Telephone: (915) 653 4576
Resident research director: Professor Carl S. Menzies
Graduate research staff: 14
Activities: Sheep and goats; range management; cattle.

Stephenville Research and Extension Center   **527**
Address: Box 292, Stephenville, TX 76401
Telephone: (817) 968 4144
Resident research director: James S. Newman
Graduate research staff: 15
Activities: Field crops, peanuts, horticulture, dairying, pecans, and fruit crops.

Uvalde Research and Extension Center   **528**
Address: PO Drawer 1051, Uvalde, TX 78801
Telephone: (512) 278 9151
Resident research director: Professor Stewart H. Fowler
Graduate research staff: 16
Activities: Crops, livestock, range and wildlife.

Weslaco Research and Extension Center   **529**
Address: 2415 East Highway 83, Weslaco, TX 78596
Telephone: (512) 968 5585
Resident research director: Professor Chan C. Connolly
Graduate research staff: 16
Activities: Citrus; vegetables; field crops.

Texas Technical University   530

Address: Box 4169, Lubbock, TX 79407
Telephone: (806) 742 2808
Status: Educational establishment with r&d capability

COLLEGE OF AGRICULTURAL   531
SCIENCES.

Dean: Dr Sam E. Curl
Associate dean for research and agricultural operations: Dr Robert C. Albin
Associate dean for industry relations: Dr J. Wayland Bennett
Departments: Agricultural economics, (interim) Dr Sujit Roy; agricultural education, Dr Jerry Stockton; agricultural engineering, Professor Marvin Dvoracek; animal science and food science, Dr Jack McCroskey; entomology, Dr Darryl Sanders; plant and soil science, Dr David Koeppe; park administration and landscape architecture, Dr James Mertes; range and wildlife management, Dr Henry Wright
Graduate research staff: 84
Activities: Agricultural economics, education and engineering; animal science; entomology; food science; plant and soil science; park administration and landscape architecture; range and wildlife management.
Publications: Three annual reports: Swine Research Report (June); Research Highlights Brush Control, Wildlife and Range Management (November); Vegetable Research Report (October).

Projects: Plant stress and water conservation (Dr David Koeppe); small ruminant production (Dr Fred Bryant); feedlot cattle nutrition and management (Dr Reed Richardson); swine production and management (Dr Leyland Tribble, Dr Don Orr); vegetable production (Dr John Downes); brush control and range management (Dr Henry Wright); wildlife management (Dr Eric Bolen); big game management (Dr David Simpson); entomology and pest management (Dr Darryl Sanders); economic evaluations (Dr Sujit Roy); meat science (Dr Boyd Ramsey).

Theracon Incorporated   532

Address: PO Box 1493, Topeka, KS 66601
Telephone: (913) 286 1451
Status: Independent research centre
Director: Dr Woodrow Nelson
Director, Research and Development: Harold E. Scheid
Graduate research staff: 3
Activities: Pet food studies; domestic animal nutrition.
Projects: Preference studies; metabolism; wear-out; nutritional; nutritional adequacy studies.

Tufts University*   533

Status: Eduational establishment with r&d capability

SCHOOL OF VETERINARY MEDICINE 534

Address: 203 Harrison Avenue, Boston, MA 02111
Telephone: (617) 956 7603
Acting dean: Dr W. Robert Cook
Departments: Medicine, Dr James Ross; surgery, Dr Anthony Schwartz; comparative medicine and laboratory animal sciences, Dr Albert Jonas; preventive medicine and epidemiology, Dr Leonard Marcus; anatomy and cellular biology, Dr Karen Hitchcock; biochemistry and pharmacology; Dr Henry Mautner; molecular biology and microbiology, Dr Moselio Schaechter; pathology, Dr Martin Flax; physiology, Dr Eunice Bloomquist
Graduate research staff: 25
Activities: Research in the following areas: animal production, including breeding, husbandry, nutrition, and veterinary medical problems, with special emphasis on dairy and beef cattle, swine, sheep, and poultry; aquaculture, including breeding, nutrition, and veterinary medical problems which arise in high production environments; sports animal medicine, primarily equine and dealing with orthopaedic and gastrointestinal disorders. Research teams encompassing basic scientists as well as the clinical faculties will be encouraged. In addition, interchange is being promoted between faculties of the veterinary, medical, dental and nutrition schools.

Publications: Annual report.
Projects: Phylogeny of immune response/immunobiology of leukaemias and lymphomes (Dr Carol Reinisch); intestinal microbial model systems (Dr Andrew Onderdonk); physiology and biomechanics of the equine larynx (Dr W. Robert Cook); swine nutrition (Dr Jeffrey Erickson); induction of bone formation in the horse (Dr Gustave Fackelman); carbon fibre implants for large animal tendon repair (Dr Henry Valdez); cellular interactions in the immune response to transplantation antigens (Dr Anthony Schwartz); comparative cardiology (Dr James Ross); livestock production in third world countries (Dr Albert Sollod); locomotion and mechanical stress on bones (Dr Lance Lanyon).

Union Camp Corporation*   535

Status: Independent research organization

WOODLANDS DIVISION*   536

Address: PO Box 1391, Savannah, GA 31402
Manager research: B.F. Malac

Woodlands Research Department, Franklin*   537

Address: Franklin, VA 23851
Telephone: (804) 569 4518
Research forester: M.J. Jones

Woodlands Research Department, Rincon*   538

Address: PO Box 216, Rincon, GA 31326
Telephone: (912) 826 5556
Activities: Applied research in southern pine silviculture, genetics, soils, and pest management.

United States Department of Agriculture   539

Status: Government department

AGRICULTURAL RESEARCH SERVICE 540
– ARS
Address: Washington, DC 20250
Telephone: (301) 344 3084
Deputy Administrator: T.J. Army

Crop Production   541

Telephone: (301) 344 3912
Acting chief: Howard J. Brooks
Sections: Fruit, nut, and speciality crops, Howard J. Brooks; sugar crops production, vacant; plant physiology, vacant; small grains, Leland W. Briggle; forage and range, Gerald E. Carlson; oilseeds, Robert C. Leffel; crop mechanization, Louis A. Liljedahl; cotton, Philip A. Miller; vegetable and florist nursery crops, Anson E. Thompson

Crop Protection   542

Telephone: (301) 344 2913
Chief: Marshall D. Levin
Sections: Fruit and vegetable insects, Merrill L. Cleveland; plant pathology and nematology, William M. Dowler; insect taxonomy, John J. Drea Jr; crops insect control, Robert D. Jackson; pest management, Dr Waldemar Klassen; cotton and tobacco industries, Richard L. Ridgeway; pest control materials, Paul H. Schwartz Jr; weed control, Warren C. Shaw

Eastern Region   543

Eastern Regional Research Centre   544
Address: 600 East Mermaid Lane, Philadelphia, PA 19118
Telephone: (215) 233 6622
Director: Dr H.L. Rothbart
Activities: The aims of research are as follows: to develop new scientific knowledge, innovation, and technology in food science, chemistry, biochemistry, chemical engineering, microbiology, and related disciplines through basic, applied, and developmental research on farm commodities including milk, meat, animal hides and skins, fruits, vegetables, and other farm products.
Research covers maintenance and improvement of a safe, nutritious, high quality food supply; acquisition of basic knowledge of agricultural products, their constituents and use in food and nonfood applications; utilization of by-products, particularly potential pollutants; soil, water, and air problems; energy conservation; and development of advanced product concepts and processing technologies, especially applied to extending the quality and shelf life of foods and the production of leather.
Projects: Engineering systems analysis (James C. Craig Jr); chemical and food engineering applications (Curtis Panzer); materials handling and equipment design (Wolfgang K. Heiland); food additives (Dr Gerhard Maerker); food contaminants (Dr Walter Fiddler); microbiological safety (Dr Robert L. Buchanan); food irradiation (Dr Eugen Wierbicki); dairy research (Dr Harold M. Farrell Jr); food quality (Dr Virginia Holsinger); food chemistry (Dr Thomas A. Foglia);

meat and meat products (Dr John H. Woychik); microbial biochemistry (Dr George A. Somkuti); hide processing (Dr David G. Bailey); hides and leather modification (Dr Peter R. Buechler); renewable resources (Dr Warner M. Linfield); separations science (Robert A. Barford); biophysics (Dr James R. Cavanaugh); surface chemistry (Dr Donald G. Cornell); spectroscopy (Dr Philip E. Pfeffer); plant biochemistry and growth regulation (Dr Marvin P. Thompson); phytochemistry and plant resistance mechanisms (Dr Stanley F. Osman); quality and technology of horticultural crops (Dr Gerald M. Sapers).

Engineering Science Laboratory   545
Chief: J.C. Craig Jr

Food Safety Laboratory   546
Chief: Dr D. Thayer

Food Science Laboratory   547
Chief: Dr J.H. Woychik

Hides and Leather Laboratory   548
Chief: Dr S.H. Feairheller

Plant Science Laboratory   549
Chief: Dr D.D. Bills

**Physical Chemistry and Instrumentation   550
Laboratory**
Chief: Dr J.R. Cavanaugh

Livestock and Veterinary Sciences   551

Telephone: (301) 344 3294
Acting Chief: Dr H.G. Purchase
Sections: Poultry production, Dr H.G. Purchase; internal parasites, vacant; rural housing, vacant; domestic animal diseases, Arthur A. Andersen; diseases affecting man/animals, Ralph A. Bram; swine, Roger J. Gerrits; dairy cattle, Charles A. Kiddy; foreign and vector borne diseases, Dyarl D. King; sheep and other animals, Clair E. Terrill

North Central Region   552

National Animal Disease Center   553
Address: PO Box 70, Ames, IA 50010
Telephone: (515) 232 0250
Director: Phillip A. O'Berry
Graduate research staff: 70
Activities: The aim of the centre is to conduct basic and applied research on the diseases of livestock and poultry that are of major economic importance to agriculture. Efforts are devoted to the study of approximately 30 diseases or disease complexes affecting cattle, swine, poultry, sheep, horses and other animals. This strong and broadly based intramural research programme is supplemented by an extensive extramural programme of research projects at 24 US universities and at veterinary research institutes in Egypt, India, and Pakistan.

Projects: Bovine mastitis (J.S. McDonald); bovine brucellosis (B.L. Deyoe); reproductive diseases of cattle (J.H. Bryner); leptospirosis of cattle (A.B. Thiermann); infectious bovine keratoconjunctivitis (pinkeye) of cattle (G.W. Pugh); paratuberculosis of cattle; mycobacteriosis of swine (R.S. Merka); swine erysipelas; streptococcic lympadenitis of swine (jowl abscesses) (R.L. Wood); swine dysentery (L.A. Joens); mycoses and mycotoxises (A.C. Pier); pasteurellosis (fowl cholera); turkey airsacculitis (K.R. Rhoades); calf scours (H.W. Moon); respiratory diseases of sheep (R.C. Cutlip); respiratory and septicemic diseases of poultry (N.P. Cheville); bovine lymphosarcoma (M.J. Van Der Maaten); respiratory and reproductive diseases of swine (W.L. Mengeling); viral enteric diseases of swine (L.J. Memeny); ornithosis in poultry (J. Tessler); pseudorabies (E.C. Pirtle); respiratory diseases of cattle (K.G. Gillette); gastroenteric physiology (S.C. Whipp); mineral metabolism studies (E.T. Littledike); gastrointestinal microbiology (M.J. Allison); development of radio telemetry system for transmitting physiologic data; biological laboratory safety research (J.F. Sullivan).

Bacteriological and Mycological Research   **554**
Laboratory
Chief: Dr A.C. Pier

Pathological Research Laboratory   **555**
Chief: Dr N.F. Cheville

Physiopathological Research Laboratory   **556**
Chief: Dr A.C. Pier

Virological Research Laboratory   **557**
Chief: Dr W.L. Mengeling

Northeastern Region   558

Address: Building 003, Barc-West, Beltsville, MD 20705
Telephone: (301) 344 3418
Regional Administrator: Steven C. King

Agricultural Research Centre   **559**
Address: Building 003, Barc-West, Beltsville, MD 20705
Director: Paul A. Putnam

Analytical Chemistry Laboratory   **560**
Address: Building 306, Barc-East, Beltsville, MD 20705
Chief: Kenneth R. Hill

Biological Waste Management and Organic   **561**
Resources Laboratory
Address: Building 007, Barc-West, Beltsville, MD 20705
Chief: James F. Parr

Biologically Active Products Laboratory   **562**
Address: Building 306, Barc-East, Beltsville, MD 20705
Chief: Martin Jacobson

Insect Reproduction Laboratory   **563**
Address: Building 306, Barc-East, Beltsville, MD 20705
Chief: Alexei B. Borkovec

Livestock Insects Laboratory   **564**
Address: Building 307, Barc-East, Beltsville, MD 20705
Chief: Dora K. Hayes

Organic Chemical Synthesis Laboratory   **565**
Address: Building 307, Barc-East, Beltsville, MD 20705
Chief: Jack R. Plimmer

Soil Nitrogen and Environmental Chemistry   **566**
Laboratory
Address: Building 007, Barc-West, Beltsville, MD 20705
Chief: Alan W. Taylor

Pesticide Degradation Laboratory   **567**
Address: Building 050, Barc-West, Beltsville, MD 20705
Chief: Philip C. Kearney

Weed Science Laboratory   **568**
Address: Building 001, Barc-West, Beltsville, MD 20705
Chief: Dayton L. Klingman

Animal Parasitology Institute   **569**
Address: Building 1040, Barc-East, Beltsville, MD 20705
Chairman: Harry Herlich

Hemoparasitic Diseases Clinic   **570**
Address: Building 1072, Barc-East, Beltsville, MD 20705
Chief: Kenneth L. Kuttler

Non-Ruminant Parasitic Diseases   **571**
Laboratory
Address: Building 1040, Barc-East, Beltsville, MD 20705
Head: Kenneth D. Murrell

Parasite Classification and Distribution Unit   **572**
Address: Building 1180, Barc-East, Beltsville, MD 20705
Head: J.R. Lichtenfels

Poultry Parasitic Diseases Laboratory   **573**
Address: Building 1040, Barc-East, Beltsville, MD 20705
Chief: Michael D. Ruff

Ruminant Parasitic Diseases Laboratory   574
Address: Building 1040, Barc-East, Beltsville, MD 20705
Chief: Ronald Fayer

Animal Science Institute   575
Address: Building 200, Barc-East, Beltsville, MD 20705
Laboratory Chief: Lewis W. Smith

**Animal Improvement Programmes   576
Laboratory**
Address: Building 263, Barc-East, Beltsville, MD 20705
Chief: Frank N. Dickinson

Avian Physiology Laboratory   577
Address: Building 262, Barc-East, Beltsville, MD 20705
Chief: Aelene Cecil

Meat Science Research Laboratory   578
Address: Building 201, Barc-East, Beltsville, MD 20705
Chief: Anthony Kotula

Milk Secretion and Mastitis Laboratory   579
Address: Building 173, Barc-East, Beltsville, MD 20705
Chief: Robert H. Miller

Non-Ruminant Animal Nutrition Laboratory   580
Address: Building 200, Barc-East, Beltsville, MD 20705
Technical Advisor: Ben Bereskin

Reproduction Laboratory   581
Address: Building 200, Barc-East, Beltsville, MD 20705
Chief: Harold Hawk

Ruminant Nutrition Laboratory   582
Address: Building 200, Barc-East, Beltsville, MD 20705
Chief: Paul Moe

Horticultural Science Institute   583
Address: Building 003, Barc-West, Beltsville, MD 20705
Chairman: Albert A. Piringer Jr

Florist and Nursery Crops Laboratory   584
Address: Building 004, Barc-West, Beltsville, MD 20705
Chief: Henry M. Cathey

Fruit Laboratory   585
Address: Building 004, Barc-West, Beltsville, MD 20705
Chief: Miklos Faust

Horticulture Crops Quality Laboratory   586
Address: Building 002, Barc-West, Beltsville, MD 20705
Chief: Robert E. Hardenhurg

Instrumentation Research Laboratory   587
Address: Building 002, Barc-West, Beltsville, MD 20705
Chief: Karl H. Norris

Postharvest Physiology Laboratory   588
Address: Building 002, Barc-West, Beltsville, MD 20705
Chief: Morris Liebermann

Vegetable Laboratory   589
Address: Barc-West, Beltsville, MD 20705
Chief: R.E. Webb

**Insect Identification and Beneficial Insect   590
Introduction Institute**
Address: Building 003, Barc-West, Beltsville, MD 20705
Chairman: Lloyd V. Knutson

Beneficial Insect Introduction Laboratory   591
Address: Building 417, Barc-East, Beltsville, MD 20705
Chief: Jack R. Coulson

Glenn Dale Plant Materials Research   592
Address: Plant Introduction Station, PO Box 88, Glenn Dale, MD 20769
Research horticulturalist: B.J. Parilman

Systematic Entomology Laboratory   593
Address: Building 003, Barc-West, Beltsville, MD 20705
Chief: Paul M. Marsh

Plant Genetics and Germplasm Institute   594
Address: Barc-West, Beltsville, MD 20705
Chairman: John G. Moseman

Economic Botany Laboratory   595
Address: Barc-East, Beltsville, MD 20705
Chief: James A. Duke

Field Crops Laboratory   596
Address: Barc-West, Beltsville, MD 20705

Germplasm Resources Laboratory   597
Address: Barc-West, Beltsville, MD 20705
Chief: Howard E. Waterworth

Plant Taxonomy Laboratory   598
Address: Barc-West, Beltsville, MD 20705
Chief: Robert E. Perdue Jr

Seed Research Laboratory   599
Address: Building 006, Barc-West, Beltsville, MD 20705
Chief: Aref A. Abdul-Baki

Tobacco Laboratory   **600**
Address: Barc-West, Beltsville, MD 20705
Chief: Tien C. Tso

Plant Physiology Institute   **601**
Address: Building 001, Barc-West, Beltsville, MD 20705
Laboratory Chief: M.N. Christiansen

Agricultural Equipment Laboratory   **602**
Address: Building 303, Barc-East, Beltsville, MD 20705
Chief: Lowell E. Campbell

Cell Culture and Nitrogen Fixation   **603**
Laboratory
Address: Building 011A, Barc-West, Beltsville, MD 20705
Chief: Gideon W. Schaeffer

Hydrology Laboratory   **604**
Address: Building 007, Barc-West, Beltsville, MD 20705

Light and Plant Growth Laboratory   **605**
Address: Building 046A, Barc-West, Beltsville, MD 20705
Chief: N. Jerry Chatterton

Plant Hormone and Regulators Laboratory   **606**
Address: Building 050, Barc-West, Beltsville, MD 20705
Chief: George L. Steffens

Plant Stress Laboratory   **607**
Address: Building 001, Barc-West, Beltsville, MD 20705
Chief: William Wergin

Water Data Laboratory   **608**
Address: Building 007, Barc-West, Beltsville, MD 20705
Head: James B. Burford

Plant Protection Institute   **609**
Address: Building 011A, Barc-West, Beltsville, MD 20705
Chairman: Burton Endo

Applied Plant Pathology Laboratory   **610**
Address: Building 004, Barc-West, Beltsville, MD 20705
Chief: Athey G. Gillaspie Jr

Bioenvironmental Bee Laboratory   **611**
Address: Building 476, Barc-East, Beltsville, MD 20705
Chief: H. Shimanuki

Insect Pathology Laboratory   **612**
Chief: James Vaughn

Insect Physiology Laboratory   **613**
Chief: James Svoboda

Mycology Laboratory   **614**
Chief: Paul Lentz

Nematode Laboratory   **615**
Chief: Raymond Rebois

Plant Virology Laboratory   **616**
Chief: Russell Steere

Soilborne Diseases Laboratory   **617**
Chief: George Papavizas

United States Regional Pasture Research   **618**
Laboratory
Address: Curtin Road, University Park, PA 16802
Location leader: William C. Templeton Jr

Northern Region   619

Northern Regional Research Centre   **620**
Address: 1815 North University Street, Peoria, IL 61604
Telephone: (309) 685 4011
Director: William H. Tallent
Graduate research staff: 160
Activities: Chemical, physical, engineering, biochemical, and microbiological research on corn, wheat, sorghum, oats, soya beans, and horticultural and special crops, to provide information leading to improvements in production, processing, and uses of these and other commodities. Among specific projects are ones designed to increase yields and decrease losses (notably by insect damage) both before and after harvest, to reduce processing costs and energy consumption and enhance food safety and quality, and to find uses for agricultural by-products and residues.
Publications: Annual report; list of publications and patents, annually.
Projects: Biochemical engineering (Dr R.J. Bothast); chemical engineering (G.E. Hamerstrand); field crops post-harvest research (R.A. Anderson); hydrocarbon plants and biomass (M.O. Bagby); physical chemistry (Dr W.M. Doane); renewable resources (F.H. Otey); cereal and food biochemistry (Dr F.R. Dintzis); cereal proteins research (Dr J.S. Wall); food physical chemistry (Dr E.B. Bagley); agricultural microbiology (Dr R.W. Detroy); Agricultural Research Service culture collection research (Dr C.P. Kurtzman); microbiological biochemistry (Dr M.E. Slodki); mycotoxin analytical and chemical research (Dr O.L. Shotwell); mycotoxin microbiology and biochemistry (M.D. Grove); composition and characterization (Dr C.R. Smith Jr); instrumental analysis (R. Kleiman); natural toxicants (Dr H.L. Tookey); plant biochemistry and photosynthesis (Dr J.A. Rothfus); biochemical and biophysical properties (Dr E.A. Emken); exploratory organic reactions (Dr E.H. Pryde); lipid chemistry (Dr E.N. Frankel); oilseed meal products (Dr W.J. Wolf).

Biomaterials Conversion Laboratory   **621**
Chief: Dr W.M. Doane

Cereal Science and Foods Laboratory   **622**
Chief: Dr G.E. Inglett

Fermentation Laboratory   **623**
Chief: Dr C.W. Hesseltine

Horticultural and Special Crops Laboratory   **624**
Chief: Dr L.H. Princen

Northern Agricultural Energy Centre   **625**
Manager: M.O. Bagby

Oilseed Crops Laboratory   **626**
Chief: T.L. Mounts

Post Harvest Science and Technology   627

Telephone: (301) 344 4075
Acting Chief: Andrew M. Cowan
Sections: Animal products, Richard H. Alsmeyer; processing fibres, Nelson F. Getchell; processing field crops, Wilda H. Martinez; transportation and facility, Roy E. McDonald; health and safety, Jane F. Robens

Soil, Water, and Air Sciences   628

Telephone: (301) 344 3648
Chief: Harold L. Barrows
Sections: Watershed hydrology, David A. Farrell; soil fertility and plant nutrition, Ronald F. Follett; water management, Dr Marvin E. Jensen; environmental quality, Jesse Lunin; energy, Steve Rawlins; aerospace technology, Jerry C. Ritche

Southern Region   629

Southern Regional Research Centre   **630**
Address: PO Box 19687, 1100 Robert E. Lee Boulevard, New Orleans, LA 70179
Telephone: (504) 589 7511
Telex: 810-951-5112
Acting Director: Dr Ivan W. Kirk
Sections: Cotton quality research, Dr H.H. Ramey Jr
Graduate research staff: 119
Activities: Research to extend basic knowledge of cotton fibre structure and elucidate mechanisms of cellulose reactions; investigation of ultrastructure and cytology of oilseeds, characterization of mechanisms of oilseed maturation, and determination of physicochemical characteristics of oilseed protein; development of energy-efficient and environmentally sound methods for processing cotton, cottonseed, peanuts, rice and sugarcane; quality, safety and nutritive value protection of food and feed; analysis of mycological flora of southern crops, interactions between microorganisms and host plants and/or seeds such as corn and cottonseed, and mode of contamination of southern crops by Fusarium

and other fungi; development of methods for improving catfish quality and recovery and conversion of catfish processing wastes; and studies of mechanisms of weed seed dormancy, germination, and control.
Publications: Abstract, biennial.
Projects: Cytochemical research (Dr T.J. Jacks); materials research (Dr R.J. Berni); plant products analysis research (Dr F.G. Carpenter); protective finishes research (G.L. Drake); special products research (Dr R.J. Harper Jr); polymer finishes research (Dr S.L. Vail); fabric engineering and development research (R.M.H. Kullman); agri-particulates analysis and control research (Dr J. Montalvo); fibre and yarn technology research (G.L. Louis); advance systems and technologies research (Dr D.P. Thibodeaux); textile chemical engineering research (Dr J.P. Neumeyer); food products research (J.J. Spadaro); oilseed products research (S.P. Koltun); natural polymer structure research (Dr S.P. Rowland); polymer products research (Dr C.M. Welch); physical-chemical research (Dr Ruth R. Benerito); food flavour research (Dr A.J. St Angelo); oils and carbohydrates research (Dr F. Parrish); biochemistry research (Dr R.L. Ory); protein products research (Dr J.P. Cherry); food and feed safety research (Dr A. Ciegler); cotton quality research (Dr H.H. Ramey).

Composition and Properties Laboratory   **631**
Chief: Dr R.J. Berni

Cotton Textile Chemistry Laboratory   **632**
Chief: Dr S.L. Vail

Cotton Textile Properties Laboratory   **633**
Chief: A. Baril Jr

Engineering and Development Laboratory   **634**
Chief: A. Graci

Natural Polymers Laboratory   **635**
Acting Chief: Dr R.R. Benerito

Oilseed and Food Laboratory   **636**
Acting Chief: Dr A. Ciegler

Western Region   637

AGRICULTURAL RESEARCH SERVICE   **638**
Address: 800 Buchanan Street, Berkeley, CA 94710
Telephone: (415) 486 3661
Director: Arthur I. Morgan Jr
Associate director: R.P. Murrmann
Sections: Cereals, R.M. Saunders; engineering, A.I. Morgan Jr (acting head); food proteins, D.D. Kasarda; food quality, R. Teranish; wool, A. Pittman (acting head); food technology, C.C. Huxsoll; plant physiology and chemistry, G. Fuller; chemical and structural analysis, R. Lundin (acting head); nutrients, A.A. Betschart; natural products chemistry, L. Jurd; plant protection phytochemistry, A.C. Waiss; toxicology, J.C.

Smith*Graduate research staff:* 120
Activities: Major problem areas of emphasis range from basic research supporting agriculture production to research on post-harvest technology. Example areas of emphasis are plant protection, plant bioscience, nutrition, land safety energy conservation and production, food processing technology, wool processing technology. The programme is national in scope, but concentrates on commodities and problems unique to western agriculture in the USA.

United States Food and Drug Administration 639

– FDA
Address: Washington, DC

BUREAU OF FOODS 640

Address: 200 C Street Southwest, Washington, DC
Telephone: (202) 245 8850
Telex: 7108229530 FDA-WSH
Status: Official research centre
Director: Dr Sanford A. Miller
Sections: Planning and operations, Bradley Rosenthal; toxicological sciences, Dr Albert C. Kolbye; nutrition and food sciences, Dr Allan L. Forbes; compliance, Taylor M. Quinn; physical sciences, Dr Robert M. Schaffner
Graduate research staff: 450
Activities: Regulatory science including development of analytical and toxicological methods; development of food guidelines and standards; risk assessment methods.
Projects: Food additives (Gerad McCowin); chemical contaminants (Dr Charles F. Jelinek); food sanitation (Dr R.B. Real Jr); nutrition (Dr John Vanderveen); consumer sciences and food economics (Dr Raymond Schuckar); cosmetics (Heinz J. Eiermann); animal drugs (Dr Robert Scheuplein).

United States Sugar Corporation Research Department* 641

Address: PO Drawer 1207, Clewiston, FL 33440
Telephone: (813) 983 8121
Status: Official research centre
Research director: Dr Edwin H. Todd
Activities: Studies on sugarcane varieties; soils and fertilizers for sugarcane production.

Universal Foods Corporation 642

Address: 433 East Michigan Street, Milwaukee, WI 53202
Telephone: (414) 271 6755
Publications: Annual report.

ROGERS FOODS 643

Address: 2020 West Main, Turlock, CA 95380
Telephone: (209) 667 2777
Status: Research centre within an industrial company
Director of seed research and development: Robert M. Rice
Graduate research staff: 7
Activities: Plant production; horticulture (garlic, onion, chilli) and processing of these products; seed research and development.
Projects: Onion breeding (Kieth W. Trammell); chilli pepper breeding (Dr Steven Czaplewski).

TECHNICAL CENTER 644

Address: 6143 North 60 Street, Milwaukee, WI 53218
Telephone: (414) 271 6755
Telex: 26842
Status: Research centre within an industrial company
Vice-president, research: Dr Gary W. Sanderson
Sections: Microbial and biochemical research, Dr Tilak W. Nagodawithana; engineering development, Robert F. Dale; cereal technology, Elmer J. Cooper; analytical laboratory, William Luth; cheese research, Dr Richard E. Willits
Graduate research staff: 22
Activities: Food science: basic and applied research in fermentation sciences, bakers yeast, wine yeast, distillers yeast, nutritional yeast and invertase production; applied research in natural cheese production, new cheese substitute products; applied research in chilli, onion and garlic horticulture; fermentation processes for use in the manufacture of fuel alcohol.
Projects: Yeast for alcohol production (Dr S.L. Chen); instant active dry yeast (Dr Lawrence Skogerson); substitute cheese products and processes (Dr Richard E. Willits); vegetable proteins (Dr Vijay Kumar Sood).

University of Alaska* 645

Status: Educational establishment with r&d capability

ALASKA AGRICULTURAL EXPERIMENT STATION* 646

Address: Fairbanks, AK 99701
Activities: Red meats production from swine and cattle, dairy science, and soils and climatic resources. Research units are maintained in Homer and Palmer.

University of Arizona 647

Address: Tucson, AZ 85721
Status: Educational establishment with r&d capability

COLLEGE OF AGRICULTURE 648

Telephone: (602) 626-2711
Dean: Dr B.P. Cardon

Arizona Agricultural Experiment Station 649

Director: Professor L.W. Dewhirst
Graduate research staff: 85
Activities: Research in the following areas: soil science; drainage and irrigation; land use; plant production - cotton, small grains, alfalfa; plant development - guayule, guar, jojoba; plant improvement - safflower, sorghum, soya beans; animal production, breeding and nutrition, dairy beef; veterinary medicine; agricultural engineering and building; food science; wildlife and fisheries; range management.
Projects: The station has 125 full-time scientists with 224 projects on budgeted funds. The overriding concern is maintaining an agricultural productivity and income base in an arid climate with a diminishing water supply. Research is aimed at development of varieties of traditional crops that are drought and salt tolerant. In addition, considerable effort is given to the domestication of alternative crops which are low water consuming while producing a useful agricultural product.

Agricultural Economics Department 650
Head: Professor Jimmye S. Hillman

Agricultural Education Department 651
Head: Professor Floyd G. McCormick Jr

Animal Sciences Department 652
Head: Professor C. Brent Theurer

Entomology Department 653
Head: Professor George W. Ware

Nutrition and Food Science Department 654
Head: Professor Darrel E. Goll

Plant Pathology Department 655
Head: Professor Merritt R. Nelson

Plant Sciences Department 656
Head: Professor LeMoyne Hogan

School of Home Economics 657
Head: Professor Robert R. Rice

School of Renewable Natural Resources 658
Head: Professor Erven H. Zube

Soils, Water, and Engineering Department 659
Head: Dr Wilford R. Gardner

Veterinary Science Department 660
Head: Professor John C. Mare

University of Arkansas 661

Address: Fayetteville, AR 72701
Status: Educational establishment with r&d capability

DIVISION OF AGRICULTURE 662

Agriculture Experiment Station 663

Address: Room 217, Agriculture Building, Fayetteville, AR 72701
Telephone: (501) 575 4446
Director: Dr L.O. Warren
Activities: Research in the following areas: soil science; drainage and irrigation; plant production including rice, soya beans, wheat, cotton, forages, horticultural and forestry; animal production including poultry, beef, dairy swine; agricultural engineering and building; freshwater fisheries; food science; forestry and forest products.
Projects: The station has approximately 350 research projects in the relevant areas of interest to the state of Arkansas, where the major crops and livestock are as follows: soya beans, cotton, rice, wheat, forage and other grain; tomatoes, various types of beans, spinach and other vegetable crops; apples, peaches and other horticultural crops; poultry, turkeys, beef, dairy cattle, and swine

Department of Agricultural Economics and Rural Sociology 664
Head: Dr H.J. Meenen

Department of Agricultural Engineering 665
Head: B.B. Bryan

Department of Agronomy 666
Acting Head: Dr J.T. Gilmour

Department of Animal Sciences 667
Head: Dr F.W. Kellogg

Department of Entomology 668
Head: Dr G.J. Musick

Department of Forestry   **669**
Head: Dr B.G. Blackmon

Department of Home Economics   **670**
Head: Dr D.A. Larery

Department of Horticulture and Forestry   **671**
Head: Dr G.A. Bradley

Department of Horticultural Food Science   **672**
Head: Dr A.A. Kattan

Department of Plant Pathology   **673**
Head: Dr D.A. Slack

Northeast Research and Extension Centre   **674**
Director: Dr A.M. Simpson Jr

Rice Research and Extension Centre   **675**
Resident Director: Francis J. Williams

Southeast Research and Extension Centre   **676**
Director: Dr Gerald W. Brown

Southwest Research and Extension Centre   **677**
Resident Director: Dr W.C. Loe

University of California*   678

– UCLA
Address: Los Angeles
Status: Educational establishment with r&d capability

UNIVERSITY OF CALIFORNIA,   679
BERKELEY*

Status: Educational establishment with r&d capability

California Agricultural Experiment   680
Station

Address: 317 University Hall, Berkeley, CA 94720
Telephone: (415) 642 9300
Director: Lowell N. Lewis
Assistant to Director: Paul Casamajor
Graduate research staff: over 800
Activities: Nearly 1 200 research projects are carried out by the station, which has 60 departments on three campuses of the university. Subjects covered include: natural resources - soil science, drainage and irrigation, land use, water; plant production (over 200 types of crops) - breeding, protection, crop husbandry, plant pathology; animal production (beef cattle, dairy cattle, and sheep) - breeding, husbandry and nutrition, veterinary medicine, pathology; agricultural engineering and building; fisheries; food science and nutrition; forestry and forest products; entomology.
Publications: California Agriculture, monthly.

Forest Products Laboratory   681

Address: 47th and Hoffman Streets, Richmond, CA 94804
Telephone: (415) 231 9456
Director: Donald G. Arganbright
Graduate research staff: 24
Activities: Research, education and public service to achieve the maximum benefit to the economy from the harvest and use of forest crops through: developing processes and designs that will permit the application and use of wood at minimum cost to the user; helping the forest products industry in its efforts to meet environmental responsibilities during harvesting, processing, and marketing of wood and in the disposal of residues; mobilizing the knowledge and experience of disciplines in the university other than wood science and technology on the problems of wood processing and use; providing training for young men and women entering careers in the forest products industries; adding to the sum of knowledge about wood; transmitting findings to the public and interested users through an efficient system of communication.
Publications: Annual report.
Projects: Mechanical properties (Dr Arno P. Schniewind); wood and extractive chemistry (Dr Eugene Zavarin); physical processing and properties (Dr Donald G. Arganbright); pulp and paper chemistry (Dr David L. Brink); forest products pathology (Dr W. Wayne Wilcox); composite wood products (Dr George Grozidits); mechanical processing (Dr Barney Klamecki); wood quality; wood in use (William A. Dost).

UNIVERSITY OF CALIFORNIA, DAVIS   682

Status: Educational establishment with r&d capability

International Programs   683

Address: 275 Mark Hall, Davis, CA 95616
Telephone: (916) 752 7071
Telex: 910-531-0785
Associate Dean: David W. Robinson
Graduate research staff: 200
Activities: Agricultural research and development projects in cooperation with lesser developed countries.
Publications: Annual project reports.
Projects: USAID/Agricultural Development Systems Project Egypt - agricultural development, systems and analysis, technical information delivery (Frank C. Child); rice research and training project (Maurice Peterson); small ruminants collaborative research support programme - animal production, health, reproduction of sheep and goats in Brazil, Indonesia, Kenya, Morocco, and Peru (David W. Robinson).

UNITED STATES OF CALIFORNIA, RIVERSIDE*   684

Status: Educational establishment with r&d capability

Dry Lands Research Institute   685

Address: Riverside, CA 92521
Telephone: (714) 787 4773
Director: Stahrl W. Edmunds
Graduate research staff: 8
Activities: Research on regional economic development, especially with relation to natural, agricultural, and human resources. The studies have a broad inter-disciplinary base and research teams in agricultural production, resource economics, and management.
Publications: Annual report.
Projects: Regional economic development and employment planning for Riverside and San Bernardino counties; strengthening the administrative capability of governments to manage development projects in developing countries (Stahrl W. Edmunds); revisions and updating of 1977 national, state and county input-output tables (Everard Lofting).

University of Connecticut   686

Address: U-136 Storrs, CT 06268
Status: Educational establishment with r&d capability

COLLEGE OF AGRICULTURE AND NATURAL RESOURCES   687

Dean: E.J. Kersting

Storrs Agricultural Experiment Station   688

Telephone: (203) 486 2917
Director: E.J. Kersting
Graduate research staff: 69
Publications: Annual research report.

Department of Agricultural Engineering   689
Head: Dr R.A. Aldrich
Activities: Research into controlled environmental food systems.
Projects: Bio-physical factors affecting energy requirements for poultry production (R.P. Prince, P.E. Stake); systems design for controlled environment plant growth (J.W. Bartok, Dr R.A. Aldrich, R.A. Ashley).

Department of Agricultural Engineering and Rural Sociology   690
Acting Head: Dr M.W. Kottke
Activities: Research into land and water use.
Projects: Population trends in Connecticut and the Northeast (T.E. Steahr, W.H. Groff, K.P. Hadden); outdoor recreation, tourism and public interest (M.W. Kottke); forest management, marketing, and small industry development pilot programme (J.R. Bentley); econometric submodel for New England Agriculture (T.C. Lee, S.K. Seaver, R.O.P. Farrish); socioeconomic factors and rural land use (R.L. Leonard, I.F. Fellows); economic impact of 200-mile limit on rural northeast coastal communities (M.A. Altobello, R.O.P. Farrish); improving the distribution of socioeconomic resources in rural areas (K.P. Haddea); economic analysis of contract production and marketing of broilers and eggs in the northeast; spatial organization of the dairy industry in Northeast (D.G. Stitts); socioeconomic perspectives of small farmers (W.H. Graff).

Department of Animal Genetics   691
Head: Dr L.J. Pierro
Activities: Research in teratology.
Projects: Genetic control of cadmium susceptibility (Dr L.J. Pierro); stress response compounds in adverse biological activity and improved resistance in potatoes; serum proteins in development of embryo culture in the surveillance for congenital abnormalities (Dr N.W. Klein); genetic control of connective tissue metabolism (Dr P.F. Goetinck).

Department of Animal Industries   692
Head: Dr W.A. Cowan
Activities: Research in the areas of dairy food and technology and animal husbandry.
Projects: More efficient methods of beef production; endocrine relationships in the reproductive performance of sheep; control of reproduction in the bovine female (C.O. Woody Jr, Dr W.A. Cowan); development of puberty in males and control of spermatogenesis (C.O. Woody Jr, J.W. Riesen).

Department of Nutritional Sciences   693
Head: Dr K.L. Knox
Activities: Research into high protein diets; vitamin studies.
Projects: Mode of inheritance and gene actions of poultry mutations (Dr R.G. Somes Jr); improving the assurance of quality and safety of consumers' food (P.E. Stake); concentrations of prostaglandins, serum triglyceride, chloresterol and fatty acid profile (E.J. McCosh-Lilie); genetic-physiological mechanisms associated with nutrient utilization in poultry (Dr R.G. Somes Jr); human nutrition improvement (Dr R.G. Jensen, A.M. Ferris); vitamin A-nutriture and nuclear binding of 14C and 3H - labelled retinoids in rat testes (Dr K.L. Knox); role of insulin in renal calcium reabsorption (L.H. Allen); influence of excess vitamin E and

other compounds on the absorption of fat-soluble vitamins (W.J. Pudelkiewicz).

Department of Pathobiology   694

Acting Head: Dr L.J. Pierro
Activities: Research into mastitis and poultry viruses.
Projects: Brucellosis; pullorum disease control; control of infectious mastitis in dairy cattle (Dr L.J. Pierro); animal diseases survey (E.S. Bryant, D.S. Wyand, L. van der Heide); infectious diseases affecting reproduction in dairy cattle (Dr M.E. Tourtellotte); gastroenteric diseases of dogs and other animals; enteric diseases in adult cattle (H.J. Van Kruiningen); avian adenoviruses (Dr C.N. Burke, L. van der Heide); mycoplasmosis in calves (Dr M.E. Tourtellotte, Dr S.W. Nielsen); hematological diseases (Dr T.N. Fredrickson); cells involved in spontaneous regression of tumours; resistance to mastitis in dairy cattle Dr T.J. Yang).

Department of Plant Science   695

Acting Head: Dr R.W. Wengel
Activities: Research in agronomy, environmental horticulture, and landscape design.
Projects: Use of herbicides in forage crop management; weed-crop competition (Dr R.A. Peters); selection and propagation of new and unique forms of native conifers from witches'-broom progeny (Dr S. Waxman); breeding cool season forage species for improved feed value and productivity (Dr W.W. Washko); breeding and evaluation of new potato clones and varieties (R.A. Ashley); forage-livestock systems for land with soil and site limitations (D.W. Allinson); cytogenetics, hybridization behaviour, and horticultural improvement of vinca (R.D. Parker); origins, transformation and management of nitrogen in soils, waters, and plants (G.F. Griffin); pesticide impact assessment (Dr M.G. Savos); environmental planning (Dr W.C. Kennard, M.W. Lefor); characterization of soils formed on glacial tills (H.D. Luce); impact of climatic variability on agriculture (Dr W.C. Kennard).

Department of Renewable Natural Resources   696

Head: Dr J.E. Bethune
Activities: Research in forest economics.
Projects: Local economic impacts of forestry activities; wood removal impacts on small private nonindustrial forests (Dr J.E. Bethune, D.R. Miller, W.R. Bentley); forest microclimate and water use characteristics; dynamics and energetics of soil-plant-atmosphere continuum (D.R. Miller).

University of Delaware*   697

Address: Newark, DE 19711
Status: Educational establishment with r&d capability

COLLEGE OF AGRICULTURE   698

Dean: Dr Donald F. Crossan

Delaware Agricultural Experiment   699
Station

Telephone: (302) 738 2501
Director: Dr Donald F. Crossan
Graduate research staff: 56
Activities: Research in the following areas: soil science; drainage and irrigation; land use; corn and soya bean production - plant breeding, crop husbandry, plant protection; poultry, dairy, and swine production - husbandry and nutrition; agricultural engineering and building; forestry; food economics.
Publications: Annual report.

Department of Agricultural and Food   700
Economics

Head: Dr Raymond Smith

Department of Agricultural   701
Engineering

Head: Dr Norman Collins

Department of Animal Science and   702
Agricultural Biochemistry

Head: Dr John Rosenberger

Department of Entomology and   703
Applied Ecology

Head: Dr Dewey Caron

Department of Plant Science   704

Head: Dr Charles Curtis

University of Georgia   705

Address: Athens, GA 30602
Status: Educational establishment with r&d capability

COLLEGE OF AGRICULTURE   706

Dean: William P. Flatt
Director, Experiment Stations: E.B. Browne
Activities: The five main aims of research in the College's experiment stations are as follows: more efficient use of energy in agriculture; more efficient use of water resources; more efficient production, processing, and marketing of crops, livestock, and poultry; improved

environmental quality with less soil erosion and more economic control of insects, diseases, and wheats; improved safety and nutritional value of foods.

Publications: Georgia Agricultural Experiment Stations Annual Report.

Projects: The main research priorities of the College's experiment stations are as follows:

Agronomy - improved efficiency of crop management; development of new crop varieties; more efficient land use and management; improvement of feed and forage production; new programme for testing new varieties.

Horticulture - improved yield of high-quality fruits and nuts; development of new varieties of sweet potatoes; reduction of peach tree losses; more efficient production of vegetable crops; increased profit potential of the floriculture - ornamental plant industry.

Forestry - timber management; forest resource management; hydrology, watershed management, and micrometeorology; impact of forest environment on fishes and wildlife; improved wood products development.

Poultry science - improved efficiency of meat and egg production; improved poultry product quality; improved energy utilization; disease control; decrease economic losses from mycotoxins in feed.

Animal science - evaluation of forage crops; improved utilization of forages; mutation of brood sows; improved efficiency of livestock production.

Entomology - biology of important insects, ticks, and mites; utilization of natural or biological control agents, development of precise methods for detecting and predicting insect occurrence, movement, and damage; precision in selecting and timing applications and dosages of insecticides; pesticide resistance. Plant pathology - information on plant disease organisms; genetic disease resistance; economical and safe usage of chemicals for disease and nematode control; basic biology of plant pathogens; biological control of plant diseases.

Food science - quality improvement by identifying and removing hazardous and toxic agents; reduction of food cost by energy-efficient methods of processing and preservation; increased knowledge of food production; factors influencing food consumption.

Nutrition - factors influencing consumer acceptance of foods; nutritional factors affecting human illnesses; nutrition research in the areas of lipid and protein synthesis; metabolic relationships between protein intake and kidney disease; genetic basis of carbohydrate-induced lipaernia and glycaemia.

Agricultural engineering - reducing petroleum consumption in agriculture; irrigated multi-cropping systems development; agricultural structures for livestock and poultry production systems; technology, designs, and equipment for harvesting, handling, and processing woody plants, fruits, vegetables, and nuts for energy production; on-farm production of power alcohol.

Agricultural economics - improved profitability through economic and organizational efficiency; affect of socioeconomic changes on food demand and consumption patterns; changes in rural and urban institutions; effectiveness of coordination and exchange arrangements for agricultural commodities.

Coastal Plain Station 707

Address: Tifton, GA 31794
Telephone: (912) 386 3338
Director: vacant

Southeast Georgia Branch Station 708
Address: Midville, GA 30441
Telephone: (912) 589 7472
Superintendent: Charles E. Perry

Southwest Georgia Branch Station 709
Address: Plains, GA 31780
Telephone: (912) 824 4375
Superintendent: Robert B. Moss

College Experiment Station 710

Address: 107 Conner Hall, Athens, GA 30602

Animal Sciences Department 711
Senior staff: M.G. McCartney; animal/dairy science, D.G. Spruill

Foods and Economics Department 712
Senior staff: Agricultural economics, S.J. Brannen; food science, vacant; home economics, Emily Q. Pou

Plant Protection and Engineering Deparment 713
Senior staff: Plant pathology, W.N. Garrett; entomology, T.D. Canerday; agricultural engineering, R.H. Brown

Plant Sciences Department 714
Senior staff: Agronomy, W.L. Colville; horticulture, C.H. Hendershott; forestry, L.A. Hargreaves

Georgia Experiment Station * 715

Address: Experiment, 30212
Telephone: (404) 228-7263
Associate Director: Curtis R. Jackson
Graduate research staff: 62
Activities: Agronomic and horticultural plant production and harvesting; food and feed processing; swine and beef cattle production; production and marketing economics.
Projects: The College is involved in approximately 86 funded projects. These are in the areas of small grains, food demand, weed control, control of disease of forage and feed grain crops, nutritional studies on apples, minimum-tillage multiple-cropping systems, turfgrass management.

Agricultural Economics Department*   **716**
Head: J.C. Purcell

Agricultural Engineering Department*   **717**
Head: B.P. Verma

Agronomy Department*   **718**
Head: O.E. Anderson

Animal Science Department*   **719**
Head: M.E. McCullough

Central Georgia Branch Station   **720**
Address: Eatonton, GA 31024
Telephone: (404) 845 6015
Superintendent: Grady V. Calvert

Entomology Department   **721**
Head: J.O. Howell

Food Science Department   **722**
Head: T. Nakayama

Georgia Mountain Branch Station   **723**
Address: Blairsville, GA 30512
Telephone: (404) 745 2655
Superintendent: James W. Dobson

Horticulture Department   **724**
Head: B.B. Brantley

Northwest Georgia Branch Station   **725**
Address: Calhoun, GA 30701
Telephone: (404) 629 2696
Superintendent: Edward G. Worley

Plant Pathology Department   **726**
Head: J.T. Walker

University of Guam   727

Status: Educational establishment with r&d capability

COLLEGE OF AGRICULTURE AND   728
LIFE SCIENCES

Agricultural Experiment Station   729

Address: UOG Station, Mangilao, GU 96913
Telephone: 734 3113; 734 2579; 734 2575
Telex: 6275
Dean and Director: Wilfred P. Leon Guerrero
Associate Dean: Dr R. Muniappan
Activities: Research in the following areas: winged
bean potential as a crop; varietal performance studies on
major vegetable crops; evaluation of different methods
for production of ornamental plants; selection, introduction, and breeding of tropical fruit crops; determination
of plant diseases; soil factors and soil-crop interactions

to suppress diseases caused by soil-borne plant
pathogens; development of integrated pest management
systems; fate of aided pesticides; testing the efficacy of
Sevin and other alternate insecticides for control of
sweet potato weevils; impact of insecticides on pole bean
yield and natural enemies of Liriomyza sativae; selection, introduction and breeding of tropical fruit crops;
use of local feedstuffs available for swine production; soil
science; irrigation; soil fertility.
Projects: Varietal performance studies on major vegetable crops, winged bean potential as a crop; selection,
introduction, and breeding of tropical fruit crops; trickle
irrigation to improve crop production and management
(Dr Chin-Tiam Lee); varying plant populations on
growth, yield, and quality of vegetable crops; evaluation
of cultural methods for production of ornamental plants
(Dr Syamal K. Sengupta); soil fertility survey (Dr Jefren
L. Demeterio); determination of plant diseases (Dr
Richard G. Beaver); biological control in pest management systems (Dr James R. Nechols); development of
integrated pest management systems (Dr James R.
Nechols); soil factors and soil-crop interaction to suppress diseases (Dr Richard G. Beaver); fate of added
pesticides in Guam (Dr Jefren L. Demeterio); testing the
efficacy of Sevin and other alternate insecticides for
control of sweet potato weevils (Mr Edwin Pickop);
impact of insecticides on pole bean yield and natural
enemies of Liriomyza sativae (Dr Donald M. Nafus, Dr
Ilse H. Schreiner); nitrogen metabolism of fresh-water
prawns in relation to diet (Dr Stephen G. Nelson); use of
local feedstuffs available for swine production (Dr
Anastacio Palafox); genetic assessment of
Macrobrachium (Dr Daniel B. Matlock).

University of Hawaii at   730
Manoa

Status: Educational establishment with r&d capability

COLLEGE OF TROPICAL   731
AGRICULTURE AND HUMAN
RESOURCES

Hawaii Institute of Tropical Agriculture and Human Resources 732

Address: 3050 Maile Way, Honolulu, HI 96822
Telephone: (808) 8131
Telex: UNIHAW
Director: Dr N.P. Kefford
Departments: Agricultural and resource economics, Chauncey Ching; agricultural biochemistry, John Hylin; agricultural engineering, M. Ray Smith; agronomy and soil science, Peter Rotar; animal sciences, Richard Stanley; botany, Bruce Cooil; entomology, John Beardsley; horticulture, Roy Nishimoto; human resources, Betsy Bergen; plant pathology, Oliver Holtzmann
Activities: Research in the following areas: natural resources - soil science, land use, water and watersheds; plant production (sugar, pineapple, corn, vegetables, fruits and nuts, horticultural plants, forage, range and pasture) plant breeding and genetics, crop husbandry and nutrition, plant protection, harvesting and processing, waste management, marketing; animal production (beef, dairy, swine, poultry) - livestock breeding and genetics, livestock husbandry and nutrition, veterinary medicine, waste management, marketing; agricultural engineering; fresh-water aquaculture; food science and human nutrition; human resources - home economics, textiles and clothing, human development.
Projects: Soil-plant water and nutrition relations (G. Uehara, S. El-Swaify); control of insects in - fruit and vegetable crops (J. Beardsley), field crops and range (R. Namba), horticultural crops (R. Mau); control of disease in fruit and vegetable crops (A. Alvarez), field crops and range (W. Apt), horticultural plants (M. Aragaki); improved biological efficiency of fruit and vegetable crops (M. Awada), field crops and range (J. Bowen), animal production (J. Carpenter); new and improved fruit and vegetable crops (H. Nakasone), horticultural plants (H. Kamemoto), field crops (J. Brewbaker); mechanization in production and harvesting of fruit and vegetable crops (J.K. Wang), field crops (T. Leung), horticultural plants (R. Smith); production management systems for crops (J. Tamini), animals (C. Campbell); environmental stress in animal production (J. Nolan); reproductive performance of animals (R. Stanley); structural changes in agriculture (C. Ching); agricultural marketing (F. Scott); human nutrition (R. Van Reen).

UNIVERSITY ATTACHED INSTITUTES 733

Harold L. Lyon Arboretum 734

Address: 3860 Manoa Road, Honolulu, HI 96822
Telephone: (808) 988-3177
Director: Yoneo Sagawa
Graduate research staff: 5
Activities: Research in the following areas: Natural resources - drainage and irrigation; soil science, land use; plant production - plant introduction, plant breeding, preservation of germplasm, preservation of native and endemic plants; forestry; botany - taxonomy, ethnobotany.
Publications: Harold L. Lyon Arboretum Lecture, annually.
Projects: To promote, facilitate, and execute research, instruction, and service especially related to its unique resources.

Hawaii Institute of Marine Biology 735

Address: PO Box 1346 Coconut Island, Kaneohe, HI 96744
Telephone: (808) 247 6631
Director: Dr Philip Helfrich
Graduate research staff: 16
Activities: Research in marine fisheries, ecology, and pollution.
Projects: Coral reef studies (Paul Jokiel); northwestern Hawaiian island studies (Richard Grigg); aquaculture (Arlo Fast); mesopelagic fish ecology (Thomas Clarke); marine toxins (Nancy Withers).

University of Idaho 736

Address: Moscow, ID 83843
Status: Educational establishment with r&d capability

COLLEGE OF AGRICULTURE 737

Dean: Raymond J. Miller

Idaho Agricultural Experiment Station 738

Telephone: (208) 885 6681
Director: Raymond J. Miller
Graduate research staff: 120

Agricultural Economics Department 739
Head: Dr R.W. Schermerhorn
Projects: Alternative energy sources for agriculture (S.M. Smith, C.L. Peterson); utilization of social and health services in Idaho (J.E. Carlson); economic evaluation of public investment in research (A.A. Araji); statistical design and interpretation of agricultural research data (D.O. Everson); economic interrelationships within Idaho (R.B. Long).

Agricultural Engineering Department   **740**

Head: Dr D.W. Fitzsimmons

Projects: Hydrology of agriculturally related areas of Idaho (Myron Molnan); interseeding and planting steep slopes on rough rangelands with a minimum of soil disturbance; design and development of research equipment, devices, and processes (W.L. Moden); equipment for producing and maintaining quality potatoes (James L. Halderson, Walter C. Sparks); subsurface irrigation with intermittent aeration (T.S. Longley); maintaining quality of stored grain in Idaho (J.L. Halderson, L.E. Sandvol).

Agricultural Information Department   **741**

Head: M.W. Stellmon

Animal Sciences Department   **742**

Head: C.F. Petersen

Projects: Reproductive performance in cattle and sheep (R.C. Bull, E.H. Stauber, R.G. Saser, D.G. Waldhalm); improved poultry house environment (C.F. Petersen, J.E. Dixon, E.A. Sauter); evaluation of agricultural and food processing by-products as feed for ruminants; barley selection and evaluation for beef cattle (D.D. Hinman).

Bacteriology and Biochemistry Department   **743**

Head: Dr W.E. Magee

Projects: Vitamin analysis in selected foods (J.A.L. Augustin); evaluation of protein quality (K.R. Davis); mechanism and control of clycolate synthesis and metabolism in higher plants (David J. Oliver).

Entomology Department   **744**

Head: Dr W.F. Barr

Projects: Idaho insect survey (W.F. Barr); to establish, improve, and evaluate biological control in pest management systems (G.W. Bishop); influence of natural and altered watersheds on aquatic insect populations and community structure (M.A. Brusven); damage by corn earworm populations on sweet corn seed (D.R. Scott); maximizing the effectiveness of bees as pollinators of agricultural crops (Norman D. Waters); integrated management of the leafcutting bee in the Pacific northwest (Leslie P. Kish); release of genetically selected clones to reduce the pest status of the green peach aphid (G.W. Bishop); recognition, evaluation, and manipulation of insects on Idaho grasslands; controlling mortality of alfalfa leafcutter bee larva (N.D. Waters, L.P. Kish).

Plant and Soil Sciences Department   **745**

Head: Dr G.A. Lee

Projects: Development of improved disease resistant potato varieties (J.J. Parek); testing and evaluating agronomic and horticultural crops for Idaho agriculture (A.A. Boe); soil characterization and genesis studies in support of Idaho soil surveys, their interpretation and use (M.A. Forsberg); characteristics and source of the loess deposits of southern Idaho (G.C. Lewis); improvement of Idaho fescue, Festuca idahoensis Elmer, by genotypic recurrent selection (R.D. Ensign); epidemiology and control of soilborne plant pathogens of potato (J.R Davis); erosion research for northern Idaho (R.W. Harder, C.L. Peterson, E. Michalson, J. Carlson, M. Molnan); breeding Austrian winter peas (Dick L. Auld); crop yield potential as affected by the rhizosphere, soil, and other environmental factors; water properties, spatial variability and implications in soil management (C.B. Holder); seed production, horticultural crops (G.F. Stallknecht); sorption of ions on soil and uptake of plants as influenced by organic residues (D.V. Naylor); development of the basic parameters for nematode pest management decisions (A.M. Finley); importance and control of perennial and range weeds in Idaho (D.C. Thill); host affinities and interactions of nematode and fungal pathogens of field crops in Idaho (R.R. Romanko); maintaining and improving seed potato quality (D.F. Hammond); control and competitive effects of selected annual weeds on irrigated land of Idaho (R.G. Brenchley).

Veterinary Science Department   **746**

Head: Dr Floyd W. Frank

Projects: Changing incidences of animal diseases (F.W. Frank); biological protection of livestock against internal parasites (R.F. Hall, Bruce Z. Lang); limiting stress of food producing animals to increase efficiency (D.P. Olson, R.C. Bull); control of F. Hepatica by vaccination and/or chemotherapeutic methods and parasite effect in cattle (R.F. Hall, B.Z. Lang, L.F. Woodard).

COLLEGE OF FORESTRY   747

University of Illinois   748

Status: Educational establishment with r&d capability

COLLEGE OF AGRICULTURE   749

Illinois Agricultural Experiment Station   750

Address: 211 Mumford Hall, 1301 West Gregory Drive, Urbana, IL 61801

Telephone: (217) 333 0240

Associate Director: Dr Benjamin A. Jones

Graduate research staff: 150

Publications: Biennial report; *Illinois Research*, quarterly.

Agricultural Economics Department   **751**

Head: W.D. Seitz

Agricultural Engineering Department 752
Head: R.R. Yoerger
Projects: Automatic monitoring and analysis of physiological parameters of lactating dairy cows (E.F. Olver); tillage systems for corn and soya beans (J.C. Siemens); improved equipment for pesticide application in soya beans (B.J. Butler); home sewage systems for areas with soils unsuitable for subsurface seepage fields (D.H. Vanderholm); sediment yield from small agricultural areas (J.K. Mitchell); nitrogen as an environmental quality factor (W.D. Lembke); technology of alcohol manufacture on the farm; soya bean quality (E.D. Rodda); diesel fuel research - ethanol (C.E. Goering); simulation and optimization of flat plate solar collectors for grain drying (G.C. Shove); soil and water control systems (P.N. Walker); off-road equipment operator vibration; improving soya bean production equipment (D.L. Hoag); dynamics of tillage, earthmoving, and traction (C.E. Goering); time and energy reductions for field machine operations (D.R. Hunt); reducing exhaust ventilation odours in swine housing (J.N. Scarborough); solar drying of high speed hay accumulator packages (G.C. Shove); potential contribution of sediment to surface waters from the erosion of surface mined lands (J.K. Mitchell); marketing and delivery of quality cereals and oilseeds in domestic and foreign markets; breakage susceptibility of corn (M.R. Paulsen); improved analysis and design of farm buildings (J.O. Curtis); drainability of the high clay content soils with a restrictive layer; feasibility of greenhouses heated with surface application of power plant cooling water (P.N. Walker); electrochemical conversion of animal wastes into protein and hydrogen (D.L. Day).

Agricultural Entomology Department 753
Head: W.H. Luckmann
Projects: Biology and impact of entomophilic microsporidia (J.V. Maddox); control tactics and management systems for arthropod pests of soya beans (M. Kogan); environmental implication and interactions of pesticide usage (A. Felsot); genetics and ecology of the European corn borer (W.H. Luckmann); forecasting the economic impact of black cutworm and corn rootworms (W.G. Ruesink); pest management strategies for leafhoppers, spittle-bugs, and aphids on alfalfa (E.J. Armbrust); development of microbial agents for use in integrated pest management systems (J.V. Maddox); pest losses (W.G. Ruesink); control of insects and insect vectors of brittleroot on horseradish (C.E. Eastman); arthropod management and economic losses of insects on livestock (S. Moore); biosystematics of insects (W.E. LaBerge); insects of ornamental plants (J.E. Appleby); evaluation of insecticides (E. Levine); bionomics and management of soil arthropods (W.G. Ruesink); pesticide residues (A. Felsot); parasites and predators (C.E. White).

Agronomy Department 754
Head: R.W. Howell
Projects: Crops testing (D.W. Graffis); mitochondrial configuration and efficiency, plant energy status and yield (D.E. Koeppe); cellular photosynthetic processes and the regulation of photosynthesis (C.J. Arntzen); soil and crop management (J.F. Welch); symbiotic N. fixation in soya beans (R.H. Hageman); microbial formation and degradation of nitrosated herbicides (M.A. Cole); breeding corn for high oil content (D.E. Alexander); soil survey (J.D. Alexander); improvement of soya bean production in the tropics, sub-tropics, and temperate zones of the world (C.N. Hittle); corn germplasm improvement (J.W. Dudley); new plants (J.R. Harlan); properties of clinoptilolite related to soil fertility (R.L. Jones); herbicide reaction with soils (W.L. Banwart); reclamation of surface-mined land (I.J. Jansen); improvement and testing of spring and winter small grains (C.M. Brown); biosystematics of some weedy grasses (J.M.J. de Wet); structural chemistry of soil humic substances (F.J. Stevenson); soyabean breeding and genetics (C.D. Nickell); genetics of plasmids in Rhizobium japonicum (M.A. Cole); plant cell cultures as biochemical and genetic tools (J.M. Widholm); optimum irrigation regime for vegetable crops on sandy land (M.D. Thorne); nutrient uptake characteristics of plant roots (R.H. Beck); mechanisms of yield depression by water deficits in soya beans (J.S. Boyer); use of genic male sterility in commercial hybrid corn production (E.B. Patterson); glycine germplasm data storage and retrieval system (T. Hymowitz); soil landscape characteristics affecting land use planning and rural development (J.B. Fehrenbacher); fat synthesis in immature soya bean seeds (R.W. Rinne); methods in the spectrographic analysis of soils and plants (T.R. Peck); agricultural benefits and environmental changes from the use of organic waste on field crops (T.D. Hinesly); comparative effects of weed competition, herbicides, and other weed control practices on plant response (E.L. Knake); electrophoretic separation of soil and reference clays into different mobility groups (A.H. Beavers); soil and crop management practices for southwest Illinois (L.F. Welch); cytological and genetical studies in Glycine max (H.H. Hadley, T. Hymowitz); single crop and double crop no-tillage grain production systems (G.E. McKibben, E.L. Knake, J.C. Siemens, L.L. Getz); the occurrence and activity of enzymes in corn (R.H. Hageman); soil maps for relating soil productivity to value of agricultural land (J.D. Alexander, F.J. Reiss); foliar fertilization of agronomic crops (L.F. Welch, C.M. Brown); efficiency of nutrient uptake and utilization by plants (R.G. Hoef); varietal improvement and genetic behaviour of forage legumes (D.A. Miller); improving seasonal distribution of forage yield and quality on soils of southern Illinois (J.J. Faix); energetics of nonpolar adsorption by soils and sediments (J.J. Hassett, W.L. Banwart); prediction of the phosphate pollution hazard of Illinois soils (J.J. Hassett); improving management

and design of cropping systems (R.R. Johnson); responses of forage and pasture plants to cultural and environmental factors (J.A. Jackobs, D.A. Miller, D.W. Graffis); dynamics and energetics of the soil-plant atmosphere continuum (D.E. Koeppe); new screening method for finding soya bean plants which lack photorespiration (J.M. Widholm, W.L. Ogren); effects of iron on crystal structure and colloidal properties of clay mineral (J.W. Stucki); use of special genetic stocks in breeding sorghums and soya beans (H.H. Hadley); studies of chromosome translocations in maize and their use as research tools (E.B. Patterson); physiology and biochemistry of herbicide action in plants (F.W. Slife).

Animal Science Department   755
Head: D.E. Becker
Projects: Reestablishment of ovarian cycles in postpartum cows (D.J. Kesler); mechanisms of gonadotropin action in the hen's ovary (J.M. Bahr); prenatal factor - foetal survival (P.J. Dziuk); fertility of lactating sows (L.H. Thompson); genetics of litter size in beef cattle (H.W. Norton); blood types and productivity in pigs and sheep (B.A. Rasmusen); genetic evaluation and monitoring of performance-tested beef herds (D. Gianola); metabolic interactions of mineral ions in animal tissues (R.M. Forbes); factors affecting chemically induced acute pulmonary disease (J.E. Garst); metabolic regulation in the avian (S.P. Mistry); efficiency of lamb production (U.S. Garrigus); protein supplements for wintering beef cows (B.A. Weichenthal); metabolic effects of cell wall fractions in livestock (G.C. Fahey); evaluation of alfalfa hay, haylage, and silage for beef production (L.L. Berger); factors affecting rumen microbial ecology (J.B. Russell); effects of energy level and monensin on reproduction performance and lactation (G.C. Fahey); forage production and utilization systems as a base for livestock production (F.C. Hinds); nutritional factors influencing equine growth and productivity (W.W. Albert); sectioned and formed meats (G.R. Schmidt); pulmonary status in swine as a function of environment (S.E. Curtis); cardiovascular responses of domestic animals to hot environments (P.C. Harrison); management, environment, physiological factors of swine production; animal waste management systems for the 1980's - nutritive value of oxidation ditch residues (A.H. Jensen); biological availability of nutrients in feed ingredients (D.H. Baker).

Dairy Science Department   756
Head: W.R. Gomes
Projects: Improving dairy cattle through breeding (M. Grossman); fatty and aromatic acid catabolizing bacteria in methanogenic ecosystems (M.P. Bryant); survival capability: anaerobic microbial cohabitants (R.B. Hespell); transport mechanism for IgGl across the mammary secretory cell, milk protein formation (B.L. Larson); nitrogen metabolism in ruminants (J.H. Clark); calcium transport and distribution in mammary cells (C.R. Baumrucker); factors affecting reproductive per-

formance and cytogenetics of newborn calves (J.R. Lodge); factors affecting embryogenesis in mammals (C.N. Graves); change in rumen and tissue metabolism associated with the low-milk-fat syndrome (C.L. Davis); physiology and biochemistry of mammalian sperm acrosomal enzymes (C.N. Graves); improving large dairy herd management practices; animal wastes; resistance to mastitis (S.L. Spahr); regulation of glycolysis and pyrimidine metabolism in the bovine mammary gland (J.L. Robinson); optimizing the nutritional utilization of forages by dairy cattle (E.H. Jaster); automation in feeding dairy cattle (K.E. Harshbarger).

Forestry Department   757
Head: G.L Rolfe
Projects: Flakeboard house siding from flakes treated with glycol and phenolic chemicals (J.K. Guiher); chemical changes in atmospheric deposition (A.R. Gilmore); nutrient dynamics in bottomland forests in Illinois (G.L. Rolfe); nitrogen fixation in black alder (J.O. Dawson); role of mitochondria in the growth of trees in the Midwest (D.E. Koeppe); study of the insects of trees in the Midwest (J.E. Appleby); sampling methods based on the spatial distribution of trees (G.Z. Gertner); rheological behaviour of a reconstituted hardwood composite (P. Chow); solar energy through woody biomass production; water quality from forested watersheds in southern Illinois; disposal of liquid waste from a cattle feedlot on a forested watershed (G.L. Rolfe); ethanol tree-seedling screening for tolerance to low soil oxygen conditions (J.O. Dawson); energy analysis of Illinois forest silvicultural and management practices (S.L. Brown); trends in wilderness use and preferences (R.A. Young); dynamics and energetics of the soil-plant-atmosphere continuum (F.A. Bazzaz); forest tree improvement through selection and breeding (J.J. Jokela).

Horticulture Department   758
Head: W.L. George Jr
Projects: Genetics and physiology of sweet corn quality and biological efficiency (A.M. Rhodes); fertilization and nutrient utilization by vegetable crops (J.M. Swiader); chlorophyll biosynthesis and chloroplast biogenesis in vitro (C.A. Rebeiz); the role of amino acids in seed germination; biological and chemical systems for control of vegetable yield and quality (W.E. Splittstoesser); productivity of horticultural crops as related to cell wall biogenesis (D.B. Dickinson); new plants (M.A. Dirr); optimization of crop production in the ornamental plant industry (J.N. Hubbell); plant water stress and container plant-soil-water relations (L.A. Spomer); thatch development and its impact in turfgrass ecosystems (A.J. Turgeon); breeding greenhouse chrysanthemums (J.R. Culbert); trial and experimental garden for outdoor ornamentals (G.M. Fosler); testing and evaluation of woody plant materials; tree nut and persimmon breeding (J.C. McDaniel); propagation and growth and development of perennial plants by in vitro

culture (M.M. Meyer); nitrogen metabolism and translocation in perennial fruit trees (J.S. Titus); scion/rootstock and interstem effects on apple growth and fruiting (R.K. Simmons); climate and peach tree development (J.B. Mowry); behaviour of worker bees (E.R. Jaycox); analysis of the impact of pesticides on commercial horticulture (H.J. Hopen); development and use of asexual techniques in plant improvement (R.M. Skirvin); breeding and genetics of commercially important characters in the apple (D.F. Dayton); completion and testing of microecosystems for investigating the fate of RPAR pesticides in turf (A.F. Turgeon); evaluation of the herbicides cacodylic acid, paraquat, and methazole for use in the commercial production of nursery crops (D.J. Williams).

Human Resources and Family Studies Department 759
Head: M.M. Dunsing
Projects: Mineral bioavailability, functionality, and organoleptic properties of processed soya foods (B.P. Klein); quality and safety of foods in households and institutions (B.P. Klein, M.A. Khan, F.O. Van Duyne); lipid oxidizing enzyme systems in seed vegetables (B.P. Klein); the effect of iron deficiency on selenium metabolism and requirement (D.K. Layman); iron status during infancy as influenced by food iron sources (M.F. Picciano, D. Stastney, J.S. Buck).

Plant Pathology Department 760
Head: R.E. Ford
Projects: Pesticide impact assessment research and data analysis (M.C. Shurtleff); persistent transmission of plant viruses by aphid vectors - mechanisms of vector specificity (C.J. D'Arcy); detection of seed-borne soya bean viruses by serological tests (G.M. Milbrath); effect of herbicides and other pesticides on root rot diseases of pulses (B.J. Jacobsen); replication of a single-stranded DNA plant virus (R.M. Goodman); seed-borne bacterial and fungal pathogens of soya bean and seed quality (J.B. Sinclair); integrated approaches to management of disease caused by soil-borne plant pathogens on field crops (J.W. Gerdemann); genetics and physiology of plant parasite interactions (A.L. Hooker); occurrence and effects of mycotoxins in feeds and foods (D.G. White); use of soil factors and soil crop interactions to suppress diseases caused by soil-borne plant pathogens (J.D. Paxton); viruses and mycoplasma-like organisms causing diseases of corn and sorghum (R.E. Ford); physiology and biochemistry of plant pathogenic fungi (P.D. Shaw); research on fungus physiology and antibiotics mechanism of action (D. Gottlieb);biology of endomycorrhizal (vesicular-arbuscular VA) fungi (J.W. Gerdemann); control of diseases of apples, peaches, nectarines, cherries, strawberries, and blueberries; detection, survival and control of plant pathogenic bacteria on seeds and plant propagative material (S.M. Ries); biological significance and ecology of stylet-bearing nematodes; reduction of corn losses due to nematodes in

the North Central Region (R.B. Male); aetiology, epidemiology, and control of vascular diseases of trees and shrubs (D. Neely).

Veterinary Research Department 761
Head: R.E. Dierkes
Projects: Development of cell lines for propagation of TGE virus of pigs (A.M. Watrach); resistance to mastitis in dairy cattle (J. Simon); the induction of interferon in the nasal secretions of calves (J.M. Cummings); effects of Zearalanol implants on reproductive function in bulls (R.S. Ott); prevention and control of enteric diseases of swine (M. Ristic); eperythrozoonosis of swine (A.R. Smith); avian respiratory disease control for consumer protection and quality (D.N. Tripathy); immunological factors associated with haemonchiasis in lambs (M.E. Mansfield); differential diagnosis of respiratory diseases of swine involving mycoplasma species (G.T. Woods); occurrence and effects of mycotoxins in feeds and foods (R.F. Bevill); bovine anaplasmosis: anti-platelet activity and mechanism of anaemia (C.A. Carson); bovine respiratory diseases (G.T. Woods)

University of Iowa 762

Address: Iowa City, IA 52240
Status: Educational establishment with r&d capability

COLLEGE OF MEDICINE 763

Institute of Agricultural Medicine 764

Address: Oakdale, IA 52319
Telephone: (319) 353 4872
Head: Dr Keith R. Long
Sections: Comparative medicine, Dr Kelley J. Donham; toxicology, Dr Donald P. Morgan; industrial hygiene, Dr Clyde M. Berry; environmental chemistry, Dr Clyde W. Frank; accident prevention and international programmes, L.W. Knapp
Graduate research staff: 25
Activities: To determine the occupational health and safety problems that effect the health of farmers and their families.
Projects: Pesticide studies (K.W. Kirby); bovine leukemia (Dr K.J. Donham); worker health in confinement housing structure for swine (Dr K.J. Donham); low back pain of farmer (L.W. Knapp).

University of Kentucky 765

Address: Lexington, KY 40506
Status: Educational establishment with r&d capability

COLLEGE OF AGRICULTURE   766

Agricultural Experiment Station   767
Address: Lexington, KY 40546

Agricultural Engineering Department   768
Activities: Subjects of research include: simulation models to describe growth of forage crops, their utilization by beef animals, and animal growth; economics of corn irrigation; performance of sediment ponds; efficiency of tobacco harvesting and marketing; performance of solar energy systems.

Agronomy Department   769
Activities: Areas of research include: soya bean seed production; tobacco quality, leaf curing, production of low-tar tobacco; forage crops; reproductive performance of ewes related to diet; pasture renovation.

Animal Nutrition Department   770
Projects: Preweaning performance of cows and calves as influenced by breed type and kind of pasture (N.W. Bradley, C.T. Dougherty, J.A. Boling, N. Gay); availability of phorphams and other minerals, in pigs feedstuffs (C.S. Stober, G.L. Cromwell, T.S. Stahly); systems of finishing weaned calves (N.W. Bradley, J.A. Boling, N. Gay, C.T. Dougherty, R.M. Stone); systems of producing beef using different combinations of forage and grain (N. Gay, W.G. Moody, J.A. Boling, N.W. Bradley, J.R. Overfield).

Animal Science Department   771
Activities: The main aims of research are: nutrition - to improve the efficiency of utilizing the resources available for production of animals and their products and to improve the quality and shelf life of the products by the following means - evaluation of forages; developing methods for harvesting and storing forages; supplements to forages; determination of minimum nutrient levels that will result in maximum or most economical performances at various stages of production; evaluation of the efficacy and safety of non-nutritive additives to improve performance of animals; maximizing reproductive efficiency of beef, dairy, sheep, and swine - feeding, management, and selection programmes to increase conception rate and reduce embryo and offspring losses, studies of hormone patterns, predetermining sex of offspring; new and improved methods for increasing cheese yield; improved methods of processing and storing pork.

Entomology Department   772
Activities: Areas of research include: physiological and biochemical functions of insect - pests insecticide resistance, susceptibility to unfavourable environmental conditions, reproductive capabilities, selection of host plants, impact of parasites and predators, effect of water balance on target insects, parasitization; development of techniques of insect control - use of pathogens to reduce reproduction potential or general vigour of insects.

Food Science Department   773
Projects: Yield losses in cheddar cheese from stored milk (C.L. Hicks, J. O'Leary); characteristics of dry-cured hams held for different times prior to curing (J.D. Martin, J.D. Kemp, B.E. Langlois).

Forestry Department   774
Activities: Research in the following areas; techniques for promoting rapid growth of pulp wood species through intensive cultural practices, and the effect of these practices on the physical and chemical properties of wood; revegetation of surface-mined land; use of hardwood bark for erosion control and soil improvement; biological and economic feasibility of growing black locust alder and other tree species in 'biomass plantations'; ecological research in virgin and second-growth hardwood forests; computer modelling to develop silviculture techniques and management systems; new techniques and products for replanting forest land; improved methods of drying lumber; factors affecting recreational use of forest land.

Genetics and Physiology Department   775
Projects: Swine embryos growth in culture (W.M. Graves, D. Olds).

Horticulture Department   776
Activities: Research and development in the following areas; plant breeding for vegetable and fruit improvement; development of dwarf vine types of cucumber, muskmelons, honeydew melons, watermelons and squash for home garden use; chemical factors affecting quality in cucurbits; superior rootstock in peaches; varietal testing of fruit and vegetables; ornamentals research; energy-related greenhouse production; computerized land use planning.

Plant Breeding and Genetics Department   777
Projects: Burley tobacco breeding (M.T. Nielsen, C.C. Litton, G.B. Collins, J.H. Smiley); pest control and disease resistance in tobacco; soya bean breeding and genetics (J.H. Orf); winter oat breeding and barley breeding genetics (V.C. Finkner); tall fesane improvement; breeding turf grasses (R.C. Buckner, P.B. Burrins); genetic studies of red clover (N.L. Taylor, G.B. Collins, P.L. Cornelius).

Plant Pathology Department   778
Activities: Subjects of research include: biochemistry of disease reaction in plants; studies on aetiology and ecology of fungal diseases of tobacco, soya beans, and trees; virus diseases of soya beans, red clover, and tobacco; soya bean cyst nematode; development of strategies for use of fungicides, nematicides, crop rotation, and resistant varieties; fungicidal control of blue mold, black shank, and damping-off in tobacco; effective and rapid detection of seedborne viruses.

Soil Science Department **779**
Projects: Classification and characterization of soils (H.H. Bailey, R.L. Blevins); soil water studies (G.W. Thomas, R.E. Phillips); irrigation (J.H. Herbeck, L.W. Murdock, M. Rasnate); long-term tillage effect on soil properties (R.L. Blevins, G.W. Thomas, M.S. Smith, W.W. Frye); grasses and legumes in revegetation of surface-mined spoils (R.I. Barnsihel); utilization of municipal sewage sludge (R.I. Barnsihel); aluminium in soil organic matter (W.L. Aargrove G.W. Thomas).

Weed Science Department **780**
Projects: Control of Nimbleweed, Johnson grass, and black nightshade (J.M. Rosemond, A.J. Powell, W.W. Wilt, S.R. Martin, L.G. Rodrigue); control of woody perennials including multiflora rose (R.K. Mann, W.W. Wilt); effects of dinitroaniline herbicides on sorghum root growth (P.P.C. Lin).

Veterinary Science Department **781**
Activities: Subjects of research include: methods of prevention and control of animal disease; more efficient management of reproductive functions; vaccine control of 'virus abortion'; control of contagious equine metritis; molecular biology and virology of herpes viruses; parasite control in sheep, cattle, and horses; eyeworm in cattle and horses; reproductive physiology of broodmares; pharmacology of the horse.

University of Maine, Orono* **782**

Status: Educational establishment with r&d capability

COLLEGE OF LIFE SCIENCES AND AGRICULTURE EXPERIMENT STATION **783**

Address: Office of the Associate Director, 103 Winslow Hall, University of Maine, Orono, ME 04469
Dean and director: Kenneth E. Wing
Associate director: Dr David E. Leonard
Graduate research staff: 121
Activities: Research in the following areas: agriculture; forestry - wildlife; biological sciences; resource economics and development; environmental studies; social sciences; marine resources; food science and nutrition; animal sciences - livestock production, husbandry and nutrition; plant production; plant breeding; crop husbandry; soil science; land use; plant protection; agricultural engineering.
Publications: Technical bulletins; annual report.

Agricultural and Resource Economics Department **784**
Head: Dr Wallace Dunham

Agricultural Engineering Department **785**
Head: Dr Norman Smith

Animal and Veterinary Science Department **786**
Head: Dr Stanley Musgrave

Biochemistry Department **787**
Head: Dr Joseph Lerner

Botany and Plant Pathology Department **788**
Head: Dr Douglas Gelinas

Entomology Department **789**
Head: Dr Howard Y. Forsythe

Food Science Department **790**
Head: Professor Gordon Ramsdell

Microbiology Department **791**
Head: Dr Bruce Nicholson

Plant and Soil Science Department **792**
Head: Dr James Swasey

School of Forest Resources **793**
Head: Dr Fred B. Knight

University of Maryland* **794**

Status: Educational establishment with r&d capability

MARYLAND AGRICULTURAL 795
EXPERIMENT STATION

Address: University of Maryland, College Park, MD 20742
Telephone: (301) 454 3707
Director: W.L. Harris
Sections: Food, nutrition and administration: Dr E.S. Prather
Graduate research staff: 82
Activities: Research is conducted in five major broad areas: natural resources and forestry; plants and crops; animals and poultry; economics and rural life; and general resource technology. Work in these areas is directed toward three main goals: to conduct research that promotes a sound and prosperous Maryland agriculture that will insure a continuous supply of nutritional foods, useful fibre, and natural products in adequate amounts at reasonable cost; to develop knowledge that will enable producers and users of natural resources to conserve and manage them wisely and to help insure a physical environment of high quality; to contribute, through research and investigation, to the improvement of human health and economic and social surroundings as related to agriculture, thereby strengthening human resources to enjoy life more fully and to participate in a complex society.
Publications: Annual report.

Agricultural and Extension Education 796
Departments

Head: Dr C.L. Nelson
Activities: Socioenconomic studies and agriculture.

Agricultural and Resource Economics 797
Departments

Head: Dr V.J. Norton
Activities: Broilers; sewage treatment; socioeconomic factors.

Agricultural Engineering Department 798

Head: Dr L.E. Stewart
Activities: Energy requirements for poultry production; energy sources in farming; water quality and supply; pollution; food processing.

Agronomy Department 799

Head: Dr J.R. Miller
Activities: Crop improvements; forage crops; weeds and herbicides; soil properties.

Animal Science Department 800

Head: Dr E.P. Young
Activities: Beef cattle production.

Botany Department 801

Head: Dr G.W. Patterson
Activities: Plant diseases; parasites; nematodes; plant development and herbicides; production of ornamental plants.

Centre for Environmental and 802
Estuarine Studies

Dairy Science Department 803

Head: Dr D.C. Westhoff
Activities: Improved production of dairy cows; milk and milk products.

Entomology Department 804

Head: Dr A.L. Steinhauer
Activities: Pest control.

Food, Nutrition, and Institution 805
Administration

Head: Dr E.S. Prather

Horticulture Department 806

Head: Dr B.A. Twigg
Activities: Ornamental plants; fruit plants; forest trees; disease control.

Poultry Science Department 807

Head: Dr O.P. Thomas
Activities: Broilers and laying hens.

Veterinary Science Department 808

Head: Dr R.C. Hammond
Activities: Virus diseases in cattle.

University of 809
Massachusetts *

Status: Educational establishment with r&d capability

COLLEGE OF FOOD AND NATURAL RESOURCES 810

Address: 217 Stockbridge Hall, Amherst, MA 01003
Director: Dr Daniel I. Padberg
Graduate research staff: 100

Massachusetts Agricultural Experiment Station 811

Cranberry Station 812
Projects: Mechanization of cranberry production operations (J.S. Norton); integrated pest management of cranberry arthoropod pests (C.F. Brodel); management practices to increase cranberry fruit and pigment production (K.H. Deubert).

Entomology Department 813
Acting head: Dr John Edman
Projects: Integrated control of hylema species of Massachusetts vegetable crops (D.N. Ferro); host-plant interaction in natural regulation of forest insects; effects of Massachusetts release of Comperia mercei on the brown banded cockroach, Supella longipalpa (J.S. Elkinton); hormonal control of insect diapause (C.M. Yin); biological control of housefly Musca domesticus, in poultry house (J.G. Stoffolano); integrated management techniques of arthropod apple pests (R.J. Prokopy); biting fly-vertebrate host interrelationships (J.D. Edman); biology and control of aphids and viruses of green peppers in northwestern US (D.N. Ferro).

Environmental Sciences Department 814
Head: Dr Warren Litsky
Projects: Limnological impact of acid rain (R.A. Coler).

Food Engineering Department 815
Head: Dr Joe T. Clayton
Projects: Engineering properties of food and their role in processing operations (M. Peleg).

Food and Resource Economics Department 816
Head: Dr John F. Foster
Projects: Economic impacts of the 200-mile limit on northeastern coastal communities (D.A. Storey); effects of integrated pest management on apple grading and packing (R.W. Christensen); identification of economically viable farm enterprises in Massachusetts (B.C. Field); economic analysis of contract production and marketing of broilers and eggs in the northeast (E.E. Dorr); cooperative marketing of wood products in Massachusetts (B.J. Morzuch); economic feasibility of alternative service delivery systems in rural communities (G.R. McDowell); determinants of early enrollment in the special supplemental food programme for women and children (C. Willis).

Food Science and Nutrition Department 817
Head: Dr R. Glenn Brown
Projects: Processing induced interactions of the major food nutrients (W.W. Nawar); relationship between food borne and human clinical isolates of Psedomonas putrefactiens (R.E. Levin); influence of dietary survey methodology upon malnutrition (K.W. Samonds); human nutrition improvement (V. Beal); spore lytic enzymes of Clostridium perfringens (R.G. Labbe); enzymatic control of quality factors in marine foods (H.O. Hultin); manufacture of processed meat and/or fish products with underutilized marine species (E.M. Buck); stable lactase of microbiological origin (R.R. Mahoney).

Forestry and Wildlife Management Department 818
Head: Dr Joseph S. Larson
Projects: Management of mixed hardwood-conifer forest resources in Massachusetts; beech bark disease (W.A. Patterson); factors affecting beaver disposal patterns (J.J. Kennelly); development of wind turbine blades from wood (R.B. Hoadley); energy requirements for drying northeastern lumber (W.W. Rice); apical control of cambial activity in hardwood branches (B.F. Wilson); methodology and quantification of timber supply in Massachusetts (P.A. Harou); effects of forest practices, soil, and precipitation on water resources in Massachusetts (D.L. Mader).

Plant and Soil Sciences Department 819
Head: Dr Allen V. Barker
Projects: Growth of vegetable crops in nutritional regimes caused by soil acidity; increased resistance in sweet corn to maize dwarf mosaic virus; plant germplasm (G.J. Hochmuth); spodozol soils (P.L.M. Veneman); adaptation, culture, persistance of perennial ryegrass in New England (K.A. Hurto); alternatives in feed crops for Massachusetts (S.J. Herbert); maximizing vegetable yields on small farms; new potato clones and varieties (R.J. Precheur); hydrologic properties and hydrologic behaviour of Massachusetts soils (D. Hillel); accelerated growth of nursery crops (J.R. Havis); apple production and quality as influenced by orchard practices (M. Drake); sulphur dioxide and acid rain (L.E. Craker); postharvest physiology of fruits (W.J. Bramlage); efficient turfgrass culture with limited inputs of water and energy (W.A. Torello).

Plant Pathology Department   **820**
Head: Dr Mark S. Mount
Projects: Recycled waste products: effects on plant diseases, symbiotic associations and reproductive processes (W.J. Manning); impact of nematode population dynamics on crop yield and growth (R.A. Rohde); nematode population dynamics and crop growth and yield, fungal and viral diseases of cranberries (B.M. Zuckerman); biology and management of vascular diseases of trees (T.A. Tattar); biology and control of viruses affecting vegetable crops (G.N. Agrios); fungal and viral diseases of cranberries (B.M. Zuckerman).

Suburban Experiment Station   **821**
Head: Professor Gordon W. Fellows
Projects: Cytogenetic manipulation of germplasm for corn improvement; genetics and physiology of sweet corn quality, pest resistance and yield (W.C. Galinat); epidemiology and control of vegetable and floricultural crop diseases (G. Moorman); breeding solanacae crops for resistance to biotic and abiotic illness (C. Nicklow); recycled waste products; chemical changes in atmospheric deposition and effects on land and surface waters (W.A. Feder).

Veterinary and Animal Sciences Department   **822**
Head: Dr James B. Marcum
Projects: Genetic bases for economically important mutant phenotypes of domestic fowl (J.R. Smyth); genetic bases for resistance to avian diseases; immunity of susceptible and resistant chickens vaccinated with Marek's disease tumour cell antigens; serological techniques for detection EIAV antigen and antibody; immune response of cattle to Brucellae (M. Sevoian); biophysical factors affecting energy requirements for poultry production (D. Anderson); optimizing intake and utilization of forage and protein by ruminants (M.J. Poos); increasing natural resistance to mastitis during the dry period (S.P. Oliver); control and improvement of reproduction in the bovine female (D.L. Black); immune mechanisms in the diagnosis and control of economically important poultry diseases (O.M. Weinack).

University of Minnesota*   823

Status: Educational establishment with r&d capability

COLLEGE OF AGRICULTURE   824
Address: 229 Agricultural Administration Building
Dean: Dr J.S. Tammen, University Park, PA 16802
Telephone: (814) 865 5419
Director: Dr Samuel H. Smith
Assistant director: Dr R.F. Hutton
Sections: Dairy and animal science, Dr P.J. Wangsness; veterinary science, Dr C.S. Card; entomology and pesticide research laboratory, Dr C.W. Pitts; poultry science, Dr K. Goodwin; horticulture, Dr F.H. Witham; food science, Dr P.G. Keeney; agricultural and extension education, Dr S. Curtis; agricultural economics and rural sociology, Dr J.W. Malone; plant pathology, Dr P. Nelson; agronomy, Dr J.L. Starling; agricultural engineering, Dr H.V. Walton; school of forest resources, Dr R.S. Bond
Graduate research staff: 300
Activities: Natural resources: soil science, drainage and irrigation, land use; plant production: plant breeding, crop husbandry, plant protection; animal production: livestock breeding, livestock husbandry and nutrition, veterinary medicine; agricultural engineering and building; fresh-water fisheries; food science; forestry and forest products; agricultural, fresh-water fisheries, forestry and food economics; soil sciences.

COLLEGE OF BIOLOGICAL SCIENCES   825
Dean: Dr R.S. Caldecott
Activities: Research in the following areas: comparative biochemistry of mammal milks; nucleic acid metabolism; nuclear magnetic resonance spectroscopy; gas chromatography; daily oscillations in sensitivity of plants to mechanical and chemical treatments; genetic coding and organization of maize cells.

COLLEGE OF VETERINARY MEDICINE   826
Dean: R.H. Dunlop
Activities: Research in the following areas: diseases of important food animals (pigs, cattle, sheep and poultry), particularly respiratory and enteric diseases; pathogenesis, immunology, and disease prevention; reproductive physiology.

INSTITUTE OF FORESTRY AND HOME ECONOMICS   827
Dean: Dr W.F. Hueg
Activities: Research in forest management and remote sensing.

MINNESOTA AGRICULTURAL EXPERIMENT STATION 828

– MAES

Address: 220 Coffey Hall, 1420 Eckles Avenue, St Paul, MN 55108
Telephone: (612) 373 0751
Director: Dr Richard J. Sauer
Graduate research staff: 536
Activities: The mission of the experiment station is to organize and support scientists from the colleges and departments of the University of Minnesota, who conduct research to improve the production, processing, quality, and marketing of food and other agricultural products, forests and forest products; also research on human nutrition, family and community life, recreation and tourism and overall environmental quality. The station represents an integral part of the triumvirate mission of a land grant university. Thus, the mission is strongly linked with the aims of the Agricultural Extension Service and those of the College of Agriculture, Forestry and Home Economics.
Publications: Minnesota Science.

Agricultural and Applied Economics Department 829

Head: Dr G. Edward Schuh
Activities: Economic analyses of Minnesota agriculture (swine, crop, dairy and beef production); and of US agriculture, including food supply, demand, transportation and consumption; and of world agriculture, including export practices, competition, and productivity growth.

Agricultural Engineering Department 830

Address: University of Minnesota, 1390 Eckles Avenue, St Paul, MN 55108
Telephone: (612) 373-1305
Head: Dr A.M. Flikke
Graduate research staff: 21
Activities: Teaching, research and extension, drainage and irrigation; land use; agricultural buildings; food processing; energy; power and machinery, power and processing.
Projects: Marketing quality of cereals and oilseeds (R.J. Gustafson); biomass conversion to energy (R.V. Morey); harvesting of biomass (C.E. Schertz); wild rice processing studies (J. Strait); tillage studies (J.A. True); thermal processes in food (D.R. Thompson); irrigation practices (E.R. Allred); plant stress in irrigation (D.C. Slack); watershed hydrology (C.L. Larson); farm waste management (P.R. Goodrich); livestock environment (K.A. Janni); animal health studies (K.A. Jordan).

Agronomy and Plant Genetics Department 831

Head: Dr H.W. Johnson
Activities: Research in the following areas: plant breeding and genetics (wheat, oats, flax, corn, soyabean, alfalfa, barley, and wild rice); cytogenetics; physiology of crop productivity; forage production and microbial-host plant specificity in forage crops; weed control; seed production.

Animal Science Department 832

Head: Dr R.W. Touchberry
Activities: Research in the following areas: animal nutrition and management (sheep, pigs, turkeys, beef and dairy cattle, horses, chickens); reproductive physiology; neurobiology and behavioural stress; physiology of tissue and cell growth.

Entomology, Fisheries, and Wildlife Department 833

Head: Dr M.W. Weller
Activities: Research in the following areas: systematics, ecology, and management of insect pests (Mallophaga, Aphididae, European corn borer, leafhoppers, and cereal, alfalfa and sunflower pests); chemical and biological pesticides; ecology and management of beneficial insects in forestry and farming; bee production; ecology and management of lake and river fisheries; wildlife in Minnesota, including moose, deer, bald eagle and waterfowl protection; habitat conservation, particularly wetlands.

Food Science and Nutrition Department 834

Head: Dr E.F. Caldwell
Activities: Research in the following areas: cheese and milk product technology; microbial and fermented foods; microorganisms in food systems; nutritional aspects of ageing and disease, food studies by pyrolysis gas chromatography, acoustic analysis, computer modelling; flavour research; heat treatment for processed foods.

Forest Products Department 835

Head: Dr J.G. Haygreen
Activities: Research in the following areas: wood processing technology; lumber and fuelwood drying; material properties of wood products; energy conservation.

Forest Resources Department   836

Head: Dr G.N. Brown
Activities: Research in the following areas: reafforestation; wood production; forest management; recreation facilities; ecology; forest tree diseases.

Horticultural Science and Landscape   837
Architecture Department

Address: 305 Alderman Hall, 1970 Folwell Avenue, St Paul, MN 55108
Telephone: (612) 373-1026
Acting head: Professor Jane Price McKinnon
Departments: Landscape arboretum, Francis deVos; landscape architecture, Professor Peter Olin
Graduate research staff: 27
Activities: Modification of environment; land use; plant breeding; plant protection; plant breeding - vegetables and fruits.
Projects: Plant breeding (Peter Ascher, D. Davis); plant physiology (P. Li, J. Carter, A. Markhart, S. Desborough, E. Stadelmann, M. Brenner); floriculture (R. Widemer, H. Wilkins, D. Koranski); vegetable crops (L. Waters, D. Davis, P. Read, F. Lauer, J. Sowokinos); woody plant materials (H. Pellett, B. Swanson, F. deVos, M. Eisel); fruit crops (L. Hertz, S. Munson); turf management (D. White, D. Taylor); landscape architecture (P. Olin, V. Cline).

Plant Pathology Department   838

Head: Dr D.W. French
Activities: Research in the following areas: identification and control of fungal, virus, bacterial, rust, and pollution-related diseases of plants and crops; microbiology, including electron microscope virus studies, and cell surface recognition in plant-pathogen relations.

Rural Sociology Department   839

Head: Dr G.A. Donohue
Activities: Research in the following areas: social structure and demographic change in rural Minnesota; mass media and the dissemination of technical knowledge.

Soil Science Department   840

Head: Dr W.P. Martin
Activities: Research in the following areas: Minnesota soil survey; nitrification, and soil fertility in relation to agricultural practices; water and landscape management.

University of Missouri -   841
Columbia*

Status: Educational establishment with r&d capability

MISSOURI AGRICULTURAL   842
EXPERIMENTAL STATION

Address: 2-44 Agriculture Building, Columbia, MO 65211
Telephone: (314) 882 7488
Dean and director: Dr A. Max Lennon
Associate dean and director: W.H. Pfander
Graduate research staff: 115
Activities: Research in all areas of agriculture, food science, and forestry, including crops production research on corn, wheat, barley, oats, cotton, soyabeans, sorghum, tobacco, and sunflowers; and also animal production research on beef and dairy cattle, swine, turkeys, chicken, and sheep.
Projects: Forage utilization in beef production systems (G.B. Garner, T.E. Fairbrother, W.G. Hires, R.E. Morrow, A.G. Matches, V.E. Jacobs).

Agricultural Economics Department   843

Chairman: Dr Charles L. Cramer
Projects: Agribusiness risk management (G.T. Devino); economic implications of land and water management problems relating to agriculture (D.E. Ervin).

Agricultural Education Department   844

Chairman: Dr Curtis R. Weston

Agricultural Engineering Department   845

Chairman: Dr C. LeRoy Day
Projects: Calorimetry and environmental requirements for animal shelters (M.D. Shanklin); energy and by-product recovery from livestock processing residues (D.M. Sievers); engineering services (N.F. Meador); microbiology of mesophilic, anaerobic digestion of organic matter (E.L. Iannotti).

Agronomy Department   846

Chairman: Dr Roger L. Mitchell
Projects: Mechanism of heredity in corn (M.G. Neuffer); maize quality improvement (A.L. Karr); weed control systems (O.H. Fletchall); breeding and genetics of forage grasses, primarily Festuca and Dactylis (D.A. Sleper); rhizosphere chemistry (R.W. Blanchar); soil fertility - nutrition and testing (J.R. Brown); farm seeds (L.E. Cavanah); soyabean breeding (S.C. Anand); influence of cultural practices and environmental factors

on crop performance (H.C. Minor); herbicidal cycles in conservation tillage systems (H.D. Kerr); biological efficiency of soyabeans (V.D. Luedders); forage-livestock systems and plant-animal research techniques (A.G. Matches); biochemistry and physiology of forage crops (C.J. Nelson); weed control in forage and agronomic crops (E.J. Peters); development of short-season cotton, Gossypium hirsutum (W.P. Sappenfield); soil genesis, classification and interpretation (C.L. Scrivner); breeding and genetics of wheat and oats (D.T. Sechler); losses of pesticides, sediments and nutrients from farmland producing soyabeans (G.E. Smith); maize breeding and genetics (M.S. Zuber); physiological genetics with arabidopsis (G.P. Redei); cytogenetics of wheat (G. Kimber); physiology and growth of the soyabean plant (D.G. Blevins); soil fertility - maximizing plant nutrient utilization (R.G. Hanson); macro and micro nutrients in cropping systems in the northern cotton belts (J.A. Roth); effects of weather on corn growth (M.E. Keener); organic pesticides in soil and their effects on soil microorganisms (L.B. Hughes).

Animal Health and Disease Department 847

Activities: Annual projects in the field of animal pathology; diseases in birds, cattle, pigs, sheep; fungal diseases in animals.

Animal Science Department 848

Chairman: Dr Bobby D. Moser
Projects: Nitrogen assimilation by ungulates and associated microorganisms (J.M. Asplund); reproductive efficiency (B.N. Day); swine nutrition (T.L. Veum); post-weaning management and breeding of cattle (J.A. Paterson); producing beef cows and calves on forages (R.E. Morrow); swine management (G.W. Jesse); beef cattle breeding (D.W. Vogt).

Atmospheric Science Department 849

Chairman: Dr Wayne L. Decker
Projects: Energetics and climate dynamics of the atmospheric general circulation (E.C. Kung); characterization of Missouri's climatic resources (W.L. Decker); energy interactions in mesoscale systems (G.L. Darkow).

Biochemistry Department 850

Chairman: Dr Milton S. Feather
Projects: Trace element analysis of agricultural materials (E.E. Pickett); micronutrients: metabolism and metabolic function (B.L. O'Dell); physiological response of animals to forage constituents (G.B. Garner); chromatographic characterization and automated analysis of biologically important molecules (C.W. Gehrke); genetics and regulation of nitrogen fixation in photosynthetic bacteria (Judy D. Wall); amine-assisted carbohydrate dehydration reactions (M.S. Feather); physiology and genetics of the soyabean ureolytic enzymes (J.C. Polacco).

Dairy Science Department 851

Chairman: Dr Fredric A. Martz
Projects: Endocrine factors in mammary gland growth and lactation (R.R. Anderson); forage nitrogen utilization in the dairy cow (R.L. Belyea); dairy farm enterprise, and utilization of forages by dairy cattle (F.A. Martz); limiting environmental stress on growth, reproduction and lactation physiology of cattle (H.D. Johnson); endocrine mechanisms controlling bovine reproduction (H.A. Garverick); physiological and biophysical mechanisms in cell preservation (C.P. Merilan); non-surgical transfer and low temperature storage of bovine embryos (J.D. Sikes).

Entomology Department 852

Chairman: Dr Thomas R. Yonke
Projects: Metabolism and mode of action of acaricides and insecticides (C.O. Knowles); bionomics and control of soil inhabiting arthoropods (A.J. Keaster); development physiology of plant-feeding insects (G.M. Chippendale); management of grain and soyabean insect pests (A.J. Keaster); biology and control of arthropod pests of livestock and humans (R.D. Hall); biology and systematics of Hemiptera and Homoptera (T.R. Yonke); biological control of insect pests and weeds (R.L. Kirkland); systems science and mathematical modelling for insect pest management in cropping systems (R.H. Ward).

Extension and Education Department 853

Chairman: Dr John G. Gross

Food Science and Nutrition Department 854

Chairman: Dr H. Donald Naumann
Projects: Function, nutritive composition, quality, stability and efficient production of poultry products (O.J. Cotterill); psychrotrophic bacteria: their enzymes and substrates in foods and food processing (R.T. Marshall); processing and distribution factors that influence meat utilization (W.C. Stringer); controlling energy usage during heat processing (Nan F. Unklesbay); effect of food processing and dietary component characteristics on nutrient bio-availability (D.T. Gordon); nutritional quality and safety of corn and soyabeans and corn-soyabean mixtures by fermentation (M.L. Fields).

Home Economics Department 855

Projects: Quality of life as influenced by area of residence (E.J. Metzen); metabolic interactions among iron, copper, and molybdenum: effect of iron nutriture (R.P. Dowdy); family resource management (I.F. Beutler); effects of exercise on trace mineral status and requirements of women (Donna M. Jeffery); nitrogen, amino acid and energy requirements of guinea pigs and humans (Helen L. Anderson); trace mineral nutrition of pregnancy, lactation and infancy (J.T. Typpo); body composition, blood lipids, hypertension and electrocardiographic changes with ageing in humans (Margaret A. Flynn).

Horticulture Department 856

Chairman: Dr Donald A. Hegwood
Projects: Vegetable environmental-hereditary interrelationships (V.N. Lambeth); greenhouse-grown potted plants (M.N. Rogers); turfgrass genotypes for stress resistance and energy and water conservation (J.H. Dunn); toxic trace substances in the human environment (D.D. Hemphill); promotion of horticulture (D.A. Hegwood); rooting and establishment of cuttings of woody ornamental plants (C.J. Starbuck); increased yield and quality of small fruit crops (J.M. Williams); energy conservation in greenhouses (D.H. Trinklein); increasing the productivity, quality, and nutritive value of fruit and nut crops (J.A. Hopfinger); nutrient testing procedures for soilless media (N.J. Natarella).

Plant Pathology Department 857

Chairman: Dr Victor H. Dropkin
Projects: Phytonematodes: soyabean cyst and pine wilt (V.H. Dropkin); virus-like and mycoplasma disorders of forest, nursery and orchard crops (D.F. Millikan); role of a genome-linked protein in viral pathogenesis and structure (O.P. Sehgal); ecology of soil-borne fungi (T.D. Wyllie); response of maize and sunflower to fungal pathogens and aflatoxin producers (O.H. Calvert); control of the soyabean cyst nematode and other root diseases (C.H. Baldwin); biochemistry and biophysics of host-pathogen interactions (A.L. Karr); fruit disease control and pest management (P.W. Steiner); bacterial disease-related alterations in membrane function (A. Novacky); aetiology, disease losses and control methods in soyabean and wheat diseases (E.W. Palm); ultrastructure of vesicular-arbuscular mycorrhizae and plant pathogenic fungi (M.F. Brown).

Poultry Science Department 858

Chairman: Dr James E. Savage
Projects: Nutrient utilization in growth and reproduction of poultry (J.E. Savage); layer production and management practices (J.M. Vandepopuliere).

Rural Sociology Department 859

Chairman: Dr Michael F. Nolan
Projects: Community development (J.K. Benson); sociological analysis of agricultural structures and organizations (W.D. Heffernan); population change in nonmetropolitan Missouri 1970-1985 (R.R. Campbell); assessing technology impacts (J.L. Gilles).

School of Forestry, Fisheries and Wildlife 860

Chairman: Dr Donald P. Duncan
Projects: Hardwood ecology (G.S. Cox); forest hydrology of small karst watersheds in the Missouri Ozarks (C.D. Settergren); influence of environmental factors on wood formation and properties (E.A. McGinnes); walnut multicropping management (H.E. Garrett); ecology and management studies of stream fishes (T.R. Finger); ecological and management studies of furbearing mammals (E.K. Fritzell); freezing avoidance by deep supercooling of water in woody plants (M.F. George); drying behaviour and utilization of woods and residues (B.E. Cutter); ashland wildlife research area (E.K. Fritzell); wildlife and selected plant species in lowland hardwood wetlands (L.H. Fredrickson); terrestrial wildlife habitat (T.S. Baskett).

Veterinary Medicine Department 861

University of Nebraska - Lincoln* 862

Status: Educational establishment with r&d capability
Chancellor: Roy A. Young

INSTITUTE OF AGRICULTURE AND NATURAL RESOURCES* 863

Address: Lincoln, NE 68588-0422
Vice Chancellor: Martin A. Massengale

Nebraska Agricultural Experiment Station 864

Address: University of Nebraska - Lincoln, 221 Agricultural Hall, Lincoln, NE 68583-0704
Telephone: (402) 472 2045
Dean and Director: Roy G. Arnold
Graduate research staff: 435
Activities: The station conducts research work in the following subject areas: natural resources, including soil science, water resources, conservation systems, environmental programmes, climatology and meteorology, irrigation, land use and economics, and wildlife; plant production (corn, sorghum, soya beans, wheat, range and forage crops, dry edible beans, sugarbeets; turf grass, ornamentals, vegetables) including plant breeding, crop husbandry, and plant protection; animal production (beef, swine, dairy, poultry) including livestock breeding, livestock husbandry and nutrition, and animal health and disease; agricultural engineering, including buildings and facilities (livestock, grain storage), water and irrigation, energy, power and machinery; food science; forestry - shelterbelts, tree improvement; agricultural economics; human nutrition; human development.
Publications: Nebraska Farm, Ranch and Home Quarterly; Annual report.

Agricultural Biochemistry Department 865
Head: Dr H.W. Knoche
Senior staff: Nutritional biochemistry, Dr R.L. Borchers; plant disease biochemistry, Dr J.M. Daly; enzymes, Dr F.W. Wagner

Agricultural Communications Department 866
Head: Dr R.L. Fleming

Agricultural Economics Department 867
Head: Dr G.J. Vollmar
Senior staff: Marketing, Dr G.D. Anderson, Dr J.B. Hassler, Dr J.G. Kendrick, Dr M.S. Turner; resource economics, Dr M.E. Baker, Dr L.K. Fischer; farm management, Dr L.L. Bitney; production economics, Dr G.A. Helmers

Agricultural Engineering Department 868
Head: Dr W.E. Splinter
Senior staff: Livestock - insect control, Dr I.L. Berry; livestock environment, J.A. Deshazer; irrigation engineering, Dr W.F. Kroutil; tractor testing, Dr L.I. Leviticus; product processing and systems analysis, Dr T.L. Thompson, systems engineering, Dr K.L. Von Bargen

Agricultural Meteorology and Climatology Centre 869
Head: Dr N.J. Rosenberg

Agronomy Department 870
Head: Dr R.G. Gast
Senior staff: Weed science, Dr O.C. Burnside; sorghum physiology, Dr R.B. Clark; crop breeding, Dr W.A. Compton; crop physiology, Dr J.D. Eastin, Dr C.Y. Sullivan; crop production, Dr A.D. Flowerday, Dr D.G. Hanway; statistics, genetics, Dr C.O. Gardner; forage genetics, Dr H.J. Gorz, Dr F.A. Haskins; winter wheat, Dr V.A. Johnson; alfalfa, Dr W.R. Kerr; soil microbiology, Dr T.M. McCalla; soil fertility, Dr G.A. Peterson, Dr J. Power, Dr R.C. Sorensen, Dr R.A. Wiese; noncropland weed control, Dr M.K. McCarty; cytogenetics, Dr M.R. Morris; forage physiology, Dr L.E. Moser; water resources and crop modelling, Dr J.M. Norman; sorghum breeding, Dr W.M. Ross; small grain breeding, Dr J.W. Schmidt; soil physics, Dr D. Schwartzendruber; soya bean breeding, Dr J.H. Williams

Animal Science Department 871
Head: Dr I.T. Omtvedt
Senior staff: Ruminant nutrition, Dr T.J. Klopfenstein; dairy production, Dr F.E. Eldridge; poultry products, Dr G.W. Froning; poultry production, Dr E.W. Gleaves; beef nutrition, Dr P.Q. Guyer, Dr J.K. Ward; meats, Dr R.W. Mandigo; dairy nutrition, Dr F.G. Owen; swine nutrition, Dr E.R. Peo; poultry nutrition, Dr T.W. Sullivan; swine physiology, Dr D.R. Zimmerman
Publications: Nebraska Beef Report; Nebraska Swine Report; Nebraska Dairy Report (annuals).

Biometrics and Information Systems Centre 872
Head: Dr W.M. Schutz
Senior staff: Statistics, Dr R.F. Mumm, Dr W.W. Stroup

Education and Family Resources Department 873
Head: Dr G. Newkirk

Entomology Department 874
Head: Dr E.A. Dickason
Senior staff: Insect physiology, Dr H.J. Ball; forage insect investigations, Dr G.R. Manglitz; biology and ecology of crop insects, Dr K.P. Pruess; veterinary entomology, Dr W.M. Rogoff; insect transmission of plant pathogens, Dr R. Staples

Environmental Programmes Division 875
Head: Dr R.E. Gold

Food Science and Technology Department 876
Head: Dr A.L. Branen
Senior staff: Food quality, Dr T.E. Hartung; food microbiology, Dr R.B. Maxcy; protein chemistry, Dr L.D. Satterlee; food chemistry, Dr K.M. Shahani

Forestry, Fisheries and Wildlife Department 877
Head: Dr G.L. Hergenrader

Horticulture Department 878
Head: Dr R.D. Uhlinger
Senior staff: Dr D.P. Coyne, Dr E.J. Kinbacher, Dr R.E. Neild, Dr M.L. Schusler

Human Development and the Family 879
Department
Head: Dr N. Stinnett

Human Nutrition and Food Service 880
Management Department
Head: Dr H. Fox
Senior staff: Nutrition, Dr C. Kies; food service management, Dr M.E. Knickrehm

North Platte Station 881
Head: Dr L.J. Sumption
Senior staff: Entomology (livestock insects), Dr J.B. Campbell; beef, Dr D.C. Canton; swine, Dr D.M. Danielson; range management, Dr J.T. Nichols

Northeast Station 882
Head: Dr Cal J. Ward
Senior staff: Agronomy, Dr G.W. Rehm

Panhandle Station 883
Head: Dr J.L. Weihing
Senior staff: Animal science, Dr D.B. Hudman; plant pathology, Dr E.D. Kerr; horticulture, Dr R.B. O'Keefe

Plant Pathology Department 884
Head: Dr M.G. Boosalis
Senior staff: Virus diseases, Dr M.K. Brakke; tree diseases, Dr G.W. Peterson; microbial physiology, Dr J.L. Van Etten; bacterial diseases, Dr A.K. Vidaver

Roman L. Hruska US Meat Animal Research 885
Centre
Head: Dr R. Oltjen
Senior staff: Agricultural engineering, Dr Y.R. Chen, Dr G.L. Hahn; physiology research, Dr R.K. Christenson, Dr J.D. Crouse, Dr S.E. Echternkamp, Dr J.J. Ford, Dr D.D. Lunstra, Dr R.R. Maurer, Dr J.C. Pekas, Dr B.D. Schanbacher, Dr R.T. Stone, Dr C.W. Weems; food technology, Dr H.R. Cross; genetics, Dr L.D. Young; chemistry, Dr C.L. Ferrell, Dr H.J. Mersmann, Dr R.L. Prior, Dr L.L. Richer; agricultural engineering, Dr G.L. Hahn, Dr A.G. Hashimoto; agricultural economy, Dr V.L. Harrison; animal science, Dr J.L. Koong, Dr W.G. Pond, Dr W.E. Wheeler, Dr J.T. Yen; microbiology, Dr R.C. Manak, Dr V.W. Varel

South Central Station 886
Head: Dr C. Stonecipher
Senior staff: Plant pathology, Dr B.L. Doupnik

Southeast Extension and Research Centre 887
Head: Dr L. Young

Textiles, Clothing, and Design Department 888
Head: Dr A. Newton
Activities: Flame-retardant fabrics; burn injury studies; clothing comfort.

University Field Laboratory 889
Head: Dr Warren W. Sahs

Veterinary Science Department 890
Head: Dr E.O. Dickinson
Sections: Parasitology, Dr D.L. Ferguson; microbiology, Dr M.L. Frey

Nebraska Water Resources Centre 891

Address: 310 Agricultural Hall, University of Nebraska, Lincoln, NE 68583-0710
Telephone: (402) 472 3305
Director: William L. Powers
Activities: The Nebraska Water Resources Centre is one of the divisions of the University of Nebraska Institute of Agriculture and Natural Resources. The primary function of the centre is to bring together water researchers, users and funding sources. It also functions as a centre for integrating university training and water research programmes with the needs and efforts of federal, state and local agencies. The staff are also involved at various times in contract and grant research, teaching water resources courses and providing water information to various publics.

Five categories have been developed as focal points for the centre's research programme: water quantity management; water use efficiency and conservation (supply and demand management); water quality; natural disaster prediction response; legal, institutional, economic and social aspects. Along with a comprehensive research programme, the centre has engaged in many aspects of scientific information dissemination.

Projects: Reduction in development of bloomforming blue-green algae by nutrient enrichment to maintain desirable pre-bloom dominants (James Rosowski); enhancement of water quality in Nebraska farm ponds by control of eutrophication through biomanipulation (Gary L. Hergenrader); parasite communities as indicator systems for predicting the effects of surface water management options on the biota of prairie rivers (John Janovy); increased water conservation and percolation through improved tillage practices (Howard Wittmuss); conservation of soil, water and energy through reduced tillage systems (Elbert C. Dickey); water conservation through limited irrigation of corn and grain sorghum in the Great Plains (Darrell G. Watts); tillage practice effects on water conservation and the efficiency and management of surface irrigation systems (Dean Eisenhauer); high Plains-Ogallala aquifer study (Raymond Supalla).

University of Nevada* 892

Status: Educational establishment with r&d capability

NEVADA AGRICULTURAL 893
EXPERIMENT STATION*

Address: Reno, NV 89557
Telephone: (702) 784 6611
Dean and Director: Dr Dale W. Bohmont
Activities: Conducts basic and applied research on agricultural and natural resource problems of the state. Maintains several field research stations in Nevada state.

University of Southwestern 894
Louisiana

– USL
Status: Educational establishment with r&d capability

CENTER FOR GREENHOUSE 895
RESEARCH

Address: PO Box 40847, USL Station, Lafayette, LA 70504
Telephone: (318) 264 6486
Director: Dr J. Robert Barry
Graduate research staff: 3
Activities: The aim of this research centre is to provide information relating to production of greenhouse crops for commercial growers. Areas of research are: variety testing (tomatoes, cucumbers and lettuce); growth medium studies (rice-hulls, pine bark, sand, hadite and mixtures of these materials are under comparison as media for tomato production); cultural practice evaluations (such as tomato fruit pruning and stage of maturity for harvesting tomato fruits); greenhouse pest control (control for disease and insect pests of greenhouse vegetable crops).
Projects: Greenhouse vegetable research (Dr J. Robert Barry); growth medium studies (Dr J. Robert Barry); tomato and cucumber variety trials (Gregory Davis); cultural practice research (Gregory Davis).

CRAWFISH RESEARCH CENTER 896

Address: USL Box 44650, Lafayette, LA 70504
Telephone: (318) 264 6647
Director: Don Gooch
Graduate research staff: 2
Activities: Research into crawfish production.
Projects: In crawfish (Procambarus clarkii) ponds: water quality; forages; carp culture for bait; water recirculation and turbidity; population structure; brood stock counts.

University of Tennessee* 897

Status: Educational establishment with r&d capability

INSTITUTE OF AGRICULTURE* 898

Tennessee Agricultural Experiment 899
Station

Address: PO Box 1071, Knoxville, TN 37901
Telephone: (615) 974 7121
Dean: D.N. Gossett
Departments: Agricultural economics, Joe A. Martin; agricultural engineering, Houston Luttrell; animal science, (acting) D.O. Richardson; entomology and plant pathology, C.J. Southards; food technology and science, J.T. Miles; forestry, wildlife and fisheries, Gary Schneider; ornamental horticulture and landscape design, D.B. Williams; plant and soil science, L.F. Seatz
Publications: Annual CRIS reports.

University of Vermont 900

– UVM
Status: Educational establishment with r&d capability

VERMONT AGRICULTURAL 901
EXPERIMENT STATION

Address: Morrill Hall, Burlington, VT 05404-0106
Telephone: (802) 656 2980
Dean and Director: Dr Robert O. Sinclair
Assistant director: R.J. Hopp
Graduate research staff: 41
Activities: Dairy: reproduction, diseases, forage crops, milk and dairy products, marketing; poultry: nutrition, physiology, management; maple: maple trees, maple sap collection, maple syrup production; human nutrition and consumerism: nutritional status, cancer and nutrition, diet and nutrition; crops and soils: culture, disease

and pest control, winter hardiness, soils; environment and ecology: water resources and pollution, forest management; recreation: resources and control; rural development: planning, land use, socio-economic development; basic research: physiology, biochemistry, and bacteriology of plant growth.

Publications: Vermont Science (Annual).

Agricultural and Resource Economics Department   902

Head: Dr Fred C. Webster

Animal Sciences Department   903

Head: Dr Leonard S. Bull

Basic Research Section   904

Projects: Ultrastructural comparison of the cytoplasm in normal and male - sterile wheat (B.B. Hyde); chemotaxis and root nodulation by rhizobia (W.W. Currier); membrane potentials and the mechanism of amino acids transport in plants (B. Etherton); biochemistry of ribosomes and selected proteins (D.L. Weller); biochemistry of plant glycoproteins (D.W. Racusen, M.W. Foote); responses of plants to visible and ultraviolet radiation (R.M. Klein); plant tissue culture under conditions of anisotropic mechanical stress (P.M. Lintilhac); genes regulating the life cycles and plant pathogenicity in fungi: a molecular approach (R.C. Ullrich); precursors in maple sap affecting syrup quality (M.F. Morselli).

Botany Department   905

Head: Dr Hubert W. Vogelmann

Crops and Soils Research Section   906

Projects: Evaluation and adaptive studies of important horticultural and agronomic crops (S.C. Wiggans, B.R. Boyce, W.M. Murphy); chemical control of major insect pests affecting apples (G.B. MacCollom); virus and mycoplasma diseases of deciduous tree fruits and grapevines (B.R. Boyce); pesticide effects on two insect predators in fruit-growing areas (B.L. Parker); physiological and genetic functions of rhizomorphs in the root-rotting fungus Armillaria (R.C. Ullrich); scion/rootstock and interstem effects on apple-tree growth and fruiting; use of growth regulators to hasten development and improve the performance of apple trees (J.F. Costante); insects and diseases adversely affecting agronomic and horticultural crops in Vermont (G.B. MacCollom, A.R. Gotlieb); pesticide impact assessment for Vermont (G.B. MacCollom, R.E. Desrosiers); development of a fast, efficient, serological assay for Scleroderris canker (A.R. Gothleb); influence of low winter temperatures on several northern-grown fruit plants (B.R. Boyce); adaptation of nursery propagation and overwintering practices for northern New England (N.E. Pellett); soil properties affecting sorption of heavy metals from wastes; chemical identification of deficient and excessive nutrients in soil (R.J. Bartlett).

Dairy Research Section   907

Projects: Development of a controlled ovulation and pregnancy programme in cattle; identification of a pheromone involved in mating behaviour in cattle (K.R. Simmons); infectious diseases affecting reproduction in cattle (W.D. Bolton, R.W. Murray); systematic anthelmintic treatment of dairy replacement heifers (J.R. Kunkel, W.D. Bolton); development of tests for early detection of Brucella abortus and other infections in cattle (J.R. Kunkel); economic impact of internal parasitism control measures in dairy heifers (J.R. Kunkel); relative profitability of dairy cows and its relationship to selection traits (J.A. Gilmore); factors affecting voluntary intake of minerals (J.G. Welch); management practices and profitability in dairy and livestock production (A.M. Smith, J.R. Kunkel); feasibility study of solar heating systems for Vermont dairy production (G.D. Wells); feasibility of perennial ryegrass and hybrid derivatives as northern forage crops (G.M. Wood, J.G. Welch); optimizing intake and digestibility of forages by ruminants (J.G. Welch, A.M. Smith); increasing the use of legumes in crop rotations (W.M. Murphy); quality changes in milk production and processing from farm to consumer (H.V. Atherton, A.H. Duthie); cheese manufacturing and whey utilization (K.M. Nilson); quantitative extraction and analysis of free fatty acids from milk; quantitative extraction and analysis of fatty acids from Mozzarella cheese (A.H. Duthie); analysis of the spatial organization of the Northeast dairy industry; economic analysis of multiple component pricing of milk; milk hauling in New England: cost and economic impact (F.C. Webster, C.L. Fife).

Environmental and Ecology Research Station   908

Projects: Antibiotic and heavy metal-resistant bacteria in Lake Champlain water; biodegradation in a freshwater ecosystem (R.E. Sjogren); effects of hydrology, geomorphology, climate, and production upon peatland development in Vermont; preparation of a classification scheme for the peatlands of New England (I.A. Worley); effects of accumulation of heavy metals on forest and alpine ecosystems in the Green Mountains of Vermont (H.W. Vogelmann); land disposal of sludge from small towns in rural districts (F.R. Magdoff, R.E. Sjogren, G.D. Wells, R.G. Fritz, E.A. Cassell, C.A. Phillips); production of biomass for energy on abandoned farmland in the Northeast (F.M. Laing).

Human Nutrition and Consumerism Research Section 909

Projects: Human nutrition improvement (S.M. Hopp, E.D. Schlenker, R.P. Clarke); faecal steroid excretion and 7-dehydroxylase and B-glucuronidase activities in colon cancer patients (J.K. Ross); nutrient intake analysis of 400 University of Vermont students (L.B. Carew, Jr, W. Helstowski); evaluation of nutritional techniques to lower egg cholesterol (L.B. Carew, Jr, M. Henault, Joline Dion); nutritional status of selected participants in Title VII nutrition programme in Vermont (E.D. Schlenker, S.M. Hopp, R.P. Clarke).

Human Nutrition and Foods Department 910

Head: Dr Eleanor D. Schlenker

Maple Research Station 911

Projects: Growth, vigour, and sap sugar content of young sugar maple trees (F.M. Laing); maple-processing equipment and related problems; vacuum techniques for maple sap collection: effect on product and tree (F.M. Laing, M.F. Morselli).

Microbiology and Biochemistry Department 912

Head: Dr Robert E. Sjogren

Plant and Soil Science Department 913

Head: Dr Alan R. Gotlieb

Poultry Research Station 914

Projects: Mechanism of action of essential fatty acids as medicated by the pituitary gland (L.B. Carew, Jr, D.C. Foss); effect of restricted feeding time on performance of brown-egg caged layers (D.C. Foss, L.S. Mercia, L.B. Carew, Jr); nutritional research of international value: animal nutrition, primarily poultry, that will be useful in developing countries (L.B. Carew, Jr); biophysical factors affecting energy requirements for poultry production; poultry and nutrition (D.C. Foss).

Recreation Research Section 915

Projects: Breeding and evaluation of Kentucky bluegrass and associated species for turf (G.M. Wood); recreation marketing adjustments in the Northeast (M.I. Bevins).

Rural Development Research Section 916

Projects: Economic analysis to facilitate land use planning decision making in small towns; socio-economic analysis of development and decline of small towns (F.O. Sargent); value transfer associated with the private-public land interface in Vermont (A.H. Gilbert); economic analysis of producing and marketing selected commodities in Vermont; enterprise selection for alternative agriculture in Vermont (N.H. Pelsue, Jr); availability of educational services for the rural craftsperson; training needs for rural craftsmen (T.K. Bloom); impact of in and out migration and population redistribution in the Northeast (M.I. Bevins, R.O. Sinclair); potential for improving incomes of small farms in Vermont; socio-economic factors and rural land use (R.O. Sinclair, R.H. Tremblay, N.H. Pelsue, Jr); improving the distribution of socio-economic resources in rural areas; Vermont community data bank (F.E. Schmidt); service delivery arrangements for older people in the Northeast (R.T. Coward).

Textiles, Merchandising and Consumer Studies Department 917

Head: Sylvia J. Emanuel

Vocational Education and Technology Department 918

Head: Gerald R. Fuller

University of Washington* 919

Address: Seattle, WA 98195
Status: Educational establishment with r&d capability

COLLEGE OF FOREST RESOURCES 920

Address: University of Washington AR-10, Seattle, WA 98195
Telephone: (206) 543 2730
Dean: David B. Thorud
Associate dean, research: Dr Stanley P. Gessel
Associate dean, instruction: Dr Thomas R. Waggener

Institute of Forest Resources   921

Director: David B. Thorud
Senior staff: G. Graham Allan, Mr Bare, James S. Bethel, Caroline S. Bledsoe, Gordon Bradley, Linda B. Brubaker, Ben S. Bryant, Mr Burke, Dale W. Cole, Mr Dawson, Mr Dowdle, Charles H. Driver, Robert L. Edmonds, Donald R. Field, Leo J. Fritschen, Robert I. Gara, Dr Stanley P. Gessel, Francis Grenlich, Charles C. Grier, Mr Hatheway, Bjorn F. Hrutfiord, Jens Jorgensen, Lawrence Leney, Mr Leopold, David A. Manuwal, Mr Morison, Chadwick D. Oliver, Steward G. Pickford, Mr Rustagi, Kyosti V. Sarkanen, Peter Schiess, Gerard F. Schreuder, David R.M. Scott, Mr Sharpe, Reinhard Stettler, Richard D. Taber, Mr Thomas, Fiorenzo C. Ugolini, Mr van Klaveren, Kristiina A. Vogt, Dr Thomas R. Waggener, Mr Witt, David D. Wooldridge, Robert J. Zasoki
Graduate research staff: 54
Activities: The institute is the research, continuing education, and information branch of the College of Forest Resources. Besides administering federally funded and state-supported programmes in research, it coordinates cooperatively sponsored research and teaching programmes with federal, state, and private agencies.

Center for Forest Ecosystem Studies   922

Director: Dr Dale W. Cole
Activities: The centre provides administrative supervision for all lands of the college, including both the arboretum and forest properties. In addition, it is responsible for the college research programmes in the biological areas. The interests of the faculty working in the biological-based investigations are diverse, ranging from basic considerations of plant growth to the application of such information to the analysis of forest ecosystems. Research projects within the centre include both individual studies concerned with the many aspects of forest ecosystems and interdisciplinary programmes such as Ecosystems Studies.
Projects: Dynamics of soil forming processes in arctic Alaska; palaeoclimate, palaeoecology, and biogeographical significance of a glacial refuginim in Southeast Alaska (Fiorenzo C. Ugolini); effects of Mount Saint Helen's tephra on belowground biological processes in Abies amabilis zone forests of Washington (Charles C. Grier); effects of acid rain on nutrient status of forest ecosystems (Dale W. Cole); nutritional and toxic effects of sewage sludge in forest ecosystems (Stephen D. West, Robert J. Zasoski).

Center for International Forest Resources Studies   923

Director: Dr James S. Bethel
Activities: Development of and assistance with programmes of study of forest resources in other lands and their products with respect to their biology, management, economics, manufacture, legislation, and administration. Specific programmes now active include studies of tropical forest ecosystems in Latin America and Thailand, analysis of alternatives in the utilization of tropical forest in Honduras, solution of problems in forest utilization in Thailand, foreign log supply and the domestic market, national parks in Central America, and control of insect pests of mahogany.

Center for Resource Management Studies   924

Director: Dr Gerard Schreuder
Activities: Land-use planning and decision making in forest management and forest industry; public policies as they influence land use, resource management, outdoor recreation, and the forest industry; goods and services and environmental protection in resource management, harvesting, and wood processing; improving the yield on the utilization of forest resources.
Projects: Effect of delayed application of a controlled release herbicide and fertilizer on the growth of Douglas-fir seedlings (G. Graham Allan); cooperative forest lands monitoring project around Northwest Alloys, Incorporated, Plant Site, Addy, Washington (Gerard F. Schreuder); duration of fire stages (David V. Sandberg); colour from polyphenolic extractives in pulping and bleaching (Bjorn F. Hrutfiord); polymer-optical interactions (G. Graham Allan); demonstration of feasibility of using agricultural residues for low-cut roofing panels (Ben S. Bryant); investigation of sulphur deficiency and growth response to sulphur in Douglas-fir stands (Stanley P. Gessel).

SCHOOL OF FISHERIES   925

Address: Seattle, WA 98195
Telephone: (206) 543 4270
Director: Donald E. Bevan
Graduate research staff: 85

Division of Fisheries Science and   926
Aquatic Ecology

Chairman: Dr Bruce Miller
Activities: Biology of marine fish; life histories, stocks physiology, biochemistry and general ecology; stock assessment and fisheries management; pollution and food chain research.

Division of Food Science and Technology 927

Chairman: Dr John Liston
Activities: General food science; seafood technology; food microbiology; food chemistry; food engineering.

Division of Quantitative Science in Fisheries 928

Chairman: Dr Douglas Chapman
Activities: Mathematics of wild and cultured fish populations; population dynamics and quantitative aspects of fisheries management; statistics and general biomathematics.

Fisheries Research Institute 929

Director: Dr Robert L. Burgner

University of Wisconsin - Madison* 930

Address: 1500 Johnson Drive, Madison WI 53706
Telephone: (608) 263 2197
Status: Educational establishment with r&d capability
Chairman, materials science programme: Professor R.A. Dodd
Sections: Surface science, Professor M.G. Lagally; phase transformations, Professor J.H. Perepezko; superconductivity, Professor D.L. Larbalesbier; nuclear materials, Professor G.L. Kulcinski

FOOD RESEARCH LABORATORY* 931

Address: 1925 Willow Drive, Madison, WI 53706
Telephone: (608) 263 7777
Director: Dr E.M. Foster
Activities: Food microbiology and toxicology.

WISCONSIN AGRICULTURAL EXPERIMENT STATION* 932

Address: 136 Agricultural Hall, Madison, WI 53706
Telephone: (608) 262 1254
Director: Dr Leo M. Walsh

University of Wisconsin - River Falls* 933

Status: Educational establishment with r&d capability

COLLEGE OF AGRICULTURE 934

Address: River Falls, WI 54022
Telephone: (715) 425 3841
Director: Roger A. Swanson
Departments: Agricultural economics, Dr R. Vern Elefson; agricultural education, Dr Marvin D. Thompson; animal and food science, Dr Dean Henderson; agricultural engineering and technology, Dr Charles Jones; plant and earth science, Dr Sam Huffman
Graduate research staff: 52
Activities: Basic and applied research in the agricultural sciences: maize, forage crops, fruit crops, pigs, sheep, dairy management, dairy foods.
Projects: Projected growth of irrigated acreages in the Wisconsin central sands (Dr J. Shatava); evaluation of progress from recurrent selection for grain yield and stalk quality in maize (Dr S.K. Carlson); high-fat feeding in late gestation and lactation-survival and growth of baby pigs (Dr P.B. George); improvement of quackgrass for use as a forage species (Dr S. Carlson, Dr L. Greub); use of antimicrobial peptides to extend the shelf life of dairy foods (Dr P.C. Vasavada); combined gas-exchange/modified-atmosphere systems for the preservation of processed raw vegetables (Dr S.C. Ridley); evaluation of the understanding and use of the DHI somatic cell counting programme in dairy management (Dr James Schwalm); influence of feeding time and feed quality on parturition in ewes (Dr A. Jilek, R. Erickson); adaptability and culture of selected fruit crops in western Wisconsin (Dr R. Tomesh).

RESEARCH CENTER* 935

Telephone: (715) 425 3841
Dean: Dr Roger A. Swanson

University of Wyoming* 936

Status: Educational establishment with r&d capability

WYOMING AGRICULTURAL EXPERIMENT STATION 937

Address: Box 3354, University of Wyoming, Laramie, WY 82071
Telephone: (307) 766 4133
Dean and director: Harold J. Tuma
Associate director, research: C.C. Kaltenbach
Graduate research staff: 75
Activities: Most aspects of crop and livestock production in the arid west are covered. Areas of emphasis include range use and management; water use and management; physiology, nutrition, disease and management aspects of sheep and cattle production;

quality and safety of meat and meat products; feed and forage crop production as influenced by insects, diseases; economics of producing and marketing agricultural products and basic biochemical mechanisms of plant and animal production.

Agricultural Economics Department   938

Head: Andrew Vanvig
Projects: Farm and ranch real estate values in Wyoming (Andrew Vanvig, J.L. Stephenson).

Agricultural Engineering Department   939

Head: James L. Smith

Agricultural Extension Department   940

Head: I. Skelton

Animal Science Department   941

Head: Frank C. Hinds
Projects: Reproductive performance in domestic ruminants (T.G. Dunn, C.C. Kaltenbach, T.R. Varnell); optimization of the use of range and complementary forages for red meat production (J.W. Waggoner, H.D. Radloff, C.C. Kaltenbach).

Biochemistry Department   942

Head: Ivan Kaiser

Home Economics Department   943

Head: Margaret Boyd

Microbiology and Veterinary Medicine 944
Department

Head: Alan C. Pier
Projects: Protection of livestock against internal parasites by management methods (R.C. Bergstrom, N. Kingston, W.R. Jolley); exploratory biological studies (J.O. Tucker).

Plant Science Department   945

Head: Lee Painter
Projects: Biology and control of field crop insects (C.C. Burkhardt); potato selection and production (K.E. Bohnenblust, L.H. Paules); physiological criteria for forage plant breeding (R.H. Delaney, K.E. Bohnenblust); biology and control of insects affecting livestock; development of integrated strategies for management of mosquito populations (J.E. Lloyd); weed control in agronomic crops (N.E. Humburg, H.P. Alley, A.F. Gale); phenology and productivity of arid lands;

biotic communities of forests and adjacent grazing lands (H.G. Fisser); evaluation of grasses and legumes for forage and seed production in Wyoming (R.H. Abernethy, H. Radloff); evaluation of field crops in Wyoming (L.R. Richardson); physiological effects of chemical and biological agents on insect pests (F.A. Larson); soil management for crop nutrition in Wyoming (H.W. Hough, P.C. Singleton); adaptation and production evaluations of small grain varieties (B.J. Kolp); basic research on biology and behaviour of Wyoming insects; identity and distribution of Wyoming insects (R.J. Lavigne).

Upjohn Company *   946

ASGROW SEED COMPANY *   947

Address: 7000 Portage Road, Kalamazoo, MI 49001
Telephone: (616) 323 4000
Status: Research centre within an industrial company
Research director: G. Chicco
Activities: Development of superior varieties and hybrids of vegetable and agronomic crops; applied research in seed production and seed technology.

Utah State University   948

Address: Logan, UT 84322
Status: Educational establishment with r&d capability

UTAH AGRICULTURAL EXPERIMENT   949
STATION

Address: Utah State University (UMC 48), Logan, UT 84322
Telephone: (801) 750 2206
Director: Doyle J. Matthews
Graduate research staff: 150
Activities: Investigations basic to the problems of agriculture in its broadest aspects including development and improvement of the rural home and rural life and the maximum contribution of agriculture to the welfare of consumers: soil and water conservation and use; plant and animal production and protection; plant and animal health; processing, distributing, marketing and utilization of food and agricultural products; forestry, including range management and range products, multiple use of forest and range lands; human nutrition and family life; rural and community development; economics in agriculture; irrigation and drainage; basic biology related to agriculture; agricultural engineering and building.

Publications: Annual progress reports, each of 175 projects.

Agricultural Irrigation and Engineering Department 950

Affiliation: College of Engineering
Head: Jack Keller

Animal, Dairy and Veterinary Sciences Department 951

Affiliation: College of Agriculture
Head: S.J. Kleinschuster

Applied Statistics and Computer Science Department 952

Affiliation: College of Science
Head: R.L. Hurst

Biology Department 953

Affiliation: College of Science
Head: G.W. Miller

Chemistry Department 954

Affiliation: College of Science
Head: K. Morse

Economics Department 955

Affiliation: College of Agriculture and Business
Head: W. Cris Lewis

Nutrition and Food Science Department 956

Affiliation: Colleges of Agriculture and Family Life
Head: C.A. Ernstrom

Plant Sciences Department 957

Affiliation: College of Agriculture
Head: Keith R. Allred

Range Science Department 958

Affiliation: College of Natural Resources
Head: D. Dwyer

Soil Science and Biometeorology Department 959

Affiliation: College of Agriculture
Head: J.J. Jurinak

Wildlife Sciences Department 960

Affiliation: College of Natural Resources
Head: G.S. Innis

UTAH COOPERATIVE WILDLIFE RESEARCH UNIT 961

Address: Utah State University (UMC-52), Logan, UT 84322
Telephone: (801) 750 2466
Affiliation: College of Natural Resources
Leader: Dr David R. Anderson
Graduate research staff: 14
Activities: Natural resources; wildlife.
Projects: Remote sensing of large ungulates; line transect sampling; density estimation; capture - recapture sampling (Dr D.R. Anderson); black bear ecology; mule deer ecology; cougar ecology (F. Lindzey).

Washington State University* 962

Address: Pullman, WA 99164
Status: Educational establishment with r&d capability

AGRICULTURAL RESEARCH CENTER 963

Telephone: (509) 335 4563
Director: Dr Landis L. Boyd
Assistant director: D.L. Lee
Graduate research staff: 225
Activities: The four primary areas of research are: maintaining an adequate supply of food and fibre products; natural resource management and conservation; human resources and community development; maintaining a quality environment in rural areas.
Projects: Approximately 320 research projects on all facets of agriculture and rural living.

Agricultural Economics Department 964

Head: L.F. Rogers

Agricultural Engineering Department 965

Head: L.G. King

Agricultural Information Department 966

Head: H.E. Cameron

Agronomy and Soils Department 967

Head: J.C. Engibous

Animal Sciences Department   968

Head: R.L. Preston

Entomology Department   969

Head: E.P. Catts

Food Science and Technology   970
Department

Head: A.L. Branen

Forestry and Range Management   971
Department

Head: D.C. LeMaster

Home Economics Research Center   972

Head: M. Mitchell

Horticulture and Landscape   973
Architecture Department

Head: P.E. Kolattukudy

Irrigated Agriculture Research and 974
Extension Center

Head: L.R. Faulkner

Northwest Washington Research and   975
Extension Unit

Head: R.A. Norton

Plant Introduction Department   976

Head: S.M. Dietz

Plant Pathology Department   977

Head: A.D. Davison

Rural Sociology Department   978

Head: D.A. Dillman

Southwestern Washington Research   979
Unit

Head: C.H. Shanks, Jr

Statistical Services Department   980

Head: T.S. Russell

Tree Fruit Research Center   981

Head: R.P. Larsen

Veterinary Science Department   982

Head: W.G. Huber

Western Washington Research and   983
Extension Center

Head: E.C. Bay

Wood Technology Department   984

Head: T.M. Maloney

Weyerhaeuser Company   985

Address: Tacoma, WA 98477
Publications: Annual report.

Weyerhaeuser Technology   986
Center

Telephone: (206) 924 2345
Telex: 327400
Affiliation: Forest Products Corporation, Weyer-
haeuser Company
Vice-president of R&D: Dr J. Laurence Kulp
Divisions: Forestry and timber products, Dr M.R.
Lembersky; wood products, Dr G.L. Comstock; fibre
products, Dr D.F. Root; energy and environment, Dr
D.R. Raymond; corporate, E.L. Soule
Graduate research staff: 491
Activities: Efforts to maximize use of solar energy for
tree growth, use total forest biomass, achieve lowest cost
manufacturing, implement technological advances
rapidly, and meet environmental and energy requirements economically.

Winston Laboratories   987
Incorporated*

Address: 25 Mount Vernon Street, Ridgefield Park, NJ
07660
Telephone: (201) 440 0022
Status: Independent research centre
Activities: Food analytical chemistry, biochemistry,
and cereal chemistry.

Yale University* 988

Status: Educational establishment with r&d capability

SCHOOL OF FORESTRY AND ENVIRONMENTAL STUDIES 989

Address: Sage Hall, 205 Prospect Street, New Haven, 06511
Acting Dean: Dr William H. Smith
Graduate research staff: 27
Research activities: Tissue culture regeneration of trees, wood formation, DNA characterization of cell and tree populations; effects of toxic and radioactive materials on tree growth, physiology and morphology of plants in alpine and stressed environments (Professor G.P. Berlyn); computer-assisted map analysis, cartography modelling of natural resource systems (Professor J.K. Berry); microeconomics in renewable resource management (Professor C.L. Binkley); structure and function of the temperate forest ecosystem with specific interest in biogeochemistry, hydrology, geology, soils, productivity, nutrient cycling, aquatic ecology, palaeoecology and animal ecology (Professor F.H. Bormann); sociological aspects of natural resources (Professor W.R. Burch); application of biometrical procedures to forestry and ecological problems (Professor G.M. Furnival); policy and management of issues relating to the interaction of people and wildlife, especially animals (Professor S.R. Kellert); effect of ionizing radiation on forest trees, vegetative propagation of trees, hybridization, progeny testing and forest tree improvement (Professor F. Mergen); competition between closely related species and population models for endangered species (Professor R.S. Miller); food resource availability and predation pressure on the social behaviour and spacing systems of mammals (P.D. Moehlman); energy budget of forest stands, meteorological modelling of air pollution dispersal, fire-weather interactions, and bioclimatic indices for outdoor recreation (Professor W.E. Reifsnyder); trace element cycling in terrestrial ecosystems (Dr T.G. Siccama); silviculture practices and structural changes in forests (Professor D.M. Smith); fungal ecology and physiology, tree root exudation and associated rhizosphere ecology and biochemistry, and the relationship between air pollution and woody plants and microorganisms (Professor W.H. Smith); soil science (Professor G.K. Voigt); comparative forest and natural resource policies of developed countries (Professor A.C. Worrell).

UPPER VOLTA

Centre Technique Forestier Tropical - Centre D'Ouagadougou*   1

[Tropical Forestry Technical Centre - Ouagadougou Centre]
Address: Boîte Postale 303, Ouagadougou
Status: Official research centre
Director: J. Piot Icgref
Activities: Research in silviculture and soil erosion.

Comité Permanent Interétats de Lutte contre la Sécheresse dans le Sahel   2

– CILSS
[Permanent Interstate Committee for Drought Control in the Sahel]
Address: BP 7049, Ouagadougou
Telephone: 342-52; 343-55
Telex: 5263 COMITER Ouga
Status: International development association
Executive Secretary: Seck Mame N'Diack
Director of Projects and Programmes: Mr Rapadem-naba
Administrative and Financial Director: Mr Madingar
Director of Documentation and Information: Mr J. Grey-Johnson
Activities: The CILSS Executive Secretariat based in Ouagadougou is the headquarters of an organization composed of eight member states (Cape-Verde Islands, Chad, Gambia, Mali, Mauritania, Niger, Senegal, and Upper Volta) dedicated to the investigation and control of drought in the Sahel region.
Publications: Report of the Council of Ministers (semi annual); *State of Progress of the First Generation Programme* (bi-annual report).

Projects: The First Generation Programme consists of 612 regional and national projects under the Director of Projects and Programmes.

CENTRE FOR OPERATIONAL HYDROMETEOROLOGY AND AGROMETEOROLOGY   3

Address: Niamey, Nigeria
Activities: The centre is responsible for training hydromet technicians for CILSS member states and other countries.

REGIONAL MANAGEMENT UNIT   4

Projects: Integrated pest management (Mr Mady-Keita).

SAHEL INSTITUTE   5

Address: Bamako, Mali
Activities: The institute is responsible for the coordination of technical and scientific research in the Mali sub-region in all areas of agricultural production.

Institut de Recherches Agronomiques Tropicales et des Cultures Viviéres en Haute-Volta* 6

– IRAT/HV
[Tropical Agriculture and Food Crops Research Institute - Upper Volta]
Address: Boîte Postale 596, Ougadougou
Affiliation: IRAT, Paris, France
Status: Official research centre
Director: M. Poulain
Activities: Applied research to rural development: intensification of agriculture for improved productivity related to food crops (particularly cereals) and certain cash crops. There are two research stations at Saria and Farako-Bo.
Publications: Annual report.

Institut de Recherches du Coton et des Textiles Exotiques* 7

– IRCT
[Cotton and Exotic Textiles Research Institute]
Address: Boîte Postale 267, Bobo Dioulasso
Affiliation: IRCT, Paris, France
Status: Official research centre
Director: H. Corre
Activities: Cotton and textile fibres research.

Institut de Recherches pour les Huiles et Oléagineux HV 8

– IRHO HV
[Research Institute for Oils and Oil Crops - Upper Volta Section]
Address: BP 1345, Ouagadougou
Telephone: 342 45
Status: Official research centre
Director: Christian Pilasso
Departments: Selection, Didier Balma; fertilization: cultural techniques, Salawu Asimi
Graduate research staff: 4
Activities: Agronomic research: soya beans, groundnuts, sesame.
Publications: Annual report.

URUGUAY

Agromax SA (Hiperfosfato)   1

Address: Avenida Uruguay 874, Montevideo
Telephone: 91 22 46
Telex: HIPER UY 992
Status: Research centre within an industrial company
Director: Nicolas Herrera MacLean
Sections: Commercial, José Luis Ruis Bracco; research, Domingo V. Luizzi Vitureira, Alfredo Castells Montes; factory, Jorge Barattin Lafourcade
Graduate research staff: 11
Activities: Crops; pastures.
Publications: Biannual report.
Projects: Crops (Juan Carlos Millot Contini); pastures (D.V. Luizzi Vitureira).

Centro de Investigaciones   2
Agrícolas Alberto Boerger

– CIAAB
[Alberto Boerger Agricultural Research Centre]
Address: 1374 Treinta y Tres, 4th Floor, Montevideo
Telephone: 90 04 48/92
Parent body: Ministry of Agriculture and Fisheries
Status: Official research centre
General director: Juan Antonio Curotto
Graduate research staff: 60
Activities: Natural resources; soil science, land use; plant production (cash crops, cereals, oleaginous vegetables, fruit trees): plant breeding, plant protection; animal production (beef, dairy, wool); livestock husbandry and nutrition.
Publications: Research activities (annual); progress report (annual).
Projects: Climate (Walter Corsi); soils (Roberto Díaz); forage crops (Mario Allegri); integrated research (Roberto Symonds); animal production (Dr Dante Geymonat); fruit trees (Rodolfo Talice); vegetables (César Maeso); plant protection (Joaquín Carbonell); citrus (Ismael Muller).

CITRUS EXPERIMENT STATION   3
Head: Ismael Muller

EASTERN AGRICULTURAL   4
EXPERIMENT STATION
Head: John Grierson

FARM ANIMALS EXPERIMENT   5
STATION
Head: Mario Mondelli

LA ESTANZUELA AGRICULTURAL   6
EXPERIMENT STATION
Head: Mario Allegri

LAS BRUJAS FARMING EXPERIMENT   7
STATION
Head: Joaquín Carbonell

NORTHERN AGRICULTURAL   8
EXPERIMENT STATION
Head: José Silva

Instituto Nacional de Pesca 9

– INAPE
[National Fisheries Institute]
Address: Casilla de Correos 1612, Montevideo
Telephone: 41 75 76
Affiliation: Ministry of Agriculture and Fisheries
Status: Official research centre
Director General: General Capitán de Navio U.W. Perez
Departments : Fish biology, Dr Herbert Nion; aquaculture, Dr Zoel Varela
Graduate research staff: 4
Activities: Development of technologies for intensive farming of local species, mainly Siluriformes (catfish).
Publications: Informes Técnicos, irregular.
Projects: Fish farming of local freshwater species (Zoel Varela).

Universidad de la Republica* 10

[University of the Republic]
Address: Avenida 18 de Julio 1968, Montevideo
Telephone: 409201
Status: Educational establishment with r&d capability

FACULTY OF AGRONOMY* 11

Head: Professor D.H. Facci

FACULTY OF VETERINARY MEDICINE* 12

Head: Professor Dr H. Lazaneo

VENEZUELA

Centro Nacional de Investigaciones Agropecuarias 1

– CENIAP
[National Centre for Agricultural Research]
Address: Apartado 4653, Maracay 2101, Estado Aragua
Telephone: (043) 22485; 22475
Telex: 43402 SIRCA VE
Parent body: Ministry of Agriculture
Status: Official research centre
Director: Dr Claudio Chicco
Sections: Projects analysis, Omar Martínez; agricultural communications, Juan Green
Activities: Natural resources: soil science, drainage and irrigation, land use; plant production: plant breeding, crop husbandry, plant protection; animal production: livestock breeding, livestock husbandry and nutrition, veterinary medicine.
Publications: Agronomía Tropical; Veterinaria Tropical (biannual).

AGRONOMY RESEARCH INSTITUTE 2

Head: Simón Ortega

ANIMAL HUSBANDRY RESEARCH INSTITUTE

Head: Gladys de Sosa

GENERAL AGRICULTURAL RESEARCH INSTITUTE 4

Head: Luis Ayala

VETERINARY RESEARCH INSTITUTE 5

Head: Manuel Vargas

Consejo Nacional de Investigaciones Agrícolas * 6

[National Council for Agricultural Research]
Address: Esquina de Cipreses, Edificio Miguel, 5° piso Apartado 12844, Caracas
Status: Official research centre
Activities: The council is the official body charged with assisting the Ministry of Agriculture in the preparation of research programmes, the coordination of such research work throughout the country, and the evaluation of the results of such programmes.

Estacion Experimental de Café * 7

[Experimental Coffee Station]
Address: Rubio, Bramón, Táchira
Status: Official research centre
Director: Alfredo Rivas Vázquez
Publications: Annual report.

Estación Experimental de Caucagua * 8

[Caucagua Research Station]
Address: Caucagua, Estado Miranda
Status: Official research centre
Activities: Cocoa breeding, agronomical research on citrus fruits, bananas, annato, yams, cassava plant and ocumo. Research is developing on avocado, mango, horticulture and mixed and multiple cropping.

Fundación para el Desarrollo de la Región Centro Occidental de Venezuela 9

– FUDECO
[Central-Western Region of Venezuela Development Foundation]
Address: Apartado 523, Barquisimeto, 3001-A, Lara
Telephone: (051) 512011; 512210
Telex: 51314 Ve
Parent body: Ministry of Planning (Cordiplan)
Status: Official research centre
Chairman: Miguel Nucete Hubner
Graduate research staff: 69
Activities: Soil science, drainage and irrigation, land use; agricultural planning for land development; freshwater and marine fisheries; forestry and agro-forestry; environmental analysis for agroecological planning.
Publications: Región (quarterly magazine); technical bulletin.

AREA DEVELOPMENT SECTION 10

Sections: Falcón state, Andrés González; Lara state, Reinaldo Mujica; Portuguese state, Amilcar Briceño; Yaracny state, Aquiles Velásquez
Projects: Quíbor-Yacambú (Reinaldo Mujica); Las Lomas; agricultural and fisheries, Falcón (Andrés González); Turén I,II; Guanare Masparro (Amilcar Briceño); Yaracuy Valley agricultural project (Aquiles Velásquez).

INFORMATION SCIENCE SECTION 11

Head: Cecilio Riera

PLANNING SECTION 12

Head: José Manuel Castillo

PROJECTS AND TECHNICAL ASSISTANCE SECTION 13

Head: Enrique Ron
Projects: Aroa Valley agricultural project (Miguel Guillory).

REGIONAL INFORMATION SYSTEMS SECTION 14

Head: Hermán Andueza

SOCIAL SCIENCE RESEARCH SECTION 15

Head: Oly de Izcaray
Projects: Environmental analysis (Edilberto Ferrer Véliz).

Fundación Servicio Para el Agricultor 16

– FUSAGRI
[Farmers' Service Foundation]
Address: Apartado 2224, Caracas, 1062
Telephone: (02) 2845521
Status: Independent research centre
President: Luis Marcano C.
National supervisor: Jesus S. Silva
Assistant to the President: Eddie A. Ramírez
Director, Cagua Station: José E. Martínez
Director, Zulia programme: Euro Bracho
Director, Guara experimental station: Sabas González
Graduate research staff: 70
Activities: Crop husbandry, plant protection - citrus, soyabean, vegetables, rice; livestock husbandry and nutrition; veterinary medicine. Programmes are developed in accordance with agreements with other institutions or companies that wholly or partly subsidize their operational costs.
Publications: Noticias Agrícolas (monthly).
Projects: Horticulture (Angel Florez); fruits (Mario Cermeli); soyabean (Pedro J. Rodríguez; Delta (Sabas González); Zulia (Euro Bracho); Falcón (Pastor Peña).

Instituto Agrario Nacional* 17

[National Agrarian Institute]
Address: Quinta Barrancas, Avenida San Carlos, Vista Alegre, Caracas 102
Status: Official research centre
President Antonio Merchan
Activities: Agrarian reform.

Instituto de Investigaciones Veterinarias* 18

[Veterinary Research Institute]
Address: Apartado 70, Maracay, Estado Aragua
Status: Official research centre
Director: Dr Manuel Vargas Díaz

Instítuto Forestal Latino-Americano   19

[Latin-American Forestry Institute]
Address: PO Box 36, Merida
Telephone: (074) 522440
Telex: 74179
Status: Independent research centre
Director: Dr Julio Cesar Centeno
Departments: Information and documentation, H. Raets; research, Dr J.C. Centeno
Graduate research staff: 4
Activities: Forestry, forest products and timber structures. An important project investigates in-grade timber properties and long-term strength properties of timber.
Publications: Latin American Forestry Journal (Spanish); Bibliographical bulletin on Latin American forestry literature (Spanish and English versions); forestry bibliography (translations into Spanish from world literature).

Instituto Nacional de Nutrición*   20

[National Institute of Nutrition]
Address: Apartado 2049, Caracas
Status: Official research centre
Director: Dr Luis Bermúdez Chauno
Publications: Archivos Latinamericanos de Nutricion.

Instituto Venezolano de Investigaciones Científicas *   21

– IVIC
[Scientific Research Institute of Venezuela]
Address: Apartado 1827, Caracas
Status: Official research centre
Director: Miguel Layrisse
Activities: Research in biology, chemistry, medicine, physics, mathematics and technology, and atomic research.

Universidad Central de Venezuela   22

– UCV
[Venezuela Central University]
Address: Ciudad Universitaria, Los Chaguaramos, Apartado Postal 104, Caracas
Telephone: 619811
Status: Educational establishment with r&d capability

FACULTY OF AGRICULTURE   23

Head: Professor Dr J.R. Rodríguez Brito

Edaphology Institute   24

Address: Maracay, Aragua
Telephone: (043) 28996
Director: Dr Antonio L. Mayorca
Sections: Ecology, Dr Anibal Rosales; chemicals, Dr Melitón Adams; soil-nutrient-plant, Dr Eduardo Casanova; soil-water-plant, Dr Omar Rodríguez
Graduate research staff: 21
Activities: Soil mineralogy; soil correlation; soil genesis; cultivation adaptability; soil microbiology; crop fertilization; soil chemistry; erosion; salinity; structural problems.
Publications: Revista de Suelos Tropicales.

Universidad de los Andes*   25

[University of the Andes]
Address: Avenida 3, Independencia, Zona Postal 802, Mérida
Telephone: 23201
Status: Educational establishment with r&d capability

FACULTY OF FORESTRY*   26

Head: Professor J.R. Corredor T.

Universidad del Zulia   27

[Zulia University]
Status: Educational establishment with r&d capability

FACULTY OF AGRICULTURE 28

Address: Apartado 526, Maracaibo, Zulia
Telephone: (061) 512208; 512248; 512197
Dean: Romer González
Graduate research staff: 11
Activities: Natural resources; soil science, land use; plant production: plant breeding (sorghum, maize, yucca, legumes), crop husbandry (sorghum, plantains, maize, yucca, legumes); animal production: livestock breeding, livestock husbandry and nutrition; food sciences. Research is principally on the improvement of agronomic practices and products better suited to the ecological conditions of the area. Other main area of research concern soil condition and utilization.
Publications: Journal; *Agro Información*; *Agro Técnico*; *Agro Jardín.*
Projects: Investigation of grain leguminae (R. Avila); market qualities and reproduction of Yorkshire pigs (E. Wilhelm); plantain diseases (L. Sosa); effect of antibiotics and chemotherapy on the growth of Landrace pigs (G. Ríos); genetic improvement of sorghum (A. Reinoso); protein concentrates from tropical foliage (A. Escoda); use of hen-dung as a protein supplement for forage (E. Golding); taxonomic census of phytophagous acaridae (G. Alvarado); pests transmitted by plantain seeds; plantain production; techno-economic study of regional viticulture (A. Nava, C. Nava); maize improvement (F. Tong); economic damage to sorghum caused by Spodoptera frugiperda (Oscar Domínguez); cacao diseases (F. León); management, conservation and storage of pepper (Livia de Ramadan, Elivia Trocóniz); effect of nitrogen, phosphorus and potassium on yields of Sorghum bicolor (O. Andrade); acid soils and graminae growth (J.R. Paredes); milking patterns (E. Rincón); effect of micro-climate and sun on yucca production (C.F. Quintero); maturation of different seeds (traditional and potential) in the area (F. Oropeza); growth characteristics, breeding and improvement of cross-bred cattle (J. Ríos, A. Ocando); behaviour and adaptability of different species in soils with a high aluminium content (I. Urdaneta); biology and control of insect pests (J.R. Labrador, E. Rubio); cultivation of acid soils (D. Mata); laboratory analysis and soil fertility (A. Atencio); forage conservation (A. del Villar); food technology for agro-industrial purpos es (O. Ferrer); genetic improvement of yucca (J.R. Tineo); different forms of applying phosphorus P32 to sorghum (P. Santiago, E. Martínez).

Agricultural Engineering Department 29

Head: Maximiano Valbuena

Agronomy Department 30

Head: Francisca Tong

Chemistry Department 31

Head: Robinson Arrieta

Edaphology Department 32

Head: Idelmo Villalobos

Phytosanitary Department 33

Head: Guillermo Alvarado

Social Sciences Department 34

Head: Jesus Pazos

Statistics Department 35

Head: J.J. Villasmil

Zootechnics Department 36

Head: Alonso Fernández

VIETNAM

Forest Research Institute 1

Address: Chèm-Tù liem, Hànôi
Status: Official research centre
Director: Vǔ Biet Linh
Departments: Forest inventory and management; forest planting mechanization; forest soil; forest tree biochemistry; forest tree physiology; forest tree protection; genetics of forest tree seed; special forest products; silviculture and forest ecology (afforestation).
Graduate research staff: 200
Activities: Natural resources; soil; plant selection; plant protection against pests and diseases; plant physiology and ecology; forest plantation; forest planting implements.
Projects: Forest plantation to supply raw material for pulp and paper industries; exploitation, rehabilitation and improvement of natural forests to supply logs for construction and export; forest plantation for watershed protection, sand dune fixation and soil improvement; raising, exploitation of special forest products; agro-forestry in forest plantation and forest management.

Institute of Agricultural Research* 2

Address: 121 Nguyen-Binh-Kheim, Ho-Chi-Minh City
Status: Official research centre
Activities: Research into general agronomy, applied pedology; phytopathology, stock farming, mechanization, and conservation of products.

VIRGIN ISLANDS

College of the Virgin Islands   1

Status: Educational establishment with r&d capability

AGRICULTURAL EXPERIMENT STATION   2

– CVIAES
Address: PO Box 920, Kingshill, St Croix 00850-0920
Telephone: (809) 778 0246; 778 0050
Director: Dr D.S. Padda
Assistant director: Dr H.D. Hupp
Graduate research staff: 4
Activities: Agronomy: sorghum and tropical forage production systems and varietal trials; aquaculture: production systems for tilapia; animal science: Senepol cattle breed characterization, breeding methods, husbandry; vegetables: crop husbandry and varietal trials of root crops, tomatoes, eggplants; fruits; citrus production, papaya, mango production; irrigation: minimum water use on fruit and vegetables; pest management: pest problems in livestock, fruits, vegetables and field crops.
Publications: Annual report.
Projects: Agronomy (Dr A. Hegab); aquaculture (Dr J. Rakocy); animal science (Dr H. Hupp); vegetables (Dr A. Navarro); fruits (C. Ramcharan); irrigation (S. Buzdugan); pest management (Dr W. Knausenberger).

YUGOSLAVIA

Biološki Institut* 1

[Biological Institute]
Address: POB 39, 50000 Dubrovnik
Telephone: 050-27-937
Status: Independent research centre
Director: Dr Adam Benović
Sections: Zooplankton, Dr F. Krsinić; phytoplankton, D. Viličić; aquaculture, Dr Adam Balenović; chemistry, R. Balenović ornithology, Dr I. Tutman; ornithology, Dr I. Tutman; botany, S. Hečimović
Graduate research staff: 11
Activities: Ecology and experimental work in phytoplankton, zooplankton, molluscs (oysters) and fish (yellow teil); bird life of the South Dalmatian coast; plant cover of the Adriatic coast and islands; analytical chemistry.

Inštitut Primenu Nuklearne Energije u Poljoprivredi, Veterinarstru i Šumarstvu* 2

– INEP
[Institute for the Application of Nuclear Energy in Agriculture, Veterinary Medicine and Forestry]
Address: Banatska 316, POB 46, 11080 Bemun-Beograd
Telephone: 219 252
Status: Official research centre
Director: Dr Stevan Simić

DEPARTMENT OF PESTICIDES* 3
Head: J. Perić

DEPARTMENT OF PHYSIOLOGY AND RADIOBIOLOGY* 4
Head: M. Movsesijan

DEPARTMENT OF PLANT PHYSIOLOGY AND SOIL SCIENCES* 5
Head: D. Milivojević

LABORATORY OF BIOPHYSICAL AND ANALYTICAL CHEMISTRY* 6
Head: S. Kapor

Inštitut Rudjer Bošković* 7

[Rudjer Bošković Institute]
Address: 54 Bijenička Cesta, POB 1016, 41001 Zagreb
Telephone: 041-272 611
Telex: 21383 yu irb zg
Status: Official research centre
Director-General: V. Kundić
Departments:
Theoretical physics, Dr N. Zovko
Materials science and electronics, Dr B. Etlinger
Physics, energetics and appliances, Dr P. Thomaš
Physical chemistry, Dr M. Orhanović
Organic chemistry and biochemistry, Dr N. Ljubešić
Experimental biology and medicine, Dr D. Petrović
Technology, nuclear energy and protection, Dr I. Dvornik
Laser and atomic research, Dr A. Peršin
Graduate research staff: 350

Activities: Particle physics, low-energy nuclear physics, nuclear reactions, nuclear models, electromagnetic interactions, nuclear spectroscopy, few-nucleon problems and nuclear forces, electronic instrumentation, radiochemistry and radiation chemistry, isotope effects, experimental therapy, radiochemical characterization of microconstituents in sea, radioecology, nuclear radiation dosimetry. Nuclear methods are applied to technology, medicine, nuclear energy, isotope production, geophysics, and to other fields of fundamental and applied science. The basic major equipment consists of a cyclotron of 16 MV deuteron energy, two Cockcroft-Walton accelerators, cobalt radiation facilities, lasers, computers, different spectrometers and analytical instruments. The Institute incorporates the Centre for Marine Research, part of which is located at Rovinj. The Institute has links with the Sveučilište u Zagrebu.

CENTAR ZA ISTRAZIVANJE MORA - ROVINJ*   8

[Centre for Marine Research - Rovinj Laboratory]
Address: Paliage 5, 52210 Rovinj
Head: Dr B. Osretić
Activities: Fundamental physical, chemical, and biological investigations in the North Adriatic with special regard to: pollution; microconstituents; primary productivity; benthic communities; shellfish culture; radioecology; microbiology; ecology and systematics of marine organisms.

CENTAR ZA ISTRAZIVANJE MORA - ZAGREB*   9

[Centre for Marine Research - Zagreb Laboratory]
Address: 54 Bijenička Cesta, POB 1016, 41001 Zagreb
Head: Dr M. Branica
Activities: The same as those of the Rovinj Laboratory.

Institut za Ekonomiku Poljoprivrede   10

[Institute of Agricultural Economics]
Address: Cara Uroša, 11000 Beograd
Telephone: 011-622 957
Status: Independent research centre
Director: Svetolik Popović
Sections: Agricultural policy, Dr Radisav Badnjarević; agricultural marketing, Srboljub Ivanović; regional and mountain agriculture, Dr Milorad Lečić; social sector farms, Radovan Pajević
Graduate research staff: 23
Activities: Research into all aspects of economy in agriculture.

Projects: Long-term development of Yugoslav agriculture (Svetolik Popović); regional development project Montenegro II (Slobodan Milić); agroindustry survey (Mileta Popović); peasant holdings income survey (Radisav Badnjarević); demand and supply prospects of Yugoslav agriculture and agroindustry (Srboljub Ivanović); organization and management in Yugoslav agriculture (Radovan Pajević); development of agriculture in mountain regions of Yugoslavia (Milorad Lečić).

Institut za Mechanizaciju Poljoprivrede - Zemun   11

[Institute for Agricultural Mechanization – Zemun]
Address: POB 41, 11080 Zemun
Telephone: 011-212 403
Status: Independent research centre
Head: Radmilo Orović
Divisions: Agricultural, Raleta Savić; Mechanical Engineering, Djordje Danilović; Civil Engineering, Miloš Radosavljević
Graduate research staff: 45
Activities: Introducing mechanization into agriculture and livestock production and breeding.
Publications: Poljoprivredna Tehnika, annually.

Inštitut za Oceanografiju i Ribarstvo*   12

[Oceanography and Fisheries Institute]
Address: Moše pijade 67, POB 114, 58000 Split
Telephone: 058-46 688
Status: Official research centre
Director: Dr Špan Ante
Sections: Fisheries, Ivo Kačić; marine biology, Dr Tereza Pucher Petković; oceanography, Dr Mira Zore Armanda
Graduate research staff: 34
Activities: Hydrography, geology and ecology of marine organisms; fisheries techniques.
Publications: Acta Adriatica.

Inštitut za Slatkovodna Ribarstvo*   13

[Institute of Freshwater Fisheries]
Address: Ul Drenovocko 30, YU 41000 Zagreb
Status: Official research centre

Institut za Stočarstvo   14

– IS
[Livestock Institute]
Address: Zemun Polje, pošt fah 108, 11080 Zemun, Sribija
Telephone: 011-697 706
Status: Official research centre
Director: Dr Vitomir Kostic
Departments: Agricultural engineering and building, B. Simović; cattle, Dr L. Romčević; nutrition, Dr G. Veličković; sheep, Dr V. Ćeranicć; pig, Dr V. Anastasijević; poultry, Z. Pavloski
Graduate research staff: 30
Activities: Research in the following areas: animal production (cattle, pigs, sheep, poultry); livestock breeding; livestock husbandry and nutrition; crop husbandry; agricultural engineering and building; food science.

Institut za Šumarstvo i   15
Drvnu Industriju

[Forestry and Wood Industry Institute]
Address: PO Box 185, Kneza Višeslava 3, 11030 Beograd
Telephone: 011-553 355
Director: Jovan Djurdjević.
Departments: Forestry and hunting management, Dr S. Šmit; forest utilization, erosion control and wood processing, B. Vulović; forest management, S. Spremović
Graduate research staff: 65
Activities: The institute is engaged in fundamental and applied development research and engineering design in the fields of forestry and hunting, the wood processing industry, and land protection against erosion. Specific areas of research are: natural resources; soil science; plant breeding; crop husbandry; plant protection; forestry and forest products; forestry economics; forest ecology and phytocenology; hunting and hunting management. The main autochthonous coniferous species are Scots pine, Austrian pine, Norway spruce, and Serbian spruce (Picea omorica); Douglas fir and eastern white pine are among introduced coniferous species. The main broadleaved species are; beech (Fagus moesiaca), oaks (Quercus robur, Quercus petraea, Quercus farnetto), poplars, willows.
Publications: Zbornik radova Instituta za šumarstvo i drvnu industriju (Annals of the Institute of Forestry and Wood Industry), annually.

Projects: Investigation into the most appropriate methods of afforestation of bare lands, and improving devastated forests (Dr M. Jovanović); improvement and optimum utilization of forest in Serbia (Dr N. Jović, Dr N. Veselinović); needs and optimal solutions for permanent supply of wood as raw material (Dr Ž. Milin, Dr R. Marović); improvement of hunting management in Serbia (Dr D. Jović, Dr D. Bojvić).

Inštitut za Tutun   16

[Tobacco Institute]
Address: 9500 Prilep
Status: Official research centre
Director: Dr J. Mickovski

Institut za Voćarstvo   17

[Fruit Research Institute]
Address: Vojvode Stepe 9, 32000 Čačak, Srbija
Telephone: 032-3593
Status: Independent research centre.
Director: Vojin Bugarčić
Deputy director: Dr Milojko Ranković
Departments: Pomology and breeding, Žarko Tešović; physiology and agricultural engineering, Radoslav Janković; plant protection, Dr Milojko Ranković; fruit technology and storage, Dr Ljubica Janda
Activities: The main aim of the institute is to improve fruit growing through introducing into production high-quality, better yielding fruit cultivars. It is also concerned with work on breeding new apple, pear, plum, apricot, peach, sweet and sour cherry, quince, raspberry, and strawberry cultivars. It is engaged in the work on training systems, pruning, fruit nutrition, soil management, and control of pests and diseases. The study of fruit tree virus diseases and the production of virus-free material are of major importance. The technological value of fruits and fruit storage techniques are also studied as a part of the work on the selection and breeding of new cultivars.
Projects: Ecological and economic-biological studies of fruit tree cultivars and rootstocks (Vojin Bugarčić); breeding and selection of fruit cultivars and rootstocks having bigger biological-economic potentials (Žarko Tešović); intensification of cropping and yields increase in fruit crops (Radoslav Janković); selection of superior indigenous ecotypes of the Prunus and investigation of resistance of some selections and cultivars to Sharka (plum pox) virus (Dr Asen Stančević).

Institut za Zaštitu Bilja   18

[Plant Protection Institute]
Address: Teodora Drajzera 9, PO Box 936, 11001 Beograd
Telephone: 011-640 049
Status: Official research centre
Director: Dr Ljubiša Vasiljević
Departments: Plant diseases, Dr M. Jordović; plant pests, Dr. M. Injac; phytopharmacy, Dr K. Mijatović; disease, insect, and rodent control, B. Manjolović
Graduate research staff: 17
Activities: Research into the biology of viruses, fungus, nematodes, insects, mites, rodents, and weeds, and their control in agricultural crops.
Publications: Plant Protection (quarterly journal).

Institut za Zemjodelstvo   19

[Agriculture Institute]
Address: Goce Delčev 27, Strumica, SR Macedonia
Telephone: 092-71 130; 71 069
Telex: 53684
Affiliation: Industrial Agriculture Consortium
Status: Research centre within an industrial company
Director: Dragi Pirikliev
Departments: Plant breeding and seed production, Toma Spasov; plant protection, Ristov Tasev
Graduate research staff: 12
Activities: Plant production - cotton, tomatoes, french beans, peppers, watermelons; plant breeding - cotton, tomatoes, peppers; plant protection - many crops, including cotton, tomatoes, french beans, peppers, and watermelons.
Projects: Breeding early maturing cotton varieties; evaluation of new cotton varieties as part of the regional experiments (Dragi Pirikliev); the influence of environment on yield and quality of cotton in relation to genotype (Toma Spasov); problems relating to cultural practices, plant nutrition, application of irrigation, and chemical weed control, as well as the control of pests and diseases. (Milan Gorgiev, Vasil Kocevski, Dobri Jakimov, Risto Vučkov, Snežana Piperevska).

Inštitut za Zemljište *   20

[Soil Science Institute]
Address: Teodora Drajzera 7, Beograd
Telephone: 011-688 821
Status: Educational establishment with r&d capability
Director: Dr Petar Irović
Graduate research staff: 25

SOIL FERTILITY SECTION *   21
Head: Dr N. Marković

SOIL MICROBIOLOGY SECTION *   22
Head: Dr Z. Vojinović

Jugoslavenska Akademija Znanosti i Umjetnosti *   23

[Yugoslav Academy of Sciences and Arts]
Address: Zrinski trg 11, 41000 Zagreb
Status: Educational establishment with r&d capability
President: Academician Jakov Sirotković
Vice-Presidents: Academician Branko Kesić, Academician Andre Mohorovičić
General Secretary: Academician Hrvoje Požar

DEPARTMENT OF NATURAL SCIENCES *   24
Head: Academician Zoran Bujas

SCIENTIFIC UNITS BELONGING TO THE ACADEMY: 25

Institute for the History of Natural, Mathematical and Medical Sciences, Zagreb *   26

Institute of Ornithology, Zagreb *   27

Jugoslovenski Inštitut za Tehnologiju Mesa   28

[Yugoslav Institute of Meat Technology]
Address: Kaćanskog 13, PO Box 548, Beograd
Telephone: 011-650 655
Telex: YU INMES 12300
Director: Dr Radoljub Tadić
Departments: Meat hygiene and technology; bacteriology; chemistry; biological residues; organoleptics; poultry meat technology; technological designing; metal packaging; paper and plastic packaging; publishing

Activities: Modern technology and hygiene of meat, meat products, fats of animal origin, poultry meat and eggs, venison and fish, animal food, additives, packaging raw materials connected with meat products; solution of scientific problems within these fields; improvement of Yugoslav industry within above mentioned fields; laboratory analyses and expert opinions for the above products (in the country and abroad); mechanical-technological designs for meat industry; scientific-professional assistance to various organizations from the said fields.
Graduate research staff: 50
Projects: Environmental pollution - significance of the finding of toxic elements from the standpoint of soundness of footstuffs of animal origin. (Dr Veselinka Djordjević); degradation of pesticides in meat as influenced by useful microflora. (Dr Vera Višacki).

Kmetijski Inštitut Slovenije   29

[Agricultural Institute of Slovenia]
Address: Hacquetova 2, 61000 Ljubljana
Telephone: 061-323 064
Status: Official research centre
Director: Professor Vilko Štern
Graduate research staff: 43
Publications: Annual report.

AGRICULTURAL ECONIMICS   30

Head: Slavko Gliha
Activities: Economics of animal husbandry, farm management, mountain farms, large social sector farms, and private farms.
Projects: Formation of socioeconomic production systems for the rational use of agricultural land (Slavko Gliha); bases of agribusiness development (Matija Kovačič).

AGRICULTURAL MECHANIZATION   31

Head: Marjan Mrhar
Activities: Mechanization of mountain farms; harvesting; solar energy studies.

ANIMAL HUSBANDRY   32

Head: France Goršič
Activities: Nutrition; genetic improvement.
Projects: Aetiology, ecology, and environmental protection in animal husbandry (France Goršič); genetics and selection of animals; animal nutrition (Janez Verbič); systems of agricultural production (France Goršič).

CENTRAL LABORATORY   33

Head: Dušan Trčelj
Activities: Chemical analysis of wines, soil, forage, and fertilizers; selection and breeding of new varieties of field crops and grasses.
Technological parameters of food, of plant and animal origin.

FARM MECHANIZATION*   34

Head: Raoul Jenčić
Activities: Mechanization of mountain farms; harvesting; solar energy studies.

FIELD CROP PRODUCTION   35

Head: Jože Silc
Activities: Plant production; selection.
Projects: Ecology, pedology, and nutrition of agricultural plant (Lojze Briški); biological bases of agricultural plants.

FRUIT AND WINE PRODUCTION   36

Head: Milena Lekšan
Activities: Selection of new varieties of apples, peaches, small fruits, vines.

PLANT PROTECTION   37

Head: Vukadin Šišakovič
Activities: Protection of potato, maize, cereals, grasses, clovers, and vegetables.
Projects: Pathology and protection of agricultural plants (Hržič, Šišakovič).

Kmetijski Zavod, Maribor,   38

[Maribor Agricultural Station]
Address: Vinarska 14, 62101 Maribor
Telephone: 062-21 191
Status: Independent research centre
Director: Jože Protner
Sections: Fruit growing, France Lombergar; viticulture, Boris Beloglavec; soil science, Dušan Šrok; plant protection, Stojan Vrabl; agriculture, Bže Miklavc; seed control, Jožica Plakolm; hail defence, Tone Dvoršak
Graduate research staff: 17
Activities: Applicational research and advisory work - fruit cultures; viticulture; soil science; plant protection; land use; agricultural engineering and building; labour analysis.
Publications: Annual report.

LABORATORY 39

Head: Sonja Ciglenečki
Activities: Wine analysis; soil and plant analysis; manure analysis; food analysis.

Makedonska Akademija na Naukite i Umjetnostite* 40

[Macedonian Academy of Sciences and Arts]
Address: POB 428, 91000 Skopje
Telephone: 091-235 311
President: Academician Mihailo Apostolski
Vice-president: Academician Blagoj Popov
Secretary: Academician Evgeni Dimitrov

BIOLOGICAL AND MEDICAL SCIENCES DEPARTMENT* 41

Secretary: Kiril Mičevski

Pasterov Zavod* 42

[Pasteur Institute]
Address: Hajduk Veljkova 1, POB 341, 21000 Novi Sad
Telephone: 021-611 003
Status: Official research centre
Director: Dr Miloš Petrović
Sections: Rabies, Dr Miloš Petrović; human virology, Dr Miroslav Petrović; insect virology, Dr Čiril Sidor; plant virology, Dejan Stakić
Graduate research staff: 7

Poljoprivredri Fakultet 43

[Faculty of Agriculture]
Status: Educational establishment with r&d capability

OOUR INSTITUT ZA STOČARSTVO 44

[Livestock Institute]
Address: Veljka Vlahovića 2, PO Box 171, Novi Sad, SAP Vojvodina
Telephone: 021-58 504
Director Professor Slobodan Živković
Departments: Cattle production, Dr S. Bačvanski; swine production, Dr M. Nikolić; poultry, Dr Č. Višnjić; animal nutrition, Dr S. Kovčin
Graduate research staff: 57

Activities: Research to improve genetic potential of animals for production of meat, milk, and eggs, and also to increase efficiency of feed utilization in animal production.
Publications: Annual report.
Projects: Silage as dominant roughage for milking cows (S. Bačvanski); possibilities of using dry sugar beet pulp as a basis or the only source of energy in feeding cows and fattening bulls. (R. Jovanović); effect of the individual cattle genotypes and their morphological and production characteristics (M. Nenadović); effect of hybridization on swine poduction of SAP Vojvodina with special reference to reproductive and productive characteristics of the hybrids obtained (D. Mančic); yield and characteristics of meat in non-castrated fattening pigs (T. Bokorov); improvement of the genetic capability of domestic light type of fowl (M. Milovanović); additives in animal nutrition (S. Živković).

Slovenska Akademija Znanosti in Umetnosti* 45

[Slovene Academy of Sciences and Arts]
Address: Novi trg 3, 61000 Ljubljana
Telephone: 061-23 961
Status: Educational establishment with r&d capability
President: Professor Dr Janez Milčinski
Senior staff: Dr Fran Zwitter, Dr Janko Jurančić, Dr Dušan Hadži, Dr Ivan Rakovec, Anton Ingolić, Dr Franc Novak
Graduate research staff: 105
Activities: Natural and cultural heritage of the Slovene nation; archaeology; botany; climatology; earth sciences; ecology; medical science; microbiology; zoology.
Publications: Yearbook.

Sour Hercegovina RO IRI Mostar 46

[Hercegovina Research and Development Centre]
Address: 79201 Buna, Mostar, Bosna and Hercegovina
Telephone: 088-71-228
Telex: 46-262 YU HEPOK
Status: Official research centre
Director: Dr Vladimir Trninić
Departments: Floriculture, Pehar Jakov; oenology, Kovačina Rade; viticulture, Pediša Hivzo; cattle breeding, Antunović Ivica; fruit-growing, Buljko Muhamed; plant patology, Nadaždin Vojislav; soil science, Hanić Elvedin
Graduate research staff: 40

Activities: Research into: soil science; drainage and irrigation; land use; plant breeding; crop husbandry; plant protection; livestock breeding; sheep and cattle husbandry and nutrition; veterinary medicine; forestry and forest production.
Publications: Godišnji izvještaja.

Srpska Akademija Nauka Umetnosti*   47

[Serbian Academy of Sciences and Arts]
Address: Knez Mihailova 35, 11001 Beograd
Status: Educational establishment with r&d capability
President: Academician Pavle Savić
Vice Presidents: Academician Dušan Kanazir, Academician Vojislar Djurić
General Secretary: Academician Milutin Garašanin

NATURAL AND MATHEMATICAL   48
SCIENCES DEPARTMENT*

Secretary: Academician Jovan Belić

Sveučilište u Zagrebu*   49

[Zagreb University]
Address: Trg Maršala Tita 14, POB 815, 4100 Zagreb
Telephone: 041-272 411
Rector: Dr Ivan Jirleović

FACULTY OF AGRICULTURE   50

Address: Šimunska 25, Zagreb
Telephone: (041) 216 777
Dean: Professor Dr Ante Gole
Graduate research staff: 131
Activities: Soil science; drainage and irrigation; land use; plant breeding; crop husbandry; plant protection; horticulture; livestock breeding; livestock husbandry and nutrition; veterinary medicine; agricultural engineering and building; freshwater fisheries; food science.

Department of Agricultural Botany*   51
Head: Professor N. Plavšić

**Department of Agricultural Exchange   52
and Agricultural Economics***
Head: Professor B. Štancl

**Department of Agricultural Machinery   53
and Storage***
Heads: Professor D. Komunjer, Professor J. Brčić

**Department of Agricultural   54
Phytoenology***
Head: Professor I. Kovačević

**Department of Agricultural Product   55
Technology***
Head: vacant

**Department of Bases of Special   56
Agriculture***
Head: Professor J. Čižek

**Department of Cattle Breeding and   57
Horse Breeding***
Head: Professor M. Car

Department of Dairy Manufacturing*   58
Head: Professor S. Miletić

**Department of Domestic Animals   59
Nutrition, Cattle Breeding and
Nutrition***
Head: Professor H. Zlatić

Department of Elements of Biology*   60
Head: Professor A. Šarić

**Department of Entomology and Plant   61
Protection***
Heads: Professor I. Milatović, Professor M. Maceljski

**Department of General Fruit   62
Cultivation***
Head: Professor I. Miljković

**Department of General Plant   63
Production***
Head: Professor V. Mihalić

**Department of Genetics and Research   64
Techniques***
Head: Professor M. Kump

**Department of Industrial Plants, 65
Agricultural Crops and Bases of
Special Agriculture***

Head: Professor J. Gotlin

Department of Microbiology* 66

Head: Professor M. Prsa

**Department of Organization and 67
Economics of Fruit, Vine, Vegetable
Growing and of Agricultural Estates***

Heads: Professor Z. Vincek, Professor P. Karoglav

**Department of Organization and 68
Management of a Cattle Breeding
Enterprise***

Head: Professor D. Dokmanović

Department of Pedology* 69

Heads: Professor A. Škorić, Professor Z. Racz

**Department of Physiology, 70
Embryology and Anatomy of Domestic
Animals***

Head: Professor Z. Stilinović

Department of Phytopathology* 71

Heads: Professor J. Kišpatić, Professor I. Milatović

**Department of Pig, Sheep and General 72
Cattle Breeding***

Head: Professor S. Jančić

Department of Plant Nutrition* 73

Head: Professor J. Anić

**Department of Social Studies and 74
Agricultural Sociology***

Head: vacant

**Department of Special Fruit 75
Cultivation***

Head: Professor R. Gliha

**Department of Statistical Methods and 76
Research Techniques in Cattle
Breeding***

Head: Professor S. Barić

**Department of Technology and 77
Quality of Milk and Milk Products***

Head: Professor D. Sabadoš

Department of Vegetable Cultivation* 78

Head: Professor P. Pavlek

**Department of Veterinary Medicine 79
with Zoohygiene***

Head: vacant

Department of Vine Growing* 80

Heads: Professor R. Licul, Professor D. Premužić

**Department of Zoology for Cattle 81
Breeders and Beekeepers***

Head: Professor L. Schmidt

FACULTY OF FORESTRY* 82

Address: Šimunska 25, Zagreb
Dean: Dr Ivo Dekanić

**Department of Chemical Wood 83
Processing and Organic Chemistry***

Head: Professor I. Opačić

**Department of Descriptive Geometry 84
and Forest Photogrammetry***

Head: Professor Z. Tomašegović

**Department of Electrotechnics and 85
Wood Machinery***

Head: Professor D. Hamm

Department of Forest Entomology* 86

Head: Professor M. Androić

**Department of Forest Exploitation and 87
Organization of Forestry Work***

Head: Professor R. Benić

Department of Forest Genetics and Dendrology* 88

Head: Professor M. Vidaković

Department of Forest Management* 89

Head: Professor D. Klepac

Department of Forestry and Reclaiming of Degraded Forest Lands* 90

Head: Professor I. Dekanić

Department of Industrial Timber Building* 91

Head: Professor N. Lovrić

Department of Machinery, Lifting and Transport Machines* 92

Head: Professor M. Biljan

Department of Organization and Rationalization in Forestry* 93

Head: Professor B. Kraljić

Department of Wood Anatomy and Protection* 94

Head: Professor I. Spaić

Department of Wood Technology* 95

Head: Professor S. Badjun

FACULTY OF NATURAL SCIENCES AND MATHEMATICS* 96

Address: Socijalističke Revolucije 8, Zagreb
Dean: Dr Ivan Crkvenčić

Department of Botany* 97

Head: Professor Z. Devidé
Senior staff: Professor R. Domac, Professor D. Miličić, Professor Lj. Ilijanić, Professor Z. Pavletić

Department of Zoology* 98

Head: Professor I. Matoničkin
Senior staff: Professor B. Djulić, Professor H. Gamulin-Brida, Professor M. Meštrov, Professor B. Rodé, Professor A. Kaštelan

FACULTY OF TECHNOLOGY* 99

Address: Pierottijeva 6, Zagreb
Dean: Dr Mladen Bravar

Department of Fermentation* 100

Head: Professor S. Ban

Department of Food Analysis* 101

Heads: Professor M. Filajdić, Professor F. Mihelić

Department of Food Technology* 102

Heads: Professor N. Jurković, Professor Z. Gerić

Department of Fruit and Vegetable Technology* 103

Head: Professor T. Lovrić

Department of Milk Technology* 104

Head: Professor A. Petričić

FACULTY OF VETERINARY SCIENCE* 105

Address: Heinzelova 55, Zagreb
Dean: Dr V. Mitin

Department of Anatomy, Histology and Embryology* 106

Heads: Professor U. Bego, Professor A. Frank

Department of Biology* 107

Head: Professor I. Ehrlich

Department of Domestic Animals Nutrition* 108

Head: Professor M. Kalivodai

Department of Forensic and Administrative Veterinary Medicine* 109

Head: Professor J. Perić

Department of Internal Diseases of Domestic Animals* 110

Head: Professor S. Forenbacher

Department of Microbiology and 111
Immunology*

Heads: Professor E. Topolnik, Professor S. Cvetinić

Department of Parasitology and 112
Invasive Diseases*

Heads: Professor M. Žuković, Professor T. Wikerhauser

Department of Pathological Anatomy* 113

Head: Professor B. Maržan

Department of Pathological 114
Physiology*

Heads: Professor S. Krvavica, Professor T. Martinčić

Department of Pharmacology and 115
Toxicology*

Heads: Professor M. Delak, Professor V. Srebočan

Department of Physiology* 116

Head: Professor D. Timet
Senior staff: Professor I. Valpotić, Professor V. Mitini

Department of Poultry Pathology* 117

Head: Professor M. Kralj

Department of Surgery* 118

Head: Professor K. Čermak

Department of Zoohygiene* 119

Heads: Professor J. Ivoš, Professor A. Asaj

Univerza Edvarda Kardelja 120
v Ljubljani*

[Edward Kardelj University of Ljubljana]
Address: Trg Osvoboditve 11, Ljubljana
Telephone: 061-22052
Status: Educational establishment with r&d capability

FACULTY OF BIOTECHNICS* 121

Address: POB 25, 61115 Ljubljana
Dean: Professor Dr F. Bucar

Division of Veterinary Medicine* 122

Address: POB 25, 61115 Ljubljana
Telephone: 061-261664
Dean: Professor Dr D. Šabec
Research head: Professor Dr V. Simčič
Sections: Diagnostics, prevention and eradication of animal infectious diseases; animal health; domestic animal reproduction, Professor Dr Ivo Vomer; physiology and pathology of animal nutrition, Professor Dr Nestor Klemenc; food hygiene and bromatology, Professor Dr Marjan Milohnoja
Graduate research staff: 40
Activities: The research programme for 1981-85 includes the following studies carried out on behalf of the territory of SR Slovenia and activities in veterinary medicine required by producers and the State administration: morphology, physiology, pharmacology and toxicology; public health and food hygiene; epizoology; animal nutrition; domestic animal reproduction; intensive animal production studies; health control and breeding of horses and carnivora; veterinary surgery; game, bees and fishes research.
Publications: Research reports, twice yearly.

FACULTY OF NATURAL SCIENCES 123
AND TECHNOLOGY*

Dean: Professor Dr J. Sovinc

Chemical Institute* 124

Address: Murnikova 6, POB 537, 61001 Ljubljana
Telephone: 061-22 625
Acting director: Professor Savo Lapanje
Graduate research staff: 50
Activities: Inorganic and organic synthesis; crystallography; spectroscopy; electrolysis; thermal analysis; mineral water analysis and investigation; environmental analysis; enzyme, protein and alkaloid studies; industrial chemistry and chemical engineering; biochemistry; pharmaceutical and agricultural chemistry.

Universitet Kiril i Metódi 125
vo Skopje*

[University of Skopje]
Address: Bulevar Krste, Misirkov BB, 9100 Skopje
Status: Educational establishment with r&d capability

FACULTY OF AGRICULTURE* 126

Dean: Professor R. Lozanovski

FACULTY OF FORESTRY* 127

Dean: Professor S. Todorovski

Univerzitet u Beogradu* 128

[Belgrade University]
Address: Studentski Trg 1, 11001 Beograd, 6
Status: Educational establishment with r&d capability
Rector: Dr Dragoslav Janković
Vice-Rector: Professor Dr Slobodan Unković

FACULTY OF AGRICULTURE* 129

Address: Nemanjina 6, Zemun
Dean: Professor M. Babović

Department of Agricultural Calculations* 130

Head: Professor Lj. Stanković

Department of Agricultural Chemistry, Soil Fertility and Fertilizers* 131

Head: Professor M. Pantović
Senior staff: Professor Ž. Popović, Professor M. Petrović

Department of Agricultural Economics* 132

Head: Professor P. Marković
Senior staff: Professor B. Radovanović, Professor V. Randjelović, Professor J. Simović

Department of Agricultural Machinery* 133

Head: Professor J. Mićić
Senior staff: Professor M. Pavlović, Professor P. Milošević, Professor P. Nenić

Department of Animal Breeding* 134

Head: Professor B. Simović

Department of Animal Husbandry* 135

Head: Professor M. Milojić

Department of Animal Nutrition* 136

Head: Professor K. Šljivovački
Senior staff: Professor D. Zeremski, Professor A. Pavličević

Department of Animal Science* 137

Head: Professor N. Mitić

Department of Apiculture* 138

Head: Professor B. Konstantinović

Department of Biochemistry* 139

Heads: Professor M. Džamić, Professor D. Veličković

Department of Botany* 140

Heads: Professor M. Kojić, Professor Ž. Živanović

Department of Chemistry* 141

Head: Professor M. Djuričić
Senior staff: Professor A. Jokić, Professor M. Jakšić

Department of Dairy Technology* 142

Heads: Professor J. Djordjević, Professor R. Stefanović

Department of Domestic Animal Physiology* 143

Heads: Professor J. Radulović, Professor B. Panić

Department of Entomology* 144

Heads: Professor B. Ilić, Professor N. Tanasijević

Department of Farm Organization* 145

Heads: Professor D. Pejin, Professor D. Bajčetić

Department of Field Crops* 146

Heads: Professor B. Milojić, Professor M. Šuput

Department of Food Technology* 147

Heads: Professor G. Niketić-Aleksić, Professor M. Gugešević-Djaković

Department of Forage Crops Production* 148

Head: Professor M. Mijatović

Department of Fruit Growing* 149

Heads: Professor M. Jovanović, Professor S. Savić

Department of Fruit Selection*   150
Head: Professor B. Pejkić

Department of General Agronomy*   151
Head: Professor D. Božić

Department of General Viticulture*   152
Heads: Professor M. Milosavljević, Professor L. Avramov

Department of Genetics*   153
Heads: Professor M. Marić, Professor A. Djokić

Department of Geology and Agriculture*   154
Head: Professor D. Aleksandrović

Department of Irrigation and Drainage*   155
Heads: Professor D. Stojičević, Professor D. Marković

Department of Livestock Breeding*   156
Head: Professor V. Petrović

Department of Marketing*   157
Head: Professor A. Tomin

Department of Meat Technology*   158
Heads: Professor J. Joksimović, Professor S. Karan Djurdjić

Department of Meteorology and Climatology*   159
Heads: Professor S. Stanojević, Professor N. Todorović

Department of Microbiology*   160
Heads: Professor M. Todorović, Professor M. Šutić

Department of Pedology*   161
Head: Professor M. Živković
Senior staff: Professor S. Stojanović, Professor R. Korunović

Department of Phytopathology*   162
Heads: Professor D. Šutić, Professor M. Babović

Department of Phytopharmacy*   163
Head: Professor R. Kljajić

Department of Plant Breeding*   164
Head: Professor B. Jovanović

Department of Plant Ecology*   165
Head: Professor M. Stojanović

Department of Plant Physiology*   166
Heads: Professor Dj. Jelenić, Professor R. Džamić

Department of Pomology*   167
Head: Professor S. Bulatović

Department of Soil Chemistry*   168
Head: Professor M. Jakovljović

Department of Statistics*   169
Heads: Professor V. Erdeljan, Professor D. Ljesov

Department of Tobacco Processing*   170
Head: Professor A. Demin

Department of Veterinary Science*   171
Head: Professor R. Bešlin

Department of Vine Production and Technology*   172
Head: Professor R. Paunović

Department of Zoology and Ecology*   173
Head: Professor D. Šinzar

FACULTY OF FORESTRY*   174
Address: Kneza Viseslava br 1, Beograd

FACULTY OF MINING AND GEOLOGY*   175
Address: Djušina 7, Beograd
Dean: Professor D. Nikolić

Department of Palaeobotany*   176
Head: Professor N. Pantić

Zavod za Ribarstvo na SR Makedonija* 196

[Fisheries Institute of Macedonia]
Address: Gradskipark, PO Box 190, 9100 Skopje
Status: Official research centre
Director: Dr N. Petrovski

Zavod za Unapreduvanje na Lozarstvto i Vinarstvto na SR Makedonija* 197

[Macedonian Institute for the Advancement of Viticulture]
Address: Naselba Butel 1, Skopje
Status: Official research centre
Director: Dr D. Pemovski

Zavod za Unapreduvanje na Stučarstvoto na SR Makedonija* 198

[Macedonian Institute for the Advancement of Animal Husbandry]
Address: Avtokomanda, 91000 Skopje
Status: Official research centre
Director: Dr B. Vaskov

ZAIRE

Institut de Recherche Scientifique*   1

– IRS
[Scientific Research Institute]
Address: BP 3474, Kinshasa-Gombe
Telex: 21162
Director-general: Professor Iteke Fefe Bochoa
Activities: General research in food sciences, in addition to other non-agricultural topics.

Institut National pour l'Étude et la Recherche Agronomique*   2

– INERA
[National Institute for Agronomic Research and Study]
Address: BP 1513, Kisangani
Director: Ngondo-Mojungwo
Activities: The aim of the institute is to promote the scientific development of agriculture.

ZAMBIA

Central Fisheries Research Institute 1

Address: Box 100, Chilanga, Lusaka
Telephone: (01) 278096
Affiliation: Department of Fisheries
Status: Official research centre
Chief fisheries research officer: P.M. Chipungu
Field station research units: Lake Tanganyika, M. Pearce; Lake Mweru-Wa-Ntipa, D. Richardson; Lake Mweru, vacant; Lake Bangweulu, vacant; Lake Kariba, vacant; Kafue, H. Mudenda
Graduate research staff: 8
Activities: Fresh-water fisheries.
Projects: fish processing; fish biology; fishing gear and craft trials; lake limnology.

Central Veterinary Research Institute 2

Address: PO Box 33980, Lusaka
Affiliation: Department of Veterinary Medicine and Tsetse Control Services
Status: Official research centre
Assistant research director: Dr H.G.B. Chizyuka
Graduate research staff: 20
Activities: Veterinary medicine (livestock).
Publications: Annual report.
Projects: Diagnosis, surveillance and research in livestock diseases (Dr H.G.B. Chizyuka); FAO/UNDP animal disease control project (H.F. Schels).

Luapula Regional Research Station 3

Address: Box 129, Mansa, Luapula
Telephone: 233/259
Status: Official research centre
Officer-in-charge: D.H. Mudenda
Departments: Agronomy (general), H. Masole
Graduate research staff: 5
Activities: Plant production carrying out trials on a wide range of crops such as maize, sunflowers, groundnuts, soyabeans, rice, cassava, and sweet potatoes. A coffee research programme is planned, which will be centred on irrigated robusta coffee.
Publications: Annual report.
Projects: Root and tuber research (M.S.C. Simwambana).

Mount Makulu Central Agricultural Station 4

Address: Private Bag 7, Chilanga
Telephone: 278655
Affiliation: Department of Agriculture
Status: Official research centre
Assistant research director: W. Chibasa
Sections: Seed services, I. Kaliangile; national oil development programme, Dr Ravagnan; plant protection, M. Kabuswe; weed research, D. Vernon; food conservation and storage unit, E. Sakufiwa; agriculture/chemistry, K. Munyinda; soil science (physics), A.M. Bunyolo; microbiology, R. Nyemba; soil advisory, J.K. McPhillips; adaptive research planning unit, S. Kean; cereals (maize, wheat, sorghum, rice), A.J. Prior; tree crops, Abou El-Neel; root and tuber crops (cassava, sweet potatoes), M.S.C. Simwambana; grain legumes (beans, bambara groundnuts, cowpeas), Dr D.M. Naik; soil productivity, Mr Svads; animal husbandry and pas-

ture ecology, vacant; vegetables, Mr Kurup
Graduate research staff: 27
Activities: Grain, tuber and vegetable crops research, directed at development of varieties adapted to local ecology and biological conditions; soils research, concentrating on management of acidic and low fertility soils of high rainfall areas for higher productivity; soil microbiological work, seeking local or imported strains of modulating bacteria.
Publications: Annual report of research branch; annual coordination report.

KABWE REGIONAL RESEARCH STATION   5

Address: PO Box 80908, Kabwe, Central Province
Telephone: (052) 3201-2
Affiliation: Ministry of Agriculture and Water Development
Officer-in-charge: F.J.C. Sikazwe
Sections: Tobacco agronomy, J.W. Ogonowski; tobacco breeding, vacant; cereals and legumes agronomy, vacant; plant pathology, vacant; chemistry laboratory, W. N'Ambi; adaptive research team, P. Makungu, K. Chanda
Graduate research staff: 5
Activities: Tobacco, maize, sunflowers, soyabeans, groundnuts, sweet potatoes: plant breeding, crop husbandry, plant protection, chemical analysis of crop samples, seed certification. The general aim is to find and recommend to farmers the most suitable varieties, fertilization levels, chemicals effective against pests and diseases, herbicides, nematocides, time of planting and spacing, and other improved measures.
Publications: Annual report.
Projects: Virginia tobacco variety trial; Burley tobacco variety trial; tobacco lime trial (J.W. Ogonowski); maize variety trial; long-term maize fertilizer trial; sunflower variety trial; adaptive research trials on farmers' fields (P. Makungu, K. Chanda); seed certification (P.M. Jones).

Msekera Agricultural Research Station   6

Address: Box 89, Chipatas, Eastern Province
Telephone: (062) 21725
Status: Official research centre
Officer-in-charge: Boniface A. Mulyate
Departments: Agronomy, E. Hohmann; groundnut breeding, Dr R.S. Sandhu; soil survey, D.B. Clayton
Graduate research staff: 6
Activities: Field assessments of various crops, fertilizer response, resistance to disease and pests; breeding for high-yield oil, disease resistance and uniformity (constriction) in groundnuts.
Publications: Annual report.

National Council for Scientific Research   7

Affiliation: National Commission for Development Planning

TREE IMPROVEMENT RESEARCH CENTRE   8

– TIRC/NCSR
Address: PO Box 21210, Kitwe
Telephone: (02) 216734
Status: Official research centre
Head: Dr A.S. Hans
Departments: Breeding and genetics, Dr A.S. Hans; physiology, vacant; pathology, vacant; ecology, C.K. Mwamba; silviculture, vacant
Graduate research staff: 5
Activities: Genetic, physiological and silvicultural research to improve forest trees, shrubs, and herbs of economic value; study of ecology, distribution, variations within species of selected genera of trees, shrubs, herbs showing potential for timber, food, and medicinal, chemical and industrial values.
Publications: Annual report.
Projects: Breeding, improvement and domestication of wild forest fruit trees; physiology of propagation and silvi-ecological research.

National Food and Nutrition Commission 9

Address: Box 32669, Lusaka
Telephone: (01) 211426
Status: Official research centre
Executive secretary: A.P. Vamoer
Units: Food policy planning, H.N. Siulanda; food science and technology, D. Muntemba; nutrition education and training, S. Bleecker; public health nutrition, H. Patel; mass communications (acting), I. Gordon
Graduate research staff: 8
Activities: To reduce the wastage of manpower through death, disease and disability due directly or indirectly to malnutrition; to increase the learning capacity of schoolchildren through better nutrition; to improve the productivity and working efficiency of adults.
Publications: Annual report.
Projects: Food and nutrition study (Dr S. Kumar); nutrition surveillance system (Dr S.L. Nyaywa, C.Y. Chikamba); improving nutrition through agricultural and rural development (Dr K. Kwofie, H.N. Siulanda).

National Irrigation Research Station 10

– NIRS
Address: PO Box 68, Mazabuka, Southern Province
Telephone: (342) 30405/30116
Affiliation: Ministry of Agriculture and Water Development
Officer-in-charge and principal research officer: G.C.H. Hill
Sections: Irrigation agronomy, A.M. Mwaipaya; irrigation engineering, vacant; soil science, A.M. Mwaipaya; horizontal resistance breeding programme for wheat, P. Groot; vegetable research programme, K.A. Kurup; cropping systems, A.M. Mwaipaya; crop physiology, vacant
Graduate research staff: 9
Activities: Investigations into irrigation systems and layout, drainage and soil science, irrigation agronomy and consumptive use of water; research is also extended to suitable rotations for all year round cropping, leading to cropping systems research and varietal testing. Irrigation facilities are also provided for other research activities, such as cereals and oil seeds breeding, the vegetable research programme, irrigated pastures; fodder crops research, and plant protection.
Publications: Annual report.
Projects: Irrigation research and development (UNDP/FAO sponsored) - a national project (FAO senior technical adviser, M.A. Qasem; NIRS irrigation development, W. Haarthoorn).

University of Zambia 11

Status: Educational establishment with r&d capability

SCHOOL OF AGRICULTURAL SCIENCES 12

Address: PO Box 32379, Lusaka
Dean: Dr R. Nadaraja
Graduate research staff: 32
Activities: Soil sciences; irrigation; land use; plant breeding; plant protection; crop husbandry; livestock nutrition and reproduction.
Publications: Annual research progress report.

Agricultural Engineering Department 13

Head: C.E. Klooster

Animal Science Department 14

Head: A. Chimwano
Projects: Physiological, nutritional, management and health factors which affect productivity of national rural herds of cattle (Dr R. Nadaraja); performance of poultry on various diet combinations with local ingredients; studies on nutritive value of Zambian energy feeds and plant protein sources in pig rations (A. Chimwano).

Crop Sciences Department 15

Head: Dr M. Macfarlane
Projects: Wheat research breeding and selection (N.S. Sisodia, I. Javaid); varietal testing of vegetables (G. Chibiliti); soyabean research (N.S. Sisodia).

Rural Economy and Extension Education Department 16

Head: L.M. Malambo

Soil Science Department 17

Head: Dr K.S. Gill
Projects: Soil-water-plant relations and root-shoot-growth patterns of some important crops (Dr K.S. Gill, J.C. Patel); residual effects of lime application (J.A. Toogood); physical characterisation of UNZA farm soils (J.A. Toogood, Dr K.S. Gill).

ZIMBABWE

Agricultural Research Council

1

– ARC
Address: PO Box 8108, Harare
Telephone: (01) 704531
Status: Official research centre
Chairman: K.D. Kiskman
Executive Secretary: R.C. Smith
Divisions: Crop research, W.R. Mills; livestock and pastures research, Dr P. Chigura; research services, Dr P. Grant; education and executive branch, J.W. Walsh
Activities: Through its Department of Research and Specialist Services and its divisions, the Agricultural Research Council organizes research covering most areas of agriculture, including the following: soil sciences; plant protection, entomology, pathology, and nematology; plant breeding and crop husbandry (maize, wheat cotton, sorghum, and millet); livestock breeding (sheep, beef and dairy cattle); livestock husbandry and nutrition, including investigation of pasture and legumes as nutritional bases; agricultural engineering and intermediate technology.
Publications: Annual report; Division of Livestock and Pastures annual report; other technical and divisional reports.
Projects: Beef cattle cross breeding (H. Ward); animal nutrition (Dr R. Richardson); cotton research (J. Gledhill); horticulture and coffee research (M. Dale); crop breeding (R.J. Tattersfield); weed research (Dr P. Thomas); legumes and pastures (Dr J. Clatworthy).

Agronomy Institute

2

Address: PO Box 8100, Causeway, Harare
Telephone: 704531
Telex: AGRISEARCH
Affiliation: Agricultural Research Council
Status: Official research centre
Acting Head: E.E. Whingwiri
Departments: Weed research, P.E.L. Thomas; crop physiology, E.E. Whingwiri; general agronomy, I.K. Mariga
Graduate research staff: 8
Activities: Crop husbandry and weed control research in maize, wheat, sorghum, finger and bullrush millet, cassava, field beans, groundnuts, and soyabeans.
Publications: Annual report, Agronomy Institute; Annual report, Agronomy Institute, Weed Research Team.
Projects: Weed control with herbicides in various crops (P.V.M. Richards); weed seed behaviour studies (P.E.L. Thomas); maize and wheat physiology studies (E.E. Whingwiri); agronomic studies on maize, wheat, cassava, field beans, finger millet and other crops (I.K. Mariga); agronomic studies on peasant farms (A.N. Mashiringwani); farming systems research (E.M. Shumba).

Aquatic Ecology and Fisheries Branch 3

Address: PO Box 8365, Causeway, Harare
Telephone: (01) 790816
Affiliation: Department of National Parks and Wildlife Management
Status: Official research centre
Chief Ecologist: M.I. van der Lingen
Graduate research staff: 16
Activities: Research and development in fresh-water fisheries, aquaculture, and capture fisheries.
Publications: Annual report.
Projects: Population dynamics and stock assessment of Limnothrissa mioden (B. Marshall); population dynamics of Hydrocynus vittatus (J. Langerman); aquaculture investigations and hybridization (R. Burne, P.J. Thomson); trout culture (J. English); productivity of large irrigation lakes (S. Chimbuyu).

HYDROBIOLOGICAL RESEARCH UNIT 4

Head: vacant

INYANGA TROUT RESEARCH CENTRE 5

Head: J. English

LAKE KARIBA FISHERIES RESEARCH INSTITUTE 6

Head: F.J.R. Junos

LAKE KYLE FISHERIES RESEARCH STATION 7

Head: P.J. Thompson

LAKE MCILWAINE HIGHVELD FISHERIES RESEARCH STATION 8

Head: R. Burne

MATEBELELAND FISHERIES RESEARCH UNIT 9

Head: S. Chimbuya

Chibero College of Agriculture 10

Address: PO Box 901, Norton
Telephone: (172) 230
Telex: CHIBCOL
Affiliation: Agricultural Research Council
Status: Educational establishment with r&d capability
College Principal: Lovegot Tendengu
Graduate research staff: 7
Publications: Annual report.

Chipinga Research Station 11

– CRS
Address: PO Box 61, Chipinga
Telephone: (137) 476
Affiliation: Agricultural Research Council
Status: Official research centre
Officer-in-Charge: M.O. Dale
Activities: Coffee trials.
Projects: Coffee intensive systems trial; coffee nutrition trials; coffee variety trials; coffee ethrel trials; coffee vegetative propagation trials (R. Bester).

Chiredzi Research Station 12

Address: PO Box 97, Chiredzi
Telephone: 397-398
Affiliation: Agricultural Research Council
Status: Official research centre
Director: vacant
Head of Cotton Pest Research: Dr J.H. Brettell
Head of Horticulture: M. Dale
Sections: Maize research, R.C. Olver; soya beans breeding, J.R. Tattersfield; groundnut breeding, C.L. Hildebrand
Graduate research staff: 2
Activities: Research in the following areas: plant production, including cotton, maize, groundnuts, soya beans, wheat, rice, vegetables and cassava; horticultural production, including tropical and subtropical fruit and nut trees; plant breeding of cotton, maize, groundnuts and soya beans; plant protection and cotton pest research.
Through lack of professional staff this station is run as an offstation unit, with the heads of department appointed variously from the Cotton Research Institute, the Horticultural and Coffee Research Institute, and the Crop Breeding Institute (see separate entries).
Publications: Biannual reports.
Projects: Investigations into why commercial cotton yields are significantly lower than those obtained on the Research Station (A.W.M. te Braake).

Cotton Research Institute   13

Address: PO Box 765, Gatooma
Telephone: (158) 2210; 2311
Affiliation: Agricultural Research Council
Status: Official research centre
Head: J.A. Gledhill
Departments: Cotton pest research, Dr J.H. Brettell; cotton agronomy and physiology, G.G. Rabey; cotton breeding, R.F.E. Jarvis
Graduate research staff: 9
Activities: Research into: cotton breeding; cotton crop husbandry; and cotton crop protection.
Publications: Annual report.
Projects: Chemical and biological methods of boll-worm control; chrysopid biology, and insecticide screening (Dr J.H. Brettell); sucking pest and red spider mite control; whitefly biology and control (A.C.Z. Musuna); pesticides spray application methods (J.A. Gledhill); medium and long staple cotton breeding; cotton breeding stocks development (R.F.E. Jarvis); verticillium wilt resistance breeding (R. Chinodya); cotton plant development studies, soil fertility and crop nutrition trials, and cotton crop production trials (G.G. Rabey).

Crop Breeding Institute   14

Address: PO Box 8100, Causeway, Harare
Telephone: (10) 704531
Affiliation: Agricultural Research Council
Status: Official research centre
Head: J.R. Tattersfield
Graduate research staff: 9
Activities: Research work is concentrated entirely on plant breeding, in order to develop improved varieties of the major food crops grown in Zimbabwe.
Publications: Annual report.
Projects: Maize breeding (R.C. Olver); pearl millet and sorghum breeding (J.N. Mushonga); wheat and barley breeding (C.J.J. Badenhast); groundnuts breeding (G.L. Hildebrand); soya bean breeding (J.R. Tattersfield); sunflower breeding (D.E. Roberts); potato breeding (M.J. Joyce).

Department of Veterinary Services *   15

Affiliation: Agricultural Research Council
Status: Official research centre

TSETSE AND TRYPANOSOMIASIS   16
CONTROL BRANCH, RESEARCH
SECTION

Address: PO Box 8283, Causeway, Harare
Telephone: (10) 707381
Telex: Tsetse
Chief Glossinologist: G.A. Vale
Graduate research staff: 5
Activities: Veterinary medicine.
Publications: Animal report of the Branch of Tsetse and Trypanosomiasis Control.
Projects: Development and application of stationary baits to control and sample populations of Tsetse flies.

VETERINARY RESEARCH   17
LABORATORY

Address: PO Box 8101, Causeway, Harare
Telephone: (10) 705885
Assistant Director of Veterinary Services (Research): J.A. Lawrence
Departments: Bacteriology, I. Macadam; biochemistry, U.H. Ushewokunze-Obatolu; pathology, C.M. Foggin; poultry, A.H. J. Colley; protozoology, B.H. Fivaz; tick research, R.A.I. Norval; virology, N.K. Blackburn; wild life research, J.B. Condy
Graduate research staff: 14
Activities: The laboratory functions primarily as a diagnostic centre but also carries out research into economically-important diseases of domestic livestock, with particular reference to their epidemiology and control in Zimbabwe.
Publications: Annual report.
Projects: Epidemiology of rabies (C.M. Foggin); ecology of ticks of domestic animals (R.A.I. Norval); epidemiology of tick-borne diseases of cattle (J.A. Lawrence); arthropod-borne viruses of farm livestock (N.K. Blackburn); diseases of wild life transmissible to domestic animals (J.B. Condy).

Forestry Commission   18

Address: PO Box HG 595, Highlands, Harare
Telephone: (10) 46878
Telex: 4-386 RH
Status: Official research centre
Divisional Manager: P.F. Banks
Publications: Annual report.

FOREST RESEARCH CENTRE 19

Divisional Manager, Research: L.J. Mullin
Graduate research staff: 8
Activities: Research in the following areas: introduction, provenance testing, and breeding of forest trees, mainly pines and eucalyptus; silvicultural research, with special emphasis on spacing trials, thinning regimes and plantation inventories; production, collection, processing, and storage of forest tree seeds; forest pest management; agroforestry research and arid area species/provenance trials.
Projects: Pine improvement programme; agroforestry research (L.J. Mullin); Eucalyptus improvement programme (D.R. Quaile); silvicultural research (J.S. Brouard); seed production (B.R.T. Seward); entomology (Y. Katerere).

Grasslands Research Station 20

Address: PO Box 3701, Marandellas
Telephone: (128) 3527
Affiliation: Agricultural Research Countil
Status: Official research centre
Head: O.T. Mufandaedza
Departments: Animal production, J.L. Grant; pastures, R.D. Kelly
Graduate research staff: 9
Activities: Animal production and livestock husbandry and nutrition (beef cattle and sheep); plant production and crop husbandry (legume-based pastures).
Projects: Beef production systems (J.L. Grant); nutrition of beef steers (J. de W. Tiffin); sheep production systems (P. Retzlaff); legume introduction and screening (J.N. Clatworthy); legume establishment and nutrition (P.J. Grant); grazing trials on legume pastures (R.D. Kelly); pasture legume seed production (A.D. Irvine).

Henderson Research Station 21

Address: PO Box 222A, Harare
Telephone: (Mazoe) 481
Telex: AGROZOE
Affiliation: Agricultural Research Council
Status: Official research centre
Head: Dr M.G.W. Rodel
Activities: Animal reproduction physiology research is mainly associated with problems in beef cattle, but also to some extent with dairy cows, with some work on nutrition. Pasture research is concerned with nitrogen grass pastures for beef production, although the emphasis is changing to grass/legume pastures for beef cattle; a certain amount of grass breeding work is being

done. Poultry research is concerned with diets for broilers and layers using locally available feeds. Weed research (conducted on behalf of the Crop Research Institute) is concerned with control of weeds in major crops using chemical means, weed biology studies, and manipulating growth of crops using growth regulators. A certain amount of agronomic work, and variety testing is conducted.
Projects: Animal reproductive physiology research (Dr C.T. McCabe); tropical and sub-tropical pasture research (Sir Peter Mills); poultry research (P. Keen); weed research team (Dr P.E.L. Thomas).

Horticulture and Coffee Research Institute 22

– HCRI
Address: PO Box 3701, Marandellas
Telephone: (128) 4122
Affiliation: Agricultural Research Council
Status: Official research centre
Director: M.O. Dale
Graduate research staff: 3
Activities: Research work consists mostly of trials dealing with collections of suitably adapted cultivars; but also with vegetative propagation of the stone fruit and intensive planting systems, a small amount of breeding is done with the pyrethrum: macadamias, guayule and jojoba are cultivated locally in an attempt to find the most suitable area for their production.
Publications: Report (biannual).
Projects: Stone fruit cultivar collection, stone fruit intensive systems, and grape cultivar collection (M.O. Dale, G.D. Rose, P.A. Spencer); vegetable cultivar trials (P.A. Spencer); vegetable fertilizer trials, Macadamia variety local trial (M.O. Dale); Guayule and Jojoba trials (G.D. Rose and M.O. Dale); Pyrethrum trials (G. D. Rose).

Makoholi Experiment Station 23

Address: PO Box 9182, Fort Victoria, 16 Victoria
Telephone: (139) 3255
Affiliation: Agricultural Research Council
Status: Official research centre
Head: Oneas Tichafa Mufandaedza
Departments: Livestock husbandry and nutrition, I.A. Dude; veld and pasture, D. Sibanda
Graduate research staff: 3
Activities: The station is one of four main centres in Zimbabwe where research work is being carried out in

livestock and pasture management: It is situated in the drier part of the country characterized by poor granite derived sands. Research in animal production is based on the use of the indigenous type animal (Mashona) most suited to the environment, whilst the introduction of tropical pasture legumes into natural grazing forms the basis for pasture research.

Projects: Nutrition pre- and post first calving on the second conception of heifers first calving at two years of age (J. de W. Tiffin); assessment of the effects on animal production of introducing Siratro or Fine Stem Stylo into veld in a medium rainfall area; animal bodymass changes, plant yield and botanical composition in Siratro reinforced veld and unimproved veld grazed in different ways (R.D. Kelly).

Matopos Research Station   24

Address: PO Box K5137, Bulawayo
Telephone: 2/18
Affiliation: Agricultural Research Council
Status: Official research centre
Head: H.K. Ward
Activities: The station is primarily concerned with extensive and semiextensive land use in a semi-arid environment. Emphasis in investigation work is therefore given to means of increasing sustained livestock production from natural vegetation (veld). To this end, the ecology, management and utilization of veld, and the breeding and nutrition of grazing animals, especially beef cattle, are studied. Further understanding of the components of the veld-animal complex, together with the provision of basic information for the development of models of production systems, are likely to continue as major research objectives in future years. In particular, information is required on the interplay of rainfall, veld stability, fluctuations in availability and nutrient composition of herbage, intake, stocking rate, growth and productivity of different classes and genotypes of cattle and of animal species complementary to cattle. The long-term aim is to achieve a synthesis of the various facets of component research so that sound predictions of productivity and related effects can be made when changes to production systems are considered.

LIVESTOCK BREEDING SECTION   25

Head: H.K. Ward
Projects: Improvement of Tuli and Nkone cattle (J.W.I. Brownlee); selection response and genotype/environment interaction in Afrikaner cattle; assessment of the productivity of a wide range of straight- and crossbred beef cattle from representative Bos indicus and Bos taurus breeds (H.K. Ward).

RANGE ECOLOGY AND   26
MANAGEMENT SECTION

Head: R.P. Denny
Projects: Effects of burning and mowing; bush-grass interrelations; bush control using chemicals (P.T. Spear); improvement of plant cover on sodic soils; simulation of grass production on red clay soils from soil water balance, plant water relations and dry matter production (P.J. Dye); studies of the harvester termite Hodotermes mossambicus (J.L. Bissett); continuous versus rotational grazing on sand soils; simulation of beef production on red clay soils: relations between stocking pressure and body mass change of steers (R.P. Denny).

RANGE LIVESTOCK NUTRITION   27
SECTION

Head: F.D. Richardson
Projects: Performance of beef cows and their progeny in relation to stocking rate and amount of dry season protein; use of biochemical parameters to monitor nutritional status of grazing beef cattle; modelling of metabolism in beef cattle.

Ministry for Agriculture*   28

Address: P. Bag 7701, Causeway, Harare
Status: Geovernment department

DEPARTMENT OF AGRICULTURAL,   29
TECHNICAL AND EXTENSION
SERVICES*

Address: PO Box 8117, Causeway, Harare

DEPARTMENT OF RESEARCH AND   30
SPECIALIST SERVICES*

Address: PO Box 8108, Causeway, Harare

DEPARTMENT OF VETERINARY   31
SERVICES*

Address: PO Box 8012, Causeway, Harare

National Herbarium and Botanic Garden* 32

Address: PO Box 8100, Causeway, Harare
Status: Official research centre
Officer-in-Charge: T. Muller (Botanic Garden)
Keeper: R.B. Drummand (Herbarium)
Activities: Botanical and ecological research. The centre provides an identification service from its collection of approximately 250,000 specimens.

Pig Industry Board Experimental Farm 33

Address: PO Box HG 297, Highlands, Harare
Telephone: (174) 25724
Affiliation: Pig Industry Board
Status: Independent research centre
Director: Dr D.H. Holness
Activities: The farm is largely funded and controlled by the pig producers of Zimbabwe. Its major function is research and development in relation to the problems faced by pig producers. In addition, new findings from research elsewhere are tested and integrated into the pig industry in Zimbabwe.
Publications: Annual Pig Testing Report.
Projects: Nutrition of the breeding sow and the pig during the growing/fattening phase; genetic improvement of the Zimbabwe national pig herd; aspects of pig management appropriate to Zimbabwe.

Tobacco Research Board 34

– TRB
Address: PO Box 1909, Harare
Telephone: (10) 50411
Status: Association funding research
Director: Dr I. McDonald
Departments: Agronomy, L.T.V. Cousins; productivity, D.A. Baxter; liaison, J.L. Stocks; agricultural engineering, vacant; analytical chemistry, L. Toet; entomology, Dr M.J.P. Shaw; nematology, J.A. Shepherd; plant breeding, R.A.F. Ternouth; plant pathology, Dr J.S. Cole; plant physiology, Dr H.D. Papenfus; soil chemistry, Dr W.W. Ryding; statistics, R.A.J. Wixley
Graduate research staff: 23
Activities: The Tobacco Research Board's experimental centre was set up by the tobacco growers' associations of Zimbabwe, with substantial financial support from the government, in order to provide guides to tobacco growing techniques on farms at all systems levels, to improve the efficiency and economy of all operations, and to solve short and long-term problems in

the following fields: agricultural engineering, analytical chemistry, crop physiology, agronomy, plant breeding, plant protection, soil chemistry, production techniques and extension.
Publications: Annual report.
Projects: Control of weeds in tobacco growing; oriental tobacco production (L.T.V. Cousins); ecology and control of Cyperus esculentus (J. Lapham); control of insect pests of tobacco; economics and control of aphid-transmitted virus infections of tobacco (M.J.P. Shaw); control of Meloidogyne species in tobacco (J.A. Shepherd); techniques in studying soil-borne nematodes (J.A. Way); incorporating disease resistance and improved quality into tobacco cultivars (R.A.F. Ternouth); the haploid technique in tobacco breeding (K. Mtindi); incorporating resistance to Meloidogyne javanica into tobacco cultivars (J. Mackenzie); epidemiology and control of bacterial diseases of tobacco; spraying techniques in tobacco growing (M.W. Deall); ecology and control of storage moulds in tobacco (C.R. Fisher); ecology and control of soil-borne diseases of tobacco (J.S. Cole, Z. Zvenyika); seed germination and seedling production; a study of flowering in tobacco; irrigation techniques for tobacco production (H.D. Papenfus); fertilizer rates and application methods; soil nitrogen and moisture effects on tobacco leaf biochemistry (W.W. Ryding); soil texture and land preparation; farm storage methods for tobacco (D.A. Baxter); tobacco production methods - efficiencies (R.T. Garvin); air-cured tobacco production (M.M.M. Chimbumu); pesticide residue analysis in tobacco (L. Toet); dissemination of findings and recommendations (J.L. Stocks).

Zimbabwe Sugar Association Experiment Station 35

Address: PO Box 7006, Chiredzi
Telephone: (133-8) 514/515
Telex: 3218 EXPO ZW
Status: Independent research centre
Director: K.E. Cackett
Graduate research staff: 9
Activities: The station exists to investigate problems associated with the production of sugarcane in the south-eastern lowveld of Zimbabwe. Main areas of research include the following: plant breeding; pests and diseases; soil chemistry; crop physiology; irrigation research; general agronomy and weed control; fertilizers; sucrose and ethanol production. The station maintains 50-100 research projects at any one time.
Publications: Main Report (biennial).

How to find the full entry using the indexes

Throughout the directory each entry in each chapter has been given a separate reference number, commencing with 1 (one) and proceeding sequentially to the end of that chapter. The chapters commence with 'International establishments', and proceed through countries in alphabetical order to the end of the directory section. Each chapter has been given an abbreviation so that unique references can be made in the indexes to a particular establishment entry. The abbreviations used are given on the Contents and index abbreviations page in the prelims of this publication, and may be found in the running heads.

Thus the index entry 'gfr 183' refers uniquely to entry number 183 in the German Federal Republic chapter.

Headings in the Subject index have largely been made as detailed as the information will allow. The reader should therefore choose as specific a keyword as possible for the initial search, as broader terms have been used only where detail is not available or where activities are genuinely comprehensive over a given field.

TITLES OF ESTABLISHMENTS INDEX

TITLES OF ESTABLISHMENTS INDEX

TITLES OF ESTABLISHMENTS INDEX

TITLES OF ESTABLISHMENTS INDEX

TITLES OF ESTABLISHMENTS INDEX

TITLES OF ESTABLISHMENTS INDEX

TITLES OF ESTABLISHMENTS INDEX

TITLES OF ESTABLISHMENTS INDEX

TITLES OF ESTABLISHMENTS INDEX

SUBJECT INDEX

orchids   nze 83; tai 37
organic chemistry   aus 196, 201; ind 10; usa 446, 561, 565
ornamental plants   aus 40, 72; bel 148, 185; bra 56; can 22, 30, 31, 37, 42, 55, 75, 137; egy 51; fra 22, 162, 285; gfr 379, 381; hun 246; isr 37; jap 83, 107, 228; mus 1; net 34, 40, 124; nie 99; nze 73; phi 119, 121; saf 19; tha 49; tur 40; uni 121, 255, 433; usa 100, 218, 256, 453, 706, 729, 758, 776, 801, 806, 856, 864
   diseases   gfr 52, 62, 68; tha 35; usa 457
   pests   ber 1; usa 753
ornamental plants, agricultural facilities   swe 128
oysters   bra 78; usa 456

packaging   aus 72; uni 272; usa 267, 317
   food   aus 128, 260; bra 88; can 117; isr 7; nor 14, 72; rom 35; saf 28; usa 217, 377; yug 28
packaging materials   isr 51; net 25; uni 272
palaeobotany   uni 337, 561, 563, 669
palaeontology   uni 585, 699
paleozoology   uni 734
palm kernel oil   ben 9; bra 3, 14; ecu 9; gfr 84; hon 3, 9; ino 40, 128; ivc 7; may 39; net 67; nie 26, 118, 139; uga 8
paper   bra 81; ind 385; isr 51; jap 176; nze 3; saf 40; usa 478, 681
paper products   isr 51
paper pulp   jap 176
paper technology   cam 2; col 12; uni 458; usa 488, 490
parasites   ind 102; mex 13, 15; nie 26, 67 see also animal parasites; plant parasites
parrots   aus 180; pue 2
passion fruit   ken 6
pasta   aus 42; aut 18; fra 42; gfr 120
patents   ind 164
pathology   ind 275; nie 67, 75, 126; uni 792; usa 104, 351
   animal   uni 358
   comparative   fra 322
   plant   aus 76; usa 457
pathology (animal)   int 30, 41, 53, 117; arg 9; aus 64, 67, 71, 214, 331, 332, 333, 334, 335; aut 10, 11; bel 92; bra 45, 63, 66; can 3, 4, 5, 144, 171; car 4; chi 35; col 6, 11; cub 12; cyp 14; den 118; ecu 4; fra 94; gam 1; gfr 556, 559, 569; hun 37, 52, 57; ind 27, 105, 107, 143, 327, 351; ino 3, 7, 43; ira 30; ire 2, 77, 96; isr 62, 75; jap 91; ken 15; kir 1, 2; may 66; mex 31, 34; mor 12; net 194, 195, 204; nie 112, 126; nor 121, 156, 160; nze 79, 100; per 1, 9; phi 33, 39; pol 154; por 41; rom 36; saf 65; sen 11; spa 4; sri 51; swe 97; tan 16, 28, 29; tha 49, 53; uga 1; uni 26, 27, 34, 82, 274, 358, 391, 602, 623, 628, 673, 800, 801; usa 43, 111, 265, 310, 317, 382, 419, 445, 553, 555, 556, 680, 746, 755, 781, 847, 864; zam 2 see also animal parasites; animal viruses; fungal diseases (animal)
pathology (plant)   int 18; arg 4, 26, 44; aus 5, 36, 37, 54, 79, 165, 227, 287, 297, 299, 318, 326; ban 11, 20, 38; bel 96, 102, 122, 146, 153, 185; bra 53, 60, 65, 106, 126, 164; bru 5; cam 5; can 13, 19, 27, 28, 30, 31, 47, 101, 135, 159; chi 25; col 6; con 3; cyp 6, 7, 7; den 34, 88, 170; ecu 22; els 1; fij 4; fra 16, 21, 48, 311; frg 1; gfr 49, 51, 54, 60, 62; gre 17, 18, 20, 56; hng 2; hon 11; hun 233, 244; ind 7, 104, 118, 147, 155, 167, 181, 192, 228, 231, 292, 316, 347, 379; ino 17; ira 27, 66; ire 10; isr 38; ivc 7; jap 29, 30, 35, 36, 65, 114, 133, 191, 221; kor 2; may 42; mus 1; nca 12; net 53, 68, 71, 72, 114, 125, 149, 154, 157; nie 3, 137; nze 13, 41, 73; pak 23; per 30; phi 6, 107; por 17; rom 32; saf 7, 44; sri 9, 16, 35, 40; sud 6; swe 157; swi 28; tai 18, 38; tan 7, 15; tha 2, 35, 44; tri 1; tur 22, 24, 25, 39, 99; uga 21; uni 105, 159, 231, 283, 342, 373, 457, 503, 509, 560, 635, 662, 695, 769, 823; usa 29, 85, 88,

106, 158, 248, 279, 304, 382, 393, 402, 515, 610, 655, 673, 680, 706, 726, 729, 760, 778, 838, 883, 884, 886, 906, 977; vie 2; zam 5, 6; zim 1 see also fungal diseases (plants); plant parasites; plant viruses
pathology (plant)   chn 39
pawpaws   guy 1; nie 95; sri 31
peaches   aus 2, 73; bra 50; ind 168; reu 2; saf 19; usa 457, 758, 776
   pests   usa 744
peanuts   int 5; aus 80; ban 36; bol 1; bra 50; can 101; car 10; ino 17; jap 51; saf 23, 37; sol 1; tai 26, 30, 37; tha 24, 51, 55; usa 527, 630; zim 14
pearls   jap 135
pears   arg 50; can 44; ita 100; nor 122; uni 247
   pests   jap 107
peas   int 38; aus 10, 68, 74, 315; ban 40; bra 8, 84; cam 5; can 50, 53, 160; ita 8; mal 15; nie 64, 139, 145; nze 72; rom 28; tha 70; uni 203; usa 70, 745
   diseases   can 55
peat   gfr 379, 456; ino 135; nze 98; uni 83, 414; usa 908
pectins   bul 21; nie 18
pedology   see soil science
peppercorns   bra 3; can 50; may 17; mus 1; sri 31; ven 28
peppers (vegetables)   cub 14; ita 8; tur 68; usa 643; yug 19
   pests   usa 813
perennial plants   gut 1; tun 3; uga 5
perfumes   ind 163; tur 97; usa 228
pest control   int 22, 24, 41, 59, 112, 115; arg 26, 36, 39, 50; aus 4, 76, 92, 137, 149, 165, 167, 180, 227, 290, 296; ban 19, 41, 44; bra 59, 139; bru 3, 5; can 27, 28, 37, 60, 64, 76, 99, 135; col 6; cub 2, 14; cze 71; egy 46; fij 5; fra 126, 141; gre 1; gut 1; hng 2; hon 7; ind 7, 40, 88, 91, 108, 167, 191, 208, 224, 300, 324, 375; ira 20, 23; ivc 7, 12; jam 4; jap 29, 30, 116, 119, 189; nca 1; net 67, 128; nie 3, 126, 137, 145; nig 1; nze 7, 13, 14, 21, 33, 57, 65, 70, 73, 77, 81, 96; pak 23; phi 39, 58, 107; por 3; reu 2; saf 16, 44, 67; sen 6; sol 1; sri 9, 14, 19, 24; swe 154; swi 68; syr 10; tai 3, 22, 26, 30, 34, 37, 43, 55; tan 15, 17, 19; tha 2, 70; tri 1; tun 8; tur 22, 23, 24, 25, 39; uni 1, 105, 201, 202, 248, 305, 373, 434, 458, 493, 560, 578, 635, 663, 675, 762, 778, 821; upv 4; usa 2, 51, 66, 99, 124, 171, 193, 201, 216, 242, 317, 338, 347, 382, 446, 478, 515, 516, 706, 729, 744, 753, 772, 804, 812, 813, 833, 852, 895, 945; ven 28; via 2; zam 5; zim 1, 12, 19
   biological methods   aus 143; fra 25; ice 19; isr 41; uni 799
   tropical   ino 54
pest control, pathology plant   cyp 7
pest-resistance tests   int 41; can 105; ecu 4; gfr 49; ind 181; kor 18; mus 3; tai 37, 55; tan 7; zam 6
pesticides   arg 10; aus 5, 22, 30, 99, 196, 246, 292; bel 84, 87, 123, 171; bra 60; bru 3; can 28, 36, 71, 105, 166; cub 15; egy 46; els 1; fin 103; fra 129; gfr 49; hun 196; ind 40, 190, 375; ira 20; isr 1, 9, 83; ita 112; jap 190; ken 5; mex 13; rom 32; saf 44; sri 9; swe 154; swi 68; tan 3, 17; tha 2, 3, 10; tur 39; uni 1, 372, 619, 662; usa 107, 327, 667, 752, 758, 764, 833, 846, 906; zim 13 see also acaricides; insecticides; nematocides; rodenticides
   residues   int 24; aus 318; aut 18, 44; bel 83; bra 65, 88; can 14, 43, 96; chn 32; fra 20; gfr 121, 135, 281, 454; ind 7, 108, 208, 300; ita 20; net 160; nze 62; swe 82; tur 14; uni 65, 79, 166, 216, 797; usa 241, 446, 448, 729, 753
pet foods   uni 98; usa 427, 532
pharmaceutical technology   usa 418
pharmaceutical technology (veterinary)   uni 262
pharmaceuticals   see drugs; drugs (veterinary)
pharmacology   nca 11

salmon   can 70; fra 80; ire 35; nor 23; uni 166; usa 929

salt soils   aus 73, 164; can 59; ind 71; net 103; uni 743

sandy soils   ira 6; nie 140; nze 71; saf 23, 67

sawing machines   png 1

sawn timber   aus 20, 54, 90; ban 30; can 72; gfr 73; gre 88; mal 6; may 8, 39; net 99; nie 32; phi 48, 68; saf 40, 46; tha 58; tri 5; uga 26; usa 774

sea fishing   aus 340; bel 1, 148; chi 39; phi 9; uni 149, 422, 679

seals (mammals)   uni 424, 685

seedlings   kor 8; may 32; nie 99

seeds   int 22, 61; arg 50; aus 1, 14, 54, 68, 72, 77, 259; aut 9, 30, 68; ban 3; bez 1; bol 1; bra 22, 34, 44, 50, 53; can 170; col 6; den 174; els 4; fin 217; gha 7; hun 248; ind 176, 296, 302, 311, 313, 314, 315, 333, 361; ino 41, 128; ira 51; isr 2; ivc 7; jam 2; jap 54; ken 4, 6; kor 8; mex 1; net 117; nie 14, 98, 99; nze 29, 63; phi 8, 47, 72, 107; png 5; rom 28; saf 59; sen 7; sri 9; tai 2; tha 2, 24, 30, 31, 34, 53; tur 40; uni 281, 417, 532, 616; usa 3, 51, 248, 599, 643, 947; ven 28; zam 5; zim 19

   diseases   bra 64

   storage   may 42; phi 58

sensory analysis (food)   arg 11; ind 48; uni 343, 742, 789

sequoia   usa 185

serums   aus 56; isr 72; jap 35, 88; nie 112; nor 157

sesame oil   bol 1; mex 1; upv 8

sewage treatment   bel 89; jap 107; uni 274, 457, 675, 806; usa 480, 779, 797

sheep   int 5; alb 1; aus 1, 4, 6, 10, 66, 67, 75, 78, 104, 221, 223, 234, 256, 285, 291, 295, 306, 308, 317, 326, 329, 337; bel 111, 113, 149; bra 164; can 9, 18, 26, 27, 39, 122, 133, 158; con 2; cze 79; fra 109, 141, 291; gam 1; gfr 43; gre 41; hun 188; ice 19; ind 60, 62, 64, 67, 88; ira 44; ire 2, 73; irq 1, 12; isr 62; ita 115, 117, 118; kor 20; mal 18; mex 1; net 5, 87, 159, 165; nie 45, 80; nor 92; nze 7, 8, 9, 11, 21, 28, 30, 33, 59, 78, 79, 85; per 12; png 8; por 30, 34; saf 2, 17, 20, 22, 26, 42, 61; sen 11; she 1; spa 29; sri 49; tan 1; tun 2, 4; tur 9, 27; uni 7, 58, 111, 157, 176, 187, 188, 242, 339, 356, 365, 370, 374, 400, 454, 460, 505, 673, 681, 721, 800, 801, 819; usa 68, 97, 342, 466, 526, 680, 683, 692, 742, 755, 769, 771, 842, 934, 937; yug 46; zim 20

   diseases   aus 60, 65, 71, 105, 217, 341; can 125; gfr 565; ind 66; ira 32, 35; uni 573; usa 761, 826, 847

   parasites   aus 39

sheep-farming   arg 9; aus 33, 44, 74; can 95; ire 16; mex 31, 51; nie 139; tan 2; usa 463

sheep-shearing   aus 33, 99, 203

   chemical methods   aus 196

shellfish   aus 257, 305, 341; bel 157; bra 34, 46; can 70; gfr 99; gre 34; ind 160, 161; jap 12, 13, 18, 19; mus 2; net 22, 178; uni 148, 149, 259, 276, 422, 472, 621, 622, 784; usa 729

   diseases   net 182

shelterbelts (forestry)   nie 36, 44; uni 412, 514; usa 192, 441

shrubs   usa 157

silage   gfr 132; net 165; uni 123, 242, 411

silk   jap 138; tur 55

silk farming   chn 32, 40; ita 112; jap 138, 192, 194, 212; ken 10, 11; mus 1; tha 2, 37

silkworms   jap 193, 196, 197, 212

silos   can 8, 113, 161

silt   bot 4; nie 140

silviculture   arg 2, 4; aus 73, 90, 92, 232, 283; aut 28; ban 30; bel 186; bra 81, 129; can 74, 78, 98, 99; chi 34; cos 4; cub 4; cyp 12; cze 58; den 154; ecu 1; eth 1; fin 66; fra 31, 72, 112; frg 4; gfr 277; 278, 280; gre 87, 89; hun 128; ice 17; ind 242, 246; ira 4; ita 3, 67; ivc 2; jap 76, 79, 116, 129, 186,

208; ken 3; kor 3, 25; mal 6; may 1, 37; mor 14; nep 2; net 120; nie 38; nze 4; pak 4, 17; phi 37, 45, 47, 48, 50, 58, 64, 68, 71, 72, 75, 76, 77, 78, 79, 80; png 5; pol 43; saf 59; sol 2; sri 27, 36; sur 1; swe 24; syr 9; tai 41, 45, 50; tan 18; tha 49, 58, 61; tri 5; tun 9; uga 26; uni 159; upv 1; usa 150, 151, 154, 181, 193, 200, 207, 243, 438, 478, 538, 695, 757, 924, 989; vie 1; yug 15; zam 8; zim 19

sisal   ind 232; tan 5

slaughtering   aus 65, 341; gfr 87, 89; ino 62; net 197; nor 54

smallholdings   can 2; may 35; png 8; sri 7; uni 741; usa 124, 472; zim 2

smoking (food processing)   isr 23; nie 17; uni 449

snakes   ecu 4

soaps   ind 160, 161; spa 51

social sciences and humanities   nca 1; ven 34

sociology   ind 148; phi 38; usa 397, 739, 759, 824, 873, 879

   rural   int 26, 45; arg 9; aus 261; ban 1; bel 165; bra 30, 127; can 130, 132; chi 3; cos 4; cze 66; egy 45; fra 144, 194; frg 1; gua 1; ino 104, 105, 128, 135; jap 37, 119, 122; kor 23; may 18; mex 18, 22; nie 46, 67, 68, 79, 82, 83, 126, 138, 139, 148, 152; nze 20; pak 23; phi 33, 59, 85, 113, 118; png 8; por 4; saf 14; spa 4; sri 1, 8; tan 21; tha 55; tur 10; uni 701, 731, 735; usa 9, 24, 232, 267, 306, 444, 463, 472, 664, 690, 796, 839, 859, 916, 949, 963, 978; ven 15

   total   net 169

   urban   nie 138

soft drinks   bra 88; rom 35; usa 76

soil conservation   int 5; arg 4, 9; aus 190, 224, 258, 274, 275, 276, 338; bel 187; bra 34, 59, 167; can 101; chn 14; den 120; dom 1; hun 130, 240; ind 70; ira 6, 56; isr 84; nze 36, 37, 41; phi 60; png 8; pol 171; saf 12, 14; tai 21; tun 1; usa 874

soil erosion   aus 30; can 130; chn 18; fra 110; frg 3; gfr 150; ice 14; jap 75, 182, 226; ken 14; mex 26; nor 49; nze 12, 37, 39, 40, 44, 46; phi 61; sud 9; uni 95, 430; upv 1; usa 706, 774; ven 24; yug 15

soil improvement   chn 3, 19; cub 21; gfr 41; gui 3; hun 88, 199; ind 186, 317; isr 40; jap 140, 157; net 35; nze 85; phi 102; swe 82; tur 14, 40; usa 331, 332, 937; zam 4

soil mechanics   aus 219; bra 113; fra 106

soil pollution   aus 225; bel 87, 122, 167; can 166; den 179; gfr 509; ind 317; may 21; nie 140; rom 31; spa 5; tur 14; uni 578, 771

soil profile   aus 38; cub 21; usa 458

soil science   int 8; alb 1; arg 18, 39, 45; aus 5, 10, 11, 12, 73, 74, 78, 90, 94, 185, 188, 238, 246, 260, 297, 302, 315, 326, 330; ban 2, 12, 21, 34, 40, 42, 48; bel 1, 85, 102, 150, 163, 166; ben 6; bol 6; bra 3, 12, 50, 58, 99, 103, 117, 138, 139, 164; cam 4; can 14, 19, 24, 26, 27, 33, 53, 77, 80, 82, 91, 115, 120, 122, 147, 158, 162, 165, 166; chi 8, 15, 20, 25; chn 32, 41; col 12, 15; con 2; cos 3; cze 14, 80; den 166, 181; dom 1; ecu 9, 23; egy 10, 52; fin 191; frg 1; gdr 47; gfr 41, 133, 274, 282, 379, 456; gha 13; gre 33, 36; gua 1; guy 3; hng 2; hon 2; hun 65, 89, 188; ice 14; ind 32, 54, 71, 74, 89, 147, 157, 185, 188, 217, 231, 247, 292, 317, 336, 339, 367; ino 2, 10, 11, 36, 108, 128, 137; ira 16, 44, 52, 66, 82; irq 1, 20, 23; isr 39, 88; ita 20; ivc 10; jap 52, 110, 125, 129, 167, 171, 172, 230; ken 3; kor 20; kuw 1; lib 3; liy 2; may 16, 23, 46; mex 11, 23, 26; mor 1; mus 1; nca 10; net 18, 30, 67, 104, 169; nie 3, 45, 51, 55, 83, 95, 145, 150, 152, 159; nor 95, 98; nze 2, 18, 70, 81; pak 23; per 33; phi 9, 16, 34, 41, 88, 94, 146; png 7; pol 153, 185; por 5; rom 31; saf 7, 17, 22, 42, 48, 49, 58, 61; she 1; sin 1; spa 4, 5, 28, 32, 39, 78; sri 40; sud 21; sur 1; swa 1; swe 29, 82; swi 24, 75; syr 11; tai 19, 30, 40, 51; tan 15, 21; tha 3, 24, 32, 38, 45, 46, 55, 62, 67; tri 14; tun 12; tur 9, 14, 42, 61, 75; uga 11, 19; uni 15, 42, 114, 173, 176, 188, 233, 283, 434,

viruses   uni 437 *see also* animal viruses; insect viruses; plant viruses

vitamins   jap 49; uni 742; usa 693, 743

viticulture   int 73; arg 9, 11, 27, 28; aus 17, 35, 79, 161, 227, 261, 272, 312; bel 218; bra 56, 139; can 37, 45, 95, 101; cyp 6; cze 14; fra 43, 140, 265, 328; gfr 151, 202, 282, 512; hun 188, 247; ira 66; isr 61; nze 16, 33, 84; rom 33; saf 16, 41, 66; swi 61; tha 34; tur 13, 15, 16, 67; uni 121; ven 28; yug 38, 197

volcanic soils   gua 1; jap 52, 55

walnut   usa 162

walnuts   usa 860

waste disposal   uni 542; usa 908

waste-disposal engineering   hun 130; jap 37

waste handling   can 107, 122, 126, 140, 147; net 29; nze 52

waste recovery   aus 164, 197, 321; can 9, 89, 167; cyp 4; dom 2; ind 181, 188, 295, 325; isr 51; net 31; nze 67; phi 102, 104, 119; tha 58; uni 522; usa 284, 357, 503, 621, 752
   food   nie 18
   wood   may 8

waste treatment   nze 31, 61

water-chestnuts   usa 70

water features   mex 20

water pollution   aus 84; bel 84, 87, 167, 190; can 70, 73, 84; gfr 371; ind 317, 395; jap 19, 125; mex 26; net 37; nze 46, 48, 50, 51; spa 5; tai 48; tha 12; uga 27; uni 161, 578, 622, 688, 725, 782, 797; usa 91, 478, 757

water power   nze 47; phi 119; sri 6

water purification   fra 171

water resources   arg 27, 58; aus 10, 29, 90, 184, 189, 219, 225, 238, 275, 276, 285; ban 48; bel 187; ben 6; bra 34; bru 2; can 102, 162; car 9; chn 14; cze 72; den 44; egy 10; fra 105, 108, 142, 304; gre 63; gua 1; gut 1; hun 90, 130; ind 70, 88, 157, 203, 336; irq 6; isr 17, 39; jap 133; may 16, 52; mex 20, 26; mor 3; nie 138; nze 12, 15, 36, 38; pak 18, 23; per 26; phi 61, 62, 88, 94, 113; sen 8; sri 1, 6; sur 1; tai 21, 52; tha 38, 45, 53; tri 4, 14; uni 161, 235, 585; upv 2; usa 2, 64, 80, 163, 171, 182, 187, 200, 202, 226, 237, 240, 280, 310, 333, 341, 434, 518, 522, 523, 604, 628, 659, 690, 706, 740, 798, 818, 840, 860, 891, 901, 949

water retention and flow works   bel 162; fra 104, 108, 245, 304; jap 160; mex 20; mor 6; net 18, 102; phi 135; rom 23; spa 75; swe 29; uni 488, 743; usa 659 *see also* dams; irrigation works; land drainage works

water supply   els 1; ira 6; sri 28; sud 2; tan 16; usa 732

water supply engineering   gfr 379; hun 240; ind 297; isr 87; phi 61; tun 1; usa 864

water treatment   aus 164, 199; can 147; por 15
   for aquaculture   uni 524

waterfowl   aus 180

watermelons   usa 74

waterways   phi 9

weed control   arg 36; aus 1, 4, 11, 20, 68, 78, 86, 92, 137, 138, 150, 165, 290; bel 83, 177; bru 2; can 8, 10, 12, 13, 19, 24, 27, 29, 30, 33, 34, 39, 49, 60, 101, 122, 134, 158, 170; cyp 7; dom 1; fij 4; gfr 458, 486; gre 22, 37, 50, 59, 90; hun 233, 262; ind 186, 226, 299, 390, 397; ira 20, 44, 46; ire 10; jam 4; jap 133; ken 1, 5; lib 3; mal 14; mar 1; may 34; mex 26; mus 1, 3; net 67, 72, 96; nie 137, 140; nze 7, 7, 17, 21, 70, 73, 77, 81; phi 3, 150; png 2; saf 38, 44, 59; sri 9; sur 1; swe 157; tha 38, 47; tur 22, 23, 24, 39, 70; uni 20, 42, 58, 292, 353, 408, 560, 615, 675, 738, 821; usa 109, 205, 248, 286, 330, 331, 334, 335, 336, 336, 346, 458, 464, 466, 517, 568,

630, 695, 715, 745, 780, 846, 852, 945; yug 18; zim 1, 2, 21 *se also* herbicides
   waterweeds   aus 145; sud 24

weeding equipment   gfr 69

whales   uni 424

wheat   int 10; arg 9, 39; aus 5, 10, 41, 42, 68, 78, 135, 225, 232, 261, 265, 310; ban 40; bol 1, 2; bra 6, 17, 22, 50, 59, 164; cam 5; can 43, 46, 48, 108, 110, 160; chn 32; den 172; fin 224; fra 42, 330; gfr 486; hun 65; ind 71, 107, 127, 170, 171, 172, 174, 175, 193, 202, 257, 292, 302, 305, 307, 309, 371; ire 42; ita 92, 94; jor 1; mal 14; mex 1; nie 1, 64, 75; nze 22, 71; per 6; saf 38; sen 2, 6; sud 9; tan 5; tur 63, 68, 94, 98; uni 342, 372, 799; usa 80, 253, 346, 463, 516, 663, 706, 842, 846, 864; zam 15; zim 2, 14
   diseases   ind 316; usa 857

whey   aus 70

wildlife   arg 9; aus 10, 84, 90, 180, 339; can 122, 163; chi 25; cos 4; ind 238, 249; jap 186; kir 4; mex 19; nie 27, 69, 75; nze 5, 14; pak 4, 19; phi 45, 54, 56; pue 2; sud 14; tri 5, 6; uni 168, 590, 817; usa 92, 103, 153, 158, 160, 167, 171, 180, 187, 191, 193, 200, 227, 252, 317, 349, 370, 434, 437, 440, 463, 478, 528, 531, 649, 818, 833, 860, 877, 960, 961, 989

wind-driven generators   may 39

wind power   chn 5; ind 203; nze 12; phi 119

wines   int 73; arg 11, 29; aus 17, 31, 271, 305; bra 88; chi 15, 19, 25; cze 14; fra 62, 265, 327, 328; gfr 151, 370; isr 61; nze 62; pol 143; rom 33; saf 12, 41, 66; swi 61; tur 16, 67; uni 249; usa 76, 323

wood   int 49; arg 4, 18; aus 54, 90, 92, 254, 283, 289; ban 30; bra 81, 117; cam 2; can 97, 103, 120, 156, 165; chi 25; chn 31; con 1; cos 4; dom 1; fin 148; frg 1, 2; gfr 193, 278, 369; ind 238, 250, 301; ino 1, 135; ira 2; ita 3, 66; jap 75, 110, 115, 125, 175, 177, 178; may 7; mex 1, 3, 11, 26; net 20; nie 27, 32; nze 2, 3, 68; phi 37; saf 46, 49; swi 14; tai 41; tha 49, 60; tri 5; tun 9; tur 69; uga 26; uni 305; usa 134, 148, 166, 167, 197, 466, 496, 663, 680, 681; ven 19; vie 1; zam 8
   as fuel   phi 47, 58, 79
   properties   ecu 1; may 8

wood based sheet materials   gfr 114; tur 69; jap 177, 178; usa 16, 757

wood defects   int 59; arg 4; phi 50

wood fibres   kor 25

wood preservation   int 59; arg 23; bel 189; ecu 1; gfr 113; ind 250; jap 179; may 37; net 20; nze 3; saf 40; sri 27; uga 26; uni 71, 305, 491

wood products   arg 4; ban 32; can 72; gfr 115; isr 51; ivc 2; jap 78, 80; ken 3; kor 20; mal 6; may 2, 9, 49; mex 19; net 99; pak 11; per 12; phi 44, 111; saf 40; she 1; sri 27; tai 41, 47; uni 71, 305; usa 200, 478, 489, 681, 706, 835, 986

wood pulp   aus 20, 197; bra 81; gfr 114; jap 80, 129, 176; kor 25; nze 3; saf 40; tai 49; usa 478, 490, 681, 774; vie 1

wood technology   int 59; arg 4, 19, 23; aus 90; bul 19; cam 2; can 72, 78; chi 36; chn 12, 31; cub 2, 5; ecu 1; frg 5; gfr 113, 193; gre 88; ita 66; ivc 2; jap 20, 80, 177, 180; may 8, 37; mex 11; mor 14; net 20, 67; nor 73; nze 3; phi 44; pol 43; saf 40; sri 27; swi 25; tai 47; tha 58; tun 9; usa 134, 148, 162, 171, 193, 201, 310, 319, 466, 681, 818, 824, 835, 860, 924, 984, 986; yug 15

wood working machines   jap 180; may 37; net 20; nze 3

wool   aus 1, 10, 32, 34, 115, 202, 203, 232, 306; gfr 197; ind 59, 60, 61, 62, 64, 65; nze 19, 30, 67, 85, 102; saf 3, 20, 26, 42; uru 2; usa 252, 638

working animals   tan 21

working environment   *see* environment (working)